T0361336

Molecular Docking for Computer-Aided Drug Design

Molecular Docking for Computer-Aided Drug Design

Fundamentals, Techniques, Resources and Applications

Edited by

MOHANE S. COUMAR
Centre for Bioinformatics
School of Life Sciences
Pondicherry University
Kalapet, Pondicherry, India

ACADEMIC PRESS

An imprint of Elsevier

ELSEVIER

Academic Press is an imprint of Elsevier
125 London Wall, London EC2Y 5AS, United Kingdom
525 B Street, Suite 1650, San Diego, CA 92101, United States
50 Hampshire Street, 5th Floor, Cambridge, MA 02139, United States
The Boulevard, Langford Lane, Kidlington, Oxford OX5 1GB, United Kingdom

Notices
Knowledge and best practice in this field are constantly changing. As new research and experience broaden
our understanding, changes in research methods, professional practices, or medical treatment may become
necessary.

Practitioners and researchers must always rely on their own experience and knowledge in evaluating and
using any information, methods, compounds, or experiments described herein. In using such information or
methods they should be mindful of their own safety and the safety of others, including parties for whom
they have a professional responsibility.

To the fullest extent of the law, neither the Publisher nor the authors, contributors, or editors, assume
any liability for any injury and/or damage to persons or property as a matter of products liability, negligence or
otherwise, or from any use or operation of any methods, products, instructions, or ideas contained in
the material herein.

Library of Congress Cataloging-in-Publication Data
A catalog record for this book is available from the Library of Congress

British Library Cataloguing-in-Publication Data
A catalogue record for this book is available from the British Library

ISBN: 978-0-12-822312-3

For information on all Academic Press publications visit our website at
https://www.elsevier.com/books-and-journals

Publisher: Andre Gerhard Wolff
Editorial Project Manager: Tracy I. Tufagaa
Production Project Manager: Kiruthika Govindaraju
Cover Designer: Alan Studholme

Typeset by TNQ Technologies

"கற்றது கை மண் அளவு கல்லாதது உலகளவு"

"Katrathu Kai Mann Alavu, Kallathathu Ulagalavu"

(What you have learnt is a mere handful; what you haven't learnt is the size of the world)
                                                                    -Thirukkural

This book is dedicated to my parents Selvaraj and Gunasundari, my mentors Dharam Paul Jindal (late) and Hsing-Pang Hsieh, my wife Vasundhara Devi, and my daughters Iniya and Lishitha.

List of Contributors

Azizeh Abdolmaleki, PhD
Department of Chemistry
Tuyserkan Branch
Islamic Azad University
Tuyserkan, Iran

Imlimaong Aier, M.Tech
Department of Bioinformatics & Applied Sciences
Indian Institute of Information Technology —
    Allahabad
Allahabad, Uttar Pradesh, India

Carolina Horta Andrade, PhD
Laboratory for Molecular Modeling and Drug Design
Universidade Federal de Goiás
Faculdade de Farmácia
Goiânia, GO, Brazil

Tamanna Anwar, PhD
Centre of Bioinformatics Research and Technology
Aligarh, India

Hemant Arya, PhD
Department of Biotechnology
Central University of Rajasthan
Bandar Sindri, Rajasthan, India

Tarun Kumar Bhatt, PhD
Department of Biotechnology
Central University of Rajasthan
Bandar Sindri, Rajasthan, India

Andrzej J. Bojarski, PhD
Maj Institute of Pharmacology
Polish Academy of Sciences
Kraków, Poland

Francesca Cavaliere, PhD student
Molecular Modelling Lab
Department of Food and Drug
University of Parma
Parma, Italy

Sibani Sen Chakraborty, PhD
Department of Microbiology
West Bengal State University
Barasat, West Bengal, India

Jeyaraj Pandian Chitra, PhD
Department of Biotechnology
Dr. Umayal Ramanathan College for Women
Alagappa University
Karaikudi, Tamil Nadu, India

José Correa-Basurto, PhD
Laboratorio de Diseño y Desarrollo de Nuevos
    Fármacos e Innovación Biotecnológica (Laboratory
    for the Design and Development of New Drugs and
    Biotechnological Innovation)
Escuela Superior de Medicina
Instituto Politécnico Nacional
Plan de San Luis y Díaz Mirón
Ciudad de México, Mexico

Mohane Selvaraj Coumar, M.Pharm, PhD
Centre for Bioinformatics
School of Life Sciences
Pondicherry University
Kalapet, Pondicherry, India

Pietro Cozzini, PhD
Molecular Modelling Lab
Department of Food and Drug
University of Parma
Parma, Italy

R. Vasundhara Devi, M.Tech, PhD
Departrment of Computer Science
School of Engineering & Technology
Pondicherry University
Kalapet, Pondicherry, India

Teodora Djikic, PhD
University of Belgrade - Faculty of Pharmacy
Department of Pharmaceutical Chemistry
Belgrade, Serbia

Zarko Gagic, PhD PharmD
University of Banja Luka - Faculty of Medicine
Department of Pharmaceutical Chemistry
Banja Luka, Bosnia and Herzegovina

Jahan B. Ghasemi, PhD
Chemistry Faculty
Drug Design in Silico Laboratory
University of Tehran
Tehran, Iran

Divya Gupta, M.Tech, PhD scholar
Interdisciplinary Biotechnology Unit
Aligarh Muslim University
Aligarh, Uttar Pradesh, India

Department of Life Sciences
Uttarakhand Technical University
Dehradun, Uttarakhand, India

Ravi Guru Raj Rao, PhD student
Structural Biology and Bio-Computing Lab
Department of Bioinformatics
Science Campus
Alagappa University
Karaikudi, Tamil Nadu, India

Sapna Jain, PhD
School of Engineering
University of Petroleum and Energy Studies (UPES)
Dehradun, Uttarakhand, India

Jeyaraman Jeyakanthan, MSc, MPhil, PhD
Structural Biology and Bio-Computing Lab
Department of Bioinformatics
Science Campus
Alagappa University
Karaikudi, Tamil Nadu, India

Asad U. Khan, PhD
Interdisciplinary Biotechnology Unit
Aligarh Muslim University
Aligarh, Uttar Pradesh, India

Sree Karani Kondapuram, PhD scholar
Centre for Bioinformatics
School of Life Sciences
Pondicherry University
Kalapet, Pondicherry, India

Pawan Kumar, PhD
National Institute of Immunology
New Delhi, India

Nathalie Lagarde, PharmD, PhD
Laboratoire GBCM
Conservatoire National des Arts et Métiers
HESAM Université
Paris, France

Florent Langenfeld, PharmD, PhD
Laboratoire GBCM
Conservatoire National des Arts et Métiers
HESAM Université
Paris, France

Neeraj Mahindroo, PhD
School of Health Sciences
University of Petroleum and Energy Studies (UPES)
Dehradun, Uttarakhand, India

Sabrina Silva Mendonca, MSc
Faculdade de Farmácia
Laboratory for Molecular Modeling and Drug Design
Universidade Federal de Goiás
Goiânia, GO, Brazil

Matthieu Montes, PhD
Laboratoire GBCM
Conservatoire National des Arts et Métiers
HESAM Université
Paris, France

José Teofilo Moreira-Filho, PhD
Faculdade de Farmácia
Laboratory for Molecular Modeling and Drug Design
Universidade Federal de Goiás
Goiânia, GO, Brazil

Melina Mottin, PhD
Faculdade de Farmácia
Laboratory for Molecular Modeling and Drug Design
Universidade Federal de Goiás
Goiânia, Brazil

Ayaluru Murali, PhD
Centre for Bioinformatics
Pondicherry University
Kalapet, Pondicherry, India

Mutharasappan Nachiappan, MSc, PhD
Structural Biology and Bio-Computing Lab
Department of Bioinformatics
Science Campus
Alagappa University
Karaikudi, Tamil Nadu, India

Bruno Junior Neves, PhD
Laboratory for Molecular Modeling and Drug Design
Faculdade de Farmácia
Universidade Federal de Goiás
Goiânia, GO, Brazil

Katarina Nikolic, PhD PharmD
University of Belgrade - Faculty of Pharmacy
Department of Pharmaceutical Chemistry
Belgrade, Serbia

Sourav Pal, M.Pharm
Department of Organic and Medicinal Chemistry
CSIR-Indian Institute of Chemical Biology
Kolkata, West Bengal, India

Academy of Scientific and Innovative Research
Ghaziabad, Uttar Pradesh, India

Archana Pan, PhD
Centre for Bioinformatics
School of Life Sciences
Pondicherry University
Kalapet, Pondicherry, India

Jeevan Patra, M.Pharm
School of Health Sciences
University of Petroleum and Energy Studies (UPES)
Dehradun, Uttarakhand, India

Sabina Podlewska, PhD
Department of Technology and Biotechnology
    of Drugs
Jagiellonian University Medical College
Kraków, Poland

Maj Institute of Pharmacology
Polish Academy of Sciences
Kraków, Poland

Dhamodharan Prabhu, MSc, PhD
Structural Biology and Bio-Computing Lab
Department of Bioinformatics
Science Campus
Alagappa University
Karaikudi, Tamil Nadu, India

G. Pranavathiyani, MSc
Centre for Bioinformatics
School of Life Sciences
Pondicherry University
Kalapet, Pondicherry, India

Sundarraj Rajamanikandan, PhD
Structural Biology and Bio-Computing Lab
Departmentof Bioinformatics
Science Campus
Alagappa University
Karaikudi, Tamil Nadu, India

Muthukumaran Rajagopalan, PhD
Centre for Bioinformatics
Pondicherry University
Kalapet, Pondicherry, India

Amutha Ramaswamy, PhD
Centre for Bioinformatics
Pondicherry University
Pondicherry, India

Manon Réau, PhD
Laboratoire GBCM
Conservatoire National des Arts et Métiers
HESAM Université
Paris, France

Mariadasse Richard, PhD student
Structural Biology and Bio-Computing Lab
Department of Bioinformatics
Science Campus
Alagappa University
Karaikudi, Tamil Nadu, India

Patricia Saenz-Méndez, PhD
Facultad de Química
Computational Chemistry and Biology Group
UdelaR
Montevideo, Uruguay

Faculty of Health, Science and Technology
Department of Engineering and Chemical Sciences
Karlstad University
Karlstad, Sweden

Balasubramanian Sangeetha, PhD
Centre for Bioinformatics
Pondicherry University
Kalapet, Pondicherry, India

Sailu Sarvagalla, PhD
Division of Biology
Indian Institute of Science Education and Research
    (IISER), Tirupati
Tirupati, Andhra Pradesh, India

Daniela Schuster, Univ.-Prof. Dr.
Institute of Pharmacy
Department of Pharmaceutical and Medicinal
    Chemistry
Paracelsus Medical University
Salzburg, Austria

Thomas Scior, PhD
Faculty of Chemical Sciences
Laboratory of Computational Molecular Simulations
BUAP
Puebla, Mexico

Asma Sellami, Pharm D
Laboratoire GBCM
Conservatoire National des Arts et Métiers
HESAM Université
Paris, France

Fereshteh Shiri, PhD
Associate Professor
Analytical Chemistry
University of Zabol
Zabol, Iran

Deepanmol Singh, M.Pharm
School of Health Sciences
University of Petroleum and Energy Studies (UPES)
Dehradun, Uttarakhand, India

Yudibeth Sixto-López, PhD
Laboratorio de Diseño y Desarrollo de Nuevos
    Fármacos e Innovación Biotecnológica (Laboratory
    for the Design and Development of New Drugs and
    Biotechnological Innovation)
Escuela Superior de Medicina
Instituto Politécnico Nacional
Plan de San Luis y Díaz Mirón
Ciudad de México, Mexico

Bruna Katiele de Paula Sousa, MSc
Faculdade de Farmácia
Laboratory for Molecular Modeling and Drug Design
Universidade Federal de Goiás
Goiânia, GO, Brazil

Giulia Spaggiari, PhD student
Molecular Modelling Lab
Department of Food and Drug
University of Parma
Parma, Italy

Arindam Talukdar, M.Pharm, PhD
Department of Organic and Medicinal Chemistry
CSIR-Indian Institute of Chemical Biology
Kolkata, West Bengal, India

Academy of Scientific and Innovative Research
Ghaziabad, Uttar Pradesh, India

Veronika Temml, PhD
Institute of Pharmacy
Department of Pharmaceutical and Medicinal
    Chemistry
Paracelsus Medical University
Salzburg, Austria

Pritish Kumar Varadwaj, PhD
Department of Bioinformatics & Applied Sciences
Indian Institute of Information Technology, Allahabad
Allahabad, Uttar Pradesh, India

Preface

Human health status is of paramount importance in terms of economic productivity, as well as social and mental well-being. However, guaranteeing a good health condition during our life span is surreal. Developments over the past five decades have provided better health care/medicine and increased human life expectancy. Even though there is a huge discord in the average life expectancy of humans residing in different parts of the world, there is an overall trend toward betterment, which is mainly due to better living conditions and the availability of effective and quality medicines. In this respect, one of the most impressive 21st century developments is the decoding of the human genome. This is now coupled with a gigantic leap in our ability to carry out computational work in real time with the help of supercomputers and cloud-based computing. For example, a distributed computing project achieved a speed of 2.43 exaflops (1 exaflop is 10^{18} floating points) during April 2020 for helping to understand COVID-19 related drug targets (data from Folding@home). Such computational power was out of scientist's reach just a few years back.

With a molecular-level understanding of many human diseases, the development of drugs that specifically target and perturb the disease protein is on the rise. However, the process of discovering a drug, even now, relies more on trial and error experimental testing, resulting in long development cycles and expenditure in the range of billions of US dollar. Saving both time and money for discovering drugs could be achieved by incorporating computational approaches in a few of the drug discovery stages. Computer-aided drug design (CADD) is a term applied to a group of techniques associated primarily with the early stages of drug discovery for lead identification and lead optimization; they can speed up the process of identifying molecules for testing in animal models and moving them to clinical trials. Nowadays, CADD techniques are integrated into the iterative process of design, build (synthesize), and experimental testing of the molecules. The most widely used CADD technique is docking, which aims to predict the interaction between two molecules (e.g., a drug and a protein target) in 3D. The predicted interaction between a molecule/drug and a target protein could aid in the identification and development of newer molecules with better interaction to the target in shorter periods. This book brings in experts' from all over the world to discuss their point of view and recent findings in the fundamentals, resources, and application of docking, with a focus on the discovery of new drugs.

I invite students as well as the research community to read and benefit from the book and apply the knowledge to develop better drugs.

January 2021
Pondicherry, India
Mohane S. Coumar

Acknowledgments

I take this opportunity to thank all the authors of the book chapters for their wonderful contribution and support in reviewing the chapters. Also, I thank the Elsevier editorial team members Ms. Mary, Kathy, Tracy, Kiruthika, Kavitha, Sajana, and others for patiently working, guiding, and encouraging me for the past 1 year for this project. Last but not the least, I thank my wife Vasundhara Devi and my daughters Iniya and Lishitha for their understanding and support during my long hours spent with the computer, instead of them.

January 2021
Pondicherry, India
Mohane S. Coumar

Contents

xv

CHAPTER 1

Modern Tools and Techniques in Computer-Aided Drug Design

TAMANNA ANWAR • PAWAN KUMAR • ASAD U. KHAN

1 OVERVIEW OF COMPUTER-AIDED DRUG DESIGN

The approaches applied in drug development in the present time are very expensive and slow irrespective of the tremendous technological advancements in drug discovery approaches. In such situation of rising pressure of reducing time and cost for safe and effective drug discovery, the focus has moved toward the initial phases of drug discovery and development. Computer-aided drug design (CADD) approaches are now immensely used in the discovery of drug more efficiently and accurately. The cost of discovery and development of drugs can be reduced by 50% with the use of CADD (Xiang et al., 2012).

For more than three decades, CADD approaches have been applied in various stages of drug discovery (Fig. 1.1). Several of the marketed drugs discovered till date have been developed with the help of CADD techniques (Table 1.1). Furthermore, CADD also helps in predicting the novel therapeutic uses of the FDA (Food and Drug Administration) approved drugs; this strategy is termed as "drug repurposing" and will be discussed later in the chapter.

The aim of using CADD approaches is to predict a promising compound that brings a desired effect after binding to the particular biological target. Conventionally, high-throughput screening is used for testing large number of compounds on automated assays to achieve the required effects. In this case, the drug development procedure is not only time-consuming but requires extensive investment. Therefore, to reduce this burden, CADD approaches are applied so that the chemical compounds can be virtually screened first, which will significantly reduce the number of compounds going for experimental screening (Yu & Mackerell, 2017). With the advancement in the information technology (IT), computational power, and availability of big data, recently new approaches have been applied in CADD, which includes machine learning (ML), deep learning (DL), artificial intelligence techniques, and data mining to further enhance the speed and accuracy of drug discovery. In future, drug discovery strategies will very much rely on these advanced IT techniques, which will help in the selection of features (drug and receptor features), image processing, clustering of compounds, etc. For example, to see the drug's impact on patients, ML approaches are used which benefits in the development of drugs that are safe and effective and take less time in the development than the conventional methods. The importance of ML in CADD is well recognized and there are several reports on its successful applications (Khamis & Gomaa, 2015; Vamathevan et al., 2019). In the ML-based approach, large data sets are trained with the help of mathematical framework, which is then applied for the prediction or classification of a new data set (Deo, 2015).

Advancement in the different aspects of computational approaches aid in CADD such as ML approaches help in modeling complex systems that will provide insight into the designing and essential knowledge of molecules. However, DL approaches help in quickly selecting compounds based on pattern recognition, as well as it can be used for early detection of disease and management of the disease. Traditional CADD approaches can be broadly divided into two groups depending upon the availability of the target protein structure: (1) structure-based drug design (SBDD) and (2) ligand-based drug design (LBDD). Availability of the target protein structure provides additional edge in the direct hit to lead optimization process. SBDD includes approaches such as molecular docking, virtual screening (VS), structure-based pharmacophore modeling, and de novo drug design, whereas LBDD approaches include similarity-based screening, quantitative structure–activity relationship (QSAR) modeling, ligand-based pharmacophore modeling, and scaffold hopping (Fig. 1.2).

Molecular Docking for Computer-Aided Drug Design. https://doi.org/10.1016/B978-0-12-822312-3.00011-4

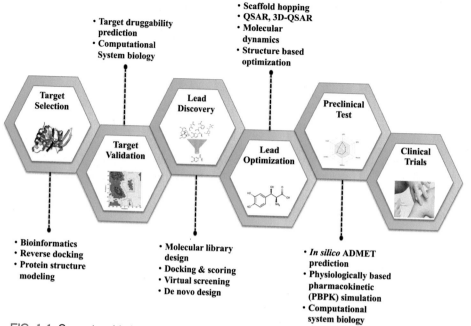

FIG. 1.1 Computer-aided drug design approaches applied in various stages of drug discovery.

TABLE 1.1
List of Drugs Developed with Computer-Aided Drug Design (CADD) Approaches.

Drug	Indication	CADD Approach	Status	References
Saquinavir	Inhibitor of HIV proteases	Structure-based drug design	Approved 1995	Drie (2007)
Nelfinavir	Inhibitor of HIV	Structure-based drug design	Approved 1997	Fischer & Robin Ganellin (2006)
Norfloxacin	Bacterial DNA gyrase Inhibitor	Quantitative structure–activity relationship	Approved 1998	Roy (2015)
Zanamivir	Antiviral (influenza A and B)	Modeling de novo design	Approved 1999	Clark (2006)
Amprenavir	HIV	Protein modeling and molecular dynamics	Approved 1999	Wlodawer & Vondrasek (1998)
Zolmitriptan	Migraine	Pharmacophore modeling	Approved 2003	Clark (2006), Glen et al. (1995)
Dorzolamide	Glaucoma and ocular hypertension	Fragment-based screening	Approved 2012	Grover et al. (2006)

2 CHEMICAL LIBRARIES

Traditionally, for finding a hit against any target in drug discovery, the structure of compounds that can act as inhibitor or activator is required for docking/VS. A high-throughput virtual screening (HTVS) method utilizes chemical databases having millions of compounds to shortlist potential compounds for synthesis. A large number of databases offer structures of chemical compounds, biological targets, and data pertaining to bioactivity for drug discovery. These databases are an

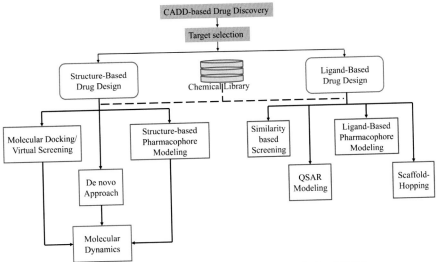

FIG. 1.2 Classification of Computer aided drug design (CADD).

exclusive source for identifying new chemical structures against biological targets. Apart from being a conventional diverse database of chemical structures, considerable attention is given on annotating chemical libraries with a view to provide information on the correlation among the chemical compound and its biological function. Several public and commercial repositories of chemical compounds essential for CADD are highlighted. The information of the drug-like compounds and their physiochemical properties can be retrieved from various databases that are available freely, e.g., PubChem, ZINC, ChEMBL, DrugBank, etc. (Table 1.2). Many resources are also available commercially such as Jubilant BioSys, GVK Bio, and Aureus Pharma. These are large databases of target-centric compounds, focusing mainly on kinases, G protein—coupled receptors, nuclear hormone receptors, or ion channels. The major source of chemical data in these databases comes from patents.

3 STRUCTURE-BASED APPROACHES AND SCREENING

SBDD method utilizes the knowledge of 3D structure of the receptor or target for VS and lead optimization. Thus, for receptors/targets having their crystal structure or modeled structure available, this method can be applied. Types of SBDD methods include molecular docking, structure-based 3D pharmacophore modeling, and de novo drug design methods. It is imperative to check whether the selected target is "druggable," i.e., its biological behavior can be altered by binding small molecule. A

target with a very deep, large, and/or highly charged binding pocket is considered unsuitable for SBDD (Fauman et al., 2011). Generally, a structure with high resolution (1.5 Å) and a large ligand binding in its active site is preferred (Rueda, Bottegoni, & Abagyan, 2010).

3.1 Target Structure and Validation

The most extensively used resources of 3D structure determined either by X-ray crystallographic method or nuclear magnetic resonance (NMR) is the Protein Data Bank (PDB) database available at http://www.rcsb.org/pdb. The current version contains 162,529 structures, which is largely determined by X-ray crystallography (88.9%); the fraction of NMR spectroscopy and electron microscopy (EM) determined structures is very low (https://www.rcsb.org/stats/summary) (Berman et al., 2002). In cases where the protein structure is not determined experimentally, again computational approaches can be applied to model the protein structure by homology modeling. The homologous structure is modeled with the help of sequence similarity to the experimentally determined structure of a similar protein. One of the most frequently used software for homology modeling which is freely available is MODELLER (Andrej Šali, 1993). There are several other homology modeling tools/servers available freely for, e.g., Swiss Model, Phyre2, LOMETS, CPHmodels 3.2, I-TASSER, etc.

Among the available solved structures in PDB, X-ray—based crystal structures are still dominating over the other experimental approaches such as NMR and cryo-EM (Cooper et al., 2011). In the drug design

TABLE 1.2

General Resources for Retrieving Chemical Compounds for Docking and Virtual Screening.

Database	Description	License type
ChemSpider http://www.chemspider.com	It is a free database of chemical structures that provides fast text and structure-based searches across 81 million chemical compounds gathered from 278 data sources.	Free
eMolecules Plus https://www.emolecules.com	It contains more than 8 million chemical compounds obtained from the network of global chemical suppliers. The chemicals can be ordered from the website as suppliers are directly connected.	Commercial
ACD (BIOVIA Available Chemicals Directory) https://www.3ds.com/products-services/biovia/products/scientific-informatics/biovia-databases/	It is one of the largest structure-searchable collections of commercially available chemicals in the world, having 10 million unique chemical structures.	Commercial
iResearch Library https://www.chemnavigator.com/cnc/products/iRL.asp	It consists of over 160 million commercially available chemical structures.	Commercial
PubChem https://pubchem.ncbi.nlm.nih.gov/	It is a huge collection of chemical compounds that mostly includes small molecules but macromolecules are also included. PubChem Substance (253 million), PubChem Compound (103 million), and PubChem Bioactives (268 million) are the three components of the dynamically expanding PubChem database.	Free
ZINC https://zinc.docking.org/	It is a large database of 230 million purchasable compounds along with their physicochemical properties. The molecules are available in 3D formats that are ready to dock.	Free
ChEMBL https://www.ebi.ac.uk/chembl/	This database includes bioactive molecules that have properties of drug-like compounds as well as the data of their chemical, bioactivity, and genomic properties are also included. It consists of around 2 million compounds, 13377 drug targets, and 15996368 activities.	Free
BindingDB www.bindingdb.org/bind/index.jsp	It is a publicly available database of binding affinities of small drug-like molecules with their corresponding candidate drug targets. It includes 1,854,767 binding data, for 7493 protein targets and 820,433 small molecules.	Free
PDBeChem https://www.ebi.ac.uk/pdbe-srv/pdbechem/	A database of ligands, small molecules, and monomers referred in Protein Data Bank (PDB) entries. It is consisting of 30899 ligands data.	Free
SuperNatural II http://bioinf-applied.charite.de/supernatural_new/index.php	This database consists of naturally occurring products. It consists of 325,508 natural compounds.	Free
NPACT http://crdd.osdd.net/raghava/npact/	This is a database of 1574 phytochemicals with anticancerous activity.	Free

TABLE 1.2
General Resources for Retrieving Chemical Compounds for Docking and Virtual Screening.—cont'd

Database	Description	License type
DrugBank https://www.drugbank.ca/	The latest version 5.1.5 contains 13548 chemical compounds including 2628 FDA-approved molecules, 1372 approved biologics, 131 nutraceuticals, and over 6363 experimental drugs.	Free
SuperDRUG2 http://cheminfo.charite.de/superdrug2/index.html	This is a database of marketed drugs that consists of 4600 active pharmaceutical ingredients.	Free
GDB-17 http://gdb.unibe.ch	This database consists of 166.4 billion molecules, which are up to 17 atoms of C, N, O, S, and halogens.	Free
KEGG Drug Database https://www.genome.jp/kegg/drug/	It is a compressive database of drugs approved in Japan, the United States, and Europe. It consists of 11,274 drug entries.	Free
SPECS https://www.specs.net/	It contains more than 3,50,000 compounds suitable for synthesis.	Commercial
Maybridge https://www.maybridge.com	It consists of over 53,000 hit-like and lead-like organic compounds.	Free

pipeline, crystallography has gained more importance as this technique is at the heart of SBDD and fragment-based drug design approaches (Cooper et al., 2011). As per the study published by Westbrook et al., 210 new molecular entries (NMEs) are approved by the FDA between 2010 and 2016, and for these NMEs, around 94% of molecular targets are available in the PDB database (Westbrook & Burley, 2019). Very recently, the wwPDB OneDep system has been set up as a single channel for deposition, validation, and biocuration of all incoming structures (Young et al., 2017). OneDep will ensure consistency in the process at the data deposition as well as internal biocuration level.

As the starting structure influences the outcomes in drug designing process, several quality checks are now introduced apart from the structural resolution and R-factor to assess the quality of the experimental structure (Table 1.3). To maintain the data accuracy of the PDB structure, several measures have been taken such as no theoretical structure is now considered from 2006 onwards, structure factor amplitudes/intensities for crystal structures are required with each structural deposition, and each submitted structure should be published in the journal (Kirchmair et al., 2008). At the structure level, the validation matrix is provided to show the accuracy at the structural, geometric, and electron density (ED) level (Fig. 1.3).

ED maps are now provided for all deposited structures and can be used by both experts and novice to assess more about the quality and characteristic of the protein under consideration. Understanding of the user from the ED maps also ruled out the possible biases incorporated by the used modeling procedure, crystallographer expertise, and familiarity. Though ED maps have given the flexibility to the user to analyze the experimental structure carefully, however, the correct representation of the small ligand molecules at the binding site is still a matter of concern. Interpretation of the ligand position binding partly or full, with or without water from the available ED maps, is a laborious task (Smart et al., 2018). Low-resolution structures especially below 3 Å tend to be trickier where water-based interactions play a crucial role between ligand and protein. To emphasize the critical challenges associated with protein–ligand complex crystallography, Smart et al. (2018) have analyzed the PDB ligand and assess the validation report in detail and examined the geometric and ED fit for the same (Smart et al., 2018).

3.2 Molecular Docking and Virtual Screening

One of the most extensively used computational tools in CADD is molecular docking, which is used for determining the complex structure produced by two or more

TABLE 1.3
Tools/Web Server Generally Used for the Protein Structure Validation and Quality Assessment.

Program	Description	Stand-alone/Web server
PQS	Analyze the quaternary protein structures deposited in the Protein Data Bank	Web server
WHAT IF	Tool for protein structure quality checks	Both
Prosa-web	Assess the quality score with respect to known protein structures.	Web server
PROCHECK	Tool to check the stereochemical quality of the protein structure	Stand-alone
PROCHECK—nuclear magnetic resonance (NMR)	Tool to check the stereochemical quality of the NMR protein structure	Stand-alone
MolProbity	Validate the protein structure at different levels	Web server
NQ Flipper	Erroneous Asn and Gln rotamer detection	Web server
PSVS	Protein structure assessment suite	Web server

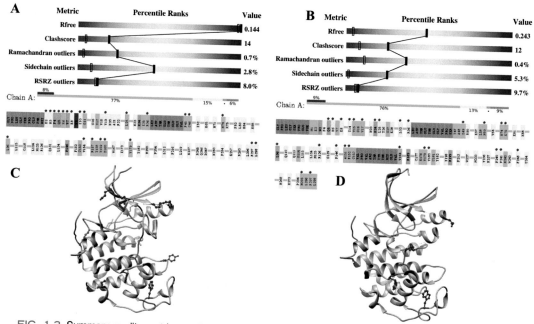

FIG. 1.3 Summary quality metrics available in the wwPDB validation reports. PDB-ID 6GUK (**A** & **C**), and 6Q3C (**B** & **D**) Residues showing the deviation from the experimantal Electron Density Map are shown in red colour (**C** & **D**).

interacting molecules. The docking process involves predicting the 3D conformation of the hit or ligand inside the binding cavity of the target. Several possible ligand poses are generated through molecular docking which are then ranked on the basis of scoring function (SF). The process of simulating the ligand and the receptor to form a stable complex can be considered as a "lock-and-key model," where the position of key (ligand) is optimized to accommodate into the lock (target binding pocket). The three vital components of molecular docking include the "receptor," the "ligand," and the docking program. The prediction of binding interaction among the protein target and the ligand, the orientation of the ligand in the target's binding pocket, and the scoring of the interaction are achieved by docking programs. The conformational search algorithm explores the poses inside a particular conformational space, while the role of SF is to score each pose that shows its relative binding affinity (Meng et al., 2012). Considerably, the docking program will generate a group of poses for each ligand such that every pose has its own docking score. Generally, the pose that is ranked at the top is considered the best pose of docking; however, the selection of the final pose should not only depend upon the docking score but also on the chemical knowledge and experimental data, if available. The docking program generates the poses by treating the ligand molecule as flexible, and the conformational search algorithm is used for sampling the ligand's torsional degrees of freedom and keeping the target rigid. The accuracy of docking relies on the conformational sampling coverage as well as the SF. Structure-based virtual screening (SBVS) can be done to identify the potential activities available in a large chemical compound database by carrying out docking (Clark, 2008; Schneider, 2010).

3.2.1 Sampling algorithm

The mode of ligand and target binding is possible in several ways as they have six degrees of translational and rotational freedom in addition to the freedom of conformational degrees. Generating all the possible conformations computationally would be highly expensive. Thus, several sampling algorithms were proposed and extensively applied in molecular docking tools (Table 1.4). The ligand is mapped into the active

TABLE 1.4
Docking Programs Used in Computer-Aided Drug Design (CADD) and Their Features.

Docking Program	Characteristic	Sampling Algorithm	Scoring Function	License	References
AutoDock	It is an automated tool for docking consisting of an autogrid, which is used to compute grid, and an autodock, which is used for docking ligands on the grid created by autogrid.	Genetic algorithms, Monte Carlo	Force field based	Open source	Forli et al. (2016)
DOCK	The latest release is built with an improved algorithm to predict binding poses by adding new features like force field scoring enhanced by solvation and receptor flexibility.	Incremental construction, Energy minimization	Force field based	Academic	Ewing et al. (2001)
FRED	An exhaustive search (ES) algorithm is used to identify the ligand's best binding pose in the receptor binding site.	Exhaustive search	Knowledge based	Academic	McGann (2011)
FlexX	It is a tool provided by BioSolveIT for flexible ligand docking. It is fully automated and docking is performed with an incremental construction algorithm.	Incremental construction	Empirical	Commercial	Kramer et al. (1999)

Continued

TABLE 1.4
Docking Programs Used in Computer-Aided Drug Design (CADD) and Their Features.—cont'd

Docking Program	Characteristic	Sampling Algorithm	Scoring Function	License	References
Glide	Glide is a molecular docking suite of software provided by Schrödinger. It offers several modes for virtual screening such as high-throughput virtual screening, standard precision, and extra precision.	Exhaustive search, energy minimization, Monte Carlo	Empirical	Commercial	Friesner et al. (2004)
GOLD	It applies a genetic algorithm for predicting poses of the ligand. It can be configured.	Genetic algorithms	Empirical, knowledge based	Commercial	Verdonk et al. (2003)
ICM	This is an easy-to-use software provided by Molsoft, LLC. The software can be used for chemical clustering, chemical similarity searching, molecular modeling, virtual screening of ligands, fully flexible docking, etc.	Monte Carlo	Empirical	Commercial	Neves et al. (2012)
Surflex-Dock	In Surflex-Dock, the active site ligand is used to produce putative poses, and a combination of similarity searches methods is applied to predict the probable pose of ligand in the binding site.	Incremental construction	Empirical	Commercial	Jain (2007)

site of the target with the help of matching algorithms, on the basis of its shape features and chemical properties. The benefit of matching algorithms is its speed; therefore, active compounds enrichment from vast libraries can be done using this method (Moitessier et al., 2008). This algorithm was used in the older versions of DOCK (Kuntz et al., 1982). Incremental construction algorithm utilizes fragmental and incremental method to place the ligand in the active site. The ligand is fragmented along the rotatable bonds, and then at first the largest fragment is docked inside the binding pocket leading to the addition of rest of the fragments incrementally (Rarey et al., 1996). Other fragment-based algorithms include multiple copy simultaneous search (Eisen et al., 1994) and LUDI (Böhm, 1992a). Programs that implement fragment-based methods comprise DOCK 4.0 (Ewing et al., 2001), FlexX (Rarey et al., 1996), and Surflex (Jain, 2003).

Exhaustive search (ES) is a type of systematic search algorithm, which is used for flexible ligand docking. To perform ES, the ligand's rotatable bonds are systematically rotated at a certain interval, which results in a huge number of ligand conformations. Thus, for initial screening, geometric/chemical constraints are applied after which more accurate refinement procedures are used. FRED (McGann et al., 2003) and Glide (Friesner et al., 2004) are examples of programs that use ES algorithm. Monte Carlo (MC) and genetic algorithms (GA) belong to the class of stochastic methods. In this class, the conformational space is searched by randomly changing the conformation of the ligand. Both of these algorithms produce a series of random modifications to a ligand or an ensemble of ligands, which is further evaluated on the basis of probability or fitness function. Due to the randomness of conformational sampling, docking is run several times to confirm that the convergence is reached. The programs that apply the MC methods include an earlier version of AutoDock (Goodsell & Olson, 1990), ICM (Abagyan et al., 1994), and Glide (Friesner et al., 2004). GA have been applied in programs such as AutoDock (Morris et al., 1998) and GOLD (Verdonk et al., 2003). Molecular dynamics (MD) (Cornell et al., 1995; Weiner et al., 1984) and energy minimization Mare powerful simulation methods used extensively in MD. These methods are computationally expensive; thus, these methods are applied for refining or rescoring ligand poses produced by other methods. The simulation method is used by the programs DOCK (Kuntz et al., 1982) and Glide (Friesner et al., 2004).

3.2.2 Scoring function

The SF is applied to evaluate the docking poses generated by docking programs to quantitatively measure the quality of the fit (Rajamani & Good, 2007). Along with the evaluation of ligand poses, the SF also evaluates the ligand binding energy and ranks them accordingly to select the best binding ligand. The two main components of any SF are its speed and accuracy. There are three classical categories of SF, i.e., force field (FF)-, empirical-, and knowledge-based SFs. The SF based on FF is calculated on physical atomic interactions like van der Waals (VDW) and electrostatic interactions as well as on bond lengths, bond angles, and dihedrals (Aqvist et al., 2002; Kollman, 1993). The disadvantage with the FF-based SF is its computational speed, which is very slow. Extensions of FF-based SFs include the hydrogen bonds, solvations, and entropy contributions. Further refinement of the result of FF-based docking can be done by applying techniques like linear interaction energy and free energy perturbation (FEP) methods. Empirical SFs are applied to measure the binding free energy (FE) by utilizing various aspects of a protein–ligand complex, for example, hydrogen bond, VDW energy, ionic interaction, hydrophobic effect, binding entropy, etc. (Guedes et al., 2018). Knowledge-based SFs use the experimentally determined structures to get the information of frequencies as well as distance of interatomic contacts

in the ligand-protein complex. To improve the accuracy of docking prediction, two or more SFs are applied in some programs, which is referred to as "Consensus Scoring" (Huang et al., 2010). The docking programs applying different SFs are cited in Table 1.4.

Recently, ML-based SFs trained on the complex structures of protein and ligand have gained much attention. This model does not work on predetermined functional forms but is rather developed by supervised learning algorithms (Li et al., 2020). By using the SFs based on ML, the intermolecular binding interactions can be captured implicitly that are difficult to model explicitly. ML-based applications have speedup the inhibitor designing process with desired pharmacodynamics and pharmacokinetic properties compared with the rational *in silico* approaches (Mak & Pichika, 2019). Due to the enormous possibility from the available chemical, genomic, and structural data, its applications are now ranging from the VS-based inhibitor identification, target protein prediction (Kaushik et al., 2020; Zheng et al., 2020), improved consensus docking score development (Ericksen et al., 2017), protein structure prediction (Torrisi et al., 2020), protein–protein interaction prediction (Du et al., 2017), de novo molecule design (Kadurin et al., 2017; Olivecrona et al., 2017), and many more.

ML-based SF used for the prediction of binding affinity performed better than several classical SFs (Ain et al., 2015; Ballester et al., 2014; Khamis et al., 2015). In a very recent study, Su et al. in 2020 have related the performance of six different ML-based SF models to nullify the assumption of overlapping training and test set. The study reports that the performance of the ML models is mostly dependent on the size of the training set used as well as on the content of the training set (Su et al., 2020). However, the docking software does not implement ML-based SFs directly, rather these are generally used for rescoring as these SFs are dependent on training data sets (Zhang, Ai, et al., 2017). The ML-based SFs help in improving the precision of docking done by classical methods by rescoring.

3.2.2.1 Support vector machine.
The application of support vector machine (SVM) in SBVS is often done to separate active and inactive ligand poses, and regression model of SVM is applied to predict the binding affinities (Zhang, Ai et al., 2017). A study was done where SVM was combined with the empirical function on the basis of energy terms; as a result, there was an increase in the accuracy of prediction in VS, as well as a correlation among SVM-based and experimental binding affinities was reported (Brylinski, 2013; Kinnings et al., 2011).

Analysis of the HIV protease by ML-based SF SVM-SP performed better than Glide, ChemScore, GoldScore, and X-Score (Li et al., 2011). In another study on 40 DUD2 targets, MIEC-SVM proved to be better than Glide and X-Score (Ding et al., 2013).

3.2.2.2 Random forest.
In this classification algorithm, learning is based on multiple decision trees, which is used for classification, regression, etc. The randomness of features is used while building each tree to produce uncorrelated forest with multiple trees, the prediction accuracy of the ensemble of trees is much more than any of the individual trees. Random forests (RFs) have been shown to increase the accuracy of conventional SF by replacing multiple linear regression (Afifi & Al-Sadek, 2018; Wang & Zhang, 2017). In a recent study, RF-based score was developed and compared with five classical SFs. ML-based SF has achieved a very high hit rate at 1% level (55.6%) compared to Vina, which only showed the 16.2% hit rate. Compared to Vina-based predicted activity correlation (Pearson correlation −0.18), RF score has gained Pearson correlation of 0.56 (Wójcikowski et al., 2017).

3.2.2.3 Artificial neural network.
Recently, artificial neural network (ANN) has been used extensively in CADD. It is a computational model inspired by biological neural networks. ANN is generally used for QSAR modeling (Cang et al., 2018), but often it is also used to predict binding affinities. An ANN-based SF "NNScore 2.0" predicts binding affinity, as the latest version considers more of binding properties (Durrant & McCammon, 2011). Moreover, NNScore rescoring function can be applied to increase the performance of scoring (Durrant et al., 2013). The prediction accuracy of the classical ANN-based SF can be greatly increased by incorporating techniques such as boosting or bagging (Ashtawy & Mahapatra, 2018). Despite the high precision in the prediction of binding affinity, the ANN-based SFs are incapable of working fine with high dimension data, limiting their application in commercial docking tools.

3.2.2.4 Deep learning.
The DL-based SF can extract features from unsupervised data, which is unstructured or unlabeled along with model fitting. The most common model of DL-based SF is convolutional neural network (Ragoza et al., 2017; Wallach et al., 2015), which can be applied for classification of drug binding and prediction of binding affinity (Gomes, Ramsundar, et al., 2017;

Stepniewska-Dziubinska et al., 2018). It has been revealed that convolutional neural network models perform better when compared with classical ML models (Bengio et al., 2013), but it is more time-consuming due to the increase in the network complexity of model.

Given a set of training data consisting of an active and inactive compounds, the data can be trained by applying ML-based SFs such as RF-Score (Ballester & Mitchell, 2010), NNScore (Durrant & McCammon, 2011) and SFCscore (Sotriffer et al., 2008; Zilian & Sotriffer, 2013) to find out the known ligands by potency with high accuracy (Wójcikowski et al., 2017). As mentioned earlier, the SF's accuracy can be further improved by applying a hybrid SF that is an integration of different SFs. However, the hybrid SFs are more efficient but more time taking.

3.3 De Novo Drug Design
De novo drug design approach is another most promising SBDD method which allows the generation of the chemical compounds from scratch in the receptor binding site with desirable drug-like properties (Mauser & Guba, 2008; Schneider & Fechner, 2005). Though this approach of novel molecular design is nearly two decades old, its contribution in the drug discovery projects is recently increasing due to its sound applicability and availability of the de novo designing computational program (Schneider & Fechner, 2005). This approach of the drug design process attempts to explore the virtually infinite chemical search space and only captures the building blocks, which is necessary for filling the available interaction space in the substrate binding site (Schneider et al., 2009). So, in the de novo approach, virtual compound generation protocol attempts to imitate the medicinal/synthetic chemist way of designing the virtual compound, while applied SF preform as a virtual assay (Lameijer et al., 2007).

To facilitate the de novo drug designing process, many different tools are published to adapt the multi-objective optimization process (Devi et al., 2015; Nicolaou et al., 2012) and so this approach comes up with many solutions depending upon the initial parameters chosen. Ludi (Böhm, 1992b), LEGEND (Honma et al., 2001), LigBuilder (Wang et al., 2000), BIBuilder (Teodoro & Muegge, 2011), and LiGen (Beccari et al., 2013) are some programs which are developed to assist the de novo drug designing process. As this approach uses all possible combinations to link the available blocks in the respective protein substrate binding site, different sets of rules are formulated to reduce the generated chemical space to a very feasible number of compounds. Following rules can be implemented to select the de novo chemical hit compound.

(1) Compound should be synthetically accessible
(2) Compound should follow the drug-like/lead-like properties
(3) Generated compounds should be diverse in scaffold

4 LIGAND-BASED APPROACHES AND SCREENING

Contrary to SBDD, LBDD does not require the target 3D structure information, rather the minimum information critical for LBDD method is the knowledge about at least one active compound, which is then utilized for ligand-based virtual screening (LBVS) to pull out similar compounds from databases. This method collects information from the set of reference compounds that are reported in different studies to interact with the target of interest or possess the desired activity. The compounds are represented such that the physiochemical properties relevant to the preferred interaction are retained, while other irrelevant information is excluded. LBDD method for drug discovery is based on "similar property principle" according to which compounds having structural similarity (structure, pharmacophoric features, molecular fields, etc.) will have similar properties. The fundamental approaches for LBDD to identify known actives are either based on chemical similarity or building a model to predict biological activity from chemical structures. LBDD techniques include ligand-based pharmacophore, fingerprint-based similarity methods, and QSAR. The techniques used in LBVS such as substructure mining and fingerprint searches are faster in comparison to SBVS methods like molecular docking. The LBVS technique has helped in finding several promising compounds on the basis of properties such as physiochemical or thermodynamic properties (Forli, 2015). However, the SBVS approach of VS is considered better than LBVS when the target's 3D structure is available (Lyne, 2002). In some cases, where both the target and ligand are known, a hybrid method is used that combines both SBVS and LBVS for achieving better results.

LBVS methods represent compounds with a set of features/descriptors; these descriptors could be either structural or physiochemical and generated with tools based on mechanisms like knowledge-based, molecular mechanics, or quantum mechanics. The molecular descriptors are classified as 1D, 2D, 3D, 4D, etc., according to the chemical structure's dimensionality it

is computed from. Several tools are available for computing molecular descriptors which will be discussed later in the chapter. Molecular fingerprint and similarity searches, pharmacophore modeling, and QSAR are the popular approaches of LBDD (Acharya et al., 2010).

4.1 Molecular Fingerprint and Similarity Searches

In this technique, compound libraries are screened based on the molecular fingerprint taken from the known ligands of a particular target to search compounds with similar fingerprint (Vogt & Bajorath, 2011). The theory behind this approach is that the molecules having chemical or physicochemical similarity ought to possess similarity in binding properties (Gomes, Muratov, et al., 2017; Yu & Mackerell, 2017). This approach does not consider the biological activity of the known ligands. Similarity searches are simple but effective and computationally less expensive than pharmacophore modeling and QSAR. In VS, similarity search method is advantageous when only few distinct ligands are known to inhibit a particular target and other methods as pharmacophore screening or structure-based design cannot be applied. The most widely used tool for similarity searching is molecular fingerprint, in which the molecular structure and properties are represented as bit strings. The bit string helps in the identification of presence or absence of molecular features (Xue et al., 2003), which is represented in a quantifiable manner. Every bit in the bit string denotes one molecular substructure/fragment or feature. The bit is fixed to 1 if the fragment is present and 0 if the fragment is absent (Fig. 1.4). The fingerprint-based methods include substructure key—based fingerprints, topological or hashed fingerprints, and circular fingerprints (Cereto-Massagué et al., 2015). The basic difference in these approaches is in the method of translating structural information into the bit string. Each bit represents a certain descriptor or value in substructure key—based fingerprints (Fig. 1.4a) (James et al., 2011). In topological fingerprints, analysis of all the fragments of a molecule is done. Generally, a path is created up to a predefined number of bonds and next all the paths are hashed to build fingerprints. It is likely that the same bit is set by multiple fragments in this method (Fig. 1.4b). The circular fingerprints are also hashed, but here in place of considering paths in the molecule, each atoms environment is documented up to a defined radius. This method is widely applied in VS on the basis of full structure similarity (Fig. 1.4c) (Cereto-Massagué et al., 2015).

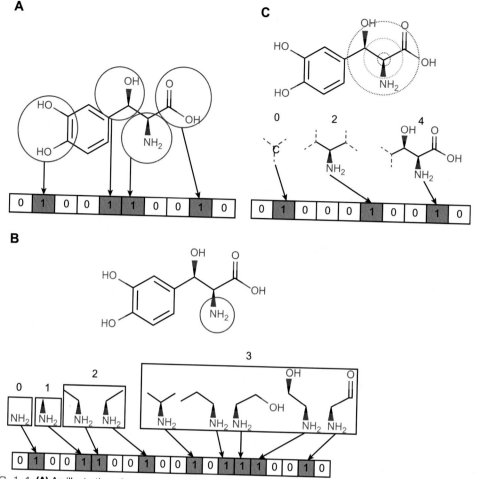

FIG. 1.4 **(A)** An illustration of a substructure key–based fingerprint; molecular substructures represented by bits that are present in the molecule (encircled) are set to 1 and those absent are set to 0. **(B)** Representation of a topological fingerprint. All atoms starting from the amino group of the molecule are shown; the fragment length and subsequent bit in the fingerprint are denoted. Different linear pathway fragments are generated based on the preset number of bonds that are translated into bit strings. **(C)** Representation of a circular fingerprint in which fragment generation starts from a central atom and considers the fragments within a preset radius (e.g., two or four bonds); these fragments are then transformed into bit strings.

Apart from the substructure fingerprint, properties of molecules can also be defined as fingerprint; these property-based fingerprints include functional class fingerprints, pharmacophore fingerprints, reaction fingerprints, etc. The pharmacophore models can also be used as a type of molecular fingerprint. The fragments of the molecule can be transformed into pharmacophoric features; the existence or nonexistence of these features aids in fingerprint creation. However, 3D pharmacophore models are frequently applied to detect chemical functionalities necessary for biological activity as well as for searching large databases of 3D compounds (Cereto-Massagué et al., 2015).

The bit string once created using any of the individual approaches described that the similarity within two molecules is quantified. The molecular similarity can be accessed in different ways; several similarities and distance-based metrics used with fingerprints are mentioned in Table 1.4. Generally, euclidean distance is used for this purpose, but as per the industry standards for molecular fingerprint, Tanimoto coefficient is usually used (Bajusz et al., 2015), which can be evaluated by the formula given in Table 1.5. Tanimoto coefficient lies between the range of 0 and 1; however, sometimes it is also represented in percent. A value ≥0.85 of the Tanimoto coefficient represents two compounds that are reasonably similar (Martin et al., 2002).

It has been observed that the longer bit strings perform better in similarity searching as they have a greater amount of stored information (Sastry et al., 2010). Fingerprint similarity search has been

TABLE 1.5
List of Similarity Coefficients and Distances Used for Fingerprint Search.

Similarity/Distance coefficient	Expression	Range
Tanimoto/Jaccard coefficient	$N_c/N_a + N_b - N_c$	0–1
Dice coefficient	$2N_c/N_a + N_b$	0–1
Cosine similarity	$N_c/\sqrt{(N_aN_b)}$	0–1
Euclidean distance	$\sqrt{(N_a + N_b - 2N_c)}$	0–N
Hamming distance	$N_a + N_b - 2N_c$	0–N
Russell–RAO coefficient	N_c/m	0–1
Forbes coefficient	N_cm/N_aN_b	0–1
Soergel distance	$N_a + N_b - 2N_c/ N_a + N_b - N_c$	0–1

Note: For the fingerprint of two compounds a and b, N_a represents the total number of bits set to 1 in compound a, N_b is the total number of bits set to 1 in compound b, N_c is the number of bits set to 1 in both a and b, and m represents the total number of bits present in the fingerprint.

implemented in various chemical databases for searching similar compounds within a range of defined Tanimoto coefficient, for example, PubChem (Wang, Bryant, et al., 2017), ChEMBL (Bento et al., 2014), ZINC (Irwin & Shoichet, 2005), ChemSpider (Pence & Williams, 2010; Royal Society of Chemistry, 2015), etc. The fingerprint method can be used to study the databases for compound diversity by grouping similar compounds. The software and web servers used for fingerprint-based VS are listed in Table 1.6.

The latest approach in fingerprint-based similarity searching is to use a combination of different VS methods (either fingerprint-based or other VS methods), specifically combining molecular fingerprint similarity method with SBVS (Ahmed et al., 2014; Broccatelli & Brown, 2014; Willett, 2013). As a result of applying a combination of approaches, the compounds performing best will be those that are ranked highest by different methods, leading to an increase in the performance of the VS. Fingerprint-based methods are very extensively used for activity predictions because of their speed, particularly in the area of target fishing, where the query compound is compared with millions of compounds having known activities.

4.2 Pharmacophore Modeling

Most of the biological structures such as proteins or DNA respond to the binding of small chemical molecules, and this response modulates the biological outcomes. How compounds interact with respective protein receptors depends upon the combination of interaction patterns available between protein and ligand molecules. Chemical interactions or chemical features such as hydrophobic, hydrogen bond acceptor, hydrogen bond donor, and ring are the major driving forces in defining the protein—ligand interactions. In the computational drug discovery pipeline, encoding the chemical features in high degree of abstraction is known as 3D pharmacophore features. The term "3D pharmacophore" came into the picture at the starting of the 19th century; however, the concept gradually progressed through many stages, and around the late 80s and early 90s, VS experiments were performed with the help of computational programs (Table 1.7). With time, the pharmacophore concept has evolved from ligand-based approach and receptor-ligand based approach to ab initio receptor-based approach (Kumar et al., 2017; Yang, 2010). With the help of this approach, many successful applications of lead optimization and finding of active molecules have been achieved (Neves et al., 2009; Schuster et al., 2008). Apart from the drug discovery—based application, pharmacophore features are also now in use to design focused chemical library and for scaffold hopping (Shin & Seong, 2013). Apart from the ligand-based pharmacophore modeling, protein—ligand complex—based pharmacophore features are also found to be very valuable in finding the novel inhibitors (Salam et al., 2009; Yang et al., 2009). Apart from the ligand and protein—ligand interaction—based pharmacophore approach, many other pharmacophores perceiving approaches are reported in the literature, and some are detailed below.

4.2.1 Water pharmacophore approach

Water molecules occupied at the unliganded protein binding site are mostly engaged with directional forces or with hydrophobic forces, and over 85% of the protein—ligand complexes have been identified to have one or more bridging water interacting with both protein and ligand (Lu et al., 2007). Most of the time, water-mediated interactions are found to affect the thermodynamic signature of the binding affinity of the ligand (Duan et al., 2017; Spyrakis et al., 2017). Incoming ligand displaces the ordered water molecules from the receptor binding site and consequently disturbs the hydrogen bond network between water and protein. This displacement of the water to the bulk solvent affects the entropy-driven thermodynamic properties of the system (Dunitz, 1994). It thereby

TABLE 1.6
Software and Web Resources for Fingerprint-Based Virtual Screening.

Software/Web server	License Type	Web Address
Instant JChem	Free	https://chemaxon.com/products/instant-jchem
Open Babel	Free	http://openbabel.org
RDKit	Free	http://www.rdkit.org
Chemistry Development Kit	Free	http://sourceforge.net/projects/cdk/
Indigo Toolkit	Free	http://ggasoftware.com/opensource/indigo
ChemFP	Free	http://chemfp.com
DecoyFinder	Free	http://urvnutrigenomica-ctns.github.io/DecoyFinder/
FLAP	Free	http://www.moldiscovery.com/soft_flap.php
jCompoundMapper	Free	http://jcompoundmapper.sourceforge.net/
MayaChemTools	Free	http://www.mayachemtools.org/
OEChem TK	Commercial	https://www.eyesopen.com/oechem-tk
Canvas from Schrödinger	Commercial	http://www.schrodinger.com/Canvas/
Molecular Operating Environment (MOE)	Commercial	https://www.chemcomp.com/Products.htm
SYBYL-X	Commercial	http://www.tripos.com/
Pipeline Pilot	Commercial	http://accelrys.com/
PubChem	Free	http://pubchem.ncbi.nlm.nih.gov/
AURAmol	Free	https://www.cs.york.ac.uk/auramol/
ChemSpider	Free	http://www.chemspider.com/
ZINCPharmer	Free	http://zincpharmer.csb.pitt.edu/
ChemDes	Free	http://www.scbdd.com/chemdes
wwLigCSRre	Free	https://bioserv.rpbs.univ-paris-diderot.fr/services/wwLigCSRre/
SwissSimilarity	Free	http://www.swisssimilarity.ch/

renders the water molecule as a crucial mediator to decipher the protein—ligand binding (Cappel et al., 2017; Wong & Lightstone, 2011). Studies suggest that not all binding site water is displaceable, and so some unsuccessful attempts during lead optimization are also reported (Clarke et al., 2001; Kadirvelraj et al., 2008). Strongly bound conserved water molecules are found to be difficult to replace so weakly bound waters are considered as a good choice for lead optimization.

In the last decade, several computational approaches have been employed to estimate the thermodynamic components of the FE and to assess accurate binding FE (Bucher et al., 2018). These methods have been used to pinpoint the potential hotspots for novel inhibitor designing as well as existing inhibitor optimization. Apart from the thermodynamics-based approach, many crystal structures and simulation-based strategies were also established to identify and characterize the binding

TABLE 1.7
List of 3D Pharmacophore Generation Tools/Web Interfaces Commonly Employed for Pharmacophore Features Extraction From the Ligand, Apo protein Binding Site, or the Protein–Ligand Complex.

Program Name	Input Type	Scoring Method	References
Pharao	Ligand	Overlay	Taminau et al. (2008)
Pharmagist	Ligand	Overlay	Schneidman-Duhovny et al. (2008)
Pharmer	Ligand, complex	RMSD	Koes & Camacho (2011)
Phase	Ligand, apo structure, complex	RMSD	Dixon et al. (2006)
MOE	Ligand, apo structure, complex	RMSD	Inc. (2015)
Catalyst	Ligand, apo structure, complex	Overlay	BIOVIA Discovery Studio | Pharmacophore and Ligand-Based Design (n.d.)
LigandScout	Ligand, apo structure, complex	Overlay	Wolber & Langer (2005)

TABLE 1.8
Advanced Pharmacophore Perceiving Approaches Based on the Apo Protein Ensemble Structures.

Category	Approach Used	References
Molecular dynamics–based approach	Hydration site–restricted pharmacophore	Hu & Lill (2012)
	SLICS-pharmacophore with multiprobe molecules	Yu et al. (2015)
	MixMD—Mixed solvent simulations–based hotspot mapping	Lexa & Carlson (2011)
	Water pharmacophore	Jung et al. (2018)
Molecular interaction field grid–based approach	CliquePharm—A clique-based *Ab initio* multiclass pharmacophore generation program	Kaalia et al. (2016)
	WaterMap	Cappel et al. (2017)

hotspots for novel inhibitor designing (Table 1.8). Very recently, Jung et al. (2018) have demonstrated a water-based ab initio pharmacophore modeling from the receptor binding site only. In this approach, protein conformations were sampled using the MD, and water at the binding pocket was explored for hydration site analysis and pharmacophore features annotation. This approach is successfully implemented on the seven different proteins (Jung et al., 2018).

4.2.2 Dynamic pharmacophore approach

Proteins are the main apparatus for biological function, and flexibility is the key that determines its function (Teague, 2003). SBDD methods mainly neglect this factor (Lexa & Carlson, 2011). Many reports have shown that incorporation of protein dynamics improves the accuracy of the predicted hit molecules (Lexa & Carlson, 2011). Many ensemble-based

approaches have been demonstrated to identify the pharmacophore models by utilizing many crystal structures (Wieder et al., 2016), NMR structures (Hornak & Simmerling, 2007), or MD (Carlson et al., 2000). These approaches were successfully implemented against diverse proteins such as HIV integrase (Carlson et al., 2000), HIV protease (Hornak & Simmerling, 2007), and DHFR proteins (Lerner et al., 2007).

4.2.3 Ab initio specificity/selectivity pharmacophore approach

Interaction complementarity is the necessity that is mostly utilized by the pharmacophore features. Ligand- and protein–ligand–based pharmacophore modeling are mostly limited by the availability of the appropriate bioactive assays and diverse scaffold complexes, respectively. So, to circumvent this limitation, many new methods have emerged in the last decade utilizing

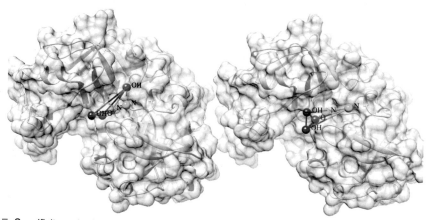

FIG. 1.5 Specificity and selectivity pharmacophore models for malarial protease (plasmepsin class of protein) generated by the CliquePharm approach. The left side figure shows the five-point specificity pharmacophore model carrying two hydroxyl probes (OH), two amides (N), and one carbonyl probe (O); while on the right side, five-point selectivity pharmacophore model is shown having the same size and feature types; however, this model is selective for malarial aspartic protease, not for human aspartic protease. So, the selective model will exclude the features that are common in both malarial and human protease and only design the model from the features that are only available to the malarial class of protease.

only receptor binding site information itself, and many new methods have been developed to encompass the possible combinations of available interactions to design and improve the inhibitor/lead molecules (Schaller et al., 2020). Molecular interaction field (MIF) based methods dominated in this field, and FLAP program (Baroni et al., 2007) from the molecular discovery was developed to elucidate the pharmacophore features from the apoprotein structure. Very recently, one more MIF-based method named CliquePharm is developed to design the specificity and selectivity ab initio pharmacophore models for the aspartic protease class of proteins (Kaalia et al., 2016). In this approach, a clique-based method is employed to identify the most frequent cliques across the selected protease followed by rule-based selectivity pharmacophore modeling having features selective to malarial aspartic protease over human protease (Fig. 1.5). The study has reported an ensemble of pharmacophore models of different sizes and combination types which can aid in the fragment to lead molecule generation (Kumar et al., 2017).

4.3 Quantitative Structure–Activity Relationship

The QSAR approach is based on the fact that the biological activity of a ligand depends on the arrangement of atoms in its molecular structure. Putting it differently, making changes in the structure of the molecule will result in the modification of its activity. QSAR is a widely used technique in LBDD.

The structural information is denoted as molecular descriptors, and the biological activity in QSAR is estimated in terms of the function of molecular descriptors (biological activity $= f$ (molecular descriptors)). A large amount of training data set is required in this method to extract descriptors or molecular features. The model developed based on the biological activities of the similar known ligands is used for predicting new compounds. Some of the techniques used in model generation include multiple linear regression, principal component regression, partial least squares (PLS) regression, ML, neural networks, etc. The main difference in the pharmacophore model and QSAR approaches is that the pharmacophore model only considers the features of the ligand while in QSAR ligand features as well as the features correlated with the biological activity among the ligand and the receptor is also considered. One of the advantages of QSAR method over pharmacophore model is that it can recognize whether a particular feature of a drug is influencing positively or negatively to its activity (Leelananda & Lindert, 2016). The QSAR method is generally classified either on the basis of dimensionality of the descriptors or on the basis of biological activity, which include chemical measurements and biological assays. The dimensionality of QSAR descriptors ranges from 0D to 6D, but generally 2D QSAR and 3D QSAR methods are used for model generation. A 2D QSAR method includes geometric and topological properties, molecular fingerprints, and polar surface area, but it excludes 3D orientation of the molecule. However, in 3D QSAR method, conformation of the molecule and alignment are used for the descriptor calculation (Dudek et al., 2006; Todeschini & Consonni, 2010). A 3D QSAR utilizes different methods including principal component analysis, ANNs, PLS method, cluster analysis, etc., and 3D GRID-based methods, such as comparative molecular field analysis, comparative molecular similarity indices analysis, hypothetical active site lattice, etc., are used for predicting the desired properties quantitatively (Arulsudar, Subramanian, & Murthy, 2005; Hemmateenejad et al., 2004; Oprea, 2003).

In drug discovery, QSAR approaches aid in pulling out hits from large molecular compound libraries. The strength of QSAR methods can be further improved by incorporating varying chemical and biological data using ANNs. This technique was found to be very effective in the development of antistreptococcal drugs (Speck-Planche et al., 2013). Several software packages are available for ANN analysis such as Matlab, Mathematica, Stuttgart Neural Network Simulator, etc. Molecular descriptors can be calculated using the tools like PaDEL-Descriptor, Dragon, etc. PaDEL-Descriptor is an open source software for the generation of descriptors and fingerprints developed using Chemistry Development Kit. It can generate 1875 descriptors including 1D, 2D, and 3D and 12 types of fingerprints (Yap, 2011). Dragon is capable of generating more than 4000 descriptors for a single molecule. It also has a web-based version which is available freely but only for limited number of compounds and features are also restricted (Tetko et al., 2005).

There are abundant of tools available for performing QSAR, apart from the available software packages, and workflow automation tools such as Taverna, Pipeline Pilot, Galaxy, KNIME, etc., are used to develop complete QSAR workflows. It is a convenient and more proficient method to manage large chemical data sets, automate lengthy process, and assist in data analysis. The automated QSAR modeling workflow of KNIME integrates all the tools required to perform various steps in QSAR analysis. The advantage of these workflows is that the QSAR models can be easily built by directly accessing the online or private chemical databases without having proficiency in ML or programming. Some of the widely used tools used for building and analysis of QSAR models are listed in Table 1.9.

The success of QSAR has been reported in a great number of researches. A QSAR model was built consisting of 3133 compounds that were either active or inactive against the malaria-causing parasite *Plasmodium falciparum*. The models were developed using descriptors 0D, 1D, 2D, ISIDA-2D fragments

descriptors, and SVM method. VS of these compounds resulted in shortlisting of 176 possible antimalarial compounds, while after validation 25 hits showed antimalarial activities having very less cytotoxicity to mammalian cells (Zhang et al., 2013). In another study, different ML methods were combined with molecular fingerprints to develop QSAR models which were used to identify potential hits against *Mycobacterium tuberculosis* (causative agent of tuberculosis). The model was used for screening all the chalcone compounds retrieved, and the results indicate that designed heteroaryl chalcones identified can be promising lead candidates against tuberculosis (Gomes, Braga, et al., 2017). In another study, adamantane-based inhibitors of Ebola virus envelope glycoprotein were identified using docking, pharmacophore and 3D-QSAR approaches. The predicted molecule performed better than the standard drugs oseltamivir and zanamivir (Mali & Chaudhari, 2019). In a recent study, QSAR approach combined with scaffold hopping was used to identify three potent inhibitors of FABP4 (adipocyte fatty acid binding protein 4), which was later identified as a molecular target for treating certain type of cancers, type 2 diabetes, and other metabolic diseases (Floresta et al., 2019).

5 APPLICATIONS OF CADD IN DRUG DISCOVERY

5.1 Virtual Screening

It is an integral part of CADD to screen novel active compounds from chemical libraries. VS is routinely used by scientists and pharmaceutical companies as one of the methods in the process of drug discovery (Lavecchia & Giovanni, 2013). Presently, the methods used in VS have immensely improved in terms of performance, utility, and user-friendliness, leading to the extensive use of VS in drug discovery. With the advent of supercomputing and cloud computing, it is now possible to narrow down huge chemical space within few hours. Both structure-based and ligand-based approaches discussed above are used for VS to discover lead compound.

5.2 Lead Optimization

The role of lead optimization is to preserve or improve the required characteristics of the main components of the drug, at the same time minimizing its toxicity. The lead compounds identified after VS can be refined to increase selectivity and specificity for a given target. Prior to synthesis of the lead compounds, properties such as binding affinity, selectivity, physiochemical and adsorption, distribution, metabolism, excretion (ADME),

TABLE 1.9

Tools Used for Quantitative Structure—Activity Relationship (QSAR) Modeling in Drug Design.

Tool	Description	License Type
AutoQSAR https://www.schrodinger.com/autoqsar	Fully automated creation and application of QSAR models	Commercial
QSARpro https://www.vlifesciences.com/products/QSARPro/Product_QSARpro.php	Used for QSAR modeling and activity prediction	Commercial
PharmQSAR https://new.pharmacelera.com/pharmqsar/	Software package for automated QSAR model development	Commercial
eTOXlab http://phi.imim.es/envoy/	Automated QSAR model development and validation	Free
OCHEM https://www.eyesopen.com/molecular-modeling	A web-based platform for fully automated QSAR modeling	Free
DELPHOS http://lidecc.cs.uns.edu.ar/index.php/sw/delphos	It is used for development of QSAR models	Free
AutoWeka https://www.cs.ubc.ca/labs/beta/Projects/autoweka/	Software used for data mining for QSAR and model development	Free
3D-QSAR https://www.3d-qsar.com/	Used for the development of QSAR models	Free
AZOrange http://github.com/AZcompTox/AZOrange	It is a QSAR modeling package based on machine learning	Free
GUSAR http://www.way2drug.com/gusar/	Web-based platform for QSAR modeling	Free
Taverna http://cdk.sourceforge.net/cdk-taverna/	A chemoinformatics workflow	Free
Pipeline Pilot https://www.3ds.com/products-services/biovia/products/data-science/pipeline-pilot/	Tool for workflow automation	Commercial
Galaxy https://galaxyproject.org/	Tool for workflow automation	Free
KNIME https://www.knime.org/	Tool for workflow automation	Free

and toxicity (T) are optimized (Cheng et al., 2011). The approaches used in VS such as QSAR and pharmacophore modeling are significantly used in lead optimization (John et al., 2011; Pirhadi et al., 2013). Hopfinger et al. applied 4D-QSAR modeling to develop a virtual screen for glycogen phosphorylase inhibitors (Hopfinger et al., 1999). Singh et al. combined 3D-QSAR with 3D pharmacophore searching for screening and optimizing specific integrin antagonists (Singh et al., 2002).

ML tools, such as ANN, hidden Markov models (HMM), SVM, decision tree learning, RF, Naive Bayes, and belief networks, are also employed in lead optimization (Byvatov et al., 2003; Olivecrona et al., 2017). Finally, the analysis of ADMET properties is carried out in the lead optimization phase (Macalino et al., 2015). Several computational tools as well as web servers are available for the prediction of ADMET properties (Table 1.10).

TABLE 1.10
Software and Web Resources for ADMET Prediction.

Software/Web server	License Type	Web Address
ADMET Predictor	Commercial	http://www.simulations-plus.com/software/admet-property-prediction-qsar/
QikProp	Commercial	https://www.schrodinger.com/
ADMET and predictive toxicology	Commercial	https://www.3ds.com/products-services/biovia/products/molecular-modeling-simulation/biovia-discovery-studio/qsar-admet-and-predictive-toxicology/
PreADMET PC version 2.0	Commercial	https://preadmet.bmdrc.kr/preadmet-pc-version-2-0/
PreADMET	Free	https://preadmet.bmdrc.kr/
SwissADME	Free	http://www.swissadme.ch/
ADMETlab	Free	http://admet.scbdd.com/calcpre/index/
admetSAR	Free	http://lmmd.ecust.edu.cn/admetsar1/
FAF-Drugs4	Free	http://fafdrugs4.mti.univ-paris-diderot.fr/
ALOGPS	Free	http://www.vcclab.org/lab/alogps/

5.3 Scaffold Hopping

Drug discovery pipeline is intended to identify the novel chemical entities, which is not only unique in its mechanism of action against a particular target protein but also novel in the chemical structural core and not available in the known drug chemical space. A high-throughput screening approach can identify the initial lead molecules having the potential to inhibit the function of the target protein; however, computational approaches can be further used for assessing the chemical diversity as well as the lead optimization process. Among the different computational approaches, scaffold hopping or lead hopping is used to design the "isofunctional" chemical entity, i.e., two compounds having the nearly same activity but carrying the different chemical scaffolds (Nakano et al., 2020). Scaffold hopping, in a general sense, is viewed to find the diverse chemical entities from the computational screening (Chen et al., 2014); however, more systematic application includes the scaffold replacement by step-by-step modification of the core scaffold of the compound series (Fig. 1.6). Modifications that might generally be employed in this case include heterocyclic replacements, ring closure or opening, peptidomimetics, and chemical topology—based modifications. Among the different reasons to carry out the scaffold hopping, replacing a chemically complex natural product with a synthetically accessible molecule and improving the pharmacological properties of known actives are the main applications for a medicinal chemist (Hu et al., 2017).

Depending upon the case study and availability of the background information, different methodologies are adapted for scaffold hopping and employed (Bajorath, 2017; Sun et al., 2012). In general, all available scaffold hopping methodology can be classified as ligand-based hoping and protein structure—based hopping. In the former category are pharmacophore-based (Lauri & Bartlett, 1994), topological pharmacophore graph-based (Nakano et al., 2020), lead molecule shape-based (Rush et al., 2005), and selected candidate compound-based hopping (Vogt et al., 2010). Scaffold hopping by 2D fingerprint was thoroughly analyzed by the Bajorath group and pointed out the possible limitation and applicability of these fingerprints in this regard and based on the findings; the paper has reported the guidelines useful for scaffold hopping using the 2D fingerprint (Vogt et al., 2010). In the latter case, protein structural information is integrated in a different way to achieve the scaffold hopping. In one case study, predocked fragment database is used to analyze the receptor binding site by scoring the different sets of fragments at a particular site followed by proposing the new ligand molecules having the maximum interaction score (Lin & Tseng, 2011). In one more study of the same line, Silverman and co-workers have reported that improved selectivity and drug-like properties can be achieved using the fragment-based scaffold hopping (Ji et al., 2009). Calculation of the protein—ligand binding FEs is a regular task to estimate the strength of the interaction between two molecules (protein and bound ligand) and can be further used to estimate

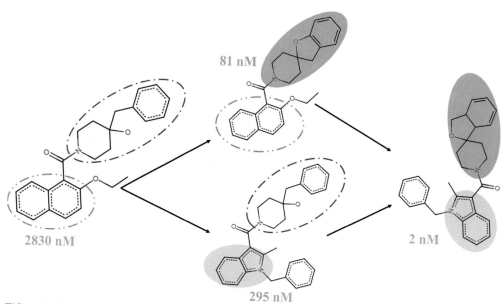

FIG. 1.6 Scaffold hopping to optimize the initial lead identified from the high-throughput screening against human vasopressin 1a receptor (Ratni et al., 2015).

the change of FEs while changing the particular R-groups; however, it is challenging to estimate FE for evaluating the scaffold hopping modification. Wang et al. have come up with an FEP-based method to pursue FEP for scaffold hopping related modification in computationally feasible manner (Wang, Deng, et al., 2017). Wang et al. have implemented proposed method for six pharmaceutically important proteins and showed that predicted binding affinities for each modification have good correlation with the experimental affinity.

5.4 Multitargeted Approaches/ Polypharmacology

Over the years, the drug discovery pipeline has incorporated many new features such as in vivo models to a single protein target drug for a single mechanism. This single target—based drug design process primarily supported by the lock-and-key model proposed more than a century ago by E. Fischer (1894). Through over the many decades, multidimensional biological data have piled up to understand the underlying mechanism of particular drug targets; however, drug discovery efforts continued to identify the selective drugs (keys) to inhibit single mechanism—based target (lock). This approach of drug design is also supported mainly by the reductionist view of systems biology (Maggiora, 2011) and by an ever-growing number of crystal structures to understand the mechanistic view at the

molecular level (e.g., understanding the interaction pattern critical for selective ligand designing).

Conventional single target—based drug discovery is found to be blinded, which neglects the other processes directly/indirectly connected through complex metabolic/signaling networks (Maggiora, 2011). Drug target—based analysis has shown that some drugs can simultaneously bind to many protein targets, thereby eliciting either biological activities or adverse effects (Frantz, 2005). One of the example drugs of this type is aspirin, which has shown many different processes of mechanisms along with cyclooxygenase inhibition (Koeberle & Werz, 2014). Large-scale multidimensional experimental biological data have demonstrated that biological processes are arranged in higher levels of hierarchical nature and single perturbation affects the whole complex network. This introduced a multitargeted drug design paradigm (Hopkins, 2008). Less than decades-old polypharmacology (interaction of single drugs to multiple targets) (Paolini et al., 2006), approach is now shifting the central dogma of the drug discovery process as this approach has found many encouraging results over former drug discovery approaches (Hopkins, 2008). This approach is identified as a capable solution for the treatment of the complex diseases like cancer (Raghavendra et al., 2018), neurological disease (Stephenson et al., 2005; Więckowska et al., 2016), inflammation (Hwang et al., 2013), and infectious diseases (Li et al., 2014).

TABLE 1.11
List of Databases of Repurposed Drugs.

Database	Drugs	Web Address
Repurposed Drug Database	300	http://www.drugrepurposingportal.com/repurposed-drug-database.php
RepoDb	268	http://apps.chiragjpgroup.org/repoDB/
Drug Repurposing Hub	13,553	http://www.broadinstitute.org/repurposing
ReDO (Repurposing Drugs in Oncology)	~270	http://www.redo-project.org/db/
COVID-19 Drug Repurposing Database	128	https://www.excelra.com/covid-19-drug-repurposing-database/

Drug repurposing is a direct way of analyzing the polypharmacological indication other than an indication they were initially being directed. Among the in silico methods, docking-based protocol is well known for polypharmacological inhibitor design. One study reported by Pinzi et al. (2018) has discussed the role of proper protein conformations selection for the docking-based protocol to achieve a polypharmacological profile. The author has suggested that for proteins having dynamics and different states (active/inactive), conformation assessment is required, as a multitarget drug is accommodated by diverse binding pockets (Pinzi et al., 2018). In the same line, many articles have reported different chemoinformatic/in silico methods/techniques which can be utilized for multitargeted drug designing (Koutsoukas et al., 2011; Antonio Lavecchia & Cerchia, 2016; Zhang, Pei, & Lai, 2017).

Using the chemical data mining approach, Bajorath group has analyzed the crystal complexes data deposited in the PDB for finding a template for multitarget ligand design. After performing the systematic search of available protein complexes, 702 ligands were identified that bound to different protein families and so-called multitargeted ligands. From the multitargeted ligands, analog-based scaffolds were isolated as a template for further multitargeted ligand design (Gilberg et al., 2018).

5.5 Drug Repurposing/Reprofiling

One of the approaches that became increasingly popular recently in the field of discovery and development of drugs is to identify novel uses of already approved drugs, known as "drug repurposing" or repositioning. The term drug repurposing was greatly highlighted recently as COVID-19 pandemic prevailed throughout the world, putting drug repurposing techniques on fast track. Several US FDA-approved drugs have shown activity against multiple targets (also called promiscuity) implying that it can be repurposed for therapeutic benefit to tackle other diseases. In comparison to the discovery of novel medicine, the advantages of drug repurposing are that the drug can move to the trial fast, reduced cost, and less risk of unfavorable results. Drug repurposing has also changed the failures in drug discovery into breakthroughs by discovering their new therapeutic uses. Presently, several hundred drugs are utilized for the disease they were originally developed for as well as for repurposed indications. Acetylsalicylic acid (aspirin) was launched in 1897 as a nonsteroidal antiinflammatory drug that was later in 1956 used as an antithrombotic drug (Desborough & Keeling, 2017). There are several repositories of repurposed drugs that are freely accessible (Table 1.11).

CADD approaches greatly help in fast-tracking the repositioning process by revealing the unknown targets of the approved or failed drugs. Repurposing of several drugs has been done in the past, one of the classical examples is Sildenafil. The pill was originally developed for treating high blood pressure and angina but later developed as a treatment for erectile dysfunction and now repurposed for pulmonary hypertension. The novel coronavirus (nCoV-2019) that is the causative agent of the disease COVID-19 has quickly caused pandemic. Irrespective of the severity of the disease, pathogen-specific antivirals are missing. Hence, for a short-term response to fight nCoV-2019, computational drug repurposing techniques stand out as a potential approach. Several approved drugs have been identified which show anti-nCoV-2019 activity, some of these compounds including chloroquine, tetrandrine, umifenovir, carrimycin, damageprevir, lopinavir are in phase 4 of clinical trials (Lima et al., 2020).

Computational drug repurposing approaches can be broadly classified into phenotypic or blind screening method, knowledge-based method, and data-based drug repurposing method. Furthermore, knowledge and data-based drug repurposing method is classified into target-based approach, signature-based approach, network-based approach, and targeted mechanism-based approach. Phenotypic or blind screening method including biological activity, structural, or pharmaceutical information about the mode of action of drug is not included, rather it depends on unexpected identification from experiments performed for certain drugs and diseases (Ma et al., 2013). This method is useful when the target structure information is very little or unknown. As this approach does not require prior structural information of the target, it can be apt for numerous diseases. Target-based drug repurposing method employs either target or ligand structure for HTVS of compound libraries through molecular docking/pharmacophore modeling. Here the tools and techniques described for molecular docking and VS are applied. This is one of the most common drug repurposing techniques used by scientists (Jin & Wong, 2014; Li et al., 2016). Jin et al. (2020) used a target-based approach for screening more than 10,000 molecules against the target protease (Mpro) of COVID-19 virus and identified N3 and ebselen as possible candidates to treat COVID-19 (Jin et al., 2020). Knowledge and data-based drug repurposing method uses bioinformatics or cheminformatics approaches to apply existing knowledge of drugs such as the chemical structure of the drug and target, drug target networks, FDA approval labels, pathways information, clinical trial information, adverse effect, etc., to drug repurposing method, improving the accuracy of prediction. Zhou et al. (2020) implemented the network proximity analyses of drug targets and host interactions of human coronavirus (HCoV) in the human interactome and identified 16 repurposable drugs aginst HCoV; the study also provides a powerful network-based approach to quickly identify repurposable drugs or drug combinations against novel coronavirus (Zhou et al., 2020).

Genome-wide association studies (GWAS) based method is used to identify the single-nucleotide polymorphisms (SNPs) linked to specific disease for identification of the genes that could be potential drug targets. Thousands of SNPs can be detected simultaneously with the help of GWAS; these data are then used to distinguish genes related with the specific disease and to understand the drug response to these variations (Sanseau et al., 2012). ML-based approaches require a considerably large amount of data for building predictive models. Currently, abundance of data pertaining to the information on the protein target, structure of small molecules, side effect profiles, gene expression data, etc., is available, which helps in studying the mechanism of the disease or mode of action of the drug as well as in identifying the new indications for prevailing drugs. ML-based methods have excelled greatly in the last few years, and several approaches based on this method have been suggested (Li et al., 2016). The ML approaches that are applied in drug repurposing include logistic regression, DL, RF, SVM, and neural network. For an efficient drug repurposing, any of the in silico drug repurposing methods or their combination can be applied depending upon the objectives and information availability.

6 CONCLUSIONS

CADD is a powerful tool in modern drug discovery for the search of potential therapeutic compounds. It has now become the most suitable alternative for high-throughput screening, which is used routinely in drug discovery and development. The techniques/tools used in CADD are applied in almost all the stages of drug discovery pipeline, as it has the ability to fast-track the process of hit identification, hit to lead, and lead optimization (binding affinity, ADME and toxicity, etc.). In the last two decades, advancements in the computational drug designing protocol at the level of techniques/tools coupled with the progress in computational power have enabled the scientific community to generate disease-oriented quick and reliable solutions at low cost in manageable time space. The current pandemic state emerged due to the coronavirus can be understood as the best example of this type where, in the 2−3 months time scale, many reports have come up across the globe against different coronavirus target proteins. Modern computational techniques have helped in achieving molecular-level understanding and also predicting the promising inhibitors. Apart from the advancements in the rational drug designing approaches, large-scale generation of the multidimensional biological data has geared up the ML/DL/AI-based model development, which has shown improvement in the prediction accuracy and in time complexity compared to traditional approaches. Though these data-driven models are highly dependent upon the quality and quantity of the data, as more and more data will come up, the prediction accuracy will also scale up. At the same time, there is a continuous need of further improvements in prediction algorithms to discover promising new drugs and predict new indications for existing drugs.

REFERENCES

Abagyan, R., Totrov, M., & Kuznetsov, D. (1994). ICM—a new method for protein modeling and design: Applications to docking and structure prediction from the distorted native conformation. *Journal of Computational Chemistry*. https://doi.org/10.1002/jcc.540150503

Acharya, C., Coop, A., Polli, J. E., & MacKerell, A. D. (2010). Recent advances in ligand-based drug design: Relevance and utility of the conformationally sampled pharmacophore approach. *Current Computer-Aided Drug Design*. https://doi.org/10.2174/157340911793743547

Afifi, K., & Al-Sadek, A. F. (2018). Improving classical scoring functions using random forest: The non-additivity of free energy terms' contributions in binding. *Chemical Biology and Drug Design*. https://doi.org/10.1111/cbdd.13206

Ahmed, A., Saeed, F., Salim, N., & Abdo, A. (2014). Condorcet and borda count fusion method for ligand-based virtual screening. *Journal of Cheminformatics*. https://doi.org/10.1186/1758-2946-6-19

Ain, Q. U., Aleksandrova, A., Roessler, F. D., & Ballester, P. J. (2015). Machine-learning scoring functions to improve structure-based binding affinity prediction and virtual screening. *Wiley Interdisciplinary Reviews: Computational Molecular Science*. https://doi.org/10.1002/wcms.1225

Aqvist, J., Luzhkov, V. B., & Brandsdal, B. O. (2002). Ligand binding affinities from MD simulations. *Accounts of Chemical Research*. https://doi.org/10.1021/ar010014p

Arulsudar, N., Subramanian, N., & Murthy, R. S. R. (2005). Comparison of artificial neural network and multiple linear regression in the optimization of formulation parameters of leuprolide acetate loaded liposomes. *Journal of Pharmacy and Pharmaceutical Sciences*, 8(2), 243–258.

Ashtawy, H. M., & Mahapatra, N. R. (2018). Boosted neural networks scoring functions for accurate ligand docking and ranking. *Journal of Bioinformatics and Computational Biology*. https://doi.org/10.1142/S021972001850004X

Bajorath, J. (2017). Computational scaffold hopping: Cornerstone for the future of drug design? *Future Medicinal Chemistry*, 9(7), 629–631. https://doi.org/10.4155/fmc-2017-0043

Bajusz, D., Rácz, A., & Héberger, K. (2015). Why is Tanimoto index an appropriate choice for fingerprint-based similarity calculations? *Journal of Cheminformatics*. https://doi.org/10.1186/s13321-015-0069-3

Ballester, P. J., & Mitchell, J. B. O. (2010). A machine learning approach to predicting protein-ligand binding affinity with applications to molecular docking. *Bioinformatics*. https://doi.org/10.1093/bioinformatics/btq112

Ballester, P. J., Schreyer, A., & Blundell, T. L. (2014). Does a more precise chemical description of protein-ligand complexes lead to more accurate prediction of binding affinity? *Journal of Chemical Information and Modeling*. https://doi.org/10.1021/ci500091r

Baroni, M., Cruciani, G., Sciabola, S., Perruccio, F., & Mason, J. S. (2007). A common reference framework for analyzing/comparing proteins and ligands. Fingerprints for ligands and proteins (FLAP): Theory and application. *Journal of Chemical Information and Modeling*, 47(2), 279–294. https://doi.org/10.1021/ci600253e

Beccari, A. R., Cavazzoni, C., Beato, C., & Costantino, G. (2013). LiGen: A high performance workflow for chemistry driven de Novo design. *Journal of Chemical Information and Modeling*, 53(6), 1518–1527. https://doi.org/10.1021/ci400078g

Bengio, Y., Courville, A., & Vincent, P. (2013). Representation learning: A review and new perspectives. *IEEE Transactions on Pattern Analysis and Machine Intelligence*. https://doi.org/10.1109/TPAMI.2013.50

Bento, A. P., Gaulton, A., Hersey, A., Bellis, L. J., Chambers, J., Davies, M., Krüger, F. A., Light, Y., Mak, L., McGlinchey, S., Nowotka, M., Papadatos, G., Santos, R., & Overington, J. P. (2014). The ChEMBL bioactivity database: An update. *Nucleic Acids Research*. https://doi.org/10.1093/nar/gkt1031

Berman, H. M., Battistuz, T., Bhat, T. N., Bluhm, W. F., Bourne, P. E., Burkhardt, K., Feng, Z., Gilliland, G. L., Iype, L., Jain, S., Fagan, P., Marvin, J., Padilla, D., Ravichandran, V., Schneider, B., Thanki, N., Weissig, H., Westbrook, J. D., & Zardecki, C. (2002). The protein data bank. *Acta Crystallographica Section D Biological Crystallography*. https://doi.org/10.1107/S0907444902003451

BIOVIA Discovery Studio | Pharmacophore and Ligand-Based Design. (n.d.). Retrieved June 17, 2020, from https://www.3dsbiovia.com/products/collaborative-science/biovia-discovery-studio/pharmacophore-and-ligand-based-design.html.

Böhm, H. J. (1992a). Ludi: Rule-based automatic design of new substituents for enzyme inhibitor leads. *Journal of Computer-Aided Molecular Design*. https://doi.org/10.1007/BF00126217

Böhm, H. J. (1992b). The computer program Ludi: A new method for the de novo design of enzyme inhibitors. *Journal of Computer-Aided Molecular Design*, 6(1), 61–78. https://doi.org/10.1007/BF00124387

Broccatelli, F., & Brown, N. (2014). Best of both worlds: On the complementarity of ligand-based and structure-based virtual screening. *Journal of Chemical Information and Modeling*. https://doi.org/10.1021/ci5001604

Brylinski, M. (2013). Nonlinear scoring functions for similarity-based ligand docking and binding affinity prediction. *Journal of Chemical Information and Modeling*. https://doi.org/10.1021/ci400510e

Bucher, D., Stouten, P., & Triballeau, N. (2018). Shedding light on important waters for drug design: Simulations versus grid-based methods. *Journal of Chemical Information and Modeling*. https://doi.org/10.1021/acs.jcim.7b00642

Byvatov, E., Fechner, U., Sadowski, J., & Schneider, G. (2003). Comparison of support vector machine and artificial neural network systems for drug/nondrug classification. *Journal of Chemical Information and Computer Sciences*. https://doi.org/10.1021/ci0341161

Cang, Z., Mu, L., & Wei, G. W. (2018). Representability of algebraic topology for biomolecules in machine learning based scoring and virtual screening. *PLoS Computational Biology*. https://doi.org/10.1371/journal.pcbi.1005929

Cappel, D., Sherman, W., & Beuming, T. (2017). Calculating water thermodynamics in the binding site of proteins —

applications of WaterMap to drug discovery. *Current Topics in Medicinal Chemistry.* https://doi.org/10.2174/1568026617666170414141452

Carlson, H. A., Masukawa, K. M., Rubins, K., Bushman, F. D., Jorgensen, W. L., Lins, R. D., Briggs, J. M., & McCammon, J. A. (2000). Developing a dynamic pharmacophore model for HIV-1 integrase. *Journal of Medicinal Chemistry.* https://doi.org/10.1021/jm990322h

Cereto-Massagué, A., Ojeda, M. J., Valls, C., Mulero, M., Garcia-Vallvé, S., & Pujadas, G. (2015). Molecular fingerprint similarity search in virtual screening. *Methods.* https://doi.org/10.1016/j.ymeth.2014.08.005

Cheng, K., Korfmacher, W., White, R., & Njoroge, F. (2011). Lead optimization in discovery drug metabolism and pharmacokinetics/case study: The Hepatitis C virus (HCV) protease inhibitor SCH 503034. *Dyes and Drugs.* https://doi.org/10.1201/b13128-15

Chen, Y.-C., Totrov, M., & Abagyan, R. (2014). Docking to multiple pockets or ligand fields for screening, activity prediction and scaffold hopping. *Future Medicinal Chemistry, 6*(16), 1741−1755. https://doi.org/10.4155/fmc.14.113

Clark, D. E. (2006). What has computer-aided molecular design ever done for drug discovery? *Expert Opinion on Drug Discovery.* https://doi.org/10.1517/17460441.1.2.103

Clarke, C., Woods, R. J., Gluska, J., Cooper, A., Nutley, M. A., & Boons, G. J. (2001). Involvement of water in carbohydrate-protein binding. *Journal of the American Chemical Society.* https://doi.org/10.1021/ja004315q

Cooper, D. R., Porebski, P. J., Chruszcz, M., & Minor, W. (2011). X-ray crystallography: Assessment and validation of protein-small molecule complexes for drug discovery. *Expert Opinion on Drug Discovery.* https://doi.org/10.1517/17460441.2011.585154

Cornell, W. D., Cieplak, P., Bayly, C. I., Gould, I. R., Merz, K. M., Ferguson, Spellmeyer, D. C., Fox, T., Caldwell, J. W., & Kollman, P. A. (1995). A 2nd Generation Force-Field for the Simulation of Proteins, Nucleic-Acids, and Organic-Molecules. *Journal of the American Chemical Society, 117*(19), 5179−5197.

Clark, D. E. (2008). What has virtual screening ever done for drug discovery? *Expert Opinion on Drug Discovery, 3*(8), 841−851.

Deo, R. C. (2015). Machine learning in medicine. *Circulation.* https://doi.org/10.1161/CIRCULATIONAHA.115.001593

Desborough, M. J. R., & Keeling, D. M. (2017). The aspirin story − from willow to wonder drug. *British Journal of Haematology.* https://doi.org/10.1111/bjh.14520

Devi, R. V., Sathya, S. S., & Coumar, M. S. (2015). Evolutionary algorithms for de novo drug design - a survey. *Applied Soft Computing Journal.* https://doi.org/10.1016/j.asoc.2014.09.042

Ding, B., Wang, J., Li, N., & Wang, W. (2013). Characterization of small molecule binding. I. Accurate identification of strong inhibitors in virtual screening. *Journal of Chemical Information and Modeling.* https://doi.org/10.1021/ci300508m

Dixon, S. L., Smondyrev, A. M., Knoll, E. H., Rao, S. N., Shaw, D. E., & Friesner, R. A. (2006). Phase: A new engine for pharmacophore perception, 3D QSAR model development, and 3D database screening: 1. Methodology and preliminary results. *Journal of Computer-Aided Molecular Design, 20*(10−11), 647−671. https://doi.org/10.1007/s10822-006-9087-6

Drie, J. H. (2007). Computer-aided drug design: The next 20 years. *Journal of Computer-Aided Molecular Design.* https://doi.org/10.1007/s10822-007-9142-y

Duan, L., Feng, G., Wang, X., Wang, L., & Zhang, Q. (2017). Effect of electrostatic polarization and bridging water on CDK2-ligand binding affinities calculated using a highly efficient interaction entropy method. *Physical Chemistry Chemical Physics.* https://doi.org/10.1039/c7cp00841d

Dudek, A., Arodz, T., & Galvez, J. (2006). Computational methods in developing quantitative structure-activity relationships (QSAR): A review. *Combinatorial Chemistry and High Throughput Screening.* https://doi.org/10.2174/138620706776055539

Dunitz, J. D. (1994). The entropic cost of bound water in crystals and biomolecules. *Science.* https://doi.org/10.1126/science.264.5159.670

Durrant, J. D., Friedman, A. J., Rogers, K. E., & McCammon, J. A. (2013). Comparing neural-network scoring functions and the state of the art: Applications to common library screening. *Journal of Chemical Information and Modeling.* https://doi.org/10.1021/ci400042y

Durrant, J. D., & McCammon, J. A. (2011). NNScore 2.0: A neural-network receptor-ligand scoring function. *Journal of Chemical Information and Modeling.* https://doi.org/10.1021/ci2003889

Du, X., Sun, S., Hu, C., Yao, Y., Yan, Y., & Zhang, Y. (2017). DeepPPI: Boosting prediction of protein-protein interactions with deep neural networks. *Journal of Chemical Information and Modeling.* https://doi.org/10.1021/acs.jcim.7b00028

Eisen, M. B., Wiley, D. C., Karplus, M., & Hubbard, R. E. (1994). HOOK: A program for finding novel molecular architectures that satisfy the chemical and steric requirements of a macromolecule binding site. *Proteins: Structure, Function, and Bioinformatics.* https://doi.org/10.1002/prot.340190305

Ericksen, S. S., Wu, H., Zhang, H., Michael, L. A., Newton, M. A., Hoffmann, F. M., & Wildman, S. A. (2017). Machine learning consensus scoring improves performance across targets in structure-based virtual screening. *Journal of Chemical Information and Modeling.* https://doi.org/10.1021/acs.jcim.7b00153

Ewing, T. J. A., Makino, S., Skillman, A. G., & Kuntz, I. D. (2001). Dock 4.0: Search strategies for automated molecular docking of flexible molecule databases. *Journal of Computer-Aided Molecular Design.* https://doi.org/10.1023/A:1011115820450

Fauman, E. B., Rai, B. K., & Huang, E. S. (2011). Structure-based druggability assessment-identifying suitable targets for small molecule therapeutics. *Current Opinion in Chemical Biology.* https://doi.org/10.1016/j.cbpa.2011.05.020

Fischer, E. (1894). Einfluss der Configuration der Wirkung der Enzyme. *Berichte Der Deutschen Chemischen Gesellschaft.* https://doi.org/10.1002/cber.18940270364

Fischer, J., & Robin Ganellin, C. (2006). Analogue-based drug discovery. *Analogue-based Drug Discovery*. https://doi.org/10.1002/3527608001

Floresta, G., Cilibrizzi, A., Abbate, V., Spampinato, A., Zagni, C., & Rescifina, A. (2019). 3D-QSAR assisted identification of FABP4 inhibitors: An effective scaffold hopping analysis/QSAR evaluation. *Bioorganic Chemistry*. https://doi.org/10.1016/j.bioorg.2018.11.045

Forli, S. (2015). Charting a path to success in virtual screening. *Molecules*. https://doi.org/10.3390/molecules201018732

Forli, S., Huey, R., Pique, M. E., Sanner, M. F., Goodsell, D. S., & Olson, A. J. (2016). Computational protein-ligand docking and virtual drug screening with the AutoDock suite. *Nature Protocols*. https://doi.org/10.1038/nprot.2016.051

Frantz, S. (2005). Playing dirty. *Nature*. https://doi.org/10.1038/437942a

Friesner, R. A., Banks, J. L., Murphy, R. B., Halgren, T. A., Klicic, J. J., Mainz, D. T., Repasky, M. P., Knoll, E. H., Shelley, M., Perry, J. K., Shaw, D. E., Francis, P., & Shenkin, P. S. (2004). Glide: A new approach for rapid, accurate docking and scoring. 1. Method and assessment of docking accuracy. *Journal of Medicinal Chemistry*. https://doi.org/10.1021/jm0306430

Gilberg, E., Stumpfe, D., & Bajorath, J. (2018). X-ray-structure-based identification of compounds with activity against targets from different families and generation of templates for multitarget ligand design. *ACS Omega*. https://doi.org/10.1021/acsomega.7b01849

Glen, R. C., Martin, G. R., Hill, A. P., Hyde, R. M., Woollard, P. M., Salmon, J. A., Buckingham, J., & Robertson, A. D. (1995). Computer-aided design and synthesis of 5-substituted tryptamines and their pharmacology at the 5-HT1d receptor: Discovery of compounds with potential anti-migraine properties. *Journal of Medicinal Chemistry*. https://doi.org/10.1021/jm00018a016

Gomes, M. N., Braga, R. C., Grzelak, E. M., Neves, B. J., Muratov, E., Ma, R., Klein, L. L., Cho, S., Oliveira, G. R., Franzblau, S. G., & Andrade, C. H. (2017). QSAR-driven design, synthesis and discovery of potent chalcone derivatives with antitubercular activity. *European Journal of Medicinal Chemistry*. https://doi.org/10.1016/j.ejmech.2017.05.026

Gomes, M. N., Muratov, E. N., Pereira, M., Peixoto, J. C., Rosseto, L. P., Cravo, P. V. L., Andrade, C. H., & Neves, B. J. (2017). Chalcone derivatives: Promising starting points for drug design. *Molecules*. https://doi.org/10.3390/molecules22081210

Gomes, J., Ramsundar, B., Feinberg, E. N., & Pande, V. S. (2017). *Atomic convolutional networks for predicting protein-ligand binding affinity* (pp. 1–17). Retrieved from http://arxiv.org/abs/1703.10603.

Goodsell, D. S., & Olson, A. J. (1990). Automated docking of substrates to proteins by simulated annealing. *Proteins: Structure, Function, and Bioinformatics*. https://doi.org/10.1002/prot.340080302

Grover, S., Apushkin, M. A., & Fishman, G. A. (2006). Topical dorzolamide for the treatment of cystoid macular edema in patients with retinitis pigmentosa. *American Journal of Ophthalmology*. https://doi.org/10.1016/j.ajo.2005.12.030

Guedes, I. A., Pereira, F. S. S., & Dardenne, L. E. (2018). Empirical scoring functions for structure-based virtual screening: Applications, critical aspects, and challenges. *Frontiers in Pharmacology*. https://doi.org/10.3389/fphar.2018.01089

Hemmateenejad, B., Safarpour, M. A., Miri, R., & Taghavi, F. (2004). Application of ab initio theory to QSAR study of 1,4-dihydropyridine-based calcium channel blockers using GA-MLR and PC-GA-ANN procedures. *Journal of Computational Chemistry*. https://doi.org/10.1002/jcc.20066

Honma, T., Hayashi, K., Aoyama, T., Hashimoto, N., Machida, T., Fukasawa, K., Iwama, T., Ikeura, C., Ikuta, M., Suzuki, T. I., Iwasawa, Y., Hayama, T., Nishimura, S., & Morishima, H. (2001). Structure-based generation of a new class of potent Cdk4 inhibitors: New de Novo design strategy and library design. *Journal of Medicinal Chemistry, 44*(26), 4615–4627. https://doi.org/10.1021/jm0103256

Hopfinger, A. J., Reaka, A., Venkatarangan, P., Duca, J. S., & Wang, S. (1999). Construction of a virtual high throughput screen by 4D-QSAR analysis: Application to a combinatorial library of glucose inhibitors of glycogen phosphorylase b. *Journal of Chemical Information and Computer Sciences*. https://doi.org/10.1021/ci990032+

Hopkins, A. L. (2008). Network pharmacology: The next paradigm in drug discovery. *Nature Chemical Biology*. https://doi.org/10.1038/nchembio.118

Hornak, V., & Simmerling, C. (2007). Targeting structural flexibility in HIV-1 protease inhibitor binding. *Drug Discovery Today, 12*(3–4), 132–138. https://doi.org/10.1016/j.drudis.2006.12.011

Huang, S. Y., Grinter, S. Z., & Zou, X. (2010). Scoring functions and their evaluation methods for protein-ligand docking: Recent advances and future directions. *Physical Chemistry Chemical Physics*. https://doi.org/10.1039/c0cp00151a

Hu, B., & Lill, M. A. (2012). Protein pharmacophore selection using hydration-site analysis. *Journal of Chemical Information and Modeling*. https://doi.org/10.1021/ci200620h

Hu, Y., Stumpfe, D., & Bajorath, J. (2017). Recent advances in scaffold hopping. *Journal of Medicinal Chemistry, 60*(4), 1238–1246. https://doi.org/10.1021/acs.jmedchem.6b01437

Hwang, S. H., Wecksler, A. T., Wagner, K., & Hammock, B. D. (2013). Rationally designed multitarget agents against inflammation and pain. *Current Medicinal Chemistry*. https://doi.org/10.2174/0929867311320130013

Inc, C. C. G. (2015). Molecular Operating Environment (MOE), 2015.01. *1010 Sherbooke St.West, suite #910, Montreal, QC, Canada, H3A 2R7*.

Irwin, J. J., & Shoichet, B. K. (2005). Zinc - a free database of commercially available compounds for virtual screening. *Journal of Chemical Information and Modeling*. https://doi.org/10.1021/ci049714+

Jain, A. N. (2003). Surflex: Fully automatic flexible molecular docking using a molecular similarity-based search engine. *Journal of Medicinal Chemistry*. https://doi.org/10.1021/jm020406h

Jain, A. N. (2007). Surflex-Dock 2.1: Robust performance from ligand energetic modeling, ring flexibility, and knowledge-based search. *Journal of Computer-Aided Molecular Design*. https://doi.org/10.1007/s10822-007-9114-2

James, C., Weininger, D., & Delaney, J. (2011). *Daylight theory manual version 4.9*. Laguna Niguel, CA: Daylight Chemical Information Systems, Inc.

Ji, H., Li, H., Martásek, P., Roman, L. J., Poulos, T. L., & Silverman, R. B. (2009). Discovery of highly potent and selective inhibitors of neuronal nitric oxide synthase by fragment hopping. *Journal of Medicinal Chemistry, 52*(3), 779–797. https://doi.org/10.1021/jm801220a

Jin, Z., Du, X., Xu, Y., Deng, Y., Liu, M., Zhao, Y., Zhang, B., Li, X., Zhang, L., Peng, C., Duan, Y., Yu, J., Wang, L., Yang, K., Liu, F., Jiang, R., Yang, X., You, T., Liu, X., … (2020). Structure of Mpro from COVID-19 virus and discovery of its inhibitors. *Nature*. https://doi.org/10.1038/s41586-020-2223-y

Jin, G., & Wong, S. T. C. (2014). Toward better drug repositioning: Prioritizing and integrating existing methods into efficient pipelines. *Drug Discovery Today*. https://doi.org/10.1016/j.drudis.2013.11.005

John, S., Thangapandian, S., Sakkiah, S., & Lee, K. W. (2011). Discovery of potential pancreatic cholesterol esterase inhibitors using pharmacophore modelling, virtual screening, and optimization studies. *Journal of Enzyme Inhibition and Medicinal Chemistry*. https://doi.org/10.3109/14756366.2010.535795

Jung, S. W., Kim, M., Ramsey, S., Kurtzman, T., & Cho, A. E. (2018). Water pharmacophore: Designing ligands using molecular dynamics simulations with water. *Scientific Reports*. https://doi.org/10.1038/s41598-018-28546-z

Kaalia, R., Srinivasan, A., Kumar, A., & Ghosh, I. (2016). ILP-assisted de novo drug design. *Machine Learning, 103*(3), 309–341. https://doi.org/10.1007/s10994-016-5556-x

Kadirvelraj, R., Foley, B. L., Dyekjær, J. D., & Woods, R. J. (2008). Involvement of water in carbohydrate-protein binding: Concanavalin A revisited. *Journal of the American Chemical Society*. https://doi.org/10.1021/ja8039663

Kadurin, A., Nikolenko, S., Khrabrov, K., Aliper, A., & Zhavoronkov, A. (2017). DruGAN: An advanced generative adversarial autoencoder model for de novo generation of new molecules with desired molecular properties in silico. *Molecular Pharmaceutics*. https://doi.org/10.1021/acs.molpharmaceut.7b00346

Kaushik, A. C., Mehmood, A., Dai, X., & Wei, D. Q. (2020). A comparative chemogenic analysis for predicting drug-target pair via machine learning approaches. *Scientific Reports*. https://doi.org/10.1038/s41598-020-63842-7

Khamis, M. A., & Gomaa, W. (2015). Comparative assessment of machine-learning scoring functions on PDBbind 2013. *Engineering Applications of Artificial Intelligence*. https://doi.org/10.1016/j.engappai.2015.06.021

Khamis, M. A., Gomaa, W., & Ahmed, W. F. (2015). Machine learning in computational docking. *Artificial Intelligence in Medicine*. https://doi.org/10.1016/j.artmed.2015.02.002

Kinnings, S. L., Liu, N., Tonge, P. J., Jackson, R. M., Xie, L., & Bourne, P. E. (2011). A machine learning-based method to improve docking scoring functions and its application to drug repurposing. *Journal of Chemical Information and Modeling*. https://doi.org/10.1021/ci100369f

Kirchmair, J., Markt, P., Distinto, S., Schuster, D., Spitzer, G. M., Liedl, K. R., Langer, T., & Wolber, G. (2008). The Protein Data Bank (PDB), its related services and software tools as key components for in silico guided drug discovery. *Journal of Medicinal Chemistry*. https://doi.org/10.1021/jm8005977

Koeberle, A., & Werz, O. (2014). Multi-target approach for natural products in inflammation. *Drug Discovery Today*. https://doi.org/10.1016/j.drudis.2014.08.006

Koes, D. R., & Camacho, C. J. (2011). Pharmer: Efficient and exact pharmacophore search. *Journal of Chemical Information and Modeling*. https://doi.org/10.1021/ci200097m

Kollman, P. (1993). Free energy calculations: Applications to chemical and biochemical phenomena. *Chemical Reviews*. https://doi.org/10.1021/cr00023a004

Koutsoukas, A., Simms, B., Kirchmair, J., Bond, P. J., Whitmore, A. V., Zimmer, S., Young, M. P., Jenkins, J. L., Glick, M., Glen, R. C., & Bender, A. (2011). From in silico target prediction to multi-target drug design: Current databases, methods and applications. *Journal of Proteomics*. https://doi.org/10.1016/j.jprot.2011.05.011

Kramer, B., Rarey, M., & Lengauer, T. (1999). Evaluation of the FLEXX incremental construction algorithm for protein–ligand docking. *Proteins: Structure, Function, and Bioinformatics, 37*(2), 228–241. https://doi.org/10.1002/(SICI)1097-0134(19991101)37:2<228::AID-PROT8>3.0.CO;2–8

Kumar, P., Kaalia, R., Srinivasan, A., & Ghosh, I. (2017). Multiple target based pharmacophore designing from active site structures. *SAR and QSAR in Environmental Research*. https://doi.org/10.1080/1062936X.2017.1401555

Kuntz, I. D., Blaney, J. M., Oatley, S. J., Langridge, R., & Ferrin, T. E. (1982). A geometric approach to macromolecule-ligand interactions. *Journal of Molecular Biology*. https://doi.org/10.1016/0022-2836(82)90153-X

Lameijer, E.-W., Tromp, R. A., Spanjersberg, R. F., Brussee, J., & Ijzerman, A. P. (2007). Designing active template molecules by combining computational de novo design and human chemist's expertise. *Journal of Medicinal Chemistry, 50*(8), 1925–1932. https://doi.org/10.1021/jm061356+

Lauri, G., & Bartlett, P. A. (1994). Caveat: A program to facilitate the design of organic molecules. *Journal of Computer-Aided Molecular Design, 8*(1), 51–66. https://doi.org/10.1007/BF00124349

Lavecchia, Antonio, & Cerchia, C. (2016). In silico methods to address polypharmacology: Current status, applications and future perspectives. *Drug Discovery Today*. https://doi.org/10.1016/j.drudis.2015.12.007

Lavecchia, A., & Giovanni, C. (2013). Virtual screening strategies in drug discovery: A critical review. *Current Medicinal Chemistry*. https://doi.org/10.2174/09298673113209990001

Leelananda, S. P., & Lindert, S. (2016). Computational methods in drug discovery. *Beilstein Journal of Organic Chemistry*. https://doi.org/10.3762/bjoc.12.267

Lerner, M. G., Bowman, A. L., & Carlson, H. A. (2007). Incorporating dynamics in *E. coli* dihydrofolate reductase enhances structure-based drug discovery. *Journal of Chemical Information and Modeling, 47*(6), 2358–2365. https://doi.org/10.1021/ci700167n

Lexa, K. W., & Carlson, H. A. (2011). Full protein flexibility is essential for proper hot-spot mapping. *Journal of the American Chemical Society, 133*(2), 200−202. https://doi.org/10.1021/ja1079332

Lima, W. G., Brito, J. C. M., Overhage, J., & Nizer, W. S. da C. (2020). The potential of drug repositioning as a short-term strategy for the control and treatment of COVID-19 (SARS-CoV-2): A systematic review. *Archives of Virology.* https://doi.org/10.1007/s00705-020-04693-5

Lin, F.-Y., & Tseng, Y. J. (2011). Structure-based fragment hopping for lead optimization using predocked fragment database. *Journal of Chemical Information and Modeling, 51*(7), 1703−1715. https://doi.org/10.1021/ci200136j

Li, K., Schurig-Briccio, L. A., Feng, X., Upadhyay, A., Pujari, V., Lechartier, B., Fontes, L., Yang, H., Rao, G., Zhu, W., Gulati, A., No, J. H., Cintra, Gi., Bogue, S., Liu, Y. L., Molohon, K., Orlean, P., Mitchell, D. A., Freitas-Junior, L., ... (2014). Multitarget drug discovery for tuberculosis and other infectious diseases. *Journal of Medicinal Chemistry.* https://doi.org/10.1021/jm500131s

Li, H., Sze, K. H., Lu, G., & Ballester, P. J. (2020). Machine-learning scoring functions for structure-based drug lead optimization. *Wiley Interdisciplinary Reviews: Computational Molecular Science.* https://doi.org/10.1002/wcms.1465

Li, L., Wang, B., & Meroueh, S. O. (2011). Support vector regression scoring of receptor-ligand complexes for rank-ordering and virtual screening of chemical libraries. *Journal of Chemical Information and Modeling.* https://doi.org/10.1021/ci200078f

Li, J., Zheng, S., Chen, B., Butte, A. J., Swamidass, S. J., & Lu, Z. (2016). A survey of current trends in computational drug repositioning. *Briefings in Bioinformatics.* https://doi.org/10.1093/bib/bbv020

Lu, Y., Wang, R., Yang, C. Y., & Wang, S. (2007). Analysis of ligand-bound water molecules in high-resolution crystal structures of protein-ligand complexes. *Journal of Chemical Information and Modeling.* https://doi.org/10.1021/ci6003527

Lyne, P. D. (2002). Structure-based virtual screening: An overview. *Drug Discovery Today.* https://doi.org/10.1016/S1359-6446(02) 02483−2

Macalino, S. J. Y., Gosu, V., Hong, S., & Choi, S. (2015). Role of computer-aided drug design in modern drug discovery. *Archives of Pharmacal Research.* https://doi.org/10.1007/s12272-015-0640-5

Ma, D. L., Chan, D. S. H., & Leung, C. H. (2013). Drug repositioning by structure-based virtual screening. *Chemical Society Reviews.* https://doi.org/10.1039/c2cs35357a

Maggiora, G. M. (2011). The reductionist paradox: Are the laws of chemistry and physics sufficient for the discovery of new drugs? *Journal of Computer-Aided Molecular Design.* https://doi.org/10.1007/s10822-011-9447-8

Mak, K. K., & Pichika, M. R. (2019). Artificial intelligence in drug development: Present status and future prospects. *Drug Discovery Today.* https://doi.org/10.1016/j.drudis.2018.11.014

Mali, S. N., & Chaudhari, H. K. (2019). Molecular modelling studies on adamantane-based Ebola virus GP-1 inhibitors

using docking, pharmacophore and 3D-QSAR. *SAR and QSAR in Environmental Research.* https://doi.org/10.1080/1062936X.2019.1573377

Martin, Y. C., Kofron, J. L., & Traphagen, L. M. (2002). Do structurally similar molecules have similar biological activity? *Journal of Medicinal Chemistry.* https://doi.org/10.1021/jm020155c

Mauser, H., & Guba, W. (2008). Recent developments in de novo design and scaffold hopping. *Current Opinion in Drug Discovery and Development, 11*(3), 365−374.

McGann, M. (2011). FRED pose prediction and virtual screening accuracy. *Journal of Chemical Information and Modeling.* https://doi.org/10.1021/ci100436p

McGann, M. R., Almond, H. R., Nicholls, A., Grant, J. A., & Brown, F. K. (2003). Gaussian docking functions. *Biopolymers.* https://doi.org/10.1002/bip.10207

Meng, X.-Y., Zhang, H.-X., Mezei, M., & Cui, M. (2012). Molecular docking: A powerful approach for structure-based drug discovery. *Current Computer-Aided Drug Design.* https://doi.org/10.2174/157340911795677602

Moitessier, N., Englebienne, P., Lee, D., Lawandi, J., & Corbeil, C. (2008). Towards the development of universal, fast and highly accurate docking/scoring methods: A long way to go. *British Journal of Pharmacology, 153*(1), S7−S26.

Morris, G. M., Goodsell, D. S., Halliday, R. S., Huey, R., Hart, W. E., Belew, R. K., & Olson, A. J. (1998). Automated docking using a Lamarckian genetic algorithm and an empirical binding free energy function. *Journal of Computational Chemistry.* https://doi.org/10.1002/(SICI)1096−1987X(19981115)19:14<1639::AID-JCC10>3.0.CO;2-B

Nakano, H., Miyao, T., & Funatsu, K. (2020). Exploring topological pharmacophore graphs for scaffold hopping. *Journal of Chemical Information and Modeling, 60*(4), 2073−2081. https://doi.org/10.1021/acs.jcim.0c00098

Neves, M. A. C., Dinis, T. C. P., Colombo, G., & Sá E Melo, M. L. (2009). Fast three dimensional pharmacophore virtual screening of new potent non-steroid aromatase inhibitors. *Journal of Medicinal Chemistry.* https://doi.org/10.1021/jm800945c

Neves, M. A. C., Totrov, M., & Abagyan, R. (2012). Docking and scoring with ICM: The benchmarking results and strategies for improvement. *Journal of Computer-Aided Molecular Design.* https://doi.org/10.1007/s10822-012-9547-0

Nicolaou, C., Kannas, C., & Loizidou, E. (2012). Multi-objective optimization methods in de novo drug design. *Mini Reviews in Medicinal Chemistry, 12*(10), 979−987. https://doi.org/10.2174/138955712802762284

Olivecrona, M., Blaschke, T., Engkvist, O., & Chen, H. (2017). Molecular de-novo design through deep reinforcement learning. *Journal of Cheminformatics.* https://doi.org/10.1186/s13321-017-0235-x

Oprea, T. (2003). 3D QSAR modeling in drug design. *Computational Medicinal Chemistry for drug Discovery.* https://doi.org/10.1201/9780203913390.ch22

Paolini, G. V., Shapland, R. H. B., Van Hoorn, W. P., Mason, J. S., & Hopkins, A. L. (2006). Global mapping of pharmacological space. *Nature Biotechnology.* https://doi.org/10.1038/nbt1228

Pence, H. E., & Williams, A. (2010). Chemspider: An online chemical information resource. *Journal of Chemical Education.* https://doi.org/10.1021/ed100697w

Pinzi, L., Caporuscio, F., & Rastelli, G. (2018). Selection of protein conformations for structure-based polypharmacology studies. *Drug Discovery Today.* https://doi.org/10.1016/j.drudis.2018.08.007

Pirhadi, S., Shiri, F., & Ghasemi, J. B. (2013). Methods and applications of structure based pharmacophores in drug discovery. *Current Topics in Medicinal Chemistry.* https://doi.org/10.2174/1568026611313090006

Raghavendra, N. M., Pingili, D., Kadasi, S., Mettu, A., & Prasad, S. V. U. M. (2018). Dual or multi-targeting inhibitors: The next generation anticancer agents. *European Journal of Medicinal Chemistry.* https://doi.org/10.1016/j.ejmech.2017.10.021

Ragoza, M., Hochuli, J., Idrobo, E., Sunseri, J., & Koes, D. R. (2017). Protein-ligand scoring with convolutional neural networks. *Journal of Chemical Information and Modeling.* https://doi.org/10.1021/acs.jcim.6b00740

Rajamani, R., & Good, A. C. (2007). Ranking poses in structure-based lead discovery and optimization: Current trends in scoring function development. *Current Opinion in Drug Discovery and Development, 10*(3), 308–315. 17554857.

Rarey, M., Kramer, B., Lengauer, T., & Klebe, G. (1996). A fast flexible docking method using an incremental construction algorithm. *Journal of Molecular Biology.* https://doi.org/10.1006/jmbi.1996.0477

Ratni, H., Rogers-Evans, M., Bissantz, C., Grundschober, C., Moreau, J., Schuler, F., Fischer, H., Alvarez Sanchez, R., & Schnider, P. (2015). Discovery of highly selective brain-penetrant vasopressin 1a antagonists for the potential treatment of autism via a chemogenomic and scaffold hopping approach. *Journal of Medicinal Chemistry, 58*(5), 2275–2289. https://doi.org/10.1021/jm501745f

Roy, K. (2015). Quantitative structure-activity relationships in drug design, predictive toxicology, and risk assessment. *Quantitative Structure-Activity Relationships in Drug Design, Predictive Toxicology, and Risk Assessment.* https://doi.org/10.4018/978-1-4666-8136-1

Royal Society of Chemistry. (2015). *ChemSpider. Search and Share Chemistry.* Royal Society of Chemistry.

Rueda, M., Bottegoni, G., & Abagyan, R. (2010). Recipes for the selection of exptl protein conformations for virtual screening. *Journal of Chemical Information and Modeling, 50*(1), 186–193.

Rush, T. S., Grant, J. A., Mosyak, L., & Nicholls, A. (2005). A shape-based 3-D scaffold hopping method and its application to a bacterial protein–protein interaction. *Journal of Medicinal Chemistry, 48*(5), 1489–1495. https://doi.org/10.1021/jm040163o

Salam, N. K., Nuti, R., & Sherman, W. (2009). Novel method for generating structure-based pharmacophores using energetic analysis. *Journal of Chemical Information and Modeling.* https://doi.org/10.1021/ci900212v

Šali, A. (1993). MODELLER A program for protein structure modeling. In *Comparative protein modelling by satisfaction of spatial restraints.*

Sanseau, P., Agarwal, P., Barnes, M. R., Pastinen, T., Richards, J. B., Cardon, L. R., & Mooser, V. (2012). Use of genome-wide association studies for drug repositioning. *Nature Biotechnology.* https://doi.org/10.1038/nbt.2151

Sastry, M., Lowrie, J. F., Dixon, S. L., & Sherman, W. (2010). Large-scale systematic analysis of 2D fingerprint methods and parameters to improve virtual screening enrichments. *Journal of Chemical Information and Modeling.* https://doi.org/10.1021/ci100062n

Schaller, D., Sribar, D., Noonan, T., Deng, L., Nguyen, T. N., Pach, S., Machalz, D., Bermudez, M., & Wolber, G. (2020). Next generation 3D pharmacophore modeling. *Wiley Interdisciplinary Reviews: Computational Molecular Science.* https://doi.org/10.1002/wcms.1468

Schneider, G. (2010). Virtual screening: An endless staircase? *Nature Reviews Drug Discovery.* https://doi.org/10.1038/nrd3139

Schneider, G., & Fechner, U. (2005). Computer-based de novo design of drug-like molecules. *Nature Reviews Drug Discovery, 4*(8), 649–663. https://doi.org/10.1038/nrd1799

Schneider, G., Hartenfeller, M., Reutlinger, M., Tanrikulu, Y., Proschak, E., & Schneider, P. (2009). Voyages to the (un)known: Adaptive design of bioactive compounds. *Trends in Biotechnology, 27*(1), 18–26. https://doi.org/10.1016/j.tibtech.2008.09.005

Schneidman-Duhovny, D., Dror, O., Inbar, Y., Nussinov, R., & Wolfson, H. J. (2008). PharmaGist: A webserver for ligand-based pharmacophore detection. *Nucleic Acids Research, 36*(Web Server), W223–W228. https://doi.org/10.1093/nar/gkn187

Schuster, D., Nashev, L. G., Kirchmair, J., Laggner, C., Wolber, G., Langer, T., & Odermatt, A. (2008). Discovery of nonsteroidal 17β-hydroxysteroid dehydrogenase 1 inhibitors by pharmacophore-based screening of virtual compound libraries. *Journal of Medicinal Chemistry.* https://doi.org/10.1021/jm800054h

Shin, W. J., & Seong, B. L. (2013). Recent advances in pharmacophore modeling and its application to anti-influenza drug discovery. *Expert Opinion on Drug Discovery.* https://doi.org/10.1517/17460441.2013.767795

Singh, J., Abraham, W. M., Adams, S. P., Van Vlijmen, H., Liao, Y., Lee, W. C., Cornebise, M., Harris, M., Shu, I. H., Gill, A., & Cuervo, J. H. (2002). Identification of potent and novel α4β1 antagonists using in silico screening. *Journal of Medicinal Chemistry.* https://doi.org/10.1021/jm020054e

Smart, O. S., Horský, V., Gore, S., Vařeková, R. S., Bendová, V., Kleywegt, G. J., & Velankar, S. (2018). Validation of ligands in macromolecular structures determined by X-ray crystallography. *Acta Crystallographica Section D: Structural Biology.* https://doi.org/10.1107/S2059798318002541

Sotriffer, C. A., Sanschagrin, P., Matter, H., & Klebe, G. (2008). SFCscore: Scoring functions for affinity prediction of protein-ligand complexes. *Proteins: Structure, Function and Genetics.* https://doi.org/10.1002/prot.22058

Speck-Planche, A., Kleandrova, V. V., & Cordeiro, M. N. D. S. (2013). Chemoinformatics for rational discovery of safe antibacterial drugs: Simultaneous predictions of biological activity against streptococci and toxicological profiles in laboratory animals. *Bioorganic and Medicinal Chemistry*. https://doi.org/10.1016/j.bmc.2013.03.015

Spyrakis, F., Ahmed, M. H., Bayden, A. S., Cozzini, P., Mozzarelli, A., & Kellogg, G. E. (2017). The roles of water in the protein matrix: A largely untapped resource for drug discovery. *Journal of Medicinal Chemistry*. https://doi.org/10.1021/acs.jmedchem.7b00057

Stephenson, V. C., Heyding, R. A., & Weaver, D. F. (2005). The "promiscuous drug concept" with applications to Alzheimer's disease. *FEBS Letters*. https://doi.org/10.1016/j.febslet.2005.01.019

Stepniewska-Dziubinska, M. M., Zielenkiewicz, P., & Siedlecki, P. (2018). Development and evaluation of a deep learning model for protein−ligand binding affinity prediction. *Bioinformatics*. https://doi.org/10.1093/bioinformatics/bty374

Su, M., Feng, G., Liu, Z., Li, Y., & Wang, R. (2020). Tapping on the black box: How is the scoring power of a machine-learning scoring function dependent on the training set? *Journal of Chemical Information and Modeling*. https://doi.org/10.1021/acs.jcim.9b00714

Sun, H., Tawa, G., & Wallqvist, A. (2012). Classification of scaffold-hopping approaches. *Drug Discovery Today*, *17*(7−8), 310−324. https://doi.org/10.1016/j.drudis.2011.10.024

Taminau, J., Thijs, G., & De Winter, H. (2008). Pharao: Pharmacophore alignment and optimization. *Journal of Molecular Graphics and Modelling*. https://doi.org/10.1016/j.jmgm.2008.04.003

Teague, S. J. (2003). Implications of protein flexibility for drug discovery. *Nature Reviews Drug Discovery*, *2*(7), 527−541. https://doi.org/10.1038/nrd1129

Teodoro, M., & Muegge, I. (2011). BIBuilder: Exhaustive searching for de novo ligands. *Molecular Informatics*, *30*(1), 63−75. https://doi.org/10.1002/minf.201000122

Tetko, I. V., Gasteiger, J., Todeschini, R., Mauri, A., Livingstone, D., Ertl, P., Palyulin, V. A., Radchenko, E. V., Zefirov, N. S., Makarenko, A. S., Tanchuk, V. Y., & Prokopenko, V. V. (2005). Virtual computational chemistry laboratory - design and description. *Journal of Computer-Aided Molecular Design*. https://doi.org/10.1007/s10822-005-8694-y

Todeschini, R., & Consonni, V. (2010). Molecular descriptors for chemoinformatics. *Molecular Descriptors for Chemoinformatics*. https://doi.org/10.1002/9783527628766

Torrisi, M., Pollastri, G., & Le, Q. (2020). Deep learning methods in protein structure prediction. *Computational and Structural Biotechnology Journal*. https://doi.org/10.1016/j.csbj.2019.12.011

Vamathevan, J., Clark, D., Czodrowski, P., Dunham, I., Ferran, E., Lee, G., Li, B., Madabhushi, A., Shah, P., Spitzer, M., & Zhao, S. (2019). Applications of machine learning in drug discovery and development. *Nature Reviews Drug Discovery*. https://doi.org/10.1038/s41573-019-0024-5

Verdonk, M. L., Cole, J. C., Hartshorn, M. J., Murray, C. W., & Taylor, R. D. (2003). Improved protein-ligand docking using GOLD. *Proteins: Structure, Function, and Genetics*. https://doi.org/10.1002/prot.10465

Vogt, M., & Bajorath, J. (2011). Predicting the performance of fingerprint similarity searching. *Methods in Molecular Biology (Clifton, N.J.)*. https://doi.org/10.1007/978-1-60761-839-3_6

Vogt, M., Stumpfe, D., Geppert, H., & Bajorath, J. (2010). Scaffold hopping using two-dimensional fingerprints: True potential, black magic, or a hopeless endeavor? Guidelines for virtual screening. *Journal of Medicinal Chemistry*, *53*(15), 5707−5715. https://doi.org/10.1021/jm100492z

Wallach, I., Dzamba, M., & Heifets, A. (2015). *AtomNet: A deep convolutional neural network for bioactivity prediction in structure-based drug discovery* (pp. 1−11). Retrieved from http://arxiv.org/abs/1510.02855.

Wang, Y., Bryant, S. H., Cheng, T., Wang, J., Gindulyte, A., Shoemaker, B. A., Thiessen, P. A., He, S., & Zhang, J. (2017). PubChem BioAssay: 2017 update. *Nucleic Acids Research*. https://doi.org/10.1093/nar/gkw1118

Wang, L., Deng, Y., Wu, Y., Kim, B., LeBard, D. N., Wandschneider, D., Beachy, M., Friesner, R. A., & Abel, R. (2017). Accurate modeling of scaffold hopping transformations in drug discovery. *Journal of Chemical Theory and Computation*, *13*(1), 42−54. https://doi.org/10.1021/acs.jctc.6b00991

Wang, R., Gao, Y., & Lai, L. (2000). LigBuilder: A multi-purpose program for structure-based drug design. *Journal of Molecular Modeling*, *6*(7−8), 498−516. https://doi.org/10.1007/s0089400060498

Wang, C., & Zhang, Y. (2017). Improving scoring-docking-screening powers of protein−ligand scoring functions using random forest. *Journal of Computational Chemistry*. https://doi.org/10.1002/jcc.24667

Weiner, S. J., Kollman, P. A., Singh, U. C., Case, D. A., Ghio, C., Alagona, G., Profeta, S., & Weiner, P. (1984). A new force field for molecular mechanical simulation of nucleic acids and proteins. *Journal of the American Chemical Society*. https://doi.org/10.1021/ja00315a051

Westbrook, J. D., & Burley, S. K. (2019). How structural biologists and the protein Data Bank contributed to recent FDA new drug approvals. *Structure*. https://doi.org/10.1016/j.str.2018.11.007

Więckowska, A., Kołaczkowski, M., Bucki, A., Godyń, J., Marcinkowska, M., Więckowski, K., Zaręba, P., Siwek, A., Kazek, G., Głuch-Lutwin, M., Mierzejewski, P., Bienkowski, P., Sienkiewicz-Jarosz, H., Knez, D., Wichur, T., Gobec, S., & Malawska, B. (2016). Novel multi-target-directed ligands for Alzheimer's disease: Combining cholinesterase inhibitors and 5−HT6 receptor antagonists. Design, synthesis and biological evaluation. *European Journal of Medicinal Chemistry*. https://doi.org/10.1016/j.ejmech.2016.08.016

Wieder, M., Perricone, U., Seidel, T., Boresch, S., & Langer, T. (2016). Comparing pharmacophore models derived from crystal structures and from molecular dynamics simulations. *Monatshefte Fur Chemie*, *147*(3), 553−563. https://doi.org/10.1007/s00706-016-1674-1

Willett, P. (2013). Fusing similarity rankings in ligand-based virtual screening. *Computational and Structural Biotechnology Journal*. https://doi.org/10.5936/csbj.201302002

Wlodawer, A., & Vondrasek, J. (1998). Inhibitors of HIV-1 protease: A major success of structure-assisted drug design. *Annual Review of Biophysics and Biomolecular Structure.* https://doi.org/10.1146/annurev.biophys.27.1.249

Wójcikowski, M., Ballester, P. J., & Siedlecki, P. (2017). Performance of machine-learning scoring functions in structure-based virtual screening. *Scientific Reports.* https://doi.org/10.1038/srep46710

Wolber, G., & Langer, T. (2005). LigandScout: 3-D pharmacophores derived from protein-bound ligands and their use as virtual screening filters. *Journal of Chemical Information and Modeling.* https://doi.org/10.1021/ci049885e

Wong, S. E., & Lightstone, F. C. (2011). Accounting for water molecules in drug design. *Expert Opinion on Drug Discovery.* https://doi.org/10.1517/17460441.2011.534452

Xiang, M., Cao, Y., Fan, W., Chen, L., & Mo, Y. (2012). Computer-aided drug design: Lead discovery and optimization. *Combinatorial Chemistry and High Throughput Screening.* https://doi.org/10.2174/138620712799361825

Xue, L., Godden, J. W., Stahura, F. L., & Bajorath, J. (2003). Design and evaluation of a molecular fingerprint involving the transformation of property descriptor values into a binary classification scheme. *Journal of Chemical Information and Computer Sciences.* https://doi.org/10.1021/ci030285+

Yang, S.-Y. (2010). Pharmacophore modeling and applications in drug discovery: Challenges and recent advances. *Drug Discovery Today, 15*(11−12), 444−450. https://doi.org/10.1016/j.drudis.2010.03.013

Yang, H., Shen, Y., Chen, J., Jiang, Q., Leng, Y., & Shen, J. (2009). Structure-based virtual screening for identification of novel 11β-HSD1 inhibitors. *European Journal of Medicinal Chemistry.* https://doi.org/10.1016/j.ejmech.2008.06.005

Yap, C. W. (2011). PaDEL-descriptor: An open source software to calculate molecular descriptors and fingerprints. *Journal of Computational Chemistry.* https://doi.org/10.1002/jcc.21707

Young, J. Y., Westbrook, J. D., Feng, Z., Sala, R., Peisach, E., Oldfield, T. J., Sen, S., Gutmanas, A., Armstrong, D. R., Berrisford, J. M., Chen, L., Chen, M., Di Costanzo, L., Dimitropoulos, D., Gao, G., Ghosh, S., Gore, S., Guranovic, V., Hendrickx, P. M. S., … (2017). OneDep: Unified wwPDB system for deposition, biocuration, and validation of macromolecular structures in the PDB archive. *Structure.* https://doi.org/10.1016/j.str.2017.01.004

Yu, W., Lakkaraju, S. K., Raman, E. P., Fang, L., & Mackerell, A. D. (2015). Pharmacophore modeling using site-identification by ligand competitive saturation (SILCS) with multiple probe molecules. *Journal of Chemical Information and Modeling.* https://doi.org/10.1021/ci500691p

Yu, W., & Mackerell, A. D. (2017). Computer-aided drug design methods. *Methods in Molecular Biology.* https://doi.org/10.1007/978-1-4939-6634-9_5

Zhang, L., Ai, H. X., Li, S. M., Qi, M. Y., Zhao, J., Zhao, Q., & Liu, H. S. (2017). Virtual screening approach to identifying influenza virus neuraminidase inhibitors using molecular docking combined with machine-learning-based scoring function. *Oncotarget.* https://doi.org/10.18632/oncotarget.20915

Zhang, L., Fourches, D., Sedykh, A., Zhu, H., Golbraikh, A., Ekins, S., Clark, J., Connelly, M. C., Sigal, M., Hodges, D., Guiguemde, A., Guy, R. K., & Tropsha, A. (2013). Discovery of novel antimalarial compounds enabled by QSAR-based virtual screening. *Journal of Chemical Information and Modeling.* https://doi.org/10.1021/ci300421n

Zhang, W., Pei, J., & Lai, L. (2017). Computational multitarget drug design. *Journal of Chemical Information and Modeling.* https://doi.org/10.1021/acs.jcim.6b00491

Zheng, S., Li, Y., Chen, S., Xu, J., & Yang, Y. (2020). Predicting drug−protein interaction using quasi-visual question answering system. *Nature Machine Intelligence.* https://doi.org/10.1038/s42256-020-0152-y

Zhou, Y., Hou, Y., Shen, J., Huang, Y., Martin, W., & Cheng, F. (2020). Network-based drug repurposing for novel coronavirus 2019-nCoV/SARS-CoV-2. *Cell Discovery.* https://doi.org/10.1038/s41421-020-0153-3

Zilian, D., & Sotriffer, C. A. (2013). SFCscoreRF: A random forest-based scoring function for improved affinity prediction of protein-ligand complexes. *Journal of Chemical Information and Modeling.* https://doi.org/10.1021/ci400120b

CHAPTER 2

Biomolecular Talks—Part 1: A Theoretical Revisit on Molecular Modeling and Docking Approaches

AMUTHA RAMASWAMY • SANGEETHA BALASUBRAMANIAN • MUTHUKUMARAN RAJAGOPALAN

1 BIOMOLECULES AND THEIR INTERACTIONS

Biomolecules are the substances produced by living cells. They vary widely in size as well as structures and form the basis for biological processes such as growth, metabolism, and development (Uzman, 2003). Cells are composed of organic and inorganic molecules called biomolecules, of which more than 70% is water. Biomolecules can be either endogenous (produced within the organism) or exogenous in origin, like the essential nutrients for survival. Chemically speaking, biomolecules are majorly made of the elements carbon, nitrogen, oxygen, phosphorous, and hydrogen, which form about 96% of the biomolecular space while other rare metals such as zinc, magnesium, manganese, and gold are found in trace quantities. Interactions between the cellular constituents are of great importance, as they control the structural, functional, and biological roles of an organism. The integration of computational approaches with biological sciences has greatly improved our understanding of these intricate communications at the atomic level. This chapter presents an ornamented vision of biological chemistry from the perspective of computational biology via biomolecular interactions.

1.1 Biomolecules

Biomolecules are the macromolecules including proteins, nucleic acids, lipids, carbohydrates, small molecules such as primary metabolites, secondary metabolites, natural products, etc.

Proteins are made of smaller units called amino acids, which are organic compounds containing amino and carboxylic acid functional groups. The primary structure of proteins is a linear arrangement of amino acids connected by peptide bonds, the order being defined by the protein encoding gene. The primary structure further folds into a secondary structure of either alpha helices (stabilized by hydrogen bonds between i^{th} and $i+4^{th}$ residues), parallel, or antiparallel beta-sheets. These secondary structures are connected by disordered or unstructured "loops" or "coils." The relative arrangement of these secondary structures gives rise to the tertiary structure, stabilized by disulfide bonds and van der Waals (vdW) interactions in addition to hydrogen bonds (Nelson, 2005). A quaternary arrangement of proteins can be observed when two or more polypeptide chains interact to form a multimeric protein. A specific tertiary structure that repeats itself in several proteins is called as a "fold." When these folds perform a specific function and are functionally independent from the remaining protein structure, these folds are called "domains." In several cases, proteins require additional small molecules (called as cofactors) to be functional, which are called as apo-enzymes. Inhibition studies of proteins or enzymes by small molecules generally follow competitive or allosteric inhibition processes and the most commonly targeted binding sites are the substrate binding site, cofactor binding site, or multimer interfaces (Bordoli & Schwede, 2006).

Deoxyribonucleic acid (DNA) and ribonucleic acid (RNA) are made of three types of molecular fragments: (1) nucleobases or nitrogenous bases (purines like adenine and guanine, and pyrimidines like cytosine, thymine, and uracil), (2) furanose sugars (ribose or deoxyribose), and (3) phosphates. The sugar moieties act as a bridge between a nitrogenous base (nucleoside) and a phosphodiester group to form a nucleotide. These nucleotides form the source of chemical energy (adenosine triphosphate and guanine triphosphate [GTP]), involve in cellular signaling, and act as cofactors in enzymatic reactions. The phosphor-diester bonds between several nucleotides result in a single strand of nucleic acid and hydrogen bonds between nitrogenous bases in the adjacent strands lead to the formation of a double helical strand of DNA. Hence, nucleotides are the building blocks of nucleic acids. The major conformation of DNA is double-stranded, while a small percentage of triple- and four-stranded structures of DNA are also seen in certain regulatory regions in the genome (Cooper, 2000). On the other hand, the major conformation of RNA is single-stranded with several tertiary structures stabilized by intrastrand hydrogen bonding. Such three-dimensional (3D) structure of RNA is seen in tRNA, ribozymes, and riboswitches and also forms an intricate part of ribosomes. These functionally important molecules are targeted by small molecules or oligonucleotide RNAs for gene regulation—related diseases. The mode of binding of these small molecule inhibitors (or stabilizers), as in case of G-quadruplex binders, is intercalation, end stacking, or groove binding.

Lipids are esters of saturated or unsaturated fatty acids and form the basic building blocks of all biological membranes. Lipids are amphiphilic in nature and generally possess a polar or hydrophilic head group and a nonpolar or hydrophobic fatty acid tail. The fatty acid tails are either saturated or unsaturated chains of hydrocarbons (varying between 14 and 24 carbons). The lipids give unique properties to biological membranes which can hold the cellular contents while allowing the entry and exit of certain molecules essential for cell survival. There are integral membrane proteins which act as transporters for the entry and exit of small molecules of which the ABC transporters, responsible for antibiotic resistance, form a particularly important class in lower eukaryotes (van Meer et al., 2008). Lipid membranes are of prime focus in studying the absorption and adsorption properties of pharmaceutically important small molecules such as curcumin, penetratin, and transportan, etc., in order to make them more effective in reaching their designed target molecules (Guidotti et al., 2017).

Carbohydrates (saccharides) are compounds containing carbon atoms (at least three), hydroxyl groups (few), and usually an aldehyde or ketone group. Their vital roles include (1) formation of nucleic acid backbone, (2) energy store house, fuels, and metabolic intermediates, (3) skeletal structures in the cell walls of bacteria and plants, etc. There are four major chemical groups of carbohydrates namely monosaccharides, disaccharides, oligosaccharides, and polysaccharides. Monosaccharides (e.g., glucose, fructose) and disaccharides (e.g., sucrose, lactose) are the smallest forms of

carbohydrates and are simply called as sugars (Berg et al., 2015; Nelson, 2005). Being a key component of RNA and DNA, the monosaccharides, ribose, and deoxyribose are important for the very basis of life. Polysaccharides are the most important form of carbohydrates because they store energy in the form of starch and glycogen, apart from forming the structural components cellulose and chitin. Glycosylation process adds carbohydrate moieties to proteins during the post-translational modification and increases the proteome diversity. The presence of different carbohydrate moieties signals different downstream processes and hence carbohydrates form an important part of protein—protein or protein—small molecule interactions, thereby mediating cellular signaling (Cooper, 2000).

Small molecules present in the biological environment form another major class of biomolecules due to their chemical, structural, and functional diversity. The weight of an organic compound has to be lower than an arbitrary cutoff value of 900 Da, in order to be considered as a small molecule. These small molecules serve as signaling molecules to regulate biological processes and are often used as probes or therapeutic agents. The basic building blocks of macromolecules such as amino acids, nucleotides, carbohydrates, and fatty acids are also considered as small molecules. Several plants, bacteria, fungi, and other organisms produce small molecular metabolic by-products (secondary metabolites/natural products) such as alkaloids, terpenes, etc. These metabolites are biologically active compounds with therapeutic potential and hence are commonly used as leads for drug discovery (Wink, 2003).

1.2 Biomolecular Interactions

All these biomolecules coordinate a complex interaction network in order to initiate any kind of communications between them and are the basis for all biological processes. These molecular interactions that exist between bonded atoms (molecules) as well as between nonbonded atoms are either attractive or repulsive and are noncovalent in nature. They are important in every aspect including biomolecule structural architecture, folding, ligand binding, design of biosensors, etc., and eventually dominate the entire life of living organisms.

Let us consider the G-protein—coupled receptor (GPCR) pathway as an example of biomolecular interaction. The G-proteins range from small single domain proteins to large heterotrimers and have the ability to bind to GTP and guanine diphosphate (GDP) molecules. When the G-protein is inactive, the $G\alpha$ subunit of a $\alpha\beta\gamma$ heterotrimer remains associated with a transmembrane protein called GPCR (Marrari et al., 2007). Upon receiving a signal from external stimulus (binding of a ligand), the GPCR undergoes conformational change and activates the Gα subunit of G-protein. While inactive, a GDP molecule remains bound to the Gα subunit. Upon activation, the GDP molecule is replaced by a GTP molecule inducing the separation of βγ subunits from the heterotrimer. These βγ subunits remain attached to the lipid membrane by lipid anchors. However, they are free from the GPCR and diffuse laterally to interact with other membrane proteins and thus passing on the signal further. Once the GTP is hydrolyzed to GDP, the Gα subunit assumes an inactive state and a heterotrimer is formed back, which reattaches to the GPCR and awaits for further signals. In this way, the G-proteins are turned on or off by external signals (Denis et al., 2012). In this process, several biomolecular interactions such as small molecule—protein, protein—protein, and protein—membrane promote significant interplays to mediate a biological process. In a similar manner, protein—nucleic acid interactions also play a key role in processes such as DNA replication, translation, and reverse transcription. Hence, understanding the mechanisms behind such biomolecular interactions has been a great deal of debate for the research communities from basic science to applied sciences including clinical sciences as well. Hence, it has become an indispensable platform promoting basic scientific outcome into the level of translational research including pharmaceutical interventions, design of biosensors, etc., that eventually improve the quality of human life (Ladbury, 2013).

1.2.1 Types of biomolecular interactions

Biomolecular interactions involve all types of interactions either within (intramolecular interactions) or between biomolecules (intermolecular interactions). Based on the chemical nature of interactions, they are classified as bonded and nonbonded interactions. Bonded interactions majorly refer to covalent linkages and disulfide bonds, while nonbonded interactions imply the formation of hydrogen bonds, vdW interactions, salt bridges, and electrostatic interactions. These interactions are the main driving forces for biomolecular stability, function, and signal communications. Studying the nature of these interactions and their modeling is important in areas such as drug design, nanotechnology, biosensor, and biomarker designing (Bordoli & Schwede, 2006).

1.2.1.1 Bonded interactions.
In proteins, disulfide bonds have structural as well as functional importance.

The covalent bond formed between the thiol groups from two cysteine amino acids is called a disulfide bond or a disulfide bridge. Structurally, the disulfide bonds help in protein folding by protecting the hydrophobic core and stabilizing the folded state. It can also help to hold multiple chains of a protein to form a quaternary structure. Functionally, these disulfide bonds can involve in redox reactions in regulatory proteins, due to their easy reversible nature as in the ferredoxin thioredoxin system during photosynthesis (Nikkanen & Rintamaki, 2014). Several bacteria contain redox proteins rich in disulfide bonds and special operons to code for unique disulfide bond forming proteins, which are used as therapeutic targets (Hogg, 2013). Compounds like Neratinib, an EGFR inhibitor and Ibrutinib, a BTK inhibitor forms covalent linkages with their target protein's thiol group of Cys residue giving rise to irreversible binding at the active site leading to strong inhibition potency (Kumalo et al., 2015).

1.2.1.2 Nonbonded interactions.
The nature of biomolecular interactions is dominated by nonbonded interactions such as hydrogen bonds, vdW interactions, salt bridges, and electrostatic interactions between charged groups. These interactions are distance-dependent inter- or intramolecular interactions, which predominate during the talks between biomolecules. Though nonbonded interactions are weaker than bonded interactions, they are quantitatively dominating the chemical space of biomolecular interactions and provide the necessary stability to biomolecular complexes.

vdW interactions: vdW interactions are weak distance-dependent forces between atoms or molecules. Though these forces vanish at longer distances, they play a fundamental role in structural biology, nanotechnology, and polymers. These forces are present in biomolecules as frequently as H-bonds and influence their solubility in polar or nonpolar solvents. As it is a distance-dependent force, there are attraction and repulsion components based on the interatomic distance and their atomic radii. When the atoms are at a higher separation distance, an attractive force acts on the atoms as they come closer; as the atoms move closer, the attractive force converts into a repulsive force when the atoms are very close, due to repulsion between the electron clouds. The distance at which the attractive force changes into repulsive force is called the vdW contact distance. vdW interactions are additive, nondirectional short-range interactions present between point charges, point charge—dipoles, dipole—dipole, or induced dipoles. Because they are short-range interactions, vdW interactions play a major role in stabilizing protein–small molecule and DNA–small molecule complexes in biological processes. The strength of vdW interactions ranges from 0.4 to 4.0 kcal/mol (Atkins & De Paula, 2014; Hermann et al., 2017).

Electrostatic interactions: Electrostatic interactions are the attractive force between two charged groups present either in a close distance (short range) or separated over a large distance (long range). At longer distances, these interactions are similar to coulombic interaction between point charges. However, at short range, an ionic bond can form between the two charged groups. An ionic bond is formed due to the transfer or sharing of electrons from a positively charged ion (cation) to a negatively charged ion (anion). These bonds vary from the covalent bonds in their high degree of electron transfer than sharing (as in covalent bonds). These interactions are stronger than other nonbonded interactions due to their partial covalent nature. The strength of an ionic bond can vary between 40 and 360 kcal/mol. The predominant electrostatic interactions observed in biomolecular complexes can be broadly classified as hydrogen bonds and salt bridge interactions.

Hydrogen bond interactions: A hydrogen bond is a special type of inter- or intramolecular dipole–dipole attractions created when a hydrogen atom bonded to a strong electronegative atom exists in the vicinity of another electronegative atom with a lone pair of electrons. One of these electronegative atoms is bonded to a hydrogen atom (called as donor, D), which is shared with another electronegative atom having a lone pair of electrons (called as an acceptor, A). The H-bond of the type N−H...O is stronger than C−H...O interactions. The geometric criteria to define an ideal hydrogen bond are >135 degrees angle between the D-H...A atoms and an H...A distance of ∼2 Å. Based on these geometries, the strength of a hydrogen bond can vary up to 40 kcal/mol, with interactions between 5 and 15 kcal/mol defined as moderate, >15–40 kcal/mol defined as strong, and <5 kcal/mol defined as weak interactions (Desiraju, 2011). Hence, these interactions are stronger than other nonbonded interactions and are the major intermolecular forces acting to stabilize biomolecules.

Few examples are given to highlight the importance and universality of H-bonds: (1) the two strands of DNA are held together by H-bonds between the nitrogenous bases of each strand, (2) the most common secondary structures of protein like alpha helices, beta-pleated sheets, etc., are held intact by the H-bonds between the amino acids from inter- and intrachains, (3) the interchain interactions between polysaccharides such as cellulose and chitin via the H-bonds between −OH groups, etc. (Atkins & De Paula, 2014; Price, 1999; Shu-Kun, 1999). Moreover, the cells contain more than 70% of aqueous environment, where H-bonding interactions expressed between the biomolecules and water are highly responsible for the entire cellular activities. Hence, the role of H-bonds in biological systems is ubiquitous. The H-bonds dictate most of the biological processes right from the information transfer mechanism of nucleic acids to the enzymatic activities, etc., via regulating the structural and functional activities of all biomolecules (Cerny & Hobza, 2007; Garand et al., 2012; Sivakova & Rowan, 2005).

Salt bridges: Salt bridge is a special type of electrostatic interaction, which is a combination of H-bonding and ionic bonding interactions (Rozas, 2007). Salt bridges or ion pairing interactions are stable because the dissociation of an ion pair requires high desolvation energy. They are found in proteins and are an important part of supramolecular chemistry. The strength of a salt bridge depends on several factors such as pH of the solvent, pKa of the ionizable groups, and distance between the interacting groups (Pylaeva et al., 2018) and typically ranges from −1 to −5 kcal/mol (Kumar & Nussinov, 1999). The charged amino acids such as lysine, arginine, aspartic acid, and glutamic acids commonly form salt bridges in proteins. Additionally, the amino acids with ionizable side chains such as histidine, tyrosine, and serine can also form salt bridges based on their pKas. Due to the high number of ionizable groups and the high probability of forming salt bridges, the pH at which a protein is studied is crucial (Atkins & De Paula, 2014).

1.3 Need for Computational Approaches

The study of biomolecular interactions is a complex process and requires sophisticated techniques. As these interactions are played at the atomic level, a highest possible resolution is necessary to gain a clear picture of the nature of these interactions. The development of large-scale high-throughput approaches has increased the amount of data available regarding biomolecular interactions. Such large data often contain noises, and the deduction of interaction networks from this noisy data often leads to misinterpretations as they are incomplete (D'Argenio, 2018). On the other hand, experimental techniques such as spectroscopic methods, surface plasmon resonance, isothermal titration calorimetry, mass spectrometry, nuclear magnetic resonance (NMR), and enzyme-linked immunosorbent assay, which are widely used to characterize biomolecular interactions, could not assess the entire interactome of an organism (Ladbury, 2013). Furthermore, experimental methods like yeast two-hybrid system can result in high false positives, while the use of more traditional methods are expensive and time-consuming in spite of their accuracy. Another major disadvantage of interpreting these experimental results is that different laboratory conditions lead to different observations and hence might not be reproducible. Hence, a uniform method that can handle large-scale interactome data was developed by computational methods (Browne et al., 2010). Computational techniques have gained popularity in this regard due to their efficiency in handling a large amount of data and provide a high resolution that is out of reach for most of the experimental techniques. The development of molecular mechanics (MM) based or quantum mechanics—based software and tools for the visualization and manipulation of structures has further eased the process of understanding biomolecular interactions by multifold.

2 COMPUTATIONAL APPROACHES TO STUDY BIOMOLECULAR INTERACTIONS

With the advent of computer technology, the scientific approach on learning biological chemistry has also got a new facet in exploring the functional roles of cellular constituents. The ultimate aim of modern biology resides mainly on understanding the biomolecular communications at the atomic level through intermolecular interactions like protein—DNA, protein—protein, protein—lipid, etc.

The current scenario of the computational approaches on understanding the biological chemistry falls into two categories such as MM-based approach and molecular dynamics (MD) based approach. The former one deals the system at the static level, while the latter one explores the time-dependent behavior of the biomolecule and the below sections deal these two approaches extensively. Force fields are the basic requirement for studying molecular systems by computational methods. For the standard amino acids and nucleic acid bases, the force field potentials have been parameterized and are available for direct use (Nerenberg & Head-Gordon, 2018). These force fields describe a biomolecule and model the biomolecular interaction by MM-based calculations such as energy minimizations, molecular docking, MD simulations, and computer-aided drug design (Vanommeslaeghe et al., 2014).

2.1 Molecular Mechanics—Based Approaches

MM follows classical mechanical principles to model molecular systems. In this approach, Born-Oppenheimer approximation is implied and hence the potential energy of a system is calculated as a function of nuclear coordinates alone (Born & Oppenheimer, 1927). According to MM, nuclei and electrons are treated together as unified atom-like particles (balls) and the bonds between them are considered as spring-like. The total energy (E_{Total}) of such "ball and spring model" emerges from both bonded and nonbonded interactions. Bonded energies arise from bond (E_{bond}), angle (E_{angle}), dihedral ($E_{dihedral}$), improper dihedral ($E_{improper}$), and cross terms (E_{cross}), whereas the nonbonded energies combine vdW (E_{vdW}), H-bond ($E_{H-bonds}$), and electrostatic ($E_{coulomb}$) interactions (Eq. 2.1). These interactions could sufficiently be modeled using simple potential functions with empirically derived parameters and are collectively termed as force field, as explained in Eq. (2.1).

$$E_{Total} = E_{bond} + E_{angle} + E_{dihedral} + E_{improper} + E_{cross} \tag{2.1}$$
$$+ E_{vdW} + E_{H-bonds} + E_{coulomb}$$

The potential functions and the associated parameters included in a force field are optional according to the need and hence vary between the available MM force fields. The few commonly used force fields are (1) MM2, MM3, and MM4 series for organic compounds (Allinger, 1977; Allinger & Chen, 1996; Allinger et al., 1989) and (2) AMBER (Cornell et al., 1995; Weiner et al., 1984), GROMOS (Hermans et al., 1984), CHARMM (MacKerell et al., 2002) and OPLS (Jorgensen & Tirado-Rives, 1988) for biomolecules. The potential energy function adopted in a general force field is

$$E_{Total} = \sum_{bonds} K_r (r - r_0)^2 + \sum_{angles} K_\theta (\theta - \theta_0)^2$$
$$+ \sum_{dihedrals} \frac{V_n}{2} [1 + \cos(n\varphi - \gamma)] + \sum_{i<j} \left[\frac{A_{ij}}{R_{ij}^{12}} - \frac{B_{ij}}{R_{ij}^6} + \frac{q_i q_j}{\varepsilon_r R_{ij}} \right] \tag{2.2}$$

This potential energy (Eq. 2.2) is a function of "n" atoms at "r" positions with several empirically derived parameters such as K_r, r_0, K_θ, θ_0, V_n, n, ϕ, γ, A_{ij}, B_{ij}, ε_r, q_i, and q_j. The first three terms in this potential energy function (eq. 2.2) represent bonded interactions while the last term governs the nonbonded vdW and electrostatic interactions. These energy terms are detailed below.

2.1.1 Bonded interaction energies

Bond energy: The first term in Eq. (2.2) describes the bond stretching energy between two covalently linked atoms modeled using a simple harmonic potential obeying Hooke's law. Here, K_r is the force constant and $(r - r_0)$ is the deviation of a given bond from its equilibrium bond length. This relation is derived from Taylor expansion and the simplest first order term is sufficient to represent the harmonic behavior of bond stretching energy. Hooke's law is ideal when a bond relaxes near the equilibrium position. If the bond stretches significantly beyond the equilibrium position, the energies increases to $\pm\infty$ and hence bond breaking events cannot be studied by assuming such simple harmonic potentials (Morse, 1929) (Fig. 2.1). The bond dissociation energy associated with bond breaking event is evaluated using Morse potential function as given in Eq. (2.3) (Morse, 1929).

$$E_{Morse} = D\left[\left(1 - e^{-\alpha(r-r_0)}\right)\right]^2 \qquad (2.3)$$

where D is the dissociation energy and α is the force constant. Assigning large force constants for each bond will prevent the occurrence of bond stretching beyond a limit.

Bond angle energy: The angular relaxation between three atoms is well treated harmonically using either simple harmonic potential or Hooke's law. Like bond stretching energy, the force constant (K_θ) and reference value (r_θ) govern angular bending. Angular bending requires lesser energy when compared to that of bond stretching and hence the force constants are significantly smaller in number. If required, the accuracy in the calculation of angle bending energy can be enhanced by adding higher order terms, if there is no priority for limited usage of computational time.

Dihedral angle energy: The majority of structural variations in a biological system arise by the relaxation of highly flexible backbone and hence the torsion angles are considered as "soft" degrees of freedom. The force field for dihedral energy is derived by considering all possible geometrical variations/states of the dihedral atoms. The Fourier cosine series derived for modeling torsional potential is shown in Eq. (2.4).

$$E_{torsion} = \sum_{n=0}^{N} \frac{V_n}{2}[1 + \cos(n\phi - \gamma)] \qquad (2.4)$$

where V_n is the relative height of rotational barrier, n is the number of minima observed in a 360 degrees rotation, ϕ is the dihedral angle, and γ (phase factor) determines the location of minima (Fig. 2.2).

In addition to the torsional interactions between four sequentially connected atoms, improper torsional angles (also known as out-of-plane bending) are also defined between four atoms, which are not sequentially connected. This improper torsional potential is included to maintain planarity and stereochemistry.

Cross terms: The bonded interactions are dependent on each other. The "cross terms" describe the coupling between various internal variables like bond lengths, angular, dihedral, etc. For example, when bond lengths are adjusted gradually to void off local short contacts, the corresponding angular values are also adjusted simultaneously. Such coupling effects are considered in MM force fields as cross terms (stretch–bend, stretch–torsion, etc.). Force fields can be categorized into three classes based on the inclusion of cross terms (Hwang et al., 1994). Class I force fields consider all interactions as individual harmonic potentials and do not include cross terms. Class II force fields are those which include some anharmonicity (Morse potential) and few explicit cross terms. A class III force field like MM2 and MM3 series spans beyond the classical space and accounts chemical effects such as electronegativity and hyperconjugation.

2.1.2 Nonbonded interaction energies

As the name suggests, nonbonded interactions are expressed between the atoms which are not connected by covalent linkages. For example, the coordination of metal ion with either the functional groups of substrate or the ligand is treated as nonbonded interactions. Exceptional cases include the iron atom in the heme group found in globins, which is treated as bonded. Such

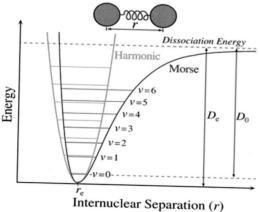

FIG. 2.1 Potential energy function of simple harmonic and Morse potential.[1]

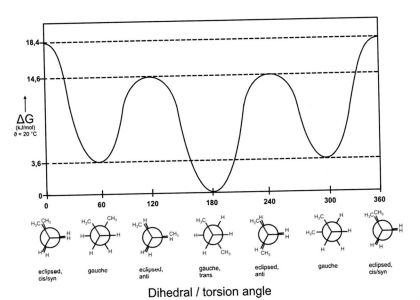

Dihedral / torsion angle

FIG. 2.2 An illustrative example for the variation in the torsional potential for sampling the dihedral angle of n-butane molecule showing periodicity.[2]

interactions between independent atoms or molecules through space fall under the category of nonbonded interactions and play a major role in stabilizing molecules. The vdW, electrostatic, H-bonding, and polar interactions are the major influencing nonbonded interactions.

Electrostatic interaction energy: The charged particles share nonbonded interactions which are electrostatic in nature and are denoted by the term E_{elec} in Eq. (2.2). The 3D distribution of electrostatic field is generally mapped using the simplest "point charge model," in which the energy between these point charges is governed by Coulomb's interactions as given in Eq. (2.5):

$$E_{elec} = \sum_{i=1}^{N_a} \sum_{j=1}^{N_b} \frac{q_i q_j}{4\pi\varepsilon_0 \varepsilon_r R_{ij}} \qquad (2.5)$$

where ε_0 is the permittivity of vacuum, ε_r is the relative permittivity of solvent medium, and q_i and q_j are partial atomic charges on atoms i and j separated by a distance R_{ij}. These charges are parameterized by fitting to an electrostatic potential obtained from electronic structure calculations.

The central multipole expansion method is also being used to calculate the electrostatic interactions. In this method, the entire electrostatic interactions are truncated into pairs of charged groups like charge–charge,

charge–dipole, dipole–dipole, dipole–quadrupole, quadrupole–quadrupole, etc., and hence become a laborious task. This approach is applicable to the molecular systems having aromatic π cloud and other systems, in which the charge distribution is not spherical. Generally, the simplest Coulomb's law is used for calculating electrostatic interactions.

Van der Waals Energy: Apart from electrostatic interactions, there are stabilizing short-range intermolecular forces termed as vdW forces, which is a weak force of attraction exerted between electrically neutral molecules that are closely interacting with each other. In fact, it is established by the temporary attractions between electron-rich regions of one molecule and electron-poor regions of another. The vdW force comprises two components: (1) an attractive term, due to dispersion or London forces (London, 1930) and induced dipoles, and (2) repulsive term, due to Pauli exclusion principle arising from nearby atoms. The Lennard-Jones 12-6 function is commonly used to model the vdW forces as in Eq. (2.6):

$$E_{vdW} = \frac{A_{ij}}{R_{ij}^{12}} - \frac{B_{ij}}{R_{ij}^6} \qquad (2.6)$$

where $A = \varepsilon r_m^{12}$, $B = 2\varepsilon r_m^6$, ε is the well depth, and r_m or σ (collision parameter) is the distance at which the Lennard-Jones potential is minimum (Fig. 2.3). Unlike other potentials (Buckingham, Hill potential), Lennard-Jones potential does not use exponential functions and hence is adopted in MM force fields.

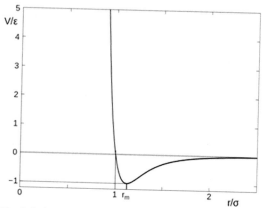

FIG. 2.3 Lennard-Jones potential curve explaining ε, r_m and σ parameters.[3]

H-bonding energy: Because H-bonding interactions are widely prevalent in nature, it has extensively been exploited in both experimental and theoretical sciences and exclusively in biological chemistry. Mathematical representation of H-bonding is also being derived similar to vdW interaction, but in a significantly shorter range. Several force fields replace the Lennard-Jones 12-6 potential by a 12-10 potential (eq. 2.7) to explicitly model electronegative atoms involved in hydrogen bond interactions.

$$E_{H-bonds} = \frac{A_{ij}}{R_{ij}^{12}} - \frac{B_{ij}}{R_{ij}^{10}} \qquad (2.7)$$

In a biological system, the nonbonded interactions are influenced by neighboring atoms and hence involve several pairwise interactions. Such many-body interactions that are not additive in nature could significantly affect the dispersion energy and hence are directly included during the parameterization of pair potentials.

2.1.3 Solvation models

In addition to the various types of interactions explained in the previous sections, incorporation of the effect of cellular aqueous environment is also inevitable to understand the physiological behavior of biomolecules. The aqueous environment, which provides suitable physiological environment for the effective solvent-mediated biomolecular activities, is mimicked by representing water molecules, counterions, etc., using theoretical models and they are referred to as solvent models.

Incorporation of solvent influence has been extensively explored using solvent models in the last few

decades. These solvent models have been classified into two classes based on the representation of solvent molecules, as implicit and explicit models, in which the former one considers the solvent as a homogeneously continuous polarizable medium and the latter one adopts an explicit geometric representation of the solvent molecule at the atomic level, and the wide network of H-bonded interactions throughout the solvent molecules reproduces the continuum effect of the medium. These two models vary largely in their speed and accuracy of parameters. The models of solvent medium have encountered a great development from implicit to explicit representations with added additional parameters for better accuracy (Zhang et al., 2017).

2.1.3.1 Explicit solvent models. Solvent molecules are explicitly included to model the solute–solvent electrostatic interactions and to reproduce the frictional drag due to solvent molecules. Water is the most commonly used solvent model while treating biological systems like proteins and nucleic acids. Generally, a water molecule is modeled either as rigid or flexible molecule with characteristic polarization and many body effects. Simple "rigid water models" such as simple point charge models (SPC and SPC/E) (Berendsen et al., 1981; Berendsen et al., 1987) and transferable intermolecular potentials (TIP, including advanced models TIP3P, TIP4P, etc.) (Jorgensen et al., 1983) are the most commonly used models. These models rely on the nonbonded interactions between water molecules, while the bonded interactions are treated by holonomic constraints in order to reduce the degrees of freedom. A water molecule is represented by three (for example, SPC and TIP3P) to five (for example, TIP5P) interaction sites of electrostatic point charges with Lennard-Jones parameters for oxygen atom. Among these models, the SPC and TIP3P models are used widely to model solvent effects at low computational cost. Both SPC and TIP3P use similar potentials and the only difference is that SPC uses a perfect tetrahedral geometry (HOH angle of 109.47 degrees) to represent a water molecule, whereas TIP3P uses 104.52 degrees as of ideal water geometry.

2.1.3.2 Implicit solvent models. The implicit solvation models assume the water medium as an infinitely continuum medium having dielectric and hydrophobic properties as that of the water molecule. The absence of explicit representation of water molecules reduces the degrees of freedom to a significant extent, thereby increasing the speed of calculations. In gas phase, the force field that describes the molecular energy is a function of molecular configuration, and the forces acting on these atoms, in that particular configuration, are determined by the gradients with respect to their positions. The question arises when the estimation of the same molecular energy is computed in the presence of solvent medium. It is the time to think about the transfer of a molecular geometry from a gaseous state to a solvent environment which requires a thermodynamically feasible space to accommodate the solute molecule. Hence, the calculation of energy (E_{tot}) of a solvated molecule requires two components (eq. 2.8): energy at the gas phase ($E_{gas\ phase}$) and free energy of solvation (ΔG_{solv}), i.e.,

$$E_{tot} = E_{gas\ phase} + \Delta G_{solv} \tag{2.8}$$

Practically, the calculation of gas phase potential energy is straightforward (Schlick, 2010) and the solvation free energy, ΔG_{solv}, is calculated from electrostatic as well as nonpolar contributions (eq. 2.9).

$$\Delta G_{solv} = \Delta G_{el} + \Delta G_{nonpolar} \tag{2.9}$$

The contribution from the hydrophobic interactions ($\Delta G_{nonpolar}$) emerges from (1) the vdW favorable attraction between the biomolecule and solvent molecules and (2) the unfavorable cost of cavity formation (to accommodate the solute) by breaking the structure of the solvent around the solute. These hydrophobic interactions are assumed to express linear dependency with the overall solvent accessible surface area (SASA) of the solute, i.e., $\Delta G_{nonpolar} \approx \sigma \times SASA$. Here, the constant σ is derived from the experimental solvation energies of small nonpolar molecules (Case et al., 2005; Gallicchio & Levy, 2004). Besides the nonpolar component, the long-range interactions contributing to the electrostatic component (ΔG_{el}) of solvation free energy are crucial in determining the molecular stability and function. Hence, significant efforts have been made to predict the ΔG_{el} more accurate and faster. MD simulations, mainly, use the Poisson–Boltzmann and generalized Born models to account for the contribution of ΔG_{el} toward solvation and are briefed in the below section.

Poisson–Boltzmann model: Computation of the electrostatic potential field $\phi(r)$ generated by the distribution of homogeneously continuous polarizable solvent medium $\rho(r)$ is carried out using the Poisson equation, $\nabla^2 \phi(r)(r) = -4\pi\rho(r)$, and the ionic distribution is taken care of by Boltzmann distribution. Hence, the electrostatic properties of the macromolecules in the physiological medium can be well explored using Poisson–Boltzmann (PB) Eq. (2.10) (Baker, 2005) as given below.

$$\nabla[\varepsilon(r)\nabla\phi(r)] = -4\pi\rho(r) + \kappa^2\varepsilon(r)\phi(r) \tag{2.10}$$

Here, κ is the Debye–Hückel parameter introducing the screening effects of the ionic distribution, and $\varepsilon(r)$ is the distance-dependent dielectric constant. Numerical formulation of PB approach has been extensively adopted in biomolecular simulations (Grant et al., 2001; Luo et al., 2002; Prabhu et al., 2004; Totrov & Abagyan, 2001).

Generalized Born model: Generalized Born model (Onufriev & Case, 2019) evaluates the electrostatic component of the solvation free energy from the pairwise interactions between the atomic charges, which is relatively simpler and computationally efficient than PB approach. For an aqueous solvation of molecules with interior dielectric value of 1, the ΔG_{el} is approximated as given below (eq. 2.11) (Still et al., 1990)

$$\Delta G_{el} \approx -\frac{1}{2}\left(1 - \frac{1}{\in_w}\right)\sum_{i,j}\frac{q_i q_j}{\sqrt{r_{ij}^2 + R_i R_j \exp\left(-\frac{r_{ij}^2}{4R_i R_j}\right)}} \tag{2.11}$$

where R_i and R_j are the Born radii of i and j atoms and r_{ij} is the distance between them. The variables q_i and q_j are the partial charges and the dielectric constant of the medium is $\in_w \gg 1$. Most of the modeling software are integrated with this model due to its simplicity and reasonable accuracy.

2.1.4 Binding energy

Interactions between biomolecules are the basis for all biological processes. For free molecules, the free energy measures the stability at thermal equilibrium and denotes the amount of energy released while forming a complex. Free energy of binding is the change in free energy when two molecules (A and B) combine to form a complex and can be calculated as the difference between the chemical potentials of complex and that of the individual molecules A and B. During such complex formation, the conformational entropy is being compensated according to the feasibility of complexation.

The binding free energy of complexes can be calculated by various computational methods using conformational sampling. These methods follow two different approaches: (1) pathway methods such as thermodynamic integration and free energy perturbation and (2) end-point methods such as molecular mechanics Poisson–Boltzman surface area (MMPBSA) and molecular mechanics generalized Born surface area (MMGBSA) (Gilson & Zhou, 2007). The pathway methods compute every infinite changes in a multistep pathway connecting the initial (unbound) and final (bound or complex) states, whereas the end-point methods generate both bound and unbound conformations of molecules and compute the binding free energy as their difference. Both the pathway and end-point methods can be performed using MD simulations in explicit solvation. Unlike explicit solvation, the application of implicit approaches could reduce the complexity in free energy analysis when used with end-point methods (Massova & Kollman, 2000). MMGBSA method has been applied successfully in the recent years, for the estimation of binding free energy in cases of biomolecular complex formation. Binding energy calculation methods go hand-in-hand with computational drug design methods to establish the energetically favorable complexes for successful therapeutics developments (Gilson & Zhou, 2007).

MMGBSA or MMPBSA: The binding energy using MMGBSA or MMPBSA method can be calculated as a difference in energy of the bound and unbound states as shown in Eq. (2.12).

$$\Delta G_{bind} = \Delta G_{complex} - (\Delta G_{receptor} + \Delta G_{ligand}) \qquad (2.12)$$

where $\Delta G = \Delta G_{gas} + \Delta G_{sol}$

$$\Delta G_{gas} = \Delta E_{bond} + \Delta E_{angle} + \Delta E_{dihedral} + \Delta E_{ele} + \Delta E_{vdW}$$

$$\Delta G_{sol} = \Delta E_{pol} + \Delta E_{nonpol}; \quad \Delta E_{nonpol} = \gamma SASA + \beta$$

The terms $\Delta G_{complex}$, $\Delta G_{receptor}$, and ΔG_{ligand} represent the average free energies of the complex, receptor, and ligand, respectively. The molecular mechanical energy in gas phase (ΔG_{gas}) is calculated as the sum of internal (bond, angle, dihedral), electrostatic, and vdW energies. The solvation energy, ΔG_{sol}, includes polar (ΔE_{pol}) and nonpolar (ΔE_{nonpol}) contributions. The polar component of binding free energy can be calculated using the Poisson–Boltzmann (MMPBSA) or the generalized Born approximation (MMGBSA) models and defining the appropriate external and internal dielectric constants to mimic the required solvent, generally, water medium. The nonpolar component of binding free energy requires approximating the SASA of our molecules of interest using algorithms like LCPO (linear combination of pairwise overlaps) approximation. By combining the internal and solvation energies and obtaining the difference between the bound and unbound states, the binding energy of a complex can be calculated. Due to the large size of molecules used in computational studies, it is feasible to split the entropy of a molecule into rotational, translational, and vibrational parts. The rotation and translational entropy can be eliminated by processing the trajectory after simulations using algorithms like rigid-rotor/harmonic-oscillator (Pearlman et al., 1995) and hence only the

vibrational entropy is assumed to contribute majorly toward the computationally derived binding energy.

2.2 Molecular Dynamics–Based Approaches

MD is a computational technique where the time-dependent behavior and dynamics of molecular systems can be simulated. This method considers all the molecules in a simulation box as flexible and hence vastly increases the conformational sampling (Karplus & Petsko, 1990). MD simulations have evolved as a promising tool to tackle many of the complex issues from theoretical physics to biological sciences. The conformational evolution of biological systems studied using MD simulations is applied right from the structure determination to enzyme inhibition by including conformational changes, stability, protein folding, molecular recognition, biomolecular interactions, ion/membrane transport mechanisms, etc. (Karplus, 2002; Karplus & McCammon, 2002). In MD, atomic motions are extracted by integrating the Newton's equation of motion, where molecular mechanics force fields are used to define the interatomic forces and the associated potential energy. The instantaneous forces acting in a system can be derived by the equation of motion ($F = ma$) and the same can also be expressed as the negative gradient of potential energy ($F = -\Delta V(r)$) in force field models (Leach, 2001). Hence, the potential energy derivative of the simulation system is related to the changes in position as a function of time as in Eq. (2.13):

$$-\frac{dV}{dr} = m \frac{d^2 r(t)}{dt^2} = a(t) \qquad (2.13)$$

where V is the potential energy of the system, F is the force exerted on any given atom at time t, $r(t)$ is the position vector of any given atom at time t, and m is the mass of the atom.

The propagation of system over time is performed by splitting the time into smaller timesteps of δt. The integration of Newton's equation of motion over several δt until achieving a desired total time t is done using numerical algorithms like Verlet and leapfrog algorithms following the finite difference principle. As an example, the time propagation using Verlet integrator can be explained as following.

For a set of particles with position r_i, the motion of the atoms in time δt can be given by Taylor's expansion as in Eq. (2.14):

$$r_i(t + \delta t) = r_i(t) + v_i(t)\delta t + \frac{1}{2} a_i(t)\delta t^2 \qquad (2.14)$$

where v_i and a_i are the first and second derivatives of the position with respect to time. Similarly, the velocities can be obtained from time t by Taylor's expansion as in Eq. (2.15):

$$v_i(t + \delta t) = v_i(t) + \frac{1}{2}[a_i(t) + a_i(t + \delta t)]\delta t \qquad (2.15)$$

where r_i, v_i and a_i are the position, velocity, and acceleration of ith atom at times t and $t + \delta t$. The acceleration acting on i^{th} atom can be derived from the Newton's equation of motion and MM potential energy as per Eq. (2.15). The different algorithms for time propagation have their own advantages and disadvantages and the choice of these algorithms requires careful considerations (Leach, 2001).

MD simulations are mainly used as a conformational sampling method to enhance the available states of biomolecules and to understand their physiological behavior. X-ray crystallographic methods could only capture the static image of a protein at any one of its functional states; however, proteins and waters are highly dynamic in nature (Karplus, 2002; Karplus & McCammon, 2002). Capturing the water molecules responsible for biomolecular interactions by experimental methods is another challenging task. In several cases, as in HIV-1 integrase, a water molecule acts as a nucleophile for enzyme catalysis (Ribeiro et al., 2012) or as in the case of HIV-1 protease, it is required to hold the protein structure together (Lawal et al., 2019). The inclusion of explicit water molecules during simulations opens up the hidden binding sites or paths and also highlights these hotspots for active water molecules. Moreover, by allowing the side chains of amino acids to move freely, the dynamic changes in binding sites can be captured and several such conformations can be used for further docking studies to identify a ligand that can bind to a majority of protein conformations.

Also, in several cases, MD simulations of unliganded proteins have revealed allosteric sites for ligand binding (Hertig et al., 2016). The relaxed complex scheme method implemented by McCammon et al. in the year 1998 includes docking of small libraries of compounds to an ensemble of protein conformations obtained by simulating the unliganded proteins conformations (Amaro et al., 2008). By this method, one can eliminate the bias of designing a ligand based on a static binding site and allow the protein to choose its own suitable inhibitor, a conformational selection binding model. A successful implementation of this method led to the identification of about 14 low-micromolar inhibitors against UDP-galactose 4′-epimerase (Durrant et al., 2010). Another method utilizing the conformational sampling schema in virtual screening is the MD-FLAP method, where MD trajectories are clustered to identify the most probable conformations, generating fingerprints for the active site in these clusters and using these fingerprints to screen databases for possible ligands (Spyrakis & Cavasotto, 2015; Spyrakis et al., 2015). Apart from conformational sampling before docking, MD simulation postdocking can be used as an efficient tool to refine and analyze the binding affinity and rescore the complexes (Salmaso & Moro, 2018).

2.3 Applications of Molecular Mechanics—Based Force Fields in Drug Design

Computer-aided drug design, began in early 1970s, is a process where new molecules are designed/identified to bind with a biological target of known (or predictable) 3D structure and express significant affinity/specificity (Cole et al., 2019). The main purpose of drug design methods is to utilize the receptor/ligand tertiary structures, for speeding up drug discovery process, or to enhance the inhibition properties of a ligand. Two strategies can be adopted for performing computer-aided drug design: (1) structure based (target-based) or (2) ligand based (analog based). All these methods use MM force fields to represent atomic level interactions and to calculate molecular geometries, energies, and motion (March-Vila et al., 2017).

Structure-based drug design methods depend on the availability of 3D structure of a protein target (Anderson, 2003). A 3D structure of protein can be obtained by X-ray crystallography, NMR, or in silico homology-based prediction methods. Once the 3D structure is known, the binding site or active site of the protein is identified. Structure-based drug design methods identify/design an inhibitor with functional properties that are complementary to the protein binding site. These methods include molecular docking and de novo ligand design. Molecular docking methods evaluate the most feasible binding geometries of a ligand at the binding site of a target in the 3D space. These binding geometries are called binding poses and include both configurational (position of the ligand in receptor binding site) and conformational sampling. These binding geometries are scored using MM and ranked based on the strength of interaction with the receptor. This process can be performed on large databases in a high-throughput manner (virtual screening), facilitating fast screening of ligands to identify good inhibitors. De novo design methods constitute the construction of molecules that are not previously synthesized. In this method, functional groups responsible for interactions with receptor are placed in 3D space complementary to the protein binding site and are connected with linker scaffolds. This method works on the assumption that only the functional groups of a ligand are responsible for its activity and not the scaffold (Batool et al., 2019).

Ligand-based drug design methods like quantitative structure–activity relationship (QSAR) and pharmacophore modeling have proven their efficiency in (1) designing/predicting the activity of new compounds and (2) searching chemical databases to identify novel lead scaffolds in the absence of receptor 3D structure (Kundaikar et al., 2014; Tawari & Degani, 2010; Telvekar & Patel, 2011). QSAR and quantitative structure–property relationship methods derive a mathematical model for biological activity using various structural and functional properties (Ravichandran & Agrawal, 2007; Srivani & Sastry, 2009; Srivani et al., 2008). This model relating activity (dependent quantity) and property (independent quantity) can be used to predict the activity of new compounds as inhibitors without the knowledge of receptor 3D structure. These relationships can be derived using statistical procedures such as regression methods, neural networks, principal component analysis (PCA), partial least squares (PLS), etc. Among these, multiple linear regression is the most commonly used method to derive a relationship between activity and more than one structural properties. When a large number of structural properties should be considered (for example, grid-based methods in 3D QSAR), linear regression fails and specialized methods such as PCA or PLS are used.

The pharmacophore model or hypothesis is a collection of chemical features (like H-bond acceptors and donors, charged or ionizable groups, hydrophobic or aromatic rings along with their spatial arrangement in terms of distance, angles, and dihedrals) which are common to a set of compounds with similar inhibition mechanism and are necessary for their inhibitory activity against a particular target. This pharmacophore hypothesis can be used as a query while screening large chemical databases to identify new compounds possessing all the necessary chemical features with a defined spatial arrangement to express biological activity against a receptor (Khedkar, Malde, & Coutinho, 2007, Khedkar, Malde, Coutinho, & Srivastava, 2007; Marriott et al., 1999).

3 MOLECULAR DOCKING

Molecular docking is a process in molecular modeling, where one can predict the favored binding orientation of one biomolecule with another resulting in a stable complex. The knowledge about a biomolecular complex might be useful for analyzing the association energies or binding energy using specific scoring functions. The relative binding orientation of the biomolecules induces various signals during biological processes resulting in agonistic or antagonistic phenomena. Hence, studying the binding affinity and binding orientation of biomolecules is of key relevance in understanding the strength and signal types. Also, under disease conditions, the knowledge about biomolecular complexes is necessary for regulating them by rational designing of small molecule inhibitors as drugs (Ferreira et al., 2015; Pinzi & Rastelli, 2019).

Molecular docking problem can be defined as an optimization problem, where one tries to identify the best fitting orientation of a ligand to bind with the target of interest. The binding of ligand to protein can be related to a "lock-and-key" model, where the right key (inhibitor) contains the right pattern (conformation) to fit into a lock (receptor site). However, in physiological conditions, both the lock and key are not static entities. The ligand conformation changes according to the protein binding site, and similarly, the protein binding site residues adjust themselves to accommodate the ligand. The structural changes in binding site residues is a minimal response for ligand-induced changes in proteins, because in many cases, ligand-induced large-scale domain rearrangements can also occur. Hence, the best docking protocol has to define the docking problem as an "induced-fit" model (Tripathi & Bankaitis, 2017).

In the early 1980s, association studies of two rigid biomolecules were initiated by docking protocols. Later, the development of searching and scoring algorithms has allowed the introduction of ligand flexibility into these docking procedures, so called "semiflexible" docking. Furthermore, the importance of conformational sampling of receptors was established as more and more ligand-induced large-scale changes were observed in biomolecules leading to the development of flexible docking algorithms in mid-1990s. During this period of development, several new applications for docking protocols were identified, ranging from simple protein–ligand docking to drug repurposing and predicting the side effects (Kitchen et al., 2004) (Figs. 2.4 and 2.5).

For a molecular docking procedure, the basic requirements are the structure of receptor and ligands (macromolecule or small molecule), conformational search procedures for sampling the receptor and ligand conformations, and some ranking or scoring scheme to

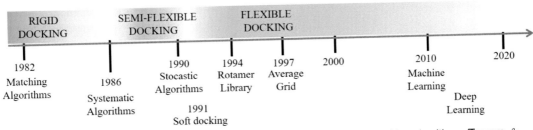

FIG. 2.4 Evolutionary timeline depicting the development of molecular docking algorithms (Tessaro & Scapozza, 2020).

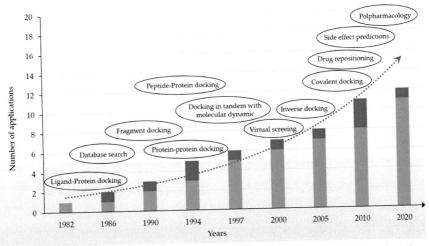

FIG. 2.5 Increase in the number of docking tools for specific purposes. The *dark blue boxes* show the number of tools added to the preexisting ones (Tessaro & Scapozza, 2020).

find out the best docking orientation. The 3D structure of receptor can be obtained by experimental methods like X-ray crystallography (Smyth & Martin, 2000) and NMR (Wider, 2000) or by computational structure prediction methods like homology modeling (Deng et al., 2018). Similarly, the 3D structure of ligands can be obtained from chemical databases, or if the compounds are small enough, the structures can be drawn using software like ChemDraw (Cousins, 2005) or ISIS-Draw. These structures are static conformations which have to be sampled for their minimum energy, physiologically stable conformations using search algorithms (Salmaso & Moro, 2018).

To dock a ligand, their binding site in the receptor structure should be known. The ligand binding sites are usually the substrate binding site, or in few cases, it can be an allosteric site. The docking protocols make an energy map (also called as "grid") of the shape, size, and properties of the binding site which is used as a guide to search the best fitting ligand orientations. Furthermore, the ligand conformations generated

from the conformational search algorithms are docked at the receptor using the generated "grid." The interaction energy gives a basic idea on the quality of the docked pose, which can be used to rank several docked poses. The interaction energies or the functions used to rank a docked pose are called the scoring functions, which vary according to the parameters used to calculate the binding affinities (Ferreira et al., 2015; Warren et al., 2006). The different types of docking protocols as well as the search and scoring algorithms used in these methods are discussed in the following sections.

3.1 Conformational Landscape Search

The potential energy function shown in Eq. (2.1) is a function of coordinates, which means any change in coordinates will change the energy of the system. The 3D way of representing the correlation between change in coordinates and change in energy is called the conformational landscape or potential energy surface (PES). It is a multidimensional surface (also called hypersurface) representing the energy of N atoms as a function of 3N Cartesian coordinates or 3N-6 internal coordinates. Representing a surface of such high dimensions is not possible except for simple cases with two or three coordinates. A simple example of one coordinate PES is the change in vdW energy as a function of interatomic distance (Lewars, 2003). The complexity of representing a total PES can be simplified by restricting to a coordinate of interest, for example, the Ramachandran plot, scanning the phi and psi angles of an amino acid backbone. The energy of each conformation with a particular combination of phi and psi torsional angles can be represented as a contoured PES as a function of only two coordinates. In simpler terms, the coordinates of interest are also called as reaction coordinates, which can be as simple as interatomic distance to a complex one like principal components (also called as collective variables). The PES is characterized by various features such as minima (least energy points), maxima (high energy points), and saddle points (linking two minima), which pose additional complexity in exploring a PES. The minima on a PES correspond to low energy state of a molecule. There can be multiple minima (local minima) on a given PES, while there is only one least energy state called the "global minima." A biomolecule is considered to be in an energetically stable state in the minima wells where the atomic arrangements are in their minimum energy conformations. Any change in these conformations will lead the molecule up into high energy conformations. Exploring the PES could reveal the properties of molecular conformational behavior like the structural changes between different minima and the varying atomic positions catalyzing enzyme reaction. Scanning the PES by introducing stepwise variations on the reaction coordinate of interest and plotting the various energy levels increases the conformational sampling of molecules of interest (Truhlar, 2003).

The process of molecular docking includes the structures of receptor as well as ligands. The receptor flexibility is handled by MD-based conformational sampling; however, the small molecules also require conformational search to find the best orientation to bind with the protein. Hence, to generate multiple conformers for small molecules, a conformational search procedure is adopted. The simplest protocol for conformational search is the grid search method. The basic steps involved in this procedure are (1) identifying all possible rotatable bonds in the molecule, (2) fixing the bond lengths and angles, (3) incrementing each rotatable bond by a fixed value to scan 360 degrees, and (4) optimizing the structure of each generated conformer until all possible conformers are generated.

A major drawback of grid search method is that the number of conformers generated increases proportionally with the increase in number of rotatable bonds (Bai et al., 2010; Wang et al., 2020). One way to handle this problem is to eliminate the high energy structures even before the optimization stage. Grid search method is a basic and simple method which follows the incremental search algorithm. Apart from this, there are several other methods like random walk algorithms (e.g., CONFGEN), MD simulations (MOE), distance geometry based (RDK$_{IT}$), and knowledge-based (CORINA) algorithms for efficient conformational sampling of small molecules (Labute, 2010; Riniker & Landrum, 2015; Watts et al., 2010). All these methods can adopt either a fragment-based or a non−fragment-based (or whole molecule) search strategy. Because a lot of fragments are common in most of the small molecules, similar to the side chains of standard amino acids, all the possible conformers (rotamers) that these molecules can adopt are predefined and stored in publicly available rotamer libraries like BCL:: Conf (Kothiwale et al., 2015). There is a vast amount of data on the structures of small molecules available in chemical databases, which is gathered to build this knowledge-based rotamer libraries. Similar to small molecule rotamer libraries, there are rotamer libraries available for amino acid side chain conformational flexibilities like dynameomics rotamer library (Scouras & Daggett, 2011) and Dunbrack rotamer library (Shapovalov & Dunbrack, 2011). The implementation of these libraries in docking protocols allows additional flexibility to the receptor conformations during the docking process.

3.2 Types of Docking
3.2.1 Rigid docking
Rigid docking is one of the fast docking protocols where both receptor and ligand structures are not allowed any vibrational degrees of freedom. The most common assumption for rigid docking is that the binding of ligand with receptor is due to their complementary nature. The complementarity can be based on the structure, as in molecular surface or shape complementarity, or based on the chemical properties of interacting groups, as in "hot-spot" technique. To account for the shape complementarity, the molecular surface of receptor and ligand is used as a descriptor defined based on their SASA. Similarly, the chemical nature of the interacting groups in receptor and ligand is considered to be complementary in the hot-spot technique. Though geometric complementarity schemes are fast and reliable, these methods cannot account for any flexibility in both receptor and ligands (Taylor et al., 2002). Protein−protein interactions are

usually considered to be based on shape complementarity and hence it is implemented in several protein—protein interaction docking tools like PATCHDOCK (Schneidman-Duhovny et al., 2005) and ZDOCK (Pierce et al., 2014). Due to less number of degrees of freedom handled in these methods, rigid docking protocols can be used to screen millions of compounds in very less time, when compared with other flexible docking methods. When combined with conformational sampling schemes, rigid docking protocols give best results. Because the chemical features are also considered in the complementarity criterion, rigid docking can also be combined with pharmacophore-based approaches for large-scale screening studies.

3.2.2 Semiflexible docking

These methods work on the assumption that the protein conformation used in the docking process is the bioactive conformation to which the ligand will bind. Hence, only the ligand is allowed to be flexible with all conformational degrees of freedom. The major disadvantage with this method is that the assumption of using one protein conformation as the bioactive one is not true always and "induced fit" binding is not possible. There are several docking algorithms following the semiflexible model, which can be broadly classified based on their conformational search schemes as follows (Huang & Zou, 2010; Kitchen et al., 2004).

3.2.2.1 Systematic search techniques. In these methods, the ligand conformations are sampled systematically using a uniform increment for each degree of freedom and then handling all values for all the degrees of freedom in a combinatorial way. The complexity can be explained using the equation calculating the number of torsional conformations sampled by systematic search given by Eq. (2.16)

$$\prod_{i=1}^{N} \prod_{j=1}^{n_{inc}} 360^{\circ}/\theta_{i,j} \qquad (2.16)$$

where N is the number of rotatable bonds, n_{inc} is the number of increments to be performed, and $\theta_{i,j}$ is the increment angle j to be added to the ith rotatable bond. The systematic search technique samples the entire conformational landscape until a minimum energy structure is reached. In order to attain the global minima, the conformational sampling has to be performed using several starting structures from different points on the conformational landscape. These methods require sampling a huge number of conformations and are further classified based on the adopted

sampling technique to improve the speed and accuracy (Brooijmans & Kuntz, 2003).

Exhaustive search: This search technique is a strict systematic search, where all possible conformers generated by combining all the degrees of freedom are searched in a systematic way. This method will lead to a combinatorial explosion if not checked by introducing certain termination criteria to limit the search space. The Glide algorithm (Friesner et al., 2006) uses an exhaustive search in its docking pipeline. Glide precomputes a grid representation of the properties and shape of the receptor active site along with an initial library of predicted minimum energy ligand conformations. By limiting the ligand conformational space using the grid representation as a guiding factor, the number of sampled conformations is greatly reduced. Following this, a high-resolution search is implemented which involves ligand optimization with MM force fields and a Monte Carlo—based search for minima.

Fragment-based search: The ligand of interest is divided into small fragments, followed by rigid docking of these fragments with the receptor site and building the fragment linkages. A partial flexibility is introduced in the process of joining the fragments. Fragment-based search is the first implemented search technique for a docking protocol in the year 1986 (DesJarlais et al., 1986). Since then, the fragment-based search techniques have been developed based on the protocol used to link the fragments. Incremental construction is a two-step fragment search which partitions a molecule into a rigid (core) backbone and flexible (side chains or functional groups) parts. Other fragment-based search methods implement distance geometry criteria (e.g., DOCK) (Allen et al., 2015) or fast shape matching algorithm (ZDOCK and FlexX) (Pierce et al., 2014; Rarey et al., 1996).

Conformational ensemble: This search technique amplifies the utility of rigid docking protocol for fast and reliable docking process. The ligand flexibility is introduced by creating an ensemble of conformers varying in all the degrees of freedom. All these conformers can then be docked with a receptor using the rigid docking protocol as implemented in tools like EUDOC (Pang et al., 2001) and MS-DOCK (Sauton et al., 2008).

3.2.2.2 Stochastic search techniques. Monte Carlo methods: Monte Carlo methods work by generating random initial configurations of ligands and scoring them in the receptor binding site. For any given ligand, a random configuration for all the degrees of freedom is applied using a random number generator algorithm. Each ligand conformation is placed in the receptor binding site and scored using a scoring function (like energy). In the next step, a small change is introduced to this conformation and scored using the same scoring function. If the score of the new conformation is improved, then the new pose is accepted. Else, its acceptance or rejection as new minima is based on a Metropolis criteria defined by the Boltzmann-based probability function as shown in Eq. (2.17) below.

$$P \sim \exp\left[\frac{-(E_1 - E_0)}{k_B T}\right] \qquad (2.17)$$

where k_B is the Boltzmann constant, T is the temperature of the system, and E_1 and E_0 are the scores of ligand conformations before and after the modifications. If the new conformation passes this criterion, then it is accepted as new minima. This entire procedure is repeated until a desired number of ligand conformations are generated. A major advantage of using the Monte Carlo search method is that the ligands never remain trapped in local minima because conformers are generated randomly irrespective of the previous states. However, a ligand conformation visited before in a search can occur again because they are generated in random and the previous states are not remembered. The docking tools like ICM (Abagyan et al., 1994) and AUTODOCK VINA (Trott & Olson, 2010) implement the Monte Carlo—based search algorithm.

Tabu search method: Tabu search method is a heuristic method, which provides approximate solutions to highly difficult optimization problems. Initially, a set of random conformations are generated and ranked using a fitness function. The random conformations are generated using a local search method by modifying a few degrees of freedom in each conformation. The changes made in the conformers rejected by the fitness function (or "tabus") are recorded for further selection process. Any new conformation with a fitness score better than the previously accepted conformations will be accepted even if it is in the list of tabus. Or, the best non-Tabu conformation will be accepted. This Tabu—non-Tabu search is repeated until a minima is found. The Tabu search methods show high accuracy, and like any other stochastic search method, the conformational search is not trapped in local minima. As all generated conformations are recorded in a short-term memory, previously visited conformations are not visited again. PSI-DOCK (Pei et al., 2006) and PRO_LEADS (Baxter et al., 1998) are two such docking tools using the Tabu search method.

Evolutionary algorithms: These algorithms are based on the principles of evolution. In biology, evolution is driven by mutations and crossing over of chromosomes and the associated genes. In this analogy, each degree of freedom in a ligand is encoded by a gene, and a chromosome is the collection of all genes that describes a ligand conformer. The most commonly used evolutionary algorithm is the genetic algorithm (GA). In GA, each chromosome is assigned a fitness score and this score is evaluated to decide whether a chromosome is viable or not. In docking protocols, the fitness score is generally calculated as the total interaction energy of ligand with the protein. Crossover events occur within chromosomes giving rise to a new generation or new ligand conformations whose fitness scores are again evaluated. Based on the fitness scores of parents, the offspring inherit the structural features and hence only those changes which are healthy for a ligand are inherited. Additionally, random mutations occur in the parents creating an entirely new gene in the offspring. After all these changes, the offspring are accepted only if their fitness value is better than their parents. Because several parameters such as mutation rates, crossover rates, and number of offspring generated are involved in these algorithms, their scoring functions are slightly complex. These algorithms can be entirely based on the ligand structure alone; however, by including the knowledge of receptor binding site, the efficiency and accuracy of these algorithms can be greatly improved. The docking program GOLD requires additional knowledge of the size and location of the ligand binding site in a receptor (Jones et al., 1997). A modification of GA is the Lamarckian GA implemented in AutoDock 3 and 4 (Morris et al., 1998). In addition to the genotypic effects, the Lamarckian GA includes the phenotypic effects in the search algorithm. Mutations and crossovers are handled in the genotypes, while the phenotype depends on the energy function after ligand optimization. The structural changes (phenotypic changes) observed after ligand optimization are mapped back onto the genes to be passed on to the next generations (Salmaso & Moro, 2018).

Swarm optimization methods: Swarm optimization algorithms are inspired by the swarm behavior of animals like a flock of birds or an ant colony to find optimal solutions. These are nature-inspired search algorithms simulating the social behavior of a swarm of animals looking for food. In molecular docking terms, each degree of freedom of a ligand is considered as an independent entity in the swarm of animals. Initially, these entities are assigned random values and, in each iterations, they vary according to other entities, while simultaneously exploring its own space. One global best solution is remembered by all the entities while also keeping track of its own best solution.

The algorithm terminates when a global best solution is found or exceeds the maximum number of required iterations. A few docking tools like SODOCK (Chen et al., 2007), PSO@AUTODOCK (Namasivayam & Gunther, 2007), and PSOVINA (Ng et al., 2015) utilize the swarm optimization methods.

3.2.3 Flexible docking

The most important drawback of docking methods is the exclusion of protein flexibility. Several attempts have been made to include protein flexibility and a few have been successful in incorporating varying degrees of protein conformational flexibility. These methods make use of either a single protein conformation integrating side chain flexibility or complex algorithms using multiple protein conformations (Alonso et al., 2006).

3.2.3.1 Single protein conformation.
These methods utilize only a single protein conformation for docking and incorporate flexibility on various levels.

Soft docking: This is the simplest method of incorporating an implicit and rather vague flexibility in protein side chains. To account for the flexibility, the vdW repulsion term in the force field—based scoring function is scaled down in order to allow close packing of ligands. This method simulates the effect of an induced fit docking, however, accounting only for a very small degree of flexibility. Also, the binding poses generated using this method are unrealistic because the refinement of docked complexes would lead to atomic clashes between the protein and ligand (Ganser et al., 2018; Jiang & Kim, 1991).

Side chain flexibility: This method accounts for slightly higher level of flexibility than the soft docking protocol. The side chains of residues in the active site are allowed flexibility by introducing alternative conformations, generally exploiting the available rotamer libraries. In certain docking tools like GOLD, a local search protocol is used to sample few degrees of freedom. By considering only side chain flexibilities, the main drawback of this method is the exclusion of large-scale protein conformational changes (Leach, 1994).

3.2.3.2 Multiple protein conformations.
Few slightly complex docking protocols might utilize more than one conformation of protein of interest to model protein flexibilities. The various protein conformations can be obtained from experimental methods or computational techniques like MD or Monte Carlo simulations. These methods follow different strategies to incorporate the protein structural diversity.

Average grid: The ensemble of protein conformations obtained from various sampling methods is concised into a single grid by averaging over the ensemble. The averaging can be a simple average or a weighted average biasing over a few conformations in the ensemble (Knegtel et al., 1997).

United description: By comparing the ensemble of protein conformations, structurally conserved (or rigid) and dynamic (or flexible) regions are identified. The rigid region is converged into a united atom description, which is then merged with the flexible parts in a combinatorial manner to generate a pool of "chimeric" protein conformations. The docking tool FlexE (Rarey et al., 1996) performs flexible docking by generating such "chimeric" pool to which all the ligands are docked.

Individual conformations: This method is the most detailed flexible docking protocol, where ligands are docked with each and every protein conformation provided as input (Huang & Zou, 2007). The results obtained from these flexible docking studies are entirely based on the protein conformations chosen as inputs. Hence, the quality of conformational sampling and the choice of input structures are very important. There are several cases where incorporating protein flexibility does not majorly improve the docking results. In particular, the use of MD sampled conformations does not improve the quality of docking as much as multiple crystal structures are expected to improve (Ganser et al., 2018; Osguthorpe et al., 2012). Several proteins, such as renin, EGFR, etc., express ligand-induced conformational changes, which need to be accounted properly while choosing the ensemble of protein structures. Sampling from a ligand bound structure, rather than from an apo structure will incorporate the ligand-induced large-scale structure changes and hence such an ensemble might provide better results.

3.3 Scoring Function and Types

Scoring functions are mathematical functions used in molecular docking methods to rank the docked complexes. These functions act as a pose selector and provide idea on the binding affinity between the two docked molecules. These functions help to discriminate between binders and nonbinders from the large number of docked conformations generated in a single docking run. The two interacting partners can be any pair of biomolecules like protein—small molecules (Jain, 2006), protein—protein (Lensink et al., 2007), or even protein—nucleic acid complexes.

The available scoring functions fall under four broad classifications. The scoring function used in a few docking tools is classified and shown in Table 2.1.

TABLE 2.1

Examples of Docking Tools Classified Based on Their Scoring Functions (Ferreira et al., 2015).

Force Field Based	Empirical Based	Knowledge Based
DOCK (Ewing et al., 2001)	AutoDock (Morris et al., 2009)	SMoG (Debroise et al., 2017)
AutoDock (Morris et al., 2009)	GlideScore (Friesner et al., 2004)	DrugScore (Gohlke et al., 2000)
GoldScore (Jones et al., 1997)	ChemScore (Eldridge et al., 1997)	PMF_Score (Muegge, 2006)
ICM (Abagyan et al., 1994)	LUDI (Bohm, 1992)	MotifScore (Xie & Hwang, 2010)
LigandFit (Venkatachalam et al., 2003)	LigScore (Krammer et al., 2005)	RF_Score (Ballester & Mitchell, 2010)
Molegro Virtual Docker (Bitencourt-Ferreira & de Azevedo, 2019)	HYDE (Reulecke et al., 2008)	PESD_SVM (Das et al., 2010)

3.3.1 Force field–based scoring functions

The energy of a system can be calculated using MM-based force fields as described in the previous sections (eq. 2.2). The bonded or intramolecular energy terms describe the vibrational energy of individual molecules, whereas the nonbonded or intermolecular energy terms are the main driving force for receptor–ligand interactions. By applying MM-based force fields, these intermolecular energies can be calculated and used to score the docked poses. In addition to the vdW, electrostatic, and H-bond interactions, few scoring functions also include the intramolecular energy of ligands. Because the ligand conformations vary considerably between the bound and unbound states, the ligands usually acquire a strain energy after docking. Several scoring functions account for this strain by including the strain energy as a penalty in the scoring function, so that ligand conformations which are too constrained to bind in the active site will be eliminated. The desolvation energies of ligand and receptor also play an important role in defining the binding affinity, which is also accounted for in certain scoring functions by using an implicit solvent model like GBSA or PBSA (Cournia et al., 2020; Foloppe & Hubbard, 2006). Few examples of softwares using force field–

based scoring functions are DOCK (Allen et al., 2015), GoldScore in GOLD (Verdonk et al., 2003), and AutoDock (Morris et al., 1998).

3.3.2 Empirical scoring functions

As the name suggests, these scoring functions are trained using the binding affinities of a known data set. These functions quantify the different types of interactions and apply a coefficient to weigh these interactions, the coefficients being optimized using a training set. The various empirical energy terms used in these scoring functions can range from the favorable nonbonded interactions like vdW, electrostatic, and H-bond interactions to the nonfavorable energy terms like desolvation and strain energies. These scoring functions are fast and hence they are commonly used in large-scale screening studies like virtual screening and de novo ligand design (Eldridge et al., 1997; Guedes et al., 2018). As these scoring functions are optimized based on data sets with known binding affinities, they can be modified and refined more easily than other scoring functions. The increasing availability of biomolecular complexes with their binding affinity data and the development in modern technologies have given a large scope for improving the accuracy of empirical scoring functions. The scoring functions like LUDI (Bohm, 1992), ChemScore (Eldridge et al., 1997), ID-Score (Li et al., 2013), and GlideScore (Friesner et al., 2006) were all developed in the regime of empirical scoring functions. The terms used in obtaining a ChemScore is shown below in Eq. (2.18).

$$\Delta G_{bind} = \Delta G_{H-bond} \sum_{H-bond} f(\Delta R, \Delta \alpha)$$
$$+ \Delta G_{metal} \sum_{metal} f(\Delta R, \Delta \alpha) + \Delta G_{lipo} \sum_{lipo} f(\Delta R)$$
$$+ \Delta G_{rot} \sum_{rot} f(P_{nl}, P'_{nl}) + \Delta G_0$$

$$(2.18)$$

3.3.3 Knowledge-based scoring functions

These functions work on the assumption that an interaction pair seen more frequently in known biomolecular complexes must be energetically favorable. By analyzing known 3D structures of biomolecular complexes available in databases like Protein Data Bank or Cambridge Structural Data Bank, the frequency of interacting atom pairs can be computed and an energy component or "potentials of mean force" can be derived. For scoring a docked complex, the energies for each interacting atom pair are taken from these precalculated and tabulated energies (or potentials of mean force) and are summed up to give the score (Gohlke et al., 2000). A few examples for knowledge-based scoring functions are DrugScore (Gohlke et al., 2000; Velec et al., 2005), PMF (Muegge, 2006), and GOLD/ASP (Mooij & Verdonk, 2005).

3.3.4 Machine learning–based scoring functions

The latest boom in technological and information sector has led to the development of machine learning–based and artificial intelligence–based scoring functions. These are regression-based models which are trained to predict either the binding affinities or classify active/inactive ligands or do both. Linear and nonlinear regression methods applied on KNN clustered data or data chosen from a random forest algorithm were shown to classify active/inactive complexes with good accuracy (Pason & Sotriffer, 2016). Recently, self-learning algorithms like support vector machines and neural networks are applied in generating target-specific drug models and interaction partner predictions (Rifaioglu et al., 2020).

Consensus scoring functions are designed by combining any of the above-said scoring functions to make stringent predictions by including a lot more interaction parameters. A few examples of consensus scoring functions are SMoG2016 (empirical and knowledge-based) (Debroise et al., 2017), DockTScore from the DockThor program (empirical and force field–based) (de Magalhães et al., 2014), and Galaxy-Dock BP2 Score (empirical, knowledge-based, and force field–based) (Baek et al., 2017).

4 CONCLUSION

Recent developments in computer technology have led to ample opportunities in every aspect of life and the same is true in scientific developments as well. Computational biology and bioinformatics are the key examples emerged from the basic sciences by the integration of physical as well as biological sciences under the computational platform. Despite the limitations of experimental techniques in chemical and biological space, it is now possible to look into the complex cellular components as a fully atomistic structural model and the interactions within the molecular network can be well analyzed at the atomic scale. This chapter focuses on the theoretical formulation of molecular modeling, simulations, and docking approaches.

Particularly, we tried to cover up the biomolecular communications/interactions from the perspective of computational structural biology. The content is outlined in such a way that the reader finds a smooth transition from, while reading, the introduction of biomolecules to the formulation of docking principles designed for the molecular interactions by sequentially highlighting the topics such as structural aspects of biomolecules, various types of biologically feasible interactions coordinating the biomolecular activities, the MM, and the MD approaches that portray the biomolecular interactions at the atomic level in a featured computational space.

The emergence of advanced computational approaches in the field of drug discovery has reshaped the molecular docking approach from static to dynamic nature of binding event, which is more appropriate in choosing a promising ligand conformation from its structural ensemble. Even though the last two decades have witnessed rapid advances in the concepts of molecular docking, certain issues like "representation of more accurate scoring function," in which the solvent models take the lead, still remains challenging. The overview of solvent models reveals the need for further developments mainly in the representation of continuum solvent medium as it poses restrictions on various observations like (1) the nature of water-mediated ligand binding, (2) water molecules forming solvation shells (first, second) around the solute, (3) inhomogeneity in calculating the forces at solute–solvent boundary due to distinct dielectric values for solute and solvent, etc. In conclusion, despite the need for still more accurate approaches, molecular docking approaches will continue to play essential roles in exploring cellular activities via biomolecular interactions and would strongly complement even the upcoming experimental developments in biological chemistry.

REFERENCES

Abagyan, R., Totrov, M., & Kuznetsov, D. (1994). ICM—a new method for protein modeling and design: Applications to docking and structure prediction from the distorted native conformation. *Journal of Computational Chemistry, 15*(5), 488–506. https://doi.org/10.1002/jcc.540150503.

Allen, W. J., Balius, T. E., Mukherjee, S., Brozell, S. R., Moustakas, D. T., Lang, P. T., Case, D. A., Kuntz, I. D., & Rizzo, R. C. (2015). DOCK 6: Impact of new features and current docking performance. *Journal of Computational Chemistry, 36*(15), 1132–1156. https://doi.org/10.1002/jcc.23905.

Allinger, N. L. (1977). Conformational analysis. 130. MM2. A hydrocarbon force field utilizing V1 and V2 torsional terms. *Journal of the American Chemical Society, 99*(25), 8127–8134. https://doi.org/10.1021/ja00467a001.

Allinger, N. L., & Chen, K. (1996). An improved force field (MM4) for saturated hydrocarbons. *Journal of Computaional Chemistry, 17*(5–6), 642–668.

Allinger, N. L., Yuh, Y. H., & Lii, J. H. (1989). Molecular mechanics. The MM3 force field for hydrocarbons. 1. *Journal of the American Chemical Society, 111*(23), 8551–8566. https://doi.org/10.1021/ja00205a001.

Alonso, H., Bliznyuk, A. A., & Gready, J. E. (2006). Combining docking and molecular dynamic simulations in drug design. *Medicinal Research Reviews, 26*(5), 531–568. https://doi.org/10.1002/med.20067.

Amaro, R. E., Baron, R., & McCammon, J. A. (2008). An improved relaxed complex scheme for receptor flexibility in computer-aided drug design. *Journal of Computer-Aided Molecular Design, 22*(9), 693–705. https://doi.org/10.1007/s10822-007-9159-2.

Anderson, A. C. (2003). The process of structure-based drug design. *Chemical Biology, 10*(9), 787–797. https://doi.org/10.1016/j.chembiol.2003.09.002.

Atkins, P. W., & De Paula, J. (2014). *Atkins' physical chemistry.*

Baek, M., Shin, W. H., Chung, H. W., & Seok, C. (2017). GalaxyDock BP2 score: A hybrid scoring function for accurate protein-ligand docking. *Journal of Computer-Aided Molecular Design, 31*(7), 653–666. https://doi.org/10.1007/s10822-017-0030-9.

Bai, F., Liu, X., Li, J., Zhang, H., Jiang, H., Wang, X., & Li, H. (2010). Bioactive conformational generation of small molecules: A comparative analysis between force-field and multiple empirical criteria based methods. *BMC Bioinformatics, 11*, 545. https://doi.org/10.1186/1471-2105-11-545.

Baker, N. A. (2005). Improving implicit solvent simulations: A poisson-centric view. *Current Opinion in Structural Biology, 15*(2), 137–143. https://doi.org/10.1016/j.sbi.2005.02.001.

Ballester, P. J., & Mitchell, J. B. (2010). A machine learning approach to predicting protein-ligand binding affinity with applications to molecular docking. *Bioinformatics, 26*(9), 1169–1175. https://doi.org/10.1093/bioinformatics/btq112.

Batool, M., Ahmad, B., & Choi, S. (2019). A structure-based drug discovery paradigm. *International Journal of Molecular Sciences, 20*(11). https://doi.org/10.3390/ijms20112783.

Baxter, C. A., Murray, C. W., Clark, D. E., Westhead, D. R., & Eldridge, M. D. (1998). Flexible docking using Tabu search and an empirical estimate of binding affinity. *Proteins, 33*(3), 367–382.

Berendsen, H. J. C., Grigera, J. R., & Straatsma, T. P. (1987). The missing term in effective pair potentials. *The Journal of Physical Chemistry, 91*(24), 6269–6271. https://doi.org/10.1021/j100308a038.

Berendsen, H. J. C., Postma, J. P. M., van Gunsteren, W. F., & Hermans, J. (1981). Interaction models for water in relation to protein hydration. In B. Pullman (Ed.), *Intermolecular forces: Proceedings of the fourteenth Jerusalem symposium on quantum chemistry and biochemistry held in Jerusalem, Israel, April 13–16, 1981* (pp. 331–342). Dordrecht: Springer Netherlands.

Berg, J. M., Tymoczko, J. L., Gatto, G. J., & Stryer, L. (2015). *[Stryer's] biochemistry* (8th ed). New York: W.H. Freeman.

Bitencourt-Ferreira, G., & de Azevedo, W. F., Jr. (2019). Molegro virtual docker for docking. *Methods in Molecular Biology, 2053*, 149–167. https://doi.org/10.1007/978-1-4939-9752-7_10.

Bohm, H. J. (1992). The computer program LUDI: A new method for the de novo design of enzyme inhibitors. *Journal of Computer-Aided Molecular Design, 6*(1), 61–78. https://doi.org/10.1007/BF00124387.

Bordoli, L., & Schwede, T. (2006). Structural Bioinformatics. By Philip E. Bourne and Helge Weissig (Eds.). *Proteomics, 6*(8), 2626–2627. https://doi.org/10.1002/pmic.200690044.

Born, M., & Oppenheimer, R. (1927). Zur Quantentheorie der Molekeln. *Annalen der Physik, 389*(20), 457–484. https://doi.org/10.1002/andp.19273892002.

Brooijmans, N., & Kuntz, I. D. (2003). Molecular recognition and docking algorithms. *Annual Review of Biophysics and Biomolecular Structure, 32*, 335–373. https://doi.org/10.1146/annurev.biophys.32.110601.142532.

Browne, F., Zheng, H., Wang, H., & Azuaje, F. (2010). From experimental approaches to computational techniques: A review on the prediction of protein-protein interactions. *Advances in Artificial Intelligence, 2010*, 924529. https://doi.org/10.1155/2010/924529.

Case, D. A., Cheatham, T. E., 3rd, Darden, T., Gohlke, H., Luo, R., Merz, K. M., Jr., Onufriev, A., Simmerling, C., Wang, B., & Woods, R. J. (2005). The amber biomolecular simulation programs. *Journal of Computational Chemistry, 26*(16), 1668–1688. https://doi.org/10.1002/jcc.20290.

Cerny, J., & Hobza, P. (2007). Non-covalent interactions in biomacromolecules. *Physical Chemistry Chemical Physics, 9*(39), 5291–5303. https://doi.org/10.1039/b704781a.

Chen, H. M., Liu, B. F., Huang, H. L., Hwang, S. F., & Ho, S. Y. (2007). SODOCK: Swarm optimization for highly flexible protein-ligand docking. *Journal of Computational Chemistry, 28*(2), 612–623. https://doi.org/10.1002/jcc.20542.

Cole, D. J., Horton, J. T., Nelson, L., & Kurdekar, V. (2019). The future of force fields in computer-aided drug design. *Future Medicinal Chemistry, 11*(18), 2359–2363. https://doi.org/10.4155/fmc-2019-0196.

Cooper, G. M. (2000). *The cell: A molecular approach.* Washington, D.C.; Sunderland, Mass: ASM Press; Sinauer Associates.

Cornell, W. D., Cieplak, P., Bayly, C. I., Gould, I. R., Merz, K. M., Ferguson, D. M., Spellmeyer, D. C., Fox, T., Caldwell, J. W., & Kollman, P. A. (1995). A second generation force field for the simulation of proteins, nucleic acids, and organic molecules. *Journal of the American Chemical Society, 117*(19), 5179–5197. https://doi.org/10.1021/ja00124a002.

Cournia, Z., Allen, B. K., Beuming, T., Pearlman, D. A., Radak, B. K., & Sherman, W. (2020). Rigorous free energy simulations in virtual screening. *Journal of Chemical Information and Modeling.* https://doi.org/10.1021/acs.jcim.0c00116.

Cousins, K. R. (2005). ChemDraw ultra 9.0. CambridgeSoft, 100 CambridgePark drive, Cambridge, MA 02140. www.cambridgesoft.com. See web site for pricing options. *Journal of the American Chemical Society, 127*(11), 4115–4116. https://doi.org/10.1021/ja0410237.

D'Argenio, V. (2018). The high-throughput analyses era: Are we ready for the data struggle? *High Throughput, 7*(1). https://doi.org/10.3390/ht7010008.

Das, S., Krein, M. P., & Breneman, C. M. (2010). Binding affinity prediction with property-encoded shape distribution signatures. *Journal of Chemical Information and Modeling, 50*(2), 298–308. https://doi.org/10.1021/ci9004139.

Debroise, T., Shakhnovich, E. I., & Cheron, N. (2017). A hybrid knowledge-based and empirical scoring function for protein-ligand interaction: SMoG2016. *Journal of Chemical Information and Modeling, 57*(3), 584—593. https://doi.org/10.1021/acs.jcim.6b00610.

Deng, H., Jia, Y., & Zhang, Y. (2018). Protein structure prediction. *International Journal of Modern Physics B, 32*(18). https://doi.org/10.1142/S021797921840009X.

Denis, C., Sauliere, A., Galandrin, S., Senard, J. M., & Gales, C. (2012). Probing heterotrimeric G protein activation: Applications to biased ligands. *Current Pharmaceutical Design, 18*(2), 128—144. https://doi.org/10.2174/1381612 12799040466.

Desiraju, G. R. (2011). A bond by any other name. *Angewandte Chemie International Edition, 50*(1), 52—59. https://doi.org/10.1002/anie.201002960.

DesJarlais, R. L., Sheridan, R. P., Dixon, J. S., Kuntz, I. D., & Venkataraghavan, R. (1986). Docking flexible ligands to macromolecular receptors by molecular shape. *Journal of Medicinal Chemistry, 29*(11), 2149—2153. https://doi.org/10.1021/jm00161a004.

Durrant, J. D., Urbaniak, M. D., Ferguson, M. A., & McCammon, J. A. (2010). Computer-aided identification of trypanosoma brucei uridine diphosphate galactose 4′-epimerase inhibitors: Toward the development of novel therapies for african sleeping sickness. *Journal of Medicinal Chemistry, 53*(13), 5025—5032. https://doi.org/10.1021/jm100456a.

Eldridge, M. D., Murray, C. W., Auton, T. R., Paolini, G. V., & Mee, R. P. (1997). Empirical scoring functions: I. The development of a fast empirical scoring function to estimate the binding affinity of ligands in receptor complexes. *Journal of Computer-Aided Molecular Design, 11*(5), 425—445. https://doi.org/10.1023/a:1007996124545.

Ewing, T. J., Makino, S., Skillman, A. G., & Kuntz, I. D. (2001). DOCK 4.0: Search strategies for automated molecular docking of flexible molecule databases. *Journal of Computer-Aided Molecular Design, 15*(5), 411—428. https://doi.org/10.10 23/a:1011115820450.

Ferreira, L. G., Dos Santos, R. N., Oliva, G., & Andricopulo, A. D. (2015). Molecular docking and structure-based drug design strategies. *Molecules, 20*(7), 13384—13421. https://doi.org/10.3390/molecules200713384.

Foloppe, N., & Hubbard, R. (2006). Towards predictive ligand design with free-energy based computational methods? *Current Medicinal Chemistry, 13*(29), 3583—3608. https://doi.org/10.2174/092986706779026165.

Friesner, R. A., Banks, J. L., Murphy, R. B., Halgren, T. A., Klicic, J. J., Mainz, D. T., Repasky, M. P., Knoll, E. H., Shelley, M., Perry, J. K., Shaw, D. E., Francis, P., & Shenkin, P. S. (2004). Glide: A new approach for rapid, accurate docking and scoring. 1. Method and assessment of docking accuracy. *Journal of Medicinal Chemistry, 47*(7), 1739—1749. https://doi.org/10.1021/jm0306430.

Friesner, R. A., Murphy, R. B., Repasky, M. P., Frye, L. L., Greenwood, J. R., Halgren, T. A., Sanschagrin, P. C., & Mainz, D. T. (2006). Extra precision glide: Docking and scoring incorporating a model of hydrophobic enclosure for protein-ligand complexes. *Journal of Medicinal Chemistry, 49*(21), 6177—6196. https://doi.org/10.1021/jm051256o.

Gallicchio, E., & Levy, R. M. (2004). AGBNP: An analytic implicit solvent model suitable for molecular dynamics simulations and high-resolution modeling. *Journal of Computational Chemistry, 25*(4), 479—499. https://doi.org/10.1002/jcc.10400.

Ganser, L. R., Lee, J., Rangadurai, A., Merriman, D. K., Kelly, M. L., Kansal, A. D., Sathyamoorthy, B., & Al-Hashimi, H. M. (2018). High-performance virtual screening by targeting a high-resolution RNA dynamic ensemble. *Nature Structural and Molecular Biology, 25*(5), 425—434. https://doi.org/10.1038/s41594-018-0062-4.

Garand, E., Kamrath, M. Z., Jordan, P. A., Wolk, A. B., Leavitt, C. M., McCoy, A. B., Miller, S. J., & Johnson, M. A. (2012). Determination of noncovalent docking by infrared spectroscopy of cold gas-phase complexes. *Science, 335*(6069), 694—698. https://doi.org/10.1126/science.1214948.

Gilson, M. K., & Zhou, H. X. (2007). Calculation of protein-ligand binding affinities. *Annual Review of Biophysics and Biomolecular Structure, 36*, 21—42. https://doi.org/10.1146/annurev.biophys.36.040306.132550.

Gohlke, H., Hendlich, M., & Klebe, G. (2000). Knowledge-based scoring function to predict protein-ligand interactions. *Journal of Molecular Biology, 295*(2), 337—356. https://doi.org/10.1006/jmbi.1999.3371.

Grant, J. A., Pickup, B. T., & Nicholls, A. (2001). A smooth permittivity function for Poisson—Boltzmann solvation methods. *Journal of Computational Chemistry, 22*(6), 608—640. https://doi.org/10.1002/jcc.1032.

Guedes, I. A., Pereira, F. S. S., & Dardenne, L. E. (2018). Empirical scoring functions for structure-based virtual screening: Applications, critical aspects, and challenges. *Frontiers in Pharmacology, 9*, 1089. https://doi.org/10.3389/fphar.2018.01089.

Guidotti, G., Brambilla, L., & Rossi, D. (2017). Cell-penetrating peptides: From basic research to clinics. *Trends in Pharmacological Sciences, 38*(4), 406—424. https://doi.org/10.1016/j.tips.2017.01.003.

Hermann, J., DiStasio, R. A., Jr., & Tkatchenko, A. (2017). First-principles models for van der Waals interactions in molecules and materials: Concepts, theory, and applications. *Chemical Reviews, 117*(6), 4714—4758. https://doi.org/10.1021/acs.chemrev.6b00446.

Hermans, J., Berendsen, H. J. C., Van Gunsteren, W. F., & Postma, J. P. M. (1984). A consistent empirical potential for water—protein interactions. *Biopolymers, 23*(8), 1513—1518. https://doi.org/10.1002/bip.360230807.

Hertig, S., Latorraca, N. R., & Dror, R. O. (2016). Revealing atomic-level mechanisms of protein allostery with molecular dynamics simulations. *PLoS Computational Biology, 12*(6), e1004746. https://doi.org/10.1371/journal.pcbi.1004746.

Hogg, P. J. (2013). Targeting allosteric disulphide bonds in cancer. *Nature Reviews Cancer, 13*(6), 425—431. https://doi.org/10.1038/nrc3519.

Huang, S. Y., & Zou, X. (2007). Ensemble docking of multiple protein structures: Considering protein structural variations in molecular docking. *Proteins, 66*(2), 399—421. https://doi.org/10.1002/prot.21214.

Huang, S. Y., & Zou, X. (2010). Advances and challenges in protein-ligand docking. *International Journal of Molecular*

Sciences, *11*(8), 3016–3034. https://doi.org/10.3390/ijms11083016.

Hwang, M. J., Stockfisch, T. P., & Hagler, A. T. (1994). Derivation of class II force fields. 2. Derivation and characterization of a class II force field, CFF93, for the alkyl functional group and alkane molecules. *Journal of the American Chemical Society,* *116*(6), 2515–2525. https://doi.org/10.1021/ja00085a036.

Jain, A. N. (2006). Scoring functions for protein-ligand docking. *Current Protein and Peptide Science,* *7*(5), 407–420. https://doi.org/10.2174/138920306778559395.

Jiang, F., & Kim, S. H. (1991). "Soft docking": Matching of molecular surface cubes. *Journal of Molecular Biology,* *219*(1), 79–102. https://doi.org/10.1016/0022-2836(91)90859-5.

Jones, G., Willett, P., Glen, R. C., Leach, A. R., & Taylor, R. (1997). Development and validation of a genetic algorithm for flexible docking. *Journal of Molecular Biology,* *267*(3), 727–748. https://doi.org/10.1006/jmbi.1996.0897.

Jorgensen, W. L., Chandrasekhar, J., Madura, J. D., Impey, R. W., & Klein, M. L. (1983). Comparison of simple potential functions for simulating liquid water. *The Journal of Chemical Physics,* *79*(2), 926–935.

Jorgensen, W. L., & Tirado-Rives, J. (1988). The OPLS [optimized potentials for liquid simulations] potential functions for proteins, energy minimizations for crystals of cyclic peptides and crambin. *Journal of the American Chemical Society,* *110*(6), 1657–1666. https://doi.org/10.1021/ja00214a001.

Karplus, M. (2002). Molecular dynamics simulations of biomolecules. *Accounts of Chemical Research,* *35*(6), 321–323. https://doi.org/10.1021/ar020082r.

Karplus, M., & McCammon, J. A. (2002). Molecular dynamics simulations of biomolecules. *Nature Structural Biology,* *9*(9), 646–652. https://doi.org/10.1038/nsb0902-646.

Karplus, M., & Petsko, G. A. (1990). Molecular dynamics simulations in biology. *Nature,* *347*(6294), 631–639. https://doi.org/10.1038/347631a0.

Khedkar, S. A., Malde, A. K., & Coutinho, E. C. (2007). Design of inhibitors of the MurF enzyme of Streptococcus pneumoniae using docking, 3D-QSAR, and de novo design. *Journal of Chemical Information and Modeling,* *47*(5), 1839–1846. https://doi.org/10.1021/ci600568u.

Khedkar, S. A., Malde, A. K., Coutinho, E. C., & Srivastava, S. (2007). Pharmacophore modeling in drug discovery and development: An overview. *Medicinal Chemistry,* *3*(2), 187–197. https://doi.org/10.2174/157340607780059521.

Kitchen, D. B., Decornez, H., Furr, J. R., & Bajorath, J. (2004). Docking and scoring in virtual screening for drug discovery: Methods and applications. *Nature Reviews Drug Discovery,* *3*(11), 935–949. https://doi.org/10.1038/nrd1549.

Knegtel, R. M., Kuntz, I. D., & Oshiro, C. M. (1997). Molecular docking to ensembles of protein structures. *Journal of Molecular Biology,* *266*(2), 424–440. https://doi.org/10.1006/jmbi.1996.0776.

Kothiwale, S., Mendenhall, J. L., & Meiler, J. (2015). BCL::Conf: Small molecule conformational sampling using a knowledge based rotamer library. *Journal of Cheminformatics,* *7*, 47. https://doi.org/10.1186/s13321-015-0095-1.

Krammer, A., Kirchhoff, P. D., Jiang, X., Venkatachalam, C. M., & Waldman, M. (2005). LigScore: A novel scoring function for predicting binding affinities. *Journal of Molecular Graphics and Modelling,* *23*(5), 395–407. https://doi.org/10.1016/j.jmgm.2004.11.007.

Kumalo, H. M., Bhakat, S., & Soliman, M. E. (2015). Theory and applications of covalent docking in drug discovery: Merits and pitfalls. *Molecules,* *20*(2), 1984–2000. https://doi.org/10.3390/molecules20021984.

Kumar, S., & Nussinov, R. (1999). Salt bridge stability in monomeric proteins. *Journal of Molecular Biology,* *293*(5), 1241–1255. https://doi.org/10.1006/jmbi.1999.3218.

Kundaikar, H. S., Agre, N. P., & Degani, M. S. (2014). Pharmacophore based 3DQSAR of phenothiazines as specific human butyrylcholinesterase inhibitors for treatment of Alzheimer's disease. *Current Computer-Aided Drug Design,* *10*(4), 335–348. https://doi.org/10.2174/1573409911666150318203528.

Labute, P. (2010). LowModeMD—implicit low-mode velocity filtering applied to conformational search of macrocycles and protein loops. *Journal of Chemical Information and Modeling,* *50*(5), 792–800. https://doi.org/10.1021/ci900508k.

Ladbury, J. E. (2013). Thermodynamics of biomolecular interactions. In G. C. K. Roberts (Ed.), *Encyclopedia of biophysics* (pp. 2589–2606). Berlin, Heidelberg: Springer Berlin Heidelberg.

Lawal, M. M., Sanusi, Z. K., Govender, T., Tolufashe, G. F., Maguire, G. E. M., Honarparvar, B., & Kruger, H. G. (2019). Unraveling the concerted catalytic mechanism of the human immunodeficiency virus type 1 (HIV-1) protease: A hybrid QM/MM study. *Structural Chemistry,* *30*(1), 409–417. https://doi.org/10.1007/s11224-018-1251-9.

Leach, A. R. (1994). Ligand docking to proteins with discrete side-chain flexibility. *Journal of Molecular Biology,* *235*(1), 345–356. https://doi.org/10.1016/s0022-2836(05)80038-5.

Leach, A. R. (2001). *Molecular modelling: Principles and applications* (2nd ed.). Harlow, England; New York: Prentice Hall.

Lensink, M. F., Mendez, R., & Wodak, S. J. (2007). Docking and scoring protein complexes: CAPRI 3rd edition. *Proteins,* *69*(4), 704–718. https://doi.org/10.1002/prot.21804.

Lewars, E. (2003). *The concept of the potential energy surface computational chemistry: Introduction to the theory and applications of molecular and quantum mechanics* (pp. 9–41). Boston, MA: Springer US.

Li, G. B., Yang, L. L., Wang, W. J., Li, L. L., & Yang, S. Y. (2013). ID-score: A new empirical scoring function based on a comprehensive set of descriptors related to protein-ligand interactions. *Journal of Chemical Information and Modeling,* *53*(3), 592–600. https://doi.org/10.1021/ci300493w.

London, F. (1930). On the theory and system of Intermolecular. *Journal of Physics,* *63*(3–4), 245–279.

Luo, R., David, L., & Gilson, M. K. (2002). Accelerated Poisson-Boltzmann calculations for static and dynamic systems. *Journal of Computational Chemistry,* *23*(13), 1244–1253. https://doi.org/10.1002/jcc.10120.

MacKerell, A. D., Brooks, B., Brooks, C. L., Nilsson, L., Roux, B., Won, Y., & Karplus, M. (2002). *CHARMM: The energy function and its parameterization encyclopedia of computational chemistry.* John Wiley & Sons, Ltd.

de Magalhães, C. S., Almeida, D. M., Barbosa, H. J. C., & Dardenne, L. E. (2014). A dynamic niching genetic algorithm strategy for docking highly flexible ligands. *Information Sciences, 289,* 206–224. https://doi.org/10.1016/j.ins.2014.08.002.

March-Vila, E., Pinzi, L., Sturm, N., Tinivella, A., Engkvist, O., Chen, H., & Rastelli, G. (2017). On the integration of in silico drug design methods for drug repurposing. *Frontiers in Pharmacology, 8,* 298. https://doi.org/10.3389/fphar.2017.00298.

Marrari, Y., Crouthamel, M., Irannejad, R., & Wedegaertner, P. B. (2007). Assembly and trafficking of heterotrimeric G proteins. *Biochemistry, 46*(26), 7665–7677. https://doi.org/10.1021/bi700338m.

Marriott, D. P., Dougall, I. G., Meghani, P., Liu, Y. J., & Flower, D. R. (1999). Lead generation using pharmacophore mapping and three-dimensional database searching: Application to muscarinic M(3) receptor antagonists. *Journal of Medicinal Chemistry, 42*(17), 3210–3216. https://doi.org/10.1021/jm980409n.

Massova, I., & Kollman, P. A. (2000). Combined molecular mechanical and continuum solvent approach (MM-PBSA/GBSA) to predict ligand binding. *Perspectives in Drug Discovery and Design, 18*(1), 113–135. https://doi.org/10.1023/A:1008763014207.

van Meer, G., Voelker, D. R., & Feigenson, G. W. (2008). Membrane lipids: Where they are and how they behave. *Nature Reviews Molecular Cell Biology, 9*(2), 112–124. https://doi.org/10.1038/nrm2330.

Mooij, W., & Verdonk, M. (2005). General and targeted statistical potentials for protein-ligand interactions. *Proteins, 61,* 272–287. https://doi.org/10.1002/prot.20588.

Morris, G. M., Goodsell, D. S., Halliday, R. S., Huey, R., Hart, W. E., Belew, R. K., & Olson, A. J. (1998). Automated docking using a Lamarckian genetic algorithm and an empirical binding free energy function. *Journal of Computational Chemistry, 19*(14), 1639–1662. https://doi.org/10.1002/(sici)1096-987x(19981115)19:14<1639::aid-jcc10>3.0.co;2-b.

Morris, G. M., Huey, R., Lindstrom, W., Sanner, M. F., Belew, R. K., Goodsell, D. S., & Olson, A. J. (2009). AutoDock4 and AutoDockTools4: Automated docking with selective receptor flexibility. *Journal of Computational Chemistry, 30*(16), 2785–2791. https://doi.org/10.1002/jcc.21256.

Morse, P. M. (1929). Diatomic molecules according to the wave mechanics. II. Vibrational levels. *Physical Review, 34*(1), 57–64. https://doi.org/10.1103/PhysRev.34.57.

Muegge, I. (2006). PMF scoring revisited. *Journal of Medicinal Chemistry, 49*(20), 5895–5902. https://doi.org/10.1021/jm050038s.

Namasivayam, V., & Gunther, R. (2007). pso@autodock: A fast flexible molecular docking program based on Swarm intelligence. *Chemical Biology and Drug Design, 70*(6), 475–484. https://doi.org/10.1111/j.1747-0285.2007.00588.x.

Nelson, D. L. (2005). *Lehninger principles of biochemistry* (4th ed.). New York: W.H. Freeman.

Nerenberg, P. S., & Head-Gordon, T. (2018). New developments in force fields for biomolecular simulations. *Current Opinion in Structural Biology, 49,* 129–138. https://doi.org/10.1016/j.sbi.2018.02.002.

Ng, M. C., Fong, S., & Siu, S. W. (2015). PSOVina: The hybrid particle swarm optimization algorithm for protein-ligand docking. *Journal of Bioinformatics and Computational Biology, 13*(3), 1541007. https://doi.org/10.1142/S0219720015410073.

Nikkanen, L., & Rintamaki, E. (2014). Thioredoxin-dependent regulatory networks in chloroplasts under fluctuating light conditions. *Philosophical Transactions of the Royal Society of London B Biological Sciences, 369*(1640), 20130224. https://doi.org/10.1098/rstb.2013.0224.

Onufriev, A. V., & Case, D. A. (2019). Generalized born implicit solvent models for biomolecules. *Annual Review of Biophysics, 48,* 275–296. https://doi.org/10.1146/annurev-biophys-052118-115325.

Osguthorpe, D. J., Sherman, W., & Hagler, A. T. (2012). Exploring protein flexibility: Incorporating structural ensembles from crystal structures and simulation into virtual screening protocols. *Journal of Physical Chemistry B, 116*(23), 6952–6959. https://doi.org/10.1021/jp3003992.

Pang, Y. P., Perola, E., Xu, K., & Prendergast, F. G. (2001). EUDOC: A computer program for identification of drug interaction sites in macromolecules and drug leads from chemical databases. *Journal of Computational Chemistry, 22*(15), 1750–1771. https://doi.org/10.1002/jcc.1129.

Pason, L. P., & Sotriffer, C. A. (2016). Empirical scoring functions for affinity prediction of protein-ligand complexes. *Molecular Informatics, 35*(11–12), 541–548. https://doi.org/10.1002/minf.201600048.

Pearlman, D. A., Case, D. A., Caldwell, J. W., Ross, W. S., Cheatham, T. E., DeBolt, S., Ferguson, D., Seibel, G., & Kollman, P. (1995). AMBER, a package of computer programs for applying molecular mechanics, normal mode analysis, molecular dynamics and free energy calculations to simulate the structural and energetic properties of molecules. *Computer Physics Communications, 91*(1), 1–41. https://doi.org/10.1016/0010-4655(95)00041-D.

Pei, J., Wang, Q., Liu, Z., Li, Q., Yang, K., & Lai, L. (2006). PSI-DOCK: Towards highly efficient and accurate flexible ligand docking. *Proteins, 62*(4), 934–946. https://doi.org/10.1002/prot.20790.

Pierce, B. G., Wiehe, K., Hwang, H., Kim, B. H., Vreven, T., & Weng, Z. (2014). ZDOCK server: Interactive docking prediction of protein-protein complexes and symmetric multimers. *Bioinformatics, 30*(12), 1771–1773. https://doi.org/10.1093/bioinformatics/btu097.

Pinzi, L., & Rastelli, G. (2019). Molecular docking: Shifting paradigms in drug discovery. *International Journal of Molecular Sciences, 20*(18). https://doi.org/10.3390/ijms20184331.

Prabhu, N. V., Zhu, P., & Sharp, K. A. (2004). Implementation and testing of stable, fast implicit solvation in molecular dynamics using the smooth-permittivity finite difference Poisson-Boltzmann method. *Journal of Computational Chemistry, 25*(16), 2049–2064. https://doi.org/10.1002/jcc.20138.

Price, S. L. (1999). Theoretical approaches to the study of non-bonded interactions. In J. A. K. Howard, F. H. Allen, & G. P. Shields (Eds.), *Implications of molecular and materials structure for new technologies* (pp. 223–234). Dordrecht: Springer Netherlands.

Pylaeva, S., Brehm, M., & Sebastiani, D. (2018). Salt bridge in aqueous solution: Strong structural motifs but weak enthalpic effect. *Scientific Reports, 8*(1), 13626. https://doi.org/10.1038/s41598-018-31935-z.

Rarey, M., Kramer, B., Lengauer, T., & Klebe, G. (1996). A fast flexible docking method using an incremental construction algorithm. *Journal of Molecular Biology, 261*(3), 470–489. https://doi.org/10.1006/jmbi.1996.0477.

Ravichandran, V., & Agrawal, R. K. (2007). Predicting anti-HIV activity of PETT derivatives: CoMFA approach. *Bioorganic and Medicinal Chemistry Letters, 17*(8), 2197–2202. https://doi.org/10.1016/j.bmcl.2007.01.103.

Reulecke, I., Lange, G., Albrecht, J., Klein, R., & Rarey, M. (2008). Towards an integrated description of hydrogen bonding and dehydration: Decreasing false positives in virtual screening with the HYDE scoring function. *ChemMedChem, 3*(6), 885–897. https://doi.org/10.1002/cmdc.200700319.

Ribeiro, A. J., Ramos, M. J., & Fernandes, P. A. (2012). The catalytic mechanism of HIV-1 integrase for DNA 3′-end processing established by QM/MM calculations. *Journal of the American Chemical Society, 134*(32), 13436–13447. https://doi.org/10.1021/ja304601k.

Rifaioglu, A. S., Nalbat, E., Atalay, V., Martin, M. J., Cetin-Atalay, R., & Doğan, T. (2020). DEEPScreen: High performance drug–target interaction prediction with convolutional neural networks using 2-D structural compound representations. *Chemical Science, 11*(9), 2531–2557. https://doi.org/10.1039/C9SC03414E.

Riniker, S., & Landrum, G. A. (2015). Better informed distance geometry: Using what we know to improve conformation generation. *Journal of Chemical Information and Modeling, 55*(12), 2562–2574. https://doi.org/10.1021/acs.jcim.5b00654.

Rozas, I. (2007). On the nature of hydrogen bonds: An overview on computational studies and a word about patterns. *Physical Chemistry Chemical Physics, 9*(22), 2782–2790. https://doi.org/10.1039/b618225a.

Salmaso, V., & Moro, S. (2018). Bridging molecular docking to molecular dynamics in exploring ligand-protein recognition process: An overview. *Frontiers in Pharmacology, 9*, 923. https://doi.org/10.3389/fphar.2018.00923.

Sauton, N., Lagorce, D., Villoutreix, B. O., & Miteva, M. A. (2008). MS-DOCK: Accurate multiple conformation generator and rigid docking protocol for multi-step virtual ligand screening. *BMC Bioinformatics, 9*, 184. https://doi.org/10.1186/1471-2105-9-184.

Schlick, T. (2010). *Molecular modeling and simulation: An interdisciplinary guide*. New York: Springer.

Schneidman-Duhovny, D., Inbar, Y., Nussinov, R., & Wolfson, H. J. (2005). PatchDock and SymmDock: Servers for rigid and symmetric docking. *Nucleic Acids Research, 33*(Web Server issue), W363–W367. https://doi.org/10.1093/nar/gki481.

Scouras, A. D., & Daggett, V. (2011). The dynameomics rotamer library: Amino acid side chain conformations and dynamics from comprehensive molecular dynamics simulations in water. *Protein Science, 20*(2), 341–352. https://doi.org/10.1002/pro.565.

Shapovalov, M. V., & Dunbrack, R. L., Jr. (2011). A smoothed backbone-dependent rotamer library for proteins derived from adaptive kernel density estimates and regressions. *Structure, 19*(6), 844–858. https://doi.org/10.1016/j.str.2011.03.019.

Shu-Kun, L. (1999). The weak hydrogen bond: Applications to structural chemistry and biology (International Union of Crystallography Monographs on Crystallography, 9). By Gautam R. Desiraju. *Molecules, 4*. https://doi.org/10.3390/41000318.

Sivakova, S., & Rowan, S. J. (2005). Nucleobases as supramolecular motifs. *Chemical Society Reviews, 34*(1), 9–21. https://doi.org/10.1039/b304608g.

Smyth, M. S., & Martin, J. H. (2000). X ray crystallography. *Molecular Pathology, 53*(1), 8–14. https://doi.org/10.1136/mp.53.1.8.

Spyrakis, F., Benedetti, P., Decherchi, S., Rocchia, W., Cavalli, A., Alcaro, S., Ortuso, F., Baroni, M., & Cruciani, G. (2015). A pipeline to enhance ligand virtual screening: Integrating molecular dynamics and fingerprints for ligand and proteins. *Journal of Chemical Information and Modeling, 55*(10), 2256–2274. https://doi.org/10.1021/acs.jcim.5b00169.

Spyrakis, F., & Cavasotto, C. N. (2015). Open challenges in structure-based virtual screening: Receptor modeling, target flexibility consideration and active site water molecules description. *Archives of Biochemistry and Biophysics, 583*, 105–119. https://doi.org/10.1016/j.abb.2015.08.002.

Srivani, P., & Sastry, G. N. (2009). Potential choline kinase inhibitors: A molecular modeling study of bis-quinolinium compounds. *Journal of Molecular Graphics and Modelling, 27*(6), 676–688. https://doi.org/10.1016/j.jmgm.2008.10.010.

Srivani, P., Usharani, D., Jemmis, E. D., & Sastry, G. N. (2008). Subtype selectivity in phosphodiesterase 4 (PDE4): A bottleneck in rational drug design. *Current Pharmaceutical Design, 14*(36), 3854–3872. https://doi.org/10.2174/138161208786898653.

Still, W. C., Tempczyk, A., Hawley, R. C., & Hendrickson, T. (1990). Semianalytical treatment of solvation for molecular mechanics and dynamics. *Journal of the American Chemical Society, 112*(16), 6127–6129. https://doi.org/10.1021/ja00172a038.

Tawari, N. R., & Degani, M. S. (2010). Pharmacophore mapping and electronic feature analysis for a series of nitroaromatic compounds with antitubercular activity. *Journal of Computational Chemistry, 31*(4), 739–751. https://doi.org/10.1002/jcc.21371.

Taylor, R. D., Jewsbury, P. J., & Essex, J. W. (2002). A review of protein-small molecule docking methods. *Journal of Computer-Aided Molecular Design, 16*(3), 151–166. https://doi.org/10.1023/a:1020155510718.

Telvekar, V. N., & Patel, K. N. (2011). Pharmacophore development and docking studies of the HIV-1 integrase inhibitors derived from N-methylpyrimidones, Dihydroxypyrimidines, and bicyclic pyrimidinones. *Chemical Biology and Drug Design*, *78*(1), 150−160. https://doi.org/10.1111/j.1747-0285.2011.01130.x.

Tessaro, F., & Scapozza, L. (2020). How 'protein-docking' translates into the new emerging field of docking small molecules to nucleic acids? *Molecules, 25*(12). https://doi.org/10.3390/molecules25122749.

Totrov, M., & Abagyan, R. (2001). Rapid boundary element solvation electrostatics calculations in folding simulations: Successful folding of a 23-residue peptide. *Biopolymers, 60*(2), 124−133. https://doi.org/10.1002/1097-0282(2001)60:2<124::AID-BIP1008>3.0.CO;2-S.

Tripathi, A., & Bankaitis, V. A. (2017). Molecular docking: From lock and key to combination lock. *Journal of Molecular Medicine and Clinical Applications, 2*(1). https://doi.org/10.16966/2575-0305.106.

Trott, O., & Olson, A. J. (2010). AutoDock Vina: Improving the speed and accuracy of docking with a new scoring function, efficient optimization, and multithreading. *Journal of Computational Chemistry, 31*(2), 455−461. https://doi.org/10.1002/jcc.21334.

Truhlar, D. G. (2003). Potential energy surfaces. In R. A. Meyers (Ed.), *Encyclopedia of physical science and technology* (3rd ed., pp. 9−17). New York: Academic Press.

Uzman, A. (2003). Molecular biology of the cell (4th ed.): Alberts, B., Johnson, A., Lewis, J., raff, M., roberts, K., and walter, P. *Biochemistry and Molecular Biology Education, 31*(4), 212−214. https://doi.org/10.1002/bmb.2003.494031049999.

Vanommeslaeghe, K., Guvench, O., & MacKerell, A. D., Jr. (2014). Molecular mechanics. *Current Pharmaceutical Design, 20*(20), 3281−3292. https://doi.org/10.2174/13816128113199990600.

Velec, H. F., Gohlke, H., & Klebe, G. (2005). DrugScore(CSD)-knowledge-based scoring function derived from small molecule crystal data with superior recognition rate of near-native ligand poses and better affinity prediction. *Journal of Medicinal Chemistry, 48*(20), 6296−6303. https://doi.org/10.1021/jm050436v.

Venkatachalam, C. M., Jiang, X., Oldfield, T., & Waldman, M. (2003). LigandFit: A novel method for the shape-directed rapid docking of ligands to protein active sites. *Journal of Molecular Graphics and Modelling, 21*(4), 289−307. https://doi.org/10.1016/s1093-3263(02)00164-x.

Verdonk, M. L., Cole, J. C., Hartshorn, M. J., Murray, C. W., & Taylor, R. D. (2003). Improved protein-ligand docking using GOLD. *Proteins, 52*(4), 609−623. https://doi.org/10.1002/prot.10465.

Wang, S., Witek, J., Landrum, G. A., & Riniker, S. (2020). Improving conformer generation for small rings and macrocycles based on distance geometry and experimental torsional-angle preferences. *Journal of Chemical Information and Modeling, 60*(4), 2044−2058. https://doi.org/10.1021/acs.jcim.0c00025.

Warren, G. L., Andrews, C. W., Capelli, A. M., Clarke, B., LaLonde, J., Lambert, M. H., Lindvall, M., Nevins, N., Semus, S. F., Senger, S., Tedesco, G., Wall, I. D., Woolven, J. M., Peishoff, C. E., & Head, M. S. (2006). A critical assessment of docking programs and scoring functions. *Journal of Medicinal Chemistry, 49*(20), 5912−5931. https://doi.org/10.1021/jm050362n.

Watts, K. S., Dalal, P., Murphy, R. B., Sherman, W., Friesner, R. A., & Shelley, J. C. (2010). ConfGen: A conformational search method for efficient generation of bioactive conformers. *Journal of Chemical Information and Modeling, 50*(4), 534−546. https://doi.org/10.1021/ci100015j.

Weiner, S. J., Kollman, P. A., Case, D. A., Singh, U. C., Ghio, C., Alagona, G., Profeta, S., & Weiner, P. (1984). A new force field for molecular mechanical simulation of nucleic acids and proteins. *Journal of the American Chemical Society, 106*(3), 765−784. https://doi.org/10.1021/ja00315a051.

Wider, G. (2000). Structure determination of biological macro-molecules in solution using nuclear magnetic resonance spectroscopy. *Biotechniques, 29*(6). https://doi.org/10.2144/00296ra01, 1278−1282, 1284−1290, 1292 passim.

Wink, M. (2003). Evolution of secondary metabolites from an ecological and molecular phylogenetic perspective. *Phytochemistry, 64*(1), 3−19. https://doi.org/10.1016/S0031-9422(03)00300-5.

Xie, Z. R., & Hwang, M. J. (2010). An interaction-motif-based scoring function for protein-ligand docking. *BMC Bioinformatics, 11*, 298. https://doi.org/10.1186/1471-2105-11-298.

Zhang, J., Zhang, H., Wu, T., Wang, Q., & van der Spoel, D. (2017). Comparison of implicit and explicit solvent models for the calculation of solvation free energy in organic solvents. *Journal of Chemical Theory and Computation, 13*(3), 1034−1043. https://doi.org/10.1021/acs.jctc.7b00169.

CHAPTER 3

Post-processing of Docking Results: Tools and Strategies

SABINA PODLEWSKA • ANDRZEJ J. BOJARSKI

1 INTRODUCTION

A growing number of resources gathering data on compound activity, together with the increasing computational power makes various in silico tools an indispensable part of drug design process. Computational strategies help not only in the identification of potentially active compounds and optimization of their physicochemical and pharmacokinetic properties but also are present at each stage of drug design pipeline, even after introduction of a drug to the market, e.g., in the tasks related to the data management.

The great popularity of computational tools in drug design—related tasks gave root to the field of computer-aided drug design (CADD), which deals with both the development of new in silico strategies for the search for new drugs and the usage of existing tools in drug design projects. The state-of-the-art CADD tools and techniques are described in detail in Chapter 1.

There are two main groups of methods, which can be distinguished within CADD: ligand-based and structure-based. The former one performs evaluation only on the basis of the structure and properties of existing ligands of considered receptor, whereas the latter makes use of the three-dimensional (3D) structure of the target protein (crystal structure or model and methods for determination of 3D target structures are described in Chapters 6—8), The core methodology in the structure-based path is docking, which is often the final step in the virtual screening (VS) cascades (as discussed in detail in Chapter 11).

As a result of this procedure, a series of ligand—receptor complexes is obtained (theoretical basis of this process is summarized in Chapter 2; Chapters 9 and 10 cover docking software guide). However, they are of limited use, if a proper evaluation of various ligand positions in the binding site is not applied. The choice of assessment approach depends mostly on the number of analyzed structures, person's experience and knowledge on desired interactions for a particular target. A small number of compounds can be analyzed visually; however, even in such cases, the set of poses should be first limited using some automatic tools.

2 SCORING FUNCTIONS

Evaluation of docking poses is crucial for the whole docking process. Docking algorithm, before returning final orientation of a ligand, needs to perform evaluation of proposed compound position. Among its requirements, despite being accurate in estimation of particular ligand—protein energy, there is a need for being fast, as the number of ligand—protein configurations can be unlimited. It should also be free of target bias, perform similarly for systems of various types, and display similar performance for different affinity ranges. Ideally, the scoring function should also be interpretable, so as the difference in its values can be explained from the ligand pose.

The current classification, according to Liu and Wang (2015), distinguishes four main types of scoring functions:

(1) Physics-based—assessment is made on the basis of the contributions coming from van der Waals and electrostatic interactions between all atoms within particular ligand-protein system. Some of such scoring functions also include the hydrogen bonding component. In the past, such methods relied solely on force fields and were computed in the gas phase (therefore, such evaluations were long classified as force field scoring functions). Later, to force field scoring functions, there started to be added also non—force field components, such as solvation models (Gilson et al., 1997; Zou et al., 1999).

Molecular Docking for Computer-Aided Drug Design. https://doi.org/10.1016/B978-0-12-822312-3.00004-7

(2) Empirical—sums up various energetic contributions, such as hydrophilic, hydrophobic contacts, hydrogen bonds, ionic interactions, steric clashes, entropic effects, etc. The final prediction is based either on the multivariate linear regression (MLR) or partial least squares (PLS; these methods are used to determine weights, which are related to particular energetic component).

A training set of ligand—protein pairs with known affinity is used in docking to develop a scoring function in its final form. Therefore, empirical scoring functions can be limited by the accuracy of the training data used. When developing empirical scoring function, one faces two problems: (1) preparation and usage of training set of large size and quality (for example, dealing with inconsistency in experimental conditions when determining ligand affinity in a particular complex) and (2) determination of energy terms to be used in the function. Example of such empirical scoring is offered by Glide (Friesner et al., 2006), whose scoring adopts the following formula:

$$\Delta G_{bind} = C_{lipo-lipo} \sum f(r_{lr}) + C_{hbond-neut-neut} \sum g(\Delta r) h(\Delta \alpha)$$
$$+ C_{hbond-neut-charged} \sum g(\Delta r) h(\Delta \alpha)$$
$$+ C_{hbond-charged-charged} \sum g(\Delta r) h(\Delta \alpha)$$
$$+ C_{max-metal-ion} \sum f(r_{lm}) + C_{rotb} H_{rotb}$$
$$+ C_{polar-phob} V_{polar-phob}$$
$$+ C_{coul} E_{coul} + C_{vdW} E_{vdW} + solvation\ terms,$$

where

$C_{lipo-lipo} \sum f(r_{lr})$—lipophilic—lipophilic term (defined as in the ChemScore),

$C_{hbond-neut-neut} \sum g(\Delta r) h(\Delta \alpha)$,

$C_{hbond-neut-charged} \sum g(\Delta r) h(\Delta \alpha)$,

$C_{hbond-charged-charged} \sum g(\Delta r) h(\Delta \alpha)$ —hydrogen bonding terms separated for cases where acceptor and donor are neutral or charged,

$C_{max-metal-ion} \sum f(r_{lm})$—metal—ligand interaction term,

$C_{rotb} H_{rotb}$—internal rotation term,

$C_{polar-phob} V_{polar-phob}$—contact of polar but non—hydrogen-bonding atom, which is found in a hydrophobic region,

$C_{coul} E_{coul}$—ligand—protein Coulomb interaction,

$C_{vdW} E_{vdW}$—ligand—protein van der Waals interaction,

solvation terms—solvent-related effects.

On the other hand, ChemScore performing poses evaluation in GOLD (Verdonk et al., 2003) has the following form:

$$ChemScore = S_{H-bond} + S_{metal} + S_{lipophilic} + P_{rotor} + P_{strain} + P_{clash}$$
$$+ [P_{covalent} + P_{constraint}]$$

with P representing various types of physical contributions to binding.

(3) Knowledge-based potential—often described in the literature also as "knowledge-based scoring function." Here, the protein—ligand complexes of known structure (e.g., ligands from ligand—protein complexes present in the Cambridge Structural Database (Groom, Bruno, Lightfoot, & Ward, 2016) or complexes itself in Protein Data Bank, PDB (Berman et al., 2000)) are used to derive pairwise potentials based on the inverse Boltzmann statistic principle (Chandler, 1987):

$$A = \sum_{i}^{lig} \sum_{j}^{prot} \omega_{ij}(r),$$

where $\omega_{ij}(r)$ is the potential between two atoms i and j, described by the following equation:

$$\omega_{ij}(r) = -k_B T \ln[g_{ij}(r)] = -k_B T \ln\left[\frac{\rho_{ij}(r)}{\rho_{ij}^*}\right]$$

where $\rho_{ij}(r)$ is the density of atom pair $i-j$ (with distance r between them), and ρ_{ij}^* is the density of atom pair $i-j$ in a state, in which the interatomic interactions are assumed to be zero (it is considered to be a reference state). The assumption underlying the knowledge-based potential approach is that the frequency of intermolecular contacts reflects the energetic preferences (the more frequent they occur, the more energetically favorable they are).

(4) Machine learning (ML) (descriptor-based scoring)—scoring using ML approaches. It is usually not incorporated directly by docking programs, but rather they are applied for rescoring previously obtained poses. It involves description of compounds with the use of various descriptors or interaction fingerprints (IFPs) and the analysis using ML methods. This type of scoring, together with examples of particular learning approaches will be discussed separately.

Table 3.1 lists different commonly used docking software and their scoring functions.

3 REPRESENTATION OF LIGAND—RECEPTOR COMPLEXES

3.1 Interaction Fingerprints

Ligand—receptor complexes returned by docking programs can be variously depicted, which facilitates

TABLE 3.1
Types of Scoring Functions in Different Docking Software.

Physics-Based	Empirical	Knowledge-Based
DOCK (Meng et al., 1992)	LigandFit (Venkatachalam et al., 2003)	DrugScore (Gohlke et al., 2000)
GOLD (Verdonk et al., 2003)	LUDI (Böhm, 1994)	SMoG (DeWitte et al., 1997)
HADDOCK (de Vries et al., 2010)	FRED (McGann, 2012)	PMF_Score (Muegge et al., 1999)
AutoDock (Goodsell et al., 1996)	Glide (Friesner et al., 2006)	MotifScore (Xie & Hwang, 2010)
ICM (Abagyan et al., 1994)	Surflex (Jain, 2003)	
Molegro Virtual Docker (Bitencourt-Ferreira & de Azevedo, 2019)	eHiTS (Zsoldos et al., 2007)	
	HYDE (Reulecke et al., 2008)	
	LigScore (Krammer et al., 2005)	
	PLP (Gehlhaar et al., 1995)	

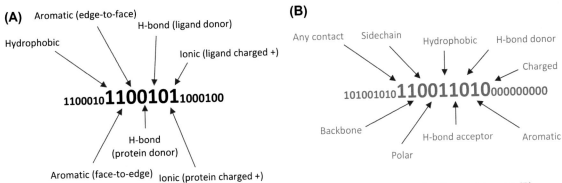

FIG. 3.1 Description of particular bits in ligand–protein contacts in **(A)** interaction fingerprints and **(B)** structural interaction fingerprints.

their storage and automatic analysis of contacts. For the sake of statistical analysis, the most useful is its representation by IFPs. In its most typical form, they transform ligand–protein contacts into the form of zero-one strings. During their generation, the target protein is screened amino acid by amino acid, and the occurrence of contacts between respective residues and ligands is indicated. Two most widely used approaches include structural interaction fingerprints (SIFts) (Deng, Chuaqui, & Singh, 2004) and IFPs (Marcou & Rognan, 2007), differentiating in terms of bits definition and set of described contacts (Fig. 3.1).

Examples of IFPs generated for co-crystallized ligand in the serotonin receptor 5-HT$_{2A}$ complex with zotepine (PDB code: 6A94 (Kimura et al., 2019)) are presented in Fig. 3.2.

Fig. 3.2 indicates differences in those two forms of ligand–receptor complexes depiction. First of all, SIFts inform about the presence of any contact between entities forming an examined complex. Therefore, if the interaction does not fall into any category of depicted by subsequent positions, the information about the contact presence is preserved thanks to the first bit. On the other hand, IFPs provide more detailed information about ligand–protein contacts, as within aromatic interactions face-to-edge and edge-to-face contacts are differentiated, additional information about ionic interaction is also given (information about the charged entity is included). Moreover, IFPs contain also information about the contacts from the side of ligands (each interaction is supplied with information about atoms from the ligand that take part in the respective contact).

A different way of depicting contacts in biological assemblies is offered by protein–ligand interaction fingerprint (PLIF) (Huang et al., 2012), implemented in the molecular operating environment (Chemical

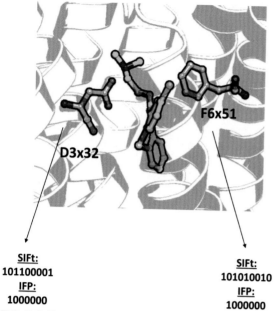

SIFt:
101100001
IFP:
1000000

SIFt:
101010010
IFP:
1000000

FIG. 3.2 Examples of structural interaction fingerprint (SIFt) and interaction fingerprint (IFP) strings generated for two selected residues (D3x32 and F6x51), numbered according to GPCRdb (Isberg et al., 2016) from the crystal structure of the 5-HT$_{2A}$R—zotepine complex.

Computing Group Inc., 2010). The fingerprint captures the presence of intermolecular interactions of the following types: hydrogen bond with side chain donor, hydrogen bond with side chain acceptor, hydrogen bond with backbone donor, hydrogen bond with backbone acceptor, ionic attraction, and surface contact. The strength of interaction also counts and strong and weak contacts are coded separately. As a result, an interaction of a ligand with each residue of a protein is encoded by 12 bits.

The extension of the PLIF approach is the structural protein—ligand interaction fingerprint (SPLIF) (Da & Kireev, 2014). An algorithm for SPLIF calculation is as follows: at first, the ligand—receptor complex is screened in terms of contacts occurrence (two atoms are considered as interacting if the distance between them is within a particular threshold). Then, each atom pair (one atom belongs to ligand and the other one to protein) is expanded to circular fragments around the two atoms, up to the specified distance, and such obtained circular fragments are annotated with an identifier using extended connectivity fingerprint ECFP2 (Rogers & Hahn, 2010). Finally, for all atoms from such obtained fragments, 3D coordinates are collected. Such a determination of interacting fragments and implicit coding of

interactions enables coding of all types of ligand—protein contacts (also those which are not captured by "traditional" SIFt and IFP approaches). The SPLIF benchmarking was performed on the focal adhesion kinase 1 (FAK1), serine/threonine protein kinase (AKT1), angiotensin converting enzyme (ACE), tryptase beta-1 (TRYB1), HMG-CoA reductase (HMDH), beta-1 adrenergic receptor (ADRB1), mineralocorticoid receptor (MCR), progesterone receptor (PRGR), glutamate receptor ionotropic kainate (GRIK1), and cyclooxygenase-2 (PGH2).

PLIF representation also constituted a basis for the development of the pharmacophore-based interaction fingerprint (Pharm-IF) (Sato et al., 2010, 34). Pharm-IF was used to detect interactions occurring in ligand—protein complex, which were classified into the following types: hydrogen bond with ligand acceptor, hydrogen bond with ligand donor, hydrogen bond in which the roles of ligand and protein atoms could not be determined, ionic interaction with ligand cation, ionic interaction with ligand anion, and hydrophobic interaction (the threshold of weak interaction was used). All possible combinations of interaction pairs from detected ligand—protein contacts were enumerated, and then interaction pairs were characterized by the pharmacophore features of ligand atoms and their distances, before the final assignment to the corresponding bin. The final matrix was produced by the summation of values of all interaction pairs.

Pérez-Nueno et al. developed atom pairs—based interaction fingerprint (APIF) (Pérez-Nueno et al., 2009). It was calculated in such a way that after determination of ligand—protein contacts, the ligand and protein atoms were labeled as hydrogen bond donor, hydrogen bond acceptor, or hydrophobic. For such determined interaction pairs, distances between two ligand and protein atoms were measured and assigned to a particular bin (bin ranges are as follows: 0—2.5, 2.5—4, 4—6, 6—9, 9—13, 13—18, >18 Å). The total length of the fingerprint is equal to 294 bits.

3.2 Ligand-based Approaches

Information about ligand orientations in the binding pockets can be encoded also in the form of descriptors generated solely from the 3D ligand conformation. Examples of such descriptors include Spectrophores (Bultinck et al., 2002). They are one-dimensional descriptors that provide information about molecules in terms of their surface properties or fields. They are calculated by surrounding the spatial structure of the compound by a set of points followed by calculation of interactions between these points and atoms of the

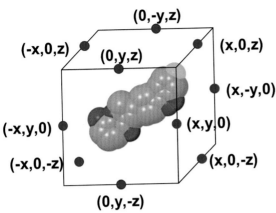

FIG. 3.3 An "artificial cage" surrounding the molecule generated for Spectrophores calculations.

considered molecule in terms of atomic properties (Fig. 3.3) (Smusz et al., 2015). The output of their generation consists of 48 positions that include values calculated from the atomic partial charges, atomic lipophilicity properties, atomic shape deviations, and atomic electrophilicity properties. During calculations, the minimization of the total interaction between the artificial points and the molecule for a given property is provided and stereospecificity might also be taken into account.

Another type of conformation-dependent compound representation are 3D MoRSE (Schuur et al., 1996) descriptors (160 descriptors allowing the representation of 3D compound structure by a fixed number of values). Their generation is based on electron diffraction equations and obtained output is then supplied by various atomic properties.

Autocorrelation 3D descriptors (Sliwoski et al., 2016) code the relative atom positions via the calculation of separation between pairs of atoms in terms of Euclidean distance (such calculated atom–pair distances are stored in the form of a histogram). Similarly, EigenSpectrum Shape Fingerprints (ESHAPE3D) (Ballester & Richards, 2007a) calculate pairwise distances between heavy atoms, which are stored in a matrix. The matrix is then diagonalized and eigenvalues computed after diagonalization characterize compound shape.

On the other hand, WHIM (Verma et al., 2010) representation focuses on size, shape, symmetry, and atom distribution. WHIM descriptors can be divided into two groups: directional (related to molecular size, molecular shape, symmetry, and atom distribution and density around the origin and along the principal axes) and global (nondirectional, analogous to directional one but the information that is related to the principal

axes is not present; therefore, the obtained description is connected only with the global view of the molecule).

Atom distances in three dimensional space also constitute basis for other 3D representations: Ultrafast Shape Recognition (USR) (Ballester & Richards, 2007b) measures all possible distances between the following parts of a molecule: compound center, the closest atom from the center, the farthest atom from the center, the farthest atom from the center (ftc), and the farthest atom from ftc. The set of distances is then supplied with some distribution data: mean distance, distribution variance, and distribution skewness.

The aforementioned 3D ligand depictions provide information about its shape, but neglect its physico-chemical features. The information about compound shape is extended by the information about detection of selected pharmacophore features (hydrophobic, aromatic, hydrogen bond donor, hydrogen bond acceptor) within the Ultrafast Shape Recognition with CREDO Atom Types (USRCAT) (Schreyer & Blundell, 2012).

There also exist approaches that compare the spatial similarity of two molecules. They usually base on the measurement of the overlap of molecules volume after their superposition. One of the most widely known tools of such a type is the rapid overlay of chemical structures (Hawkins et al., 2007). It is designed to perform superposition of two molecules in such a way that the overlapped volume is as high as possible. An alternative approach of such a type is MolShaCS (Vaz de Lima & Nascimento, 2013), which maximize not only the volume overlap but also electrostatic potential.

An extension of molecular shape comparison by pharmacophore features is provided by the shape-feature similarity (SHAFTS) (Liu et al., 2011). It uses hydrophobic center, positive charge center, negative charge center, hydrogen bond acceptor, hydrogen bond donor, and aromatic ring definitions.

Wide range of methods that consider molecular surfaces was also introduced. Such an approach is used, for example, by MolPrint3D representation (Bender et al., 2004), which computes the Interacting Surface Point Type (ISPT) descriptors by transforming interaction energies at the molecular surface. ISPT descriptors are also used by the eHiTS LASSO approach (Reid et al., 2008), which uses an extended set of ISPTs (feature vector composed of counts of particular surface points of a ligand). As ISPTs are based neither on surface properties nor topology, therefore, the similarity in terms of ISPTs is not correlated with the similarity in topological sense.

Interactions occurring within ligand–receptor complexes can also be used for tuning two-dimensional

(2D) representations, as it was done, for example, in the work of Tan and Bajorath (2009), Tan et al. (2009), where frequencies of interactions occurring in crystal structures were used to develop a weighing scheme for particular bits of 2D representation.

3.3 Interaction Fingerprint Databases

The significant growth of the number of available crystal structures, on the one hand, and the development of new approaches for translating ligand—receptor complexes in very simple forms (such as bit strings in the case of SIFts and IFPs), on the other, triggered the development of databases storing information about interactions occurring for particular targets.

KLIFs database (Kooistra et al., 2016) focuses on kinases. It is a weekly updated set of over 5000 entries from the PDB database, covering over 300 different kinases with over 3200 different co-crystallized ligands. For each PDB entry, it comprehensively analyzes interactions occurring within the ligand—protein complex and offers an engine for searching structures with similar interaction pattern, which is based on the IFP calculation. The database is available at www.klifs.vu-compmedchem.nl. A portion of an output produced by KLIFS is presented in Fig. 3.4.

Another database of interaction patterns is CREDO (Schreyer & Blundell, 2013). It is a database covering 86,903 entries from the PDB database (release 2013.2.1). The CREDO database gathers all interactions between surface-exposed residues (generated by OpenEye) in biological assemblies. Pairwise contacts are stored in the form of SIFt, with the addition of information about halogen bonds and carbonyl interactions.

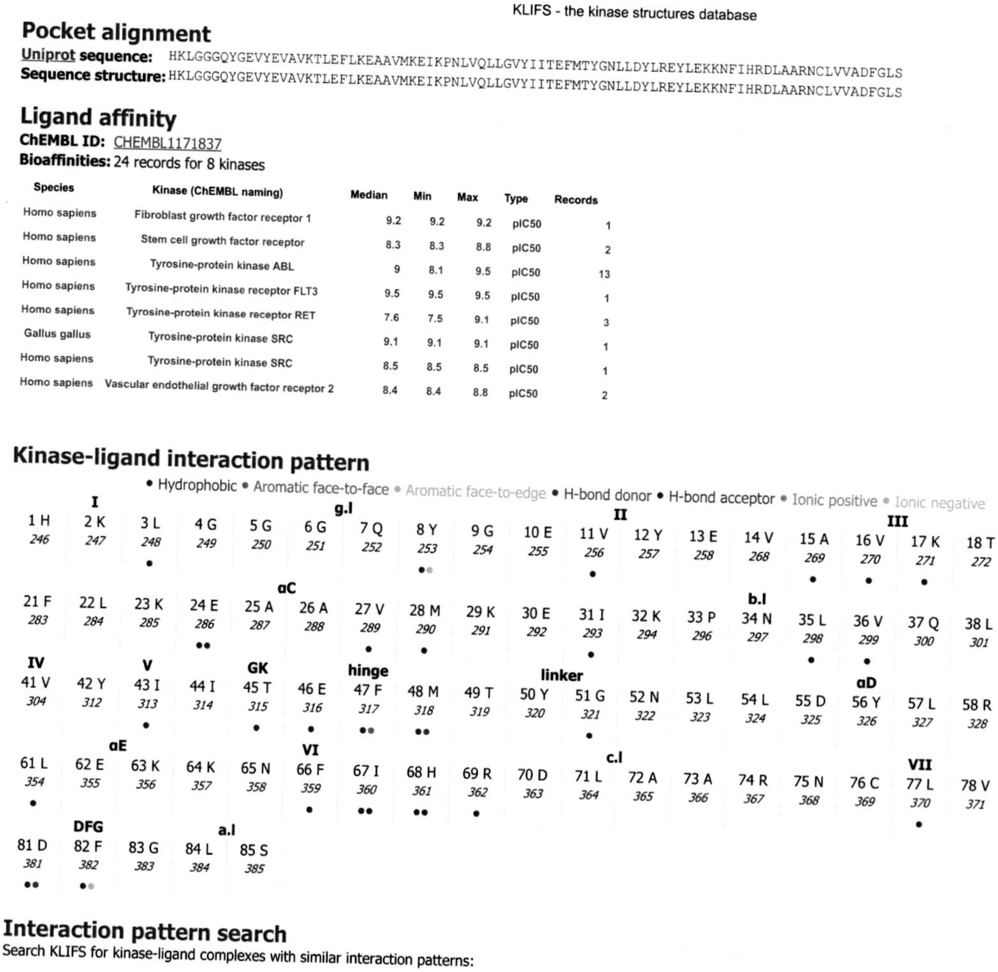

KLIFS - the kinase structures database

Pocket alignment

Uniprot sequence: HKLGGGQYGEVYEVAVKTLEFLKEAAVMKEIKPNLVQLLGVYIITEFMTYGNLLDYLREYLEKKNFIHRDLAARNCLVVADFGLS
Sequence structure: HKLGGGQYGEVYEVAVKTLEFLKEAAVMKEIKPNLVQLLGVYIITEFMTYGNLLDYLREYLEKKNFIHRDLAARNCLVVADFGLS

Ligand affinity

ChEMBL ID: CHEMBL1171837
Bioaffinities: 24 records for 8 kinases

Species	Kinase (ChEMBL naming)	Median	Min	Max	Type	Records
Homo sapiens	Fibroblast growth factor receptor 1	9.2	9.2	9.2	pIC50	1
Homo sapiens	Stem cell growth factor receptor	8.3	8.3	8.8	pIC50	2
Homo sapiens	Tyrosine-protein kinase ABL	9	8.1	9.5	pIC50	13
Homo sapiens	Tyrosine-protein kinase receptor FLT3	9.5	9.5	9.5	pIC50	1
Homo sapiens	Tyrosine-protein kinase receptor RET	7.6	7.5	9.1	pIC50	3
Gallus gallus	Tyrosine-protein kinase SRC	9.1	9.1	9.1	pIC50	1
Homo sapiens	Tyrosine-protein kinase SRC	8.5	8.5	8.5	pIC50	1
Homo sapiens	Vascular endothelial growth factor receptor 2	8.4	8.4	8.8	pIC50	2

Kinase-ligand interaction pattern

• Hydrophobic • Aromatic face-to-face • Aromatic face-to-edge • H-bond donor • H-bond acceptor • Ionic positive • Ionic negative

Interaction pattern search

Search KLIFS for kinase-ligand complexes with similar interaction patterns:

FIG. 3.4 A portion of the fragment of KLIFS output generated for nonreceptor tyrosine kinase C-abl oncogene 1. It contains pocket alignment, a summary of ligand activity, and a summary of ligand—protein contacts.

Analogous storage of protein—protein interactions is offered by PICCOLO (Bickerton et al., 2011), whereas BIPA (Lee & Blundell, 2009) stores protein—nucleic acids interactions.

Pocketome (Kufareva et al., 2012), in addition to ligand—protein interactions annotated to structural data, offers possibility to separate positions on assembly interfaces and examine compatibility of binding pockets and ligands similarity.

Sc-PDB database (Desaphy et al., 2015) gathers proteins, ligands, binding sites, and binding modes on the basis of PDB data, covering almost 10,000 of binding sites from 3678 unique proteins and 5608 ligands. It offers wide range of possibilities in terms of data analysis and searches of various types, such as querying the database for the search of similar binding pockets or complexes with similar interaction patterns. A portion of example database output is presented in Fig. 3.5.

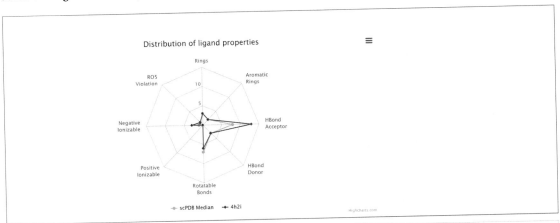

Binding mode :

What is Poseview ?

4h2i:A12

Image generated by PoseView

Ligand Atom	Protein Atom	Residue	Distance (Å)	Angle (°)	Type
O3B	ND2	ASN- 117	2.58	149.89	H-Bond (Protein Donor)
O2B	NE2	HIS- 118	2.6	172.41	H-Bond (Protein Donor)
O1A	ND2	ASN- 245	3.05	159.91	H-Bond (Protein Donor)
O2A	CZ	ARG- 354	3.73	0	Ionic (Protein Cationic)
O2A	NH1	ARG- 354	2.86	158.99	H-Bond (Protein Donor)
O5'	NH2	ARG- 354	3.25	153.62	H-Bond (Protein Donor)
O2'	ND2	ASN- 390	3.05	169.69	H-Bond (Protein Donor)
N3	ND2	ASN- 390	3.16	138.39	H-Bond (Protein Donor)
O1B	CZ	ARG- 395	3.94	0	Ionic (Protein Cationic)
O2B	CZ	ARG- 395	3.87	0	Ionic (Protein Cationic)
O1B	NH1	ARG- 395	3.01	171.98	H-Bond (Protein Donor)
O2B	NH2	ARG- 395	3.31	139.95	H-Bond (Protein Donor)
O5'	NH2	ARG- 395	2.95	140.96	H-Bond (Protein Donor)
DuAr	DuAr	PHE- 417	3.98	0	Aromatic Face/Face
C2'	CZ	PHE- 500	3.94	0	Hydrophobic
DuAr	DuAr	PHE- 500	3.8	0	Aromatic Face/Face
O3'	OD1	ASP- 506	2.55	155.17	H-Bond (Ligand Donor)
O2'	OD1	ASP- 506	3.35	127.81	H-Bond (Ligand Donor)
O1B	ZN	ZN- 601	2.06	0	Metal Acceptor
O3B	ZN	ZN- 602	2.09	0	Metal Acceptor

EXPORT

FIG. 3.5 A portion of the fragment of the sc-PDB output generated for human ecto-5′-nucleotidase (CD73), PDBID: 4H2I (Knapp et al., 2012). It provides a characterization of the ligand properties and a detailed summary of each ligand—protein interaction in the form of ligand and protein atoms that participate in each contact together with their mutual positions.

Such interaction databases can be of great use for medicinal chemists, when preparing VS experiments, to develop prefiltering rules for rejecting compounds that do not meet particular interaction criteria. Moreover, they can help in designing new ligands, as well as to detect targets with similar binding pockets to examine possible off-target effects.

4 POST-PROCESSING STRATEGIES

4.1 Visual Inspection

Visual evaluation of docking poses is an approach that can outperform any other post-processing strategy; however, it is definitely a very time-consuming strategy and can be very subjective (especially when examining a larger number of compounds). Moreover, to correctly evaluate ligand—protein complexes by visual analysis, one should possess experience and target-specific knowledge. Nevertheless, even in such cases, the results evaluation can be biased, as it also depends on the available structural data. Example analysis of docking results toward two crystal structures of serotonin receptor 5-HT$_{1B}$ is presented in Fig. 3.6. The figure presents two compounds exhibiting different activities toward 5-HT$_{1B}$R: CHEMBL66310, which is inactive toward this receptor (IC$_{50}$ > 300 000 nM), and CHEMBL387545, which is characterized by relatively good affinity to 5-HT$_{1B}$R (K_i = 51 nM). The docking outputs (obtained via Schrödinger's Glide, in extra precision) presented in Fig. 3.6 indicate a strong dependence of the obtained results on the crystal structures used for docking. 4IAR refers to the activated 5-HT$_{1B}$R

structure, with agonist co-crystallized, whereas 5V54 is co-crystallized with the 5-HT$_{1B}$R antagonist. In the former case, the occupation of the binding site by the two analyzed compounds is a little bit different; however, for 5V54, the discrimination between active and inactive entity might be more difficult.

4.2 Automatic Post-processing Protocols

As relationships between particular ligand—receptor complex features might be difficult to be captured by a human, various ML-based protocols have been developed for evaluation of the activity of docked compounds.

4.2.1 Consensus scoring

Due to the fact that each scoring function has its limitations, one of the strategies applied to improve the accuracy of docking poses evaluation is to combine assessment provided by various scoring functions into a single answer—consensus scoring. Charifson et al. (1999) proved that such a strategy improves the proper discrimination between active and inactive inhibitors of three enzymes: P38, IMPDH, and HIV protease. Thirteen scoring functions were used in the study (Table 3.2), and two docking programs (DOCK 4.0.1 (Meng et al., 1992) and GAMBLER) (based on the genetic algorithm, developed within the study) produced the docking poses. In each case, the best compound orientation according to one selected scoring scheme was selected and 12 other scoring strategies were applied for rescoring the pose. Then, separate compound ranking for each scoring strategy was

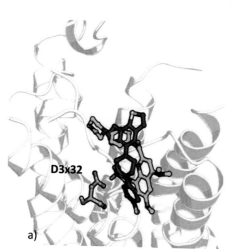
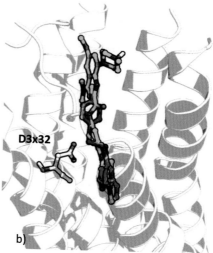

FIG. 3.6 Docking results of CHEMBL66310 (red) and CHEMBL387545 (green) to crystal structures of 5-HT$_{1B}$R: **(A)** 4IAR (Wang et al., 2013) and **(B)** 5V54 (Yin et al., 2018).

prepared and the compounds common for particular scoring strategies considered were determined. The method performance was measured by the percentage of active compounds returned (after taking into account top-scored compounds, different cutoffs were tested). Scheme of the study is presented in Fig. 3.7.

It was proved that the consensus scoring lowers the number of false positives although some scoring strategies applied individually were not very effective in discrimination between active and inactive compounds. Furthermore, Charifson et al. proposed a "typical" consensus approach, where only the best compounds occurring within each scoring strategy are further considered. Improvement of the success rate using

such an approach was also indicated by other researchers (Cheng et al., 2009; Huang et al., 2006; Kukol, 2011; Oda et al., 2006).

Recently, Palacio-Rodríguez et al. (2019) indicated that better performance is obtained when the sum of sets is taken (it provides getting rid of problems when a poor strategy is within the set of used methods); in the study, an exponential distribution was used to combine the ranks from individual docking programs. Fig. 3.8 compares "AND" (typical consensus strategy) and "OR" (presented by Palacio-Rodríguez) approaches of compounds selection.

4.2.2 Artificial intelligence in the serve of docking poses rescoring

Tools based on artificial intelligence (AI) are extensively used in various fields and everyday in the current form would not exist without them. The CADD domain also intensively makes use of various AI algorithms. Their popularity is connected mainly with the ability to deal with large amount of information in a relatively short time and detection of complex relationships occurring in data, which would be impossible to identify by human. AI-based strategies also support the assessment of docking poses, with both "standard" ML tools and deep learning (DL) strategies used.

ML models for scoring: One of the first studies using ML as a support for docking results evaluation was presented by Deng, Breneman, and Embrechts (2004) who applied Kernel-partial least squares (K-PLS) to construct a predictive model for binding affinity prediction for 61 and 105 ligand—protein complexes (two data sets were used). Ligand—protein complexes were described in the form of occurrence of ligand—protein atom pairs within a particular distance (1—6Å).

Atom pairs—based representation of ligand—protein complexes was also used by Artemenko (2008).

TABLE 3.2
Scoring Functions Used in the Study of Charifson et al. in the Consensus Scoring Strategy.

Empirical	Physics-Based	Other
Böhm (Böhm, 1992)	Merck molecular force field nonbond energy (Halgren, 1996)	Poisson—Boltzmann (Honig & Nicholls, 1995)
ChemScore (Eldridge et al., 1997)	DOCK energy score (Meng et al., 1992)	Buried lipophilic surface area (Flower, 1997)
SCORE (Wang et al., 1998)	DOCK chemical score (Meng et al., 1992)	DOCK constant score (Shoichet et al., 1992)
Piecewise linear potential (Gehlhaar et al., 1995)	Flexible ligands on a grid (Miller et al., 1994)	Volume overlap (Stouch & Jurs, 1986)
	Strain energy	

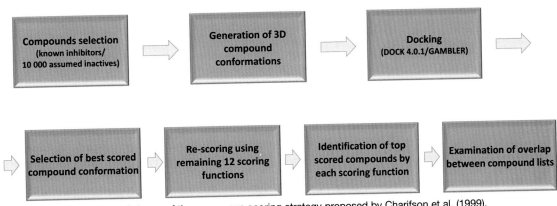

FIG. 3.7 Scheme of the consensus scoring strategy proposed by Charifson et al. (1999).

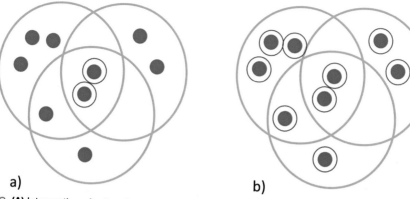

FIG. 3.8 **(A)** Intersection of sets referring to the "AND" (typical) consensus strategy, and **(B)** sum of the sets referring to the "OR" consensus strategy.

In addition, docked complexes were further characterized by physicochemical descriptors (van der Waals, electrostatic, metal–atom interactions) and complexes were evaluated with the use of neural networks.

Example of ML-based study is an improvement of the scoring provided by AutoDock Vina, by applying random forest (RF) algorithm (Li et al., 2015). RF belongs to the class of supervised algorithms. It is composed of a series of decision trees working as an ensemble (separate trees constructed on the randomly selected subset of features provide predictions and the output with the highest number of votes is taken as a final answer (Breiman, 2001)).

The original Vina (Morris et al., 2009) scoring has the following form:

$$e_k = \frac{e_{k,inter} + e_{k,intra} - e_{1,intra}}{1 + w_6 N_{rot}},$$

Where e_k is the energy for kth conformer, and $e_{k,inter}$ and $e_{k,intra}$ refer to intermolecular and intramolecular contributions, respectively. N_{rot} is the number of computed rotatable bonds.

Detailed form of $e_{k,inter}$ is as follows:

$$e_{k,inter} = w_1 \cdot Gauss1_k + w_2 \cdot Gauss2_k + w_3 \cdot Repulsion_k + w_4 \cdot Hydrophobic_k + w_5 \cdot HBonding_k$$

with the following values of weights for subsequent contributors:

$w_1 = -0.035579$
$w_2 = -0.005156$
$w_3 = 0.840245$
$w_4 = -0.035069$
$w_5 = -0.587439$
$w_6 = 0.05846$

In the study, there was only one conformer per molecule ($k = 1$), and therefore the Vina's score referring to the free energy of binding was simplified to

$$e_1 = \frac{e_{1,inter}}{1 + w_6 N_{rot}}$$

The assumption of a functional form of Vina scoring is a source of limitations of its predictions. Therefore, the RF model using six Vina features was constructed to derive relationships between the structural features and binding affinity. The presented approach was proved to greatly improve the Vina-based predictions of binding affinity.

Nondirect capturing of interactions supported by ML algorithms was used for the development of the new scoring function—RF-Score, based on the application of RF algorithm (Ballester & Mitchell, 2010). The ligand–protein depiction was based on the number of occurrences of a particular protein–ligand atom type pair, which interact within a given distance, using a minimal set of atom types (C, N, O, F, P, S, Cl, Br, I). RF-based regression was used to fit binding output to such represented data. RF-Score was proved to provide higher accuracy of binding prediction than 16 scoring functions, with which it was challenged.

A scoring function provided by docking program (eHiTS) was also improved with the use of ML methods by Kinnings et al. (2011). Support vector machines (SVMs) (Cortes & Vapnik, 1995) was used in this study to associate individual energy terms returned by eHiTS with binding affinities. SVM also belongs to the class of supervised learning algorithms dedicated to finding a hyperplane, which separates data into groups (classes) with the highest margin (Fig. 3.9). If the data cannot be separated using a linear function, the kernel trick

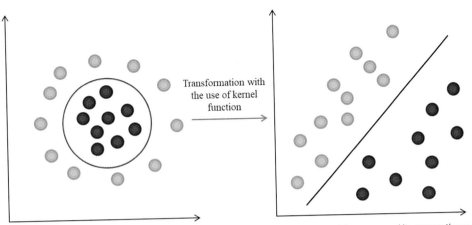

FIG. 3.9 Transformation of data using the kernel function into linearly separable space and its separation with the hyperplane found by support vector machine.

is applied to transform the data into the linearly separated space.

Kinnings et al. used a regression SVM model to predict IC_{50} values from BindingDB (Chen et al., 2001) data of inhibitors of enoyl acyl carrier protein reductase (InhA). As features, energy terms from eHiTS were taken (H_bond, Lipophil, pi_stack, solvent, steric, strain, Coulomb, entropy, Rcharge, Rshape, RlogD, Rcover, Lcharge, LlogD, Lcover, metal, lig_int, Family, depth and other) in different combinations. In addition, the classification model was developed. In this case, the model was trained on the InhA Directory of Useful Decoys (DUD) data set.

SVM is very popular algorithm in CADD studies (Melville et al., 2009; Mitchell, 2014). Das et al. (2010) predicted binding affinities with the use of SVM on the basis of the PDBbind data and using property-encoded shape distribution as docked compounds descriptions.

Interactions represented in the form of Pharm-IF constituted an input for ML methods training in the work of Sato et al. (2010). The basis for the calculation of Pharm-IF is set of distances of pharmacophore features interacting with protein atoms. The efficiency of such an approach was tested in screening for inhibitors of serine/threonine kinase PKA, nonreceptor tyrosine kinase SRC, carbonic anhydrase II, cathepsin K, and HIV-1 protease. ML algorithms applied for the predictions were Naïve Bayes classifier, RF, SVM, and artificial neural network; and Pharm-IF was challenged against PLIF. The obtained results indicated the superiority of the SVM–Pharm-IF combination over other approaches (averaged enrichment factor at 10% equal to 5.7, for Glide-based scoring it was equal to 4.2, and for PLIF-based scoring it was 4.3).

The general use of ML models in post-processing of docking studies is that the ligand–receptor complexes are transformed into descriptor- or fingerprint-based representations, and in such form, they constitute an input for ML-based activity prediction. However, there are also examples, in which ML is combined with a scoring function of another type (energy- or knowledge-based), and an iterative process of compound poses evaluation is applied. Wright et al. (2013) combined energy-based scoring function with MLR to fit experimentally determined compound affinities. After such scoring, top-ranked poses were selected, and they were then used in MLR-based fitting procedure. It was indicated, that after 4−7 repetitions of such procedure, the exponential convergence is obtained. The approach of Wright et al. is based on the previously reported concept by Jain et al. (1995), which uses the Compass algorithm (Jain, Dietterich, et al., 1994; Jain, Koile, & Chapman, 1994). Its task is to predict the compound bioactive conformation and describe it in the form of a quantitative binding site model to detect structural requirements for binding.

The iterative scoring approach was also developed by Martin and Sullivan (2008a, 2008b), whose methodology was captured within the Autoshim software. It is based on the addition of point pharmacophores to the binding site depiction of a target (the so-called "shims"). Then, weights are added to such "shims" using PLS algorithm for adjustment of the original scoring function to best fit IC_{50} values of analyzed compounds.

4.2.3 Multistep ML-based protocols for docking results evaluation

The above-presented approaches in general use one receptor conformation and one ML method in a particular experiment of compound activity prediction. Here, there are presented examples of some more complex evaluation protocols, which combine information from various sources.

Hsin et al. (2013) combined several docking programs (eHiTS, GOLD, AutoDock VINA) and used ML models in two different tasks: at first, RF-Score was adopted to rescore docking poses, and then, ML algorithm (multinomial logistic regression) was used to identify the most predictive binding modes (Fig. 3.10).

Several ways of considering output from multiple docking programs were proposed by Plewczynski et al. (2011). The output resulting from the application of various docking programs was clustered to identify poses similar to the respective native conformation (extracted from the appropriate crystal structure). The clustering is then followed by the consensus procedure with the use of the MLR to choose positions related to strong binding. The docking programs used were of various types (AutoDock 4.2.1, Glide 4.5, GOLD 3.2, Surflex 2.2, FlexX 2.2.1, eHiTS 9.0, and LigandFit2.3) to cover different approaches of obtaining ligand poses. 1300 protein—ligand pairs were used for evaluation of the applied strategies ("refined" set from the PDBbind 2007 database).

The first conclusion coming from the study was that when the docking programs were evaluated individually, no approach could be considered as being significantly better than the remaining ones (in terms of the general ability to predict ligand docking pose, assessed by RMSD). Then, using MLR, MetaScore function was developed to combine assessments provided by various scoring approaches:

$$
\begin{aligned}
MetsScore = {}& -0.378 * eHiTSSCore + 0.015 * FlexXScore \\
& -0.358 * GlideSPScore + 0.014 * GoldScore \\
& +0.004 * LigScore + 0.15 * SurflexScore
\end{aligned}
$$

The output of MetaScore corresponds to the $-\log K_d$ of the respective ligand—protein complex.

In addition, an algorithm of docking pose selection, which is closest to the native conformation, was developed (MetaPose).

The VoteDock algorithm, being the main result of the discussed work, combines both prediction of correct ligand pose and evaluation of its binding strength to the given target. It was proved that the method overcomes the limitations of individual docking software and that it is not biased by the ligand type (small/large, hydrophobic/hydrophilic, and peptides). In addition, its performance does not depend on the protein family.

A series of consensus approaches were also tested by Ericksen et al. (2017). They covered not only "traditional" consensus methods, such as minimum, maximum, median, and mean scores of individual docking scores, but also mean—variance consensus and newly developed boosting consensus scoring (BCS). The approach trains individual tree ensemble classifiers (one per target) by gradient boosting (data set composed of active compounds and decoys from DUD-E database; features docking scores from eight docking programs). Then, for each target, the BCS was determined for each compound (using leave-one-out procedure) by averaging the score from the 20 "off-target" models. The method enables construction of a boosting consensus model for each considered target without the use of experimental data.

Smusz et al. (2015) presented a multistep protocol for docking results evaluation, which takes into account two different representations of ligand—receptor complexes, predictions of 5 ML algorithms, different compound orientations in the binding site, and various receptor conformations used for docking.

The protocol was tested on the task of compound activity prediction toward two serotonin receptors 5-HT$_6$ and 5-HT$_7$. For both targets, the crystal structures are not available; therefore, their homology models were constructed. For each target, five models with the highest AUROC values obtained in retrospective docking studies were considered. Then, compounds with experimentally confirmed activity toward the considered targets, together with DUD sets, were docked in Glide to the respective receptors conformations. The obtained ligand—receptor complexes (a maximum of five poses per ligand was set) were represented with the use of SIFts and Spectrophores (both representations were used separately and combined; SIFts were

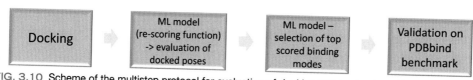

FIG. 3.10 Scheme of the multistep protocol for evaluation of docking results developed by Hsin et al. ML, machine learning.

also used in their raw form and after filtering in terms of attribute importance). The compound activity evaluation (classification study, discrimination between compounds with K_i below 100 nM and K_i above 1000 nM) was performed with the use of five ML algorithms: Naïve Bayes, sequential minimal optimization, k-nearest neighbor (IBk), decision tree J48, and RF. Then, the predictions obtained for a given compound were combined in such a way that weight average of various ML algorithms was applied (weights being the Matthews correlation coefficient values obtained for particular ML algorithm), weight average of various compounds' conformations was determined (weights being values of Glide docking score), and weight average of results obtained for different models was calculated (weights being the AUROC values obtained for particular homology model). Final predictions obtained after applying of such a weighting protocol were proved to significantly outperform any individual predictions schemes. The protocol scheme is presented in Fig. 3.11.

DL-based scoring: The wide range of possibilities and great efficiency in various types of tasks have constantly increased the application of DL methods in CADD tasks. The wide use of DL in CADD is not only limited to generative models and enumeration of combinatorial libraries containing compounds with desired activity and physicochemical profile (Ekins, 2016; Kim & Oh, 2017; Koutsoukas et al., 2017; Lenselink et al., 2017; Lusci et al., 2013; Ma et al., 2015; Popova et al., 2018; Unterthiner et al., 2015; Xu et al., 2015). Here, we summarize the DL usage of scoring ligand–receptor complexes.

Stepniewska-Dziubinska et al. (2018) developed a deep neural network for the assessment of binding affinity of ligands on the basis of ligand–receptor complexes with known affinity. The system is called Pafnucy, and it uses ligand–protein complexes from the PDBbind database v. 2016 (Liu et al., 2017). They are represented in the form of four-dimensional tensor, where each point is defined by Cartesian coordinates and vector of features (encoding atom types, hybridization, number of bonds, pharmacophore-like features: hydrophobic, aromatic, acceptor, donor, and ring; charge, information whether its ligand or protein).

Such an input is first processed by a block of 3D convolutional layers combined with a max-pooling layer; then, a block of fully connected layers is applied to finally predict the binding affinity. For the evaluation of performance, the CASF-2013 benchmark (Li et al., 2014) for the assessment of scoring power was used revealing great accuracy of Pafnucy (it was outperformed only by RF-Score v3 as reported by Wójcikowski et al. (2015)).

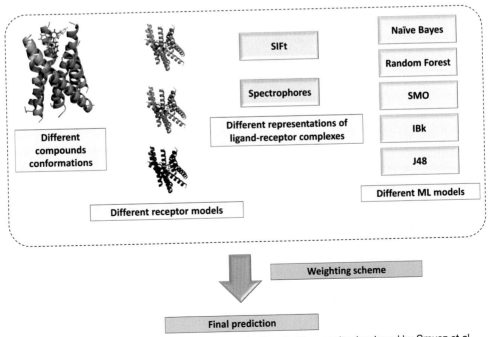

FIG. 3.11 Scheme of the multistep protocol for evaluating docking results developed by Smusz et al. *ML*, machine learning; *SIFt*, structural interaction fingerprint; *SMO*, sequential minimal optimization.

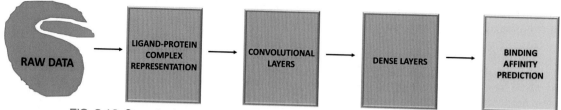

FIG. 3.12 General scheme of binding affinity predictions by convolutional neural networks.

Wojcikowski et al. (2017) also used similar approach; however, the DUD-E data set (Mysinger et al., 2012) was used for the study, including 102 protein targets and performing training on 3D conformations generated by three docking programs: AutoDock Vina, Dock 3.6, and Dock 6.6. The obtained ligand conformations were scored using several classical scoring functions. Then, the RF-Score—VS was trained on the ligand pose with the lowest docking score. As an external validation set, the DEKOIS 2.0 (Bauer et al., 2013) database (consisting of 76 targets) was used. The RF-Score—VS was proved to improve the VS performance by significantly higher hit rate when the top 1% is considered (55.6% hit rate in comparison to 16.2% obtained for Vina).

Binding affinity predictions via DL methods are also considered in KDEEP (Jiménez et al., 2018), using 3D convolutional neural networks in its basis, trained on the PDBbind v. 2016 data. Similar to the previously mentioned studies, ligand—receptor complexes were depicted in the form of 3D voxels combined with the set of properties: hydrophobic, hydrogen bond donor, hydrogen bond acceptor, aromatic, positive ionizable, negative ionizable, metallic, and total excluded volume.

Another example of the application of prediction of compound binding affinity via convolutional networks is the OnionNet (Zheng et al., 2019). In this system, the 3D interaction information is converted into a 2D form using the following procedure: at first, the "shell-like" division of the protein around the ligand is performed, and the contact numbers occurring within each interaction shell is determined, taking into account eight element types—C, N, O, H, P, S, HAX (F, Cl, Br, I), and Du (referring to all remaining elements). In the OnionNet, the number of shells was equal to 60.

The general scheme of the binding affinity predictions using convolutional neural networks is presented in Fig. 3.12.

5 CONCLUSIONS

The chapter summarized various strategies of evaluation of ligand-protein complexes obtained in the docking procedure. It covers wide range of methods—from visual inspection, via docking scoring functions included in the docking software, to IFPs and automatic analysis using ML tools. In addition, examples of hybrid methods, which uses various tools to provide final compound evaluation, are provided. The wide range of methods presented indicates the almost unlimited possibilities of further development in the field of docking and post-processing of its results. Although the docking is computationally expensive, it provides a comprehensive information about the possible ligand—protein contacts. Unfortunately, it is not free of bias, as results can be prone to the receptor conformation, which actually can depend a lot on the co-crystallized ligand or protein model selected for a study. Nevertheless, the post-processing strategies can help in dealing with such cases, as they enable using of multiple receptor conformations and also consider various ligand orientations in the binding site. Due to the high need for computational resources by docking, the future directions of computational strategies in this field might be oriented at prediction of docking output by ML, rather than using an output of docking software.

The chapter is intended to be a resource for post-processing strategies both for the beginners and the experienced users of docking.

ACKNOWLEDGMENTS

Supported by the grant OPUS 2018/31/B/NZ2/00165 financed by the National Science Center, Poland (www.ncn.gov.pl). SP was supported by the Foundation for Polish Science (FNP) within the START scholarship.

REFERENCES

Abagyan, R., Totrov, M., & Kuznetsov, D. (1994). ICM—a new method for protein modeling and design: Applications to docking and structure prediction from the distorted native conformation. *Journal of Computational Chemistry, 15*(5), 488—506.

Artemenko, N. (2008). Distance dependent scoring function for describing protein-ligand intermolecular interactions. *Journal of Chemical Information and Modeling, 48*(3), 569—574.

Ballester, P. J., & Mitchell, J. B. (2010). A machine learning approach to predicting protein-ligand binding affinity with applications to molecular docking. *Bioinformatics, 26*(9), 1169–1175.

Ballester, P. J., & Richards, W. G. (2007a). Ultrafast shape recognition for similarity search in molecular databases. *Proceedings of the Royal Society A: Mathematical, Physical & Engineering Sciences, 463*(2081), 1307–1321.

Ballester, P. J., & Richards, W. G. (2007b). Ultrafast shape recognition to search compound databases for similar molecular shapes. *Journal of Computational Chemistry, 28*(10), 1711–1723.

Bauer, M. R., Ibrahim, T. M., Vogel, S. M., & Boeckler, F. M. (2013). Evaluation and optimization of virtual screening workflows with DEKOIS 2.0–a public library of challenging docking benchmark sets. *Journal of Chemical Information and Modeling, 53*(6), 1447–1462.

Bender, A., Mussa, H. Y., Gill, G. S., & Glen, R. C. (2004). Molecular surface point environments for virtual screening and the elucidation of binding patterns (MOLPRINT 3D). *Journal of Medicinal Chemistry, 47*(26), 6569–6583.

Berman, H. M., Westbrook, J., Feng, Z., Gilliland, G., Bhat, T. N., Weissig, H., Shindyalov, I. N., & Bourne, P. E. (2000). The protein data bank. *Nucleic Acids Research, 28*(1), 235–242.

Bickerton, G. R., Higueruelo, A. P., & Blundell, T. L. (2011). Comprehensive, atomic-level characterization of structurally characterized protein-protein interactions: The PICCOLO database. *BMC Bioinformatics, 12*, 313.

Bitencourt-Ferreira, G., & de Azevedo, W. F., Jr. (2019). Molegro virtual docker for docking. *Methods in Molecular Biology. 2053*, pp. 149–167.

Böhm, H. J. (1992). The computer program LUDI: A new method for the de novo design of enzyme inhibitors. *Journal of Computer-Aided Molecular Design, 6*(1), 61–78.

Böhm, H. J. (1994). The development of a simple empirical scoring function to estimate the binding constant for a protein-ligand complex of known three-dimensional structure. *Journal of Computer-Aided Molecular Design, 8*(3), 243–256.

Breiman, L. (2001). Random forests. *Machine Learning, 45*(1), 5–32.

Bultinck, P., Langenaeker, W., Lahorte, P., De Proft, F., Geerlings, P., Van Alsenoy, C., & Tollenaere, J. P. (2002). The electronegativity equalization method II: Applicability of different atomic charge schemes. *The Journal of Physical Chemistry A, 106*(34), 7895–7901.

Chandler, D. (1987). *Introduction to modern statistical mechanics.* New York: Oxford University Press.

Charifson, P. S., Corkery, J. J., Murcko, M. A., & Walters, W. P. (1999). Consensus scoring: A method for obtaining improved hit rates from docking databases of three-dimensional structures into proteins. *Journal of Medicinal Chemistry, 42*(25), 5100–5109.

Chemical Computing Group Inc. (2010). *Molecular operating environment (MOE) (Version v.2010.10).*

Cheng, T., Li, X., Li, Y., Liu, Z., & Wang, R. (2009). Comparative assessment of scoring functions on a diverse test set. *Journal of Chemical Information and Modeling, 49*(4), 1079–1093.

Chen, X., Lin, Y., & Gilson, M. K. (2001). The binding database: Overview and user's guide. *Biopolymers, 61*(2), 127–141.

Cortes, C., & Vapnik, V. (1995). Support-vector networks. *Machine Learning, 20*(3), 273–297.

Da, C., & Kireev, D. (2014). Structural protein-ligand interaction fingerprints (SPLIF) for structure-based virtual screening: Method and benchmark study. *Journal of Chemical Information and Modeling, 54*(9), 2555–2561.

Das, S., Krein, M. P., & Breneman, C. M. (2010). Binding affinity prediction with property-encoded shape distribution signatures. *Journal of Chemical Information and Modeling, 50*(2), 298–308.

Deng, W., Breneman, C., & Embrechts, M. J. (2004). Predicting protein-ligand binding affinities using novel geometrical descriptors and machine-learning methods. *Journal of Chemical Information and Computer Sciences, 44*(2), 699–703.

Deng, Z., Chuaqui, C., & Singh, J. (2004). Structural interaction fingerprint (SIFt): A novel method for analyzing three-dimensional protein-ligand binding interactions. *Journal of Medicinal Chemistry, 47*(2), 337–344.

Desaphy, J., Bret, G., Rognan, D., & Kellenberger, E. (2015). sc-PDB: a 3D-database of ligandable binding sites–10 years on. *Nucleic Acids Research, 43*, D399–D404.

DeWitte, R. S., Ishchenko, A. V., & Shakhnovich, E. I. (1997). SMoG: de Novo design method based on simple, fast, and accurate free energy estimates. 2. Case studies in molecular design. *Journal of the American Chemical Society, 119*(20), 4608–4617.

Ekins, S. (2016). The next era: Deep learning in pharmaceutical research. *Pharmaceutical Research, 33*(11), 2594–2603.

Eldridge, M. D., Murray, C. W., Auton, T. R., Paolini, G. V., & Mee, R. P. (1997). Empirical scoring functions: I. The development of a fast empirical scoring function to estimate the binding affinity of ligands in receptor complexes. *Journal of Computer-Aided Molecular Design, 11*(5), 425–445.

Ericksen, S. S., Wu, H., Zhang, H., Michael, L. A., Newton, M. A., Hoffmann, F. M., & Wildman, S. A. (2017). Machine learning consensus scoring improves performance across targets in structure-based virtual screening. *Journal of Chemical Information and Modeling, 57*(7), 1579–1590.

Flower, D. R. (1997). SERF: A program for accessible surface area calculations. *Journal of Molecular Graphics and Modelling, 15*(4), 238–244.

Friesner, R. A., Murphy, R. B., Repasky, M. P., Frye, L. L., Greenwood, J. R., Halgren, T. A., Sanschagrin, P. C., & Mainz, D. T. (2006). Extra precision glide: Docking and scoring incorporating a model of hydrophobic enclosure for protein-ligand complexes. *Journal of Medicinal Chemistry, 49*(21), 6177–6196.

Gehlhaar, D. K., Verkhivker, G. M., Rejto, P. A., Sherman, C. J., Fogel, D. B., Fogel, L. J., & Freer, S. T. (1995). Molecular recognition of the inhibitor AG-1343 by HIV-1 protease: Conformationally flexible docking by evolutionary programming. *Chemistry & Biology, 2*(5), 317–324.

Gilson, M. K., Given, J. A., & Head, M. S. (1997). A new class of models for computing receptor-ligand binding affinities. *Chemistry & Biology, 4*(2), 87–92.

Gohlke, H., Hendlich, M., & Klebe, G. (2000). Knowledge-based scoring function to predict protein-ligand interactions. *Journal of Molecular Biology, 295*(2), 337–356.

Goodsell, D. S., Morris, G. M., & Olson, A. J. (1996). Automated docking of flexible ligands: Applications of AutoDock. *Journal of Molecular Recognition, 9*(1), 1–5.

Groom, C. R., Bruno, I. J., Lightfoot, M. P., & Ward, S. C. (2016). The Cambridge structural database. *Acta Crystallographica Section B: Structural Science, Crystal Engineering and Materials, 72*(Pt 2), 171–179.

Halgren, T. A. (1996). Merck molecular force field. III. Molecular geometries and vibrational frequencies for MMFF94. *Journal of Computational Chemistry, 17*(5-6), 553–586.

Hawkins, P. C., Skillman, A. G., & Nicholls, A. (2007). Comparison of shape-matching and docking as virtual screening tools. *Journal of Medicinal Chemistry, 50*(1), 74–82.

Honig, B., & Nicholls, A. (1995). Classical electrostatics in biology and chemistry. *Science, 268*(5214), 1144–1149.

Hsin, K. Y., Ghosh, S., & Kitano, H. (2013). Combining machine learning systems and multiple docking simulation packages to improve docking prediction reliability for network pharmacology. *PLoS One, 8*(12), e83922.

Huang, Q., Jin, H., Liu, Q., Wu, Q., Kang, H., Cao, Z., & Zhu, R. (2012). Proteochemometric modeling of the bioactivity spectra of HIV-1 protease inhibitors by introducing protein-ligand interaction fingerprint. *PLoS One, 7*(7), e41698.

Huang, N., Shoichet, B. K., & Irwin, J. J. (2006). Benchmarking sets for molecular docking. *Journal of Medicinal Chemistry, 49*(23), 6789–6801.

Isberg, V., Mordalski, S., Munk, C., Rataj, K., Harpsøe, K., Hauser, A. S., Vroling, B., Bojarski, A. J., Vriend, G., & Gloriam, D. E. (2016). GPCRdb: An information system for G protein-coupled receptors. *Nucleic Acids Research, 44*(D1), D356–D364.

Jain, A. N. (2003). Surflex: Fully automatic flexible molecular docking using a molecular similarity-based search engine. *Journal of Medicinal Chemistry, 46*(4), 499–511.

Jain, A. N., Dietterich, T. G., Lathrop, R. H., Chapman, D., Critchlow, R. E., Bauer, B. E., Webster, T. A., & Lozano-Perez, T. (1994). A shape-based machine learning tool for drug design. *Journal of Computer-Aided Molecular Design, 8*(6), 635–652.

Jain, A. N., Koile, K., & Chapman, D. (1994). Compass: Predicting biological activities from molecular surface properties. Performance comparisons on a steroid benchmark. *Journal of Medicinal Chemistry, 37*(15), 2315–2327.

Jain, A. N., Harris, N. L., & Park, J. Y. (1995). Quantitative binding site model generation: Compass applied to multiple chemotypes targeting the 5-ht1a receptor. *Journal of Medicinal Chemistry, 38*(8), 1295–1308.

Jiménez, J., Škalič, M., Martínez-Rosell, G., & De Fabritiis, G. (2018). K(DEEP): Protein-Ligand absolute binding affinity prediction via 3D-convolutional neural networks. *Journal of Chemical Information and Modeling, 58*(2), 287–296.

Kim, I.-W., & Oh, J. M. (2017). Deep learning: From chemoinformatics to precision medicine. *J Pharm Investig, 47*(4), 317–323.

Kimura, K. T., Asada, H., Inoue, A., Kadji, F. M. N., Im, D., Mori, C., Arakawa, T., Hirata, K., Nomura, N., Aoki, J., Iwata, S., & Shimamura, T. (2019). Structures of the 5-HT(2A) receptor in complex with the antipsychotics risperidone and zotepine. *Nature Structural & Molecular Biology, 26*(2), 121–128.

Kinnings, S. L., Liu, N., Tonge, P. J., Jackson, R. M., Xie, L., & Bourne, P. E. (2011). A machine learning-based method to improve docking scoring functions and its application to drug repurposing. *Journal of Chemical Information and Modeling, 51*(2), 408–419.

Knapp, K., Zebisch, M., Pippel, J., El-Tayeb, A., Müller, C. E., & Sträter, N. (2012). Crystal structure of the human ecto-5′-nucleotidase (CD73): Insights into the regulation of purinergic signaling. *Structure, 20*(12), 2161–2173.

Kooistra, A. J., Kanev, G. K., van Linden, O. P., Leurs, R., de Esch, I. J., & de Graaf, C. (2016). KLIFS: A structural kinase-ligand interaction database. *Nucleic Acids Research, 44*(D1), D365–D371.

Koutsoukas, A., Monaghan, K. J., Li, X., & Huan, J. (2017). Deep-learning: Investigating deep neural networks hyper-parameters and comparison of performance to shallow methods for modeling bioactivity data. *Journal of Cheminformatics, 9*(1), 42.

Krammer, A., Kirchhoff, P. D., Jiang, X., Venkatachalam, C. M., & Waldman, M. (2005). LigScore: A novel scoring function for predicting binding affinities. *Journal of Molecular Graphics and Modelling, 23*(5), 395–407.

Kufareva, I., Ilatovskiy, A. V., & Abagyan, R. (2012). Pocketome: An encyclopedia of small-molecule binding sites in 4D. *Nucleic Acids Research, 40*, D535–D540.

Kukol, A. (2011). Consensus virtual screening approaches to predict protein ligands. *European Journal of Medicinal Chemistry, 46*(9), 4661–4664.

Lee, S., & Blundell, T. L. (2009). BIPA: A database for protein-nucleic acid interaction in 3D structures. *Bioinformatics, 25*(12), 1559–1560.

Lenselink, E. B., Ten Dijke, N., Bongers, B., Papadatos, G., van Vlijmen, H. W. T., Kowalczyk, W., Ijzerman, A. P., & van Westen, G. J. P. (2017). Beyond the hype: Deep neural networks outperform established methods using a ChEMBL bioactivity benchmark set. *Journal of Cheminformatics, 9*(1), 45.

Li, Y., Han, L., Liu, Z., & Wang, R. (2014). Comparative assessment of scoring functions on an updated benchmark: 2. Evaluation methods and general results. *Journal of Chemical Information and Modeling, 54*(6), 1717–1736.

Li, H., Leung, K. S., Wong, M. H., & Ballester, P. J. (2015). Improving AutoDock Vina using random forest: The growing accuracy of binding affinity prediction by the effective exploitation of larger data sets. *Molecular Informatics, 34*(2–3), 115–126.

Liu, X., Jiang, H., & Li, H. (2011). SHAFTS: A hybrid approach for 3D molecular similarity calculation. 1. Method and assessment of virtual screening. *Journal of Chemical Information and Modeling, 51*(9), 2372–2385.

Liu, Z., Su, M., Han, L., Liu, J., Yang, Q., Li, Y., & Wang, R. (2017). Forging the basis for developing protein-ligand interaction scoring functions. *Accounts of Chemical Research*, 50(2), 302–309.

Liu, J., & Wang, R. (2015). Classification of current scoring functions. *Journal of Chemical Information and Modeling*, 55(3), 475–482.

Lusci, A., Pollastri, G., & Baldi, P. (2013). Deep architectures and deep learning in chemoinformatics: The prediction of aqueous solubility for drug-like molecules. *Journal of Chemical Information and Modeling, 53*(7), 1563–1575.

Marcou, G., & Rognan, D. (2007). Optimizing fragment and scaffold docking by use of molecular interaction fingerprints. *Journal of Chemical Information and Modeling, 47*(1), 195–207.

Martin, E. J., & Sullivan, D. C. (2008a). AutoShim: Empirically corrected scoring functions for quantitative docking with a crystal structure and IC50 training data. *Journal of Chemical Information and Modeling, 48*(4), 861–872.

Martin, E. J., & Sullivan, D. C. (2008b). Surrogate AutoShim: Predocking into a universal ensemble kinase receptor for three dimensional activity prediction, very quickly, without a crystal structure. *Journal of Chemical Information and Modeling, 48*(4), 873–881.

Ma, J., Sheridan, R. P., Liaw, A., Dahl, G. E., & Svetnik, V. (2015). Deep neural nets as a method for quantitative structure-activity relationships. *Journal of Chemical Information and Modeling, 55*(2), 263–274.

McGann, M. (2012). FRED and HYBRID docking performance on standardized datasets. *Journal of Computer-Aided Molecular Design, 26*(8), 897–906.

Melville, J. L., Burke, E. K., & Hirst, J. D. (2009). Machine learning in virtual screening. *Combinatorial Chemistry & High Throughput Screening, 12*(4), 332–343.

Meng, E. C., Shoichet, B. K., & Kuntz, I. D. (1992). Automated docking with grid-based energy evaluation. *Journal of Computational Chemistry, 13*(4), 505–524.

Miller, M. D., Kearsley, S. K., Underwood, D. J., & Sheridan, R. P. (1994). FLOG: A system to select 'quasi-flexible' ligands complementary to a receptor of known three-dimensional structure. *Journal of Computer-Aided Molecular Design, 8*(2), 153–174.

Mitchell, J. B. (2014). Machine learning methods in chemoinformatics. *Wiley Interdisciplinary Reviews: Computational Molecular Science, 4*(5), 468–481.

Morris, G. M., Huey, R., Lindstrom, W., Sanner, M. F., Belew, R. K., Goodsell, D. S., & Olson, A. J. (2009). AutoDock4 and AutoDockTools4: Automated docking with selective receptor flexibility. *Journal of Computational Chemistry, 30*(16), 2785–2791.

Muegge, I., Martin, Y. C., Hajduk, P. J., & Fesik, S. W. (1999). Evaluation of PMF scoring in docking weak ligands to the FK506 binding protein. *Journal of Medicinal Chemistry, 42*(14), 2498–2503.

Mysinger, M. M., Carchia, M., Irwin, J. J., & Shoichet, B. K. (2012). Directory of useful decoys, enhanced (DUD-E): Better ligands and decoys for better benchmarking. *Journal of Medicinal Chemistry, 55*(14), 6582–6594.

Oda, A., Tsuchida, K., Takakura, T., Yamaotsu, N., & Hirono, S. (2006). Comparison of consensus scoring strategies for evaluating computational models of Protein–Ligand complexes. *Journal of Chemical Information and Modeling, 46*(1), 380–391.

Palacio-Rodríguez, K., Lans, I., Cavasotto, C. N., & Cossio, P. (2019). Exponential consensus ranking improves the outcome in docking and receptor ensemble docking. *Scientific Reports, 9*(1), 5142.

Pérez-Nueno, V. I., Rabal, O., Borrell, J. I., & Teixidó, J. (2009). APIF: A new interaction fingerprint based on atom pairs and its application to virtual screening. *Journal of Chemical Information and Modeling, 49*(5), 1245–1260.

Plewczynski, D., Łaźniewski, M., von Grotthuss, M., Rychlewski, L., & Ginalski, K. (2011). VoteDock: Consensus docking method for prediction of protein-ligand interactions. *Journal of Computational Chemistry, 32*(4), 568–581.

Popova, M., Isayev, O., & Tropsha, A. (2018). Deep reinforcement learning for de novo drug design. *Science Advances, 4*(7), eaap7885.

Reid, D., Sadjad, B. S., Zsoldos, Z., & Simon, A. (2008). LASSO-ligand activity by surface similarity order: A new tool for ligand based virtual screening. *Journal of Computer-Aided Molecular Design, 22*(6–7), 479–487.

Reulecke, I., Lange, G., Albrecht, J., Klein, R., & Rarey, M. (2008). Towards an integrated description of hydrogen bonding and dehydration: Decreasing false positives in virtual screening with the HYDE scoring function. *ChemMedChem, 3*(6), 885–897.

Rogers, D., & Hahn, M. (2010). Extended-connectivity fingerprints. *Journal of Chemical Information and Modeling, 50*(5), 742–754.

Sato, T., Honma, T., & Yokoyama, S. (2010). Combining machine learning and pharmacophore-based interaction fingerprint for in silico screening. *Journal of Chemical Information and Modeling, 50*(1), 170–185.

Schreyer, A. M., & Blundell, T. (2012). USRCAT: Real-time ultrafast shape recognition with pharmacophoric constraints. *Journal of Cheminformatics, 4*(1), 27.

Schreyer, A. M., & Blundell, T. L. (2013). CREDO: A structural interactomics database for drug discovery. *Database (Oxford), 2013*, bat049.

Schuur, J. H., Selzer, P., & Gasteiger, J. (1996). The coding of the three-dimensional structure of molecules by molecular transforms and its application to structure-spectra correlations and studies of biological activity. *Journal of Chemical Information and Computer Sciences, 36*(2), 334–344.

Shoichet, B. K., Kuntz, I. D., & Bodian, D. L. (1992). Molecular docking using shape descriptors. *Journal of Computational Chemistry, 13*(3), 380–397.

Sliwoski, G., Mendenhall, J., & Meiler, J. (2016). Autocorrelation descriptor improvements for QSAR: 2DA_Sign and 3DA_Sign. *Journal of Computer-Aided Molecular Design, 30*(3), 209–217.

Smusz, S., Mordalski, S., Witek, J., Rataj, K., Kafel, R., & Bojarski, A. J. (2015). Multi-step protocol for automatic evaluation of docking results based on machine learning methods—A case study of serotonin receptors 5-HT(6) and 5-HT(7). *Journal of Chemical Information and Modeling, 55*(4), 823–832.

Stepniewska-Dziubinska, M. M., Zielenkiewicz, P., & Siedlecki, P. (2018). Development and evaluation of a deep learning model for protein-ligand binding affinity prediction. *Bioinformatics, 34*(21), 3666–3674.

Stouch, T. R., & Jurs, P. C. (1986). A simple method for the representation, quantification, and comparison of the volumes and shapes of chemical compounds. *Journal of Chemical Information and Computer Sciences, 26*(1), 4–12.

Tan, L., & Bajorath, J. (2009). Utilizing target-ligand interaction information in fingerprint searching for ligands of related targets. *Chemical Biology & Drug Design, 74*(1), 25–32.

Tan, L., Vogt, M., & Bajorath, J. (2009). Three-dimensional protein-ligand interaction scaling of two-dimensional fingerprints. *Chemical Biology & Drug Design, 74*(5), 449–456.

Unterthiner, T., Mayr, A., & Wegner, J. K. (2015). Deep learning as an opportunity in virtual screening. In *Paper presented at the NIPS work 2014.*

Vaz de Lima, L. A., & Nascimento, A. S. (2013). MolShaCS: A free and open source tool for ligand similarity identification based on Gaussian descriptors. *European Journal of Medicinal Chemistry, 59*, 296–303.

Venkatachalam, C. M., Jiang, X., Oldfield, T., & Waldman, M. (2003). LigandFit: A novel method for the shape-directed rapid docking of ligands to protein active sites. *Journal of Molecular Graphics and Modelling, 21*(4), 289–307.

Verdonk, M. L., Cole, J. C., Hartshorn, M. J., Murray, C. W., & Taylor, R. D. (2003). Improved protein-ligand docking using GOLD. *Proteins, 52*(4), 609–623.

Verma, J., Khedkar, V. M., & Coutinho, E. C. (2010). 3D-QSAR in drug design–a review. *Current Topics in Medicinal Chemistry, 10*(1), 95–115.

de Vries, S. J., van Dijk, M., & Bonvin, A. M. (2010). The HADDOCK web server for data-driven biomolecular docking. *Nature Protocols, 5*(5), 883–897.

Wang, C., Jiang, Y., Ma, J., Wu, H., Wacker, D., Katritch, V., Han, G. W., Liu, W., Huang, X. P., Vardy, E., McCorvy, J. D., Gao, X., Zhou, X. E., Melcher, K., Zhang, C., Bai, F., Yang, H., Yang, L.Jiang, H., … (2013). Structural basis for molecular recognition at serotonin receptors. *Science, 340*(6132), 610–614.

Wang, R., Liu, L., Lai, L., & Tang, Y. (1998). Score: A new empirical method for estimating the binding affinity of a protein-ligand complex. *Molecular Modeling Annual, 4*(12), 379–394.

Wójcikowski, M., Ballester, P. J., & Siedlecki, P. (2017). Performance of machine-learning scoring functions in structure-based virtual screening. *Scientific Reports, 7*, 46710.

Wójcikowski, M., Zielenkiewicz, P., & Siedlecki, P. (2015). Open Drug Discovery Toolkit (ODDT): A new open-source player in the drug discovery field. *Journal of Cheminformatics, 7*, 26.

Wright, J. S., Anderson, J. M., Shadnia, H., Durst, T., & Katzenellenbogen, J. A. (2013). Experimental versus predicted affinities for ligand binding to estrogen receptor: Iterative selection and rescoring of docked poses systematically improves the correlation. *Journal of Computer-Aided Molecular Design, 27*(8), 707–721.

Xie, Z. R., & Hwang, M. J. (2010). An interaction-motif-based scoring function for protein-ligand docking. *BMC Bioinformatics, 11*, 298.

Xu, Y., Dai, Z., Chen, F., Gao, S., Pei, J., & Lai, L. (2015). Deep learning for drug-induced liver injury. *Journal of Chemical Information and Modeling, 55*(10), 2085–2093.

Yin, W., Zhou, X. E., Yang, D., de Waal, P. W., Wang, M., Dai, A., Cai, X., Huang, C. Y., Liu, P., Wang, X., Yin, Y., Liu, B., Zhou, Y., Wang, J., Liu, H., Caffrey, M., Melcher, K., Xu, Y., Wang, M. W., … (2018). Crystal structure of the human 5-HT(1B) serotonin receptor bound to an inverse agonist. *Cell Discovery, 4*, 12.

Zheng, L., Fan, J., & Mu, Y. (2019). OnionNet: A multiple-layer intermolecular-contact-based convolutional neural network for protein-ligand binding affinity prediction. *ACS Omega, 4*(14), 15956–15965.

Zou, X., Yaxiong, & Kuntz, I. D. (1999). Inclusion of solvation in ligand binding free energy calculations using the generalized-born model. *Journal of the American Chemical Society, 121*(35), 8033–8043.

Zsoldos, Z., Reid, D., Simon, A., Sadjad, S. B., & Johnson, A. P. (2007). eHiTS: a new fast, exhaustive flexible ligand docking system. *Journal of Molecular Graphics and Modelling, 26*(1), 198–212.

CHAPTER 4

Best Practices for Docking-Based Virtual Screening

BRUNO JUNIOR NEVES • MELINA MOTTIN • JOSÉ TEOFILO MOREIRA-FILHO •
BRUNA KATIELE DE PAULA SOUSA • SABRINA SILVA MENDONCA •
CAROLINA HORTA ANDRADE

1 INTRODUCTION

Molecular docking is one of most extensively used tool of computer-assisted drug design that simulates molecular interactions between two molecules and usually predicts the binding mode and affinity between them. In other words, docking describes the fitting of two molecules in three-dimensional (3D) space (Kitchen et al., 2004). Since its first appearance in the mid-1980s (Kuntz et al., 1982), docking has proved to be an important component of drug discovery projects, whether in academy or industry. The three main motivations are the advances in structural biology, the impressive development of computers, and the growing availability of protein 3D structures and compound databases (Morris & Lim-Wilby, 2008). Consequently, more than 60 docking tools and programs have been developed for both commercial and academic use, such as DockThor (Magalhães et al., 2014; Santos et al., 2020), DOCK (Venkatachalam et al., 2003), Auto-Dock (Osterberg et al., 2002), FlexX (Rarey et al., 1996), GOLD (Jones et al., 1997), Glide (Friesner et al., 2004), FRED (McGann, 2011; McGann et al., 2003), MOE-Dock (Corbeil et al., 2012), AutoDock Vina (Oleg & Olson, 2010), and many others.

Fig. 4.1 shows the key steps in molecular docking. The system for a ligand–protein docking tools requires the following information: the 3D structure of a macromolecular target and the designed or virtual compound(s) of interest. Ideally, the structure of macromolecular target shall be experimentally resolved, usually by nuclear magnetic resonance (NMR), X-ray crystallography, or cryogenic electron microscopy (cryo-EM). Or it can be computationally predicted using any of the 3D protein structure prediction approaches (Pinzi & Rastelli, 2019).

Before performing the molecular docking, the 3D structure of macromolecular target and the ligands should be prepared as follows: missing side chains and loops of protein are added, protonation states are assigned, and simplified molecular charge schemes are calculated. Then, a user-defined docking search space (known as grid box) is calculated based on the coordinates of the binding site (e.g., catalytic or allosteric sites). After defining the grid box, molecular docking is practiced by an initial prediction of the **complementarity** (binding mode) of a ligand inside a grid box using a **search algorithm** and then predicting the affinity by means of a **scoring function** (Kitchen et al., 2004; Pinzi & Rastelli, 2019). Search algorithms fall into three major categories: systematic, stochastic, and deterministic (Guedes et al., 2014). Most of them take the macromolecule as a rigid body, while ligands are treated as flexible. The classical scoring functions (e.g., empirical, force field–based, or knowledge-based) usually allocate a common set of energy terms that linearly lead to the score (Guedes et al., 2014). As a result, the top orientation/conformation (pose), especially the one along the minimum energy score, may be assigned as the preferable binding mode.

Despite the opportunities offered by molecular docking, its applicability is limited by low accuracy of scoring functions and a reduced sampling of both macromolecule and ligand conformations in pose prediction (Coupez & Lewis, 2006; Elokely & Doerksen, 2013; Pantsar & Poso, 2018). Such inherent weaknesses are exacerbated by a lack of user experience when performing important steps, such as the preparation of ligands and protein, enrichment analysis, rescoring of poses, and many others. These sequences of errors during docking negatively impact the hit rates in virtual

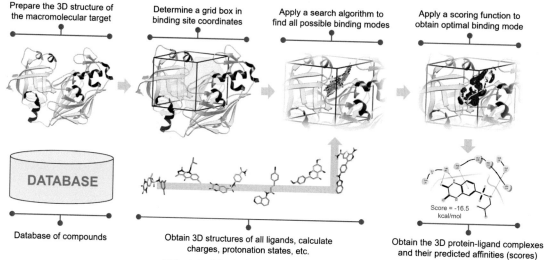

FIG. 4.1 A typical molecular docking workflow.

screening (VS) campaigns, leading to the random selection and acquisition of inactives for experimental validation. In this chapter, we intend to highlight the strengths and pitfalls of main steps of molecular docking and suggest possible roadmaps, methods, and strategies, which can increase hit rate in VS and advance to finding new drug candidates.

2 DOCKING-BASED VIRTUAL SCREENING

Since its first introduction in 1997 (Horvath, 1997), docking-based virtual screening (DBVS) has been considered a more efficient, faster, and less expensive approach when compared with experimental high-throughput screening (HTS) assays. The approach consists in identifying a subset of potentially bioactive compounds (i.e., 10^1 to 10^2 chemical structures) that bind to a specific target from a large library of compounds (i.e., 10^3 to 10^9 chemical structures) (Braga et al., 2014; Lionta et al., 2014). The general scheme of a typical DBVS is shown in Fig. 4.2.

DBVS is usually represented in the form of a funnel that incorporates different filters that remove undesirable compounds (Braga et al., 2014; Lionta et al., 2014). Generally, chemical similarity filters and rule-based approaches (e.g., Lipinski rule of five (Lipinski et al., 2001), Veber' rules (Veber et al., 2002), and pan-assay interference compounds (Baell & Holloway, 2010; Jasial et al., 2018)) are typically incorporated in DBVS workflows, as well as docking and rescoring strategies. In

addition to the filters mentioned above, ligand-based drug discovery tools, such as pharmacophore models (Vuorinen & Schuster, 2015) and quantitative structure–activity relationship (QSAR) models (Neves et al., 2018), can be efficiently incorporated into the DBVS workflow. This combination allows to predict other biological endpoints (e.g., physicochemical, pharmacokinetics, and toxicological properties), to minimize the limitations of docking, and to increase reliability in predictions (Braga et al., 2014). The final processing step in a DBVS consists in the final analysis and visual inspection of the predicted poses, raking the compounds and prioritizing the best hits for experimental validation. At this stage, the main noncovalent interactions of the ligand alongside the amino acids of the binding site, such as hydrophobic, hydrogen bonds, π-stacking, cation-π, and salt bridges, are evaluated, as well as docking scores (Freitas & Schapira, 2017). Besides this manual inspection of the docking poses, Fassio et al. (2019) developed the nAPOLI (analysis of protein–ligand interactions), a web server for analyzing conserved protein–ligand interactions and docking poses in large scale (Fassio et al., 2019). Unlike most servers that perform the ligand–protein interaction analysis of each docking pose, nAPOLI server allows the inspection of several docking poses at atomic level; comparion with ligands co-crystallized in proteins; and comprehensive graph representations and reports of the interacting ligand residues, to detect and explore conserved noncovalent interactions.

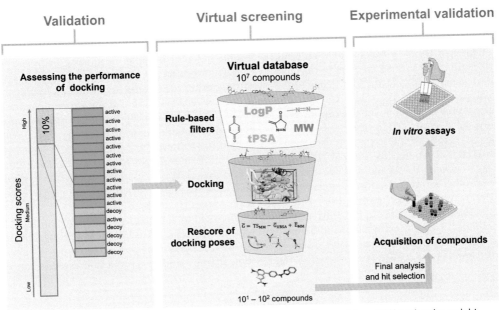

FIG. 4.2 An example of typical docking-based virtual screening workflow. *MW*, molecular weight.

Although docking appears to be an easy-to-use approach, there is a large complexity in the level of elaborateness in the protocols used to implement VS workflows around the world. In general, 3D structures of compounds and macromolecules must be carefully prepared to mimic the experimental conditions of the corresponding in vitro/in vivo assays or biophase (Sastry et al., 2013). Essentially, virtual compounds should be prepared to generate 3D geometries, designate appropriate bond orders, and create accessible tautomer and protonation states before VS. In turn, 3D structures of macromolecule need to be prepared in order to add all missing side chains and amino acids, add hydrogen atoms, eliminate atomic clashes, and optimize hydrogen bonds that are not solved in X-ray crystallography (Sastry et al., 2013). Furthermore, the accuracy of the docking protocol must be estimated using benchmark datasets and standard metrics (Braga & Andrade, 2013; Torres et al., 2019). Details of ligand and macromolecule preparation, as well as docking validation and enrichment are presented in the sections below.

Once carried out with due care, hit rates from DBVS usually reach 40%, while the hit rates of HTS campaigns range between 0.01% and 0.1% (Zhu et al., 2013). Despite this, the expectation that molecular docking can completely replace experimental assays is overoptimistic. So, in vitro experimental evaluation of virtual hits must be considered as the final and most important step of any DBVS study (Neves et al., 2016).

3 BEST PRACTICES IN MOLECULAR DOCKING FOR VIRTUAL SCREENING

Although docking of chemical databases is widely used as molecular filter in VS campaigns, this method retains important weaknesses that compromise the hit rate during experimental evaluation. The most important failures are related to (1) low resolution of the 3D structure of macromolecule; (2) inadequate choice of bioactive conformation of the ligands and macromolecule; (3) incorrect setting of structural water molecules and protonation states of amino acid residues; (4) low correlation between scoring functions and experimental affinity; and (5) the absence of an evaluation of enrichment using a benchmark dataset and descriptors. Further down in this chapter, we outline some guidelines for best practices in docking-based VS.

3.1 Calculation of Protonation States and Structural Waters

The majority of solved 3D protein structures does not provide the position of hydrogen atoms (Blakeley et al., 2015; Fisher et al., 2012); thus, the protonation states should be calculated on a theoretical base (Brink & Exner, 2009). From the perspective of the ligand, several chemical groups present many protonation states and tautomeric forms (Bax et al., 2017; Milletti et al., 2009). Different protonation states either of the receptor or the ligand led to variability in conformation, functional groups, molecular surface, and intermolecular interaction patterns (Milletti et al., 2009). Thus, wrong prediction of protonation states has a serious impact in DBVS, as the binding mode and docking scores can be strongly influenced (Brink & Exner, 2009; ten Brink & Exner, 2010; Mitra et al., 2008). In order to calculate the protonation states accurately, the knowledge of the environmental pH and the pK_a values of ionizable groups at this specific pH are key factors (Garcia-Moreno, 2009; Mitra et al., 2011). Thus, it is feasible to determine the charge distribution within each molecule (Petukh et al., 2013), and the ionized groups of opposite charge between the ligand and the receptor will attract each other forming electrostatic interactions (Bertrand, 1997; Kundrotas & Alexov, 2006; Petukh et al., 2013). These electrostatic interactions play a crucial role for ligand—receptor recognition and binding affinity (Fig. 4.3) (Freitas & Schapira, 2017; Kukić & Nielsen, 2010; Smith et al., 2012).

Another crucial step preceding docking calculation is the selection of essential active site water molecules (Lie et al., 2011). Water molecules play an important role in ligand—receptor binding, facilitating hydrogen bonds and contributing to entropic and enthalpic changes (Cheng et al., 2012). Structural water molecules are present in several proteins that are experimentally determined. A study analyzing protein—ligand complexes from the PDBbind database (Cheng et al., 2009; Li et al., 2014) showed that the number of ligand-bound water molecules ranges from 0 to 21. The study also found that 76% of these interfacial waters are considered to be bridging interactions among ligand and protein residues (Lu et al., 2007).

The rigid treatment of structural waters at the binding site represents a feasible strategy. However, treating the possible positions and orientations of them in a static manner could lead to a docking bias concerning

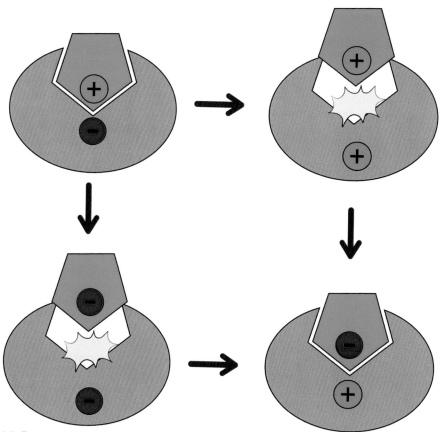

FIG. 4.3 Representation of electrostatic interactions between charged groups of ligands (green) and receptors (gray). Attractive interactions occur between the opposite-sign poles of the ligand and the receptor (+/−). Like-sign poles (+/+ and −/−) of the ligand and the receptor result in repulsive interactions.

pose prediction and score accuracy, with the direct consequence of increasing rates of false positives in VS hit lists (Huggins & Tidor, 2011; Kroemer, 2007; Santos et al., 2010). On the other hand, the presence of water moieties narrows the binding site volume to be occupied by the incoming ligand, and the ligand could be biased toward the correct binding mode with a subsequent increase of enrichment (Lie et al., 2011).

One way to assess the important water moieties to ligand–receptor binding is through the redocking of experimentally obtained ligands in the presence and absence of waters, and use the most accurate mode to replicate experimental results for DBVS (Berry et al., 2015; Huang & Shoichet, 2008). Water molecules making three hydrogen bonds with the protein are probably less displaced by ligands and could be maintained in docking studies (Yang et al., 2006; Hornak et al., 2006). In addition, water molecules involved in hydrogen bond bridges between the ligand and receptor in experimentally solved structures are also deemed to be important and should be kept in docking calculations (Huang & Shoichet, 2008).

3.2 Pose Prediction

Molecular docking exhaustively covers the space of intermolecular interactions as possible to calculate appropriate ligand-binding conformation or corresponding free energy of binding (Berry et al., 2015). One of the limitations of this technique is that similar scores can be obtained from significantly different poses (Houston & Walkinshaw, 2013). In this sense, incorrect poses can result in low-accuracy affinity predictions and reduced hit rates during experimental validation (Houston & Walkinshaw, 2013).

Frequently, the precision of docking poses is estimated by redocking and cross-docking approaches. Redocking, sometimes also called back or self-docking, deals with the ability to reproduce experimental binding of co-crystalized ligand given a rigid macromolecule state. Cross-docking involves the use of different co-crystalized ligands retrieved from multiple PDB structures of the same target to dock with a specific macromolecule state. The accuracy of the ligands that are docked may be calculated by root mean square deviation (RMSD) (Bell & Zhang, 2019). The RMSD measures the average distance between docking pose and the respective experimental conformation through Eq. (4.1):

$$RMSD = \sqrt{\frac{1}{N} \sum_{i=1}^{N} d_i^2} \qquad (4.1)$$

where d_i is the Euclidean distance between the ith pair of corresponding atoms, and N is the number of atoms in the ligand (Bell & Zhang, 2019). In general, it is well accepted that docking can reproduce the experimental binding mode when RMSD is lower than 2.0 Å, but values closest to zero are optimal. Although important, a simple RMSD is not sufficient to validate docking-based VS studies. Of course, thousands to millions of compounds are likely being scored in VS, but validating each chemical group of ligand would be impossible (Berry et al., 2015).

3.3 Assessing the Performance of DBVS

The progress of DBVS is associated with its capacity to rank a larger fraction of actives at higher locations of the hit list than a random compound selection. Despite this convincing and useful concept, the assessment of its performance is often neglected, either due to the limitations of the method or due to the researcher's lack of knowledge. A docking algorithm that randomly ranks numerous compounds from a virtual database is of limited use. Therefore, it is crucial to evaluate the performance of the method, using known actives and inactives or decoys (Kirchmair et al., 2008; Novoa et al., 2012).

While docking scores can contribute to the prediction of binding affinity, they are extensively recognized to weakly correspond with the experimental property (Mackey & Melville, 2009). Therefore, a routine task in DBVS studies should be the estimative of the enrichment (i.e., ranking) using a benchmark dataset of known actives and inactives. Currently, millions of compounds with bioactivity data for thousands of macromolecular targets are openly accessible in databases, such as PubChem Bioassay (Wang et al., 2012), BindingDB (Liu et al., 2007), and ChEMBL (Gaulton et al., 2012). The unavailability of activity data for the macromolecular target under study does not exclude the docking study. However, for those macromolecular targets with known actives and inactives, it is extremely important to use them as benchmark datasets to guide the docking of query compounds.

The composition and size of benchmark datasets is critical to limit the detrimental effect of bias on docking performance (Réau et al., 2018). In typical HTS collections, usually only a small percentage of compounds had been registered with biological activity data (Neves et al., 2016). On the other hand, structure–activity relationship (SAR) studies involving a specific scaffold have artificial distribution, with few inactives for each active. In order to avoid this artificial distribution, putatively inactives, so-called decoys, have been used generally for benchmarking datasets (Huang et al., 2006). The physicochemical properties of the decoys in the database should resemble those of the real actives sufficiently enough so that enrichment rate is not an overoptimistic separation of irrelevant properties (Huang et al., 2006). The decoys nevertheless should be structurally different from the actives, so that they are likely to be inactives (Mysinger et al., 2012). Generally, typical ratios of benchmark sets have >36 decoys for each 1 active (Huang et al., 2006) in order to reproduce the distribution of active and inactives in a randomized or HTS screening.

Following the report of the first benchmark decoys database (Huang et al., 2006), the composition of decoy sets has been pointed out to impact docking performance outcomes either positively or negatively (Réau et al., 2018). For example, the potential presence of actives in the decoy set may undervalue the enrichment rate because decoys are usually assumed to be inactives (Good & Oprea, 2008; Verdonk et al., 2004). On the other hand, the physicochemical dissimilarity of actives and decoys may overestimate the enrichment rate (Bissantz et al., 2000; Vogel et al., 2011). In view of this conundrum, several decoy databases have been developed to minimize bias in decoy sets or at least establish a common basis or reference with a constant composition. The Directory of Useful Decoys Enhanced (DUD-E) is accepted as such an all-purpose reference, nowadays (Huang et al., 2006; Mysinger et al., 2012), Demanding Evaluation Kits for Objective in Silico Screening (DEKOIS) (Ibrahim et al., 2015; Vogel et al., 2011) and Maximum Unbiased Validation (MUV) (Rohrer & Baumann, 2009) are other useful decoy sets.

Fig. 4.4 shows three hypothetical distributions for actives and decoys in a VS according to their docking scores. In an ideal situation (Fig. 4.4A), docking scores

all actives (TP, true positives) before decoys (TN, true negatives) in top x% ordered list (e.g., 0.1%, 1%, 5%, 10%). In contrast, if part of decoys are scored higher than actives, an overlap between the distribution of docking scores is observed (Fig. 4.4B), which will drive to the prediction of false positives (FP) and false negatives (FN), respectively. Finally, the docking scores of actives and decoys randomly distributed in top x% of the list and cause an overlapping distribution (Fig. 4.4C) that corresponds to the worst situation in a VS (Kirchmair et al., 2008; Novoa et al., 2012).

Accordingly, several enrichment descriptors (metrics) have been used to estimate the "early recognition" in VS practice (Braga & Andrade, 2013; Novoa et al., 2012). In this section, we grouped these metrics into classic and advanced. The most used classic descriptors are receiver operating characteristic (ROC) curve (Fawcett, 2006), area under the ROC curve (AUC) (Fawcett, 2006), and enrichment factor (EF) (Mackey & Melville, 2009).

The ROC curve provides a visual summary of TP rate (TPR, or sensitivity) and FP rate (FPR, or 1—specificity) to show how often the inactives (or decoys) are classified as actives (Fawcett, 2006). Fig. 4.5 shows a theoretical ROC curve.

The visualization of the high (a), good (b), and acceptable (c) for the actives and decoys enables a straightforward interpretation of the enrichment. Thus, an optimal ROC curve will remain shifted to the upper-right corner of the graph, where all actives are retrieved before decoys, being consisted with TPR of 1 and FPR of 0. In contrast to that, the randomly (d) and worsened (e) curves tend toward the TPR lower than 0.5 and FPR upper than 0.5 with an increasing number of actives and decoys (Fawcett, 2006).

The AUC determines the entire 2D area underneath the entire ROC curve to investigate how better a

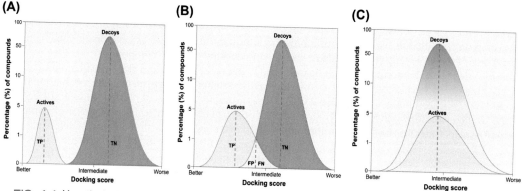

FIG. 4.4 Hypothetical distributions for docking scores of actives compounds and inactives/decoys under ideal **(A)**, acceptable **(B)**, and bad **(C)** enrichment rates. *FN*, false negatives; *FP*, false positives; *TN*, true negatives; *TP*, true positives.

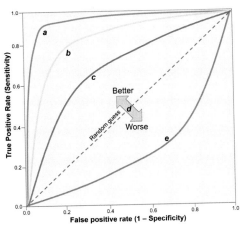

FIG. 4.5 Schematic diagram relating docking performance to the receiver operating characteristic (ROC) curve. The classification rate in curve *a* is better than *b*, *c*, and *e*, respectively. The ROC curve for a randomly docking ranking is represented as a *dotted line* in *d*.

predicted active compound is ranked compared to a randomly picked decoy (Fawcett, 2006; Mackey & Melville, 2009). The AUC is typically calculated for the whole of the ROC curve using Eq. (4.2):

$$AUC = \frac{1}{n} \sum_{i=1}^{n} (1 - f_1) \qquad (4.2)$$

where f_1 is the fraction of decoys ranked higher than the *i*th active. A general guide for classifying the AUC results is shown as follows: AUC < 0.5 is a failure; $0.5 \leq$ AUC < 0.7 is poor; $0.7 \leq$ AUC < 0.8 is acceptable; $0.8 \leq$ AUC < 0.9 is good; and $0.9 \leq$ AUC ≤ 1 is excellent (Braga & Andrade, 2013). The AUC value is an effective and quickly feasible assessment for demonstrating the global performance of VS. However, the AUC is not suitable for the investigation of early recognition of actives in the top *x*% ranked list.

Despite that AUC is easily interpreted, many practitioners of VS are, rightly, most concerned about early performance (top-ranked compounds) of the methods they use (Mackey & Melville, 2009). So, the EF was introduced to measure how much better a method is than a randomly ordered list, by obtaining the portion of actives scored in the top *x*% of the ordered list, contrasted with the ratio among actives and decoys in the total benchmarking dataset (Braga & Andrade, 2013; Mackey & Melville, 2009). This metric is usually represented for a given percentage of the filtered dataset. For example, $EF_{1\%}$ would represent the EF value obtained from 1% of the top-ranked compounds. The EF is defined according to Eq. (4.3):

$$EF_{x\%} = \frac{n_{x\%}/N_{x\%}}{n/N} \qquad (4.3)$$

where *n* represents the total number of actives, *N* represents the total number of compounds (actives and decoys) in the database, $n_{x\%}$ represents the number of actives, and $N_{x\%}$ represents the total number of compounds in the *x*% ordered list, respectively. At top 10% of the ordered list, the maximum EF value is 10, at 5% of the ordered list, the maximum is 20, and at 1% of the ordered list, the maximum value obtainable is 100. The main disadvantages of the EF are the high dependency on the proportion of actives in the benchmarking dataset and the inability to estimate the distribution of actives at the top *x*% rank-ordered list (Braga & Andrade, 2013).

Although extremely important, the AUC and EF, they are not able to discern the order of actives and decoys in top *x*% list. To overpower the "early recognition" issue, advanced metrics, such as the robust initial enhancement (RIE) (Sheridan et al., 2001), Boltzmann-enhanced discrimination of ROC (BEDROC) (Truchon & Bayly, 2007), and partial area under ROC curve (pAUC) (McClish, 1989), were developed.

The RIE metric has a similar meaning to EF as it shows how often the frequency of the actives in top *x*% ranked list is superior to an arbitrary ranking. Unlike EF, the RIE metric has the advantage of including the contributions of actives into the final score (Sheridan et al., 2001; Truchon & Bayly, 2007). Another advantage of the RIE metric over EF is its ability to estimate the distribution of actives at the top of the ordered list. The RIE is defined according to Eq. (4.4):

$$RIE = \frac{\frac{1}{n} \sum_{i=n}^{n} e^{-\alpha x_i}}{\frac{1}{n} \left(\frac{1 - e^{-\alpha}}{e^{\frac{\alpha}{n}} - 1} \right)} \qquad (4.4)$$

where x_i is the relative rank of the *i*th active and α is a tuning parameter. Although the RIE addresses the "early recognition" problem, it is sensitive to the rate of active compounds in the benchmarking dataset (Braga & Andrade, 2013; Truchon & Bayly, 2007).

BEDROC is a metric that undertakes the probability of an active compound to be ranked preceding a randomly chosen compound. This descriptor uses a decreasing exponential weighting function that focuses on active compounds listed at the top of the ordered list (Braga & Andrade, 2013; Truchon & Bayly, 2007). Moreover, BEDROC has a linear relationship with RIE (Sheridan et al., 2001; Truchon & Bayly, 2007) and is defined by Eq. (4.5):

$$\begin{aligned} \text{BEDROC} = \text{RIE} \; x \; & \frac{R_a \sinh\left(\frac{\alpha}{2}\right)}{\cosh\left(\frac{\alpha}{2}\right) - \cosh\left(\frac{\alpha}{2} - \alpha R_\alpha\right)} \\ & + \frac{1}{1 - \exp(\alpha(1 - R_\alpha))} \end{aligned} \quad (4.5)$$

where α is an exponential weighting factor which controls the emphasis given to early recognition, and R_α is the proportion of actives in the benchmarking set. The BEDROC is obtained by linearly scaling the RIE metric to [0,1]. However, the ratio of actives and decoys can significantly influence the BEDROC score. More specifically, if the ratio of decoys is pretty low, the BEDROC will be unable to properly estimate enrichment rate (Braga & Andrade, 2013; Truchon & Bayly, 2007).

The pAUC is a relative approach that focuses only on a part of the ROC curve. It allows the selection of models with high TPR and low FPR in the top $x\%$ ranked list, rather than models with better average AUC performance (McClish, 1989). The value of the pAUC varies between 0 and 1, where 0.5 coincides to a random ranking while 1 means a perfect ranking (all actives are ranked above the decoys). The pAUC is defined according to Eq. (4.6).

$$\text{pAUC}_{x\%} = \frac{1}{2}\left[1 + \frac{\text{AUC}_{x\%} - \text{AUC}_{\text{random at } x\%}}{\text{AUC}_{\text{perfect at } x\%} - \text{AUC}_{\text{random at } x\%}}\right] \quad (4.6)$$

3.4 Post-Docking Processing

Although remarkable improvements in docking algorithms have been obtained over the years, resulting scores and experimental binding affinities generally do not correlate (Rastelli & Pinzi, 2019). As a consequence of this low correlation, a variable number of FP populates the top-ranked lists in VS campaigns. Hence, docking results have been improved using rigorous post-docking processing strategies to provide higher hit rates in VS, as described below (Rastelli & Pinzi, 2019).

3.4.1 Binding free energy estimations
Within hierarchical DBVS workflows, the prioritization of virtual hits is generally based on a simple and fast scoring function. Subsequently, a more rigorous and time-consuming scoring scheme that usually attempts to include solvation and entropic effects is used to rescore the top hits (Genheden & Ryde, 2015; Rastelli & Pinzi, 2019).

Several rescoring methods have been developed for optimized DBVS outputs. Notably, approaches based on binding free energy calculation usually correlate better with experimental data and, consequently, are more suitable to ranking compounds in VS (Genheden & Ryde, 2015; Gohlke et al., 2000; Pearlman & Charifson, 2001; Pu et al., 2017). The most widely used binding free energy estimators are molecular mechanics energies combined with the Poisson–Boltzmann surface area continuum solvation (MM/PBSA) (Kollman et al., 2000) or generalized Born surface area continuum solvation (MM/GBSA) models (Srinivasan et al., 1998). They are typically intermediate in both computational cost and accuracy between empirical docking scoring and strict alchemical perturbation methods (Genheden & Ryde, 2015; Kuhn et al., 2005; Thompson et al., 2008). The MM/PBSA and MM/GBSA calculates the free energy according to Eq. (4.7):

$$\overline{G} = \text{TS}_{\text{MM}} - \overline{G}_{\text{XBSA}} + \overline{E}_{\text{MM}} \quad (4.7)$$

where $\overline{G}$ is the predicted average free energy, TS_{MM} is the solute entropy that can be calculated by quasi harmonic analysis of the trajectory or normal mode analysis (Srinivasan et al., 1998), $\overline{G}_{\text{XBSA}}$ is the the polar and nonpolar contributions to the solvation free energies typically obtained by solving the PB equation or by using the GB model. The $\overline{E}_{\text{MM}}$ is the average molecular mechanical energy:

$$\overline{E}_{\text{MM}} = \overline{E}_{\text{bond}} + \overline{E}_{\text{angle}} + \overline{E}_{\text{torsion}} + \overline{E}_{\text{vdW}} + \overline{E}_{\text{elec}} \quad (4.8)$$

where these correspond to the bond, angle, torsion, van der Waals, and electrostatic energies in the molecular mechanics force field (Kollman et al., 2000). The ΔG for ligand–protein association can be calculated using Eq. (4.9):

$$\Delta G = \overline{G}_{\text{complex}} - \overline{G}_{\text{ligand}} - \overline{G}_{\text{protein}} \quad (4.9)$$

Although the MM/PB(GB)SA methods may be useful to rescore docking poses, it ignores the entropy of water molecules in the binding site before and after ligand binding and structural changes in the macromolecule and ligand upon binding. In addition, the accuracy of ΔG results strongly depends on details in the method, especially the charges, the continuum solvation method, the dielectric constant, the sampling method, and the entropies (Genheden & Ryde, 2015).

3.4.2 Score normalization
The energy-based scoring functions usually compute the sum of intermolecular interactions. Therefore, smaller compounds usually receive less favorable scores than larger ones (Carta et al., 2007; Zhong et al., 2010). Usually, hit-and-fragment selection methods from HTS include statistical analyses such as ligand efficiency

(LE), which normalizes the experimental activity to molecular weight (MW), heavy atom count, or other molecular size quantifiers (Kenny, 2019).

To optimize DBVS outputs, Carta and coauthors developed a normalization procedure to attenuate the influence of MW in docking scores (Carta et al., 2007). Notably, this method does not increase the scoring function in terms of physical relevance or enrichment but favors the selection of compounds with more intermolecular interactions per atom. The higher ratio of intermolecular interactions may increase the possibility of selected virtual hits progressing successfully through the experimental in vitro assays and hit-to-lead optimization (Carta et al., 2007; Pan et al., 2003).

First, the scores resulting from each scoring function and the corresponding MW of each were normalized according to Eq. (4.10):

$$\text{normValue} = \frac{X - \min}{\max - \min} \quad (4.10)$$

where X represents either the MW or the actual score for each compound, and min and max are the minimum and maximum values of the screened set, respectively (Carta et al., 2007). Then, the ranked list must be rescored to reduce the dependence of the scoring function on MW using Eq. (4.11):

$$\text{corrScore} = \text{normScore} \, (1 - \text{normMW})^{1/4} \quad (4.11)$$

where *normScore* is the normalized score and *normMW* is the normalized MW according to the previous formula. The power function $(1 - \text{normMW})^{1/4}$ penalizes only those molecules which exhibit high MW and leave the remaining relatively unchanged (Carta et al., 2007).

3.4.3 Consensus scoring

Although docking programs share the principles of pose prediction and binding affinity estimation, they use somewhat different scoring algorithms, as well as different degrees of protein and ligand flexibility. In addition to the different scoring functions, each program has distinct atom typing, methods for obtaining the atomic partial charges and different datasets of ligand–protein were used to train the program (Kukol, 2011). Therefore, each program can estimate different corresponding binding affinity. One approach to deal with these issues—instead of choosing a single "correct" function—is to merge data obtained from different scoring functions in a single value, called consensus scoring (Houston & Walkinshaw, 2013; Palacio-Rodríguez et al., 2019).

The consensus score has been successfully explored in the field of QSAR (Ajay, 1994; Feher, 2006; So & Karplus, 1996) and molecular similarity (Feher, 2006; Ginn et al., 1997). Consequently, its applicability has been extended to the docking field. The consensus of docking scores combine results of different docking programs (Houston & Walkinshaw, 2013; Palacio-Rodríguez et al., 2019). Usually, the consensus scores have led to superior accuracy when compared with single scoring results (Houston & Walkinshaw, 2013), leading early enrichment of known actives in VS (Charifson et al., 1999; Feher, 2006; Kukol, 2011). The limitation of consensus docking relies/depends mainly on the performance of docking programs. Having poor performance could be due to the scoring function parameterization and training set dependencies (Palacio-Rodríguez et al., 2019).

The combination of docking scores can be performed as a sum ranks method, simply adding up the individual scores and averaging them (Clark et al., 2002; Houston & Walkinshaw, 2013; Palacio-Rodríguez et al., 2019). However, despite the fact that the values of scoring functions are always given as energy values, they are not commensurate. Thus, several consensus scoring methods have been used to combine the scoring functions values, such as worst-best ranks (Clark et al., 2002); deprecated sum ranks (Clark et al., 2002); coarse quantiles voting (Feher, 2006); and multivariate methods (Feher, 2006).

A nonlinear consensus score is the worst-best ranks approach. In this method, the worst rank for each compound is rejected and the remainder entries are sorted (Clark et al., 2002). On the other hand, in the deprecated sum ranks approach, for each compound, only the best ranks are summed (Clark et al., 2002).

In coarse quantiles voting, the values obtained by each scoring function rely on coarse positions or quantiles in range rather than ranks (Clark et al., 2002; Feher, 2006). If the score of a candidate configuration falls into the best half of range of values, it receives one favorable vote from each scoring function. The total number of votes received is called CScore (Clark et al., 2002; Feher, 2006), and its values may produce tied scores, if calculated using only one criterion (best half), hindering the rank analysis.

Clark et al. (2002) compared CScore, deprecated sum ranks, worst-best ranks, and sum ranks methods, combining results from GOLD, DOCK, and FlexX docking programs (Clark et al., 2002). They showed that sum ranks and CScore were more reliable and robust than the individual scoring functions for the considered datasets.

The multivariate methods for combining scores include principal component analysis (PCA), projection to latent structures (PLS), and discriminant analysis (Livingstone, 1995). Terp et al. (2001) combined eight different scoring functions and analyzed using multivariate statistical methods PCA and PLS to select the best pose between different docked conformations and to quantify the binding affinity (Terp et al., 2001). PCA model successfully predicts ligand binding modes and was superior of using simple CScore, as CScore latter does not distinguish among tied scores. The PLS model was obtained from the regression of the five most predictive scoring functions, and the correlation with the experimental pK_i was pretty superior to the performance of a single scoring function (Terp et al., 2001).

3.4.4 Machine learning–based scoring functions

By the last decades, several efforts have been devoted to improve classical scoring functions (i.e., knowledge-based, force field, and empirical). Despite the progress made, these classical scoring functions remain weak predictors of binding affinity of ligand–protein complexes. The main reason for that is the inability of the scoring function to reliably determine the correct binding affinity of the docking pose (Khamis et al., 2015; Shen et al., 2020). Furthermore, many scoring functions incorrectly assume that each interaction contributes toward the binding affinity in an additive and linear manner. These limitations highlight the need for new scoring functions based on nonlinear models (Khamis et al., 2015; Shen et al., 2020).

Since 2010, several ML-based scoring functions have been successfully introduced in docking to learn from the nonlinear dependency of the intermolecular interactions and more correlation with the experimental binding affinity (Khamis et al., 2015; Shen et al., 2020). ML is seen as a subfield of artificial intelligence that uses statistical techniques to build mathematical models from various data types. Once developed, the ML model can be used to make predictions or decisions without being explicitly programmed to do so (Ekins et al., 2019). Fig. 4.6 schematizes the main steps involved in the development of ML-based scoring functions.

Currently, PDBbind (Cheng et al., 2009; Li et al., 2014) and Community Structure-Activity Resource (CSAR) (Smith et al., 2011) are the most consolidated benchmark repositories to develop global scoring functions, as they provide structural data for a pool of experimentally solved protein–ligand structures with determined K_d or K_i information. For developing target-specific scoring functions, the experimental data can be retrieved from PubChem Bioassay (Wang et al., 2012), BindingDB (Liu et al., 2007), and ChEMBL (Gaulton et al., 2012) and then submitted to pose generation. After data collection, structures are represented as a set of protein–ligand complex features that are relevant in predicting the binding affinity (Ain et al., 2015; Khamis et al., 2015). Traditionally, these features are structure-based fingerprints, energy terms used in classical scoring functions, algebraic topology approaches, or basic structural information, such as atoms and bonds (Ain et al., 2015; Khamis et al., 2015).

Like most other applications of ML, the division of benchmark dataset in training and validation (test)

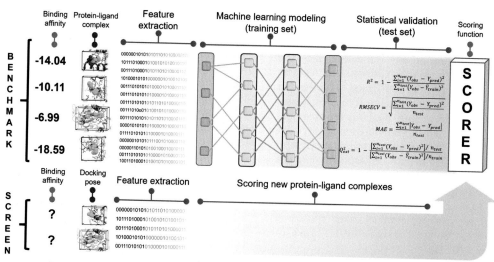

FIG. 4.6 Typical workflow for generation of machine learning–based scoring functions.

sets is typically required to establish an ML-based scoring function. After division, an ML algorithm is applied to the training set in order to discover an empirical function that can achieve an optimal relationship between the protein—ligand complex features and binding affinities. Once generated and validated using test sets and standard metrics, established models may be used as standard scoring function in docking programs or used as rescoring filter in DBVS studies (Ain et al., 2015; Khamis et al., 2015; Shen et al., 2020).

Table 4.1 shows some examples of different scoring functions assessed on the PDBbind database and correlated then with their scoring power (Cheng et al., 2009; Wang et al., 2004, 2005). Analyzed with classical scoring functions, ML-based approaches always achieved a remarkable improvement on the prediction power. Generally, the Spearman's correlation (Rp) between the ML-based scores and experimental binding data is higher than > 0.75, whereas this correlation does not reach 0.50 using classical scoring functions. Ensemble methods such as random forest (RF) (Breiman, 2001) and gradient boosting trees (Friedman, 2001) are the most popular and effective ML tools to develop scoring functions. Support vector machine (SVM) was elected to be used for target-specific VS (Ashtawy & Mahapatra, 2015; Shen et al., 2020). However, due to the powerful descriptor extraction capacity of the deep learning methods, convolutional neural networks (Chen et al., 2018) have been recently implemented, as they just need simple atomic features such as types, distances, and partial charges as the input.

4 INCORPORATING PROTEIN FLEXIBILITY

In physiological environment, proteins present dynamic behavior, with subtle movements of their backbones or side chains, extensive loop movements, and flexibility at the binding sites, acquiring multiple conformations. Moreover, proteins can undergo conformational changes upon ligand binding, phenomenon referred as induced fit. Both protein and ligand can modify their conformations and adjust their complementarity to enhance favorable contacts and diminish unfavorable ones, augmenting the total binding free energy (Alonso et al., 2006). Lack of consideration of such flexibility in docking studies can lead to inaccurate docking poses and poor correlation with binding affinity energies (Alonso et al., 2006; Rao et al., 2008).

Several approaches have been developed to incorporate protein flexibility in docking studies (Fig. 4.7). They can be categorized into (1) single-input strategy and (2) multiple-input strategy. In the single-input strategy, the conformational space of protein binding site is explored in terms of side chain and/or backbone flexibility, mainly using soft docking and induced fit docking (IFD) approaches. On the other hand, the multiple-input strategy uses a pool of protein conformations obtained from crystallographic structures, NMR conformations, Monte Carlo (MC), or molecular dynamics (MD) conformations, mainly using ensemble docking. The determinant and limiting factors to select the approach to be used in each study case will depend on the time and computational resources available. Also, it depends on how flexible a protein is, to decide if whole protein or only binding site residues will be considered in flexible approaches.

4.1 Soft Docking Approach

Soft docking approach considers and allows small degrees of local flexibility for proteins through the weakening of the repulsive term in the Lennard-Jones potential between ligand and receptor, generating a "soft" interaction, despite treating the receptor as a rigid macromolecule (Alonso et al., 2006; Ferrari et al., 2004; Nabuurs et al., 2007). The attenuation allows protein surface penetration by the ligand as well as small and localized changes in the environment (Alonso et al., 2006). The advantage of soft docking is the computational efficiency and low cost, but its applications are limited to the local flexibility at the binding site.

Some examples of programs that also use soft docking methods to allow protein flexibility are DOCK (Venkatachalam et al., 2003) and DockThor (Magalhães et al., 2014; Santos et al., 2020). Ferrari et al. (2004) confronted the approaches of multiple protein conformations and soft docking in VS. They suggested that soft docking was superior in the identification of known ligands to the hard scoring function (i.e., when only a single protein conformation was used), but it was worse when multiple conformations were used (Ferrari et al., 2004).

4.2 Side Chain and Backbone Flexibility Approaches

Side chain flexibility can be considered (1) before docking calculations, using alternative residues conformations; (2) during docking process, using dynamic exploration of rotamer libraries; and (3) after docking calculations, performing final optimization of contacts of amino acids. Some approaches explicitly consider side chain flexibility, particularly those in the binding site, during docking calculations or after ligand binding, using rotamer libraries (set of amino acid side chain conformations) to inspect the conformational space

TABLE 4.1

Performance Comparisons of Some Machine Learning (ML)—Based and Classical Scoring Functions.

Scoring Function	Method	Feature Representation	R_p	RMSE	Year	References
ML-BASED SCORING FUNCTIONS						
EIC-Score	GBT	Element interactive curvatures	0.83	1.77	2019	Nguyen and Wei (2019)
TopBP	GBT and CNN	Topological features	0.86	1.65	2018	Cang et al. (2018)
K_{DEEP}	CNN	Features based on 3D grid	0.82	1.27	2018	Jiménez et al. (2018)
Pafnucy	CNN	Atomic coordinates and some features associated with basic atoms, bonds, or partial charges	0.78	1.48	2018	Stepniewska-Dziubinska et al. (2018)
TNet-BP	CNN	Element-specific persistent homology	0.83	1.37	2017	Cang and Wei (2017)
$\Delta_{vina}RF_{20}$	RF	AutoDock Vina and solvent-accessible surface area terms	0.82	—	2017	Wang and Zhang (2017)
FFT-BP	GBT	Free energy binding terms	0.75	1.99	2017	Wang et al. (2017)
RI-Score	RF	Element-specific rigidity index	0.81	1.85	2017	Nguyen et al. (2017)
RF-Score-v3	RF	Terms from RF-Score and Vina	0.80	1.51	2015	Li et al. (2015)
BgN-Score	ENN	X-score, AffiScore, GOLD, and RF-Score terms	0.82		2015	Ashtawy and Mahapatra (2015)
RF-Score-v2	RF	Counts of elemental atom pairs	0.80	1.53	2014	Ballester et al. (2014)
RF-Score-v1	RF	Counts of elemental atom pairs	0.78	1.58	2012	Ballester and Mitchell (2010)
CScore	ANN	Elemental atom pairs transformed by two fuzzy membership functions	0.77	1.5	2011	Ouyang et al. (2011)
CLASSICAL SCORING FUNCTIONS						
LigScore2	Knowledge-based	Polar, van der Waals (vdW), and desolvation terms	0.46	2.12	2005	Krammer et al. (2005)
GlideScore-XP	Empirical	Electrostatic, vdW, and desolvation interaction terms	0.46	2.14	2004	Friesner et al. (2004)
D-Score	Force field—based	Coulombic and Lennard-Jones potentials	0.39	2.19	2001	Ewing et al. (2001)
ChemScore	Empirical	Rotable bonds, lipophilic, H-bond, polar, and metal interaction terms	0.44	2.15	1997	Eldridge et al. (1997)
GoldScore	Force field—based	H-bond, vdW interactions, ligand internal vdW, and ligand torsional strain terms	0.29	2.29	1997	Jones et al. (1997)
PLP	Empirical	H-bond and hydrophobic terms	0.54	2.00	1995	Gehlhaar et al. (1995)

ANN, artificial neural network; *CNN*, convolutional neural network; *ENN*, ensemble neural network; *GBT*, gradient boosting trees; *RF*, random forest; *RMSE*, root mean square error; R_p, Spearman's correlation between scores and experimental binding data. Statistical performances of classical scoring functions were retrieved from reference (Ashtawy & Mahapatra, 2015).

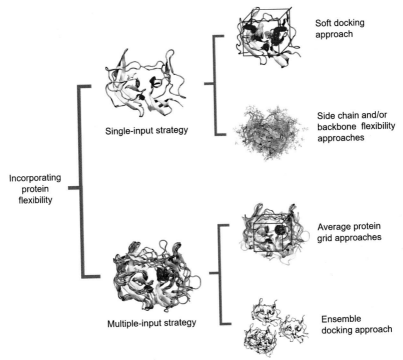

FIG. 4.7 Different approaches used to incorporate protein flexibility in docking studies.

(Alonso et al., 2006; Nabuurs et al., 2007). Several docking programs use side chain flexibility to incorporate protein flexibility in docking results. The RosettaLigand program employs MC minimization in ligand position/orientation and also in full protein side chain conformations, simultaneously, during docking calculations (Meiler & Baker, 2006). This program increases the protein conformation number obtained from the Dunbrack's rotamer library (Shapovalov & Dunbrack, 2011) using MC minimization procedure. The FLIPDock program uses genetic algorithm to optimize ligand geometry, as well as the rotamer indices for moving side chain that generates the new protein conformations during the docking process (Zhao & Sanner, 2008).

Moreover, minor backbone movements are also considered in some approaches, in addition to side chain flexibility on docking calculations. IFD protocols (Friesner et al., 2006; Halgren et al., 2004; Sherman et al., 2006) consider protein side chain degrees of freedom and also sample minor backbone movements (Sherman et al., 2006). Side chain conformations are sampled in dihedral angle space, while backbone movements are calculated through minimization process. The four critical aspects of protein sampling algorithm involves (1)

removal of conformations which have steric clashes; (2) effective minimization process; (3) sampling of only energetically appropriate side chain conformations (using rotamer libraries); and (4) use of precise energy models for scoring the protein conformations (Sherman et al., 2006). Zhong et al. (2009) performed IFD of the active and inactive states of tyrosine kinases epidermal growth factor receptor (EGFR) and Abelson leukemia virus protein kinase (ABL). They showed that IFD successfully reproduced native poses of ligands and consistently predicted binding interactions and docking poses because the results were in keeping with the available experimental data (Zhong et al., 2009).

4.3 Average Protein Grid Approaches

Average protein grid approaches combine alternative structures/conformations of the protein into a single representation, using a simple average or a differential weighting scheme that facilitates the contribution of specific conformations over the others (Alonso et al., 2006). The averaging conformations result on a final protein structure computed from the average over all atom coordinates used (if all receptor conformations are considered) or on a grid box (if only the grid representation is considered) (Alonso et al., 2006).

Average protein grid approaches presented superior results than grids build from a unique structure (Alonso et al., 2006; Carlson, 2002). Osterberg et al. (2002) tested four methods to combine multiple protein structures within a unique grid-based lookup table containing the interaction energies, using AutoDock program (Osterberg et al., 2002). They tested the (1) mean grid, (2) minimum grid, (3) energy-weighted grid, and (4) clamped grid on 21 HIV-1 protease structures. HIV-1 protease shows high flexibility mainly in the active site, enabling the protein to identify and cleave several types of peptide substrates, which can represent significant problems in docking calculation. A minimum grid takes the lowest value across all the grids, and mean grid takes a simple point-by-point average across all the grids (Osterberg et al., 2002). In clamped grid, at each point, a weighted average of energies over all structures is calculated (Knegtel et al., 1997). The energy-weighted grid uses a Boltzmann assumption for calculating the weight (W), according to the interaction energy (ΔG), as shown in Eq. (4.12):

$$W = e^{-(\Delta G/RT)} \qquad (4.12)$$

Among the four tested methods, energy-weighted grid and clamped grid demonstrated higher success rate for the correct prediction of the docked conformations with precise binding free energies. Also, both were successfully applied in the inclusion of limited protein flexibility (Osterberg et al., 2002).

4.4 Ensemble Docking Approach

Different protein conformation structures can be used in docking simulations. They are obtained either from experimental data such as crystallographic structures (Abagyan, Rueda, & Bottegoni, 2010; Craig, Essex, & Spiegel, 2010; Park, Kufareva, & Abagyan, 2010), cryo-EM, or NMR structures (Damm & Carlson, 2007; Hritz, De Ruiter, & Oostenbrink, 2008) or computationally from MD (Park & Li, 2010; Paulsen & Anderson, 2009) as well as MC conformations. The use of an ensemble of protein conformations in docking has been shown to be better than performing calculations in to a single structure of the target (Falcon, Ellingson, Smith, & Baudry, 2019; Yan & Zou, 2015).

Ensemble docking has been used to add protein conformational dynamics (Agarwal et al., 2020; Bhattarai et al., 2020; De Paris et al., 2018) and aims to reproduce a mechanism in which the ligand selects particular protein conformations to form a protein—ligand complex that is thermodynamically favored. Several docking programs can be used or adapted to perform ensemble docking or DBVS, such as GOLD (Verdonk et al., 2003), AutoDockVina (Oleg and Olson,

2010), MDock (Yan & Zou, 2015), and VinaMPI (Ellingson et al., 2013). The latter constitutes a high-throughput docking version of the AutoDockVina program developed to enable substantially large VS, applying a great number of computer cores to reduce the computational time for calculations. MDock is an academic-free docking suite, which incorporates ensemble docking method (Yan & Zou, 2015) and allows users to dock a ligand or dataset, in case of VS, simultaneously, to up to 99 protein structures. The multiple structures firstly need to be superimposed together (Yan & Zou, 2015).

Falcon, Ellingson, Smith, & Baudry (2019) performed coarse-grained MD simulations of four G-protein—coupled receptors and ensemble docking, using VinaMPI (Falcon, Ellingson, Smith, & Baudry, 2019). After obtaining conformations, a clusterization of conformations and selection of a representative structure for each cluster can be performed. Moreover, using known ligands for each protein studied can identify promising conformations scanned in an MD trajectory for ligand binding. The authors highlighted the difficulties of selecting promising snapshots from MD simulations and suggested that clustering of MD conformations do not guarantee the identification of protein conformations that are most frequently selected by the ligands, if protein movement as a whole is considered. However, if clustering is focused in a binding site region, better results are achieved (Falcon, Ellingson, Smith, & Baudry, 2019).

Additional approaches, such as ML, have been combined with ensemble docking to circumvent MD issues and facilitate the selection of relevant protein conformations (Chandak, Mayginnes, Mayes, & Wong, 2020; Wong, 2019). MD suffers from the key drawbacks of force field errors and insufficient sampling of target configurational space, related to the gap between the timescales reachable by MD and slow protein conformational changes (Amaro et al., 2018). Chandak, Mayginnes, Mayes, & Wong (2020) used several ML algorithms, such as k-nearest neighbor, logistic regression, SVM, and RF to show the improvement in ensemble docking of EGFR against 832 actives and 35,442 decoys (Chandak, Mayginnes, Mayes, & Wong, 2020). Fan et al. (2020) also applied ML to select the most relevant protein structures for ensemble docking (Fan et al., 2020). The authors employed an integrated machine learning and docking approach (ALADDIN) using RF classifiers to select—singly for every compound of interest—from an ensemble of protein structures the unique most appropriate protein structure for docking (Fan et al., 2020).

Therefore, ensemble docking can improve docking performance if compared to the docking using a unique crystal structure (Korb et al., 2012), but the performance of ensemble docking depends on the scoring function and the target (Korb et al., 2012).

5 MOLECULAR DOCKING IN FRAGMENT-BASED DRUG DISCOVERY

Although fragment-based drug discovery (FBDD) presents some advantages compared with HTS, it also faces some limitations. Despite the fact that FBDD covers a larger fraction of the chemical space, a wider exploration is still needed (Sheng & Zhang, 2013). The experimental biophysical techniques used for fragment screening have low-to-medium throughput (Keseru et al., 2016), and only hundreds to thousands of fragments are commonly screened, leaving a large portion of commercially available fragments untested (Barelier et al., 2014; Hall et al., 2014). To overcome these drawbacks, molecular docking could be used in several parts of FBDD workflow as an alternative or complementary approach, with its benefits of low cost, fastness, and no solubility concerns (Grove et al., 2016; Sheng & Zhang, 2013).

In the early stages of FBDD, DBVS allows the analysis of fragments and prioritizes the most promising for experimental testing, augmenting the exploration of the chemical space (Bian & Xie, 2018; Kumar et al., 2012; Rudling et al., 2017). Fragment-based VS campaigns can improve the screening process and achieve 10%–40% of hit rate in experimental validation (Marchand & Caflisch, 2018). In the fragment-to-lead (F2L) optimization stage, the resulting amount of optimized compounds can often exceed the throughput of experimental approaches, and molecular docking can be used to assess virtual analogs and testing experimentally those with better results (de Souza Neto et al., 2020; Grove et al., 2016). Another common scenario during F2L is the failure of the co-crystallization of a fragment, resulting in the unavailability of structural information (Chen & Shoichet, 2009; Chevillard et al., 2018; Erlanson et al., 2019). In these cases, molecular docking is an alternative tool to predict the binding mode of the fragments on apo or holo high-quality protein structures or also on a homology model, providing structural information for the F2L (de Souza Neto et al., 2020; Erlanson et al., 2019).

Considerations about fragment docking are often present in the scientific community: first, as a consequence of the low complexity, fragments can adopt several orientations in the binding site, and the poor sampling implies in the incorrect prediction of the binding modes. Second, the scoring functions are parametrized for drug-like compounds and are inappropriate for fragments (Barelier et al., 2014; Chen &

Shoichet, 2009; Wang et al., 2015). To address these concerns, some studies were performed (Vass et al., 2015). Researchers from AstraZeneca tested the Glide program (Friesner et al., 2004) by exploring several docking protocols and considering the RMSD cutoff of 2 Å; they were able to dock correctly five from a total of seven fragments (71%) using Glide standard precision protocol (Kawatkar et al., 2009). In a study by researchers from Schrödinger with 12 protein–fragment complexes, considering the RMSD cutoff of 2 Å, all fragments were docked correctly (Loving et al., 2009). In another study, researchers from Astex compared the performance of the GOLD (Jones et al., 1997) program for drug-like and fragment-like compounds and showed that GOLD performs similarly on both (Verdonk et al., 2011). In the same study, the researchers found that LE, when defined according to Eq. (4.13) (Hopkins et al., 2014; Reynolds et al., 2007; Sándor et al., 2010), is more important than MW for GOLD performance (Verdonk et al., 2011).

$$LE = \frac{\Delta G}{HAC} \qquad (4.13)$$

where ΔG is the binding free energy and HAC is the heavy atoms count of a compound. Sándor et al. (2010) investigated the docking performance of Glide using 78 targets and 190 complexes of fragments to proteins. It also showed similar accuracy for both fragments and drug-like compounds, and most of docking errors were due to scoring problems (Sándor et al., 2010; Vass et al., 2015). Thus, to enhance the outputs of fragment docking during VS campaigns, there is a need for the advancement of scoring functions more adequate to fragments (Rognan, 2011).

6 CASE STUDIES OF DBVS

Some recent successful examples of DBVS are discussed below. Lyu et al. (2019) explored the power of docking and advances in computers' hardware for discovering new AmpC β-lactamase (AmpC) and the D$_4$ dopamine receptor ligands from Real Database (2020). Initially, 99 million virtual compounds were docked into the active site of the AmpC, using the program DOCK3.7. An average of 4054 orientations per compound were used and 280 conformations for each orientation. Then, 5 out of 51 compounds selected for experimental testing were active hits. Next, 90 top-ranked analogs of the actives were also tested. Among those analogs, it was possible to identify a potent AmpC inhibitor (compound **1**, $K_i = 0.077 \ \mu M$), one of the most potent non-covalent inhibitor for AmpC ever reported

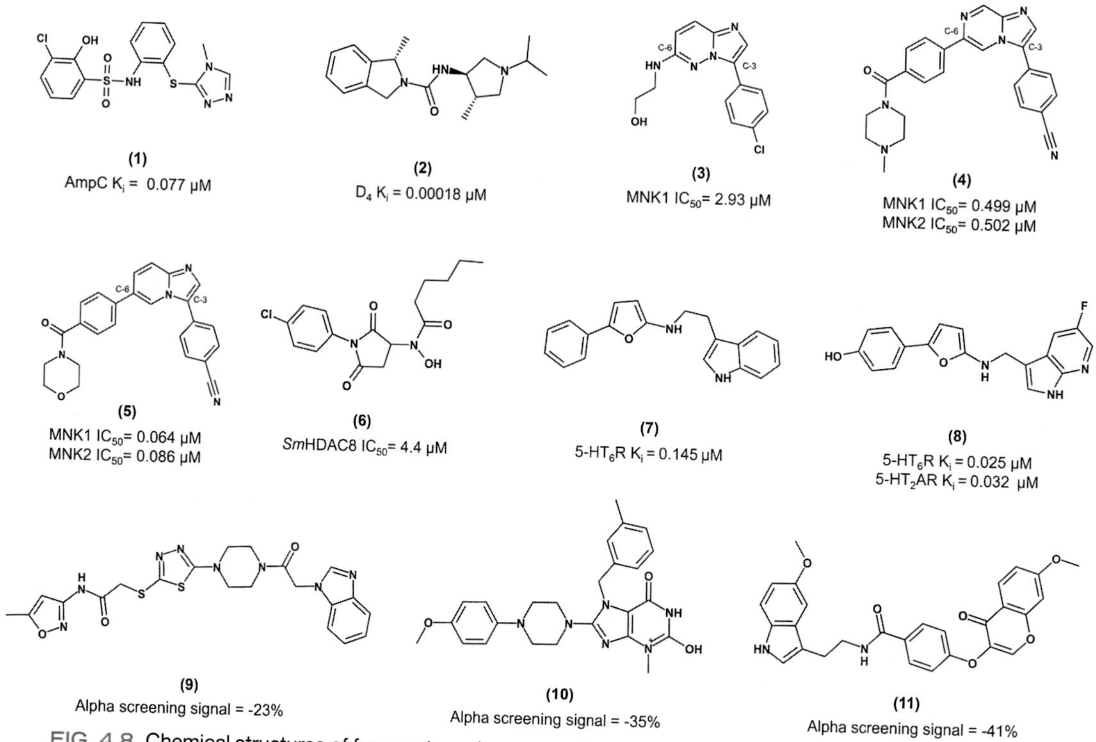

FIG. 4.8 Chemical structures of fragments and compounds identified/optimized using docking-based virtual screening.

(Fig. 4.8) (Lyu et al., 2019). In the same work, researchers docked 138 million compounds into the binding site of D_4 dopamine receptor, sampling 70 trillion complexes in a process of only 1.2 days using 1500 CPUs. The authors were able to identify 30 compounds with submicromolar activity, including one of the most potent selective full agonists for the D_4 receptor, with picomolar activity (compound **2**, $K_i = 0.00018$ μM, Fig. 4.8) (Lyu et al., 2019).

In 2010, Oyarzabal et al. (2010) performed a VS of fragments against mitogen-activated protein kinase–interacting kinase 1 and 2 (MNK1 and MNK2, respectively) using structure- and ligand-based methods and identified 26 hits (Oyarzabal et al., 2010). From this computational approach, a molecular fragment (**3**, Fig. 4.8) with MNK1 $IC_{50} = 2.93$ μM was selected for structural optimization (Yang et al., 2018). The designed compounds were subjected to molecular docking studies using the Glide program (Friesner et al., 2004) and those with good score were synthesized. From synthesized compounds, hit **4** (Fig. 4.8) presented high inhibitory activity against MNK-1 ($IC_{50} = 0.499$ μM) and MNK-2 ($IC_{50} = 0.502$ μM), as

well as good cell permeability and ADME (absorption, distribution, metabolism, and excretion) properties. The subsequent structural optimization of compound **4** led to discovery of compound **5** (Fig. 4.8), which presents activity in nanomolar range (MNK1 $IC_{50} = 0.064$ μM and MNK2 $IC_{50} = 0.086$ μM), good ADME properties, oral bioavailability, and selectivity. Compound **5** is under phase I clinical trial for the blast crisis chronic myeloid leukemia (Gagic et al., 2020; Yang et al., 2018).

Simoben et al. (2018) performed a substructure search of compounds with hydroxamate moieties in InterBioScreen database (~550,000 compounds) aiming to discover inhibitors of *Schistosoma mansoni* Zn-dependent histone deacetylase (*Sm*HDAC8) (Simoben et al., 2018). Subsequently, the hydroxamate compounds submitted to DBVS against *Sm*HDAC8 binding pocket, using Glide program. At the end of VS, nine compounds were prioritized to in vitro assays with *Sm*HDAC8. As a result, eight compounds showed inhibitory activity against *Sm*HDAC8 at low micromolar concentrations ($IC_{50} = 4.4–20.3$ μM). Among them, compound **6** (Fig. 4.8) showed the most promising results with $IC_{50} = 4.4\ \mu M$. Complementary ex vivo studies also showed that compound **6** kills 67% of the infective larval stage (schistosomula) of *S. mansoni*, after 3 days at a 100 μM concentration, and induced dose-dependent apoptosis in the larvae.

Staroń et al. (2020) developed an integrated VS approach (2D pharmacophore-based fingerprints, structural fingerprints filters, 3D pharmacophore models, and IFD) to screen potential ligands of subtype 6 of 5-hydroxytryptamine receptor ($5\text{-}HT_6R$) from Princeton BM and ChemBridge databases (Staroń et al., 2020). As a result, 92 virtual hits were prioritized for experimental evaluation. From active hits confirmed in vitro, compound **7** ($K_i = 0.145\ \mu M$, Fig. 4.8) was selected for further structural optimization. A series of 43 derivative compounds were synthesized and tested against $5\text{-}HT_6R$, $5\text{-}HT_{2a}R$, $5\text{-}HT_{1A}R$, $5\text{-}HT_7R$, and dopaminergic D_2R. The most promising compounds were optimized, supported by SAR studies. The IFD analysis showed that the designed compounds assume an atypical binding mode compared with the binding modes of previously published $5\text{-}HT_6R$ ligands. The analogue compound **8** (Fig. 4.8) was the most promising compound, showing favorable properties in terms of target affinity and selectivity ($5\text{-}HT_6R$ $K_i = 0.025\ \mu M$ and $5\text{-}HT_{2a}R = 0.032\ \mu M$) and ADMET profile (Staroń et al., 2020).

Agarwal et al. (2020) used MD simulations, ensemble docking, and IFD to VS new compounds against the wall-anchored serine-rich repeat (SRR) protein, the main mediator of the adhesion of *Streptococcus gordonii* and *Streptococcus sanguinis* in sialoglycans (Agarwal et al., 2020). The glycoprotein GspB and human serum albumin (Has) can adhere to various glycans from human platelets (Sullam & Sande, 1992). SRR adhesion proteins (Hsa_{BR}, $GspB_{BR}$, $10,712_{BR}$, $SK150_{BR}$, $SrpA_{BR}$, $SK678_{BR}$) were simulated through MD, and four representative structures were obtained for ensemble docking. The Vanderbilt small molecule database was screened against Hsa_{BR}, $GspB_{BR}$, 10712_{BR}, $SK150_{BR}$, $SrpA_{BR}$, and $SK678_{BR}$ using VinaMPI (Ellingson et al., 2013). The top 1% docking poses for Hsa_{BR}, but not within the top 1% of the other five BRs ($GspB_{BR}$, 10712_{BR}, $SK150_{BR}$, $SrpA_{BR}$, $SK678_{BR}$), to remove the sialic acid mimetics and promiscuous compounds, were experimentally tested. Approximately 250 compounds were then docked via IFD in molecular operating environment docking program (Vilar et al., 2008). Through consensus scoring, the top 1% (50 compounds) present in all the scoring were experimentally validated. Nine compounds showed activity in AlphaScreen assay, which measure ligand displacement. The compounds **9**, **10**, and **11** (Fig. 4.8) had the best results in decreasing the signal in AlphaScreen assays with Hsa_{BR}, 23%, 35%, and 41%, respectively. Integrating ensemble, IFD, and consensus docking approaches, there was a hit rate of ~20%.

7 CONCLUDING REMARKS AND FUTURE DIRECTIONS

DBVS has found its place in modern drug discovery and is widely applied with many success cases by both pharmaceutical companies and academic groups. In this chapter, we emphasized that the recent advances in scoring functions, search algorithms, consensus scoring, protein flexibility, and enrichment represent a new era of docking approaches. This chapter also highlighted the importance of performance evaluation of docking protocol to distinguish between actives and inactives, using several metrics from classic enrichment descriptors to advanced ones, as well as to compare if some methods and score functions achieve better results than others and in what conditions some metrics are more appropriate than others.

Overall, we highlight the relevance of integrating computational approaches, including docking and machine learning, with experimental techniques to accelerate drug discovery. We proposed a typical DBVS scheme, integrating all the best practices outlined here (see Fig. 4.2), and demonstrated a potential guideline to integrate such approaches. Finally, additional molecular properties, such as ADME and toxicological properties, should be considered in initial stages of drug discovery projects and, therefore, be included as filters in DBVS to help the decision-making on selecting which compounds should be experimentally evaluated.

REFERENCES

Abagyan, R., Rueda, M., & Bottegoni, G. (2010). Recipes for the selection of experimental protein conformations for virtual screening. *Journal of Chemical Information and Modeling, 50*(1), 186–193.

Agarwal, R., Bensing, B. A., Mi, D., Vinson, P. N., Baudry, J., Iverson, T. M., & Smith, J. C. (2020). Structure based virtual screening identifies novel competitive inhibitors for the sialoglycan binding protein Hsa. *Biochemical Journal, 477*(19), 3695–3707.

Ain, Q. U., Aleksandrova, A., Roessler, F. D., & Ballester, P. J. (2015). Machine-learning scoring functions to improve structure-based binding affinity prediction and virtual screening. *Wiley Interdisciplinary Reviews: Computational Molecular Science, 5*(6), 405–424.

Ajay. (1994). On better generalization by combining two or more models: A quantitative structure—activity relationship example using neural networks. *Chemometrics and Intelligent Laboratory Systems, 24*(1), 19–30.

Alonso, H., Bliznyuk, A. A., & Gready, J. E. (2006). Combining docking and molecular dynamic simulations in drug design. *Medicinal Research Reviews, 26*(5), 531–568.

Amaro, R. E., Baudry, J., Chodera, J., Demir, Ö., McCammon, J. A., Miao, Y., & Smith, J. C. (2018). Ensemble docking in drug discovery. *Biophysical Journal, 114*(10), 2271–2278.

Ashtawy, H. M., & Mahapatra, N. R. (2015). BgN-Score and BsN-Score: Bagging and boosting based ensemble neural networks scoring functions for accurate binding affinity prediction of protein-ligand complexes. *BMC Bioinformatics, 16*(S4), S8.

Baell, J. B., & Holloway, G. a (2010). New substructure filters for removal of pan assay interference compounds (PAINS) from screening libraries and for their exclusion in bioassays. *Journal of Medicinal Chemistry, 53*(7), 2719–2740.

Ballester, P. J., & Mitchell, J. B. O. (2010). A machine learning approach to predicting protein–ligand binding affinity with applications to molecular docking. *Bioinformatics, 26*(9), 1169–1175.

Ballester, P. J., Schreyer, A., & Blundell, T. L. (2014). Does a more precise chemical description of protein–ligand complexes lead to more accurate prediction of binding affinity? *Journal of Chemical Information and Modeling, 54*(3), 944–955.

Barelier, S., Eidam, O., Fish, I., Hollander, J., Figaroa, F., Nachane, R., … Siegal, G. (2014). Increasing chemical space coverage by combining empirical and computational fragment screens. *ACS Chemical Biology, 9*(7), 1528–1535.

Bax, B., Chung, C. W., & Edge, C. (2017). Getting the chemistry right: Protonation, tautomers and the importance of H atoms in biological chemistry. *Acta Crystallographica Section D: Structural Biology, 73*(2), 131–140.

Bell, E. W., & Zhang, Y. (2019). DockRMSD: An open-source tool for atom mapping and RMSD calculation of symmetric molecules through graph isomorphism. *Journal of Cheminformatics, 11*(1), 40.

Berry, M., Fielding, B., & Gamieldien, J. (2015). Practical considerations in virtual screening and molecular docking. In M. Kaufmann (Ed.), *Emerging trends in computational biology, bioinformatics, and systems biology* (1st ed., pp. 487–502). Elsevier Inc.

Bertrand, G. M. E. (1997). Electrostatics and hydrogen bonding. *Advances in Molecular and Cell Biology, 22*(C), 109–132.

Bhattarai, A., Wang, J., & Miao, Y. (2020). Retrospective ensemble docking of allosteric modulators in an adenosine G-protein-coupled receptor. *Biochimica et Biophysica Acta (BBA) - General Subjects, 1864*(8), 129615.

Bian, Y., & Xie, X.-Q. (2018). Computational fragment-based drug design: Current trends, strategies, and applications. *The AAPS Journal, 20*(3), 59.

Bissantz, C., Folkers, G., & Rognan, D. (2000). Protein-based virtual screening of chemical databases. 1. Evaluation of different docking/scoring combinations. *Journal of Medicinal Chemistry, 43*(25), 4759–4767.

Blakeley, M. P., Hasnain, S. S., & Antonyuk, S. V. (2015). Sub-atomic resolution X-ray crystallography and neutron crystallography: Promise, challenges and potential. *IUCrJ, 2*(4), 464–474.

Braga, R. C., Alves, V. M., Silva, A. C., Nascimento, M. N., Silva, F. C., Liao, L. M., & Andrade, C. H. (2014). Virtual screening strategies in medicinal chemistry: The state of the art and current challenges. *Current Topics in Medicinal Chemistry, 14*(16), 1899–1912.

Braga, R. C., & Andrade, C. H. (2013). Assessing the performance of 3D pharmacophore models in virtual screening: How good are they? *Current Topics in Medicinal Chemistry, 13*(9), 1127–1138.

Breiman, L. (2001). Random forests. *Machine Learning, 45*, 5–32.

Brink, T. T., & Exner, T. E. (2009). Influence of protonation, tautomeric, and stereoisomeric states on protein-ligand docking results. *Journal of Chemical Information and Modeling, 49*(6), 1535–1546.

Brink, T., & Exner, T. E. (2010). pK(a) based protonation states and microspecies for protein-ligand docking. *Journal of Computer-Aided Molecular Design, 24*(11), 935–942.

Cang, Z., Mu, L., & Wei, G.-W. (2018). Representability of algebraic topology for biomolecules in machine learning based scoring and virtual screening. Peng, J. (Ed.) *PLoS Computational Biology, 14*(1), e1005929.

Cang, Z., & Wei, G.-W. (2017). TopologyNet: Topology based deep convolutional and multi-task neural networks for biomolecular property predictions. Dunbrack, R. L. (Ed.) *PLoS Computational Biology, 13*(7), e1005690.

Carlson, H. A. (2002). Protein flexibility and drug design: How to hit a moving target. *Current Opinion in Chemical Biology, 6*(4), 447–452.

Carta, G., Knox, A. J. S., & Lloyd, D. G. (2007). Unbiasing scoring functions: A new normalization and rescoring strategy. *Journal of Chemical Information and Modeling, 47*(4), 1564–1571.

Chandak, T., Mayginnes, J. P., Mayes, H., & Wong, C. F. (2020). Using machine learning to improve ensemble docking for drug discovery. *Proteins: Structure, Function and Bioinformatics, 88*(10), 1263–1270.

Charifson, P. S., Corkery, J. J., Murcko, M. A., & Walters, W. P. (1999). Consensus scoring: A method for obtaining improved hit rates from docking databases of three-dimensional structures into proteins. *Journal of Medicinal Chemistry, 42*(25), 5100–5109.

Chen, H., Engkvist, O., Wang, Y., Olivecrona, M., & Blaschke, T. (2018). The rise of deep learning in drug discovery. *Drug Discovery Today, 23*(6), 1241–1250.

Cheng, T., Li, X., Li, Y., Liu, Z., & Wang, R. (2009). Comparative assessment of scoring functions on a diverse test set. *Journal of Chemical Information and Modeling, 49*(4), 1079–1093.

Cheng, T., Li, Q., Zhou, Z., Wang, Y., & Bryant, S. H. (2012). Structure-based virtual screening for drug discovery: A problem-centric review. *The AAPS Journal, 14*(1), 133–141.

Chen, Y., & Shoichet, B. K. (2009). Molecular docking and ligand specificity in fragment-based inhibitor discovery. *Nature Chemical Biology, 5*(5), 358–364.

Chevillard, F., Rimmer, H., Betti, C., Pardon, E., Ballet, S., van Hilten, N., … Kolb, P. (2018). Binding-site compatible fragment growing applied to the design of β_2-adrenergic receptor ligands. *Journal of Medicinal Chemistry, 61*(3), 1118–1129.

Clark, R. D., Strizhev, A., Leonard, J. M., Blake, J. F., & Matthew, J. B. (2002). Consensus scoring for ligand/protein interactions. *Journal of Molecular Graphics and Modelling, 20*(4), 281–295.

Corbeil, C. R., Williams, C. I., & Labute, P. (2012). Variability in docking success rates due to dataset preparation. *Journal of Computer-Aided Molecular Design, 26*(6), 775–786.

Coupez, B., & Lewis, R. A. (2006). Docking and scoring—theoretically easy, practically impossible? *Current Medicinal Chemistry, 13*(25), 2995–3003.

Craig, I. R., Essex, J. W., & Spiegel, K. (2010). Ensemble docking into multiple crystallographically derived protein structures: An evaluation based on the statistical analysis of enrichments. *Journal of Chemical Information and Modeling, 50*(4), 511–524.

Damm, K. L., & Carlson, H. A. (2007). Exploring experimental sources of multiple protein conformations in structure-based drug design. *Journal of the American Chemical Society, 129*(26), 8225–8235.

De Paris, R., Vahl Quevedo, C., Ruiz, D. D., Gargano, F., & de Souza, O. N. (2018). A selective method for optimizing ensemble docking-based experiments on an InhA Fully-Flexible receptor model. *BMC Bioinformatics, 19*(1), 235.

de Souza Neto, L. R., Moreira-Filho, J. T., Neves, B. J., Maidana, R. L. B. R., Guimarães, A. C. R., Furnham, N., … Silva-Junior, F. P. (2020). In silico strategies to support fragment-to-lead optimization in drug discovery. *Frontiers in Chemistry, 8*, 93.

Ekins, S., Puhl, A. C., Zorn, K. M., Lane, T. R., Russo, D. P., Klein, J. J., … Clark, A. M. (2019). Exploiting machine learning for end-to-end drug discovery and development. *Nature Materials, 18*(5), 435–441.

Eldridge, M. D., Murray, C. W., Auton, T. R., Paolini, G. V., & Mee, R. P. (1997). Empirical scoring functions: I. The development of a fast empirical scoring function to estimate the binding affinity of ligands in receptor complexes. *Journal of Computer-Aided Molecular Design, 11*, 425–445.

Ellingson, S. R., Smith, J. C., & Baudry, J. (2013). VinaMPI: Facilitating multiple receptor high-throughput virtual docking on high-performance computers. *Journal of Computational Chemistry, 34*(25), 2212–2221.

Elokely, K. M., & Doerksen, R. J. (2013). Docking challenge: Protein sampling and molecular docking performance. *Journal of Chemical Information and Modeling, 53*(8), 1934–1945.

Erlanson, D. A., Davis, B. J., & Jahnke, W. (2019). Fragment-based drug discovery: Advancing fragments in the absence of crystal structures. *Cell Chemical Biology, 26*(1), 9–15.

Ewing, T. J., Makino, S., Skillman, A. G., & Kuntz, I. D. (2001). DOCK 4.0: Search strategies for automated molecular docking of flexible molecule databases. *Journal of Computer-Aided Molecular Design, 15*(5), 411–428.

Falcon, W. E., Ellingson, S. R., Smith, J. C., & Baudry, J. (2019). Ensemble docking in drug discovery: How many protein configurations from molecular dynamics simulations are needed to reproduce known ligand binding? *The Journal of Physical Chemistry B, 123*(25), 5189–5195.

Fan, N., Bauer, C. A., Stork, C., Bruyn Kops, C., & Kirchmair, J. (2020). ALADDIN: Docking approach augmented by machine learning for protein structure selection yields superior virtual screening performance. *Molecular Informatics, 39*(4), 1900103.

Fassio, A. V., Santos, L. H., Silveira, S. A., Ferreira, R. S., & de Melo-Minardi, R. C. (2019). nAPOLI: a graph-based strategy to detect and visualize conserved protein-ligand interactions in large-scale. *IEEE/ACM Transactions on Computational Biology and Bioinformatics, 1–1*.

Fawcett, T. (2006). An introduction to ROC analysis. *Pattern Recognition Letters, 27*(8), 861–874.

Feher, M. (2006). Consensus scoring for protein-ligand interactions. *Drug Discovery Today, 11*(9–10), 421–428.

Ferrari, A. M., Wei, B. Q., Costantino, L., & Shoichet, B. K. (2004). Soft docking and multiple receptor conformations in virtual screening. *Journal of Medicinal Chemistry, 47*(21), 5076–5084.

Fisher, S. J., Blakeley, M. P., Cianci, M., McSweeney, S., & Helliwell, J. R. (2012). Protonation-state determination in proteins using high-resolution X-ray crystallography: Effects of resolution and completeness. *Acta Crystallographica Section D Biological Crystallography, 68*(7), 800–809.

Freitas, R. F., & Schapira, M. (2017). A systematic analysis of atomic protein-ligand interactions in the PDB. *MedChemComm, 8*(10), 1970–1981.

Friedman, J. H. (2001). Greedy function approximation: A gradient boosting machine. *The Annals of Statistics, 29*(5), 1189–1232.

Friesner, R. A., Banks, J. L., Murphy, R. B., Halgren, T. A., Klicic, J. J., Mainz, D. T., … Shenkin, P. S. (2004). Glide: A new approach for rapid, accurate docking and scoring. 1. Method and assessment of docking accuracy. *Journal of Medicinal Chemistry, 47*(7), 1739–1749.

Friesner, R. A., Murphy, R. B., Repasky, M. P., Frye, L. L., Greenwood, J. R., Halgren, T. A., … Mainz, D. T. (2006). Extra precision Glide: Docking and scoring incorporating a model of hydrophobic enclosure for protein—ligand complexes. *Journal of Medicinal Chemistry, 49*(21), 6177–6196.

Gagic, Z., Ruzic, D., Djokovic, N., Djikic, T., & Nikolic, K. (2020). In silico methods for design of kinase inhibitors as anticancer drugs. *Frontiers in Chemistry, 7*(January), 1–25.

Garcia-Moreno, B. (2009). Adaptations of proteins to cellular and subcellular pH. *Journal of Biology, 8*(11), 98.

Gaulton, A., Bellis, L. J., Bento, A. P., Chambers, J., Davies, M., Hersey, A., … Overington, J. P. (2012). ChEMBL: A large-scale bioactivity database for drug discovery. *Nucleic Acids Research, 40*(Database issue), D1100–D1107.

Gehlhaar, D. K., Verkhivker, G. M., Rejto, P. A., Sherman, C. J., Fogel, D. B., Fogel, L. J., & Freer, S. T. (1995). Molecular recognition of the inhibitor AG-1343 by HIV-1 protease: Conformationally flexible docking by evolutionary programming. *Chemistry and Biology, 2*(5), 317–324.

Genheden, S., & Ryde, U. (2015). The MM/PBSA and MM/GBSA methods to estimate ligand-binding affinities. *Expert Opinion on Drug Discovery, 10*(5), 449–461.

Ginn, C. M. R., Turner, D. B., Willett, P., Ferguson, A. M., & Heritage, T. W. (1997). Similarity searching in files of three-dimensional chemical structures: Evaluation of the EVA descriptor and combination of rankings using data fusion. *Journal of Chemical Information and Computer Sciences, 37*(1), 23–37.

Gohlke, H., Hendlich, M., & Klebe, G. (2000). Knowledge-based scoring function to predict protein-ligand interactions. *Journal of Molecular Biology, 295*(2), 337–356.

Good, A. C., & Oprea, T. I. (2008). Optimization of CAMD techniques 3. Virtual screening enrichment studies: A help or hindrance in tool selection? *Journal of Computer-Aided Molecular Design, 22*(3–4), 169–178.

Grove, L. E., Vajda, S., & Kozakov, D. (2016). Computational methods to support fragment-based drug discovery. In D. A. Erlanson, & W. Jahnke (Eds.), *Fragment-based drug discovery: Lessons and outlook* (1st ed., pp. 197–222). Weinheim, Germany: Wiley-VCH Verlag GmbH & Co. KGaA.

Guedes, I. A., de Magalhães, C. S., & Dardenne, L. E. (2014). Receptor–ligand molecular docking. *Biophysical Reviews, 6*(1), 75–87.

Halgren, T. A., Murphy, R. B., Friesner, R. A., Beard, H. S., Frye, L. L., Pollard, W. T., & Banks, J. L. (2004). Glide: A new approach for rapid, accurate docking and scoring. 2. Enrichment factors in database screening. *Journal of Medicinal Chemistry, 47*(7), 1750–1759.

Hall, R. J., Mortenson, P. N., & Murray, C. W. (2014). Efficient exploration of chemical space by fragment-based screening. *Progress in Biophysics and Molecular Biology, 116*(2–3), 82–91.

Hopkins, A. L., Keserü, G. M., Leeson, P. D., Rees, D. C., & Reynolds, C. H. (2014). The role of ligand efficiency metrics in drug discovery. *Nature Reviews. Drug Discovery, 13*(2), 105–121.

Hornak, V., Okur, A., Rizzo, R. C., & Simmerling, C. (2006). HIV-1 protease flaps spontaneously close to the correct structure in simulations following manual placement of an inhibitor into the open state. *Journal of the American Chemical Society, 128*(9), 2812–2813.

Horvath, D. (1997). A virtual screening approach applied to the search for trypanothione reductase inhibitors. *Journal of Medicinal Chemistry, 40*(15), 2412–2423.

Houston, D. R., & Walkinshaw, M. D. (2013). Consensus docking: Improving the reliability of docking in a virtual screening context. *Journal of Chemical Information and Modeling, 53*(2), 384–390.

Hritz, J., De Ruiter, A., & Oostenbrink, C. (2008). Impact of plasticity and flexibility on docking results for cytochrome P450 2D6: A combined approach of molecular dynamics and ligand docking. *Journal of Medicinal Chemistry, 51*(23), 7469–7477.

Huang, N., & Shoichet, B. K. (2008). Exploiting ordered waters in molecular docking. *Journal of Medicinal Chemistry, 51*(16), 4862–4865.

Huang, N., Shoichet, B. K., & Irwin, J. J. (2006). Benchmarking sets for molecular docking. *Journal of Medicinal Chemistry, 49*(23), 6789–6801.

Huggins, D. J., & Tidor, B. (2011). Systematic placement of structural water molecules for improved scoring of protein-ligand interactions. *Protein Engineering Design and Selection, 24*(10), 777–789.

Ibrahim, T. M., Bauer, M. R., & Boeckler, F. M. (2015). Applying DEKOIS 2.0 in structure-based virtual screening to probe the impact of preparation procedures and score normalization. *Journal of Cheminformatics, 7*(1), 21.

Jasial, S., Gilberg, E., Blaschke, T., & Bajorath, J. (2018). Machine learning distinguishes with high accuracy between pan-assay interference compounds that are promiscuous or represent dark chemical matter. *Journal of Medicinal Chemistry, 61*(22), 10255–10264.

Jiménez, J., Škalič, M., Martínez-Rosell, G., & De Fabritiis, G. (2018). K DEEP : Protein–ligand absolute binding affinity prediction via 3D-convolutional neural networks. *Journal of Chemical Information and Modeling, 58*(2), 287–296.

Jones, G., Willett, P., Glen, R. C., Leach, A. R., & Taylor, R. (1997). Development and validation of a genetic algorithm for flexible docking. *Journal of Molecular Biology, 267*(3), 727–748.

Kawatkar, S., Wang, H., Czerminski, R., & Joseph-McCarthy, D. (2009). Virtual fragment screening: An exploration of various docking and scoring protocols for fragments using Glide. *Journal of Computer-Aided Molecular Design, 23*(8), 527–539.

Kenny, P. W. (2019). The nature of ligand efficiency. *Journal of Cheminformatics, 11*(1), 8.

Keseru, G. M., Erlanson, D. A., Ferenczy, G. G., Hann, M. M., Murray, C. W., & Pickett, S. D. (2016). Design principles for fragment libraries: Maximizing the value of learnings from pharma fragment-based drug discovery (FBDD) programs for use in academia. *Journal of Medicinal Chemistry, 59*(18), 8189–8206.

Khamis, M. A., Gomaa, W., & Ahmed, W. F. (2015). Machine learning in computational docking. *Artificial Intelligence in Medicine, 63*(3), 135–152.

Kirchmair, J., Markt, P., Distinto, S., Wolber, G., & Langer, T. (2008). Evaluation of the performance of 3D virtual screening protocols: RMSD comparisons, enrichment assessments, and decoy selection—what can we learn from earlier mistakes? *Journal of Computer-Aided Molecular Design, 22*(3–4), 213–228.

Kitchen, D. B., Decornez, H., Furr, J. R., & Bajorath, J. (2004). Docking and scoring in virtual screening for drug discovery: Methods and applications. *Nature Reviews Drug Discovery, 3*(11), 935–949.

Knegtel, R. M., Kuntz, I. D., & Oshiro, C. (1997). Molecular docking to ensembles of protein structures 1 1Edited by B. Honig. *Journal of Molecular Biology, 266*(2), 424–440.

Kollman, P. A., Massova, I., Reyes, C., Kuhn, B., Huo, S., Chong, L., … Cheatham, T. E. (2000). Calculating structures and free energies of complex molecules: Combining molecular mechanics and continuum models. *Accounts of Chemical Research, 33*(12), 889–897.

Korb, O., Olsson, T. S. G., Bowden, S. J., Hall, R. J., Verdonk, M. L., Liebeschuetz, J. W., & Cole, J. C. (2012). Potential and limitations of ensemble docking. *Journal of Chemical Information and Modeling, 52*(5), 1262–1274.

Krammer, A., Kirchhoff, P. D., Jiang, X., Venkatachalam, C. M., & Waldman, M. (2005). LigScore: A novel scoring function for predicting binding affinities. *Journal of Molecular Graphics and Modelling, 23*(5), 395−407.

Kroemer, R. T. (2007). Structure-based drug design: Docking and scoring. *Current Protein and Peptide Science, 8*(4), 312−328.

Kuhn, B., Gerber, P., Schulz-Gasch, T., & Stahl, M. (2005). Validation and use of the MM-PBSA approach for drug discovery. *Journal of Medicinal Chemistry, 48*(12), 4040−4048.

Kukić, P., & Nielsen, J. E. (2010). Electrostatics in proteins and protein−ligand complexes. *Future Medicinal Chemistry, 2*(4), 647−666.

Kukol, A. (2011). Consensus virtual screening approaches to predict protein ligands. *European Journal of Medicinal Chemistry, 46*(9), 4661−4664.

Kumar, a., Voet, A., & Zhang, K. Y. (2012). Fragment based drug design: From experimental to computational approaches. *Current Medicinal Chemistry, 19*(30), 5128−5147.

Kundrotas, P. J., & Alexov, E. (2006). Electrostatic properties of protein-protein complexes. *Biophysical Journal, 91*(5), 1724−1736.

Kuntz, I. D., Blaney, J. M., Oatley, S. J., Langridge, R., & Ferrin, T. E. (1982). A geometric approach to macromolecule-ligand interactions. *Journal of Molecular Biology, 161*(2), 269−288.

Lie, M. A., Thomsen, R., Pedersen, C. N. S., Schiøtt, B., & Christensen, M. H. (2011). Molecular docking with ligand attached water molecules. *Journal of Chemical Information and Modeling, 51*(4), 909−917.

Li, Y., Han, L., Liu, Z., & Wang, R. (2014). Comparative assessment of scoring functions on an updated benchmark: 2. Evaluation methods and general results. *Journal of Chemical Information and Modeling, 54*(6), 1717−1736.

Li, H., Leung, K.-S., Wong, M.-H., & Ballester, P. J. (2015). Improving AutoDock Vina using random forest: The growing accuracy of binding affinity prediction by the effective exploitation of larger data sets. *Molecular Informatics, 34*(2−3), 115−126.

Lionta, E., Spyrou, G., Vassilatis, D. K., & Cournia, Z. (2014). Structure-based virtual screening for drug discovery: Principles, applications and recent advances. *Current Topics in Medicinal Chemistry, 14*(16), 1923−1938.

Lipinski, C. A., Lombardo, F., Dominy, B. W., & Feeney, P. J. (2001). Experimental and computational approaches to estimate solubility and permeability in drug discovery and development settings. *Advanced Drug Delivery Reviews, 64*(1−3), 3−26.

Liu, T., Lin, Y., Wen, X., Jorissen, R. N., & Gilson, M. K. (2007). BindingDB: A web-accessible database of experimentally determined protein-ligand binding affinities. *Nucleic Acids Research, 35*(Database), D198−D201.

Livingstone, D. (1995). *Data analysis for chemists. Applications to QSAR and chemical product design.* New York: Oxford University Press.

Loving, K., Salam, N. K., & Sherman, W. (2009). Energetic analysis of fragment docking and application to structure-based pharmacophore hypothesis generation. *Journal of Computer-Aided Molecular Design, 23*(8), 541−554.

Lu, Y., Wang, R., Yang, C.-Y., & Wang, S. (2007). Analysis of ligand-bound water molecules in high-resolution crystal structures of protein−ligand complexes. *Journal of Chemical Information and Modeling, 47*(2), 668−675.

Lyu, J., Wang, S., Balius, T. E., Singh, I., Levit, A., Moroz, Y. S., … Irwin, J. J. (2019). Ultra-large library docking for discovering new chemotypes. *Nature, 566*(7743), 224−229.

Mackey, M. D., & Melville, J. L. (2009). Better than random? The chemotype enrichment problem. *Journal of Chemical Information and Modeling, 49*(5), 1154−1162.

Magalhães, C. S., Almeida, D. M., Barbosa, H. J. C., & Dardenne, L. E. (2014). A dynamic niching genetic algorithm strategy for docking highly flexible ligands. *Information Sciences, 289*, 206−224.

Marchand, J. R., & Caflisch, A. (2018). In silico fragment-based drug design with SEED. *European Journal of Medicinal Chemistry, 156*, 907−917.

McClish, D. K. (1989). Analyzing a portion of the ROC curve. *Medical Decision Making, 9*(3), 190−195.

McGann, M. (2011). FRED pose prediction and virtual screening accuracy. *Journal of Chemical Information and Modeling, 51*(3), 578−596.

McGann, M. R., Almond, H. R., Nicholls, A., Grant, J. A., & Brown, F. K. (2003). Gaussian docking functions. *Biopolymers, 68*(1), 76−90.

Meiler, J., & Baker, D. (2006). ROSETTALIGAND: Protein-small molecule docking with full side-chain flexibility. *Proteins: Structure, Function, and Bioinformatics, 65*(3), 538−548.

Milletti, F., Storchi, L., Sforna, G., Cross, S., & Cruciani, G. (2009). Tautomer enumeration and stability prediction for virtual screening on large chemical databases. *Journal of Chemical Information and Modeling, 49*(1), 68−75.

Mitra, R., Shyam, R., Mitra, I., Miteva, M., & Alexov, E. (2008). Calculating the protonation states of proteins and small molecules: Implications to ligand-receptor interactions. *Current Computer-Aided Drug Design, 4*(3), 169−179.

Mitra, R. C., Zhang, Z., & Alexov, E. (2011). In silico modeling of pH-optimum of protein-protein binding. *Proteins: Structure, Function, and Bioinformatics, 79*(3), 925−936.

Morris, G. M., & Lim-Wilby, M. (2008). Molecular docking. In A. Kukol (Ed.), *Molecular modeling of proteins* (443rd ed., pp. 365−382). Humana Press.

Mysinger, M. M., Carchia, M., Irwin, J. J., & Shoichet, B. K. (2012). Directory of useful decoys, enhanced (DUD-E): Better ligands and decoys for better benchmarking. *Journal of Medicinal Chemistry, 55*(14), 6582−6594.

Nabuurs, S. B., Wagener, M., & de Vlieg, J. (2007). A flexible approach to induced fit docking. *Journal of Medicinal Chemistry, 50*(26), 6507−6518.

Neves, B. J., Braga, R. C., Melo-Filho, C. C., Moreira-Filho, J. T., Muratov, E. N., & Andrade, C. H. (2018). QSAR-based virtual screening: Advances and applications in drug discovery. *Frontiers in Pharmacology, 9*(November), 1−7.

Neves, B. J., Muratov, E., Machado, R. B., Andrade, C. H., & Cravo, P. V. L. (2016). Modern approaches to accelerate discovery of new antischistosomal drugs. *Expert Opinion on Drug Discovery, 11*(6), 557–567.

Nguyen, D. D., & Wei, G.-W. (2019). DG-GL: Differential geometry-based geometric learning of molecular datasets. *International Journal for Numerical Methods in Biomedical Engineering, 35*(3), e3179.

Nguyen, D. D., Xiao, T., Wang, M., & Wei, G.-W. (2017). Rigidity strengthening: A mechanism for protein–ligand binding. *Journal of Chemical Information and Modeling, 57*(7), 1715–1721.

Novoa, E. M., Pouplana, L. R. de, & Orozco, M. (2012). Small molecule docking from theoretical structural models. In N. V. Dokholyan (Ed.), *Computational modeling of biological systems, biological and medical physics, biomedical engineering* (pp. 75–95). Boston, MA: Springer US.

Oleg, T., & Olson, A. J. (2010). AutoDock Vina: Improving the speed and accuracy of docking with a new scoring function, efficient optimization, and multithreading. *Journal of Computational Chemistry, 31*(2), 455–461.

Osterberg, F., Morris, G. M., Sanner, M. F., Olson, A. J., & Goodsell, D. S. (2002). Automated docking to multiple target structures: Incorporation of protein mobility and structural water heterogeneity in AutoDock. *Proteins, 46*(1), 34–40.

Ouyang, X., Handoko, S. D., & Kwoh, C. K. (2011). CScore: A simple yet effective scoring function for protein-ligand binding affinity prediction using modified CMAC learning architecture. *Journal of Bioinformatics and Computational Biology, 9*(Suppl. 1), 1–14.

Oyarzabal, J., Zarich, N., Albarran, M. I., Palacios, I., Urbano-Cuadrado, M., Mateos, G., … Bischoff, J. R. (2010). Discovery of mitogen-activated protein kinase-interacting kinase 1 inhibitors by a comprehensive fragment-oriented virtual screening approach. *Journal of Medicinal Chemistry, 53*(18), 6618–6628.

Palacio-Rodríguez, K., Lans, I., Cavasotto, C. N., & Cossio, P. (2019). Exponential consensus ranking improves the outcome in docking and receptor ensemble docking. *Scientific Reports, 9*(1), 5142.

Pan, Y., Huang, N., Cho, S., & MacKerell, A. D. (2003). Consideration of molecular weight during compound selection in virtual target-based database screening. *Journal of Chemical Information and Computer Sciences, 43*(1), 267–272.

Pantsar, T., & Poso, A. (2018). Binding affinity via docking: Fact and fiction. *Molecules, 23*(8), 1899.

Park, I.-H., & Li, C. (2010). Dynamic ligand-induced-fit simulation via enhanced conformational samplings and ensemble dockings: A survivin example. *The Journal of Physical Chemistry B, 114*(15), 5144–5153.

Park, S. J., Kufareva, I., & Abagyan, R. (2010). Improved docking, screening and selectivity prediction for small molecule nuclear receptor modulators using conformational ensembles. *Journal of Computer-Aided Molecular Design, 24*(5), 459–471.

Paulsen, J. L., & Anderson, A. C. (2009). Scoring ensembles of docked protein:Ligand interactions for virtual lead

optimization. *Journal of Chemical Information and Modeling, 49*(11), 2813–2819.

Pearlman, D. A., & Charifson, P. S. (2001). Are free energy calculations useful in practice? A comparison with rapid scoring functions for the p38 MAP kinase protein system. *Journal of Medicinal Chemistry, 44*(21), 3417–3423.

Petukh, M., Stefl, S., & Alexov, E. (2013). The role of protonation states in ligand-receptor recognition and binding. *Current Pharmaceutical Design, 19*(23), 4182–4190.

Pinzi, L., & Rastelli, G. (2019). Molecular docking: Shifting paradigms in drug discovery. *International Journal of Molecular Sciences, 20*(18), 4331.

Pu, C., Yan, G., Shi, J., & Li, R. (2017). Assessing the performance of docking scoring function, FEP, MM-GBSA, and QM/MM-GBSA approaches on a series of PLK1 inhibitors. *MedChemComm, 8*(7), 1452–1458.

Rao, S., Sanschagrin, P. C., Greenwood, J. R., Repasky, M. P., Sherman, W., & Farid, R. (2008). Improving database enrichment through ensemble docking. *Journal of Computer-Aided Molecular Design, 22*(9), 621–627.

Rarey, M., Kramer, B., Lengauer, T., & Klebe, G. (1996). A fast flexible docking method using an incremental construction algorithm. *Journal of Molecular Biology, 261*(3), 470–489.

Rastelli, G., & Pinzi, L. (2019). Refinement and rescoring of virtual screening results. *Frontiers in Chemistry, 7*, 498.

Real database: The largest enumerated database of synthetically feasible molecules. (2020). *Enamine Inc.* Retrieved July 7, 2020, from https://enamine.net/.

Réau, M., Langenfeld, F., Zagury, J.-F., Lagarde, N., & Montes, M. (2018). Decoys selection in benchmarking datasets: Overview and perspectives. *Frontiers in Pharmacology, 9*, 11.

Reynolds, C. H., Bembenek, S. D., & Tounge, B. A. (2007). The role of molecular size in ligand efficiency. *Bioorganic and Medicinal Chemistry Letters, 17*(15), 4258–4261.

Rognan, D. (2011). Fragment-based approaches and computer-aided drug discovery. In T. G. Davies, & M. Hyvönen (Eds.), *Fragment-based drug discovery and X-ray crystallography* (1st ed., Vol. 310, pp. 201–222). Berlin Heidelberg: Springer Berlin Heidelberg.

Rohrer, S. G., & Baumann, K. (2009). Maximum unbiased validation (MUV) data sets for virtual screening based on PubChem bioactivity data. *Journal of Chemical Information and Modeling, 49*(2), 169–184.

Rudling, A., Gustafsson, R., Almlöf, I., Homan, E., Scobie, M., Warpman Berglund, U., … Carlsson, J. (2017). Fragment-based discovery and optimization of enzyme inhibitors by docking of commercial chemical space. *Journal of Medicinal Chemistry, 60*(19), 8160–8169.

Santos, K. B., Guedes, I. A., Karl, A. L. M., & Dardenne, L. E. (2020). Highly flexible ligand docking: Benchmarking of the DockThor program on the LEADS-PEP protein–peptide data set. *Journal of Chemical Information and Modeling, 60*(2), 667–683.

Santos, R., Hritz, J., & Oostenbrink, C. (2010). Role of water in molecular docking simulations of cytochrome P450 2D6. *Journal of Chemical Information and Modeling, 50*(1), 146–154.

Sastry, G. M., Adzhigirey, M., Day, T., Annabhimoju, R., & Sherman, W. (2013). Protein and ligand preparation: Parameters, protocols, and influence on virtual screening enrichments. *Journal of Computer-Aided Molecular Design*, 27(3), 221–234.

Sándor, M., Kiss, R., & Keserű, G. M. (2010). Virtual fragment docking by Glide: A validation study on 190 Protein–Fragment complexes. *Journal of Chemical Information and Modeling*, 50(6), 1165–1172.

Shapovalov, M. V., & Dunbrack, R. L. (2011). A smoothed backbone-dependent rotamer library for proteins derived from adaptive Kernel density estimates and regressions. *Structure*, 19(6), 844–858.

Shen, C., Ding, J., Wang, Z., Cao, D., Ding, X., & Hou, T. (2020). From machine learning to deep learning: Advances in scoring functions for protein–ligand docking. *WIREs Computational Molecular Science*, 10(1).

Sheng, C., & Zhang, W. (2013). Fragment informatics and computational fragment-based drug design: An overview and update. *Medicinal Research Reviews*, 33(3), 554–598.

Sheridan, R. P., Singh, S. B., Fluder, E. M., & Kearsley, S. K. (2001). Protocols for bridging the peptide to nonpeptide gap in topological similarity searches. *Journal of Chemical Information and Computer Sciences*, 41(5), 1395–1406.

Sherman, W., Day, T., Jacobson, M. P., Friesner, R. A., & Farid, R. (2006). Novel procedure for modeling ligand/receptor induced fit effects. *Journal of Medicinal Chemistry*, 49(2), 534–553.

Simoben, C., Robaa, D., Chakrabarti, A., Schmidtkunz, K., Marek, M., Lancelot, J., … Sippl, W. (2018). A novel class of Schistosoma mansoni histone deacetylase 8 (HDAC8) inhibitors identified by structure-based virtual screening and in vitro testing. *Molecules*, 23(3), 566.

Smith, R. D., Dunbar, J. B., Ung, P. M.-U., Esposito, E. X., Yang, C.-Y., Wang, S., & Carlson, H. A. (2011). CSAR benchmark exercise of 2010: Combined evaluation across all submitted scoring functions. *Journal of Chemical Information and Modeling*, 51(9), 2115–2131.

Smith, R. D., Engdahl, A. L., Dunbar, J. B., & Carlson, H. A. (2012). Biophysical limits of protein-ligand binding. *Journal of Chemical Information and Modeling*, 52(8), 2098–2106.

So, S. S., & Karplus, M. (1996). Evolutionary optimization in quantitative structure-activity relationship: An application of genetic neural networks. *Journal of Medicinal Chemistry*, 39(7), 1521–1530.

Srinivasan, J., Cheatham, T. E., Cieplak, P., Kollman, P. A., & Case, D. A. (1998). Continuum solvent studies of the stability of DNA, RNA, and phosphoramidate-DNA helices. *Journal of the American Chemical Society*, 120(37), 9401–9409.

Staroń, J., Kurczab, R., Warszycki, D., Satała, G., Krawczyk, M., Bugno, R., … Bojarski, A. J. (2020). Virtual screening-driven discovery of dual 5-HT6/5-HT2A receptor ligands with procognitive properties. *European Journal of Medicinal Chemistry*, 185, 111857.

Stepniewska-Dziubinska, M. M., Zielenkiewicz, P., & Siedlecki, P. (2018). Development and evaluation of a deep learning model for protein–ligand binding affinity prediction. Valencia, A. (Ed.) *Bioinformatics*, 34(21), 3666–3674.

Sullam, P. M., & Sande, M. A. (1992). Role of platelets in endocarditis: Clues from von Willebrand disease. *The Journal of Laboratory and Clinical Medicine*, 120(4), 507–509.

Terp, G. E., Johansen, B. N., Christensen, I. T., & Jørgensen, F. S. (2001). A new concept for multidimensional selection of ligand conformations (MultiSelect) and multidimensional scoring (MultiScore) of protein–ligand binding affinities. *Journal of Medicinal Chemistry*, 44(14), 2333–2343.

Thompson, D. C., Humblet, C., & Joseph-McCarthy, D. (2008). Investigation of MM-PBSA rescoring of docking poses. *Journal of Chemical Information and Modeling*, 48(5), 1081–1091.

Torres, P. H. M., Sodero, A. C. R., Jofily, P., & Silva, F. P., Jr. (2019). Key topics in molecular docking for drug design. *International Journal of Molecular Sciences*, 20(18), 4574.

Truchon, J.-F. F., & Bayly, C. I. (2007). Evaluating virtual screening methods: Good and bad metrics for the "early recognition" problem. *Journal of Chemical Information and Modeling*, 47(2), 488–508.

Vass, M., Makara, G., & Keserű, G. (2015). Fragment-based methods in drug design. In C. N. Cavasotto (Ed.), *In Silico drug discovery and design* (1st ed., pp. 353–389). Boca Raton: CRC Press.

Veber, D. F., Johnson, S. R., Cheng, H. Y., Smith, B. R., Ward, K. W., & Kopple, K. D. (2002). Molecular properties that influence the oral bioavailability of drug candidates. *Journal of Medicinal Chemistry*, 45(12), 2615–2623.

Venkatachalam, C. M., Jiang, X., Oldfield, T., & Waldman, M. (2003). LigandFit: A novel method for the shape-directed rapid docking of ligands to protein active sites. *Journal of Molecular Graphics and Modelling*, 21(4), 289–307.

Verdonk, M. L., Berdini, V., Hartshorn, M. J., Mooij, W. T. M., Murray, C. W., Taylor, R. D., & Watson, P. (2004). Virtual screening using protein–ligand docking: Avoiding artificial enrichment. *Journal of Chemical Information and Computer Sciences*, 44(3), 793–806.

Verdonk, M. L., Cole, J. C., Hartshorn, M. J., Murray, C. W., & Taylor, R. D. (2003). Improved protein-ligand docking using GOLD. *Proteins*, 52(4), 609–623.

Verdonk, M. L., Giangreco, I., Hall, R. J., Korb, O., Mortenson, P. N., & Murray, C. W. (2011). Docking performance of fragments and druglike compounds. *Journal of Medicinal Chemistry*, 54(15), 5422–5431.

Vilar, S., Cozza, G., & Moro, S. (2008). Medicinal chemistry and the molecular operating environment (MOE): Application of QSAR and molecular docking to drug discovery. *Current Topics in Medicinal Chemistry*, 8(18), 1555–1572.

Vogel, S. M., Bauer, M. R., & Boeckler, F. M. (2011). DEKOIS: Demanding evaluation Kits for objective in silico screening — a versatile tool for benchmarking docking programs and scoring functions. *Journal of Chemical Information and Modeling*, 51(10), 2650–2665.

Vuorinen, A., & Schuster, D. (2015). Methods for generating and applying pharmacophore models as virtual screening filters and for bioactivity profiling. *Methods*, 71, 113–134.

Wang, R., Fang, X., Lu, Y., & Wang, S. (2004). The PDBbind database: Collection of binding affinities for Protein—Ligand complexes with known three-dimensional structures. *Journal of Medicinal Chemistry, 47*(12), 2977—2980.

Wang, R., Fang, X., Lu, Y., Yang, C.-Y., & Wang, S. (2005). The PDBbind database: Methodologies and updates. *Journal of Medicinal Chemistry, 48*(12), 4111—4119.

Wang, T., Wu, M.-B., Chen, Z.-J., Chen, H., Lin, J.-P., & Yang, L.-R. (2015). Fragment-based drug discovery and molecular docking in drug design. *Current Pharmaceutical Biotechnology, 16*(1), 11—25.

Wang, C., & Zhang, Y. (2017). Improving scoring-docking-screening powers of protein-ligand scoring functions using random forest. *Journal of Computational Chemistry, 38*(3), 169—177.

Wang, B., Zhao, Z., Nguyen, D. D., & Wei, G.-W. (2017). Feature functional theory—binding predictor (FFT—BP) for the blind prediction of binding free energies. *Theoretical Chemistry Accounts, 136*(4), 55.

Wang, Y., Xiao, J., Suzek, T. O., Zhang, J., Wang, J., Zhou, Z., … Bryant, S. H. (2012). PubChem's bioAssay database. *Nucleic Acids Research, 40*(Database issue), D400—D412.

Wong, C. F. (2019). Improving ensemble docking for drug discovery by machine learning. *Journal of Theoretical and Computational Chemistry, 18*(3), 1920001.

Yang, H., Bartlam, M., & Rao, Z. (2006). Drug design targeting the main protease, the achilles heel of coronaviruses. *Current Pharmaceutical Design, 12*(35), 4573—4590.

Yan, C., & Zou, X. (2015). MDock: An ensemble docking suite for molecular docking, scoring and in silico screening. *Methods in Pharmacology and Toxicology,* 153—166.

Yang, H., Chennamaneni, L. R., Ho, M. W. T., Ang, S. H., Tan, E. S. W., Jeyaraj, D. A., … Nacro, K. (2018). Optimization of selective mitogen-activated protein kinase interacting kinases 1 and 2 inhibitors for the treatment of blast crisis leukemia. *Journal of Medicinal Chemistry, 61*(10), 4348—4369.

Zhao, Y., & Sanner, M. F. (2008). Protein—ligand docking with multiple flexible side chains. *Journal of Computer-Aided Molecular Design, 22*(9), 673—679.

Zhong, H., Tran, L. M., & Stang, J. L. (2009). Induced-fit docking studies of the active and inactive states of protein tyrosine kinases. *Journal of Molecular Graphics and Modelling, 28*(4), 336—346.

Zhong, S., Zhang, Y., & Xiu, Z. (2010). Rescoring ligand docking poses. *Current Opinion in Drug Discovery and Development, 13*(3), 326—334.

Zhu, T., Cao, S., Su, P.-C., Patel, R., Shah, D., Chokshi, H. B., … Hevener, K. E. (2013). Hit identification and optimization in virtual screening: Practical recommendations based on a critical literature analysis. *Journal of Medicinal Chemistry, 56*(17), 6560—6572.

CHAPTER 5

Virtual Libraries for Docking Methods: Guidelines for the Selection and the Preparation

ASMA SELLAMI • MANON RÉAU • FLORENT LANGENFELD • NATHALIE LAGARDE[a] • MATTHIEU MONTES[a]

1 INTRODUCTION

Current drug discovery relies on massive screening of chemical libraries against various extracellular and intracellular molecular targets to identify novel chemotypes with the desired mode of action. Protein—ligand molecular docking methods are nowadays widely implemented in these drug discovery pipelines as an in silico analogy of experimental high-throughput screening of large databases of compounds (Kontoyianni, 2017).

The success of this approach in identifying new potent therapeutic compounds lies for many parts in the availability of accurate docking algorithms and scoring functions and of high-quality data on the structure of the macromolecular target. The choice of the compounds database to screen is as important as the choice of the parameters mentioned above. In fact, the quality of the input is key to ensure reliable docking results, and it is very unlikely that active compounds could be identified within poorly selected compound collections, despite the use of an ideal docking tool (Forli, 2015). Several studies have shown that the size (Lyu et al., 2019), the composition, and the preparation (Corbeil et al., 2012) of virtual databases impact the success of docking predictions.

In recent years, high-throughput technologies for combinatorial and multiparallel chemical synthesis, automation technologies for the isolation of natural products, and the availability of large compound collections from commercial sources have substantially increased the size and diversity of synthesizable compound collections available for virtual screening. However, navigating and cherry-picking compounds within this huge number of compounds (Williams et al., 2012) is a challenging task and no gold standard protocol has currently been defined (Gally et al., 2019).

This chapter aims at helping choose and construct virtual screening databases dedicated to docking methods. We will, in the first part, provide insights about the possible sources of compounds and how they can be combined to create a customized database adapted to a drug discovery campaign. In the second part, we will provide an extensive description of the steps required for compounds preparation.

2 DATA COLLECTION WITHIN THE VAST CHEMICAL SPACE

The chemical space is defined as a "comprehensive collection of all possible small molecules under some reasonable restrictions considering size and composition" (Vogt, 2020). Practically, all possibly existing compounds are mapped on a mathematical space and their positions are defined according to their properties (Arús-Pous et al., 2019; Awale et al., 2017). Even if thousands to billions of natural and synthetic compounds are listed in commercial or academic, bioactivity, and natural products databases, this represents only a very small coverage of the entire possible chemical space. Various teams are thus working toward the identification of virtual compounds belonging to areas of this chemical space not yet covered by commercially available compound collections (Kontijevskis, 2017). To do so, fragment libraries are used as a starting point for building new molecules. Known chemical reactions

---

[a]These authors contributed equally to this work.

Molecular Docking for Computer-Aided Drug Design. https://doi.org/10.1016/B978-0-12-822312-3.00017-5

are applied on known building blocks and are used to generate synthetically feasible compounds or all possible molecules according to these chemical rules.

In this section, we will present different sources for collecting compounds to build virtual screening libraries. It is to note that the purpose here is not to provide a complete enumeration of all available compound libraries. The databases will be divided into actual compound collections, i.e., physically available molecules representing the total output of the pharmaceutical field (both academic and commercial collections) and virtual compound collections, i.e., theoretically tangible and synthesizable molecules (Reymond, 2015). We will also provide guidelines to choose the appropriate database tailored for a drug design project.

2.1 Actual Compounds Collections
2.1.1 Commercial and academic databases
Commercial databases represent an important source of compounds for virtual screening. They often contain more than 1 million molecules (Table 5.1). The major

advantage of these commercial collections of compounds lies in the possibility for the customer to cherry-pick and rapidly obtain selected compounds to be used in experimental assays. The average price is estimated between US$50 and US$200 for 5 mg of a given compound but it can vary according to the supplier, the amount of product ordered, and the complexity of the molecule (Rognan & Bonnet, 2014). Despite their commercial purpose, suppliers usually offer free access to the molecule's structures, provided in various file formats (2D or 3D). These databases undergo frequent updates, new products being added while other being either removed or out of stock. Virtual screening libraries constructed from these databases should ideally be prepared when the whole virtual screening protocol is settled and ready to be used.

Academic laboratories have always been a source for innovative compounds that chemical suppliers acquire and integrate in their commercial databases (Rognan & Bonnet, 2014). Initiatives to gather these academic databases within an institution, a country, or an ensemble of countries succeed in the creation of large

TABLE 5.1
Examples of Actual Compounds Collections.

	Database Name	Access	Number of Compounds	Compounds Purchasability	Website
Commercial	ChemSpider	Free	~67 millions	Link to vendors	http://www.chemspider.com/
	Ambinter	Free	~7 millions (screening compounds)	Yes	http://www.ambinter.com/
	eMolecules	Free	~7 millions	Link to vendors	https://www.emolecules.com/
	MolPort	Free	~7 millions	Yes	https://www.molport.com/
	Enamine	Free upon registration	~2.7 millions	Yes	https://enamine.net/
	ChemDiv	Free (e-mail address requested)	~1.6 millions	Yes	https://www.chemdiv.com/
	ChemBridge	Free	~1.3 millions	Yes	https://www.chembridge.com/
	IBS	Free upon registration	~550,000	Yes	https://www.ibscreen.com/
	Life Chemicals	Free (e-mail address requested)	~490,000	Yes	https://lifechemicals.com/
	Specs	Free upon registration	~350,000	Yes	https://www.specs.net/
	Asinex	Free	~260,000	Yes	https://www.asinex.com/
	Maybridge	Free upon registration	~53,000	Yes	https://www.maybridge.com/

TABLE 5.1
Examples of Actual Compounds Collections.—cont'd

	Database Name	Access	Number of Compounds	Compounds Purchasability	Website
Bioactivity	ZINC15	Free	~750 millions	Link to vendors	https://zinc15.docking.org/
	PubChem	Free	~103 millions	Link to vendors when available	https://pubchem.ncbi.nlm.nih.gov/
	ChEMBL	Free	~1,96 millions	Link to vendors when available	https://www.ebi.ac.uk/chembl/
	BindingDB	Free	~800,000	List of purchasable compounds	https://www.bindingdb.org/bind/index.jsp
	CARLSBAD	Free	~430,000	No	http://carlsbad.health.unm.edu/carlsbad/
	GLASS	Free	~340,000	No	https://zhanglab.ccmb.med.umich.edu/GLASS/index.html
	DrugBank	Free	~22,000	No	https://www.drugbank.ca/
	NR-DBIND	Free	7593	No	http://nr-dbind.drugdesign.fr/
	DrugCentral	Free	4052	No	http://drugcentral.org/
	SuperDrug2	Free	3992	No	http://cheminfo.charite.de/superdrug2/
Natural products	Super Natural II	Free	325,508	Yes	http://bioinf-applied.charite.de/supernatural_new/
	Dictionary of Natural Products	Commercial	~230,000	No	http://dnp.chemnetbase.com/
	Reaxys	Commercial	~220,000	No	https://www.reaxys.com/
	Antibase	Commercial	~40,000	No	https://www.wiley.com/en-us/AntiBase%3A+The+Natural+Compound+Identifier-p-9783527343591
	MarinLit	Commercial	~29,000	No	http://pubs.rsc.org/marinlit/
	The Natural Product Atlas	Free	~25,500	No	https://www.npatlas.org/joomla/
	AfroDB	Free	~900	No	http://african-compounds.org/nanpdb/

public databases such as the Molecular Libraries Probe Production Centers Network (MLPCN) database (National Center for Biotechnology Information (US), 2010) or the European Chemical Biology Library (ECBL) (Horvath et al., 2014). Using compounds belonging to academic databases in virtual screening protocols presents a dual interest (Rognan & Bonnet, 2014). The first advantage is that compounds gathered from different academic databases should guarantee a certain chemical diversity because each laboratory is usually focused on specific chemical scaffolds. The second one is the possibility of settling a collaboration with the academic laboratory that synthesized a given hit compound because its expertise on this chemical series could facilitate a future hit-to-lead optimization.

2.1.2 Bioactivity databases

Bioactivity databases are particularly important for drug discovery processes not only to get knowledge on biological targets and their modulation mechanisms but also to construct benchmarking datasets to design docking protocols prior to prospective virtual screenings and to construct predictive models of activity (Huang et al., 2006; Lagarde et al., 2015; Mysinger et al., 2012; Réau et al., 2018). This category includes databases such as the ZINC (Sterling & Irwin, 2015) or PubChem (Kim et al., 2019) (Table 5.1).

ZINC (ZINC Is Not Commercial) is a freely available compound collection that was initially developed to give access to the chemical structures of molecules included in commercial vendors catalogs encoded in file formats suitable for virtual screening experiments (SMILES, mol2, 3D SDF, and DOCK flexibase formats) (Irwin & Shoichet, 2005). The ZINC database was thus particularly designed to facilitate virtual screening databases preparation (Irwin, 2008). However, following the requests of ZINC investigators, the new version of the ZINC released in 2015, and thus named ZINC15, still provides purchasable data and ready-to-dock 3D formats but now also encompasses, whenever available, biological data on the proteins and biological processes modulated for more than 230 million compounds (Sterling & Irwin, 2015).

The PubChem (Kim et al., 2019) and ChEMBL databases (Gaulton et al., 2017) are two other examples of public bioactivity databases including both biological and structural information. PubChem is an open chemistry database, hosted by the US National Center for Biotechnology Information (NCBI), that gathers data from a wide variety of sources, among which are government agencies, pharmaceutical companies, chemical vendor catalogs, and scientific literature (Kim et al., 2016). PubChem is divided into three interconnected databases named "Substance," "Compound," and "BioAssay." The PubChem Compound database includes, as on July 2020, more than 100 million unique chemical structures. ChEMBL (Gaulton et al., 2017) is a manually curated open database that receives data from public and commercial organizations and scientific literature (patents and publications). ChEMBL includes over 16 million bioactivity data for over 1.96 million distinct compounds and more than 13,000 targets.

Focused bioactivity libraries are also available. They can be dedicated to a given protein family like the nuclear receptors (NRLiSt BDB (Lagarde et al., 2014), NR-DBIND (Réau et al., 2019)), the G-protein—coupled receptor (GLL) (Gatica & Cavasotto, 2012), GLASS (Chan et al., 2015), or protein kinases (PKIDB) (Carles et al., 2018).

Finally, the databases collecting information about molecules already approved as drugs or currently evaluated in clinical trials (the "drug" subset of the ChEMBL database (Gaulton et al., 2012), DrugBank (Wishart et al., 2018), DrugCentral (Ursu et al., 2019), SuperDrug2 (Siramshetty et al., 2018), and Drugs-lib (Lagarde et al., 2018)) can also be of interest. These compounds can be used for repositioning studies, i.e., find new biological target and therapeutic indications for approved or investigational drugs (Pushpakom et al., 2019). Drug repositioning is one of the emergent strategies to overcome drug attrition rates and to speed up the drug discovery process. It was shown of particular importance for cancer (Würth et al., 2016), rare diseases applications (Delavan et al., 2018) and also in situations where an urgent need of a therapeutic solution that can be immediately administered to patients is needed, such as the recent COVID-19 pandemic (Serafin et al., 2020).

The main drawback of these bioactivity databases is that synthesized samples of compounds of interest may not be easily or directly available, unlike in commercial and academic databases because bioactivity databases encompass data from a wide range of sources, including scientific publications and patents.

2.1.3 Natural products

Natural products were the first drugs ever used and have always been a source of drugs. In a recent retrospective study, it has been reported that between January 1, 1981 and September 30, 2019, 23.5% of all new approved drugs and 33.6% of new approved small molecules drugs were natural products or derivatives of natural products (Newman & Cragg, 2020). The chemical space covered by natural products is quite dissimilar to the one occupied by synthetic drug-like compounds (Morrison & Hergenrother, 2014) and natural products are believed to constitute promising starting point for drug discovery (Rodrigues et al., 2016). Natural products databases can thus be used as a source for virtual screening libraries. Numerous databases of natural products are available (for a review, see Sorokina and Steinbeck (2020), Chen et al. (2017)) that can be commercial or freely accessible (Table 5.1). Natural products databases can be comprehensive or focused on natural products used in traditional medicine (such as the AfroDb (Ntie-Kang et al., 2013) or the TCM database@ Taiwan (Chen, 2011)). The most complete database of natural products is to date the Super Natural II with more than 320,000 compounds (Banerjee et al., 2015).

2.2 Virtual Compounds Collections

To access unexplored and intellectual property-free area of the chemical space, an emerging field is to use virtual compounds, i.e., possible molecules but with undescribed synthetic access. Virtual compounds can be designed using fragment-based docking methods by assembling known building block using known chemical reactions or by enumerating all molecules respecting defined rules with meaningful chemical structures.

2.2.1 Fragment-based databases

Fragment-based drug discovery has shown efficiency in providing new drug candidates in the last 20 years (Erlanson et al., 2016). The success of fragment-based docking methods relies on the availability and quality of fragment libraries. The size and the composition of the fragment library directly influence the outcomes of fragment-based docking and it has been shown that a diverse library with size limited to 2000 fragments should be preferred (Böttcher et al., 2019; Messick et al., 2019; Shi & von Itzstein, 2019). Fragments included in the virtual screening libraries mainly comply to the "Rule of Three" (Congreve et al., 2003), derived from "Lipinski's rule of five" (Ro5), with a molecular weight inferior to 300 Da and a value inferior to 3 for the numbers of both hydrogen bond donors and hydrogen bond acceptors and for the partition coefficient between n-octanol and water (clogP) (a number of rotatable bond inferior to 3 and a polar surface area inferior to 60 Å^2 are also recommended). Fragments from large commercial libraries can be purchased for experimental testing and chemistry synthesis as shown in Table 5.2.

More than 10 million virtual fragments complying to the "Rule of Three" are provided in the FDB-17 (Visini et al., 2017). These fragments present the advantage to cover a large chemical space with a wide range of molecular size, polarity, and complexity. However, fragments of the FDB-17 identified as hits in virtual screening protocols are not directly purchasable for experimental validation.

2.2.2 Tangible compounds

Tangible compounds (Hann & Oprea, 2004) are virtual compounds that present a high synthetic feasibility using well-known or in-house building blocks and reactions. The Screenable Chemical Universe Based on Intuitive Data Organization (SCUBIDOO) (Chevillard & Kolb, 2015) tries to enhance the synthetic feasibility criterion. The SCUBIDOO database gathers more than 21 million virtual compounds generated by exhaustively reacting 7805 building blocks with the 58 most commonly used reactions in the pharmaceutical field. Commercial databases are also providing lists of

TABLE 5.2
Examples of Purchasable Fragment-Based Databases.

Database Name	Number of Fragments	Web Interface
Enamine Fragment Collection	172,723	https://enamine.net/ fragments/ fragment-collection
Asinex's Fragments	20,117	http://www.asinex. com/fragments/
OTAVAchemicals General Fragments Library	13,685	https:// otavachemicals. com/products/ fragment-libraries/ general-fragment-library
ChemBridge Fragment Library	13,500	https://www. chembridge.com/ screening_libraries/ fragment_library/,
Maybridge Ro3 Library	2500	https://www. maybridge.com/
Prestwick Drug-Fragment Library	1456	http://www. prestwickchemical. com
Prestwick F2L Library	530	

tangible compounds. For example, Life Chemicals (n.d.) provides a database of more than 500,000 tangible compounds estimated to be able to be synthesized at 90% through in-house developed and validated synthetic procedures. The REAL Space of Enamine comprises more than 3.8 billion virtual compounds that should be successfully synthesized and delivered to the customer within approximatively 3 weeks with a probability over 80% (Hoffmann & Gastreich, 2019).

2.2.3 Possible compounds

Possible compounds are virtual compounds which structures are obtained using the rules of chemistry to guide the assembling of atoms through covalent bonds. The generated databases (GDBs) initiative (Blum & Reymond, 2009; Fink & Reymond, 2007; Reymond, 2015; Ruddigkeit et al., 2012) represents the largest collection of possible compounds. GDBs aims to enumerate all possible molecules from a chemical space defined by a set of rules among which are geometrical strain, functional group stability criteria, and simple synthetic feasibility rules. The latest version, the GDB-17 (Ruddigkeit et al., 2012), includes 166 billion compounds with a maximum of 17 heavy atoms (C, N, O, S, F, Cl, Br, and I).

2.2.4 Virtual compounds and virtual screenings
These databases of virtual compounds represent a gold mine to discover new hits, especially for difficult biological targets. However, despite the recent advances toward ultralarge virtual screening using ligand-based methods (Boehm et al., 2008; Hoffmann & Gastreich, 2019; Lessel et al., 2009) and structure-based methods (Gorgulla et al., 2020; Lyu et al., 2019), screening of these enormous databases of virtual compounds using docking methods remains marginal and poorly accessible to numerous teams. The best option to explore these virtual compounds databases remains to select smaller but representative subsets. In this way, the GDBs provide smaller subsets of compounds with less chemical complexity and thus enhanced feasibility probability (Meier et al., 2020) and the SCUBIDOO includes three representative subsets named S, M, and L that contained, respectively, 9994, 99,977, and 999,794 compounds (Chevillard & Kolb, 2015). Finally, despite the efforts put to guarantee the synthetical feasibility of the virtual compounds, the compounds availability for experimental testing could be challenging.

2.3 Library Selection and Customization
As aforementioned, there is a vast number of diverse databases, and the tricky question of the one(s) to choose to construct an appropriate virtual screening database may arise. To guide this selection, two approaches are possible. The first one is to screen all possible compounds and libraries, as it has been shown that raising the number of compounds that are screened improves the true positive rates (Lyu et al., 2019). However, as previously mentioned with virtual compounds databases, ultralarge databases screening achieved using docking methods is still an exception and is limited by computing and storage capacities. The second choice is thus to select a subset of compounds among all available. The choice of the compounds to be prioritized in the database can be made based on the information available for the targeted protein. Focused libraries can be designed by filtering the database to enrich it with compounds that present structural and physicochemical similarity with known modulators of the query target or

complementarity to the binding site of the query target. Another option is to promote diversity in the retained database by removing molecules that are too similar to those already included in the database (Yosipof & Senderowitz, 2014). This selection procedure is based on the assumption that similar compounds present similar biological activities. Thus, testing a single representative compound from a group of similar ones should allow insights on the potential activity of other compound members of the same cluster (Bayada et al., 2010).

Various clustering methods, divided into supervised and unsupervised approaches, are commonly used for data selection (Downs & Barnard, 2002; Shemetulskis et al., 1995). Clustering methods are efficient to reduce the database sizes and thus the associated docking computational times; however, a recent study demonstrated that electing a single representative of one cluster, regardless of the selection procedure, was negatively impacting the docking scores (Lyu et al., 2019). Dissimilarity-based methods can also be used to select the compounds to include in the databases (Hassan et al., 1996; Lajiness & Watson, 2008). These approaches include the application of distance metrics, such as the widely used Tanimoto coefficient on a set of appropriate descriptors (Bayada et al., 2010) and diversity algorithms among which are the maximum dissimilarity algorithms (Kennard & Stone, 1969), the Kohonen neural network (Li et al., 1993), or the sphere exclusion algorithms (Hudson et al., 1996).

3 DATABASE PREPARATION
Most of the databases presented in the previous section, with few exceptions such as the ZINC15, are not dedicated to virtual screening approaches and the chemical structures provided by these databases cannot directly be used as input for docking methods. A rigorous and careful step of ligand preparation is thus necessary to ensure the success of the virtual screening protocol. Even if there is no widely accepted protocol for virtual screening databases preparation, common guidelines are found in the scientific literature (Bologa et al., 2019; Fourches et al., 2010; Gally et al., 2019; Gorgulla et al., 2020; Lagorce, Bouslama, et al., 2017; Rognan & Bonnet, 2014; Williams et al., 2012). In this section, we will detail each step of the preparation process as seen in Fig. 5.1 and provide a summary available in Table 5.3 that lists commonly used tools for one, several, or all steps of the database preparation.

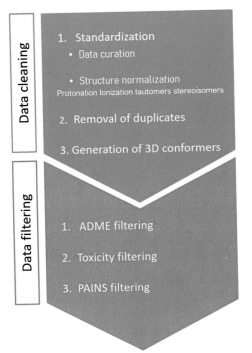

FIG. 5.1 Graphical representation of the main steps of database preparation. *ADME*, absorption, distribution, metabolism, elimination; *PAINS*, pan-assay interference compounds.

3.1 Database Cleaning
3.1.1 Standardization
3.1.1.1 Data curation. The first step of the cleaning procedure is the standardization of chemical structures, especially when the molecules are retrieved from different data sources. Compounds can be prepared with different protocols and encoded in different formats. Compact 1D and 2D formats, such as SMILES and SDF, are often privileged for large databases, but already prepared and cleaned structures in PDB or MOL2 formats can also be available. All compounds should thus be converted to the same format, and for SMILES inputs, canonical SMILES should be generated to avoid duplicates. Once the conversion is done, a major step is to identify and resolve issues such as correcting or filtering out compounds with incorrect or missing structures. Salted molecules are also problematic inputs for docking softwares and a neutralization step with elimination of the corresponding counterion is necessary. Most of the docking methods are only

able to compute one molecule at a time, and mixture components should be separated. Finally, inorganics compounds, which usually represent a small fraction of drug discovery–oriented databases, should also be removed because they often cannot be handled by docking software (Fourches et al., 2010). All these steps can rapidly be achieved with different tools (Table 5.3).

3.1.1.2 Structure normalization (protonation, ionization, tautomerization, and stereochemistry). Compounds can present different protonation or ionization states and different tautomeric forms (tautomers) according to the physiological conditions. Depending on the state and form used, the possible interactions between the ligand and the binding site may vary and impact the docking outcomes. Experimentally determined protonation, ionization, and tautomeric states are rarely provided in the databases and are mainly determined through predictions (Ten Brink & Exner, 2009). Identification of the correct protonation state is crucial for docking calculations as the presence or the absence of a specific hydrogen may drastically influence both the sampling and the scoring part of the docking. Correct placement of these atoms is important to generate and identify the right docking poses (Polgár et al., 2007) and the compounds should be protonated at the physiological pH (Knox et al., 2005) with appropriate tools (Table 5.3). Tautomers are the result of a formal migration of a hydrogen atom or proton, accompanied by a switch of a single bond and an adjacent double bond. Consequently, properties such as hydrophobicity, pKa, 3D shapes, and ability to form hydrogen bonds can vary among the different tautomers of the same molecule. Careful selection of tautomers is thus crucial for accurate docking predictions (Martin, 2009) and several methods for the enumeration of tautomers have been published (Milletti et al., 2009; Oellien et al., 2006; Sitzmann et al., 2010; Trepalin et al., 2003). The level of enumeration that should be performed depends on the objective of the docking (Martin, 2009); a reasonable number of tautomers should be selected for a large virtual screening campaign, while a full enumeration of the tautomers is required for accurate prediction of one ligand/protein complex. In the same way, stereochemistry can impact the relative binding affinity. Proper enumeration of the relevant stereoisomers and pairs of enantiomers for chiral compounds is important for databases preparation (Brooks et al., 2008).

TABLE 5.3
Examples of Tools Available for Database Preparation.

Tool	Ref[1]	Type[2]	Cleaning										Filtering			
			Standardi-zation	Inorganics	Neutraliza-tion	Mixtures	Checking structures	Duplicates	Tautomers	Stereo-isomers	Protonation	Conformers	PhysChem	Toxicity	PAINS	BBB[3]
CDK	[1]	OS	✓	✓	✓	✓	✓	✓	✓	✓	✗	✗	✓	✗	✗	✗
DataWarrior	[2]	F	✓	✓	✓	✓	✓	✓	✗	✓	✓[4]	✓	✓	✓	✗	✗
DockingServer	[3]	C	✓	✗	✗	✗	✗	✗	✗	✗	✓	✓	✗	✗	✗	✗
FAF-Drugs4	[4]	F	✓	✓	✓	✓	✓	✓	✗	✗	✓	✗	✓	✓	✓	✓
FILTER	[5]	C	✗	✗	✗	✗	✓	✗	✗	✗	✗	✗	✓	✓	✗	✗
Indigo	[6]	OS	✓	✗	✗	✗	✓	✓	✓	✓	✗	✗	✗	✗	✗	✗
Instant JChem	[7]	C(FA)	✓	✓	✓	✓	✓	✓	✓	✓	✓	✓	✓	✗	✗	✗

Name	Ref	Type[2]	1	2	3	4	5	6	7	8	9	10	11	12	13	14
LigPrep	[8]	C	✓	✓	✓	✗	✓	✓	✓	✓	✓	✓	✓	✗	✗	✗
LigQ	[9]	F	✓	✗	✗	✗	✗	✗	✓	✓	✓	✓	✗	✗	✗	✗
Open Babel	[10]	OS	✓	✗	✗	✗	✗	✓	✗	✓	✓	✓	✗	✗	✗	✗
RDKit	[11]	OS	✓	✓	✓	✓	✓	✓	✓	✓	✓	✓	✓	✗	✗	✗
Screening Assistant	[12]	F	✓	✗	✗	✗	✗	✗	✗	✗	✗	✗	✓	✓	✓	✗
Standardizer	[13]	C(FA)	✓	✓	✓	✓	✓	✓	✓	✓	✗	✓	✗	✗	✗	✗
SwissADME	[14]	F	✗	✗	✗	✗	✗	✗	✗	✗	✗	✗	✓	✗	✓	✓
UNICON	[15]	C(FA)	✓	✓	✗	✗	✓	✗	✓	✗	✓	✓	✗	✗	✗	✗
VFLP	[16]	F	✓	✗	✓	✗	✗	✗	✓	✗	✗	✗	✗	✗	✗	✗
VSPred[5]	[17]	F	✓	✓	✓	✓	✓	✓	✓	✓	✓	✓	✓	✗	✓	✗

[1] References: [1] (Willighagen et al., 2017), [2] (Sander et al., 2015),[3] (Bikadi & Hazai, 2009), [4] (Lagorce, Oliveira, et al., 2017), [5] (FILTER, n.d.), [6] (Pavlov et al., 2011), [7] (Instant JChem, n.d.), [8] (LigPrep, n.d.), [9] (Radusky et al., 2017), [10] (OpenBabel, n.d.), [11] (RDKit, n.d.), [12] (Le Guilloux et al., 2012), [13] (Standardizer, n.d.), [14] (Daina et al., 2017), [15] (Sommer et al., 2016), [16] (Gorgulla et al., 2020), [17] (Gally et al., 2019).

[2] *C*, commercial; *C(FA)*, commercial but free for academics; *F*, freely accessible; *OS*, open source.

[3] BBB (blood−brain barrier) permeability filters.

[4] Only if the ChemAxon pKa Plugin is installed.

[5] Workflow using RDKit, ChemAxon, and Indigo features.

3.1.2 Removal of duplicates

Duplicated molecules should be removed to avoid unnecessary calculations. Various sources of duplicated molecules are possible, i.e., a same compound can be incorporated from different input databases (sometimes without the same identifier) or included in the database in different mixtures or salted forms. The standardization step is thus critical to highlight duplicates that were included in different forms and not recognized as duplicated.

3.1.3 Generation of 3D conformers

Most of the time, according to data storage consideration, compound structures are collected in a 2D format. A conversion step to generate 3D conformations is often required for docking algorithms. Because the bioactive conformation is generally unknown, a common strategy is to include a representative ensemble of conformers of each compound to avoid missing potential hits (Gimeno et al., 2019). This is of primary importance when running both rigid docking and flexible docking; in the former case, the ensemble of conformers should ideally cover the binding conformation(s) of the small molecules, and in the latter case, different conformations should be enumerated to cover the different conformational orientations of nonrotatable bonds and chemical groups (Hawkins, 2017). For example, the cyclohexane can adopt either a "chair" or a "boat" conformation, and both should be considered if no experimental data support the choice of one conformation over the other. The safe number of conformers to generate depends both on the tool used and on the number of rotatable bonds of the small molecules composing the database. Different commercial (e.g., OMEGA (Hawkins et al., 2010), ConfGen (Watts et al., 2010), or iCon (Poli et al., 2018)) or free (e.g., RDKIT (RDKit, n.d.), Frog2 (Miteva et al., 2010), or Tinker (Rackers et al., 2018)) ensemble generator tools are available. They are often included within software pipelines and are then not necessarily available as stand-alone tools. They fall into two categories: the stochastic approaches that sample random values of torsion angles with respect to predefined rules (e.g., molecular dynamics or Monte Carlo simulations or distance geometry) and the systematic approaches that sample all authorized torsion angles of a molecule (e.g., brute force and rule-based enumeration). Unlike the stochastic approaches, the systematic ones always return the same lowest-score conformation but are limited to molecules with few rotatable bonds because of the combinatorial explosion of solutions. Most tools make use of knowledge-based or force field–based scoring

functions to estimate the internal energy of the generated conformers. In the latter case, the coulombic electrostatic term is generally omitted or tuned to avoid favoring conformers with intramolecular hydrogen bonds (Hawkins, 2017; Wang et al., 2020).

3.2 Database Filtering

Filtering procedures have been developed to enhance the chance that a hit identified by virtual screening will be successfully optimized as a potential drug candidate. A lot of ADME-Tox (absorption, distribution, metabolism, elimination, and toxicity) filters have been used in this purpose and will be presented here together with PAINS (pan-assay interference compounds) alerts and knowledge-based filters.

3.2.1 ADME (absorption, distribution, metabolism, elimination) filtering

In the mid-1990s, drug candidates' failure in clinical trials was mainly attributed to nonoptimal pharmacokinetics and bioavailability parameters (Kola & Landis, 2004). To overcome this issue, physicochemical filters were proposed to select compounds with pharmacokinetics values similar to actual approved drugs. The most used filters were established in 1997 by Lipinski and his colleagues from Pfizer (Lipinski et al., 1997). They selected in the World Drug Index 2245 orally available compounds that had reached at least phase II of clinical trials, hence displaying presumed adequate water solubility and intestinal permeability. The analysis of the common properties of these compounds led to the "Lipinski's Ro5," in which thresholds for molecular weight (≤ 500 Da), number of hydrogen bonds donors (≤ 5) and acceptors (≤ 10), and water/octanol partition coefficient (≤ 5) were defined. According to this rule, a compound that is not compliant to more than one of these criteria is associated with a higher risk of poor oral bioavailability.

The Ro5 has been largely used and integrated in numerous virtual screening databases preparation protocols. However, it is now of common knowledge that the Ro5 should not be considered as an absolute drug-likeness criterion because different toxic compounds are Ro5-compliant whereas some approved drugs are not (Bickerton et al., 2012) and that compounds compliant to this rule are not suitable to modulate half of the targets involved in human disease pathways (Surade & Blundell, 2012). Additionally, there is still some oral space beyond the Ro5 (Doak et al., 2014; Poongavanam et al., 2018; Tyagi et al., 2020). Virtual screening protocols aim to identify potential hits/leads for a biological target that will be modified during the hit-to-lead and lead-to-candidate drug optimization processes. Applying drug-like filters on potential hits/leads seems untimely as the subsequent optimized drug candidates may not, at the end, be Ro5 compliant. Hann and Oprea suggested to adjust the Ro5 threshold to obtain "lead-like" filters deduced from the comparison of the properties of 176 leads compounds and 532 drugs (Alvarez & Shoichet, 2005). These "lead-like" filters presented not only modified threshold values compared to the Ro5 properties but also additional criteria such as the number of aromatic rings, Caco-2 intestinal permeability, and water solubility.

Drug permeability through the blood–brain barrier (BBB) is also a major property to monitor. Drugs targeting the central nervous system should be able to cross the BBB while peripheral acting drugs should not, in order to avoid any psychotropic side effect (Di & Kerns, 2015). The two gold-standard experimental measures of BBB permeability are logBB (the concentration of drug in the brain divided by concentration in the blood) and logPS (permeability surface area product). Both values can be obtained either through experimental measures or predicted via in silico methods (Carpenter et al., 2014). Another approach to estimate the BBB penetration is to use the BBB score (Gupta et al., 2019), a simple model that is based on five physicochemical descriptors.

Numerous methods to improve "drug-like," "lead-like," and other ADME filters have been developed (Korkmaz et al., 2015; Oprea et al., 2007; Petit et al., 2012; Ridder et al., 2011; Veber et al., 2002), but the interest for "drug-like" or "lead-like" filters has considerably decreased. The control of the physicochemical properties of the compounds during the optimization process remains important (Waring et al., 2015) and databases developed for docking purposes should at least be filtered using soft physicochemical thresholds to remove compounds not suitable for docking tools (Lagarde et al., 2018).

3.2.2 Toxicity filtering

Toxicity of drug candidates can be discovered at very late stages of drug development, during clinical trials or worse after its commercialization, ruining years of efforts and billions of dollars investments. Safety and toxicity have been identified as the major sources of failure in an analysis of attrition of drug candidates from four major pharmaceutical companies (Waring et al., 2015) and represent major concerns in drug discovery processes. Many computational toxicology methods

have been developed to predict potential toxicity. These methods can be used as a prefiltering step to remove undesirable compounds from the virtual screening databases or after the virtual screening to flag predicted hits with toxicity alerts and to ensure that the flagged chemical moiety will be modified during the optimization process. Computational toxicology methods are mainly divided into three categories (Hevener, 2018): algorithms and models, chemical filters, and structural alerts tools.

In the first category, quantitative structure–activity relationship and quantitative structure–toxicity relationship models have long been the reference (Benigni & Bossa, 2008; Gini, 2016; Lapenna et al., 2010; Myshkin et al., 2012). These mathematical models can be established by correlating experimentally measured toxicity with the structure of the compounds encoded as physicochemical and geometrical descriptors. Other methods include deep learning methods (Gawehn et al., 2016; Goh et al., 2017), structure-based methods (Jing et al., 2015; Moroy et al., 2012), and toxicophore mapping (Kar & Roy, 2013; Pramanik & Roy, 2014; Singh et al., 2016).

The second category gathers a large variety of filters to identify promiscuous compounds, i.e., drug compounds that can act on multiple molecular targets, exhibiting similar or different pharmacological effects (Mei & Yang, 2018). Limited promiscuity might be a desirable property for polypharmacological drug discovery (Feldmann et al., 2019) but promiscuous compounds can also present undesirable side effect because of the modulation of unwanted targets (Mei & Yang, 2018). Filters are either rules built on physiochemical properties or structural alerts. Among the physiochemical properties evaluated (Bruns & Watson, 2012), specific attention was also given to molecular weight and lipophilicity (Bowes et al., 2012; Hughes et al., 2008; Morphy & Rankovic, 2007; Tarcsay & Keserű, 2013).

The last category includes structural alerts developed to highlight structural motifs and substructures often occurring among toxic compounds (Bruns & Watson, 2012; Pizzo et al., 2015).

A large number of freely available tools and web servers implement these computational toxicology method (ToxiPred (Mishra et al., 2014), DeepTox (Mayr et al., 2016), admetSAR (Cheng et al., 2012), ToxiM (Sharma et al., 2017), VirtualToxLab (Vedani et al., 2012), OpenTox (Tcheremenskaia et al., 2012), FAF-Drug4 (Lagorce, Bouslama, et al., 2017); ToxAlerts (Sushko et al., 2012), Screening Assistant (Le Guilloux et al., 2012)).

3.2.3 PAINS (pan-assay interference compounds) alerts

PAINS are compounds frequently identified as hits in any given assay due to the presence of common chemotypes able to interfere in biochemical assays (Baell & Nissink, 2018). PAINS behavior is linked to intrinsic redox activity, instability, reactivity, aggregation potency, covalent labeling of proteins, metal chelation, membrane disruption, fluorescence interference, and structural decomposition (Lagorce, Oliveira, et al., 2017). These compounds raised issues because numerous efforts to optimize PAINS identified as hits remained unsuccessful (Baell & Nissink, 2018). Several filters have been developed to flag PAINS (Baell & Holloway, 2010; Lagorce, Bouslama et al., 2017; Saubern et al., 2011; Sterling & Irwin, 2015) and were used to discard hits flagged with PAINS alerts prior to experimental validation. However, different studies point out that these filters are not perfect as 6%–7% of approved drugs display PAINS chemotypes (Capuzzi et al., 2017; Senger et al., 2016) and that some compounds flagged as PAINS were not confirmed by experimental assays, whereas compounds not flagged as PAINS revealed experimentally a PAINS behavior (Capuzzi et al., 2017). PAINS filters should then be used as flag alerts but not to remove compounds from virtual screening databases (Baell & Nissink, 2018; Lagorce, Oliveira, et al., 2017).

3.3 Automated Tools for Virtual Screening Databases Preparation

Numerous free or commercial methods are available for assisting the virtual screening databases preparation (Villoutreix et al., 2007). Table 5.3 lists the most popular tools with details about the corresponding monitored step of the database preparation. It is also worth mentioning that each step can be scripted by users with different programming languages. For example, the desalting procedure can be done with a simple text editing program by searching for "full stops" separating two fragments on SMILES and removing the one without any carbon atom. Finally, despite the precision of automated tools, a final manual inspection of each compound is recommended to ensure that no errors were left after the cleaning process (Fourches et al., 2010).

4 CONCLUSION

Docking methods are used in virtual screening protocols to identify hits that could be optimized into drug candidates. This can be successfully achieved by carefully selecting and preparing compounds that are included in the virtual screening compound collections.

Numerous sources of compounds and databases are available, and we described the main categories in this chapter. Due to limited resources in computer power and data storage, ultralarge databases screening using docking methods has remained poorly accessible for a long time. However, recent advances could enhance its democratization within a few years. The compounds included in a virtual screening database must be correctly prepared for docking methods. This is particularly important to avoid missing potential hits and providing unnecessary efforts to optimize inactive, toxic, or promiscuous compounds that could have been identified beforehand. No current gold standard for database preparation exists, and different docking software require different molecular file formats. We detail in this chapter the main steps of compounds cleaning filtering. Filters may be applied to enrich the database in compounds that are less likely destined to fail in the drug discovery process and to reduce the virtual screening computational times. The strategy to select compounds and to filter the database before docking depends on the virtual screening purpose and on the importance of the virtual and experimental screening facilities.

LIST OF ABBREVIATIONS

1D	One-dimensional
2D	Two-dimensional
3D	Three-dimensional
ADME/Tox	Absorption, distribution, metabolism, elimination, toxicity
Å	Ångström
BBB	Blood–brain barrier
Br	Bromine
C	Carbon
ChEMBL	Chemical database of the European Molecular Biology Laboratory
Cl	Chlorine
clogP	Partition coefficient between n-octanol and water
Da	Dalton
F	Fluorine
FDB-17	Fragment database 17
GDBs	Generated databases
GLASS	GPCR-ligand associations
GLL	G-protein–coupled receptor (GPCR) ligand library
I	Iodine
logBB	Logarithm of the ratio of the concentration of drug in the brain divided by concentration of the drug in the blood
logPS	Permeability surface area product

mg	Milligram
N	Nitrogen
NCBI	National Center for Biotechnology Information
NR-DBIND	Nuclear Receptors Database Including Negative Data
NRLiSt BDB	Nuclear Receptors Ligands and Structures Benchmarking Database
O	Oxygen
PAINS	Pan-assay interference compounds
PDB	Protein Data Bank
pH	Potential of hydrogen
pKa	Acid dissociation constant
PKIDB	Protein Kinase Inhibitor Database
S	Sulfur
SCUBIDOO	Screenable Chemical Universe Based on Intuitive Data Organization
SDF	Structure data file
SMILES	Simplified Molecular Input Line Entry Specification
TCM DB	Traditional Chinese Medicine Database
US	United States of America

REFERENCES

Alvarez, J., & Shoichet, B. (2005). *Virtual screening in drug discovery*. CRC Press.

Arús-Pous, J., Awale, M., Probst, D., & Reymond, J.-L. (2019). Exploring chemical space with machine learning. *Chimia, 73*(12), 1018−1023. https://doi.org/10.2533/chimia.2019.1018.

Awale, M., Visini, R., Probst, D., Arús-Pous, J., & Reymond, J.-L. (2017). Chemical space: Big data challenge for molecular diversity. *Chimia, 71*(10), 661−666. https://doi.org/10.2533/chimia.2017.661.

Baell, J. B., & Holloway, G. A. (2010). New substructure filters for removal of pan assay interference compounds (PAINS) from screening libraries and for their exclusion in bioassays. *Journal of Medicinal Chemistry, 53*(7), 2719−2740. https://doi.org/10.1021/jm901137j.

Baell, J. B., & Nissink, J. W. M. (2018). Seven year itch: pan-assay interference compounds (PAINS) in 2017—utility and limitations. *ACS Chemical Biology, 13*(1), 36−44. https://doi.org/10.1021/acschembio.7b00903.

Banerjee, P., Erehman, J., Gohlke, B.-O., Wilhelm, T., Preissner, R., & Dunkel, M. (2015). Super natural II—a database of natural products. *Nucleic Acids Research, 43*(D1), D935−D939. https://doi.org/10.1093/nar/gku886.

Bayada, D. M., Hamersma, H., & van Geerestein, V. J. (2010). ChemInform abstract: Molecular diversity and representativity in chemical databases. *ChemInform, 30*(17). https://doi.org/10.1002/chin.199917285.

Benigni, R., & Bossa, C. (2008). Predictivity and reliability of QSAR models: The case of mutagens and carcinogens. *Toxicology Mechanisms and Methods, 18*(2−3), 137−147. https://doi.org/10.1080/15376510701857056.

Bickerton, G. R., Paolini, G. V., Besnard, J., Muresan, S., & Hopkins, A. L. (2012). Quantifying the chemical beauty of drugs. *Nature Chemistry, 4*(2), 90−98. https://doi.org/10.1038/nchem.1243.

Bikadi, Z., & Hazai, E. (2009). Application of the PM6 semi-empirical method to modeling proteins enhances docking accuracy of AutoDock. *Journal of Cheminformatics, 1*(1), 15. https://doi.org/10.1186/1758-2946-1-15.

Blum, L. C., & Reymond, J.-L. (2009). 970 million druglike small molecules for virtual screening in the chemical universe database GDB-13. *Journal of the American Chemical Society, 131*(25), 8732−8733. https://doi.org/10.1021/ja902302h.

Boehm, M., Wu, T.-Y., Claussen, H., & Lemmen, C. (2008). Similarity searching and Scaffold Hopping in synthetically accessible combinatorial chemistry spaces. *Journal of Medicinal Chemistry, 51*(8), 2468−2480. https://doi.org/10.1021/jm0707727.

Bologa, C. G., Ursu, O., & Oprea, T. I. (2019). How to prepare a compound collection prior to virtual screening. In R. S. Larson, & T. I. Oprea (Eds.), *Bioinformatics and drug discovery* (Vol. 1939, pp. 119−138). Springer New York. https://doi.org/10.1007/978-1-4939-9089-4_7.

Böttcher, J., Dilworth, D., Reiser, U., Neumüller, R. A., Schleicher, M., Petronczki, M., Zeeb, M., Mischerikow, N., Allali-Hassani, A., Szewczyk, M. M., Li, F., Kennedy, S., Vedadi, M., Barsyte-Lovejoy, D., Brown, P. J., Huber, K. V. M., Rogers, C. M., Wells, C. I., Fedorov, O., … McConnell, D. B. (2019). Fragment-based discovery of a chemical probe for the PWWP1 domain of NSD3. *Nature Chemical Biology, 15*(8), 822−829. https://doi.org/10.1038/s41589-019-0310-x.

Bowes, J., Brown, A. J., Hamon, J., Jarolimek, W., Sridhar, A., Waldron, G., & Whitebread, S. (2012). Reducing safety-related drug attrition: The use of in vitro pharmacological profiling. *Nature Reviews Drug Discovery, 11*(12), 909−922. https://doi.org/10.1038/nrd3845.

Ten Brink, T., & Exner, T. E. (2009). Influence of protonation, tautomeric, and stereoisomeric states on Protein−Ligand docking results. *Journal of Chemical Information and Modeling, 49*(6), 1535−1546. https://doi.org/10.1021/ci800420z.

Brooks, W. H., Daniel, K. G., Sung, S.-S., & Guida, W. C. (2008). Computational validation of the importance of absolute stereochemistry in virtual screening. *Journal of Chemical Information and Modeling, 48*(3), 639−645. https://doi.org/10.1021/ci700358r.

Bruns, R. F., & Watson, I. A. (2012). Rules for identifying potentially reactive or promiscuous compounds. *Journal of Medicinal Chemistry, 55*(22), 9763−9772. https://doi.org/10.1021/jm301008n.

Capuzzi, S. J., Muratov, E. N., & Tropsha, A. (2017). Phantom PAINS: Problems with the utility of alerts for Pan-Assay INterference compoundS. *Journal of Chemical Information and Modeling, 57*(3), 417−427. https://doi.org/10.1021/acs.jcim.6b00465.

Carles, F., Bourg, S., Meyer, C., & Bonnet, P. (2018). PKIDB: A curated, annotated and updated database of protein kinase inhibitors in clinical trials. *Molecules (Basel, Switzerland)*, 23(4). https://doi.org/10.3390/molecules23040908.

Carpenter, T. S., Kirshner, D. A., Lau, E. Y., Wong, S. E., Nilmeier, J. P., & Lightstone, F. C. (2014). A method to predict blood-brain barrier permeability of drug-like compounds using molecular dynamics simulations. *Biophysical Journal*, 107(3), 630–641. https://doi.org/10.1016/j.bpj.2014.06.024.

Chan, W. K. B., Zhang, H., Yang, J., Brender, J. R., Hur, J., Özgür, A., & Zhang, Y. (2015). GLASS: A comprehensive database for experimentally validated GPCR-ligand associations. *Bioinformatics (Oxford, England)*, 31(18), 3035–3042. https://doi.org/10.1093/bioinformatics/btv302.

Chen, C. Y.-C. (2011). TCM Database@Taiwan: The world's largest traditional Chinese medicine database for drug screening in silico. *PLoS One*, 6(1), e15939. https://doi.org/10.1371/journal.pone.0015939.

Chen, Y., de Bruyn Kops, C., & Kirchmair, J. (2017). Data resources for the computer-guided discovery of bioactive natural products. *Journal of Chemical Information and Modeling*, 57(9), 2099–2111. https://doi.org/10.1021/acs.jcim.7b00341.

Cheng, F., Li, W., Zhou, Y., Shen, J., Wu, Z., Liu, G., Lee, P. W., & Tang, Y. (2012). admetSAR: A comprehensive source and free tool for assessment of chemical ADMET properties. *Journal of Chemical Information and Modeling*, 52(11), 3099–3105. https://doi.org/10.1021/ci300367a.

Chevillard, F., & Kolb, P. (2015). SCUBIDOO: A large yet screenable and easily searchable database of computationally created chemical compounds optimized toward high likelihood of synthetic tractability. *Journal of Chemical Information and Modeling*, 55(9), 1824–1835. https://doi.org/10.1021/acs.jcim.5b00203.

Congreve, M., Carr, R., Murray, C., & Jhoti, H. (2003). A 'Rule of Three' for fragment-based lead discovery? *Drug Discovery Today*, 8(19), 876–877. https://doi.org/10.1016/S1359-6446(03)02831-9.

Corbeil, C. R., Williams, C. I., & Labute, P. (2012). Variability in docking success rates due to dataset preparation. *Journal of Computer-Aided Molecular Design*, 26(6), 775–786. https://doi.org/10.1007/s10822-012-9570-1.

Daina, A., Michielin, O., & Zoete, V. (2017). SwissADME: A free web tool to evaluate pharmacokinetics, drug-likeness and medicinal chemistry friendliness of small molecules. *Scientific Reports*, 7(1), 42717. https://doi.org/10.1038/srep42717.

Delavan, B., Roberts, R., Huang, R., Bao, W., Tong, W., & Liu, Z. (2018). Computational drug repositioning for rare diseases in the era of precision medicine. *Drug Discovery Today*, 23(2), 382–394. https://doi.org/10.1016/j.drudis.2017.10.009.

Di, L., & Kerns, E. H. (2015). *Blood-brain barrier in drug discovery: Optimizing brain exposure of CNS drugs and minimizing brain side effects for peripheral drugs*. Wiley.

Doak, B. C., Over, B., Giordanetto, F., & Kihlberg, J. (2014). Oral druggable space beyond the rule of 5: Insights from drugs and clinical candidates. *Chemistry and Biology*, 21(9), 1115–1142. https://doi.org/10.1016/j.chembiol.2014.08.013.

Downs, G. M., & Barnard, J. M. (2002). Clustering methods and their uses in computational chemistry. In K. B. Lipkowitz, & D. B. Boyd (Eds.), *Reviews in computational chemistry* (Vol. 18, pp. 1–40). John Wiley & Sons, Inc. https://doi.org/10.1002/0471433519.ch1.

Erlanson, D. A., Fesik, S. W., Hubbard, R. E., Jahnke, W., & Jhoti, H. (2016). Twenty years on: The impact of fragments on drug discovery. *Nature Reviews Drug Discovery*, 15(9), 605–619. https://doi.org/10.1038/nrd.2016.109.

Feldmann, C., Miljković, F., Yonchev, D., & Bajorath, J. (2019). Identifying promiscuous compounds with activity against different target classes. *Molecules*, 24(22), 4185. https://doi.org/10.3390/molecules24224185.

FILTER. (n.d.). OpenEye Scientific Software. Retrieved July 27, 2020, from http://www.eyesopen.com.

Fink, T., & Reymond, J.-L. (2007). Virtual exploration of the chemical Universe up to 11 atoms of C, N, O, F: Assembly of 26.4 million structures (110.9 million stereoisomers) and analysis for new ring systems, stereochemistry, physicochemical properties, compound classes, and drug discovery. *Journal of Chemical Information and Modeling*, 47(2), 342–353. https://doi.org/10.1021/ci600423u.

Forli, S. (2015). Charting a path to success in virtual screening. *Molecules (Basel, Switzerland)*, 20(10), 18732–18758. https://doi.org/10.3390/molecules201018732.

Fourches, D., Muratov, E., & Tropsha, A. (2010). Trust, but verify: On the importance of chemical structure curation in cheminformatics and QSAR modeling research. *Journal of Chemical Information and Modeling*, 50(7), 1189–1204. https://doi.org/10.1021/ci100176x.

Gally, J.-M., Bourg, S., Fogha, J., Do, Q.-T., Aci-Sèche, S., & Bonnet, P. (2019). VSPrep: A KNIME workflow for the preparation of molecular databases for virtual screening. *Current Medicinal Chemistry*. https://doi.org/10.2174/0929867326666190614160451.

Gatica, E. A., & Cavasotto, C. N. (2012). Ligand and decoy sets for docking to G protein-coupled receptors. *Journal of Chemical Information and Modeling*, 52(1), 1–6. https://doi.org/10.1021/ci200412p.

Gaulton, A., Bellis, L. J., Bento, A. P., Chambers, J., Davies, M., Hersey, A., Light, Y., McGlinchey, S., Michalovich, D., Al-Lazikani, B., & Overington, J. P. (2012). ChEMBL: A large-scale bioactivity database for drug discovery. *Nucleic Acids Research*, 40(Database issue), D1100–D1107. https://doi.org/10.1093/nar/gkr777.

Gaulton, A., Hersey, A., Nowotka, M., Bento, A. P., Chambers, J., Mendez, D., Mutowo, P., Atkinson, F., Bellis, L. J., Cibrián-Uhalte, E., Davies, M., Dedman, N., Karlsson, A., Magariños, M. P., Overington, J. P., Papadatos, G., Smit, I., & Leach, A. R. (2017). The ChEMBL database in 2017. *Nucleic Acids Research*, 45(D1), D945–D954. https://doi.org/10.1093/nar/gkw1074.

Gawehn, E., Hiss, J. A., & Schneider, G. (2016). Deep learning in drug discovery. *Molecular Informatics, 35*(1), 3–14. https://doi.org/10.1002/minf.201501008.

Gimeno, A., Ojeda-Montes, M., Tomás-Hernández, S., Cereto-Massagué, A., Beltrán-Debón, R., Mulero, M., Pujadas, G., & Garcia-Vallvé, S. (2019). The light and dark sides of virtual screening: What is there to know? *International Journal of Molecular Sciences, 20*(6), 1375. https://doi.org/10.3390/ijms20061375.

Gini, G. (2016). QSAR methods. In E. Benfenati (Ed.), *In silico methods for predicting drug toxicity* (Vol. 1425, pp. 1–20). Springer New York. https://doi.org/10.1007/978-1-4939-3609-0_1.

Goh, G. B., Hodas, N. O., & Vishnu, A. (2017). Deep learning for computational chemistry. *Journal of Computational Chemistry, 38*(16), 1291–1307. https://doi.org/10.1002/jcc.24764.

Gorgulla, C., Boeszoermenyi, A., Wang, Z.-F., Fischer, P. D., Coote, P. W., Padmanabha Das, K. M., Malets, Y. S., Radchenko, D. S., Moroz, Y. S., Scott, D. A., Fackeldey, K., Hoffmann, M., Iavniuk, I., Wagner, G., & Arthanari, H. (2020). An open-source drug discovery platform enables ultra-large virtual screens. *Nature, 580*(7805), 663–668. https://doi.org/10.1038/s41586-020-2117-z.

Gupta, M., Lee, H. J., Barden, C. J., & Weaver, D. F. (2019). The blood–brain barrier (BBB) score. *Journal of Medicinal Chemistry, 62*(21), 9824–9836. https://doi.org/10.1021/acs.jmedchem.9b01220.

Hann, M. M., & Oprea, T. I. (2004). Pursuing the leadlikeness concept in pharmaceutical research. *Current Opinion in Chemical Biology, 8*(3), 255–263. https://doi.org/10.1016/j.cbpa.2004.04.003.

Hassan, M., Bielawski, J. P., Hempel, J. C., & Waldman, M. (1996). Optimization and visualization of molecular diversity of combinatorial libraries. *Molecular Diversity, 2*(1–2), 64–74. https://doi.org/10.1007/BF01718702.

Hawkins, P. C. D. (2017). Conformation generation: The state of the art. *Journal of Chemical Information and Modeling, 57*(8), 1747–1756. https://doi.org/10.1021/acs.jcim.7b00221.

Hawkins, P. C. D., Skillman, A. G., Warren, G. L., Ellingson, B. A., & Stahl, M. T. (2010). Conformer generation with OMEGA: Algorithm and validation using high quality structures from the protein databank and Cambridge structural database. *Journal of Chemical Information and Modeling, 50*(4), 572–584. https://doi.org/10.1021/ci100031x.

Hevener, K. E. (2018). Computational toxicology methods in chemical library design and high-throughput screening hit validation. In O. Nicolotti (Ed.), *Computational toxicology* (Vol. 1800, pp. 275–285). Springer New York. https://doi.org/10.1007/978-1-4939-7899-1_13.

Hoffmann, T., & Gastreich, M. (2019). The next level in chemical space navigation: Going far beyond enumerable compound libraries. *Drug Discovery Today, 24*(5), 1148–1156. https://doi.org/10.1016/j.drudis.2019.02.013.

Horvath, D., Lisurek, M., Rupp, B., Kühne, R., Specker, E., von Kries, J., Rognan, D., Andersson, C. D., Almqvist, F., Elofsson, M., Enqvist, P.-A., Gustavsson, A.-L., Remez, N.,

Mestres, J., Marcou, G., Varnek, A., Hibert, M., Quintana, J., & Frank, R. (2014). Design of a general-purpose European compound screening library for EU-OPENSCREEN. *ChemMedChem, 9*(10), 2309–2326. https://doi.org/10.1002/cmdc.201402126.

Huang, N., Shoichet, B. K., & Irwin, J. J. (2006). Benchmarking sets for molecular docking. *Journal of Medicinal Chemistry, 49*(23), 6789–6801. https://doi.org/10.1021/jm0608356.

Hudson, B. D., Hyde, R. M., Rahr, E., Wood, J., & Osman, J. (1996). Parameter based methods for compound selection from chemical databases. *Quantitative Structure-Activity Relationships, 15*(4), 285–289. https://doi.org/10.1002/qsar.19960150402.

Hughes, J. D., Blagg, J., Price, D. A., Bailey, S., DeCrescenzo, G. A., Devraj, R. V., Ellsworth, E., Fobian, Y. M., Gibbs, M. E., Gilles, R. W., Greene, N., Huang, E., Krieger-Burke, T., Loesel, J., Wager, T., Whiteley, L., & Zhang, Y. (2008). Physiochemical drug properties associated with in vivo toxicological outcomes. *Bioorganic and Medicinal Chemistry Letters, 18*(17), 4872–4875. https://doi.org/10.1016/j.bmcl.2008.07.071.

Instant JChem. ChemAxon. Retrieved July 31, 2020, from https://chemaxon.com/products/instant-jchem.

Irwin, J. J. (2008). Using ZINC to acquire a virtual screening library. *Current Protocols in Bioinformatics*. https://doi.org/10.1002/0471250953.bi1406s22 (Chapter 14), Unit 14.6.

Irwin, J. J., & Shoichet, B. K. (2005). ZINC—a free database of commercially available compounds for virtual screening. *Journal of Chemical Information and Modeling, 45*(1), 177–182. https://doi.org/10.1021/ci049714+.

Jing, Y., Easter, A., Peters, D., Kim, N., & Enyedy, I. J. (2015). In silico prediction of hERG inhibition. *Future Medicinal Chemistry, 7*(5), 571–586. https://doi.org/10.4155/fmc.15.18.

Kar, S., & Roy, K. (2013). First report on predictive chemometric modeling, 3D-toxicophore mapping and in silico screening of in vitro basal cytotoxicity of diverse organic chemicals. *Toxicology in Vitro, 27*(2), 597–608. https://doi.org/10.1016/j.tiv.2012.10.015.

Kennard, R. W., & Stone, L. A. (1969). Computer aided design of experiments. *Technometrics, 11*(1), 137–148. https://doi.org/10.1080/00401706.1969.10490666.

Kim, S., Chen, J., Cheng, T., Gindulyte, A., He, J., He, S., Li, Q., Shoemaker, B. A., Thiessen, P. A., Yu, B., Zaslavsky, L., Zhang, J., & Bolton, E. E. (2019). PubChem 2019 update: Improved access to chemical data. *Nucleic Acids Research, 47*(D1), D1102–D1109. https://doi.org/10.1093/nar/gky1033.

Kim, S., Thiessen, P. A., Bolton, E. E., Chen, J., Fu, G., Gindulyte, A., Han, L., He, J., He, S., Shoemaker, B. A., Wang, J., Yu, B., Zhang, J., & Bryant, S. H. (2016). PubChem substance and compound databases. *Nucleic Acids Research, 44*(D1), D1202–D1213. https://doi.org/10.1093/nar/gkv951.

Knox, A. J. S., Meegan, M. J., Carta, G., & Lloyd, D. G. (2005). Considerations in compound database preparation"Hidden" impact on virtual screening results. *Journal of Chemical Information and Modeling, 45*(6), 1908–1919. https://doi.org/10.1021/ci050185z.

Kola, I., & Landis, J. (2004). Can the pharmaceutical industry reduce attrition rates? *Nature Reviews Drug Discovery, 3*(8), 711−716. https://doi.org/10.1038/nrd1470.

Kontijevskis, A. (2017). Mapping of drug-like chemical Universe with reduced complexity molecular frameworks. *Journal of Chemical Information and Modeling, 57*(4), 680−699. https://doi.org/10.1021/acs.jcim.7b00006.

Kontoyianni, M. (2017). Docking and virtual screening in drug discovery. *Methods in Molecular Biology (Clifton, N.J.), 1647,* 255−266. https://doi.org/10.1007/978-1-4939-7201-2_18.

Korkmaz, S., Zararsiz, G., & Goksuluk, D. (2015). MLViS: A web tool for machine learning-based virtual screening in early-phase of drug discovery and development. *PLoS One, 10*(4), e0124600. https://doi.org/10.1371/journal.pone.0124600.

Lagarde, N., Ben Nasr, N., Jérémie, A., Guillemain, H., Laville, V., Labib, T., Zagury, J.-F., & Montes, M. (2014). NRLiSt BDB, the manually curated nuclear receptors ligands and structures benchmarking database. *Journal of Medicinal Chemistry, 57*(7), 3117−3125. https://doi.org/10.1021/jm500132p.

Lagarde, N., Rey, J., Gyulkhandanyan, A., Tufféry, P., Miteva, M. A., & Villoutreix, B. O. (2018). Online structure-based screening of purchasable approved drugs and natural compounds: Retrospective examples of drug repositioning on cancer targets. *Oncotarget, 9*(64), 32346−32361. https://doi.org/10.18632/oncotarget.25966.

Lagarde, N., Zagury, J.-F., & Montes, M. (2015). Benchmarking data sets for the evaluation of virtual ligand screening methods: Review and perspectives. *Journal of Chemical Information and Modeling, 55*(7), 1297−1307. https://doi.org/10.1021/acs.jcim.5b00090.

Lagorce, D., Bouslama, L., Becot, J., Miteva, M. A., & Villoutreix, B. O. (2017). FAF-Drugs4: Free ADME-tox filtering computations for chemical biology and early stages drug discovery. *Bioinformatics, 33*(22), 3658−3660. https://doi.org/10.1093/bioinformatics/btx491.

Lagorce, D., Oliveira, N., Miteva, M. A., & Villoutreix, B. O. (2017). Pan-assay interference compounds (PAINS) that may not be too painful for chemical biology projects. *Drug Discovery Today, 22*(8), 1131−1133. https://doi.org/10.1016/j.drudis.2017.05.017.

Lajiness, M., & Watson, I. (2008). Dissimilarity-based approaches to compound acquisition. *Current Opinion in Chemical Biology, 12*(3), 366−371. https://doi.org/10.1016/j.cbpa.2008.03.010.

Lapenna, S., Fuart Gatnik, M., & Worth, A. (2010). *Review of QSAR models and software tools for predicting acute and chronic systemic toxicity.* Publications Office of the European Union. https://op.europa.eu/en/publication-detail/-/publication/940acf32-3e4d-47cf-b4a1-eebe67cc79ae1/language-en.

Le Guilloux, V., Arrault, A., Colliandre, L., Bourg, S., Vayer, P., & Morin-Allory, L. (2012). Mining collections of compounds with screening assistant 2. *Journal of Cheminformatics, 4*(1), 20. https://doi.org/10.1186/1758-2946-4-20.

Lessel, U., Wellenzohn, B., Lilienthal, M., & Claussen, H. (2009). Searching fragment spaces with feature trees. *Journal of Chemical Information and Modeling, 49*(2), 270−279. https://doi.org/10.1021/ci800272a.

Li, X., Gasteiger, J., & Zupan, J. (1993). On the topology distortion in self-organizing feature maps. *Biological Cybernetics, 70*(2), 189−198. https://doi.org/10.1007/BF00200832.

Life Chemicals (n.d.). Leading supplier of HTS compounds, building blocks. Retrieved July 25, 2020, from https://lifechemicals.com/.

LigPrep (Version Schrödinger Release 2020-3). (n.d.). Schrödinger, LLC, New York, NY, 2020.

Lipinski, C. A., Lombardo, F., Dominy, B. W., & Feeney, P. J. (1997). Experimental and computational approaches to estimate solubility and permeability in drug discovery and development settings. *Advanced Drug Delivery Reviews, 23*(1−3), 3−25. https://doi.org/10.1016/S0169-409X(96)00423-1.

Lyu, J., Wang, S., Balius, T. E., Singh, I., Levit, A., Moroz, Y. S., O'Meara, M. J., Che, T., Algaa, E., Tolmachova, K., Tolmachev, A. A., Shoichet, B. K., Roth, B. L., & Irwin, J. J. (2019). Ultra-large library docking for discovering new chemotypes. *Nature, 566*(7743), 224−229. https://doi.org/10.1038/s41586-019-0917-9.

Martin, Y. C. (2009). Let's not forget tautomers. *Journal of Computer-Aided Molecular Design, 23*(10), 693−704. https://doi.org/10.1007/s10822-009-9303-2.

Mayr, A., Klambauer, G., Unterthiner, T., & Hochreiter, S. (2016). DeepTox: Toxicity prediction using deep learning. *Frontiers in Environmental Science, 3.* https://doi.org/10.3389/fenvs.2015.00080.

Meier, K., Bühlmann, S., Arús-Pous, J., & Reymond, J.-L. (2020). The generated databases (GDBs) as a source of 3D-shaped building blocks for use in medicinal chemistry and drug discovery. *CHIMIA International Journal for Chemistry, 74*(4), 241−246. https://doi.org/10.2533/chimia.2020.241.

Mei, Y., & Yang, B. (2018). Rational application of drug promiscuity in medicinal chemistry. *Future Medicinal Chemistry, 10*(15), 1835−1851. https://doi.org/10.4155/fmc-2018-0018.

Messick, T. E., Smith, G. R., Soldan, S. S., McDonnell, M. E., Deakyne, J. S., Malecka, K. A., Tolvinski, L., van den Heuvel, A. P. J., Gu, B.-W., Cassel, J. A., Tran, D. H., Wassermann, B. R., Zhang, Y., Velvadapu, V., Zartler, E. R., Busson, P., Reitz, A. B., & Lieberman, P. M. (2019). Structure-based design of small-molecule inhibitors of EBNA1 DNA binding blocks Epstein-Barr virus latent infection and tumor growth. *Science Translational Medicine, 11*(482), eaau5612. https://doi.org/10.1126/scitranslmed.aau5612.

Milletti, F., Storchi, L., Sforna, G., Cross, S., & Cruciani, G. (2009). Tautomer enumeration and stability prediction for virtual screening on large chemical databases. *Journal of Chemical Information and Modeling, 49*(1), 68–75. https://doi.org/10.1021/ci800340j.

Mishra, N. K., Singla, D., Agarwal, S., & Raghava, G. P. S. (2014). ToxiPred: A server for prediction of aqueous toxicity of small chemical molecules in *T. Pyriformis. Journal of Translational Toxicology, 1*(1), 21–27. https://doi.org/10.1166/jtt.2014.1005.

Miteva, M. A., Guyon, F., & Tufféry, P. (2010). Frog2: Efficient 3D conformation ensemble generator for small compounds. *Nucleic Acids Research, 38*(Web Server issue), W622–W627. https://doi.org/10.1093/nar/gkq325.

Moroy, G., Martiny, V. Y., Vayer, P., Villoutreix, B. O., & Miteva, M. A. (2012). Toward in silico structure-based ADMET prediction in drug discovery. *Drug Discovery Today, 17*(1–2), 44–55. https://doi.org/10.1016/j.drudis.2011.10.023.

Morphy, R., & Rankovic, Z. (2007). Fragments, network biology and designing multiple ligands. *Drug Discovery Today, 12*(3–4), 156–160. https://doi.org/10.1016/j.drudis.2006.12.006.

Morrison, K. C., & Hergenrother, P. J. (2014). Natural products as starting points for the synthesis of complex and diverse compounds. *Natural Product Reports, 31*(1), 6–14. https://doi.org/10.1039/c3np70063a.

Myshkin, E., Brennan, R., Khasanova, T., Sitnik, T., Serebriyskaya, T., Litvinova, E., Guryanov, A., Nikolsky, Y., Nikolskaya, T., & Bureeva, S. (2012). Prediction of organ toxicity endpoints by QSAR modeling based on precise chemical-histopathology annotations: Prediction of organ toxicity endpoints by QSAR modeling. *Chemical Biology and Drug Design, 80*(3), 406–416. https://doi.org/10.1111/j.1747-0285.2012.01411.x.

Mysinger, M. M., Carchia, M., Irwin, J. J., & Shoichet, B. K. (2012). Directory of useful decoys, enhanced (DUD-E): Better ligands and decoys for better benchmarking. *Journal of Medicinal Chemistry, 55*(14), 6582–6594. https://doi.org/10.1021/jm300687e.

National Center for Biotechnology Information (US). (2010). *Probe reports from the NIH molecular libraries program.*

Newman, D. J., & Cragg, G. M. (2020). Natural products as sources of new drugs over the nearly four decades from 01/1981 to 09/2019. *Journal of Natural Products, 83*(3), 770–803. https://doi.org/10.1021/acs.jnatprod.9b01285.

Ntie-Kang, F., Zofou, D., Babiaka, S. B., Meudom, R., Scharfe, M., Lifongo, L. L., Mbah, J. A., Mbaze, L. M., Sippl, W., & Efange, S. M. N. (2013). AfroDb: A select highly potent and diverse natural product library from African medicinal plants. *PLoS One, 8*(10), e78085. https://doi.org/10.1371/journal.pone.0078085.

Oellien, F., Cramer, J., Beyer, C., Ihlenfeldt, W.-D., & Selzer, P. M. (2006). The impact of tautomer forms on pharmacophore-based virtual screening. *Journal of Chemical Information and Modeling, 46*(6), 2342–2354. https://doi.org/10.1021/ci060109b.

OpenBabel. (n.d.). Retrieved July 27, 2020, from http://openbabel.org/wiki/Main_Page.

Oprea, T. I., Allu, T. K., Fara, D. C., Rad, R. F., Ostopovici, L., & Bologa, C. G. (2007). Lead-like, drug-like or "Pub-like": How different are they? *Journal of Computer-Aided Molecular Design, 21*(1–3), 113–119. https://doi.org/10.1007/s10822-007-9105-3.

Pavlov, D., Rybalkin, M., Karulin, B., Kozhevnikov, M., Savelyev, A., & Churinov, A. (2011). Indigo: Universal cheminformatics API. *Journal of Cheminformatics, 3*(S1), P4. https://doi.org/10.1186/1758-2946-3-S1-P4.

Petit, J., Meurice, N., Kaiser, C., & Maggiora, G. (2012). Softening the rule of five—where to draw the line? *Bioorganic and Medicinal Chemistry, 20*(18), 5343–5351. https://doi.org/10.1016/j.bmc.2011.11.064.

Pizzo, F., Gadaleta, D., Lombardo, A., Nicolotti, O., & Benfenati, E. (2015). Identification of structural alerts for liver and kidney toxicity using repeated dose toxicity data. *Chemistry Central Journal, 9*(1), 62. https://doi.org/10.1186/s13065-015-0139-7.

Polgár, T., Magyar, C., Simon, I., & Keserü, G. M. (2007). Impact of ligand protonation on virtual screening against β-secretase (BACE1). *Journal of Chemical Information and Modeling, 47*(6), 2366–2373. https://doi.org/10.1021/ci700223p.

Poli, G., Seidel, T., & Langer, T. (2018). Conformational sampling of small molecules with iCon: Performance assessment in comparison with OMEGA. *Frontiers in Chemistry, 6*, 229. https://doi.org/10.3389/fchem.2018.00229.

Poongavanam, V., Doak, B. C., & Kihlberg, J. (2018). Opportunities and guidelines for discovery of orally absorbed drugs in beyond rule of 5 space. *Current Opinion in Chemical Biology, 44*, 23–29. https://doi.org/10.1016/j.cbpa.2018.05.010.

Pramanik, S., & Roy, K. (2014). Exploring QSTR modeling and toxicophore mapping for identification of important molecular features contributing to the chemical toxicity in *Escherichia coli. Toxicology in Vitro, 28*(2), 265–272. https://doi.org/10.1016/j.tiv.2013.11.002.

Pushpakom, S., Iorio, F., Eyers, P. A., Escott, K. J., Hopper, S., Wells, A., Doig, A., Guilliams, T., Latimer, J., McNamee, C., Norris, A., Sanseau, P., Cavalla, D., & Pirmohamed, M. (2019). Drug repurposing: Progress, challenges and recommendations. *Nature Reviews Drug Discovery, 18*(1), 41–58. https://doi.org/10.1038/nrd.2018.168.

Rackers, J. A., Wang, Z., Lu, C., Laury, M. L., Lagardère, L., Schnieders, M. J., Piquemal, J.-P., Ren, P., & Ponder, J. W. (2018). Tinker 8: Software tools for molecular design. *Journal of Chemical Theory and Computation, 14*(10), 5273–5289. https://doi.org/10.1021/acs.jctc.8b00529.

Radusky, L., Ruiz-Carmona, S., Modenutti, C., Barril, X., Turjanski, A. G., & Martí, M. A. (2017). LigQ: A webserver to select and prepare ligands for virtual screening. *Journal of Chemical Information and Modeling, 57*(8), 1741–1746. https://doi.org/10.1021/acs.jcim.7b00241.

RDKit. (n.d.). Retrieved July 27, 2020, from https://www.rdkit.org/.

Réau, M., Lagarde, N., Zagury, J.-F., & Montes, M. (2019). Nuclear receptors database including negative data (NR-DBIND): A database dedicated to nuclear receptors binding data including negative data and pharmacological profile. *Journal of Medicinal Chemistry*, 62(6), 2894–2904. https://doi.org/10.1021/acs.jmedchem.8b01105.

Réau, M., Langenfeld, F., Zagury, J.-F., Lagarde, N., & Montes, M. (2018). Decoys selection in benchmarking datasets: Overview and perspectives. *Frontiers in Pharmacology*, 9, 11. https://doi.org/10.3389/fphar.2018.00011.

Reymond, J.-L. (2015). The chemical space project. *Accounts of Chemical Research*, 48(3), 722–730. https://doi.org/10.1021/ar500432k.

Ridder, L., Wang, H., de Vlieg, J., & Wagener, M. (2011). Revisiting the rule of five on the basis of pharmacokinetic data from rat. *ChemMedChem*, 6(11), 1967–1970. https://doi.org/10.1002/cmdc.201100306.

Rodrigues, T., Reker, D., Schneider, P., & Schneider, G. (2016). Counting on natural products for drug design. *Nature Chemistry*, 8(6), 531–541. https://doi.org/10.1038/nchem.2479.

Rognan, D., & Bonnet, P. (2014). Chemical databases and virtual screening. *Medecine Sciences: M/S*, 30(12), 1152–1160. https://doi.org/10.1051/medsci/20143012019.

Ruddigkeit, L., van Deursen, R., Blum, L. C., & Reymond, J.-L. (2012). Enumeration of 166 billion organic small molecules in the chemical universe database GDB-17. *Journal of Chemical Information and Modeling*, 52(11), 2864–2875. https://doi.org/10.1021/ci300415d.

Sander, T., Freyss, J., von Korff, M., & Rufener, C. (2015). Data-Warrior: An open-source program for chemistry aware data visualization and analysis. *Journal of Chemical Information and Modeling*, 55(2), 460–473. https://doi.org/10.1021/ci500588j.

Saubern, S., Guha, R., & Baell, J. B. (2011). KNIME workflow to assess PAINS filters in SMARTS format. Comparison of RDKit and Indigo cheminformatics libraries. *Molecular Informatics*, 30(10), 847–850. https://doi.org/10.1002/minf.201100076.

Senger, M. R., Fraga, C. A. M., Dantas, R. F., & Silva, F. P. (2016). Filtering promiscuous compounds in early drug discovery: Is it a good idea? *Drug Discovery Today*, 21(6), 868–872. https://doi.org/10.1016/j.drudis.2016.02.004.

Serafin, M. B., Bottega, A., Foletto, V. S., da Rosa, T. F., Hörner, A., & Hörner, R. (2020). Drug repositioning is an alternative for the treatment of coronavirus COVID-19. *International Journal of Antimicrobial Agents*, 55(6), 105969. https://doi.org/10.1016/j.ijantimicag.2020.105969.

Sharma, A. K., Srivastava, G. N., Roy, A., & Sharma, V. K. (2017). ToxiM: A toxicity prediction tool for small molecules developed using machine learning and chemoinformatics approaches. *Frontiers in Pharmacology*, 8, 880. https://doi.org/10.3389/fphar.2017.00880.

Shemetulskis, N. E., Dunbar, J. B., Dunbar, B. W., Moreland, D. W., & Humblet, C. (1995). Enhancing the diversity of a corporate database using chemical database clustering and analysis. *Journal of Computer-Aided Molecular Design*, 9(5), 407–416. https://doi.org/10.1007/BF00123998.

Shi, Y., & von Itzstein, M. (2019). How size matters: Diversity for fragment library design. *Molecules*, 24(15), 2838. https://doi.org/10.3390/molecules24152838.

Singh, P. K., Negi, A., Gupta, P. K., Chauhan, M., & Kumar, R. (2016). Toxicophore exploration as a screening technology for drug design and discovery: Techniques, scope and limitations. *Archives of Toxicology*, 90(8), 1785–1802. https://doi.org/10.1007/s00204-015-1587-5.

Siramshetty, V. B., Eckert, O. A., Gohlke, B.-O., Goede, A., Chen, Q., Devarakonda, P., Preissner, S., & Preissner, R. (2018). SuperDRUG2: A one stop resource for approved/marketed drugs. *Nucleic Acids Research*, 46(D1), D1137–D1143. https://doi.org/10.1093/nar/gkx1088.

Sitzmann, M., Ihlenfeldt, W.-D., & Nicklaus, M. C. (2010). Tautomerism in large databases. *Journal of Computer-Aided Molecular Design*, 24(6–7), 521–551. https://doi.org/10.1007/s10822-010-9346-4.

Sommer, K., Friedrich, N.-O., Bietz, S., Hilbig, M., Inhester, T., & Rarey, M. (2016). UNICON: A powerful and easy-to-use compound library converter. *Journal of Chemical Information and Modeling*, 56(6), 1105–1111. https://doi.org/10.1021/acs.jcim.6b00069.

Sorokina, M., & Steinbeck, C. (2020). Review on natural products databases: Where to find data in 2020. *Journal of Cheminformatics*, 12(1), 20. https://doi.org/10.1186/s13321-020-00424-9.

Standardizer. (n.d.). ChemAxon. https://chemaxon.com/products/chemical-structure-representation-toolkit.

Sterling, T., & Irwin, J. J. (2015). Zinc 15—ligand discovery for everyone. *Journal of Chemical Information and Modeling*, 55(11), 2324–2337. https://doi.org/10.1021/acs.jcim.5b00559.

Surade, S., & Blundell, T. L. (2012). Structural biology and drug discovery of difficult targets: The limits of ligandability. *Chemistry and Biology*, 19(1), 42–50. https://doi.org/10.1016/j.chembiol.2011.12.013.

Sushko, I., Salmina, E., Potemkin, V. A., Poda, G., & Tetko, I. V. (2012). ToxAlerts: A web server of structural alerts for toxic chemicals and compounds with potential adverse reactions. *Journal of Chemical Information and Modeling*, 52(8), 2310–2316. https://doi.org/10.1021/ci300245q.

Tarcsay, Á., & Keserű, G. M. (2013). Contributions of molecular properties to drug promiscuity: Miniperspective. *Journal of Medicinal Chemistry*, 56(5), 1789–1795. https://doi.org/10.1021/jm301514n.

Tcheremenskaia, O., Benigni, R., Nikolova, I., Jeliazkova, N., Escher, S. E., Batke, M., Baier, T., Poroikov, V., Lagunin, A., Rautenberg, M., & Hardy, B. (2012). OpenTox predictive toxicology framework: Toxicological ontology and semantic media wiki-based OpenToxipedia. *Journal of Biomedical Semantics*, 3(Suppl. 1), S7. https://doi.org/10.1186/2041-1480-3-S1-S7.

Trepalin, S. V., Skorenko, A. V., Balakin, K. V., Nasonov, A. F., Lang, S. A., Ivashchenko, A. A., & Savchuk, N. P. (2003). Advanced exact structure searching in large databases of chemical compounds. *Journal of Chemical Information and Computer Sciences*, *43*(3), 852–860. https://doi.org/10.1021/ci025582d.

Tyagi, M., Begnini, F., Poongavanam, V., Doak, B. C., & Kihlberg, J. (2020). Drug syntheses beyond the rule of 5. *Chemistry - A European Journal*, *26*(1), 49–88. https://doi.org/10.1002/chem.201902716.

Ursu, O., Holmes, J., Bologa, C. G., Yang, J. J., Mathias, S. L., Stathias, V., Nguyen, D.-T., Schürer, S., & Oprea, T. (2019). DrugCentral 2018: An update. *Nucleic Acids Research*, *47*(D1), D963–D970. https://doi.org/10.1093/nar/gky963.

Veber, D. F., Johnson, S. R., Cheng, H.-Y., Smith, B. R., Ward, K. W., & Kopple, K. D. (2002). Molecular properties that influence the oral bioavailability of drug candidates. *Journal of Medicinal Chemistry*, *45*(12), 2615–2623. https://doi.org/10.1021/jm020017n.

Vedani, A., Dobler, M., & Smieško, M. (2012). VirtualToxLab—a platform for estimating the toxic potential of drugs, chemicals and natural products. *Toxicology and Applied Pharmacology*, *261*(2), 142–153. https://doi.org/10.1016/j.taap.2012.03.018.

Villoutreix, B. O., Renault, N., Lagorce, D., Montes, M., & Miteva, M. A. (2007). Free resources to assist structure-based virtual ligand screening experiments. *Current Protein and Peptide Science*, *8*(4), 381–411. https://doi.org/10.2174/138920307781369391.

Visini, R., Awale, M., & Reymond, J.-L. (2017). Fragment database FDB-17. *Journal of Chemical Information and Modeling*, *57*(4), 700–709. https://doi.org/10.1021/acs.jcim.7b00020.

Vogt, M. (2020). How do we optimize chemical space navigation? *Expert Opinion on Drug Discovery*, *15*(5), 523–525. https://doi.org/10.1080/17460441.2020.1730324.

Wang, S., Witek, J., Landrum, G. A., & Riniker, S. (2020). Improving conformer generation for small rings and macrocycles based on distance geometry and experimental torsional-angle preferences. *Journal of Chemical Information and Modeling*, *60*(4), 2044–2058. https://doi.org/10.1021/acs.jcim.0c00025.

Waring, M. J., Arrowsmith, J., Leach, A. R., Leeson, P. D., Mandrell, S., Owen, R. M., Pairaudeau, G., Pennie, W. D., Pickett, S. D., Wang, J., Wallace, O., & Weir, A. (2015). An analysis of the attrition of drug candidates from four major pharmaceutical companies. *Nature Reviews Drug Discovery*, *14*(7), 475–486. https://doi.org/10.1038/nrd4609.

Watts, K. S., Dalal, P., Murphy, R. B., Sherman, W., Friesner, R. A., & Shelley, J. C. (2010). ConfGen: A conformational search method for efficient generation of bioactive conformers. *Journal of Chemical Information and Modeling*, *50*(4), 534–546. https://doi.org/10.1021/ci100015j.

Williams, A. J., Ekins, S., & Tkachenko, V. (2012). Towards a gold standard: Regarding quality in public domain chemistry databases and approaches to improving the situation. *Drug Discovery Today*, *17*(13–14), 685–701. https://doi.org/10.1016/j.drudis.2012.02.013.

Willighagen, E. L., Mayfield, J. W., Alvarsson, J., Berg, A., Carlsson, L., Jeliazkova, N., Kuhn, S., Pluskal, T., Rojas-Chertó, M., Spjuth, O., Torrance, G., Evelo, C. T., Guha, R., & Steinbeck, C. (2017). The Chemistry Development Kit (CDK) v2.0: Atom typing, depiction, molecular formulas, and substructure searching. *Journal of Cheminformatics*, *9*(1), 33. https://doi.org/10.1186/s13321-017-0220-4.

Wishart, D. S., Feunang, Y. D., Guo, A. C., Lo, E. J., Marcu, A., Grant, J. R., Sajed, T., Johnson, D., Li, C., Sayeeda, Z., Assempour, N., Iynkkaran, I., Liu, Y., Maciejewski, A., Gale, N., Wilson, A., Chin, L., Cummings, R., Le, D., … Wilson, M. (2018). DrugBank 5.0: A major update to the DrugBank database for 2018. *Nucleic Acids Research*, *46*(D1), D1074–D1082. https://doi.org/10.1093/nar/gkx1037.

Würth, R., Thellung, S., Bajetto, A., Mazzanti, M., Florio, T., & Barbieri, F. (2016). Drug-repositioning opportunities for cancer therapy: Novel molecular targets for known compounds. *Drug Discovery Today*, *21*(1), 190–199. https://doi.org/10.1016/j.drudis.2015.09.017.

Yosipof, A., & Senderowitz, H. (2014). Optimization of molecular representativeness. *Journal of Chemical Information and Modeling*, *54*(6), 1567–1577. https://doi.org/10.1021/ci400715n.

CHAPTER 6

3D Structural Determination of Macromolecules Using X-ray Crystallography Methods

MUTHARASAPPAN NACHIAPPAN • RAVI GURU RAJ RAO • MARIADASSE RICHARD • DHAMODHARAN PRABHU • SUNDARRAJ RAJAMANIKANDAN • JEYARAJ PANDIAN CHITRA • JEYARAMAN JEYAKANTHAN

1 INTRODUCTION

X-ray crystallography is a versatile tool in structural biology to determine the atomic and molecular structure of a crystal. The basic principle in working of X-ray crystallography is that the crystalline atoms diffract X-rays to several specific directions whose intensity and angle of the diffracted beams generate three-dimensional (3D) electron density image from which the mean position of atoms in a crystal, their chemical bonds, and disorder can be determined. X-ray crystallography still remains the most powerful method to determine the structure of macromolecular crystals including proteins, DNA, drugs, vitamins, etc. This chapter outlines all the basic procedures right from growing protein crystals to evaluating suitable optimal conditions for the growth of the crystal. Various advance methods for data collection namely X-ray, synchrotron, and X-ray free electron lasers (XFELs) are also presented. In addition, the chapter presents the processing of data in macromolecular crystallography. Solve phase, structure refinement, and model building are also revisited. The process of characterization and structure validation is also discussed. The chapter concludes by briefing the applications of solved X-ray in structure-based drug design (SBDD), computer-aided drug design, and docking studies. This chapter provides the strategies for obtaining better diffraction patterns, reduction in scattering of X-ray by water, and prevention of radiation damage to the macromolecule crystals. The overall steps involved in solving 3D structure of macromolecules using X-ray crystallography method is shown in Fig. 6.1.

2 PROTEIN PURIFICATION

A pure protein is a prerequisite for studying about its structure, physiological function, and mechanism of action. Specific proteins can be isolated from crude mixtures using appropriate purification strategies based on the size, binding affinity, biological, physical, and chemical properties. All biomolecules can be purified according to the specific properties that each protein possesses (Table 6.1).

Ion exchange chromatography (IEC) shown in Fig. 6.2A is a widely used technique in column chromatography. Based on the affinity of ions and polar solvents to the ion exchangers, they are separated in IEC process. The principle of IEC is reversible exchange of ions between the target ions and ion exchangers on the matrix. IEC is widely classified into cationic and anionic exchange chromatography. Gel filtration or size exclusion chromatography (SEC) is the simplest of all the chromatographic techniques. SEC usually separates proteins based on their molecular size or mass as they pass through resin of the chromatographic column (Fig. 6.2B). Unlike other chromatography, the molecules do not bind to the resins and the buffer composition does not affect the resolution of the eluting protein. SEC adds a polishing step to the protein molecule (Nachiappan et al., 2018). Affinity

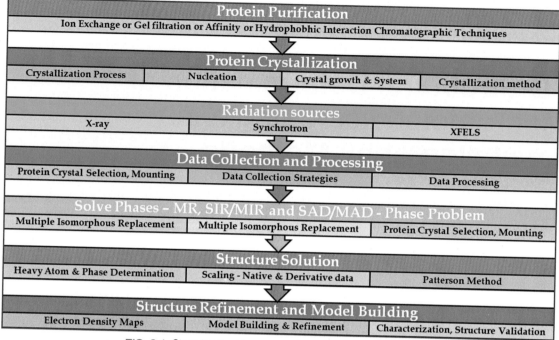

FIG. 6.1 Steps involved in macromolecular X-ray crystallography.

TABLE 6.1
Properties and Types of Chromatography.

Properties	Techniques
Charge	Ion exchange chromatography
Size	Gel filtration chromatography/size exclusion chromatography
Hydrophobicity	Hydrophobhic interaction chromatography/reversed phase chromatography
Binding domain (ligand specificity)	Affinity chromatography

chromatography (Fig. 6.2C) usually separates the protein based on the binding of protein or ligand to the chromatography matrix. This is highlighted to be the only technique that enables the purification by its biological functions and offers high resolution and selectivity. The binding of ligand and protein molecule can be mediated by hydrogen bonding, van der Waals, and hydrophobic interactions (Prabhu et al., 2020; Yamaguchi, 2015). Target molecules can be eluted from the affinity column by specific or nonspecific

reversing of the process by using competitive ligands and by altering the polarity, strength, or pH of the medium, respectively. Hydrophobhic interaction chromatography (Fig. 6.2D) is also a widely used technique for separating and purifying proteins in their native state. It is commonly referred to as "salting out" and works with the properties of hydrophobicity. Under high salt buffer conditions, proteins containing both hydrophobic and hydrophilic regions are salted out.

3 PROTEIN CRYSTALLIZATION

Crystallization of protein is the first step required for the determination of protein structure by X-ray crystallography. Because the success of macromolecular X-ray crystallography depends mainly on the quality of protein crystals, several advanced techniques have been developed to grow high-quality protein crystals. Of these, vapor diffusion, free interface diffusion, microbatch, and dialysis methods are most commonly used (Geigé & McPheson, 2001; McPherson, 1982).

3.1 Crystallization Process

Crystallization is not only a natural process but also a chemical method used for the separation of solid and liquid in which solid crystals are formed from homogeneous solution. The solution should be in supersaturated condition for crystallization process.

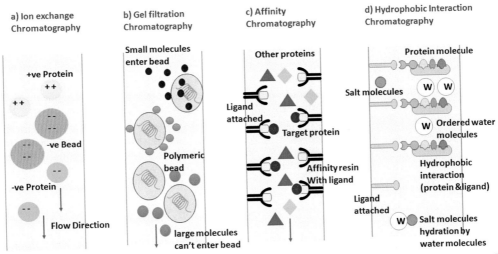

FIG. 6.2 Principle and types of chromatography- a) Ion exchange, b) Gel filtration, c) Affinity and d) Hydrophobic chromatography.

Therefore, the solution should contain higher concentration of dissolved solutes namely molecules or ions than it would be present in the equilibrium state of the saturated solution, which can be achieved by several common methods used in industries such as chemical reaction, cooling of solution, changing the pH, and addition of second solvent to decrease the solute solubility. Other techniques such as solvent evaporation are also be employed.

Nucleation and crystal growth are the two major steps in crystallization process. Nucleation involves the formation of cluster of nanometer (nm) distances by the homogeneously dispersed solute molecules in the solvent that results in increasing the concentration of solute in a small region and stabilization of the current operating conditions. Nuclei are the compact crystal structures that form stable clusters. Unstable clusters cause redissolving of crystals in the solution. Hence, to develop stable nuclei, clusters should have a critical size which depends on the working factors such as irregularities, temperature, supersaturation, etc. In nucleation stage, the periodic and definite arrangement of atoms defines the structure of crystal. "Crystal structure" specifies the internal atom arrangement and not the macroscopic properties of the crystal size and shape (Chayen & Saridakis, 2008).

The crystal growth refers to critical cluster size that can be achieved by subsequent growth of nuclei. As long as the phenomenon of supersaturation persists, nucleation and growth continue to occur simultaneously. Supersaturation is essential for the crystallization process and therefore the nucleation rate and crystal growth are determined by the obtainable solution's supersaturation condition. Different sizes and shapes of the crystals depend on the solution conditions that determine whether crystal growth or nucleation is predominant over others. Solid–liquid system reaches equilibrium after exhaustion of the supersaturation which starts completing crystallization till modifications are made to supersaturate the solution over again.

3.2 Nucleation

Nucleation event occurs in all natural as well as artificial crystallization processes involving the formation of solid crystals from a homogeneous solution. Releasing the pressure inside the container containing carbonated liquid causes the release of carbon dioxide nucleate. Nucleation process predominantly occurs at preexisting interface as in the case of boiling chips and in the strings used for making rock candy (e.g., Mentos eruptions). Nucleation can occur when the pressure is reduced and then during boiling the bulk liquid in which the superheating of liquid relies on the boiling point is dependent on the pressure of the working condition. Nucleation process is predominantly followed by heating the surface of nucleation sites. Moreover, nucleation sites are small crevices in which the maintenance or spotting occurs on free gas–liquid surface or heating surfaces with lesser wetting property, respectively.

Significant superheating of liquid is obtained after degassing the liquid and material in heating surface that can be wetted well by the liquid. Principle of nucleation is widely applied in ceramic, alloy, and polymer production. In chemistry, nucleation process also refers to the formation of phaseless multimers that are intermediates in polymerization reactions and are widely employed in amyloidogenesis and crystallization. In molecular biology, nucleation refers to a stage in assembly of polymeric structures, namely, microtubule, which is a cluster of small monomers that aggregate and arrange to speed up polymerization. Another example is the formation of actin polymers which gets stabilized by the addition of third actin molecules to the weekly bound two actin molecules. This actin trimer further adds molecules resulting in a nucleation site that enhances the lag phase of polymerization reaction.

3.3 Mechanics of Nucleation

3.3.1 Homogeneous nucleation

Homogeneous nucleation is a difficult process that occurs within the interior of the uniform substance. Supercooled liquids are the liquids that are cooled below their melting temperature (heterogenous nucleation temperature to the maximum) or cooled above the freezing temperature of pure substance (homogeneous nucleation temperature) and are useful for synthesis of metastable structures including amorphous solids. This process can also produce undesirable effects in casting or can delay the chemical processes in industries. The formation of nucleus is an energy requiring process that infers the interface formation at the boundaries of the new phase based on its surface energy. In the case of very small hypothetical nucleus, the energy is released when the volume is insufficient to create its surface and therefore the nucleation stops. Nucleation proceeds if the length of the critical nucleus decreases to a critical value. Progressive increase in the small nuclei to viable large surface occurs due to available energy in the favorable transformation phase which will eventually cause thermal activation providing energy for stable nuclei formation. This process facilitates the growth of nuclei until restoration of thermodynamic equilibrium.

3.3.2 Heterogeneous nucleation

In heterogenous nucleation, destruction of previous interface causes the release of energy. For instance, during bubbling of carbon dioxide between inner surface of bottle and water, energy is released that drives toward the formation of interfaces between bubble and water as well as between bubble and bottle. Precipitate particles can be formed from the similar effect at the solid's grain boundaries and interfere in precipitation strength which depends on homogeneous nucleation to synthesize uniformly distributed precipitate particles.

3.4 Crystal Growth

Next to nucleation stage, crystal growth is the key stage in crystallization process which occurs due to accumulation of ions, new atoms, or polymer strings into lattice of a crystal. Small nucleus comprising of newly formed crystal is produced in nucleation stage. Nucleation process occurs slowly because the initial compartments of crystal should bump in precise orientation and location that will enhance them to form crystal. Crystal growth rapidly follows after crystal nucleation stage which occurs rapidly outside the nucleation site or in prearranged system and the motif forming elements adds to growing crystal. As stated by Charles Frank in 1951, real crystals grow rapidly compared to perfect crystals because they contain dislocations that offer the necessary growth points.

Nucleation can be either heterogeneous or homogeneous with or without the influence of foreign particles, respectively, because in heterogeneous nucleation, foreign particles act as a scaffold enhancing crystal growth. Heterogeneous nucleation occurs by several methods; for instance, the crystals grow on the small inclusions or cuts in the container, scratches in the sides or bottom of the glassware and other nucleating sites like dust, or other random particles in air. Therefore, addition of foreign substances such as string or rock to the solution could provide the nucleating site to speed up the crystal growth.

The quantity of nucleating sites can also be regulated by aforementioned process. For instance, the surface of a new glass container is very smooth to permit heterogenous nucleation, while a scratched container will produce many small crystals (Fig. 6.3). Therefore, making few scratches in the container works better to obtain adequate quantity of medium-sized crystals. Similarly, addition of formerly synthesized seed crystals or small crystals to the growing crystal will provide the nucleation site in the solution. Different stages of crystal growth in crystallization process are visualized in Fig. 6.3 that differentiate the precipitate, microcrystals, and macrocrystals.

Few key features in crystal growth include the arrangement, origin of crystal growth, and the interface form into final size. Anisotropic crystals or crystals with diverse properties in different direction can be formed if the origin of crystal growth occurs unidirectional for all the crystals. The additional free energy for

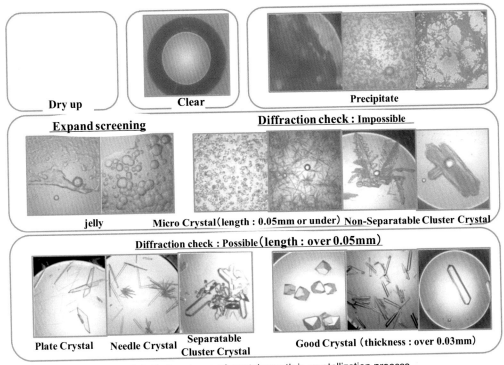

Dry up   Clear   Precipitate

Expand screening   **Diffraction check : Impossible**

jelly   Micro Crystal(length : 0.05mm or under)   Non-Separatable Cluster Crystal

Diffraction check : Possible (length : over 0.05mm)

Plate Crystal   Needle Crystal   Separatable Cluster Crystal   Good Crystal (thickness : over 0.03mm)

FIG. 6.3 Various types of crystal growth in crystallization process.

crystal growth (each volume) is determined by the interface form. Metal lattice arrangement commonly occurs in the structures like hexagonal close packed or face- or body-centered cubic. The final crystal size is significant to determine the mechanical properties of materials. Long deformation path and low internal stresses can stretch large crystals further in metals (Wiencek, 1999).

3.5 Crystallization Methods

The protein to be crystallized often dictates the method of choice. The vapor diffusion method relies on the diffusion of vaporizable species between solutions of two different concentrations in achieving supersaturation of the solute protein. Two most popular methods used to achieve this state are the hanging drop method and the sitting drop method (Fig. 6.4) (Stevens, 2000).

In the hanging drop method, the protein solution is mixed with a buffered solution containing precipitant and additives. The mixture is then placed on a siliconized glass cover slip and suspended over the well of a tissue culture plate containing the reservoir solution. The volume of the reservoir solution, which contains the precipitant and additives at a higher concentration, is much larger (few hundred microliters) than the drop

volume. The ratio of the protein solution and the precipitant solution and volume (few microliters) of the drop is varied to optimize the crystal growth. The sitting drop method involves a very similar procedure except that the protein drop is kept seated over the reservoir buffer instead of being suspended from the ceiling. Therefore, an increased volume of protein solution could be applied for crystal growth. The benefit of using the vapor diffusion method and the hanging drop method is to decrease the protein requirement and to facilitate the screening process for optimizing the growth of protein crystals.

3.6 Crystal System

Foremost constraint in X-ray crystallography is acquiring the protein crystals; most of the proteins are difficult to crystallize due to several factors that influence the formation of crystal packing such as protein solubility, purity, protein homogeneity, pH, temperature of crystal growth, crystallization type, crystallization conditions, vibration free environment, etc. Moreover, there are no straightforward procedures available for obtaining a good diffractable crystal and this has to be experimented with all possibilities (Dauter, 2017).

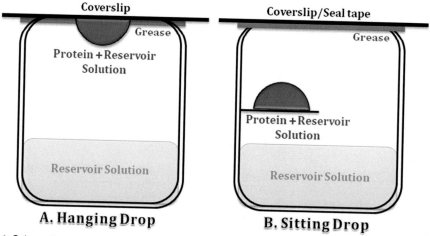

FIG. 6.4 Schematic representation of the crystallization vapor diffusion technique. **(A)** Hanging drop and **(B)** sitting drop method of crystallization.

Protein crystals are composed of molecules which are formed in a repeated and periodic pattern occurring in 3D space. This repeating unit or unit cell is defined by specifying its lengths (a, b, and c denoting its edges) and three angles between its corresponding axes (α, β, and γ). These six factors are represented as lattice parameters. Based on unit cell parameters, the crystal systems are grouped into seven different types (Table 6.2).

There are in total 32 possible combinations of symmetry operation (inversion, rotation, and reflection) in a 3D lattice termed as point groups. There are four types of crystal lattice namely primitive (P), face-centered (F), C-centered (A, B, and C), and body-centered (I), which has lattice point at every corner of unit cell, on all corner with additional point on face of each lattice, on all corner with extra points only on center of any two opposite faces, and on all corner with extra point on center of crystal lattice, respectively. The four-lattice point with seven crystal system forms 14 unique Bravais lattices (Bravais, 1849). With the described symmetry, translational, screw, and glide planes, there can be total of 230 possible crystallographic space groups. All crystals formed should come under these possible space groups; because the protein molecules are chiral, it will form only from the 65 enantiomorphic space groups. The plane of atoms developing diffraction patterns formed on exposure of protein crystal to X-ray scatter provides structural information of the molecule.

TABLE 6.2 The Crystal System.		
Crystal System	**Unit Cell Parameter**	**Symmetry**
Triclinic	$a \neq b \neq c$, $\alpha \neq \beta \neq \gamma \neq 90$ degrees	None
Monoclinic	$a \neq b \neq c$, $\alpha = \gamma = 90$ degrees $\beta \neq 90$ degrees	One twofold axis
Orthorhombic	$a \neq b \neq c$, $\alpha = \beta = \gamma = 90$ degrees	Three twofold axes
Tetragonal	$a = b \neq c$, $\alpha = \beta = \gamma = 90$ degrees	One fourfold axis
Trigonal	$a = b = c$, $\alpha = \beta = \gamma = 90$ degrees, $\gamma < 120$ degrees	One threefold axis
Hexagonal	$a = b \neq c$, $\alpha = \beta = 90$ degrees, $\gamma = 120$ degrees	One sixfold axis
Cubic	$a = b = c$, $\alpha = \beta = \gamma = 90$ degrees	Fourfold axis

4 X-RAY, SYNCHROTRON, AND XFELS

4.1 X-ray

X-ray was discovered by WC Rontgen in 1895 when he was experimenting with cathode ray tube (Röntgen, 1895). X-rays are broad range of electromagnetic radiation having variable wavelengths and energies ranging from 0.1 to 100 Å and 100 eV to 10 MeV, respectively. Fast-moving electrons when hit on the heavy metal decelerate and convert their kinetic energy into radiation to produce X-rays. X-rays are produced by first rotating anode generator where electron gun emits electrons that were accelerated by high voltage (40−50 kV), which was rapidly decelerated by heavy metal target (copper or molybdenum) (McPherson et al., 1993). When the electron hits the heavy metal, electron from low orbital is displaced and excess energy is radiated as X-ray. The efficiency of this generator is low as most of the supplied energy is lost in the form of heat. To prevent this loss of energy as heat, water cooling system is set up in both traditional cathode tube and in rotating anode such as copper or in molybdenum head that rotates at high rpm. The electron beam striking the target is constantly changed, thus dissipating heat efficiently (Souza et al., 2000). Thus, high vacuum has to be maintained to eliminate the collision between gas molecules during the production of high-intensity X-ray (5−18 kW) when compared with the traditional tubes. There are certain merits in using anode X-ray generator namely inexpensive operating and maintenance, user-friendly, efficient energy consumption, and space. Similarly, demerits of X-ray intensity are considerably less when compared with the synchrotron where fixed wavelength and crystals of small dimensions cannot be experimented.

4.2 Synchrotron Radiation

The synchrotron radiations are produced when electrons or positrons are accelerated close to relativistic speeds that are constrained to shift its direction in the effect of magnetic field. These radiations are enormously intense and emit broad array of electromagnetic spectrum from infrared rays to hard X-rays (Mills et al., 2005). Synchrotron radiation was first detected at General Electric, USA, in 1947, and it seems to be a massive energy loss problem at particle accelerator that was then found to be a huge experimental setup in late 1960s (Winick, 1998). Synchrotron radiations are incredibly large and expensive to maintain and usually need a huge city block space to construct. The arrangement mainly consists of electrons source accelerated to relativistic speed such as storage rings where in vacuum, electrons are restricted to move in a circular path under the influence of magnets along its circumference (Kirk, 2004). Electrons enter the storage ring only when it is accelerated by linear accelerator (linac) reaching numerous millions of electron volt and then moves

to booster rings where the energy is boosted to few giga volts. The first-generation radiation sources are generated by particle accelerators without using the synchrotron radiation. But then synchrotron radiation focuses mainly based on bending magnets or high magnetic field devices like wigglers that are used to produce radiation termed as second generation. With the advancement in design and magnetic devices such as both wigglers and undulators, it results in production of high-intensity radiation termed as third generation. As most electromagnetic spectrum from infrared to hard X-ray region can be obtained, synchrotron radiation has a wide range of applications in different fields such as material chemistry, magnetic materials, protein crystallography, microscopy, and surface science (Balerna & Mobilio, 2015). Currently, there are over 40 synchrotrons spread through 5 different continents available to users for various applications. One such third-generation synchrotron radiation facility is SPring-8 (Super Photon ring-8 GeV) in Japan that delivers powerful radiation capacity at 8 GeV (Fig. 6.5).

The main advantage of using synchrotron radiation is its high intensity that enables microcrystal (<10 μM dimensions) to be experimented because microcrystals have a very weak diffraction which cannot be done by conventional source. Moreover, the tunability using multiwavelength anomalous dispersion (MAD) or single-wavelength anomalous dispersion (SAD) provides the phase information of the crystal that facilitates researchers to solve the difficult unique structures. This has enabled several researchers to solve the crystal structure rapidly with high resolution of 1.2 Å or less. Microcrystals that are thought to be useless are solved using synchrotron radiations efficiently (Chen et al., 2002).

4.3 X-ray Free Electron Lasers

XFELs are advanced form of the third-generation synchrotron radiation in which improved undulator is used so that the electron could interact coherently with the emitted radiation when compared with the incoherent in third-generation synchrotrons. The main advantage over the previous generation is its full coherent X-ray in hard region and its femtosecond pulse length with high brilliance, i.e., intensity. With the high pulse, experiments that happen extremely fast like metabolic function can be validated using XFELs (Chapman, 2019). But, the construction and maintenance of XFELs are very expensive as it needs linear structure stretching to few kilometers with high power to operate.

FIG. 6.5 The view of world's largest third-generation synchrotron radiation facility SPring-8 located at Harima Science Park City, Hyogo Prefecture, Japan. (Source: Official website of SPring-8, Japan.)

5 DATA COLLECTION AND PROCESSING

Obtaining a protein crystal is a primary prerequisite for structure determination by X-ray crystallography, but not all crystals will fetch good diffraction as the crystal shape, morphology, and packing will play a key role in analyzing the diffraction quality. The basis for X-ray crystallography was laid by the Bragg's law, which states that a constructive coherent interference must arise to obtain a diffraction spot $n\lambda = 2d\sin\theta$ condition, which has to be fulfilled, where λ, d, and θ represent X-ray wavelength, interplane distance, and the angle between the incident ray and planes, respectively (Bragg, 1913).

Initially, data collections were recorded in X-ray films and are processed using laborious and time-consuming traditional photographic film where crystals are exposed for hours together with different crystals. The next advancement was the image plate that subsequently reduced the data collection time and digitalized the diffraction images that can be processed in a workstation. This was superseded by charged couple device detectors that are sensitive, accurate, and superfast so that each diffraction data can be collected within few seconds. With the latest advancements in synchrotrons, data can now be collected remotely from the leisure of lab with a simple laptop. Thus, technological innovations have streamlined the intricate process with high reliability, accuracy, and speed. Data collection is the essential experimental procedure for determination of structure, but the rest of process is done by advanced computer programs and calculations that enable novice user without much technicality to solve protein structures (Powell, 2017). A typical synchrotron radiation instrumentation used for crystallographic analysis of macromolecules is shown in Fig. 6.6.

5.1 Protein Crystal Selection and Mounting

The obtained protein crystal has to be scrutinized before data collection; it needs to be well shaped with sharp faces without reentrant angle, thus eliminating the possibilities of twining. Data collected from twined crystal will be difficult to process, as they contain independent diffraction pattern that has to be processed separately. Protein crystals will usually contain solvent that is crucial for the integrity of packing. This can be achieved by two widely used methods such as capillary mounting where crystals are closed in glass capillary with lower amount of mother liquor in it. Crystals are fished in a small fiber loop (0.025−1 mm diameter) with mother liquor flash frozen in 100 K liquid nitrogen (Hope, 1988) in flash freezing process. Each method has its own advantage over other. Moreover, the crystals that cannot be frozen could use the capillary tube to collect the data at room temperature, as freezing may affect the diffraction quality of crystals. But, flash freezing will

FIG. 6.6 A typical beamline from TPS 05A at National Synchrotron Radiation Research Centre (NSRRC), Taiwan, (1) detector with motorized platform where detector to crystal distance can be modified, (2) motorized spindle where the crystal is mount for diffraction, (3) liquid nitrogen stream knob, (4) X ray beam runs tangentially from the storage ring, (5) automated crystal mounter used to mount crystal on to the spindle. This complete setup is housed in a lead hatch to protect user from radiation.

reduce the mechanical stress and intrude and reduce the radiation damage in a crystal; thus, it will be helpful in data collection. Handling and transporting flash-frozen crystals are easier when compared to capillary tube for synchrotron trips. The disadvantage of flash freezing is mosaicity, which can be minimized by using cryo-protectant namely glycerol, PEG 400, ethylene glycol, and glucose (10%−25%) (Garman, 1999). This may affect the diffraction quality, so cryo-protectant has to be checked prior to data collection. These are the important parameters that have to be considered before data collection of protein crystals.

5.2 Data Collection Strategies

The data collection strategy will be governed by different parameters such as X-ray intensity, detectors, crystal quality, crystal symmetry, radiation damage, resolution, structure solution method, and application (Aitipamula & Vangala, 2017). Protein crystals are usually tested before data collection to obtain maximum structural information with highest resolution possible. First, the X-ray beam position is altered corresponding to the detector with the collimator and then the crystals are mounted in goniometer head spindles (Fig. 6.6), which can be rotated 360 degrees to maintain its direct exposure to X-ray beam. The distance between crystal and detector is tweaked to get highest resolution

possible; the lesser the distance, the more will be the resolution attained. The oscillation degree φ that can be set from 0.1 degree to 2 degrees swings; X-ray exposure time can be varied based on the X-ray intensity. Detector sensitivity also plays a vital role as the read-out times will be faster in synchrotrons like 10−30 min and slower in in-home sources that may take few hours to collect the entire dataset. Lower oscillation degree will significantly increase the signal-to-noise ratio that enhances better intensity data collection. Once these parameters are set, then the exposure time can be fixed by exposing the crystals to different time intervals with increased exposure. If the intensity of spots increases, then it could be fixed, but that may lead to radiation damage. Therefore, less exposure time is always better in synchrotrons and in-home source with higher exposure time can be used (Smyth & Martin, 2000).

The crystal is exposed to X-ray beam for every 90 degrees to check its diffraction quality, and based on spot intensity, the distance is adjusted. If most of the spots can be recorded within the edge, then it would be selected, but if the spots get till the edge, then they have to move closer. Lattice parameters and orientation of crystals are determined using the collected data, and with this information, the total rotation of the crystal needed for complete data collection can be determined.

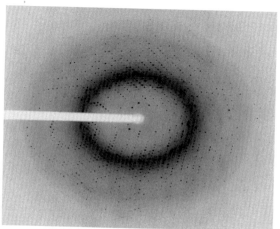

FIG. 6.7 X-ray diffraction image of insulin hormone with 1 degree oscillation at 0.1 s exposure time, 250 mm detector distance, and 1.4 Å resolution taken at the National Synchrotron Radiation Research Center, Taiwan.

reduced into a single file. The crystal lattice exposed to X-ray gives a finite diffraction spots in 3D known as reciprocal lattice. These individual reciprocal lattices are identified by Miller indices (*hkl*), which denote the number of lattice spacings from the starting point of reciprocal lattice with its direction represented as vectors. Integration of all the X-ray intensities is done by 3D profiles; it consists of information about peak size and shape (Wang, 2015). Index is assessed by fitting of the predicted diffraction pattern with that of observed diffraction or by the calculated Miller indices value. Then, the data are combined and scaled depending on their space group to provide the intensity. The scaling programs will usually provide a detailed statistics about the dataset, where completeness of data, R_{merge}, and $<I/\sigma>$ as functions of resolution are checked. The R_{merge} is given by Eq. (6.1) and the lower R_{merge} value indicates the effective resolution.

$$R_{merge} = (\Sigma hkl \ \Sigma i |Ii - <I>|)/(\Sigma hkl \ \Sigma i \ Ii) \qquad (6.1)$$

The signal-to-noise ratio (I/σ) indicates resolution and the larger I/σ value specifies better estimate of effective resolution. An effective data collection and processing can be validated by greater redundancy of data, completeness, and lower R_{merge} value as possible. Currently, there are different programs available to process X-ray data such as HKL2000 (Leslie & Powell, 2007) using Denzo program for integration of data with scalepack for scaling and merging. Mosflm (Otwinowski & Minor, 1997) utilizes aimless program for scaling and merging. XDS suite (Kabsch, 2010) and DIALS program are specific to data processing collected from synchrotron.

Most crystal with symmetry higher than P1 does not require the entire 360 degrees data to successfully solve the crystal. The availability of molecular replacement (MR) data is also an important criterion in data collection process; if not, then the crystals have to be soaked in different heavy atom to obtain its phase information using SAD or MAD methods in synchrotron with variable wavelength. Then, the above specified strategies have to be set and adjusted based on the crystal, and data collection is performed so that the reflection (Fig. 6.7) from all the lattice planes (*hkl*) with its intensities (I_{hkl}) is picked up which is followed by data processing.

5.3 Data Processing

Processing the diffraction image is very tedious prior to development of advanced software and devices. Researchers either have to determine the diffraction spots by naked eye or by digitalizing each photograph taken during data collection. The processing of data is mathematical intense, but as latest algorithms and software are developed, the process is simplified. Now most detectors have software such as Proteum, MarFLM, and Crystal-clear developed by Bruker, MarResearch, and Rigaku, respectively, to collect and process data. The first step in data processing is estimating the crystal system and its unit cell as precise as possible. Followed by indexing where all the spots in diffraction data are assigned, a three-integer *h*, *k*, and *l*, then integration of X-ray intensity and finally scaling is done. In this step, the intensity and index of each spot of entire data is

6 SOLVING PHASES IN MR, SIR/MIR, AND SAD/MAD METHODS

In the period 1912−13, there was an emerging idea in experimental science to use a small crystal as specimen and about ∼1 Å short wavelength X-ray beam from which the information about intensities of the diffraction spots and pattern can be used to infer the atomic arrangements inside the crystals. Due to extensive use of Bragg's formulation in 1929, this method was recognized. If diffraction spot phases could be availed, then phased magnitudes of the spots could be signified as Fourier coefficients and those Fourier sum facilitate to directly calculate the image of molecular structure of a crystal immediately (Sayre, 2015). To date, solving the phases is a significant theoretical problem in X-ray crystallography. But, tremendous progress in X-ray crystallography and computational efficiency has addressed the phases, even in the case of complex structures. X-rays are energetic electromagnetic waves that travel in one direction, and the electric field strength E which is described as function of distance traveled can be represented as Eq. (6.2):

$$\Delta x: E = E_0 \cos(2\pi\Delta x / \lambda) \qquad (6.2)$$

Here, the amplitude and wavelength are represented as E_0 and λ, respectively. The argument quantity of cosine function $\phi = (2\pi\Delta x/\lambda)$ is called as phase.

When X-ray passes through a crystal, it is scattered by the electron with a proportion of atomic shell. It follows different paths through the crystal and reaches the same point to the detector. The detector interferes the waves with one another and produces patterns. If the X-ray waves are in phase, crests and troughs occur in the same position, which interfere with one another and their amplitude together can produce higher amplitude waves. If X-ray waves are out of phase with one another, then destructive interference can occur and the amplitude of the waves can be reduced. This interference is referred as diffraction and it cause bright spots. The diffraction pattern of a molecule is closely related to the Fourier transform of electron density. Here, the brightness of the diffraction pattern in each point is directly proportional to the magnitude of Fourier transform coefficient of that point in Fourier space. Also, the brightness of each points in the pattern directly specifies the magnitude (scaling factor) of a Fourier transform (3D sinusoids) of electron density which can be written as sum of 3D sinusoids. But the problem is that the phase of each Fourier components cannot be directly measured. Therefore, both the magnitude and phase of Fourier component must be known to determine the electron density by entirety of the shifted and scaled sinusoids. Hence, phase is very important to determine the electron density, and without this, figuring out the electron density is quite challenging. In X-ray crystallography, phase problem refers to the inability to directly measure the phase of diffracted X-ray waves to reconstruct the electron density of a molecule (Fultz & Howe, 2013).

6.1 Phase Problem

It is obvious that electron density of each molecule in a crystal can be determined when the phases of each Fourier components (sinusoids) are identified. This technical problem referred as phasing consists of mainly two steps:

- Search of suitable structure solution to build an approximate phase of target protein referred as initial phasing.

- Adjust the phases of estimated structure with experimentally observed diffraction pattern termed as phase refinement.

MR method is widely used to determine the initial phase problem. Here, the known structure of homologous protein is considered to determine the phases. Several phasing techniques available are to get such information namely multiple isomorphous replacement (MIR), single-wavelength anomalous diffraction (SAD), and MAD methods. These methods rely on the premise of phase information that can be obtained if few atoms are replaced with some heavier atom in the protein (Taylor, 2010).

6.2 Molecular Replacement Method

In the era of structural genome project, two major components of crystallography are considered for getting rapid and efficient methods to solve the 3D structures. First one is the MAD method (explained later) and the second one is MR method which has been used to rapidly solve the homologous structure using X-ray data. MR method was first introduced in the early 1960s and the first paper mentioned about the way to solve the 3D structure using homologous structures (Scapin, 2013). MR method is employed to build the phase of unknown structure from homologous structure and the phase can be determined by the noncrystallographic symmetry (NCS). MR method became predominant and the most used method to solve the unknown structures, derived from the different sources, and it has solved around 78% of protein structure deposited in the Protein Data Bank (PDB) database. In conventional MR method, the homologous structure considered as the search model can manually or automatically edit the differed side chain, removal of loop region between the homolog and target structure to solve the known or unknown structure. Also, in this method, it is quite difficult to determine the phase, when the target protein consists of more distant homolog structures. The exponential growth of computer software and hardware results in more fast, flexible, accurate, and automated MR method that enhances the quality and quantity of initial models. Also, the conventional MR method has been improved by superimposition with distantly homologous structure to get the ensemble search model. The ensemble search model works fine with maximum-likelihood scoring method of phaser software to solve the structure (McCoy, 2006; McCoy et al., 2007). In case of imperfect correlation of sequence and structure similarity with target, it screens the sequence independently in the PDB database to overcome the issues. Recent efforts on MR method have resulted in the development of the automated phasing and structure determination with help of computer algorithms. For instance, Modeller software, I-Tasser, QUARK, AMoRe and T-Coffee, etc., basically develop the homology modeling using MR method. However, MR method relies on the availability of good quality homologous templates present in the PDB database. Further combination of iterative fragmental structure assembly and trimming of poorly modeled region enhances the success of MR method (Simpkin et al., 2019). Another approach, ab initio model, is used to predict protein structure based on the primary sequence and independent homologous structure from PDB database. Rosetta program is the first successful approach to build the structure using different fragments of unrelated proteins with the help of Monte Carlo algorithms. More recently, the structural folds of protein have been predicted by intermolecular and intramolecular interactions of residues contact from the covariance of sequence alignments. The structural bioinformatics and structural biology implications have perceived high-quality contacts of resides that predict the ab initio protein folding of large proteins. Besides everything, quality of MR method relies on the data obtained from previously refined structure of protein and rebuilding the model by borrowing the phase of existing structure.

6.3 Issues of Molecular Replacement Method

MR method consists of several major issues and is essentially independent. Failure of MR method may be due to the suboptimal conditions in any one of the following criteria (Evans & McCoy, 2007):

1. Poor quality of target model.
2. The best fit structure with template consists of different positions and orientations; in the case of different target functions, it may have discriminated the solution.
3. Slow calculation can be done with limited number of MR trials, which will give failure of solution.
4. Computational trick is very important to speed up the MR searches, although the current computational technology takes long time to calculate the searches.

MR method positions the probes within their respective unit cell of the crystal and the diffraction pattern could be close to the model matching with experimental data. Asymmetrical unit is a smallest portion of crystal, and each molecule present in asymmetrical unit needs six parameters (three rotational and three translational) to define the orientation and position of molecules (Jogl et al., 2001). This can give the better

agreement to the observed and calculated structure factor. However, this search is an exhaustive method and it demands high computational efficiency. A six-dimensional search covers $N_{rotation} = 1.5 \times 10^{12}$ points (Evans & McCoy, 2007). If the number of searches reduces the parameter, then the search will be successful. Most of the program splits the searches and selects few good solutions from the rotational to test the translational searches. Selecting and preparing an appropriate model is a very critical step in MR method. Low root mean square deviation (RMSD) in the structures has good model and compactness. Also, MR fails when the model does not match with unknown structure. The model fails to give solution for the target structure, but can solve with different space groups of target structure. Basically, low RMSD of two structures indicates the high sequence identity and builds a better model by omitting the diverse region in the structure. B-factor of an atom gives scattered information which has either high or low B-factor values (Carugo, 2019). Modification with surface-exposed residues can make low B-factor values which will give better model solution. The proteins with similar sequences are not showing same secondary structural element in the structure. Binding of ligand or crystal package of protein can alter the rigid body conformation of group or secondary structures. In this case, the protein has to be either split into two domain and can be used as MR model or the conformation of domain should be predicted in advance (Evans & McCoy, 2007).

6.4 Multiple Isomorphous Replacement

MIR is a powerful tool in X-ray crystallography to solve the phase problem of protein structures. The method derives phase information from changes in the diffraction pattern caused by the introduction of strongly scattering atoms in the crystal while keeping it identical in all other respects. In 1954, the traditional MIR method was introduced by Perutz and co-workers and became a very popular method to get solution for the phase problem, even the absence of structure closely homologous. Protein crystals comprise of an open lattice of molecules with the solvent occupation of 30%–80% of total crystal volume. MIR methods rely on the preparation and substitution of a useful derivative that requires the binding of a heavy atom such as Hg, Au, and Pt to a specific position. Ideally, selection and incorporation of heavy metal ions requires additional knowledge of crystalline structure of protein. This information cannot be found easily but can be achieved by the suggestion of sequence and mechanism of action to choose the appropriate heavy metal ions. The terminus amino acid methionine has been substituted and modified by the selenium in the sulfur atom. In some case, substrate of the protein can be substituted by the heavy atoms but this method did not give efficient result because it affects the catalytic activity of the protein. The methionine residues can be substituted by the selenomethionine and or more recently telluromethionines, which revolutionized the macromolecular crystallography through the dispersion techniques. Most of the heavy atom derivatives are produced by the soaking of crystals in the solution of heavy atoms which binds with proteins or side chain of the residues. Choosing the heavy atom appropriate for the protein is based on random and trial-and-error basis. Therefore, heavy atom data bank (HAD) is useful for determining the specificity (Carvin et al., 2012).

HAD is computer knowledge–based data retrieval system which consists of experimental and computation data driven from the successful source of existing MIR analyses determined from the protein crystal structures. The structure factor amplitude of modified (native (Fp)) and unmodified derivatives (Fph) provides the information about the heavy atom in the protein which can further be located by the Patterson or direct methods. Positions of heavy atom, information about the phases, and amplitude of the corresponding heavy atom sites can also be determined in the crystal. The observed modified (Fp) and unmodified (Fph) amplitudes are used to evaluate the phases of the entire structure. Phase ambiguity occurs for only one derivative and that can be resolved by adding more derivatives of datasets. Alternatively, phase ambiguity can be resolved by including anomalous dispersion data for each derivative (MIRAS), thereby the total number of required derivations can be reduced. The amplitude differences arise from one heavy atom that can be mentioned as isomorphous and it can produce estimated phase. In this method, a common difficulty is that it needs trial-and-error testing which can explore the compounds tightly bound with protein. Mostly, the experimental methods completely rely on the anomalous isomorphous dispersion as it intrinsically provides very accurate results though it delivers only weak signal. Therefore, this method is widely used in recent times with source of synchrotron radiation (Papiz & Winter, 2016).

6.5 Single-Wavelength Anomalous Diffraction

Unlike MAD, SAD method can be used for a single dataset at single wavelength with benefit of both speed and protection from the radiation damage of crystal while collecting data. Recently, selenium-SAD is widely used for phase determination, since the selenomethionine is frequently incorporated into the recombinant proteins. SAD experiment provides the details about the measurement of anomalous or Bijvoet, differences as shown in Eq. (6.3).

$$\Delta F^{\pm} = |F_{PH}(+)| - |F_{PH}(-)| \qquad (6.3)$$

These can be used to estimate the contribution of scattering heavy atom and to locate the heavy atom substructure in the protein by direct or Patterson method (Messerschmidt, 2007). Amplitude and phase contribution can be determined if the heavy atom substructure is known. Phase ambiguity can be overcome through the density modification procedure that has become powerful method in recent years. SAD method can efficiently utilize the anomalous scatterers such as S atom of methionine and cysteine residues or ions in macromolecules. As the SAD phasing is very effective and has benefits, it comprises of two main advantages. First one is that this method may have possibilities to evaluate the structure using a single dataset that can avoid the scaling problem and second one is the data collection. In some circumstance, it can accurately phase a dataset of anomalous differences even when present around 5% of the mean amplitude. Sometimes, the naturally occurring sulfur atoms in the cysteine and methionine residues can be used for the phasing. In synchrotron wavelengths (~1 Å), the anomalous change is very less and it requires data of high quality. Whereas in synchrotron, the wavelength is longer (~2 Å) and the signal is high that makes S-SAD phasing a reality. In case of SAD phasing problem, solvent fraction, high-resolution data, and NCS are essential to sort out. Moreover, it's known that longer synchrotron wavelength (~2 Å) gives high anomalous signal for the proteins which consist of anomalous scatterers namely Xe, P, Ca, S, Cl, or Zn. Cr Kα radiation has high wavelength of 2.29 Å. Therefore, chromium anodes are widely used for macromolecule phasing based on Se or S atoms (Xu et al., 2005).

6.6 MultiWavelength Anomalous Dispersion

MAD technique was first developed by the Wayne Hendrickson and this approach deals with solving the phase problem in structure determination. It mainly involves comparing the structure factors collected from the different wavelengths including heavy atom scatters. MAD phasing requires one or more atoms that would have X-ray absorption in the energy range and measure the X-ray intensities (around 1 Å) (Messerschmidt, 2007). Usually, the X-ray absorption edge range of atoms C, N, O, S, and H are in available energy range. The elements of periodic table atomic number roughly between 20 and 40 or <60 have accessible absorption energy edges, which include transition metal such as selenomethionine that can be artificially included in the protein structure. By adding the suitable anomalous scattering atom in the protein structure, the X-ray intensities can be measured at different wavelengths or near the absorption edge of anomalous scattering atom. Also, altering the wavelength can alter the individual reflection intensities and this intensity differences can be used directly to calculate the phase angle. Such experimental process can be performed in the synchrotron beamlines (Ealick, 2000). MAD phasing is not a trial-and-error procedure; besides, it generally gives interpretable electron density map. However, MAD phasing method gives higher quality electron density map than MIR and an initial model builds faster with very few errors.

In MAD, the first step of phase determination is to locate the anomalous scattering atoms and it can be usually studied by assessing Patterson method. However, many anomalous scattering of atoms present in the MAD structures and manual interpretation of those using Patterson method are quite complicated. Hence, two emerging approaches have been used to explain the anomalous scattering namely automated interpretation of Patterson method in combination with different Fourier transform techniques and crystallographic direct method that employees the shake and bake method. The first method can locate several Se atoms (SOLVE and CNS programs (Brunger et al., 1998; Brunger, 2007) have been used to locate the Se atoms in the structure). This method can be implemented in the computer program SnB and can be widely used for MAD structural analysis. The program SHELXD can also be used to locate the anomalous scattering substructures in direct method. Moreover, several packages are available to solve the MAD phasing and SHELXD tool can give very good results rapidly. The PHENIX automated tools provide the information of the local scaling, structure determination, phase information, and the improvements in the structure. Also, autoSHARP tools help to refine the phase when data and model errors appear (Papiz & Winter, 2016).

7 STRUCTURE SOLUTION

As dicussed in previous section, in macromolecular crystallography, structures are normally solved using any of the following four methods—direct method, MR method, single isomorphous replacement (SIR)/MIR method, or SAD/MAD method. MR method relies upon the existence of a homologous structure from which the diffraction data are derived. Isomorphous replacement method in which the crystals are soaked in a heavy atom solution and heavy atom solution and the diffraction pattern of the derivative crystal is compared to the diffraction pattern of the native crystal. This is further classified into SIR and MIR methods. Anomalous dispersion method or scattering specifies diffracting X-ray's phase changes that are exclusive from other atoms in a crystal because of strong X-ray absorbance. Thus, to obtain an anomalous scattering, a heavy atom derivative should be made. This is further classified into MAD and SAD.

7.1 Heavy Atom Analysis and Phase Determination

The technique of MIR can be augmented by the use of anomalous scattering by the heavy atoms. Accordingly, the first step in the application of the MIR method is the preparation of useful heavy atom derivatives that are isomorphous to the native crystals. The derivatives are prepared by the introduction of heavy atoms namely uranium, platinum, mercury, etc., into the crystal without damaging the protein structure or its packing within the protein crystal. Therefore, isomorphous derivative only refers to the changes in incorporation of one or more heavy atoms compared with native crystals. There are several ways of preparing heavy atom derivatives. A few of the preferred methods of introducing heavy atoms into protein crystals are (1) soaking protein crystals in solution containing the desired heavy atom, (2) co-crystallization of salts containing heavy atoms with the protein, (3) replacement of a metal ion in a metalloprotein by a suitable heavier atom, and (4) replacement of an amino acid residue by another containing a suitable heavy atom. In the soaking experiments, binding of heavy atoms to the protein is enabled by the existence of large solvent channels in crystals. Addition of one or more heavy atoms to a macromolecule shows alterations in the diffraction pattern of the derivative compared to native.

Subsequent to the preparation of isomorphous heavy atom derivative crystals, X-ray diffraction datasets are collected from them. A common data collection strategy, as described in detail in earlier section on data collection, is employed for the native as well as the derivative crystals. The isomorphous replacement method determines the amplitudes and phase angles of the heavy atom contribution to the structure factors of the derivative. This in turn requires the identification of the sites where heavy atoms are bound from the observed differences in structure factor amplitudes caused by their introduction. This is done through a series of steps, which are detailed below. The steps are used to identify and refine the heavy atom positions and finally obtain heavy atom phasing.

7.2 Scaling Between Native and Derivative Datasets

First step in heavy atom analyses involves scaling of the derivative datasets with respect to the native dataset. The scale factor can be calculated using least squares procedure (Eq. 6.2) by minimizing the summation of weighted squares of isomorphous differences relative to unknown scale factor k as shown in Eq. (6.4).

$$\sum w(k|F_{PH}| - |F_P|)^2 \qquad (6.4)$$

where F_{PH} represents the structure factor amplitudes of the derivative crystals and F_P indicates the structure factor amplitudes of native crystals.

The scaling of the derivative datasets was performed using the program SCALEIT, which is part of the CCP4 suite of programs (CCP4, 1994). In addition to the scales, the isotropic or the anisotropic temperature factors can be refined.

7.3 Isomorphous Difference Patterson Map

Locating the positions of heavy atoms is essential to obtain phase information from isomorphous replacement. After scaling the datasets, the presence of heavy atom in the derivatives is identified using isomorphous difference Patterson map. It is the most general method of identifying heavy atom position and does not require any previous information. An isomorphous difference Patterson function is a Fourier summation with amplitudes $(F_{PH} - F_P)^2$ as shown in Eq. (6.5):

$$|F_{PH} - F_P|^2 \approx |F_H \cos(\alpha_{PH} - \alpha_P)|^2 \qquad (6.5)$$

where F_H is the structure factor amplitude of the heavy atom.

Eq. (6.6) represents in the case of centric reflections

$$|F_{PH} - F_P|^2 = F_H^2. \qquad (6.6)$$

Eq. (6.7) represents in the case of acentric reflections

$$|F_{PH} - F_P|^2 \approx |F_H|^2 \cos^2(\alpha_{PH} - \alpha_H) \qquad (6.7)$$

In case of presence of anomalously scattering atoms in the crystal, an anomalous difference Patterson is used to investigate the position of anomalous scatterer.

An anomalous difference Patterson synthesis is a Fourier summation of the terms of following Eq. (6.8).

$$|F_{PH}(+) - F_{PH}(-)|^2 \approx 4/k^2 |F_H|^2 \sin^2(\alpha_{PH} - \alpha_H)$$
$$\approx F''^2_H \sin^2(\alpha_{PH} - \alpha_H) \qquad (6.8)$$

Hence, anomalous difference Patterson can also be used to confirm the position of the heavy atoms, though it is not much used. This is because, in spite of the absence of noise generated due to lack of isomorphism in an anomalous difference Patterson map, the anomalous scattering factor, F''_H, signal is called as phase is much weaker than the F_H signal. As a result, random errors in the measurements very frequently swamp the anomalous signal.

The various Patterson coefficients aforementioned were calculated from the reflection data using FFT (fast Fourier transform) program. The peaks in the map given out by FFT were searched using PEAKMAX. NPO was used for plotting selected sections of the map. The coordinates of heavy atom can subsequently be determined from this map either manually or by using the program RSPS. All these programs are a part of the CCP4 suite.

7.4 Patterson Method

The methods available to estimate the phase information uses a Patterson function P(q), which is a Fourier transform that utilizes the coefficients $|F(h)|^2$ only. Patterson is a crucial method to calculate it directly from observed data and does not require the phase information (Papiz & Winter, 2016). Moreover, they can be calculated from the model by ignoring the structure factors. The observed data driven is the vector map of contents present in the crystal which consist of intramolecular self-vectors and other vectors produced by the crystallographic and noncrystallographic symmetry. Moreover, Patterson has property to rotate the model and rotating intramolecular vectors in the same angle in radius-limited Patterson model and here the scoring is based on the matches obtained by the observed data in unrotated radius-limited Patterson. If the crystals have any rotational symmetry operators, then the Patterson comprises the cross-vectors of different atoms in the molecules related by the symmetry. Translation of molecule related to the symmetry operator can move in different directions and their respective cross-vector can also be changed. Also, the search of likelihood translation and rotation of molecule can find solution to build the structures (Evans & McCoy, 2007).

7.5 The Difference Fourier

Once one or more heavy atom sites are located, they can be utilized to determine the approximate protein phases, α_{best} (defined in the next section) that can be further employed to evaluate a difference in Fourier series with coefficients in the form of Eq. (6.9):

$$m(F_{PH} - F_P)\exp(i\ \alpha_{best}) \quad (6.9)$$

where m is the figure of merit (defined in the next section). The difference Fourier can be calculated using FFT.

8 STRUCTURE REFINEMENT AND MODEL BUILDING

Crystallographic refinement focuses to optimize the correlation between the atomic model with both chemical restraints and obtained diffraction data. Refinement of the crystal structure is the process to obtain accurate atomic parameters compared to initial model. It is an iterative process to enhance the structure model quality. Positions x, y, and z and atomic displacement parameters obtained from each atom are attuned so as to enhance the concurrence between the calculated structural model |Fc| and the observed structure factor amplitude |Fo|. REFMAC (Murshudov et al., 1996), which is supported by the CCP4 suit of programs, can be used for refinement. Both the programs can be applied for structure solution, for refinement, and for calculating the map of macromolecule structures obtained by X-ray techniques.

Structure refinement involves electron density map calculations and iteration of refinements such as positional, rigid body, restrained, unrestrained individual B-factor, and occupancy refinements. In addition, there are options in CNS to perform simulated annealing refinements both in the Cartesian and torsion angle conformational space and the option of TLS refinement in REFMAC.

8.1 Interpretation of Electron Density Maps

The electron density map is calculated using an estimate phase angles derived from the available protein structure factors. The Fourier coefficients can be determined using Eq. (6.10):

$$(2m\ F_O - D\ F_C)\exp(i\alpha C) \text{ and } (m\ F_O - D\ F_C)\exp(i\alpha C) \quad (6.10)$$

where m and D represent figure of merit and error in coordinates of the model, respectively. Electron density maps using the Fourier coefficients are considered advanced than conventional "unweighted" maps (Main, 1979; Read, 1986). Interpreting the map with poor phases or low-resolution data is the toughest process at the initial stage and generally it is easy to start with good phasing high-resolution data. There are three strategies widely implemented in the interpretation of electron density maps.

The first strategy starts at low resolution with 5–6 Å, which determines the contours of the molecules in the electron density map of the crystal. The difference in the contours of map helps to distinguish the proteins and solvent regions. At certain instances, secondary structure element α-helices that appear in cylindrical rods can also be identified in range between 4 and 6 Å, whereas β-sheets and strands at this resolution are more difficult to differentiate and sometimes are not visible. In case of second strategy, the molecule positioned in the crystal unit cell can be confidently fixed in the medium-resolution map at 3.5–2.5 Å. Medium-resolution electron density maps could facilitate the tracing of polypeptide chain, and with the known information of amino acid sequence, it is even easier to fix the polypeptide chain. However, due to medium resolution of data, there are high chances of misinterpreting secondary structural elements. Moreover, the connections between the secondary structural elements are often complicated to distinguish and the prominent position of the amino acids may be shifting from the correct position due to the flexibility of side chains. In third strategy, the higher resolution of more than 2 Å permits to identify the accurate location of amino acids and most importantly helps to overcome the misinterpretations of secondary structural elements and to predict the exact position of side chains.

The bottleneck is that high-resolution maps are not common, and in majority of the cases, high-resolution maps are derived from calculated or combined phases. After every refinement cycle, the model was inspected and manual rebuilding can be done by inspection of $2F_O$-F_C and F_O-F_C electron density maps. CNS has options to calculate σA-weighted maps where the structure-factor amplitudes are weighted in order to reduce the model bias of an incomplete or partially incorrect structure. REFMAC also produces an output file containing coefficients for σA-weighted maps. The maps are usually contoured at 1σ and 3σ, respectively, where σ refers to the RMSD of the mean density in electrons and 3 Å cutoff in the maps. Regions of poor electron density are examined with the maps contoured at a lower level. The molecular modeling package FRODO (Jones, 1978) can be used to examine and interpret the model against the electron density maps. During the final stages, omit maps can be calculated using CNS to check the correctness of the model.

8.2 Model Building in Interactive Computer Graphics Environment

In earlier days, the scope for the high-resolution maps is nearer to zero and the available low-resolution maps are also drawn on the transparent sheets for small-scale objects. The maps are highly helpful to protein crystallographers and are referred as "minimaps," which provide the overall observation of the electron density in unit cell. Sometimes, these maps are used in the preliminary tracing of polypeptides. During the early stages of crystallographic development, the complete model of the molecule is constructed with help of the homemade apparatus called as an optical comparator and it is also referred as "Richard's box" because it was invented by Richard (Richard, 1968). Advancements and new techniques in the field of crystallography and computer science have groomed the paper or clay-based model building to the current era of graphical modeling. The graphical user intervention—based interpretation of the electron density maps directly allows analyzing and fitting the atoms/amino acids in accordance with the density.

User-friendly and adaptable software packages are the prerequisite for the efficient modeling of 3D structure from X-ray crystallographic data. The major software used in crystallographic studies are given in Table 6.3.

The software is categorized into two components, namely foreground objects and background objects. The foreground objects let the user to manipulate the data through the graphical user interface and the manual intervention of the background objects is restricted. The software displays the mesh like surface representation of the electron density in the background and the atomic model can be assigned to fit inside the electron density map on the foreground. However, both the models generated using foreground and background are allowed to superimpose on each other to correct and correlate toward the building of complete structure. Advancements in the computers have resulted in the generation and development of libraries, which contains the information on the stereochemical properties of chemical groups and amino acids that generally exist in proteins and nucleic acids. The coordinates with stereochemical information in libraries facilitate the straightforward building of a polypeptide fragment or a part of the macromolecule in an ideal conformation. High-performance computers in current generations can allow the atoms of the model to be manipulated easily by movement of mouse. Even it allows shifting the angle of the entire group of atoms instead of single atom. In a similar way, the bond angles and dihedral angles are able to be rotated along with bond stretching. Trials on fitting into the electron density map during model building are minimal in case of available good phases and the suitable method to build the initial protein model has been developed by Jones and Thirup (1986). The core principle behind the method is to fully utilize the resource available in the libraries to generate the short segment of secondary structure elements for approximate placing and to suitably fit into the analyzed electron density maps. All the limitations in viewing the global molecular model on the primitive days are overcome with the technology advancements in graphic visualizer.

8.3 Refinement of the Protein Structure

The protein structure refinement process is fast and accurate in current scenario due to the performance of the high-end computers, which are efficient in handling the numerous equations with thousands of variables in a fraction of time. Moreover, the number of X-ray crystal structures available in the repositories as reference model supports well for the refinement process. The protein crystals are intrinsically not in perfect order and in most of the cases the resolution of the diffraction data ranges between 3.0 and 2.5 Å (Jeyakanthan et al., 2005; Rajakannan et al., 2002). The resolution of 2.0 Å has been possible in some protein crystals; however, data diffracted at 1.5 Å resolution are widely considered as excellent, and at certain distinctive cases, even 1.0 Å resolution data are achievable. In real space, the first attempts are initiated to refine the protein structure. The general method commonly used in the structure refinement calculates the minimal difference between the observed electron density (ρ_{obs}) and calculated model density (ρ_{calc}) as given in Eq. (6.11).

$$S = (\rho_{obs} - \rho_{calc})^2 \qquad (6.11)$$

The limitations of this method are the real-space refinement drawbacks and the instance of existing poor phases that lead to incorrect refinement. Similarly, there are diverse reasons to support reciprocal space refinement methods and various solutions have been developed to conquer the limitations of the problems, which helps to improve the observation and parameters ratio by reducing the variable numbers or by artificially enhancing the observation numbers. The earlier is referred as constrained least squares, whereas the second is referred as restrained least squares.

TABLE 6.3
Major Software Used in X-ray Crystallographic Studies.

S. No	Software	Application	Web Links	References
1	HKL-3000	A platform of software for macromolecular X-ray crystalllography	https://www.hkl-xray.com/hkl-3000	Minor, W., Cymborowski, M., Otwinowski, Z., & Chruszcz, M. (2006). HKL-3000: The integration of data reduction and structure solution — from diffraction images to an initial model in minutes. *Acta Crystallographica Section D Biological Crystallography, 62*(8), 859–866.
2	CCP4—Collaborative Computational Project No. 4 (Winn et al., 2001)	Software suite for macromolecular X-ray crystallography	http://www.ccp4.ac.uk/	Winn, M. D., Ballard, C. C., Cowtan, K. D., Dodson, E. J., Emsley, P., Evans, P. R., ... Wilson, K. S. (2011). Overview of the CCP4 suite and current developments. *Acta Crystallographica Section D Biological Crystallography, 67*(4), 235–242.
3	PHENIX	A new Python-based graphical user interface for the PHENIX suite of crystallography software	http://www.phenix-online.org	Echols, N., Grosse-Kunstleve, R. W., Afonine, P. V., Bunkóczi, G., Chen, V. B., Headd, J. J. Adams, P. D. (2012). Graphical tools for macromolecular crystallography in PHENIX. *Journal of Applied Crystallography, 45*(3), 581–586.
4	CNS-crystallography and NMR system (Brunger, 2007; Brunger et al., 1998)	Macromolecular structure determination	http://cns.csb.yale.edu/v1.1/	Brünger, A. T., Adams, P. D., Clore, G. M., DeLano, W. L., Gros, P., Grosse-Kunstleve, R. W., ... Warren, G. L. (1998). Crystallography & NMR System: A New Software Suite for Macromolecular Structure Determination. *Acta Crystallographica Section D Biological Crystallography, 54*(5), 905–921.
5	COOT—Crystallographic Object-Oriented Toolkit (Emsley and Cowtan, 2004)	A suitable software for macromolecular protein modeling using X-ray data	https://www2.mrc-lmb.cam.ac.uk/personal/pemsley/coot/	Emsley, P., & Cowtan, K. (2004). Coot: Model-building tools for molecular graphics. *Acta Crystallographica Section D Biological Crystallography, 60*(12), 2126–2132

9 CHARACTERIZATION AND CROSS-VALIDATION OF SOLVED 3D STRUCTURES

The R-factor which measures differences between the observed structure factor amplitude (F_O) and calculated structure factor amplitude (F_C) provides the quality diffraction data of the fit for a model to the diffraction data and is given by the following Eq. (6.12):

$$R = \sum \left| |F_O| - |F_C| \right| / \sum |F_O| \qquad (6.12)$$

Increasing the quantity of the adjustable parameters can make the obtained value arbitrarily low that can be used for describing the model. Statistical cross-validation method that uses free R-factor for crystallographic refinement is a reliable indicator of quality of the predicted model (Brünger, 1992; Kleywegt & Brünger, 1996). The diffraction dataset is divided into two sets namely a test set (comprising 5%−10% of randomly chosen reflections) and a complementary working set. In standard crystallographic refinement, the diffraction data are normally used as working set while the test data are not employed. The free R-factor either increases or remains constant when the model does not fit to the test set even after changes are made to improve the efficiency of the model.

9.1 Constraints and Restraints

During macromolecular refinement, few groups of atoms can be constrained or restrained to enhance the ratio of observables to parameters. The refinement programs have preferences to group atoms that can move as rigid bodies or restrain or constrain the bond angles, NCS, bond lengths, and atomic positions to specific value using suitable force constants. Once restricted freedom is provided for a parameter, restraints are given. In case of constraints, the parameters are detained to an exact value. Generally, constraint refers to restraint through infinite force constant. Asymmetric unit molecules are superimposed using superimposition of least squares and the individual atom's average coordinates (xav) are calculated in NCS symmetry restraints. Then each atom is restrained as per the following Eq. (6.13).

$$ENCS = w(x - xav)^2 \qquad (6.13)$$

The respective B-factor restraints are represented by following Eq. (6.14).

$$BNCS = (b - bav)^2 / \sigma^2 NCS \qquad (6.14)$$

where w indicates weight function, b represents individual temperature, and bav specifies average temperature factors of NCS-related atoms. B-factor restraints target deviations are represented by σNCS.

9.2 Identification of Solvent Sites and Structure Analysis

The water molecules are picked using peaks from both F_O-F_C and $2F_O-F_C$ maps contoured usually at 2.5σ and 0.8, respectively. Identification of water molecules can be performed by both manually and using automatic water picking routines in CNS. The OMIT

maps are used to verify the solvent positions. It is further verified that no solvent site can be located at a distance less than 2.3 Å from any protein atom and that the R-free values dropped on addition of water molecules to the model. An initial B-value of 30 Å is assigned to the water molecules, which can be subsequently refined. This procedure is repeated until most of the density in the maps is accounted for and the R-factor of the model is converged. The refinement programs analyze the geometrical parameters and RMSD in bond lengths, bond angles, dihedral angles, short contacts between symmetry related atoms, etc., of the refined structure. There are also options in CNS to calculate Luzzati (Luzzati, 1952) and Wilson plots (Wilson, 1942). The program also lists the energies that deviate from weights used for the refinement (Engh & Huber, 1991).

10 STRUCTURE VALIDATION

The final coordinate files and the structure factor files are used in the validation process by OMIT map. The products are removed from the coordinate file and verified with the structure factor file while generating OMIT map (Kanaujia et al., 2010). The generation of map mask over the positional areas of the products indicates the presence of ligand. Absence of map mask over the positional areas indicates the ligand in the model is not actually present and should be removed. The PROCHECK program was developed by Laskowski et al. (1993) for verifying stereochemical quality of protein structures and it is embedded in the CCP4 suite of programs for structure validations. Amino acid distribution of aminoacyl-tRNA synthetase (PDB ID: 5ZDO) in Ramachandran plot is shown in Fig. 6.8 (Mutharasappan et al., 2020).

The output will include complete residue-wise listing and graphical representation of parameters. Ramachandran plot (Ramachandran et al., 1963) is one such method used to investigate the secondary structural characteristics of proteins and stereochemistry of the modeled protein. Model quality is comparatively analyzed with other structures by the program. The output of the program includes the coordinates that incorporate the conventional numbering of atom in accordance with the guidelines provided by the International Union of Pure and Applied Chemistry (IUPAC) and the International union of Biochemistry (IUB) Commission on Biochemical Nomenclature in the year 1970. This program can be used to evaluate the model quality after each refinement cycle. Finally, the coordinates/structure is

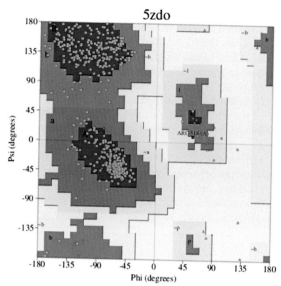

FIG. 6.8 Distribution of aminoacyl-tRNA synthetase (PDB ID: 5ZDO) amino acids in the four quadrants of Ramachandran plot.

deposited in the 3D structure repository called "PDB", which is open to the scientific community.

11 CONCLUSION

Macromolecular crystallography not only plays a vital role in accurate investigation of crystal structures of macromolecules and its complex with ligands but also is crucial in SBDD. Though there is a tremendous progress in the structural biology methods especially in X-ray crystallography, corresponding advancements in computational high-throughput screening is still on the horizon. SBDD methods use the macromolecular structural information provided by the macromolecular crystallography techniques to analyze both the key sites and significant interactions that elicit biological functions. Such information are essential in SBDD to design potent drugs against specific target of the macromolecules that interrupt the key biological pathways in disease pathologies. X-ray crystallographic macromolecular structural information is also essential for ligand-based drug design, which focuses on structure–activity relationship which guides in design of novel drugs with enhanced activity or in optimization of existing drugs.

ACKNOWLEDGMENT

JJ thank the DST INDO-TAIWAN (GITA/DST/TWN/P-86/2019 dated March 04, 2020), Board of Research in Nuclear Sciences (BRNS) (35/14/February 2018 BRNS/35009), Indian Council for Medical Research (ICMR) (No.BIC/12(07)/2015), DST-Science and Engineering Research Board (SERB) (No.EMR/2016/000498), UGC Research Award (No. F. 30−32/2016(SA-II) dated April 18, 2016), DST-Fund for Improvement of S&T Infrastructure in Universities & Higher Educational Institutions (FIST) (SR/FST/LSI-667/2016) (C), and DST-Promotion of University Research and Scientific Excellence (PURSE) (No. SR/PURSE Phase 2/38 (G), 2017). JJ and MN thank MHRD-RUSA 2.0, New Delhi (F.24−51/2014-U, Policy (TNMulti-Gen), Dept. of Edn. Govt. of India, dated October 09, 2018). DP thankfully acknowledges the Indian Council of Medical Research [No. ISRM/11/(17)/2017] for Senior Research Fellowship.

REFERENCES

Aitipamula, S., & Vangala, V. R. (2017). X-ray crystallography and its role in understanding the physicochemical properties of pharmaceutical cocrystals. *Journal of the Indian Institute of Science, 97,* 227−243.

Balerna, A., & Mobilio, S. (2015). Introduction to synchrotron radiation. In S. Mobilio, F. Boscherini, & C. Meneghini (Eds.), *Synchrotron radiation.* Berlin, Heidelberg: Springer.

Bragg, W. L. (1913). The structure of some crystals as indicated by their diffraction of x-rays. *Proceedings of the Royal Society (London), A89,* 248−277.

Bravis, A. (1849). Mémoire sur les polyèdres de forme symétrique. *Jurnal Matematika, 14,* 141−180.

Brünger, A. T. (1992). Free R value: A novel statistical quantity for assessing the accuracy of crystal structures. *Nature, 355*(6359), 472−475.

Brunger, A. T., Adams, P. D., Clore, G. M., Gros, P., Grosse-Kunstleve, R. W., Jiang, J.-S., Kuszewski, J., Nilges, N., Pannu, N. S., Read, R. J., Rice, L. M., Simonson, T., & Warren, G. L. (1998). Crystallography & NMR system (CNS)-A new software suite for macromolecular structure determination. *Acta Crystallographica Section D, 54,* 905−921.

Brunger, A. T. (2007). Version 1.2 of the crystallography and NMR system. *Nature Protocols, 2,* 2728−2733.

Carugo, O. (2019). Maximal B-factors in protein crystal structures. *Zeitschrift Fur Kristallographie - Crystalline Materials, 234*(1), 73−77.

Carvin, D., Islam, S. A., Sternberg, M. J. E., & Blundell, T. L. (2012). The preparation of heavy-atom derivatives of protein crystals for use in multiple isomorphous replacement and anomalous scattering. *International Tables for Crystallography, F*(12.1), 317−326.

Chapman, H. N. (2019). X-ray free-electron Lasers for the structure and dynamics of macromolecules. *Annual Review of Biochemistry, 88,* 35−58.

Chayen, N., & Saridakis, E. (2008). Protein crystallization: From purified protein to diffraction-quality crystal. *Nature Methods, 5,* 147−153.

Chen, C.-J., Rose, J., Newton, M., Liu, Z.-J., & Wang, B.-C. (2002). *Protein crystallography.* https://doi.org/10.1201/9781420036527.ch2.

Dauter, Z. (2017). Collection of X-ray diffraction data from macromolecular crystals. *Methods in Molecular Biology (Clifton, N.J.), 1607,* 165–184.

Ealick, S. E. (2000). Advances in multiple wavelength anomalous diffraction crystallography. *Current Opinion in Chemical Biology, 4*(5), 495–499.

Emsley, P., & Cowtan, K. (2004). Coot: Model-building tools for molecular graphics. *Acta Crystallographica Section D: Biological Crystallography, 60*(Pt 12 Pt 1), 2126–2132.

Engh, R. A., & Huber, R. (1991). Accurate bond and angle parameters for X-ray protein structure refinement. *Acta Crystallographica Section A: Foundations of Crystallography, 47*(4), 392–400.

Evans, P., & McCoy, A. (2007). An introduction to molecular replacement. *Acta Crystallographica Section D: Biological Crystallography, 64*(1), 1–10.

Fultz, B., & Howe, J. (2013). *Transmission electron microscopy and diffractometry of materials* (4th ed.).

Garman, E. (1999). Cool data: Quantity and quality. *Acta Crystallographica. Section D, Biological Crystallography, 55*(Pt 10), 1641–1653.

Geigé, R., & McPheson, A. (2001). *Crystallization: General methods.* International Tables for Crystallography (Vol. F, pp. 81–93).

Hope, H. (1988). Cryocrystallography of biological macromolecules: A generally applicable method. *Acta Crystallographica. Section B, Structural Science, 44*(Pt 1), 22–26.

Jeyakanthan, J., Inagaki, E., Kuroishi, C., & Tahirov, T. H. (2005). Structure of PIN-domain protein PH0500 from *Pyrococcus horikoshii. Acta Crystallographica Section F: Structural Biology and Crystallization Communications, 61*(5), 463–468.

Jogl, G., Tao, X., Xu, Y., & Tong, L. (2001). COMO: A program for combined molecular replacement. *Acta Crystallographica Section D: Biological Crystallography, 57*(8), 1127–1134.

Jones, T. A. (1978). A graphics model building and refinement system for macromolecules. *Journal of Applied Crystallography, 11*(4), 268–272.

Jones, T. A., & Thirup, S. (1986). Using known substructures in protein model building and crystallography. *The EMBO Journal, 5*(4), 819–822.

Kabsch, W. (2010). XDS. *Acta Crystallographica. Section D, Biological Crystallography, 66*(Pt 2), 125–132.

Kanaujia, S. P., Jeyakanthan, J., Nakagawa, N., Balasubramaniam, S., Shinkai, A., Kuramitsu, S., Yokoyama, S., & Sekar, K. (2010). Structures of apo and GTP-bound molybdenum cofactor biosynthesis protein MoaC from *Thermus thermophilus* HB8. *Acta Crystallographica Section D: Biological Crystallography, 66*(7), 821–833.

Kirk, J. G. (2004). Particle acceleration in relativistic current sheets. *Physical Review Letters, 92*(18), 181101.

Kleywegt, G. J., & Brünger, A. T. (1996). Checking your imagination: Applications of the free R value. *Structure, 4*(8), 897–904.

Laskowski, R. A., MacArthur, M. W., Moss, D. S., & Thornton, J. M. (1993). PROCHECK: A program to check the stereochemical quality of protein structures. *Journal of Applied Crystallography, 26*(2), 283–291.

Leslie, A. G. W., & Powell, H. R. (2007). Processing diffraction data with mosflm. *Evolving Methods for Macromolecular Crystallography. NATO Science Series, 245,* 41–51.

Luzzati, V. (1952). Statistical treatment of errors in the determination of crystal structures. *Acta Crystallographica, 5*(6), 802–810.

Main, P. (1979). A theoretical comparison of the β, γ′ and 2Fo−Fc syntheses. *Acta Crystallographica - Section A: Crystal Physics, Diffraction, Theoretical and General Crystallography, 35*(5), 779–785.

McCoy, A. J., Grosse-Kunstleve, R. W., Adams, P. D., Winn, M. D., Storoni, L. C., & Read, R. J. (2007). Phaser crystallographic software. *Journal of Applied Crystallography, 40*(4), 658–674.

McCoy, A. J. (2006). Solving structures of protein complexes by molecular replacement with Phaser. *Acta Crystallographica Section D: Biological Crystallography, 63*(1), 32–41.

McPherson, A. (1982). *The preparation and analysis of protein crystals* (2nd ed.). New York: John Wiley and Sons.

McPherson, A., Luk, T. S., & Thompson, B. D. (1993). Multiphoton-induced X-ray emission and amplification from clusters. *Journal of Appied Physics. B, 57,* 337–347.

Messerschmidt, A. (2007). Methods for solving the phase problem. In *X-ray crystallography of biomacromolecules* (pp. 99–139) (Chapter 5).

Mills, D. M., Helliwell, J. R., Kvick, A., Ohta, T., Robinson, I. A., & Authier, A. (2005). Report of the working group on synchrotron radiation nomenclature − brightness, spectral brightness or brilliance? *Journal of Synchrotron Radiation, 12*(Pt 3), 385.

Murshudov, G. N., Dodson, E. J., & Vagin, A. A. (1996). Application of maximum likelihood methods for macromolecular refinement. *Macromolecular Refinement, 93*–104.

Mutharasappan, N., Raj, R,.G., Mariadasse, R., Saritha, P., Amala, M., Prabhu, D., Sundarraj, R., Pandian, C., & Jeyaraman, J. (2020). Experimental and computational methods to determine protein structure and stability. In *Frontiers in protein structure, function, and dynamics.* https://doi.org/10.1007/978-981-15-5530-5_2.

Nachiappan, M., Jain, V., Sharma, A., Yogavel, M., & Jeyakanthan, J. (2018). Structural and functional analysis of Glutaminyl-tRNA synthetase (*Tt*GlnRS) from *Thermus thermophilus* HB8 and its complexes. *International Journal of Biological Macromolecules, 120*(Pt B), 1379–1386.

Otwinowski, Z., & Minor, W. (1997). Processing of X-ray diffraction data collected in oscillation mode. *Methods in Enzymology, 276,* 307–326.

Papiz, M. Z., & Winter, G. (2016). X-ray crystallography, biomolecular structure determination methods. In *Encyclopedia of spectroscopy and spectrometry* (3rd ed.). Elsevier Ltd.

Powell, H. R. (2017). X-ray data processing. *Bioscience Reports, 37*(5). BSR20170227.

Prabhu, D., Amala, M., Saritha, P., Rajamanikandan, S., Veerapandiyan, M., & Jeyakanthan, J. (2020). Functional characterization of streptomycin adenylyltransferase from *Serratia marcescens*: An experimental approach to understand the antibiotic resistance mechanism. *BMC Infectious Diseases, 20*(Suppl. 1), 324, 20.

Rajakannan, V., Yogavel, M., Poi, M.-J., Arockia Jeyaprakash, A., Jeyakanthan, J., Velmurugan, D., Tsai, M.-D., & Sekar, K. (2002). Observation of additional calcium ion in the crystal structure of the triple mutant K56, 120,121 M of bovine pancreatic phospholipase A2. *Journal of Molecular Biology, 324*(4), 755—762.

Ramachandran, G. N., Ramakrishnan, C., & Sasisekharan, V. (1963). Stereochemistry of polypeptide chain configurations. *Journal of Molecular Biology, 7*, 95—99.

Read, R. J. (1986). Improved fourier coefficients for maps using phases from partial structures with errors. *Acta Crystallographica Section A: Foundations of Crystallography, 42*(3), 140—149.

Richards, F. M. (1968). The matching of physical models to three-dimensional electron-density maps: A simple optical device. *Journal of Molecular Biology, 37*(1), 225—230.

Röntgen, W. C. (1895). *Physik Med Gesellschaft zu Würzburg, 137*, 132.

Sayre, D. (2015). X-ray crystallography: The past and present of the phase problem. *Science of Crystal Structures: Highlights in Crystallography, 13*(1), 3—16.

Scapin, G. (2013). Molecular replacement then and now. *Acta Crystallographica Section D: Biological Crystallography, 69*(11), 2266—2275.

Simpkin, A. J., Thomas, J. M. H., Simkovic, F., Keegan, R. M., & Rigden, D. J. (2019). Molecular replacement using structure predictions from databases Simpkin Adam. *J. Acta Crystallographica Section D: Structural Biology, 75*, 1051—1062.

Smyth, M. S., & Martin, J. H. (2000). x ray crystallography. *Molecular Pathology, 53*(1), 8—14.

Souza, D. H., Selistre-de-Araujo, H. S., & Garratt, R. C. (2000). Determination of the three-dimensional structure of toxins by protein crystallography. *Toxicon: Official Journal of the International Society on Toxinology, 38*(10), 1307—1353.

Stevens, R. C. (2000). High-throughput protein crystallization. *Current Opinion in Structural Biology, 10*(5), 558—563.

Taylor, G. L. (2010). Introduction to phasing. *Acta Crystallographica Section D: Biological Crystallography, 66*(4), 325—338.

Wang, J. (2015). Estimation of the quality of refined protein crystal structures. *Protein Science: A Publication of the Protein Society, 24*(5), 661—669.

Wiencek, J. M. (1999). New strategies for protein crystal growth. *Annual Review of Biomedical Engineering, 1*(1), 505—534.

Wilson, A. J. C. (1942). Determination of absolute from relative x-ray intensity data. *Nature, 150*, 151—152.

Winick, H. (1998). Synchrotron radiation sources — present capabilities and future directions. *Journal of Synchrotron Radiation, 5*(Pt 3), 168—175.

Winn, M. D., Ballard, C. C., Cowtan, K. D., Dodson, E. J., Emsley, P., Evans, P. R., Keegan, R. M., Krissinel, E. B., Leslie, A. G., McCoy, A., McNicholas, S. J., Murshudov, G. N., Pannu, N. S., Potterton, E. A., Powell, H. R., Read, R. J., Vagin, A., & Wilson, K. S. (2001). Overview of the CCP4 suite and current developments. *Acta Crystallographica Section D Biological Crystallography, 67*(Pt 4), 235—242.

Xu, H., Yang, C., Chen, L., Kataeva, I. A., Tempel, W., Lee, D., Habel, J. E., Nguyen, D., Pflugrath, J. W., Ferrara, J. D., Arendall, W. B., Richardson, J. S., Richardson, D. C., Liu, Z. J., Newton, M. G., Rose, J. P., & Wang, B. C. (2005). Away from the edge II: In-house Se-SAS phasing with chromium radiation. *Acta Crystallographica Section D: Biological Crystallography, 61*(7), 960—966.

Yamaguchi, Y. (2015). Affinity chromatography of native and recombinant proteins from receptors for Insulin and IGF-I to recombinant single chain antibodies. *Frontiers in Endocrinology, 6*, 166.

CHAPTER 7

Electron Microscopy and Single Particle Analysis for Solving Three-Dimensional Structures of Macromolecules

AYALURU MURALI

1 INTRODUCTION

Structural biology, a branch of biology that elucidates the structure of biomolecules, is one of the interesting branches of biology. Similar to other branches of sciences, the structural biology was also expanded in two directions—theoretical and experimental. The theoretical part of structural biology chiefly dealt with prediction of the models of biological macromolecules and their interaction using basic physics laws; this was integrated into bioinformatics or computational biology. These concepts were discussed in detail in Chapter 8 of this book. On the other hand, the experimental tools available for solving the structure of a macromolecule were primarily X-ray crystallography (Drenth & Mesters, 2007; Ladd et al., 2013) and nuclear magnetic resonance (NMR) spectroscopy (Rule & Hitchens, 2006; Wuthrich, 1990). However, the scenario changed with the incorporation of the transmission electron microscopy and single particle analysis (SPA) as another tool in structural biology. Extensive reviews, monographs, and books were written detailing the X-ray crystallography and NMR approaches in solving the structures of macromolecules (Drenth & Mesters, 2007; Rule & Hitchens, 2006; Wuthrich, 1990). The concept of X-ray crystallography for solving the macromolecular structure is also discussed in detail in Chapter 6 of this book. This chapter focuses on SPA and electron microscopy (EM) as a tool in structural biology and its contribution to biology.

2 THE MAJOR TECHNIQUES IN STRUCTURAL BIOLOGY

2.1 X-Ray Crystallography and Nuclear Magnetic Resonance

The X-ray crystallography basically requires a three-dimensional (3D) single crystal grown from the purified protein/protein complex. The diffraction pattern collected from this single crystal would be analyzed to generate the electron density map from which a model that fits the electron density map would be computed. Of all the structural biology tools available, X-ray crystallography is most popular and often considered as a benchmark. The resolution that one can expect from X-ray diffraction is typically of the atomic resolution (less than 1 Å in most cases). However, some consider the X-ray structure as solid structure as the entire experiment is based on crystallized protein.

On the other hand, NMR builds a model by calculating the distance constraints and suggests a model that accommodates all the distance constraints thus generated. The structure is considered as a solution structure and can give us a resolution comparable to X-ray structure (with X-ray structure giving much superior resolution). Also, the NMR structure is known to capture the dynamics of the macromolecule. However, this tool is best suited for small molecules having molecular weight typically less than about 30 kDa.

2.2 Single Particle Analysis

The newly emerging SPA combines the advantages of both X-ray structures in terms of having a wider range of molecules and NMR structures where one can expect the molecule to preserve its native conformation. On the top of that, SPA can help us to visualize various conformations of the molecule, if present.

This technique is slowly gaining popularity as seen by more and more structures being deposited (Fig. 7.1) in the EMDB database (EMDataResource, n.d.), a database that curates the structures solved using SPA technique. Also, the inclusion of a wide spectrum of molecular weights also confirms the potentiality of this tool (Fig. 7.2A). Depending on the sample

Molecular Docking for Computer-Aided Drug Design. https://doi.org/10.1016/B978-0-12-822312-3.00008-4

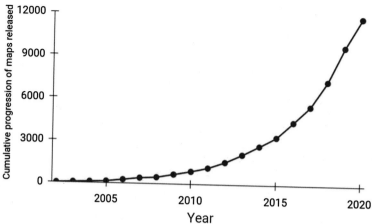

FIG. 7.1 The number of structures (cumulative) solved using single particle analysis (SPA). The monotonic growth in number of structures shows the increasing interest in use of SPA. (Data are taken from EMDB website EMDataResource. (n.d.). Retrieved September 24, 2020, from https://www.emdataresource.org/index. html.)

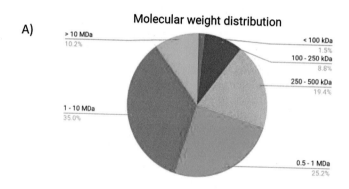

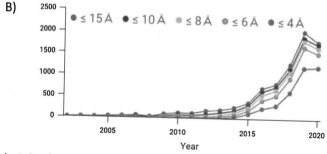

FIG. 7.2 **(A)** Pie chart showing the distribution of molecular weight of the molecules. These data include the structures solved from both negative staining and cryo-EM data. Though majority of the molecules have their molecular weights in 250–500 kDa and 1–10 MDa, structures of macromolecules with almost all ranges of molecular weights are solved using the single particle analysis (SPA) technique. **(B)** Line chart showing the cumulative number of structures solved using SPA at different resolutions. It can be seen that the high-resolution structures (with resolution < 4 Å) is at a monotonic growth. (Data are current as on September 2020 and taken from EMDB website EMDataResource. (n.d.). Retrieved September 24, 2020, from https://www.emdataresource. org/index.html.)

preparation techniques that one adopts (a detailed account on sample preparation protocols for EM experiment is given in Section 3.2), it is possible to obtain resolutions ranging from 2 Å to 15 Å as can be seen in Fig. 7.2B. Still X-ray crystallography is considered as the best tool for high-resolution structures. A separate database (EMDB) of structures was built in 2002 (EMDataResource, n.d.) and is maintained jointly by PDBe and RCSB. As on now, about 12,000 structures solved by SPA were deposited at EMDB.

3 ELECTRON MICROSCOPE

3.1 Instrumentation

The transmission electron microscope, which will be exclusively used for recording the micrographs, was discussed extensively in standard books and review articles (Bozzola & Russell, 1999; Williams & Carter, 1996). A brief outline of the instrumentation and other features is given here.

Transmission electron microscope, like any other microscope, consists of three constituent components, viz., the source, the lens system, and the detection system. The source, which basically emits high energy electrons, is called the electron gun. The lens system basically consists of electromagnets, which play a key role in electron optics. Depending on the function the particular lens perfoms, they are called condenser lens, objective lens, and projector lens (similar to conventional optical microscope). The detection system includes a fluorescent screen, a charge coupled detector (CCD) camera, or a photographic film (which was replaced by CCD cameras lately). A representative ray diagram of a typical transmission electron microscope is shown in Fig. 7.3. A brief outline of the individual components is given below.

3.1.1 Electron gun

A typical role played by an electron gun includes emission of electrons and their acceleration to high energies. The acceleration voltage influences the de Broglie's wavelength of the electron, which can control the resolution of the data. Typically, higher operating voltages result in high-resolution data. The electron gun, depending on the origin of electron emission, can be categorized into two types—the thermionic guns and field emission electron guns. The thermionic guns include the traditional tungsten filament and LaB_6 crystals, which give electrons of less coherence and hence provide low-resolution data. As the electron microscopes equipped with thermionic guns are more affordable and can stand relatively robust usage, they

are mostly used for screening the samples or for preliminary analysis. On the other hand, the field emission guns give us more resolution and are expensive. As these guns are more susceptible for contamination, higher vacuum levels are expected at the gun chamber.

Besides the type of the gun, the resolution of the microscope also depends on operating voltage. The low operating voltage (typically 100 kV) gives us better contrast at the cost of loss of resolution. On the other hand, the high operating voltage (300 kV) gives us better resolution with poor contrast. The modern electron microscopes are capable of variable operating voltages which can operate in the range of 100 −300 kV.

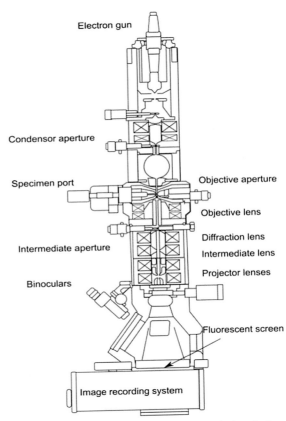

FIG. 7.3 The ray diagram of a typical transmission electron microscope. The electrons coming from the electron gun were focused on the specimen by a condenser lens. The objective lens together with projector lens plays a role in forming the image and provides magnification. The image can be viewed using a binoculars focused on the fluorescent screen (which is placed at the bottom). A provision is made in all microscopes to take the fluorescent screen off the optical path of the electron for recording the image using a charge coupled detector camara/photographic film. (Source: https://commons.wikimedia.org/wiki/File:Scheme_TEM_en.svg.)

3.1.2 Electromagnetic lenses

The electron optics is typically controlled by electromagnetic lenses. Similar to other conventional light microscopes, the electron microscope also consists of several lenses (condenser lenses, objective lenses, and projection lenses). The condenser lenses play their role in arranging a collimated beam of electrons onto the sample. The objective lens and projector lenses together contributed to magnification and image formation. Each of these lenses will be provided with apertures of variable sizes. The condenser aperture plays a role in increasing/decreasing the beam intensity, while the objective aperture controls the contrast of the image (and resolution) (see Fig. 7.3).

3.1.3 Fluorescent screen/charge coupled detector

The most common detectors in modern transmission electron microscopes are fluorescent screens and/or CCD camera. The conventional photographic film recording, which is available in few laboratories, is considered to provide high reliability and resolution. However, the photographic film approach is not favorable for automation and hence the CCD cameras have become more popular. The fluorescent screen emits green light (a characteristic of ZnS material coated on the surface of the screen), which can be viewed by the user with the help of a binocular. The user has the option to include the fluorescent or CCD/photographic film camera into the optical path for viewing or recording the image.

3.2 Sample Preparation Methods

The two popular sample preparation methods for an EM experiment are negative staining and cryo-EM. A brief outline of negative staining and cryo-EM is given in the following sections. The reviews on various sample preparation methods which will be handy are recommended for a more detailed discussion (Brenner & Horne, 1959; De Carlo & Harris, 2011; Harris et al., 1999; Horne & Pasquali Ronchetti, 1974; Lepault et al., 1983; Messaoudi et al., 2003).

3.2.1 Negative staining

The contrast required for recording the electron micrographs was obtained by the use of electron-rich heavy metal salt solutions. Unlike the conventional staining where the stain molecule binds to the sample molecule, the heavy metal stain molecule stains the background (hence the name negative stain) in the EM experiment.

In a typical negative stain experiment, a glow-discharged carbon-coated copper grid would be kept on the surface of the sample (typically 10 μL) for a short period of time (say 20–30 s) and then would be blotted quickly and moved on to the stain solution (without letting dry the sample completely). Aqueous uranyl acetate solution at a concentration of 1% (Wt/Vol) is considered as an optimal stain for routine negative staining. (Other stains which are in use for negative stain include sodium tungstate, ammonium molybdate, and lead citrate.) Optional washing with buffer solution before exposing the grid to stain solution was also practiced by few research groups. Few review articles explain other variants of the negative staining protocol for biological samples can be found (Brenner & Horne, 1959; Gelderblom et al., 1991; Harris et al., 1999; Horne & Pasquali Ronchetti, 1974; Massover & Marsh, 1997). A freshly prepared glow-discharged grid can attract dust particles with ease and hence care has to be taken to avoid the contact of the glow-discharged grids with any surface except the sample.

Because of the high electron scattering ability of the heavy metals, the negative staining data can give a good contrast; the negative staining approach will bring molecules with wider range of molecular weights ranging from as low as 65 kDa (Chinnaswamy et al., 2010) to as big ribosomal complexes to be analyzed in SPA (Sashital et al., 2014). In other words, the domain of negative stain ranges from small molecules to big molecular complexes.

However, despite a wider molecular range, the negative stain suffers from its relatively low resolution compared to cryo-EM samples. Apart from low resolution, the negatively stained samples also suffer from artifacts such as flattening (because of the presence of vacuum in the electron microscope column). A trehalose cushion protocol was suggested to overcome this flattening effect (Hirai et al., 1999).

3.2.2 Cryo-EM

The cryo-EM is an approach where the sample would be frozen rapidly without forming ice crystals. The preparation of a sample for the cryo-EM experiment requires a plunger (see Fig. 7.4). A plunger helps to freeze the sample rapidly so that the sample would be embedded in amorphous ice (rather than crystalline ice which is known to scatter electrons).

A suggested protocol for preparing the sample for cryo-EM experiment is reported by Hoenger & Aebi (1996). Typically, the sample at an appropriate concentration would be applied to a freshly glow-discharged quantifoil grid (a grid with wells of regular size at regular intervals—see the next section) placed in a chamber with controlled humidity. This helps the protein to

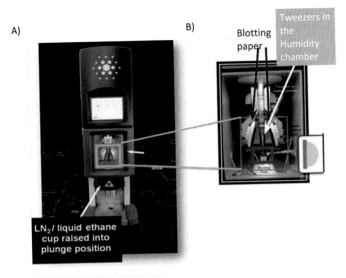

FIG. 7.4 Picture of cryo-plunger to freeze the sample for cryo-EM experiment. **(A)** Picture of Vitrobot; the grid attached to the tweezers is placed in the humid chamber (marked with blue box). **(B)** A close-up view of the humidity chamber. The sample ($\sim$10 µL) is attached through a side entry. The sample then would be blotted with blotting papers (shown with *blue arrows* in **(B)**) and plunged immediately into the liquid ethane cup (see **(A)**). The purple box shows the sample droplet on the EM grid. (Source: https://upload.wikimedia.org/wikipedia/commons/8/83/FEI_Vitrobot_%288510271991%29.jpg.)

spread uniformly on carbon film and wells. The protein which was trapped in the wells is of great importance in cryo-EM experiment, as the protein will retain its physiological conformation as compared to the protein adsorbed to the surface of the carbon film. The excess sample would be blotted and dropped into the liquid ethane (liquid ethane is preferred over liquid nitrogen in order to get amorphous ice). The thickness of the ice formed on the quantifoil grid can be controlled by changing the humidity level in the humidity chamber and the blotting angle/time (see Fig. 7.4).

However, some researchers prefer not to glow discharge the quantifoil grids as the glow discharging would promote protein adsorption on the carbon film rather than trapped in holes. However, the carbon film without glow discharge is hydrophobic in nature and will repel the protein all together. In order to see more protein trapped in the wells of a glow-discharged holey/quantifoil grids, higher concentration of protein (say 10× than that used for a regular negative stain experiment) was preferred by several research groups.

Unlike the negatively stained samples, the cryo-EM experiment generates the contrast from the scattering of the electrons by the sample itself (in the absence of heavy metal stain molecules) and hence provides poor contrast. However, the absence of external

scattering agents adds an advantage to the cryo-EM experiment in the form of high-resolution data. Also, the poor contrast sets a lower limit for the samples (typically the molecular weight of the sample should be 200 kDa or greater), thereby narrowing the domain of the molecules to only of considerable molecular size. Another precaution that one has to observe while doing the cryo-EM experiment is to minimize the exposure of the samples to the electron beam in order to minimize the possible radiation damage. This can be done by adopting few low-dosage protocols (Adrian et al., 1984; Fujiyoshi, 2013).

The choice between negatively stained sample and cryo-freeze samples depends on the nature of the sample and resolution expected. A detailed comparison of negatively stained and cryo-EM images was done by Hoenger and Aebi (1996).

3.2.3 Cryo-negative staining

Though less popular compared to negative stain and cryo-EM approaches, the cryo-negative staining (CNS) approach gives high contrast (as was seen with the negatively stained samples) and high resolution with physiologically relevant conformation (as provided by cryo-EM approach). A typical protocol (De Carlo et al., 2008; De Carlo & Harris, 2011) involves the treatment of a holey grid or a quantifoil grid (see Section 3.3 for details) with the sample similar to cryo-EM preparation, except that before blotting the specimen and freezing the grid, it will be treated with stain solution (mostly 16% w/v ammonium molybdate at neutral pH). The wells of the holey film or quantifoil film will now contain the stain molecule (ammonium molybdate), along with the protein in aqueous solution, which provides higher contrast to the protein as compared to the conventional cryo-EM preparation.

3.2.4 Two-dimensional crystallization

This is another technique that is available for elucidating the structure of the protein, though not received much attention as SPA received. Growing of two-dimensional (2D) crystals was already used to explore the structures of several membrane and non-membrane proteins (Fujiyoshi, 2013; Kimura et al., 1997; Nogales et al., 1998). For crystallizing a solvable protein, a lipid bilayer crystallization approach can be used. In a typical protocol, the protein can be incubated overnight in a humid environment with a mixture of lipids in organic solvent (chloroform/hexane). The 2D crystal thus formed can be transformed into a freshly discharged carbon-coated copper grid (see next section) and can be treated with negative stain molecules (Sun et al., 2007) or cryo-freezed (Fujiyoshi, 2013) before taking to an electron microscope. The diffraction pattern can be taken at different tilt angles of the stage and the diffraction spots can be collectively analyzed to reconstruct the 3D model of the protein (Shi et al., 2013).

3.3 EM Grids

EM grids are metallic circular discs of about 3 mm diameter with cross-bars which acts as a substrate for the specimen. Depending on the requirements of the user, a wide variety of grids are commercially available. Typically they are classified based on their mess size (ranging from 100 mesh size to 600 mesh size), the film that is applied on it (carbon, formvar, holey, or quantifoil).

The holey film grids and quantifoil film grids possess holes which can capture the proteins freely suspended, when rapidly frozen. Hence these two types of grids are specifically useful for cryo-EM experiments. Quantifoil grids possess holes of uniform size with a uniform spacing and hence are preferred for automated measurements.

4 OUTLINE OF SINGLE PARTICLE ANALYSIS
4.1 Principles of Single Particle Analysis

The basic idea of SPA lies in reconstructing the 3D model of the macromolecule from its individual projections as recorded from an electron microscope. The principles of SPA can best be understood in terms of the following steps.

4.1.1 EM data acquisition and particle selection

In order to proceed with SPA, one needs a collection of particle data. This can be generated from recording the images of well-isolated macromolecules that are clearly distinct and spatially distant from its neighboring molecule spread on the EM grid. These particles represent different orientations that the macromolecule can be found on the surface of an electron microscope grid. The higher population of isolated macromolecules can be seen at relatively lower concentrations of the macromolecule, which turned out to be an advantage over its peer techniques in structural biology.

4.1.2 Initial model generation

The particle data set thus generated from an EM experiment needs to be classified based on a 3D model, *the initial model*. The initial model can be built from the particle data set without imposing any symmetry while building the model. Alternatively, an initial model can also be built by converting the PDB structure (from RCSB server) of a homologous protein into the electron density map. However, reference-free building of the initial model from the particle data set is preferred by most research groups in order to avoid possible initial model bias.

4.1.3 Projections and classification of data set

Based on the initial model built, projections would be drawn. With these projections as reference, the particles from the EM data set would be classified into a set of classes. The resolution of the final model basically depends on number of projections made from the initial model (apart from the dependence of the resolution on sample preparation approaches and operating voltage of the microscope). The selection of number of projections should not be stretched for mere want of high resolution for inappropriate number of projections may reduce the signal to noise ratio in the class averages (see Section 4.1.4).

4.1.4 Class averages and rebuilding the model

The particles in each class after the classification of the projection matching would be averaged to generate the class averages. With the set of class averages, a new 3D model would be generated. This new model replaces the initial model and the classification of particles would be re-executed with reference to the projections drawn from the new model. This process is repeated iteratively until a stable model is resulted in two

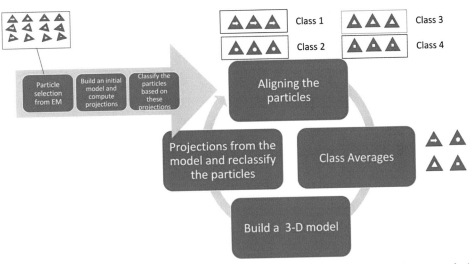

FIG. 7.5 Flowchart of single particle analysis. The particles from the electron micrographs were selected, aligned, and classified into several classes. The class averages will be used for building a three-dimensional model from which the projections will be extracted. The process is repeated iteratively until a stable structure results.

TABLE 7.1
Details of Most Popular Software Available for Performing Single Particle Analysis.

SI. No	Software	Open Source/Commercial	Website	References
1	Spider	Open source	https://spider.wadsworth.org/spider_doc/spider/docs/spider.html	Leith et al. (2012)
2	IMAGIC	Commercial (test version is available)	https://www.imagescience.de/imagic.html	van Heel & Keegstra (1981)
3	EMAN2	Open source	https://blake.bcm.edu/emanwiki/EMAN2	Tang et al. (2007)
4	Sparx	Open source	http://sparx-em.org/sparxwiki/Introduction	Hohn et al. (2007)
5	Relion	Open source	https://github.com/3dem/relion	Zivanov et al. (2018)

successive iterations. A schematic of the simplified process of SPA is shown in Fig. 7.5.

A detailed account of SPA can be found in Borkotoky et al. (2013), Tang et al. (2007).

In addition to the above steps, the SPA should also be subjected to contrast transfer function (CTF) correction. This becomes specifically important for samples prepared using cryo-EM protocol. The CTF is a function that modulates the imaging process that arises because of the electron optics (processes such as astigmatism, defocusing, etc.). A correction is necessary in order to attain high resolution in the reconstruction of the 3D images. The automated protocol for CTF correction is available for the users (Sander et al., 2003).

4.2 Available Software Tools

Though Spider (Kishchenko & Leith, 2014; Leith et al., 2012) and EMAN (Ludtke, 2016; Tang et al., 2007) software suites were among the most popular software available, numerous suites and tools are now available for reconstructing the 3D images of macromolecules using SPA. A list of various software tools available for SPA is given in Table 7.1 and is discussed elsewhere (Borkotoky et al., 2013). All these software tools are

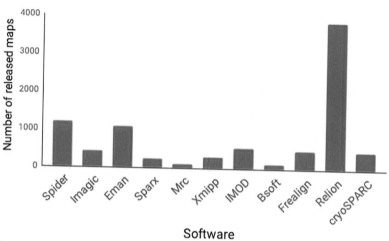

FIG. 7.6 The available software for single particle analysis along with the number of structures deposited at EMDB. (Data are taken from EMDB website EMDataResource. (n.d.). Retrieved September 24, 2020, from https://www.emdataresource.org/index.html.)

widely used by structural biologists across the globe, and structures solved by these tools are deposited in EMDB. Fig. 7.6 shows the number of structures solved by each of the popular tools available.

5 REMARKABLE RESULTS FROM SINGLE PARTICLE ANALYSIS

The novel results that were obtained using SPA are summarized here. Due to the bulk information available and every work has its novelty, only selective works were discussed here due to limitation of the space. The results are broadly divided into two categories depending on the sample preparation techniques viz. negative staining or cryo-EM with a note on the blended technique—the CNS.

5.1 Negatively Stained Samples

Though the negative staining data suffer from the low-resolution limitation, its unique advantage of having good contrast and lower limit of molecular size made it an equally important tool for biologists. Given below are few instances where low-resolution negative stain data were found in answering few biological problems.

5.1.1 Understanding the virus encapsulation mechanism

It is quite known that the nucleic acid—protein interaction plays an important role in virus encapsulation, though the mechanism is not clear yet. Cheng and Dragnea group has undertaken a series of experiments

where the nucleic acid of a brome mosaic virus (BMV) was replaced with nano gold particles of various sizes with an appropriate functional group on their surfaces. The virus-like particles (VLPs) thus obtained were imaged and their 3D images were reconstructed from negatively stained data (Sun et al., 2007). It was observed that the size of the nanoparticle played an important role in the structures of the VLPs. Smaller nano gold particles with diameter ~6 nm resulted in a virus with $T = 1$ structure, which was similar to an empty capsid (Larson et al., 2005; Lucas et al., 2001) (see Fig. 7.7) while the VLPs formed with about 9 nm nano gold particles at the core resembled the pseudo $T = 2$ virions (obtained from VIPER data base) (Natarajan et al., 2005). Also when the size of the nano gold particle approached 12 nm, a well-built $T = 3$ VLP resulted. It is interesting to note that the size of the core of a native BMV is about 14 nm confirming these results. Furthermore, by varying the charge on the nano gold particles during the capsid assembly, it was observed that the formation of VLPs was more favored after a threshold charge on the nanoparticles (Daniel et al., 2010). By mutating the capsid protein at carefully selected residues and reconstructing the 3D images of the VLPs thus formed, it was observed that the surface charge does play a role in the encapsulation of the VLPs (Hema et al., 2010).

5.1.2 Ligand-induced conformational change

In drug discovery and delivery, it is quite useful to know whether the drug really binds to the target and also the possible conformational change that happens upon

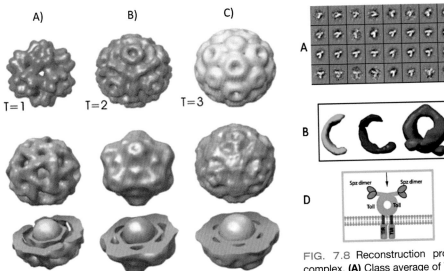

FIG. 7.7 Reconstructions of the virus-like particles (VLPs) formed with brome mosaic virus capsid proteins encapsulated around nano gold particles of different diameters (A: 6 nm, B: 9 nm, C: 12 nm). The middle row shows the reconstruction of the VLPs by using EMAN software and the bottom row represents a cross-sectional view of the reconstructions. The top row shows the representative view of the crystal structures solved (Sun et al., 2007).

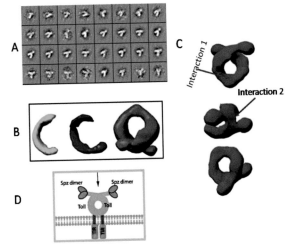

FIG. 7.8 Reconstruction process of Toll-Spatzle (Spz) complex. **(A)** Class average of Toll+Spz in 2:4 stoichiometry. **(B)** The reconstruction of three molecules Drosophila toll receptor (green), Toll-Spz complex in monomeric form (1:2 ratio—blue), and dimeric form (2:4 ratio—purple). **(C)** Different views of the Toll-Spz complex (2:4) showing two interactions one near C-terminal and the other at N-terminal (Gangloff et al., 2008). **(D)** A representative cartoon that explains the possible signaling mechanism of Toll-Spz dimer (Valanne et al., 2011).

the drug binding to the target. The SPA quite often helps the scientists to understand the drug binding mechanism. Also these data augmented with other biophysical data can increase the confidence levels of the results. The SPA of ligand-induced conformational change was well-documented and few instances are discussed below.

While analyzing the dimerization mechanism of *Drosophila* Toll receptor, it was observed that the toll receptor in its monomeric form undergoes a conformational change upon binding with its ligand, the *Spatzle* (spz) (Spatzle being in dimeric form). The reconstruction was also carried out for different receptors to ligand stoichiometric ratios. In 1:2 (toll:spz) stoichiometric ratio, a clear tightening of the receptor (more curvature of the well-known horse shoe shape of the receptor) was seen. Furthermore, in the signaling complex (in 2:4 stoichiometric ratio), it was seen that there was a huge conformational change where an interaction at N-terminal of the receptor is seen in the resultant dimeric form with the spz ligand binding at the C-terminal (Gangloff et al., 2008) (see Fig. 7.8).

Also in a separate work, the LGP2 has shown a conformational change upon binding with its double-stranded RNA molecule, and the images of LGP2 in both monomeric form and its dimeric form were reconstructed (Murali et al., 2008). The LGP2 showed a clear conformational change upon binding with a double-stranded nucleic acid as its ligand.

5.1.3 Identification of several conformations of the protein using multimodel refinement

The SPA is equipped with another feature called multimodel refinement wherein the single initial model will be replaced by multiple initial models and iteration cycle would be continued (see Section 4.1.3). This is particularly helpful in several occasions such as (1) a system where the ligand to receptor interaction is not complete and the receptors with and without ligand are coexisting in a particular dissociation constant at a given time and (2) the protein exhibits several stable conformations at a given time.

The multimodel refinement was carried out on HCV RdRp, which is known to exhibit multiple conformations (Chinnaswamy et al., 2010). As many as 50,000 particles of HCV RdRp were subjected to multimodel refinement and a range of conformations of HCV RdRp ranging from an open to a closed conformations were reconstructed using EMAN suite (Ludtke et al., 1999)

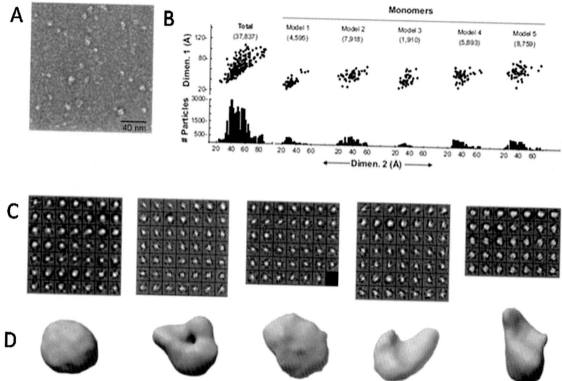

FIG. 7.9 The reconstruction of HCV RdRp to capture different conformations. **(A)** A representative micrograph for negatively stained HCV RdRp showing a heterogeneous data. **(B)** Class averages of all data set showing a range of particle diameters along with class averages obtained after performing the particle data set with multimodel refinement. **(C)** Class averages and **(D)** a representative view of the different conformations of HCV RdRp captured using single particle analysis (Chinnaswamy et al., 2010).

(Fig. 7.9). Few of the models resembled standard right-hand palm model of the HCV RdRp. Multimodel refinement was also carried out in *Drosophila* toll receptors to separate the bound and unbound receptors (see Section 5.1.2).

5.1.4 Analyzing the conformational change of a protein in different environments

Another strength of the SPA lies in its ability to monitor the conformational change of the protein as its local environment changes. It is obvious that the conformation of a protein is dependent on the pH of the buffer and also the salt concentration. This ability of the SPA was also reported with T7 RNA polymerase (T7RNAP) as a model system.

It was reported that the activity of the T7RNAP is pH-dependent and showed a considerable activity in the pH range of 7.9–9.5 with almost zero activity at highly acidic (pH < 5) and basic conditions (pH > 11) (Osumi-Davis et al., 1994). It was proposed that the

possible conformational change that the T7RNAP would undergo is to be accounted for such change in activity. Borkotoky et al. did an extensive molecular dynamics simulation of the T7RNAP in different pH conditions (Borkotoky et al., 2017). The 3D images were also reconstructed from negatively stained T7RNAP molecules under different pH conditions. The 3D models thus obtained under various pH conditions were correlated with in silico data to understand the pH-dependent activity of the T7RNAP.

5.2 Cryo-EM Samples

The cryo-EM has been the choice for researchers who aim at high-resolution structures. There are interesting results reported in the recent past and given below are a list of few remarkable results.

5.2.1 Novel SARS-CoV-2 virus

Ever since the outbreak of novel coronavirus (2019-nCoV) was reported, the pandemic condition posed a threat to the world and the scientists were actively trying to understand the structure and mechanism of the virus life cycle in order to design an effective drug for curing the COVID-19 disease.

Wrapp et al. targeted the 2019-nCoV spike glycoprotein (S) and were able to solve the structure of the spike protein in its trimeric form to a resolution of 3.5 Å (Wrapp et al., 2020) (Fig. 7.10A). Going further, they correlated their result with SARS spike protein where the receptor-binding domain (RBD) of the three monomeric spikes projecting outside and open for binding with the receptor (angiotensin-converting enzyme 2 [ACE2] receptor) (Fig. 7.10B and C) (Kirchdoerfer et al., 2018). The interaction of Spike protein and ACE2 receptor was also confirmed through surface plasmon resonance experiments. Though the observed binding of ACE2 receptor with the RBD of the nCoV is stronger than that observed with SARS—CoV, they suggested further studies for a complete understanding of the mechanism.

In a separate study, Thoms et al. worked on the nonstructural protein NSP-1, which is known to block the RIG-I—dependent innate immune response (Thoms et al., 2020). Through cryo-EM, they reported the structures of NSP1 with 40 and 80S ribosomal subunits and suggested a few structure (NSP1) based drugs for preventing the spread of the disease.

5.3 Cryo-Negative Samples

De Carlo et al. carried out the CNS approach and reported the structure of the yeast DNA-dependent RNA polymerase I (RNA PolI) to a resolution of 1.8 nm (De Carlo et al., 2003). They reported a conformational flexibility of the structures which appeared to be "open" and "closed" conformations as also observed in the heterogeneous images in the micrographs. In a separate study, by taking the GroEL as a model molecule, they showed that the CNS increased the signal to noise ratio considerably without losing the nativity of the structure (De Carlo et al., 2008). They recommended the use of

A)

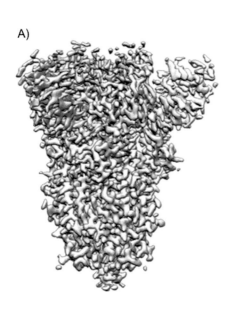

B)

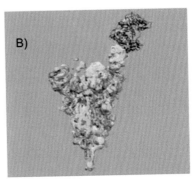

C)

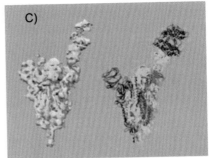

FIG. 7.10 **(A)** Cryo-EM of the 2019-nCoV spike glycoprotein (EMDB-21374). **(B and C)** Representative views of the model of SARS spike glycoprotein—human angiotensin-converting enzyme 2 (ACE2) complex (Kirchdoerfer et al., 2018). The electron density map (EMD-7582) and corresponding PDB that fits the density map (6cs2) were downloaded from EMDB and visualized using UCSF chimera software (Pettersen et al., 2004). Panel B shows the superimposition of modeled PDB into electron density map and panel C shows the PDB and density maps in juxtaposition. The three monomers were colored in blue, green, and cyan. The ACE2 was colored in red. (**(A)** The image was taken from EMDB.)

CNS wherever the EM image data suffer with a poor signal to noise ratio.

6 SINGLE PARTICLE ANALYSIS AND DRUG DISCOVERY

Scientists working in the drug design and discovery mostly look for atomic resolution images of the receptors. With the recent advancement in developing algorithms, which resulted in EM structures with high resolutions, the SPA is gaining popularity in drug design. The conformations changes in the receptor arising because of binding with the small molecule (drug) can be used as a powerful tool in drug discovery. Several reviews have discussed the use of the SPA in drug discovery (Saur et al., 2020; Scapin et al., 2018; Van Drie & Tong, 2020). Also, it is important to note that despite of attaining higher and higher resolutions, SPA and X-ray crystallography (the major tools in structural biology) are considered as complementary tools (Van Drie & Tong, 2020).

7 CONCLUSIONS AND FUTURE DIRECTIONS

The SPA is becoming a choice for more research groups as a structural biology tool to answer their biological question. With more number of protein structures being solved, this technique is gaining more popularity. In the recent pandemic situation, cryo-EM has drawn the attention of many research groups. The CNS, which has combined advantages of negative staining and cryo-EM, is yet to gain momentum. With its greater ability to provide a good signal to noise ratio yet preserving the nativity of the protein, this approach will become a choice where cryo-EM poses challenges in terms of low contrast. In a separate note, the use of 2D crystallization as a possible structural biology technique was also emphasized. The potential of the other techniques, viz., CNS staining and 2D crystallization, is yet to be tested widely. Also some prospects where the SPA can find itself as a powerful tool in the drug discovery was presented.

REFERENCES

Adrian, M., Dubochet, J., Lepault, J., & McDowall, A. W. (1984). Cryo-electron microscopy of viruses. *Nature*, *308*(5954), 32–36. https://doi.org/10.1038/308032a0

Borkotoky, S., Meena, C. K. C. K., Bhalerao, G. M. G. M., & Murali, A. (2017). An in-silico glimpse into the pH dependent structural changes of T7 RNA polymerase: A protein with simplicity. *Scientific Reports*, *7*(1), 1–12. https://doi.org/10.1038/s41598-017-06586-1

Borkotoky, S., Meena, C. K., Khan, M. W., & Murali, A. (2013). Three dimensional electron microscopy and in silico tools for macromolecular structure determination. *EXCLI Journal*, *12*, 335–346. http://www.ncbi.nlm.nih.gov/pubmed/27092033.

Bozzola, J. J., & Russell, L. D. (1999). *Electron microscopy: Principles and techniques for biologists*. Massachusetts: Jones and Bartlett Publisher. https://books.google.co.in/books?hl=en&lr=&id=zMkBAPACbEkC&oi=fnd&pg=PR21&dq=Electron+Microscopy:+Principles+and+Techniques+for+Biologists&ots=AdJV0nnJG9&sig=LlgJOxoJYoRX5Z8OJ106yy59Yp8#v=onepage&q=Electron Microscopy%3A Principles and Techniques for Biol.

Brenner, S., & Horne, R. W. (1959). A negative staining method for high resolution electron microscopy of viruses. *BBA - Biochimica et Biophysica Acta*, *34*(C), 103–110. https://doi.org/10.1016/0006-3002(59)90237-9

Chinnaswamy, S., Murali, A., Cai, H., Yi, G., Palaninathan, S., & Kao, C. C. (2010). Conformations of the monomeric hepatitis C virus RNA-dependent RNA polymerase. *Virus Adaptation and Treatment*, *2*(1), 21–39. https://doi.org/10.2147/vaat.s9101

Daniel, M.-C., Tsvetkova, I. B., Quinkert, Z. T., Murali, A., De, M., Rotello, V. M., Kao, C. C., & Dragnea, B. (2010). Role of surface charge density in nanoparticle-templated assembly of bromovirus protein cages. *ACS Nano*, *4*(7), 3853–3860. https://doi.org/10.1021/nn1005073

De Carlo, S., Boisset, N., & Hoenger, A. (2008). High-resolution single-particle 3D analysis on GroEL prepared by cryo-negative staining. *Micron*, *39*(7), 934–943. https://doi.org/10.1016/j.micron.2007.11.003

De Carlo, S., Carles, C., Riva, M., & Schultz, P. (2003). Cryo-negative staining reveals conformational flexibility within yeast RNA polymerase I. *Journal of Molecular Biology*, *329*(5), 891–902. https://doi.org/10.1016/S0022-2836(03)00510-2

De Carlo, S., & Harris, J. R. (2011). Negative staining and cryo-negative staining of macromolecules and viruses for TEM. *Micron*, *42*(2), 117–131. https://doi.org/10.1016/j.micron.2010.06.003

Drenth, J., & Mesters, J. (2007). *Principles of protein X-ray crystallography* (3rd ed.). Springer New York, ISBN 978-0-387-33334-2. https://doi.org/10.1007/0-387-33746-6

EMDataResource. (n.d.). Retrieved September 24, 2020, from https://www.emdataresource.org/index.html.

Fujiyoshi, Y. (2013). Low dose techniques and cryo-electron microscopy. *Methods in Molecular Biology*, *955*, 103–118. https://doi.org/10.1007/978-1-62703-176-9_6

Gangloff, M., Murali, A., Xiong, J., Arnot, C. J., Weber, A. N., Sandercock, A. M., Robinson, C. V., Sarisky, R., Holzenburg, A., Kao, C., & Gay, N. J. (2008). Structural insight into the mechanism of activation of the toll receptor by the dimeric ligand Spätzle. *Journal of Biological Chemistry*, *283*(21), 14629–14635. https://doi.org/10.1074/jbc.M800112200

Gelderblom, H. R., Renz, H., & Özel, M. (1991). Negative staining in diagnostic virology. *Micron and Microscopica Acta*, *22*(4), 435–447. https://doi.org/10.1016/0739-6260(91)90061-4

Harris, J. R., Roos, C., Djalali, R., Rheingans, O., Maskos, M., & Schmidt, M. (1999). Application of the negative staining technique to both aqueous and organic solvent solutions of polymer particles. *Micron, 30*(4), 289–298. https://doi.org/10.1016/S0968-4328(99)00034-7

van Heel, M., & Keegstra, W. (1981). Imagic: A fast, flexible and friendly image analysis software system. *Ultramicroscopy, 7*(2), 113–129. https://doi.org/10.1016/0304-3991(81)90001-2

Hema, M., Murali, A., Ni, P., Vaughan, R. C., Fujisaki, K., Tsvetkova, I., Dragnea, B., & Kao, C. C. (2010). Effects of amino-acid substitutions in the Brome mosaic virus capsid protein on RNA encapsidation. *Molecular Plant-Microbe Interactions, 23*(11), 1433–1447. http://apsjournals.apsnet.org/doi/10.1094/MPMI-05-10-0118.

Hirai, T., Murata, K., Mitsuoka, K., Kimura, Y., & Fujiyoshi, Y. (1999). Trehalose embedding technique for high-resolution electron crystallography: Application to structural study on bacteriorhoposin. *Journal of Electron Microscopy, 48*(5), 653–658. https://doi.org/10.1093/oxfordjournals.jmicro.a023731

Hoenger, A., & Aebi, U. (1996). 3-D reconstructions from ice-embedded and negatively stained biomacromolecular assemblies: A critical comparison. *Journal of Structural Biology, 117*(2), 99–116. https://doi.org/10.1006/jsbi.1996.0075

Hohn, M., Tang, G., Goodyear, G., Baldwin, P. R., Huang, Z., Penczek, P. A., Yang, C., Glaeser, R. M., Adams, P. D., & Ludtke, S. J. (2007). SPARX, a new environment for Cryo-EM image processing. *Journal of Structural Biology, 157*(1), 47–55. https://doi.org/10.1016/j.jsb.2006.07.003

Horne, R. W., & Pasquali Ronchetti, I. (1974). A negative staining-carbon film technique for studying viruses in the electron microscope. I. Preparative procedures for examining icasahedral and filamentous viruses. *Journal of Ultrastructure Research, 47*(3), 361–383. https://doi.org/10.1016/S0022-5320(74)90015-X

Kimura, Y., Vassylyev, D. G., Miyazawa, A., Kidera, A., Matsushima, M., Mitsuoka, K., Murata, K., Hirai, T., & Fujiyoshi, Y. (1997). Surface of bacteriorhodopsin revealed by high-resolution electron crystallography. *Nature, 389*(6647), 206–211. https://doi.org/10.1038/38323

Kirchdoerfer, R. N., Wang, N., Pallesen, J., Wrapp, D., Turner, H. L., Cottrell, C. A., Corbett, K. S., Graham, B. S., McLellan, J. S., & Ward, A. B. (2018). Stabilized coronavirus spikes are resistant to conformational changes induced by receptor recognition or proteolysis. *Scientific Reports, 8*(1), 1–11. https://doi.org/10.1038/s41598-018-34171-7

Kishchenko, G. P., & Leith, A. (2014). Spherical deconvolution improves quality of single particle reconstruction. *Journal of Structural Biology, 187*(1), 84–92. https://doi.org/10.1016/j.jsb.2014.05.002

Ladd, M., Palmer, R., Ladd, M., & Palmer, R. (2013). Proteins and macromolecular X-ray analysis. In *Structure determination by X-ray crystallography* (pp. 489–548). Springer US. https://doi.org/10.1007/978-1-4614-3954-7_10

Larson, S. B., Lucas, R. W., & McPherson, A. (2005). Crystallographic structure of the T=1 particle of brome mosaic virus. *Journal of Molecular Biology, 346*(3), 815–831. https://doi.org/10.1016/j.jmb.2004.12.015

Leith, A., Baxter, W., & Frank, J. (2012). *Use of SPIDER and SPIRE in image reconstruction* (pp. 620–623). https://doi.org/10.1107/97809553602060000874

Lepault, J., Booy, F. P., & Dubochet, J. (1983). Electron microscopy of frozen biological suspensions. *Journal of Microscopy, 129*(1), 89–102. https://doi.org/10.1111/j.1365-2818.1983.tb04163.x

Lucas, R. W., Kuznetsov, Y. G., Larson, S. B., & McPherson, A. (2001). Crystallization of brome mosaic virus and T = 1 brome mosaic virus particles following a structural transition. *Virology, 286*(2), 290–303. https://doi.org/10.1006/viro.2000.0897

Ludtke, S. J. (2016). Single-particle refinement and variability analysis in EMAN2.1. *Methods in Enzymology, 579*, 159–189. https://doi.org/10.1016/bs.mie.2016.05.001

Ludtke, S. J., Baldwin, P. R., & Chiu, W. (1999). EMAN: Semi-automated software for high-resolution single-particle reconstructions. *Journal of Structural Biology, 128*(1), 82–97. https://doi.org/10.1006/jsbi.1999.4174

Massover, W. H., & Marsh, P. (1997). Unconventional negative stains: Heavy metals are not required for negative staining. *Ultramicroscopy, 69*(2), 139–150. https://doi.org/10.1016/S0304-3991(97)00040-5

Messaoudi, C., Boudier, T., Lechaire, J. P., Rigaud, J. L., Delacroix, H., Gaill, F., & Marco, S. (2003). Use of cryo-negative staining in tomographic reconstruction of biological objects: Application to T4 bacteriophage. *Biology of the Cell, 95*(6), 393–398. https://doi.org/10.1016/S0248-4900(03)00086-8

Murali, A., Li, X., Ranjith-Kumar, C. T., Bhardwaj, K., Holzenburg, A., Li, P., & Kao, C. C. (2008). Structure and function of LGP2, a DE X (D/H) helicase that regulates the innate immunity response. *Journal of Biological Chemistry, 283*(23), 15825–15833. https://doi.org/10.1074/jbc.M800542200

Natarajan, P., Lander, G. C., Shepherd, C. M., Reddy, V. S., Brooks, C. L., & Johnson, J. E. (2005). Exploring icosahedral virus structures with VIPER. *Nature Reviews Microbiology, 3*(10), 809–817. https://doi.org/10.1038/nrmicro1283

Nogales, E., Wolf, S. G., & Downing, K. H. (1998). Structure of the αβ tubulin dimer by electron crystallography. *Nature, 391*(6663), 199–203. https://doi.org/10.1038/34465

Osumi-Davis, P. A., Sreerama, N., Volkin, D. B., Middaugh, C. R., Woody, R. W., & Woody, A. Y. M. (1994). Bacteriophage T7 RNA polymerase and its active-site mutants. Kinetic, spectroscopic and calorimetric characterization. *Journal of Molecular Biology, 237*(1), 5–19. https://doi.org/10.1006/jmbi.1994.1205

Pettersen, E. F., Goddard, T. D., Huang, C. C., Couch, G. S., Greenblatt, D. M., Meng, E. C., & Ferrin, T. E. (2004). UCSF Chimera – A visualization system for exploratory research and analysis. *Journal of Computational Chemistry, 25*(13), 1605–1612. https://doi.org/10.1002/jcc.20084

Rule, G. S., & Hitchens, T. K. (2006). *Fundamentals of protein NMR spectroscopy*. Springer-Verlag. https://doi.org/10.1007/1-4020-3500-4

Sander, B., Golas, M. M., & Stark, H. (2003). Automatic CTF correction for single particles based upon multivariate statistical analysis of individual power spectra. *Journal of Structural Biology, 142*(3), 392–401. https://doi.org/10.1016/S1047-8477(03)00072-8

Sashital, D. G., Greeman, C. A., Lyumkis, D., Potter, C. S., Carragher, B., & Williamson, J. R. (2014). A combined quantitative mass spectrometry and electron microscopy analysis of ribosomal 30S subunit assembly in *E. coli*. *ELife*, *3*. https://doi.org/10.7554/eLife.04491

Saur, M., Hartshorn, M. J., Dong, J., Reeks, J., Bunkoczi, G., Jhoti, H., & Williams, P. A. (2020). Fragment-based drug discovery using cryo-EM. *Drug Discovery Today*, *25*(3), 485–490. https://doi.org/10.1016/j.drudis.2019.12.006

Scapin, G., Potter, C. S., & Carragher, B. (2018). Cryo-EM for small molecules discovery, design, understanding, and application. *Cell Chemical Biology*, *25*(11), 1318–1325. https://doi.org/10.1016/j.chembiol.2018.07.006

Shi, D., Nannenga, B. L., Iadanza, M. G., & Gonen, T. (2013). Three-dimensional electron crystallography of protein microcrystals. *ELife*, *2013*(2). https://doi.org/10.7554/eLife.01345.001

Sun, J., DuFort, C., Daniel, M.-C., Murali, A., Chen, C., Gopinath, K., Stein, B., De, M., Rotello, V. M., Holzenburg, A., Kao, C. C., & Dragnea, B. (2007). Core-controlled polymorphism in virus-like particles. *Proceedings of the National Academy of Sciences*, *104*(4), 1354–1359. https://doi.org/10.1073/pnas.0610542104

Tang, G., Peng, L., Baldwin, P. R., Mann, D. S., Jiang, W., Rees, I., & Ludtke, S. J. (2007). EMAN2: An extensible image processing suite for electron microscopy. *Journal of Structural Biology*, *157*(1), 38–46. https://doi.org/10.1016/j.jsb.2006.05.009

Thoms, M., Buschauer, R., Ameismeier, M., Koepke, L., Denk, T., Hirschenberger, M., Kratzat, H., Hayn, M., Mackens-Kiani, T., Cheng, J., Straub, J. H., Stürzel, C. M., Fröhlich, T., Berninghausen, O., Becker, T., Kirchhoff, F., Sparrer, K. M. J., & Beckmann, R. (2020). Structural basis for translational shutdown and immune evasion by the Nsp1 protein of SARS-CoV-2. *Science*, *369*(6508), eabc8665. https://doi.org/10.1126/science.abc8665

Valanne, S., Wang, J.-H., & Rämet, M. (2011). The Drosophila toll signaling pathway. *The Journal of Immunology*, *186*(2), 649–656. https://doi.org/10.4049/jimmunol.1002302

Van Drie, J. H., & Tong, L. (2020). Cryo-EM as a powerful tool for drug discovery. *Bioorganic and Medicinal Chemistry Letters*, *30*(22), 127524. https://doi.org/10.1016/j.bmcl.2020.127524

Williams, D. B., & Carter, C. B. (1996). The transmission electron microscope BT. In D. B. Williams, & C. B. Carter (Eds.), *Transmission electron microscopy: A textbook for materials science* (pp. 3–17). Springer US. https://doi.org/10.1007/978-1-4757-2519-3_1.

Wrapp, D., Wang, N., Corbett, K. S., Goldsmith, J. A., Hsieh, C.-L., Abiona, O., Graham, B. S., & McLellan, J. S. (2020). Cryo-EM structure of the 2019-nCoV spike in the prefusion conformation. *Science*, *367*(6483), 1260–1263. https://doi.org/10.1126/science.abb2507

Wuthrich, K. (1990). Protein structure determination in solution by NMR spectroscopy. *Journal of Biological Chemistry*, *265*(36), 22059–22062. https://doi.org/10.1142/9789812795830_0001

Zivanov, J., Nakane, T., Forsberg, B. O., Kimanius, D., Hagen, W. J. H., Lindahl, E., & Scheres, S. H. W. (2018). New tools for automated high-resolution cryo-EM structure determination in RELION-3. *ELife*, *7*. https://doi.org/10.7554/eLife.42166

Computational Modeling of Protein Three-Dimensional Structure: Methods and Resources

ARCHANA PAN • G. PRANAVATHIYANI • SIBANI SEN CHAKRABORTY

1 INTRODUCTION

Proteins are biological macromolecules that perform a wide variety of functions in a living organism. They do function as enzymes, transport/storage, hormones, structural component, antibody, etc. Enzymes function as biocatalysts that carry out almost all biochemical reactions taking place in a living organism. Storage proteins act as reserves of amino acids and metal ions, utilized by the organism, whereas transport proteins facilitate to move the molecules/ions within an organism. Hormones are the messenger proteins that transmit signals to orchestrate biological processes among various cells, tissues, and organs of an organism. Cell structure and shape are usually maintained by structural proteins. The other important types of proteins are antibodies, which are produced to recognize the foreign substances (e.g., viruses and bacteria) and play crucial role in immune system in animals, including humans. These proteins of different functional classes have specific amino acid sequence compositions that can influence their structures. Solving the protein structure, particularly three-dimensional (3D) structure, often facilitates to understand the biological role of the proteins (Branden & Tooze, 2012; Laskowski et al., 2005a, 2005b; Lesk, 2014). Indeed, the protein structure determination from its primary sequence (i.e., amino acid sequence) is an active area of research in structural biology. Experimental methods such as X-ray crystallography, nuclear magnetic resonance (NMR) spectroscopy, and electron microscopy especially cryogenic electron microscopy (cryo-EM) can solve protein 3D structure with high resolution. However, the high cost involved with these methods and their labor-intensive natures are the limiting factors for their usage in protein structure prediction.

As of July 2020, 185.56 million protein sequences have been deposited in public database UniProtKB (Swiss-Prot + TrEMBL) (Bairoch, 2004). It has been possible owing to the successful completion of large-scale genome sequencing projects of several organisms, which has generated a voluminous amount of gene and protein sequences. However, currently only 1,54,918 protein 3D structures (as of July 2020) are deposited in Protein Data Bank (PDB), a repository (freely accessible) of experimentally determined protein 3D structures (Berman et al., 2000; Burley et al., 2018) (https://www.rcsb.org/). Thus, PDB library contains very less number of structures compared to the available number of protein sequences in the database $\left(\text{ratio} = \frac{\text{available protein structures}}{\text{available protein sequences}} = 0.0008 \right)$. This discrepancy is due to the difficulty associated with experiential structure determination methods. To fill the gap, researchers have been paying attention to develop several computational structure prediction methods. Currently available such methods can commonly be classified into template-based (Karplus et al., 1998; Jones, 1999; Šali & Blundell, 1993; Skolnick et al., 2004; Soding, 2005; Wu & Zhang, 2008a, 2008b; Yang et al., 2011) and template-free methods (Bradley, 2005; Bradley et al., 2005; Jauch et al., 2007; Kinch et al., 2016; Klepeis et al., 2004; Liwo et al., 2005; Xu & Zhang, 2012).

Homology modeling and threading are template-based methods, whereas ab initio modeling is known as template-free method. This chapter focuses on discussing protein 3D structure prediction methods along with relevant information and resources associated with protein structure prediction. In addition, the chapter illustrates case study using some frequently used methods.

Molecular Docking for Computer-Aided Drug Design. https://doi.org/10.1016/B978-0-12-822312-3.00023-0

2 PROTEIN STRUCTURAL ORGANIZATION

To comprehend the structure of a given protein, it is necessary to understand its hierarchical structure, i.e., its primary, secondary, tertiary, and quaternary structure (Lodish et al., 2000; Nelson et al., 2008). The simplest form of a protein structure is its primary structure. The primary structure of any protein is a linear chain consisting of 20 different amino acids, which are linked by the peptide bonds. All the amino acids have a common backbone structure; however, they vary in their side chains (i.e., R-groups) (Lodish et al., 2000). The linear chain, made up of amino acids, is known as a polypeptide chain and each amino acid along the chain is called a residue. A peptide bond is a covalent chemical bond, which joins two amino acids by removing a water molecule (H_2O) from an amino group ($-NH_2$) of one amino acid and a carboxyl group ($-COOH$) of the adjacent amino acid in a polypeptide chain. At one end of the polypeptide chain, there is a free unlinked amino group (known as N-terminus), whereas at the opposite end of the protein chain, there is a free carboxyl group (known as C-terminus). Conventionally, a polypeptide chain is represented with N-terminal amino acid on the left side and its C-terminal amino acid on the right side.

The next complex level of protein structure is the secondary structure, which arises owing to the localized organization of the segments of a polypeptide chain. When there is no stabilizing interaction, a polypeptide chain attains a random coil structure. However, the most common types of secondary structures—alpha-helix and beta-pleated sheet—arise owing to the formation of stable hydrogen bonds between atoms of the polypeptide backbone. Hydrogen bonds can be formed between carbonyl oxygen (C=O) of one amino acid and amino hydrogen (N—H) of another one, which are accountable for coiled/folded shape structure (secondary structure) of the polypeptide backbone. The alpha-helix—a curled ribbon-like structure—is a right-handed coiled helix, which is stabilized by hydrogen bonds between ith amino acid and i+4th amino acid (Rehman, Farooq, & Botelho, 2020). In a beta-pleated sheet, two (or more) different segments of a polypeptide chain lie side-by-side that are held together by hydrogen bonds and form a sheet-like structure. The beta-pleated sheets can be of two types, parallel and antiparallel beta-pleated sheets. When two beta-strands run in the same direction (e.g., both strands: N- to C-terminal direction or vice versa), it is known as parallel beta-pleated sheet. In contrast, if they are in opposite directions (strand 1: N- to C-terminal direction and strand 2: C- to N-terminal direction), then it is called antiparallel beta-pleated sheet. In addition to these two common types of secondary structures, there exists one less frequent secondary structure in nature, which is the beta-barrel. This type of secondary structure encompasses antiparallel beta-strands and that is twisted and coils to form a barrel structure, wherein the first strand and the last strand are bound by hydrogen bonds.

The 3D arrangement of all the amino acid residues in a polypeptide chain gives rise to its tertiary structure. This type of structure primarily arises owing to the interactions between side chains (i.e., R-groups) of amino acid residues of the polypeptide chain. These R-group interactions, which govern the tertiary structure of a protein, include hydrophobic interactions, hydrogen bonding, ionic bonding, dipole—dipole interactions, and London dispersion forces (such as van der Waals forces). Furthermore, in some proteins (e.g., extracellular proteins), disulfide bonds (covalent bond; much stronger compared to other types of bonds) contribute to the tertiary structure. These forces hold the secondary structure elements (α-helices, β-strands, turns, and coils) into a compact scaffold. Thus, the shape and size of a protein are dependent on its sequence and also based on its secondary structures (mainly, number, size, and arrangement). The prediction of 3D structure of a protein from its primary structure is discussed in Section 5. For the proteins (monomeric) with a single polypeptide chain, tertiary structure is the highest level of organization. Some proteins, in nature, comprise of more than one polypeptide chain, called subunits, for which quaternary structure is the highest level of organization. The subunits are held together by hydrogen bonding and London dispersion forces to give quaternary structure. The four levels of structural organization of a protein are depicted in Fig. 8.1.

3 SEQUENCE–STRUCTURE–FUNCTION RELATIONSHIP

Inferring the function of an unknown protein is considered to be the "holy grail" in biological research. Sequence (primary) similarity search against known (characterized) proteins using Basic Local Alignment Search Tool (BLAST) (Altschul et al., 1990) or phylogenetic analysis of an unknown protein with closely related proteins can provide functional information of the unknown protein. Moreover, structural similarity search between predicted structure (using computational method) of the unknown protein and solved structures present in PDB may also reveal molecular function of the unknown protein. Thus, generating the structure of an unknown protein is an important aspect to revealing its possible function. However, when one such approach fails to uncover protein function, it is wise to use the combination of sequence and structure information/features together. In this regard, an attempt has been made to amalgamate different prediction methods using sequence and structural information. For instance, ProFunc server is developed that utilizes a combination of methods such as InterProScan (Zdobnov & Apweiler, 2001) and BLAST search against PDB for predicting protein function (Laskowski et al., 2005a). The InterPro unifies several protein signature databases PROSITE, Pfam, ProDom, PRINTS, and Blocks, and using those methods it facilitates to infer function of the new/novel sequences. Thus, the

knowledge at the level of sequence and structure can be of use in understanding the biological role of unknown proteins. A pictorial representation of sequence—structure—function relationship of a protein is given in Fig. 8.2.

4 EXPERIMENTAL APPROACHES TO DETERMINE STRUCTURES

There are different widely used experiential methods to determine the protein 3D structure. They are X-ray crystallography, NMR, and cryo-EM. Every experimental method has its own merits and demerits in solving

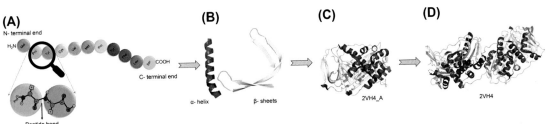

FIG. 8.1 Four levels of protein structural organization: **(A)** primary polypeptide chain of amino acids, **(B)** secondary structure elements, **(C)** tertiary structure, and **(D)** quaternary structure.

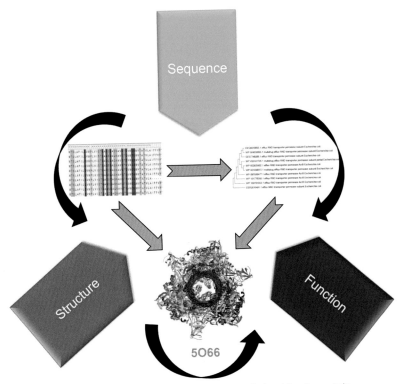

FIG. 8.2 Sequence—structure—function relationship of a protein.

structure. Currently, X-ray crystallography is the most favored method for obtaining 3D structure of biological macromolecules including proteins. In this method, the protein needs to be first crystallized, and then the crystals are placed in an X-ray beam from where X-ray diffraction datasets need to be collected. The datasets are analyzed using computational techniques for protein structure generation followed by structure refinement and validation. The X-ray crystallography method can enable to solve protein structure with high resolution and high accuracy. However, in many cases, such as in the case of membrane proteins, getting crystals is very difficult and thereby limiting its usage. Cryo-EM is often used for visualization and 3D reconstruction of comparatively larger (>~100 kDa) objects, such as individual molecules, macromolecular complexes, cellular organelles, whole cells, etc. (should be thin enough <1 µM to pass through electron beam). Unlike crystallography, this method does not demand crystallized samples. Also, cryo-EM does not require a larger amount of material, which is a plus point in comparison to crystallography and NMR methods. However, one of the pitfalls of this method is that structure resolution obtained is generally less compared to that of the other two methods. In case of NMR spectroscopy, the protein under consideration should be stable enough at room temperature for longer time duration for data acquisition. In addition, method needs a larger amount of material to begin the process. The level of obtained structure resolution is lesser than crystallography. But the method is reported to be useful when it is hard to crystallize a protein or when the dynamics of the system is required to analyze in details. As a whole, all the experiential methods are costly and labor-intensive in nature and take a longer time to solve the structure. As an alternative to experiential methods, researchers are using computational prediction methods to solve the protein structures which are comparatively faster and cost-effective.

5 COMPUTATIONAL APPROACHES FOR PREDICTING PROTEIN STRUCTURES

The two important computational approaches used for predicting protein 3D structure are template-based (homology modeling and threading) and template-free (ab initio) modeling, which are discussed below.

5.1 Template-Based Homology Modeling

Absence of sufficient experimentally determined structures prompts one to gain interest in a theoretically modeled structure by using homology modeling.

Indeed, homology modeling demands the knowledge of the known structure of a protein (template) with significant amino acid identity/similarity to the unknown protein (target). For this reason, homology modeling is known as temple-based modeling. The homology modeling method predicts the protein 3D structure following several steps: (1) selection of best template(s), (2) alignment of template and target sequence, (3) building model of target protein, (4) model optimization, and (5) model validation or evaluation (Fig. 8.3). Each step is discussed in detail below.

5.1.1 Selection of best templates

For homology modeling, a minimum of 30% amino acid identity is essential between the target and template protein sequences (Rost, 1999). The more the target similarity, the greater is the accuracy in structure prediction of the unknown protein using homology modeling. Use of more than one template structure can improve the quality of structure prediction of the target protein. If the sequence similarity is less than 25%, it is difficult to assume a common ancestral evolution and hence to predict structural similarity (Chung & Subbiah, 1996). If structural homologs are known, then structural homolog databases such as CATH, SCOP, or FSSP can be used to retrieve homologs from PDB. Among the three systems for structural data classification, FSSP is completely automated, SCOP is manually driven, and CATH uses automated process along with manual intervention (Hadley & Jones, 1999). On the contrary, if only the target sequence of the protein is known, then template proteins with sequence and structure homology can be identified using BLAST search (Altschul et al., 1990) provided on the NCBI website (https://www.ncbi.nlm.nih.gov/). Blast helps to find out regions of similarity between two aligned sequences.

5.1.2 Alignment of template and target sequence

The accuracy of the modeling process is highly influenced by the correct alignment of amino acid residues between the template and target protein. If the identity of the target protein is less than 30%, then it is better to align the sequences between the target protein and the proteins from the other family members so as to make sure the regions of structural/functional importance aligned rightly with the sequence of template. In cases with lower sequence identity, it is suggested to use alignment methods, which consider manipulation of parameters, such as gap penalties, in order to avoid errors in modeling protocol. However, for alignment

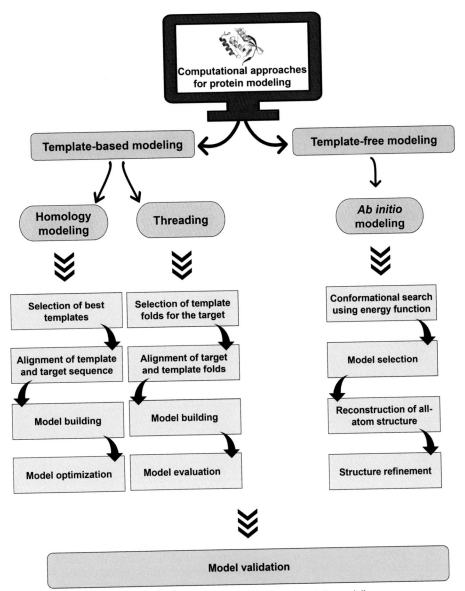

FIG. 8.3 Steps involved in computational protein modeling.

of multiple sequences as input which includes members of protein family, alignment programs such as CLUSTALX (Thompson, 1997) (http://www.clustal.org/clustal2/) and PileUP (Womble, 2000) can be used.

5.1.3 Building model of target protein
5.1.3.1 Generation of the backbone 3D structure. The alignment correction is done to generate the backbone of the target protein by making sure that functionally important and conserved residues are aligned in the correct manner. The mismatched residues and gaps due to insertions or deletions in the target sequence are scored using a scoring matrix for the pairwise alignment of two sequences. Higher score reflects better alignment. Substitution matrices such as PAM250 (Hersh & Dayhoff, 1970) and BLOSUM62 (Henikoff & Henikoff, 1992) facilitate in finding a cumulative score for the substitution of an amino acid with another in a group of proteins. The point accepted mutation (PAM) is the replacement of

an amino acid residue by another in the primary structure of a protein through natural selection. PAM matrices are based on scoring matrices for amino acid positions in sequences while BLOSUM matrices are relied on substitutions and conserved sequences in blocks. In BLOSUM62, the sequences used to create this matrix share 62% identity. The E-value is a statistical parameter, which reflects the likelihood of occurrence of an amino acid in a protein sequence by chance. Smaller the E-value, the lesser the likelihood of occurrence by chance. A sequence to structure alignment, which provides a better understanding of the similarity in 3D structure, is performed using position-specific scoring matrix (PSSM) or profile hidden Markov model (HMM) (Stojmirović et al., 2008). PSSM is used for identities greater than 35% and it assigns a score to the amino acids in a database sequence, whereas profile HMM is usually tolerated when the identity in protein sequences is less than 35%. The earlier the alignment gaps in protein backbone can be detected, the better is the chance of it being remodeled accurately, as any error in the backbone will later be conveyed in side chains.

5.1.3.2 Loop modeling to correct the folding of low-homology regions.

Loops generally are regions of low sequence conservation and connect secondary structural elements such as *helices* and *sheets*. Hence, loops adopt different conformations in homologous-related proteins. Loop modeling is thus visualized as a mini protein folding where sequence information helps in calculating loop conformation (Fiser et al., 2000). Loop modeling can be performed either by database search or conformational search. Conformational search is a part of ab initio method, which is discussed in Section 5.3.

The database contains information regarding sequence and structure of loops from all available proteins. The database is then searched to find the loop that fits in the region satisfying the geometric constraints or the similarity in sequence in loop regions of the target and template protein. After superimposition of the selected segments, the overall conformation of the loop structures requires optimization. The database search method for loop modeling is more accurate for loops of seven residues or less (Fidelis et al., 1994). For homologous proteins, insertion of loops containing more than eight residues is rare (Benner et al., 1993; Flores et al., 1993; Pascarella & Argos, 1992; Sali, 1995).

Conformational search process for loop modeling results in loops that can be determined by a scoring function. Several algorithms are available for the scoring function analysis (Contreras-Moreira et al.,

2002; Sali, 1995). The method is more applicable for modeling smaller loops than for larger loops as a greater number of loop conformations are generated, thus reducing the efficacy of the method.

5.1.3.3 Side chain modeling through conformational search.

Owing to differences between target and template protein (amino acid) sequences, side chain modeling cannot be done in the same manner as the conserved region amino acids. In such cases, rotamer databases are utilized to find the preferred conformation of the side chain amino acids different from the template structure. The rotamer databases contain preferred torsional angles of side chains for a particular side chain of preferred conformation (A1-Lazikani et al., 2001). However, if the side chain is larger than the template structure, the possibility of steric conflicts needs to be resolved during model minimization. Side chain modeling accuracy is determined by the rotamer library used, the force field used for side chain optimization, and the quality of the protein backbone, bond angle, and bond length (Xiang & Honig, 2001). Side chains in buried regions of a protein are modeled with more accuracy as more constraints are imposed on them than those which are exposed on the surface (Chakravarty et al., 2005).

5.1.4 Model optimization

5.1.4.1 Energy minimization.

Model optimization involves removal of nonfavorable contacts and idealization of bond geometry (Peitsch, 2002). Several minimization programs such as CHARMM, AMBER, and GROMOS are included in modeling programs for optimal energy minimization. Energy minimization involves removal of steric clashes between side chain residues ensuring proper covalent geometry between each atom (Contreras-Moreira et al., 2002). In energy minimization, all atoms are moved within a limited space to determine the minimum local energy. The local energy minima may be much higher than the global minima but has the advantage of allowing moderate changes to the position of the atom. Every minimization process rectifies significant errors in the stereochemistry of the model but sometimes at the cost of moving the structure much away from the original one after several cycles of minimization. Thus, current programs restrain the position of the atoms or apply few hundred cycles of minimization (Bourne & Weissig, 2003).

5.1.4.2 Molecular dynamics.

Molecular dynamics (MD) is a conformational searching protocol which allows atoms to move and collide with each other. If enough energy is provided, the molecules will be agitated and cross the local energy minima on the conformational potential energy surface of the molecule. Simulated annealing (SA) used for optimizing protein molecules simulates protein at higher temperature altering the state of the molecule and then lowering the temperature bringing back the molecule to a stable state, thereby sampling a large conformational space. The process is repeated several times obtaining multiple conformations for analysis and is carried out in presence of solvent environment. MD cycle of 10−100 ns time scale is usually run in an explicit representation of protein and solvent (Fan, 2004).

5.1.5 Model validation

While evaluation of the validity of a model, many aspects such as interacting neighboring residues and the atoms involved along with placement of residues need to be considered. The stereochemical properties of the model such as bond angles, bond lengths, proper chirality, and ring structure coupled with other geometric parameters need to be checked. Bad contacts within the model due to packing need to be eliminated.

The stereochemical features of the model can be analyzed using programs PROCHECK (Laskowski et al., 1993) and WHATCHECK (Hooft et al., 1996). The amino acid geometry can be assessed through Ramachandran plot (Ramachandran et al., 1963). Ramachandran plot is a two-dimensional (2D) plot of the torsional angles of amino acids φ (phi) and ψ (psi) in a protein sequence. The φ represents the dihedral angle between N(i-1)-C(i)-CA(i)-N(i) and ψ is the backbone dihedral angle between C(i)-CA(i)-N(i)-C(i+1). The plot was developed by G.N. Ramachandran in 1963. The φ and ψ values are plotted to obtain the conformation of the peptide and the angular spectrum lies between −180 and +180° on x-axis and y-axis. Ramachandran outlier is a representation of those amino acids which lie in the nonfavorable regions of the plot. The plot makes these evident which results from mistakes in data processing. Evaluation of spatial properties such as solvent accessibility, charged group distribution, etc., can be assessed with programs such as PROSAII (Marti-Renom et al., 2000), ANOLEA (Melo et al., 1997), and VERIFY3D (Eisenberg et al., 1997). All these programs evaluate the environment of each residue with that of the expected high-resolution X-ray structure environment. Model validation programs point out regions that require further optimization usually by manual adjustments. Several structure validation programs are available at SAVES v5.0 (Structure Analysis and Verification Server), an online server (https://servicesn.mbi.ucla.edu/SAVES/). The final model must support all biochemical data that have been predicted for the target protein.

5.2 Template-Based Threading

Protein threading, also known as "fold recognition", is used to model those proteins which do not have homologous proteins of known structures but have the same fold as unknown proteins of known structure. When no significant homology between proteins is found, protein threading can be used as a good prediction based on structural information. In this method of template-based modeling, the queried target sequence is compared against a library of several template structures of known proteins with similar fold. The target sequence is then aligned with the core of the best-fit template structure. The unknown protein is "threaded," i.e., aligned along the backbone of the template protein structure and then evaluated on the basis of the best-fit score to find how well the target fits the template structure. Once the best-fit template is selected, the target protein structure is built based on that template structure (Fig. 8.3). This method of template-based protein designing has a dual approach. First, it tries to recognize a fold in the protein corresponding to that present in the fold library generating a one-dimensional (1D) profile and aligns the target sequence to these profiles. Secondly, by assessing the interacting distances, it tries to fit in the protein in 3D structure generating the characteristic fold of the target protein. Threading works on a statistical relationship among the structures present in PDB and the sequence of the target protein to be built (Jones, 1999; Jones et al., 1992). Raptor X is an integer-based programming software for protein threading. It uses probabilistic graphical models and statistical inference to both single-template and multi-template−based protein threading (Ma et al., 2012; Peng & Xu, 2010, 2011). I-TASSER is one of the top-ranked servers in community-wide experiments (critical assessment of structure prediction [CASP]), which builds the protein structure based on iterated fragment assemblies using threading approach (Yang et al., 2015). Models generated from these servers can be further validated using SAVESv5.0 server discussed in Section 5.1.5.

5.3 Template-Free Modeling (Ab initio)

For many query proteins, template with appropriate sequence similarity (>30%) is not available in PDB. To build 3D model of such proteins, ab initio (or de novo) modeling approach can be of great use, which can generate the structure from the first principles of physics. However, the nearly all ab initio methods have some limitations and currently, to some extent, they make use of knowledge of previously known structures from PDB library for predicting the query protein's structure. Thus, the phrase "template-free" is utilized merely for naming the existing ab initio methods that do not belong to the group of homology modeling as well as threading.

The fundamental steps involved in all ab initio methods are the following: (1) conformational search using a suitable energy function, (2) model selection, (3) reconstruction of all-atom structure, (4) refinement of structure, and (5) model validation (Fig. 8.3). The principles involved in these steps are discussed below.

5.3.1 Conformational search using energy function

Theoretically, there can be numerous possible steric conformations for a specific protein, but in general, the native conformation for a typical protein is unique. Conformational search is aimed to search for all the possible (near-native) conformations of a target protein guided by an appropriate energy function. The major challenges today are (1) to design accurate energy function that can guide the query protein folding correctly and (2) to reduce the computing time for conformational search process. In ab initio modeling, two types of energy functions— physics-based energy function and knowledge-based energy function or a combination of these two types of energy function—are utilized (Lee et al., 2017). Different existing force fields, which rely on classical physics theory and contain energies associated with bond stretching, angle bending, torsion angle, nonbonded interactions (such as electrostatics and van der Waals) etc., are the physics-based energy function. The force fields, namely CHARMM (Brooks et al., 1983; MacKerell et al., 1998; Neria et al., 1996), AMBER (Cornell et al., 1995; Duan & Kollman, 1998; Weiner et al., 1984), OPLS (Jorgensen et al., 1996; Jorgensen & Tirado-Rives, 1988), and GROMOS96 (Scott et al., 1999), are utilized in MD simulations. On the other hand, when the energy function is designed based on structural information/features from experimentally solved structures deposited in PDB library, it is termed as knowledge-based energy function. For designing such energy function, structural features, such as distance-dependent atomic contact, pairwise residue contact, secondary structure propensities, hydrogen bonding, dihedral angles, side chain orientation, solvent accessibility, etc., can be utilized. The knowledge-based energy functions often show better performance in predicting protein structure in comparison to physics-based energy functions. However, they cannot explain the nature of protein folding process. Physics-based energy functions often fail to depict complicated atomic

interactions of a protein although they are based on the principle of physics. Today, the most successful ab initio methods such as ROSETTA (Das et al., 2007; Simons et al., 1997) and QUARK (Xu & Zhang, 2012) make use of the combination of these two energy functions.

Conformational search process often uses MD simulation coupled with an appropriate force field to generate the possible native structures (also known as decoy structures) for a template protein. By solving Newton's equations of motion, MD simulations, indeed, simulate protein folding and movement. It is worth to use MD simulation as it can provide the information about the folded structure as well as the process of protein folding. However, even for a typical size protein ($\sim100-300$ aa), it demands a vast amount of computational resources to complete the conformation search. Generally, the simulation time needed for folding a typical protein is in milliseconds (10^{-3} s), which is much longer (10^{12} times) than the usual incremental time step of femtoseconds (10^{-15} s) (Lee et al., 2017). Thus, it is very challenging to fold protein by MD to obtain a native/near-native structure of a protein. Conformational search relied on Monte Carlo (MC) simulation is much faster and efficient compared to that based on MD simulation. The popular conformational search methods, dependent on MC simulation, are SA and replica exchange that aid to overcome energy barriers and attain the global lower-energy state (Adcock & McCammon, 2006; Bernardi et al., 2015).

5.3.2 Model selection

Once a large number of structures (decoy structures) of a target protein are generated following the conformational search, the most challenging task is to select the models close to the native state from a pool of decoy structures. There are different "model quality assessment programs" (MQAPs) that can be used to select the best model structures from candidate structures built by different prediction methods (Fischer, 2006). Model selection methods are of two categories—energy-based and free energy–based (Lee et al., 2017). Energy-based model selection methods attempt to identify structures having lowest energy state using physics-based and knowledge-based energy function. Another popular model selection method selects the best models using scoring function (not fully relied on energy functions) that explains the compatibility of protein (target) sequence with model structures. In addition, researchers have developed a method that considers the consensus conformation from predictions done by several model selection programs. ModFold is one such consensus MQAP that uses MODCHECK, ModSSEA, and ProQ, and its performance is reported to be higher than its component programs (McGuffin, 2007). The other approach is the clustering of all decoy structures that identify the structures with lowest free energy state. The programs such as ROSETTA (Simons et al., 1997), TASSER/I-TASSER ((Roy et al., 2010; Yang et al., 2015), and QUARK (Xu & Zhang, 2012)) use this clustering technique for model selection.

5.3.3 Reconstruction of all-atom structure

Following conformational search and model selection, we only obtain the reduced model structures as most of the protein structure prediction methods consider certain kind of simplified protein representation. Thus, it is necessary to reconstruct all-atom structure for reduced model structures. The procedure of reconstruction of all-atom structure depends on the type of protein representation adopted by protein prediction methods. Several methods make use of the simple protein representation, where every atom of a protein can be depicted by "Cα-atom" and "virtual center of side chain". During conformational search, the later term facilitates to determine Cα-atom position and the resulting structure includes only Cα-atoms. In this case, structure reconstruction can be done in two steps separately: (1) rebuilding/reconstruction of backbone atoms (C, N, and O) considering the Cα-atoms position and (2) rebuilding/reconstruction of side chain for every residue (relied on side chain rotamer library).

5.3.4 Refinement of structure

The previous steps result in the complete structure of a target protein. However, sometimes the structure quality may not be good enough due to different reasons. This may happen due to the inappropriate use of energy functions or faults in conformational search process and reconstruction of all-atom structure. In conformational search process, to ensure search efficiency, the consideration of simple protein representation can affect all-atom structure quality. Thus, it is very crucial to further refine the resulted structure obtained from the previous step. Refinement of structure can be done by using MD simulation/MC simulation with an appropriate energy function. Instead of refining overall topology of the protein, the structure refinement is often done by small change in the backbone conformation. One such structure refinement method is FG-MD, which is based on MD simulation at all-atom level (Zhang et al., 2011). This method does structure refinement by collecting spatial restraints not only from global templates but also from local fragments. Another method ModRefiner uses MC simulation to rebuild and refine the structure stepwise (Xu & Zhang, 2011). Initially, the method rebuilds the backbone structure from the Cα-trace and an energy minimization simulation is performed to refine the quality of the backbone structure. Then all the side chain atoms are added to the backbone structure and again simulation is performed to minimize the energy of all-atom structure.

Finally, the model is validated using different programs such as Ramachandran plot, which is discussed in the homology model evaluation section.

6 CASE STUDY AND RESOURCES FOR MODELING PROTEIN STRUCTURES

Klebsiella pneumoniae is an emerging nosocomial pathogen that causes various diseases such as urinary tract infection, sepsis, surgical site infection, and pneumonia. Reports around the globe have indicated that this opportunistic pathogen has developed resistance to a wide range of currently used antibiotics. A recent study on the exploration of hypothetical proteins of *K. pneumoniae* identified AHS1 domain—containing protein (WP_004176857.1) as one of the potential drug targets (Pranavathiyani et al., 2020). The present case study is aimed to analyze this protein and predict its 3D structure using various available servers/tools.

The primary sequence of AHS1 domain—containing protein from *Klebsiella pneumoniae* subsp. *rhinoscleromatis* SB3432 (5-oxoprolinase subunit PxpB—*Enterobacteriaceae*) was retrieved from the UniProt database (UniProt ID: R4YD10 and NCBI Accession: WP_004176857.1) (The UniProt Consortium, 2017). The tertiary structure of this protein was generated from its amino acid sequence independently using different model prediction tools/servers, namely SWISS-MODEL, MODELLER, I-TASSER, and Robetta. Prior to 3D model prediction, the primary and the secondary structure analyses of this protein can be carried out using various bioinformatics tools to have insights into its physicochemical/functional/structural properties. The complete methodology followed in the case study is depicted in Fig. 8.4 and explained in a stepwise manner below.

6.1 Protein Sequence Analysis

In addition to UniProt database, there are several primary and secondary databases from which the primary sequence of protein of interest can be obtained. The list of sequence databases and resources is provided in Table 8.1. Each database is unique and integrated with the other databases to provide uniform service to

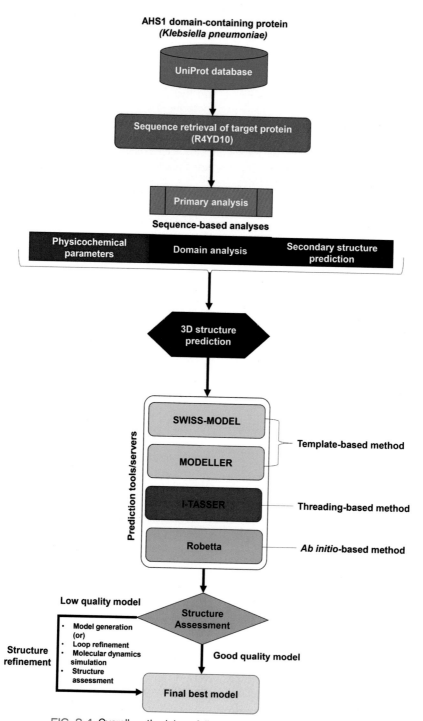

FIG. 8.4 Overall methodology followed in the present case study.

TABLE 8.1
List of Databases and Tools for Protein Sequence Retrieval, Primary Sequence Analysis, and Secondary Structure Prediction.

Database/Tool/Server	URL
SEQUENCE DATABASES	
NCBI/GenBank proteins	https://www.ncbi.nlm.nih.gov/protein/
DDBJ	https://www.ddbj.nig.ac.jp/index-e.html
EMBL-EBI	https://www.ebi.ac.uk/
UniProt	https://www.uniprot.org/
PIR	https://proteininformationresource.org/
KEGG	https://www.genome.jp/kegg/
STRUCTURAL DATABASES	
RCSB/PDB	https://www.rcsb.org/
PDBe	https://www.ebi.ac.uk/pdbe/
MMDB/NCBI Structure	https://www.ncbi.nlm.nih.gov/structure/
SCOPe	https://scop.berkeley.edu/
CATH	https://www.cathdb.info/
CDD	https://www.ncbi.nlm.nih.gov/cdd/
SEQUENCE ANALYSIS	
BLASTp	https://blast.ncbi.nlm.nih.gov/Blast.cgi?PAGE=Proteins
FASTA	https://www.ebi.ac.uk/Tools/sss/fasta/
ProtParam	https://web.expasy.org/protparam/
PredictProtein	https://www.predictprotein.org/
ExPASy	https://www.expasy.org/
DOMAIN PREDICTION/FUNCTIONAL ANALYSIS	
Pfam	https://pfam.xfam.org/
InterPro	https://www.ebi.ac.uk/interpro/
PROSITE	https://prosite.expasy.org/
SECONDARY STRUCTURE PREDICTION	
PSIPRED	http://bioinf.cs.ucl.ac.uk/psipred/
CFSSP	http://www.biogem.org/tool/chou-fasman/
GOR IV	https://npsa-prabi.ibcp.fr/cgi-bin/npsa_automat.pl?page=npsa_gor4.html
SOPMA	https://npsa-prabi.ibcp.fr/cgi-bin/npsa_automat.pl?page=/NPSA/npsa_sopma.html
Jpred	http://www.compbio.dundee.ac.uk/jpred/
Consensus secondary structure prediction	https://npsa-prabi.ibcp.fr/cgi-bin/npsa_automat.pl?page=/NPSA/npsa_seccons.html

the users. The primary sequence of AHS1 protein (UniProt ID: R4YD10) was analyzed using the ProtParam tool to understand the basic physicochemical properties of this protein. ProtParam is an ExPASy-based tool that computes various physical and chemical parameters for a given protein sequence (Gasteiger et al., 2005).

ExPASy is a bioinformatics resource portal developed and maintained by the Swiss Institute of Bioinformatics (SIB), which integrates various databases and tools that aid in the analyses of different multiomics data to serve the scientific community (Gasteiger et al., 2003). The primary analysis of AHS1 protein using ProtParam estimated that the molecular weight of the protein was 24371.97 Da, and the theoretical isoelectric point (pI) was 5.6, where the net charge of the molecule was neutral. The total length of the protein was 218 amino acids with a grand average of hydropathicity (GRAVY) of −0.230, and the computed instability index of the protein was found to be 62.15, suggesting it would be an unstable protein in *in vivo* conditions. The GRAVY value for any protein represents the ratio of the sum of hydropathy values of all amino acids to the total length of the protein sequence. The amino acid composition of AHS1 protein is depicted in Fig. 8.5A.

The AHS1 domain–containing protein retrieved from UniProt is encoded by the gene KPR_3858. However, in NCBI, the same protein is reported as "multispecies 5-oxoprolinase subunit PxpB" from *Enterobacteriaceae*. To annotate its function, a homology search was performed using BLASTp and it was observed that the protein was similar to the putative

carboxylase from *Klebsiella aerogenes* with 100% query coverage and 99.5% identity. BLAST is a widely used bioinformatics sequence similarity search tool for both nucleotide and protein sequences (Altschul et al., 1990). The tool compares the query sequence to the sequences present in the databases and may provide significantly similar hits based on the statistical measures. To validate the BLAST results, a domain search for this protein was performed using Pfam database, a large repository of protein families and domains (Finn et al., 2016). Pfam predicted that the protein contains the carboxyltransferase domain (subdomains C and D), which is a part of urea carboxylase, a catalyst of ATP- and biotin-dependent carboxylation reaction of urea. This domain is also present in other organisms such as *Bacillus subtilis* covering the whole length of KipI (kinase A inhibitor) and in *Saccharomyces cerevisiae* as urea amidolyase Dur1,2, a multifunctional biotin-dependent enzyme comprising urea carboxylase and allophanate (urea carboxylate) hydrolase domains. Several domain prediction databases and tools (e.g., Conserved Domain Database, InterPro, etc.) are developed by the researchers to facilitate the identification/prediction of domains for unknown proteins (Table 8.1).

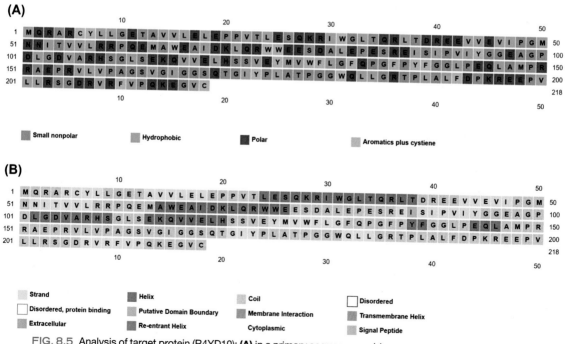

FIG. 8.5 Analysis of target protein (R4YD10): **(A)** in a primary sequence, residues are represented by different colors based on their amino acid group and **(B)** predicted secondary structure (residue-wise).

Next, the secondary structure for the protein of interest (i.e., carboxyltransferase of *K. pneumoniae*) was predicted. Various methods and tools/servers are developed by researchers to predict the secondary structure of proteins from their corresponding primary sequences (Table 8.1). These servers are mainly based on sequence-level propensity statistics or machine learning approaches (neural network and support vector machine) or HMM (Zhou et al., 2016). Herein, the secondary structure elements, namely helices, beta-strands, and coils for the query protein were predicted using PSIPRED, which is depicted in Fig. 8.5B. PSIPRED works on the principle of two feedforward neural networks, based on the results of position-specific iterated BLAST (PSI-BLAST) (Altschul et al., 1997; Altschul & Koonin, 1998). A stringent cross-validation method is used for the evaluation of PSIPRED, which is estimated to predict an average Q3 score of 81.6% (Buchan & Jones, 2019).

6.2 Protein Structure Prediction and Evaluation

Several online servers/tools that are being developed to predict/model the 3D structure of a query protein from its primary sequence are listed in Table 8.2. Each of these tools/servers builds the protein structure based on the basic principles of structure modeling, i.e., template-based or template-free approach. Procedures of 3D model generation for the protein carboxyltransferase (AHS1) using different tools/servers—SWISS-model, MODELLER, I-TASSER, and Robetta—along with tool description are discussed herein.

6.2.1 SWISS-MODEL

SWISS-MODEL is an online interactive server for protein homology modeling (https://swissmodel.expasy.org/interactive). It provides a unique integrated platform for user to build protein models in a comprehensive way. The server also provides a personal workspace for user to carry out their projects and the models can be deposited directly to ModelArchive, a public repository of predicted 3D structures (https://www.modelarchive.org/). It generates the protein model following the four key steps: selection of structural template(s), alignment of target sequence and template structure(s), model building, and model quality evaluation. These steps require specific software and integrate up-to-date protein sequence and structure databases. Each of the aforementioned steps can be repeated interactively until a satisfying modeling result is achieved (Waterhouse et al., 2018).

To begin with, the protein sequence of carboxyltransferase from *K. pneumoniae* (UniProt Accession: R4YD10) was given as input (target sequence) to the server. The sequence of target protein was searched against the PDB database to find out suitable template. The available templates (structures) were listed and one or multiple structures can be selected for building the model of the target protein. The template(s) should be selected based on sequence identity and query coverage percentage between target and template. SWISS-MODEL provides a unique score called Global Model Quality Estimation (GMQE), which is the quality estimation based on the target-template alignment and the template search method. For the given query protein, the search resulted in a total of 30 template structures. Among these, the template structure of YbgJ protein from *Escherichia coli* was selected, which was solved using X-ray diffraction method with 2.8 Å resolution. Using this template structure then the model was built and the quality was assessed for the predicted model. In fact, SWISS-MODEL server provides various structure assessment methods, including Ramachandran plot, MolProbity, quality estimate, and residue quality. Herein, the predicted model for the target protein was found to have overall 94.76% residues in favored region in the Ramachandran plot with a MolProbity score of 1.46. The steps in the model building along with quality assessment results are provided in Figs. 8.6 and 8.7.

6.2.2 MODELLER

MODELLER is a python-based tool that generates 3D structure of proteins using spatial restraints of protein assembly. It is one of the commonly used tools for protein homology modeling (comparative modeling) where user can provide a target-template sequence alignment along with the template structure, and MODELLER would automatically calculate a model having all nonhydrogen atoms (Eswar et al., 2006). This tool can also be used for multitemplate modeling, wherein several templates (structures) can be provided for a target protein, and MODELLER would compare and cluster their structural features into a single 3D model. This tool is developed and maintained at Andrej Sali Lab at the University of California, San Francisco (https://salilab.org/modeller/). The tool can be used freely by academic nonprofit organizations and is compatible with different operating systems (Windows, Linux, and Mac). Based on user needs, the tool also provides several custom features such as loop modeling, model refinement, add additional restraints, etc., for model generation. Several third-party tools/web servers

TABLE 8.2
List of Servers and Tools Available for Structure Prediction/Modeling, Validation, and Visualization.

Tool/Server	URL
STRUCTURE PREDICTION/MODELING	
SWISS-MODEL	https://swissmodel.expasy.org/
MODELLER	https://salilab.org/modeller/
RaptorX	http://raptorx.uchicago.edu/StructurePrediction/
@TOME V3	http://atome.cbs.cnrs.fr/ATOME_V3/index.html
ModWeb	https://modbase.compbio.ucsf.edu/modweb/
(PS)^2	http://ps2.life.nctu.edu.tw/index.php
I-TASSER	https://zhanglab.ccmb.med.umich.edu/I-TASSER/
Robetta	http://robetta.bakerlab.org/
Phyre^2	http://www.sbg.bio.ic.ac.uk/~phyre2/html/page.cgi?id=index
ModLoop	https://modbase.compbio.ucsf.edu/modloop/
GalaxyWEB	http://galaxy.seoklab.org/
STRUCTURE VALIDATION	
WHAT IF	https://swift.cmbi.umcn.nl/servers/html/index.html
SAVES v5.0	https://servicesn.mbi.ucla.edu/SAVES/
RAMPAGE	http://mordred.bioc.cam.ac.uk/%7Erapper/rampage.php
PDBsum	https://www.ebi.ac.uk/thornton-srv/databases/cgi-bin/pdbsum/GetPage.pl?pdbcode=index.html
ProSA-web	https://prosa.services.came.sbg.ac.at/prosa.php
ModFOLD	https://www.reading.ac.uk/bioinf/ModFOLD/ModFOLD6_form.html
MolProbity	http://molprobity.biochem.duke.edu/
VISUALIZATION TOOLS	
Pymol	https://pymol.org/
Chimera	https://www.cgl.ucsf.edu/chimera/
RasMol	http://www.openrasmol.org/
Swiss-PdbViewer	https://spdbv.vital-it.ch/
YASARA	http://www.yasara.org/

are available which incorporate MODELLER for protein modeling (e.g., EasyModeller, Modelface, BIOVIA, etc.) (Kemmish et al., 2017; Kuntal et al., 2010; Sakhteman & Zare, 2016).

To start with, a suitable template structure was identified for the query protein (i.e., carboxyltransferase of *K. pneumoniae*) using a homology search (BLASTp) against the PDB database. This resulted in eight homologous structures from different organisms, from which the best template (5DUD_B) based on the identity percentage (79.36%) with query coverage (100%) was selected for building the model (Fig. 8.8).

Essentially, to build a model using MODELLER, the following files are required:

• The target protein sequence in PIR format—the target protein in FASTA format was converted to PIR (sequence code 1YES in .pir file) using EMBOSS Seqret program (R4YD10.pir)

• The template protein structure (5DUD_B) retrieved from PDB database (5dud.pdb)

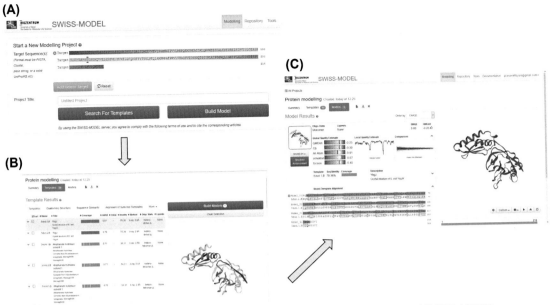

FIG. 8.6 Steps followed in protein modeling in SWISS-MODEL: **(A)** the interface of SWISS-MODEL server where the input (target sequence) is submitted for template search, **(B)** list of identified templates for the target protein with query coverage, identity, and method for solvation, sorted based on the Global Model Quality Estimation (GMQE) score. The first template is selected for three-dimensional modeling of the target protein, and **(C)** the generated model with quality assessment score and alignment between model and template structures.

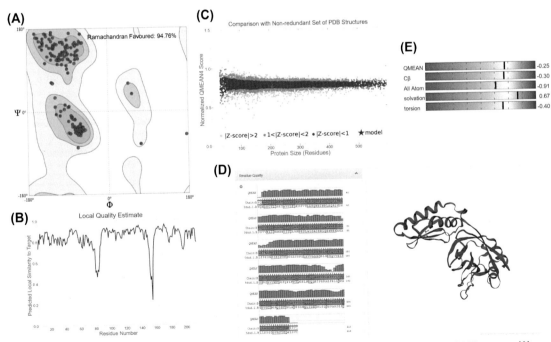

FIG. 8.7 Structure assessment results for the modeled target protein built by SWISS-MODEL server: **(A)** Ramachandran plot of all residues indicating 94.7% of the residues are in most favored region, **(B)** residue-wise quality estimation for the target protein structure based on local similarity prediction, **(C)** z-score comparison of the predicted model to the nonredundant set of structures present in Protein Data Bank (PDB), **(D)** residue-wise quality scores QMEAN along with secondary structure element comparison for the target and template structures, and **(E)** estimated quality score of the generated model.

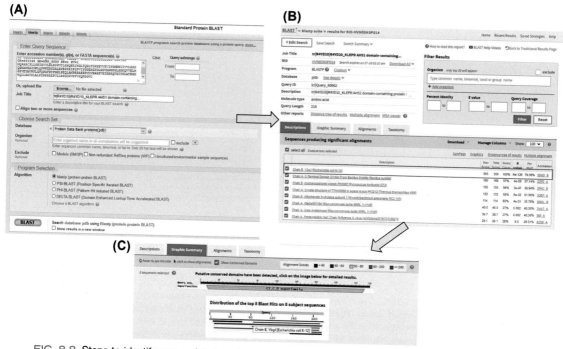

FIG. 8.8 Steps to identify appropriate template structure for the target protein: **(A)** homology search of the target protein (as input sequence) using BLASTp against the Protein Data Bank (PDB), **(B)** search results showing list of templates homologous to the target sequence, and **(C)** graphical summary of the hits obtained from BLAST search.

- "target-template.ali" sequence-structure alignment file generated using "align.py" program from MODELLER. The input file names (sequence and structure files) used in the "align.py" should be modified and the output file should be specified. Once this code align.py is run, the alignment file "target-template.ali" will be automatically generated.

Next, to generate the models for the target protein, the file "model-default.py" from MODELLER should be modified by providing the input sequence code (1YES), structure (5dud.pdb), and also the structure-sequence alignment file (target-template.ali). The number of models (e.g., 10 models) should be specified and the file "model-default.py" was renamed as "buildmodel.py". Once the "builtmodel.py" file is run on MODELLER, the 10 models were generated (as specified) along with their molpdf, DOPE, and GA341 score available in log file. All 10 models were found to have GA341 score of 1 and based on the least DOPE score (Min-yi Shen, 2006) the model 1YES.B99990002.pdb was selected. The structure assessment of the model was carried out using SAVES server. The Ramachandran plot for the predicted model revealed that the 90.9% of

the residues were present in the most favored region. The steps in the model building along with the other assessment results are depicted in Figs. 8.9 and 8.10.

6.2.3 I-TASSER

I-TASSER is an online server, which predicts the protein structure and function using threading-based assembly approach. The server requires primary sequence of the target protein as input, and then it retrieves the templates with similar folds from the PDB database using local meta-threading server (LOMETS), a meta-threading server for template-based structure prediction. A fragment-based assembly method is deployed with MC simulation and threading for generating the model. In the absence of template structure, the server builds an ab initio model for the given protein. The structure assembly is reiterated to refine the overall topology and to remove the steric clashes. From the generated assembly, the lowest energy structures are clustered and a final optimized full model for the given protein is generated (J. Yang et al., 2015). The best model can be identified by the C-score, which is a confidence score for assessing the quality of the generated models. The score is measured by calculating the significance of template

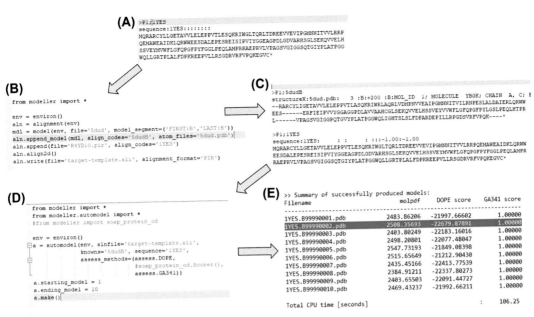

(A)
```
>P1;1YES
sequence:1YES::::::::
MQRARCYLLGETAVVLELEPFPVTLESQKRIWGLTQRLTDREEVVEVIPGMNNITVVLRRP
QEMAWEAIDKLQRWWEESDALEPESREISIPVIYGGEAGPDLGDVARHSGLSEKQVVELH
SSVEYMVWFLGFQPGFPYFGGLPEQLAMPRRAEPRVLVPAGSVGIGGSQTGIYPLATPGG
WQLLGRTPLALFDPKREEPVLLRSGDRVRFVPQKEGVC*
```

(B)
```
from modeller import *

env = environ()
aln = alignment(env)
mdl = model(env, file='5dud', model_segment=('FIRST:B','LAST:B'))
aln.append_model(mdl, align_codes='5dudB', atom_files='5dud.pdb')
aln.append(file='R4YD10.pir', align_codes='1YES')
aln.align2d()
aln.write(file='target-template.ali', alignment_format='PIR')
```

(C)
```
>P1;5dudB
structureX:5dud.pdb:   3 :B:+200 :B:MOL_ID  1; MOLECULE  YBGK; CHAIN  A, C; B
--RARCYLIGETAVVLELEPFPVTLASQKRIWRLAQRLVDMFNVVEAIPGMNNITVILRNPESLALDAIERLQRWW
EES------ERFIEIPVVYGGAGGPDLAVVAAHCGLSEKQVVELHSSVEYVVWFLGFQPGFPYLGSLPEQLHTPR
L-------VPAGSVGIGGPQTGVYPLATPGGWQLIGHTSLSLFDPARDEPILLRPGDSVRFVPQK----*

>P1;1YES
sequence:1YES:    :  : ::::-1.00:-1.00
MQRARCYLLGETAVVLELEPFPVTLESQKRIWGLTQRLTDREEVVEVIPGMNNITVVLRRPQEMAWEAIDKLQRWW
EESDALEPESREISIPVIYGGEAGPDLGDVARHSGLSEKQVVELHSSVEYMVWFLGFQPGFPYFGGLPEQLAMPR
RAEPRVLVPAGSVGIGGSQTGIYPLATPGGWQLLGRTPLALFDPKREEPVLLRSGDRVRFVPQKEGVC*
```

(D)
```
from modeller import *
from modeller.automodel import *
#from modeller import soap_protein_od

env = environ()
a = automodel(env, alnfile='target-template.ali',
              knowns='5dudB', sequence='1YES',
              assess_methods=(assess.DOPE,
                              #soap_protein_od.Scorer(),
                              assess.GA341))
a.starting_model = 1
a.ending_model = 10
a.make()
```

(E)
```
>> Summary of successfully produced models:
Filename                  molpdf        DOPE score    GA341 score
----------------------------------------------------------------
1YES.B99990001.pdb        2483.86206    -21997.66602     1.00000
1YES.B99990002.pdb        2508.35693    -22679.87891     1.00000
1YES.B99990003.pdb        2403.80249    -22183.16016     1.00000
1YES.B99990004.pdb        2498.20801    -22077.48047     1.00000
1YES.B99990005.pdb        2547.73193    -21849.08398     1.00000
1YES.B99990006.pdb        2515.65649    -21212.90430     1.00000
1YES.B99990007.pdb        2435.45166    -22413.77539     1.00000
1YES.B99990008.pdb        2384.91211    -22337.80273     1.00000
1YES.B99990009.pdb        2403.65503    -22091.44727     1.00000
1YES.B99990010.pdb        2469.43237    -21992.66211     1.00000

Total CPU time [seconds]                          :   106.25
```

FIG. 8.9 Files to create/modify for generating models with MODELLER: **(A)** the target protein sequence in PIR format with the code 1YES, **(B)** align.py script file modified with the template pdb file (5dud.pdb) and target sequence file (R4YD10.pir) to generate the "target-template.ali" file, **(C)** the generated "target-template.ali" file showing the template structure sequence alignment, **(D)** the modified script file "buildmodel.py" where the alignment file, the template structure, and sequence code are provided with required total number of models, and **(E)** the log file containing list of generated models with their modpdf, DOPE score, and GA341 score.

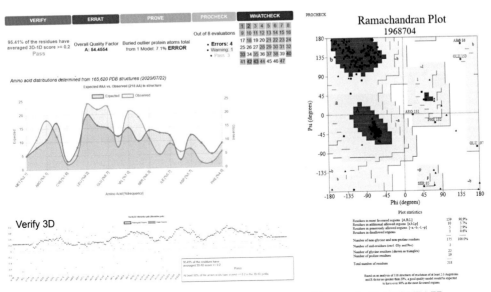

FIG. 8.10 Structure assessment results of the best model generated by MODELLER.

alignments in threading and the convergence parameters of the structure assembly simulations. This score ranges from [−5,2] where higher score represents high confidence model and low score indicates model with low confidence. It also provides TM-score and root mean square deviation, which are the standard measures for structural similarity between two structures.

To begin with, the protein sequence of the target protein (carboxyltransferase) was given as input in the I-TASSER server. Several additional options are available in I-TASSER to add additional restraints, exclude templates, or to even specify secondary structure for residues. Once submitted, the results will be sent to the user's email. For the model generation, it may take from few hours to a day or more. The results page (Fig. 8.11) contains the submitted sequence query in fasta format along with the predicted residue-wise secondary structure elements. The server also provides predicted solvent accessibility and normalized B-factor for the query protein. Solvent accessibility score ranges from 0 to 9, where 0 represents buried residue and 9 denotes highly exposed residue. B-factor is the inherent thermal mobility value of residues in proteins, which is calculated from threading protein templates obtained from PDB along with the sequence profiles generated from sequence databases. The top 10 threading templates used by I-TASSER are given in Fig. 8.11, using

which the I-TASSER server generated the model for the target protein. A total of five models were generated by the server and the best model was selected based on the C-score. In addition, I-TASSER also lists the structural analog proteins from PDB database including enzyme commission numbers for the active site and consensus gene ontology terms. The structural assessment of the model generated from I-TASSER was assessed using SAVES server, in which only 81.1% of the residues were found to be present in the most favored region in Ramachandran plot suggesting that the model needs to be refined for further use.

6.2.4 Robetta

Robetta is an automated online server for protein structure prediction. The server utilizes the Rosetta method for predicting protein structures and provides an interactive interface for the user to include multiple features such as sequence alignments for comparative (homology) modeling, constraints, local fragments, etc. The server can be used for building multichain complexes (models) and also provides options to include large-scale sampling of templates. It uses the PDB100 template database, a co-evolution-based model database, which is updated weekly, and additionally user can also provide custom templates for modeling (Kim et al., 2004).

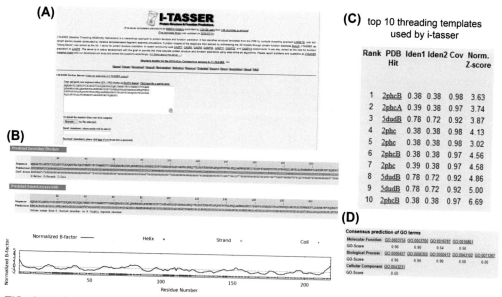

FIG. 8.11 Generation of model using I-TASSER: **(A)** the interface of I-TASSER page where the target sequence is provided, **(B)** the results of I-TASSER server showing the predicted secondary structure and solvent accessibility with confidence scores, **(C)** the top 10 chosen models by I-TASSER for threading to build the model for the target sequence, and **(D)** the consensus gene ontology terms and their scores for the target protein.

The amino acid sequence of the target protein was given as input in Robetta server for its structure prediction using ab initio method. Once the sequence is submitted, it will be in queue for structure prediction and the results will be sent to the user in their registered email. The Robetta server provides results of consensus secondary structure prediction for the target protein using various prediction methods, namely DeepCNF, PSIPRED, and SPIDER3, and it also predicts disordered or transmembrane helices, if any, in the protein. Based on the selected methods (comparative or ab initio), the server generated five models along with estimated errors. Herein, the generated model had confidence score of 0.60. The score for ab initio model is calculated based on the average of TM-score of the top 10 Rosetta scoring model. The confidence score can range from 0.0 to 1.0, where the score closer to 1.0 indicates a good model. The steps followed to generate model in Robetta server are presented in Fig. 8.12.

6.2.5 Comparison of structure assessment for the predicted models

The 3D models predicted by tools/servers—SWISS-MODEL, MODELLER, I-TASSER, and Robetta—were submitted to SAVES server to assess their quality. The structure assessment results are provided in Fig. 8.13. The SAVES server integrates various structure assessment methods, namely Verify3D, ERRAT, Prove,

PROCHECK, and WHATCHECK. Verify3D utilizes the 3D profiles of the model to the 1D sequence (of the target), assigns the structural classes based on the residue location and environment (alpha, beta, loop, polar, nonpolar, etc.), and calculates an average residue score ranging from -1 to $+1$. All the four predicted models were found to pass this assessment. The models predicted by Robetta, MODELLER, I-TASSER, and SWISS-MODEL had a total of 100%, 95.41%, 90.37%, and 82.55% of the residues, respectively, with an average 3D-1D score ≥ 0.2. ERRAT determines the overall quality factor of the predicted model by analyzing the nonbonded interactions between different atom types and plots the error function value (y-axis) against the amino acid sequence (9-residue sliding window in x-axis). The ERRAT scores are calculated based on the comparison with statistics from highly refined structures, and the model with a quality factor >50 is considered as a good quality model. All the predicted models using the four tools (SWISS-MODEL, MODELLER, I-TASSER, and Robetta) had passed this assessment (Robetta: 82.84%, SWISS-MODEL: 82.38%, I-TASSER: 81.9%, and MODELLER: 54.4%). PROVE evaluates the volume of atoms using an algorithm where each atom is assumed to be a sphere with a Z-score deviation between the predicted model and the highly resolved/refined PDB structures (resolution ≤ 2.0 Å and R-factor ≤ 0.2). If the percentage of buried atoms are

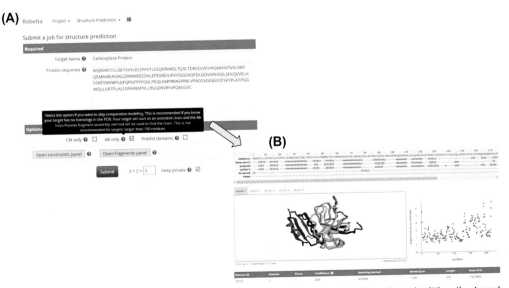

FIG. 8.12 Model generation of the target protein: **(A)** Robetta server interface for submitting the target sequence, with the option to choose the type of modeling, **(B)** the results of the ab initio modeling from Robetta server with predicted secondary structure elements and the generated models with estimated error.

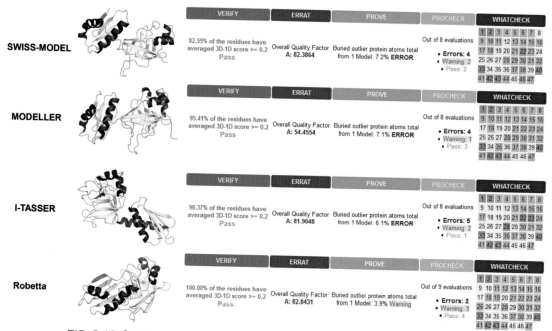

FIG. 8.13 Structure assessment scores of the models from four prediction servers/tools.

higher than 5%, it indicates "Error", whereas 1%–5% as "Warning" and <1% as "Pass". All the predicted models had more than 5% of buried residues, except the model predicted by Robetta server (3.9%). PROCHECK quantifies the stereochemical quality of a protein structure by investigating residue-wise geometry as well as overall structural geometry. It performs a number of evaluations including Ramachandran plot analysis. Similarly, WHATCHECK also performs extensive stereochemical parameter estimation for the residues in the model with the tools derived from WHATIF program. The total evaluations performed and the results (number of Error, Warning, and Pass) for all four models are depicted in Fig. 8.13. WHATCHECK provides a color code–based result for the evaluation, where green represents "pass", yellow for "warning", and red for "error" (Fig. 8.13). The Ramachandran plot from PROCHECK analysis provided the quality assessment for the four predicted models, in which the model generated by Robetta server had 90.9% of the residues in most favored region with no residues in disallowed region. The total residues in most favored region of the Ramachandran plot for the models generated by SWISS-MODEL, MODELLER, and I-TASSER were 95.3%, 90.9%, and 81.1%, respectively (Fig. 8.14). However, each of these models had residues in disallowed region

(one in SWISS-MODEL, one in MODELLER, and three in I-TASSER), suggesting that the models should be refined. The models were subjected to loop/model refinement step using web servers, but those residues still remained in the disallowed region. Thus, it is necessary to perform MD simulations to ensure the predicted model attains the possible lowest energy conformation.

7 CONCLUSIONS AND FUTURE DIRECTIONS

Solving protein 3D structures and their analyses is one of the key research areas and regarded as a great challenge in structural biology. As an alternative to experimental approaches, researchers have developed different computational methods for the prediction of protein structures. However, for many complex proteins or proteins with larger size, existing methods may not always result in structures with 100% accuracy. Structure prediction is difficult particularly when the template structure (known structure from PDB) of a target protein is not available. Homology modeling enables to predict protein structures from their amino acid sequences only when they are similar/identical to sequences of known structures available in PDB library. Threading method relies on fold library of known

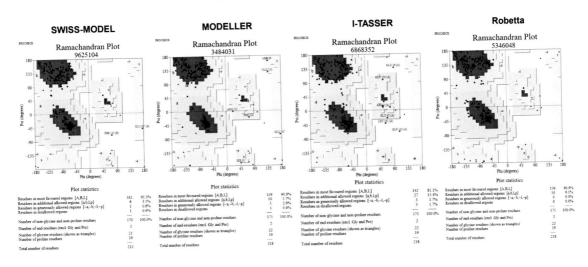

Based on an analysis of 118 structures of resolution of at least 2.0 Angstroms and R-factor no greater than 20%, a good quality model would be expected to have over 90% in the most favoured regions.

FIG. 8.14 Ramachandran plot of the four predicted models for the target protein.

structures from PDB. In case of ab initio modeling, structures are built from the principles of physics. However, still it is challenging to predict structures for larger proteins (>300 residues) by using ab initio method. This is mainly due to the intricacy involved in conformational search process and designing appropriate energy functions to explain atomic interactions that can direct protein folding simulations. Therefore, there is a continuous endeavor by researchers to develop the state-of-the-art techniques, which can facilitate researchers to predict the structures of different categories of proteins precisely.

It is worth mentioning at this point that the protein structure prediction center (https://www.predictioncenter.org/index.cgi) facilitates the advancement of the methods pertaining to protein structure prediction from primary sequence. In biannual CASP, experiments, organized by the center, indeed, carry out benchmark progress in accuracy of the current prediction methods and underline the area to be focused in future for better structure prediction. So far, 13 CASP experiments have been conducted since the starting year 1994 (CASP1 to CASP13). According to CASP, some of the high-quality structure prediction methods are ROSETTA, I-TASSER, SwissModel, MODELLER, etc. The former two are the hybrid methods that utilize physics-based and knowledge-based energy functions, while the latter two (homology modeling methods) depends on the known structures from PDB. In the last CASP meeting

(CASP13) held in 2018, AlphaFold system (Senior et al., 2020) has been introduced that can generate the accurate 3D models of proteins from primary sequence without utilizing templates. This is considered as an "unprecedented progress in the ability of computational methods to predict protein structure". AlphaFold uses deep neural networks which are trained to accurately predict physical properties of the protein. These physical properties include "distances between pairs of residues" and "angles between chemical bonds that contact those residues" in a protein. Then using this information, a protein-specific potential (representing protein structure) is constructed and can be optimized with a mathematical technique—gradient descent to obtain accurate structure predictions. It is expected that the system can be improved further to achieve more accurate prediction for sequences of unknown structure which would enable us to provide true insights into the molecular function and malfunction of proteins.

ACKNOWLEDGMENTS

The authors acknowledge the support of Centre for Bioinformatics, School of Life Sciences, Pondicherry University, Pondicherry, India, and the Department of Microbiology, West Bengal State University, Barasat, India, for providing the computational facilities. GP is indebted to the Department of Biotechnology, Govt. of India for providing DBT-BINC Fellowship.

REFERENCES

Adcock, S. A., & McCammon, J. A. (2006). Molecular dynamics: Survey of methods for simulating the activity of proteins. *Chemical Reviews, 106*(5), 1589–1615.

Al-Lazikani, B., Jung, J., Xiang, Z., & Honig, B. (2001). Protein structure prediction. *Current Opinion in Chemical Biology, 5*(1), 51–56.

Altschul, S. F., Gish, W., Miller, W., Myers, E. W., & Lipman, D. J. (1990). Basic local alignment search tool. *Journal of Molecular Biology, 215*(3).

Altschul, S. F., Madden, T. L., Schäffer, A. A., Zhang, J., Zhang, Z., Miller, W., & Lipman, D. J. (1997). Gapped BLAST and PSI-BLAST: A new generation of protein database search programs. *Nucleic Acids Research, 25*(17), 3389–3402.

Altschul, S. F., & Koonin, E. V. (1998). Iterated profile searches with PSI-BLAST—a tool for discovery in protein databases. *Trends in Biochemical Sciences, 23*(11), 444–447.

Bairoch, A. (2004). The universal protein resource (UniProt). *Nucleic Acids Research, 33*(Database issue), D154–D159.

Benner, S. A., Cohen, M. A., & Gonnet, G. H. (1993). Empirical and structural models for insertions and deletions in the divergent evolution of proteins. *Journal of Molecular Biology, 229*(4), 1065–1082.

Berman, H. M., Westbrook, J., Feng, Z., Gilliland, G., Bhat, T. N., Weissig, H., Shindyalov, I. N., & Bourne, P. E. (2000). The Protein Data Bank. *Nucleic Acids Research, 28*(1), 235–242.

Bernardi, R. C., Melo, M. C. R., & Schulten, K. (2015). Enhanced sampling techniques in molecular dynamics simulations of biological systems. *Biochimica et Biophysica Acta, 1850*(5), 872–877.

Bourne, P. E., & Weissig, H. (2003). *Structural bioinformatics.* Wiley-Liss.

Bradley, P. (2005). Toward high-resolution de novo structure prediction for small proteins. *Science, 309*(5742), 1868–1871.

Bradley, P., Malmström, L., Qian, B., Schonbrun, J., Chivian, D., Kim, D. E., Meiler, J., Misura, K. M. S., & Baker, D. (2005). Free modeling with rosetta in CASP6. *Proteins: Structure, Function, and Bioinformatics, 61*(S7), 128–134.

Branden, C. I., & Tooze, J. (2012). *Introduction to protein structure 98.*

Brooks, B. R., Bruccoleri, R. E., Olafson, B. D., States, D. J., Swaminathan, S., & Karplus, M. (1983). CHARMM: A program for macromolecular energy, minimization, and dynamics calculations. *Journal of Computational Chemistry, 4*(2), 187–217.

Buchan, D. W. A., & Jones, D. T. (2019). The PSIPRED protein analysis workbench: 20 years on. *Nucleic Acids Research, 47*(W1), W402–W407.

Burley, S. K., Berman, H. M., Christie, C., Duarte, J. M., Feng, Z., Westbrook, J., Young, J., & Zardecki, C. (2018). RCSB Protein Data Bank: Sustaining a living digital data resource that enables breakthroughs in scientific research and biomedical education. *Protein Science: A Publication of the Protein Society, 27*(1), 316–330.

Chakravarty, S., Wang, L., & Sanchez, R. (2005). Accuracy of structure-derived properties in simple comparative models of protein structures. *Nucleic Acids Research, 33*(1), 244.

Chung, S. Y., & Subbiah, S. (1996). A structural explanation for the twilight zone of protein sequence homology. *Structure, 4*(10), 1123–1127.

Contreras-Moreira, B., Fitzjohn, P. W., & Bates, P. A. (2002). Comparative modelling: An essential methodology for protein structure prediction in the post-genomic era. *Applied Bioinformatics, 1*(4), 177–190.

Cornell, W. D., Cieplak, P., Bayly, C. I., Gould, I. R., Merz, K. M., Ferguson, D. M., Spellmeyer, D. C., Fox, T., Caldwell, J. W., & Kollman, P. A. (1995). A second generation force field for the simulation of proteins, nucleic acids, and organic molecules. *Journal of the American Chemical Society, 117*(19), 5179–5197.

Das, R., Qian, B., Raman, S., Vernon, R., Thompson, J., Bradley, P., Khare, S., Tyka, Bhat, D., Chivian, D., Kim, D. E., Sheffler, W. H., Malmström, L., Wollacott, A. M., Wang, C., Andre, I., & Baker, D. (2007). Structure prediction for CASP7 targets using extensive all-atom refinement with Rosetta@home. *Proteins, 69*(Suppl. 8), 36.

Duan, Y., & Kollman, P. A. (1998). Pathways to a protein folding intermediate observed in a 1-microsecond simulation in aqueous solution. *Science, 282*(5389), 40.

Eisenberg, D., Lüthy, R., & Bowie, J. U. (1997). VERIFY3D: Assessment of protein models with three-dimensional profiles. *Methods in Enzymology, 277*, 396–404.

Eswar, N., Webb, B., Marti-Renom, M. A., Madhusudhan, M. S., Eramian, D., Shen, M.-Y., Pieper, U., & Sali, A. (2006). Comparative protein structure modeling using modeller. *Current Protocols in Bioinformatics.* Editoral Board, Andreas D. Baxevanis, et al., 0 5, Unit.

Fan, H. (2004). Refinement of homology-based protein structures by molecular dynamics simulation techniques. *Protein Science, 13*(1), 211–220.

Fidelis, K., Stern, P. S., Bacon, D., & Moult, J. (1994). Comparison of systematic search and database methods for constructing segments of protein structure. *Protein Engineering, 7*(8), 953–960.

Finn, R. D., Coggill, P., Eberhardt, R. Y., Eddy, S. R., Mistry, J., Mitchell, A. L., Potter, S. C., Punta, M., Qureshi, M., Sangrador-Vegas, A., Salazar, G. A., Tate, J., & Bateman, A. (2016). The Pfam protein families database: Towards a more sustainable future. *Nucleic Acids Research, 44*(Database issue), D279.

Fischer, D. (2006). Servers for protein structure prediction. *Current Opinion in Structural Biology, 16*(2), 178–182.

Fiser, A., Do, R. K., & Sali, A. (2000). Modeling of loops in protein structures. *Protein Science: A Publication of the Protein Society, 9*(9), 1753–1773.

Flores, T. P., Orengo, C. A., Moss, D. S., & Thornton, J. M. (1993). Comparison of conformational characteristics in structurally similar protein pairs. *Protein Science: A Publication of the Protein Society, 2*(11), 1811–1826.

Gasteiger, E., Gattiker, A., Hoogland, C., Ivanyi, I., Appel, R. D., & Bairoch, A. (2003). ExPASy: The proteomics server for in-depth protein knowledge and analysis. *Nucleic Acids Research, 31*(13), 3784.

Gasteiger, E., Hoogland, C., Gattiker, A., Duvaud, S., Wilkins, M. R., Appel, R. D., & Bairoch, A. (2005). Protein identification and analysis tools on the ExPASy server. *The Proteomics Protocols Handbook,* 571–607.

Hadley, C., & Jones, D. T. (1999). A systematic comparison of protein structure classifications: SCOP, CATH and FSSP. *Structure, 7*(9), 1099–1112.

Henikoff, S., & Henikoff, J. G. (1992). Amino acid substitution matrices from protein blocks. *Proceedings of the National Academy of Sciences of the United States of America, 89*(22), 10915–10919.

Hersh, R., & Dayhoff, M. O. (1970). Atlas of protein sequence and structure, 1969 volume 4. *Systematic Zoology, 19*(2), 112455.

Hooft, R. W. W., Vriend, G., Sander, C., & Abola, E. E. (1996). Errors in protein structures. *Nature, 381*(6580), 272.

Jauch, R., Yeo, H. C., Kolatkar, P. R., & Clarke, N. D. (2007). Assessment of CASP7 structure predictions for template free targets. *Proteins: Structure, Function, and Bioinformatics, 69*(S8), 57–67.

Jones, D. T. (1999). GenTHREADER: An efficient and reliable protein fold recognition method for genomic sequences. *Journal of Molecular Biology, 287*(4), 797–815.

Jones, D. T., Taylor, W. R., & Thornton, J. M. (1992). A new approach to protein fold recognition. *Nature, 358*(6381), 86–89.

Jorgensen, W. L., Maxwell, D. S., & Tirado-Rives, J. (1996). Development and testing of the OPLS all-atom force field on conformational energetics and properties of organic liquids. *Journal of the American Chemical Society, 118*(45), 11225.

Jorgensen, W. L., & Tirado-Rives, J. (1988). The OPLS [optimized potentials for liquid simulations] potential functions for proteins, energy minimizations for crystals of cyclic peptides and crambin. *Journal of the American Chemical Society, 110*(6), 1657.

Karplus, K., Barrett, C., & Hughey, R. (1998). Hidden Markov models for detecting remote protein homologies. *Bioinformatics, 14*(10), 846–856.

Kemmish, H., Fasnacht, M., & Yan, L. (2017). Fully automated antibody structure prediction using BIOVIA tools: Validation study. *PloS One,* 1277923.

Kim, D. E., Chivian, D., & Baker, D. (2004). Protein structure prediction and analysis using the Robetta server. *Nucleic Acids Research, 32*(Web Server issue), W526.

Kinch, L. N., Li, W., Monastyrskyy, B., Kryshtafovych, A., & Grishin, N. V. (2016). Evaluation of free modeling targets in CASP11 and ROLL. *Proteins, 84*(Suppl. 1), 51–66.

Klepeis, J. L., Wei, Y., Hecht, M. H., & Floudas, C. A. (2004). Ab initio prediction of the three-dimensional structure of a de novo designed protein: A double-blind case study. *Proteins: Structure, Function, and Bioinformatics, 58*(3), 560–570.

Kuntal, B. K., Aparoy, P., & Reddanna, P. (2010). EasyModeller: A graphical interface to MODELLER. *BMC Research Notes, 3*(1), 1–5.

Laskowski, R. A., MacArthur, M. W., Moss, D. S., & Thornton, J. M. (1993). PROCHECK: A program to check the stereochemical quality of protein structures. *Journal of Applied Crystallography, 26*(2), 283–291.

Laskowski, R. A., Watson, J. D., & Thornton, J. M. (2005a). ProFunc: A server for predicting protein function from 3D structure. *Nucleic Acids Research, 33*(Web Server), W89–W93.

Laskowski, R. A., Watson, J. D., & Thornton, J. M. (2005b). Protein function prediction using local 3D templates. *Journal of Molecular Biology, 351*(3), 614–626.

Lee, J., Freddolino, P. L., & Zhang, Y. (2017). Ab initio protein structure prediction. In *From protein structure to function with bioinformatics* (pp. 3–25).

Lesk, A. (2014). *Introduction to bioinformatics.* Oxford University Press.

Liwo, A., Khalili, M., & Scheraga, H. A. (2005). Ab initio simulations of protein-folding pathways by molecular dynamics with the united-residue model of polypeptide chains. *Proceedings of the National Academy of Sciences, 102*(7), 2362–2367.

Lodish, H., Berk, A., Lawrence Zipursky, S., Matsudaira, P., Baltimore, D., & Darnell, J. (2000). Hierarchical structure of proteins. In *Molecular cell biology* (4th ed.). W. H. Freeman.

MacKerell, A. D., Bashford, D., Bellott, M., Dunbrack, R. L., Evanseck, J. D., Field, M. J., Fischer, S., Gao, J., Guo, H., Ha, S., Joseph-McCarthy, D., Kuchnir, L., Kuczera, K., Lau, F. T. K., Mattos, C., Michnick, S., Ngo, T., Nguyen, D. T., Prodhom, B., … Karplus, M. (1998). All-atom empirical potential for molecular modeling and dynamics studies of proteins. *The Journal of Physical Chemistry B, 102*(18), 3586–3616.

Ma, J., Peng, J., Wang, S., & Xu, J. (2012). A conditional neural fields model for protein threading. *Bioinformatics, 28*(12), i59–i66.

Martí-Renom, M. A., Stuart, A. C., Fiser, A., Sánchez, R., Melo, F., & Sali, A. (2000). Comparative protein structure modeling of genes and genomes. *Annual Review of Biophysics and Biomolecular Structure, 29,* 291–325.

McGuffin, L. J. (2007). Benchmarking consensus model quality assessment for protein fold recognition. *BMC Bioinformatics, 8,* 345.

Melo, F., Devos, D., Depiereux, E., & Feytmans, E. (1997). ANOLEA: A www server to assess protein structures. In *Proceedings/international conference on intelligent systems for molecular Biology; ISMB* (p. 9322034).

Min-yi Shen, A. S. (2006). Statistical potential for assessment and prediction of protein structures. *Protein Science: A Publication of the Protein Society, 15*(11), 2507.

Nelson, D. L., Lehninger, A. L., & Cox, M. M. (2008). *Lehninger principles of biochemistry.* Macmillan.

Neria, E., Fischer, S., & Karplus, M. (1996). Simulation of activation free energies in molecular systems. *The Journal of Chemical Physics, 105*(5), 1902.

Pascarella, S., & Argos, P. (1992). Analysis of insertions/deletions in protein structures. *Journal of Molecular Biology, 224*(2), 461–471.

Peitsch, M. C. (2002). About the use of protein models. *Bioinformatics, 18*(7), 934–938.

Peng, J., & Xu, J. (2010). Low-homology protein threading. *Bioinformatics, 26*(12), i294–i300.

Peng, J., & Xu, J. (2011). RaptorX: Exploiting structure information for protein alignment by statistical inference. *Proteins, 79*(Suppl. 10), 161–171.

Pranavathiyani, G., Prava, J., Rajeev, A. C., & Pan, A. (2020). Novel target exploration from hypothetical proteins of *Klebsiella pneumoniae* MGH 78578 reveals a protein involved in host-pathogen interaction. *Frontiers in Cellular and Infection Microbiology, 10*. https://doi.org/10.3389/fcimb.2020.00109

Ramachandran, G. N., Ramakrishnan, C., & Sasisekharan, V. (1963). Stereochemistry of polypeptide chain configurations. *Journal of Molecular Biology, 7*(1), 80023–80026.

Rehman, I., Farooq, M., & Botelho, S. (2020). Biochemistry, secondary protein structure. In *StatPearls [Internet]*. StatPearls Publishing.

Rost, B. (1999). Twilight zone of protein sequence alignments. *Protein Engineering, 12*(2), 85–94.

Roy, A., Kucukural, A., & Zhang, Y. (2010). I-TASSER: A unified platform for automated protein structure and function prediction. *Nature Protocols, 5*(4), 725.

Sakhteman, A., & Zare, B. (2016). Modelface: An application programming interface (API) for homology modeling studies using modeller software. *Iranian Journal of Pharmaceutical Research : IJPR, 15*(4), 801–807.

Sali, A. (1995). Modeling mutations and homologous proteins. *Current Opinion in Biotechnology, 6*(4), 437–451.

Šali, A., & Blundell, T. L. (1993). Comparative protein modelling by satisfaction of spatial restraints. *Journal of Molecular Biology, 234*(3), 779–815.

Scott, W. R. P., Hünenberger, P. H., Tironi, I. G., Mark, A. E., Billeter, S. R., Fennen, J., Torda, A. E., Huber, T., Krüger, P., & van Gunsteren, W. F. (1999). The GROMOS biomolecular simulation program package. *The Journal of Physical Chemistry A, 103*(19), 3596–3607.

Senior, A. W., Evans, R., Jumper, J., Kirkpatrick, J., Sifre, L., Green, T., Qin, C., Žídek, A., Nelson, A. W. R., Bridgland, A., Penedones, H., Petersen, S., Simonyan, K., Crossan, S., Kohli, P., Jones, D. T., Silver, D., Kavukcuoglu, K., & Hassabis, D. (2020). Improved protein structure prediction using potentials from deep learning. *Nature, 577*(7792), 706–710.

Simons, K. T., Kooperberg, C., Huang, E., & Baker, D. (1997). Assembly of protein tertiary structures from fragments with similar local sequences using simulated annealing and bayesian scoring functions. *Journal of Molecular Biology, 268*(1), 209–225.

Skolnick, J., Kihara, D., & Zhang, Y. (2004). Development and large scale benchmark testing of the PROSPECTOR_3 threading algorithm. *Proteins: Structure, Function, and Bioinformatics, 56*(3), 502–518.

Soding, J. (2005). Protein homology detection by HMM-HMM comparison. *Bioinformatics, 21*(7), 951–960.

Stojmirović, A., Gertz, E. M., Altschul, S. F., & Yu, Y.-K. (2008). The effectiveness of position- and composition-specific gap costs for protein similarity searches. *Bioinformatics, 24*(13), i15–i23.

The UniProt Consortium. (2017). UniProt: The universal protein knowledgebase. *Nucleic Acids Research, 45*(D1), D158–D169.

Thompson, J. (1997). The CLUSTAL_X windows interface: Flexible strategies for multiple sequence alignment aided by quality analysis tools. *Nucleic Acids Research, 25*(24), 4876–4882.

Waterhouse, A., Bertoni, M., Bienert, S., Studer, G., Tauriello, G., Gumienny, R., Heer, F. T., de Beer, T. A. P., Rempfer, C., Bordoli, L., Lepore, R., & Schwede, T. (2018). SWISS-MODEL: Homology modelling of protein structures and complexes. *Nucleic Acids Research, 46*(W1), W296–W303.

Weiner, S. J., Kollman, P. A., Case, D. A., Singh, U. C., Ghio, C., Alagona, G., Profeta, S., & Weiner, P. (1984). A new force field for molecular mechanical simulation of nucleic acids and proteins. *Journal of the American Chemical Society, 106*(3), 765–784.

Womble, D. D. (2000). GCG: The Wisconsin Package of sequence analysis programs. *Methods in Molecular Biology, 132*, 3–22.

Wu, S., & Zhang, Y. (2008a). A comprehensive assessment of sequence-based and template-based methods for protein contact prediction. *Bioinformatics, 24*(7), 924.

Wu, S., & Zhang, Y. (2008b). MUSTER: Improving protein sequence profile-profile alignments by using multiple sources of structure information. *Proteins: Structure, Function, and Bioinformatics, 72*(2), 547–556.

Xiang, Z., & Honig, B. (2001). Extending the accuracy limits of prediction for side-chain conformations. *Journal of Molecular Biology, 311*(2), 421–430.

Xu, D., & Zhang, Y. (2011). Improving the physical realism and structural accuracy of protein models by a two-step atomic-level energy minimization. *Biophysical Journal, 101*(10), 2525–2534.

Xu, D., & Zhang, Y. (2012). Ab initio protein structure assembly using continuous structure fragments and optimized knowledge-based force field. *Proteins, 80*(7), 1715–1735.

Yang, Y., Faraggi, E., Zhao, H., & Zhou, Y. (2011). Improving protein fold recognition and template-based modeling by employing probabilistic-based matching between predicted one-dimensional structural properties of query and corresponding native properties of templates. *Bioinformatics, 27*(15), 2076.

Yang, J., Yan, R., Roy, A., Xu, D., Poisson, J., & Zhang, Y. (2015). The I-TASSER suite: Protein structure and function prediction. *Nature Methods, 12*(1), 7.

Zdobnov, E. M., & Apweiler, R. (2001). InterProScan - an integration platform for the signature-recognition methods in InterPro. *Bioinformatics, 17*(9), 847–848.

Zhang, J., Liang, Y., & Zhang, Y. (2011). Atomic-level protein structure refinement using fragment-guided molecular dynamics conformation sampling. *Structure, 19*(12), 1784–1795.

Zhou, Y., Kloczkowski, A., Faraggi, E., & Yang, Y. (2016). *Prediction of protein secondary structure*. Humana Press.

CHAPTER 9

Resources for Docking-Based Virtual Screening

SAILU SARVAGALLA • SREE KARANI KONDAPURAM • R. VASUNDHARA DEVI • MOHANE SELVARAJ COUMAR

1 DRUG DISCOVERY

Drug discovery is a multidisciplinary process through which a potent new therapeutic/drug molecule is identified to prevent/treat a particular disease condition (Mohsa & Greig, 2017; Silverman, 2004). It has a significant role in the practice of modern medicine and has emerged as one of the main translational activities that contribute to human health and well-being. The design and discovery of new therapeutic molecules for disease still relies on trial-and-error methods and it involves a wide range of scientific disciplines including biology, medicinal chemistry, structural biology, pharmacology, etc. Many pharmaceutical industries, academia, and scientific communities around the globe are actively involved in the search for new chemical entities (drug molecules) for a wide variety of diseases. Discovery of a drug molecule (ligand) for a particular drug target (receptor, enzyme, ion channel, transporter, protein—protein interactions, DNA, etc.) is an expensive and time-consuming procedure that normally takes an average of 10—15 years with an estimated cost of US$985—2870 million (Wouters et al., 2020).

Generally, the drug discovery process could be divided into four major steps (Ng, 2009):

i. target identification and validation,
ii. lead discovery and development (constitutes lead identification and lead optimization),
iii. preclinical studies, and
iv. clinical studies.

The discovery of a new drug molecule is required for treating a disease or illness for which no other treatment exists or it provides additional advantages over the prevailing treatments such as lesser adverse effects, higher therapeutic efficacy, improved amenability, and fewer drug—drug interactions, etc. Thus, a thorough understanding of disease biology especially the identification of impaired biomolecule/drug target function that contributes to the disease pathology is very important for the drug discovery process. Accordingly, the identification of a drug target is the first and foremost step in the drug discovery and development process. Generally, the targets are classified into two categories: one is established/reported drug targets and the other is novel drug targets that are yet to be identified. The functions and pathophysiological roles of the reported targets have already been scientifically well studied and validated, whereas the same for the novel target demands a thorough understanding of its role and has to be validated by applying various advanced research methods. Several scientific methods including phenotype screening, genetic study, transgenic animal, molecular, and functional imaging techniques are widely employed for target identification and validation (Hughes et al., 2011; Katsila et al., 2016).

Following target identification, the next step in the drug discovery process is to identify a synthetic or natural chemical molecule (commonly referred to as a lead or hit) that specifically binds to the target and thereby inhibits/induces its function. Furthermore, the potency, efficacy, safety, and other drug-like properties of the lead molecule are improved by implementing in silico, in vitro, and in vivo methods in an iterative fashion. Besides, structure—activity relationships (SARs) are developed to evaluate the pharmacokinetic and pharmacodynamic properties of lead molecules and can be further applied for the synthesis and evaluation of analogs (Guido et al., 2011; Holenz et al., 2016; Rankovic & Morphy, 2010).

The lead molecule, after several rounds of optimization, is developed into a drug molecule by passing through various preclinical and clinical studies. In preclinical studies, synthetic process, formulation, potency, and ADMET (absorption, distributions, metabolism, excretion, and toxicity) properties of drug molecules are evaluated using in vitro and in vivo experiments. Once suitable molecules are found that are potent and safe with desired ADMET properties, they are further moved to clinical testing. In the clinical study, safety, efficacy, dosage, adverse side effects, pharmacokinetics, and pharmacological properties of the drug molecule are evaluated in human volunteers and also in patients. Finally, the drug molecule is approved by regulatory agencies (e.g., US FDA, EMA, CDSCO, etc.) if it shows efficacy, safety, tolerability, and fewer side effects while administering to the patients (Ng, 2009). Thus, the drug discovery process is a highly multidisciplinary task involving biochemists, molecular biologists, structural biologists, medicinal chemists, computational biologists, pharmacologists, clinicians, statisticians, to name a few.

In this chapter, we brief on the available computer-aided drug design (CADD) techniques, tools, and resources for lead discovery and development in the drug discovery process. Particular emphasis is focused on docking tools and resources for lead identification/optimization.

1.1 CADD for Lead Discovery and Development

Lead discovery and development involves the identification of a synthetic or natural chemical molecule/peptide/antibody that specifically and efficiently binds to a drug target and thereby modulates its biological function (Guido et al., 2011; Holenz et al., 2016; Rankovic & Morphy, 2010). Lead molecules (also referred to as hit) could be considered as a prototype, from which the drug molecules are developed. The initial step in lead discovery is to identify a starting molecule that shows reasonable biological activity toward the target protein. Both experimental and computational methods are widely used for lead identification and optimization as well. Some of the extensively used experimental methods that are being used for hit identification include high-throughput screening (HTS), combinatorial library screening, knowledge-based screening, fragment-based screening, etc. (Bleicher et al., 2003; Holenz & Stoy, 2019). HTS is a well-defined and widely used experimental method in which tens and thousands of chemical entities are screened in an assay system (i.e., cell/enzyme-based assay) against the target protein to find starting chemical compounds with anticipated biological activity. In combinatorial screening (Liu et al., 2017), a large number of structurally diverse or focused library compounds are being synthesized through systematic covalent linkage of various building blocks and are subjected to screening against particular drug target to know their positive interaction with the target protein. Furthermore, the identified

positive hits are evaluated for lead development and optimization. In fragment-based screening, thousands of chemical fragments (molecular weight <250Da) are evaluated for biological activity with the desired target protein using biochemical and biophysical methods (Erlanson et al., 2020; Neto de Souza et al., 2020).

However, the hit identification using experimental methods is expensive and time-consuming as they require extensive development and validation of complex assay systems; moreover, the compounds need to be physically available for biological testing. Moreover, the hit rate for HTS and other experimental methods is extremely low and hence results in limited usage for screening a large chemical library of compounds. Alternatively, computational methods commonly known as CADD techniques such as virtual screening (VS) methods have emerged as powerful techniques for hit identification (Lionta et al., 2014; Wang et al., 2020). In VS experiments, a molecular library consisting of millions of chemical molecules is screened computationally (in silico) in a short time and the compounds which are predicted as active/positive are then subjected for further biological testing, whereas the filtered inactive/negative compounds are skipped from biological testing. This VS strategy for lead identification significantly reduces the cost and workload as compared to HTS screening (Fradera & Babaoglu, 2017; Maia et al., 2020; Scior et al., 2012).

Once a lead molecule is identified with the desired biological activity, it undergoes several rounds of chemical optimization (analog synthesis) to improve the pharmacokinetics, pharmacodynamics, and toxicity profile. Several medicinal chemistry lead optimization strategies, such as congener series synthesis, ring expansion/contraction, bioisosteric replacements, etc., are carried out to find an optimized lead (Wermuth, 2008). Such lead optimization (development) effort would provide suitable drug candidates for testing in preclinical models and clinical trials. CADD techniques could help in identifying/predict the regions in the lead molecule, where chemical modifications (analog synthesis) could be carried out to improve the potency or ADME properties of the lead molecules. Such predictions would help to minimize the number of analogs that are actually synthesized and tested experimentally and thus reduce the cost and time required for identifying optimized lead molecule. Thus, CADD techniques play an integral role in modern lead discovery and development.

Broadly, CADD methods are classified into two categories and are structure-based drug design (SBDD) and ligand-based drug design (LBDD) (Katsila et al., 2016; Leelananda & Lindert, 2016; Sliwoski et al., 2014; Yu & Jr, 2017). The main aim of these methods is to identify therapeutically effective small chemical/lead molecules that could be developed into a drug for curing/preventing disease. The SBDD techniques including molecular docking, structure-based VS, pharmacophore screening (structure-based), de novo drug design, and molecular dynamics (MD) techniques are widely used for both lead identification and optimization (Baig et al., 2016; Sliwoski et al., 2014; Yu & Jr, 2017). In molecular docking, the interaction and binding mode of the ligand molecule can be predicted with the target protein, and subsequently, the derived knowledge could be used for better lead identification and optimization process (Berry et al., 2015; Ferreira et al., 2015; Maruca et al., 2019; Pinzi & Rastelli, 2019; Ruyck et al., 2016). Docking-based virtual screening (DBVS) is one of the most widely used SBDD techniques for hit identification. In this method, millions of chemical library compounds are virtually evaluated for interaction with the biological target using molecular docking (Binkowski et al., 2014; Forli, 2015; Fradera & Babaoglu, 2017; Kumar & Zhang, 2015; Maia et al., 2020; Shoichet, 2006; Zhang et al., 2015). Additionally, the structure-based pharmacophore screening methods (i.e., e-pharmacophore) could be used to screen a large number of library compounds for hit identification (Pirhadi et al., 2013; Seidel et al., 2019). Using de novo drug design techniques, a diverse set of novel chemical or lead molecules can be designed (Devi et al., 2015; Nicolaou et al., 2012; Schneider & Clark, 2019).

The LBDD methods, including similarity search, substructure search, pharmacophore (ligand-based) screening, quantitative structure—activity relationship (QSAR), etc., are extensively used both for lead identification and lead optimization processes (Acharya et al., 2011; Banegas-Luna et al., 2018; Yan et al., 2016). In ligand-based methods, only the existing ligand molecule knowledge is used to generate a pharmacophore model that could be used for predicting new molecules with biological activity. In addition to this, the LBDD techniques can also be used for evaluating ligand molecular properties including ADMET properties. Thus, CADD techniques are extremely helpful for better lead discovery and development process. Moreover, these methods are mostly used in conjunction with experimental methods for effective lead discovery and development. Fig. 9.1 shows the various stages of drug discovery and the use of CADD methods for lead identification and optimization.

2 A BRIEF VIEW OF DOCKING AND ITS USE

Molecular docking is a computational method used to predict the mode of binding between the interacting biomolecules while forming stable complex structure, e.g., protein—ligand (Forli et al., 2016; Pinzi & Rastelli, 2019; R. N. Santos et al., 2018), protein—protein

(Gromiha et al., 2017; Rabbani et al., 2018; Soni & Madhusudhan, 2017), protein—nucleic acid (Tuszynska et al., 2015), and nucleic acid—ligand (Chen et al., 2012; Luo et al., 2019; Selvaraj & Singh, 2018) complexes. Subsequently, the derived knowledge from docking can be rationally used to assess the binding affinity and interaction mechanism of biomolecules at the atomic level to further comprehend their role in various pathophysiological conditions. Thus, the main objective of molecular docking is to obtain near-native binding conformation of the biomolecular interactions with optimal binding free energy. Accordingly, various molecular docking programs are widely used in rational drug design and discovery process. Successively, the derived information could be used to optimize the binding efficiency of the ligand with the target protein. Thus, docking plays a key role in modern rational drug design and discovery (Ferreira et al., 2015; Pinzi & Rastelli, 2019; Ruyck et al., 2016).

In principle, there are two types of docking methods that are being extensively used. One method is a shape complementary approach (Yan & Huang, 2019; Zhang et al., 2009) and the second method is a simulated approach (Goodsell & Olson, 1990; Totrov & Abagyan, 2008). In the former method, a set of surface/geometric features (i.e., descriptors) of the protein and ligand, which are complementary to each other, facilitate interactions to form a stable complex structure. In general, these features (descriptors) may include the protein—ligand solvent accessible surface area, hydrophobicity, and acidic and basic regions. The docking programs which use shape complementary methods are typically fast and robust, and they do not make any dynamic change (flexibility) in the protein/ligand structure. Although these types of docking programs are predominantly used to model the protein—protein interactions, ligand-based pharmacophore screening, and docking, few recently developed docking programs incorporate the protein—ligand flexibility as well. In the latter method, the protein and ligands are actually simulated in its characteristic biological environment to find their native binding conformation by calculating their pairwise interaction energies. In this type of docking programs, the ligand is separated from the protein by a physical distance, and later, it is allowed to move in its conformational space to find its optimal binding position at the active site of the protein. Such ligand movement is mediated by its conformational relaxation via translations, rotations, and torsional angle (conformational) changes (Ferreira et al., 2015; Meng et al., 2011). Ligand's various binding conformation energies are calculated and ranked for selecting the best fitting ligand pose.

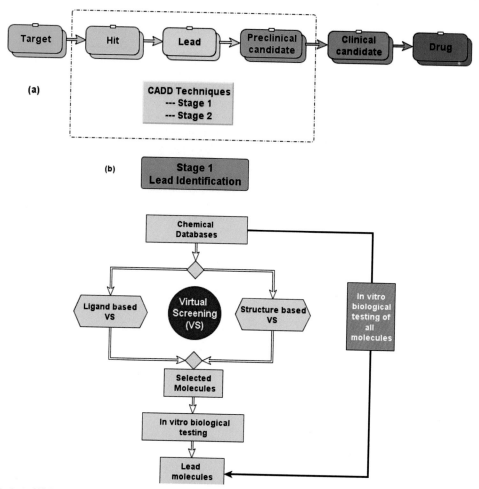

FIG. 9.1 **(A)** Application of computer-aided drug design (CADD) techniques in various drug discovery stages, **(B)** application of CADD-based virtual screening for lead identification, and **(C)** application of CADD for lead optimization.

The docking tools/programs comprise of two important components (Halperin et al., 2002; Saikia & Bordoloi, 2019; Sulimov et al., 2019):

i. search algorithm and
ii. scoring function.

The search algorithm searches or generates all the possible conformations and orientation of the ligand molecule in the search space, i.e., in the active site of the target protein. The search algorithms are classified into two types and are systematic and stochastic; different search algorithms such as genetic algorithm (GA), simulated annealing, and Monte Carlo algorithms are developed and implemented in several docking tools (Meng et al., 2011). For example, AutoDock and GOLD docking tools implement GA-based searching.

The generated poses are scored and ranked based on the protein–ligand interaction. Protein–ligand interactions that are considered in various scoring algorithms include hydrogen bond, electrostatic, and van der Waals interaction. Some of the scoring functions also consider solvation energies in their energy function terms. Scoring algorithms are classified into four major types such as force field-based, empirical-based, knowledge-based potentials, and machine learning–based scoring functions (Guedes et al., 2018; Li et al., 2019). Readers interested in more details about the search and scoring algorithms used in docking may refer to Chapters 2 and 3 in this book.

As discussed before, docking can help in understanding the molecular mode of interaction between a drug and its target. It is useful for both lead identification

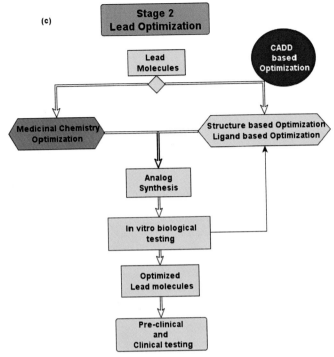

FIG. 9.1 **cont'd.**

and optimization. Also, docking can be useful for fragment-based drug design (Erlanson et al., 2004), drug repurposing (Sarvagalla et al., 2019), and also for target identification in target-fishing experiments (Jenkins et al., 2006; Kharkar et al., 2014; Lee et al., 2016). Readers interested to know more about the various uses of docking may refer to Chapters 11 and 12 in this book. Fig. 9.2 summarizes the overall requirements/inputs for docking and its use in a general sense.

3 RESOURCES FOR DOCKING

Biomolecules are dynamic in nature and they perform their function through interaction with other biomolecules. Thus, resolving the biomolecules (i.e., protein–protein or protein–ligand or protein–nucleic acids) interaction and their conformational binding mechanism at the structural (atomic) level will play a significant role in understanding various pathophysiological conditions. Experimental methods such as X-ray (Maveyraud & Mourey, 2020), nuclear magnetic resonance (NMR) (Takeuchi et al., 2019), and cryo-EM (Renaud et al., 2018) are being extensively used to resolve these interactions. However, resolving these complex interactions using experimental methods is time-consuming and expensive. Alternatively, in silico

methods such as docking can be used to predict the biomolecules' interaction and binding mechanism. Additionally, docking has an immense role in the drug discovery process where it can be used to predict the binding affinity and interaction mode of ligand (i.e., small molecule/peptide, etc.) with the target (i.e., protein/nucleic acid) molecule.

Accordingly, to perform a docking study, one needs to have the following:

i. **Target information**: a high-resolution X-ray or NMR or cryo-EM or homology modeled structures of the target protein with known ligand-binding site/cavity. So, one can use this ligand-binding site knowledge for docking other novel ligand molecules. In the absence of a known ligand-binding site, various in silico prediction tools can be used to predict the ligand-binding sites.

ii. **Ligand information**: a ligand molecule or a set of ligand molecules, whose interaction with the target we intend to predict. The resulting target–ligand interaction knowledge could be used to optimize the ligand interaction and efficiency.

iii. **Docking program/tool**: it is essential for carrying out the target–ligand docking process. These tools can be used for simple docking of protein–ligand, protein–protein, or protein–DNA complexes.

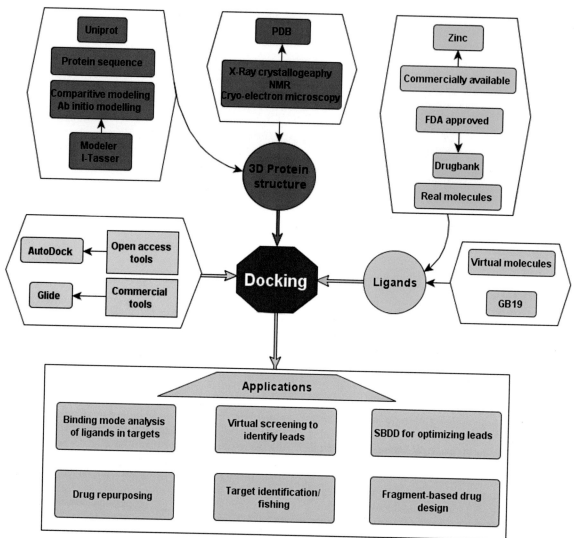

FIG. 9.2 Schematic representation of various inputs for the protein—ligand docking process, along with the applications of docking.

These tools are specific for the job at hand. Most of the tools meant for protein—ligand docking also can handle VS of a large library of compounds.

iv. **Computational facilities**: a simple desktop or high-performance computing (HPC) or cloud computing can be used for running the docking calculation depending on the amount of data that need to be processed and the availability of resources. The availability of HPC will help to carry out VS of large databases with millions of compounds in reasonable time frames; such mega screening projects are difficult to handle in simple desktop systems.

Fig. 9.2 shows the various input information and output obtained from a typical docking study. In the following subsections, we describe the available drug target/receptor databases, ligand databases, various docking tools for protein—ligand docking/VS, along with different computing facilities that can be used.

3.1 Target Information

Drug target is a biomolecule in the living organism that when altered (e.g., mutation or expression level changes) is associated with a particular disease condition. Binding of a ligand or a drug molecule to the target can lead to a change in its function that subsequently shows a therapeutic effect on the disease. Due to advancements in cell and molecular biology, genomics, and proteomics, a huge number of targets have been identified and reported in the literature. However, fewer numbers of target three-dimensional (3D) structures are resolved and reported in various databases. Thus, there is a huge gap in target identification and resolving its corresponding 3D structural information. This gap can be filled with computational modeling tools such as homology modeling/threading and ab initio methods. The reported target can be either protein or nucleic acids. However, the predominantly reported drug targets are proteins including G protein–coupled receptors, enzymes (e.g., kinases, proteases, phosphatases, etc.), ion channels (e.g., ligand-gated ion channels, voltage-gated ion channels), transporters (e.g., SGLT2), and structural proteins (e.g., tubulin membrane transport proteins) (Santos et al., 2017).

There are several databases available for storing and retrieving target sequences and 3D structural information. One such repository for storing and retrieving structural information of biomolecules is Protein Data Bank (PDB; https://www.rcsb.org/) (Burley et al., 2017), where most of the reported 3D structures with/without their ligand molecules are deposited for the benefit of the scientific community. This is a freely available database where one can retrieve and analyze the structural data of the proteins and nucleic acids. Additionally, it provides various structural analyses and visualization tools for a better understanding of these structures. Presently, PDB is having 164,174 numbers of biological macromolecule structures. Biological Magnetic Resonance Bank (BMRB; http://www.bmrb.wisc.edu/) (Ulrich et al., 2008) is another freely accessible database that stores the information of the proteins, peptides, nucleic acids, and other biomolecules derived from NMR spectroscopy.

DrugBank (Wishart et al., 2018) is a freely available online database that comprises information on drug/ligand molecules and their corresponding targets. In the absence of target 3D structural information, one can use target sequence information for modeling their structures using computational tools. UniProt (Bateman et al., 2017) is a freely available protein sequence database that comprises information about the biological function of the proteins derived from literature. The National Center for Biotechnology Information (NCBI; http://www.ncbi.nlm.nih.gov) also provides rich information and several useful tools for protein sequence analysis. For example, Protein (NCBI, 1988) is an open-access database with a collection of protein sequences derived

from annotated coding regions in GenBank, RefSeq, and TPA, as well as records from SwissProt, PDB, PIR, and PRF. A few of the important target sequence and structure databases is presented in Table 9.1.

3.2 Ligand Information

Ligand is a small chemical molecule or biomolecule (i.e., peptide/antibody) that binds to the target and thereby changes its function, which subsequently shows a therapeutic effect on the disease. Several databases store ligand information that is useful for docking studies. As aforementioned, DrugBank (Wishart et al., 2018) is a freely available online database that contains information about drugs, their mode of action, and their targets. The current release of DrugBank is version 5.1.6, which has 13,572 drug entries. Out of this, 2631 are approved small molecule drugs, 1377 approved biologics (proteins, peptides, vaccines, and allergenic), 131 nutraceuticals, and over 6374 experimental (discovery-phase) drugs. These drugs are linked to 5246 nonredundant protein (i.e., drug target/enzyme/transporter/carrier) sequences.

PubChem (Kim et al., 2016) is open access, the world's largest chemistry database that majorly contains information about small molecules and their biological activity. However, it also contains information on larger biomolecules such as nucleotides, carbohydrates, lipids, peptides, and chemically modified macromolecules information. ZINC (Irwin et al., 2012; Irwin & Shoichet, 2005) is a free database of commercially available compounds for VS. Over 230 million purchasable compounds in different file formats (including SMILES, 3D SDF, mol2, and DOCK flexibase format) are available for download and use in VS experiments. NCI (Voigt et al., 2001) is another open-access database that contains approximately 250,000 small molecule structures that can be used directly for VS and docking. There are additional small molecule/ligand databases which are presented in Table 9.2.

Many of these databases provide easy access to the chemical information in a format suitable for VS experiments, such as that of SDF format. The chemical data downloaded from databases can be cleaned or analyzed using cheminformatics data handling software such as DataWarrior (http://www.openmolecules.org/datawarrior/download.html) (Sander et al., 2015). These tools help to analyze the datasets for their physiochemical properties and calculate other descriptors. Such an analysis will help to gauge the nature of VS datasets. The databases discussed in Table 9.2 provide information on real compounds (compounds already synthesized in the lab or those naturally occurring)

TABLE 9.1
List of Available Drug—Target Databases.

Database/Server	Web Link	Description	References
DrugBank	https://www.drugbank.ca/	It provides drug target information for the drugs that are in clinical use and also for those molecules that are in advanced clinical testing. Sequences of 5246 nonredundant proteins (i.e., drug target/enzyme/transporter/carrier) are linked to these drug entries.	Wishart et al. (2018)
Protein Data Bank	https://www.rcsb.org/	PDB (Protein Data Bank) is an open-access database that provides information on three-dimensional structures of proteins, nucleic acids, and their complex structures with ligand molecules. Presently, it comprises 164,174 macromolecular structures that are useful for structure-based drug design.	Burley et al. (2017)
Therapeutic target database	http://db.idrblab.net/ttd/	Therapeutic Target Database (TTD) is a freely available database with drug target information.	Chen et al. (2002)
SuperTarget	http://insilico.charite.de/supertarget/index.php?site=home	It is an open-access database that incorporates the information about drugs, such as medical indications, adverse effects, drug metabolism, pathways, and Gene Ontology terms for the drug targets. Presently, the updated database contains 6219 drug targets that are annotated with 195,770 compounds (including approved drugs).	Hecker et al. (2012)
STITCH	http://stitch.embl.de/	STITCH (Search Tool for Interactions of CHemicals) is a freely accessible database that integrates the information about protein—ligand interaction derived from metabolic pathways, crystal structures, and binding experiments.	Kuhn et al. (2008)
UniProt	https://www.uniprot.org/	It is a freely available protein sequence database that collates literature information about the biological function of the proteins.	Bateman et al. (2017)
RefSeq	https://www.ncbi.nlm.nih.gov/refseq/	It is an open-access, annotated, and curated database of nucleotide sequences and their protein products.	Pruitt et al. (2007)
Protein	https://www.ncbi.nlm.nih.gov/protein/	Protein is an open-access database that contains a collection of protein sequences derived from annotated coding regions in GenBank, RefSeq, and TPA, as well as records from SwissProt, PDB, PIR, and PRF.	NCBI (1988)

TABLE 9.1
List of Available Drug–Target Databases.—cont'd

Database/Server	Web Link	Description	References
Biological Magnetic Resonance Bank	http://www.bmrb.wisc.edu/	It is a freely accessible database that stores NMR spectroscopy–derived structural information about the proteins, peptides, nucleic acids, and other biomolecules.	Ulrich et al. (2008)
Nucleic acid database	http://ndbserver.rutgers.edu/	Nucleic acid database is a freely available web portal that provides information about 3D structures of nucleic acid and their complexes. Additionally, it provides tools and software for nucleic acid sequence and structure analysis.	Buvaneswari et al. (2014)
GenBank	https://www.ncbi.nlm.nih.gov/genbank/	It is a freely available NIH genetic sequence database, which collects and annotates publicly available DNA sequences. Moreover, GenBank is part of the International Nucleotide Sequence Database Collaboration that includes the European Nucleotide Archive (ENA) and DNA Data Bank of Japan (DDBJ). These three databases exchange data daily.	Benson et al. (2013)

TABLE 9.2
List of Available Drug/Ligand/Chemical Databases.

Database/Server	Web Link	Description	References
DrugBank	https://www.drugbank.ca/	It is an online freely accessible database that provides information about drugs, their mode of action, and their targets. The latest release of DrugBank version 5.1.6 has information about 13,572 drugs, including that of approved small molecule drugs (2631), approved biologics (1377; proteins, peptides, vaccines, and allergenic), nutraceuticals (131), and experimental (6374; discovery-phase) drugs.	Wishart et al. (2018)
PubChem	https://pubchem.ncbi.nlm.nih.gov/	It is open access, the world's largest chemistry database that majorly contains information about small molecules and their biological activity.	Kim et al. (2016)
ZINC	https://zinc.docking.org/	It is a free database of commercially available compounds that can be downloaded and used for virtual screening (VS). It has over 230 million purchasable compounds in different file formats (including SMILES, mol2, 3D SDF, and DOCK flexibase format).	Irwin et al. (2012), Irwin and Shoichet (2005)

Continued

TABLE 9.2
List of Available Drug/Ligand/Chemical Databases.—cont'd

Database/Server	Web Link	Description	References
ChEMBL	https://www.ebi.ac.uk/chembl/	It is an open-access database that contains biological activity data of small molecules collated from several core journals in the field of medicinal chemistry.	Davies et al. (2015), Mendez et al. (2019)
NCI	https://cactus.nci.nih.gov/download/nci/	It is an open-access database that contains approximately 250,000 small molecule structures that can be used for VS and docking.	Voigt et al. (2001)
ChemDB	http://cdb.ics.uci.edu/	It is a chemical database that contains around 5 million commercially available small molecule compounds. These can be used as synthetic building blocks and as leads for drug design projects.	Chen et al. (2007)
ChemSpider	http://www.chemspider.com/About.aspx	It is a freely available chemical structure database. It offers text and chemical structure search to access over 67 million compounds from different data sources.	Little et al. (2012)
BindingDB	https://www.bindingdb.org/bind/index.jsp	It is a freely accessible public database that contains the experimentally derived binding affinities of small molecules with drug targets.	Gilson et al. (2016), Liu et al. (2007)
PDBeChem	https://www.ebi.ac.uk/pdbe-srv/pdbechem/	It is a search engine which stores information on chemical species including small molecules, ligands, metal ions, amino acids, and nucleotides found in the Protein Data Bank.	Dimitropoulos et al. (2006)
HMDB	https://hmdb.ca/	HMDB (Human Metabolome Database) is an open-access database that contains information related to small molecule metabolites found in the human body. Furthermore, these data are classified as chemical, clinical data, and molecular biology/biochemistry data.	Wishart et al. (2007)
SMPDB	https://smpdb.ca/	SMPDB (Small Molecule Pathway Database) is an open-access database with more than 30,000 small molecule pathways found in humans only.	Frolkis et al. (2009), Jewison et al. (2014)
TTD	http://db.idrblab.net/ttd/	TTD (Therapeutic Target Database) is an open-access database that integrates and provides the drug targets information and their corresponding drug molecules.	Chen et al. (2002)
SuperDRUG2	http://cheminfo.charite.de/superdrug2/index.html	It is a freely available database with more than 4600 active pharmaceutical ingredients. It provides information on drugs such as their chemical structures, side effects, their physicochemical and pharmacokinetic properties, and drug–drug interactions.	Siramshetty et al. (2018)

that are in many cases commercially available and hence purchasable. In addition to such compounds, there are synthetically feasible virtual compound libraries available for computational investigations (Hoffmann & Gastreich, 2019). For example, the GDB-17 database set (Ruddigkeit et al., 2013) has over 150 billion possible compounds made up from a maximum of 17 atoms (C, N, O, S, X). Also, the real dataset from enamine (https://enamine.net/library-synthesis/real-compounds/real-database) has over 1.36 billion of readily synthesizable molecules that are Ro5 (Lipinski's rule of five) compliant. They are available in ready-to-use VS formats (SMILES and SDF) and are readily synthesizable upon request. Such virtual compounds could also be investigated in DBVS to identify potential hits.

3.3 Docking Tools

The main aim of bioinformatics/computational biology is to better understand the living organism at the molecular level. In a way, this can be done by understanding the biomolecular interactions at the molecular level. Docking is a tool that could help us to understand the molecular-level interaction of biomolecules. Additionally, docking is widely used in the drug discovery process to predict the ligand/small molecule interaction with the receptor/target protein. Once the receptor and ligand structures are retrieved for docking purposes, then both need to be separately prepared/processed to correct their geometry/missing features using tools. Next, the ligand-binding site (active site/substrate binding site) in the target protein needs to be defined for the docking program to recognize the search area. The ligand-binding site can be detected based on the previously existing ligand/substrate in the active site of the target protein. Many protein structures in PDB are cocrystallized along with a ligand molecule, which will help in identifying the correct binding site residues for carrying out the docking. In case the ligand-binding site is not known, it can be predicted using various reported ligand-binding site prediction tools (Table 9.3).

TABLE 9.3
List of Available Binding Site Prediction Tools/Server.

Database/ Server	Web Link	Description	References
ATPbind	https://zhanglab.ccmb.med.umich.edu/ATPbind/	It is a freely available metaserver which uses a support vector machine method to predict the ATP binding site in the protein.	Hu et al. (2018)
CASTp	http://sts.bioc.uni-edu/castp/index.html?2cpk	Computed Atlas of Surface Topography of proteins (CASTp) is an online resource for predicting concave surface region on protein structures.	Binkowski et al. (2003)
COACH	https://zhanglab.ccmb.med.umich.edu/COACH/	It is a freely available tool that uses a metaserver approach to predict the protein—ligand-binding site.	Yang et al. (2013)
Despite	http://www.playmolecule.org/deepsite/	It is an open-source web-based tool that predicts the ligand-binding pockets on the protein surface using a knowledge-based convolutional neural network approach.	Jiménez et al. (2017)
eFindSite	http://brylinski.cct.lsu.edu/efindsite	It is a freely available ligand-binding site prediction tool that uses metathreading, machine learning, and auxiliary ligands to effectively forecast the ligand-binding site in protein models.	Brylinski and Feinstein (2013)
FINDSITE	http://cssb.biology.gatech.edu/findsite	It is a freely available ligand-binding site prediction and protein functional annotation tool. It uses threading-based methods to detect common ligand-binding sites in a set of evolutionarily related proteins.	Brylinski and Skolnick (2008)

Continued

TABLE 9.3
List of Available Binding Site Prediction Tools/Server.—cont'd

Database/Server	Web Link	Description	References
Fpocket	http://fpocket.sourceforge.net/	It is an open-source platform for detecting cavities/pockets on the protein surface.	Guilloux et al. (2009)
Pocketome	http://www.Pocketome.org/	It is an encyclopedia of conformational ensembles of druggable binding sites derived from cocrystal structures available in the Protein Data Bank and Uniprot Knowledgebase.	Kufareva et al. (2012)
PocketDepth	http://proline.physics.iisc.ernet.in/pocketdepth/	It is a geometry-based pocket prediction web server that uses a depth-based clustering method to identify the pocket in proteins.	Kalidas and Chandra (2008)
3DLigandsite	http://www.sbg.bio.ic.ac.uk/~3dligandsite/	It is an automated method for the prediction of ligand-binding sites in 3D protein structures.	Wass et al. (2010)

These tools use various algorithms to detect the cavity in the target protein. Typically, the largest cavity is the active site and can be considered for docking of small molecule ligand, e.g., CASTp (Computed Atlas of Surface Topography of proteins) web server that can be accessed online for locating, delineating, and measuring geometric and topological properties of protein structures. It was launched first in 2003 and the current version of CASTp 3.0 is released in 2018 (Tian et al., 2018).

Finally, the ligand molecule can be docked at the defined binding site of the target protein using various docking tools to know its binding affinity and interaction mechanism.

Over the period, various docking tools/web servers (i.e., AutoDock, Dock, GalaxyDock, ICM-Docking, GOLD, Glide, FlexX, and LigandFit, etc.; Table 9.4) were developed; each docking tool utilizes its own search algorithm and scoring function for pose generation and refinement process. Although there are a large number of docking tools/server reported in the literature, one should bear in mind that the developed docking programs are not suitable for all of the given system. Hence, one should be careful while choosing a docking tool and it is recommended to use different docking tools to correctly assess the protein–ligand interaction. For example, AutoDock (Forli et al., 2016) is a freely available docking tool that is widely used for docking of small molecule/drug candidates to the receptor protein/drug target. It uses a variety of search algorithms including Monte Carlo, simulated annealing, GA, and a hybrid local search GA.

Dock6 (Allen et al., 2015) is another open-source docking program for academia. It is a rigid body docking program that uses the protein and ligand geometry features for predicting the lowest energy binding mode. GalaxyDock (Shin & Seok, 2012) is an open-source protein–ligand docking module that allows the flexibility of preselected side chains of ligand using conformational space annealing method. ICM-Docking (Neves, 2012) is another open-source desktop docking module for performing protein–ligand, protein–protein, and protein–peptide docking. GOLD (Verdonk et al., 2003) is a commercially available docking tool that uses the GA algorithm for ligand search. GOLD is widely employed in VS, lead optimization, and identifying the correct binding mode of small molecule inhibitors/ligands. Glide (Friesner et al., 2004; Halgren et al., 2004) is another commercially available docking module that offers a full range of speed versus accuracy. It includes HTVS (high-throughput virtual screening), SP (standard precision), and XP (extra precision) docking modules, with an incremental level of accuracy. Glide uses an exhaustive search and empirical scoring function to rank the ligands (Friesner et al., 2006).

3.3.1 Virtual screening tools

As discussed above, VS is a CADD technique used to search large libraries of molecules to identify the ligand that most likely binds to the drug target and thereby alter (inhibit) the function of the target. For this

TABLE 9.4
List Available Docking/Virtual Screening Tools/Servers.

Tool/Server	Web Link	Description	References
AutoDock	http://autodock.scripps.edu/	It is a freely available open-source molecular docking software suit. It is widely used for docking of small molecule/drug candidates to the receptor protein/drug target.	Forli et al. (2016)
AutoDock Vina	http://vina.scripps.edu/	It is a new generation of docking software for performing docking/virtual screening. It is faster and significantly improved the accuracy of binding mode prediction compared with AutoDock.	Trott and Olson (2010)
UCSF Dock	http://dock.compbio.ucsf.edu/	Dock is an open-source docking program for academia. It is a rigid body docking program that uses the protein and ligand geometry features for predicting the lowest energy binding mode.	Allen et al. (2015)
DockingServer	https://www.dockingserver.com/web	It is a web-based service that performs docking of a small molecule to the receptor proteins.	Hazai et al. (2009)
FlexX	https://www.biosolveit.de/FlexX/	It is a commercially available standalone docking tool. It predicts the protein binding site and performs virtual screening.	Schellhammer and Rarey (2004)
Glide	https://www.schrodinger.com/glide	Glide is a commercially available docking module that offers a full range of speed versus accuracy. It includes high-throughput virtual screening, standard precision, and extra precision docking modules.	Friesner et al. (2004), Halgren et al. (2004)
GOLD	https://www.ccdc.cam.ac.uk/solutions/csd-discovery/components/gold/	Gold is a commercial docking software that could be used for predicting the correct binding mode of small molecule inhibitors/ligands. It is also useful for virtual screening and lead optimization.	Verdonk et al. (2003)
ICM-Docking	http://www.molsoft.com/docking.html	It is an open-source desktop docking module that offers a step-by-step GUI docking menu. Using this module, we could perform protein–ligand, protein–protein, and protein–peptide docking.	Neves (2012)
GalaxyDock	http://galaxy.seoklab.org/softwares/galaxydock.html	It is an open-source protein–ligand docking module that uses the conformational space annealing method to allow the flexibility of preselected side chains of ligands.	Shin and Seok (2012)
LigandFit	https://www.phenix-online.org/documentation/reference/ligandfit.html	LigandFit is a commercially available docking module which uses shape-based filters in combination with Monte Carlo conformational search to generate ligand poses in the active site of the protein.	Venkatachalam et al. (2003)
NPDock	http://genesilico.pl/NPDock/submit	It is a freely available docking tool for predicting RNA–protein and DNA–protein complex structures. It uses GRAMM for global macromolecular docking.	Tuszynska et al. (2015)

Continued

TABLE 9.4
List Available Docking/Virtual Screening Tools/Servers.—cont'd

Tool/Server	Web Link	Description	References
MCDock	https://github.com/andersx/mcdock	It is a freely available standalone docking module that uses the Monte Carlo simulation approach to perform molecular docking of two biomolecules.	Liu and Wang (1999)
Surflex	https://www.biopharmics.com/downloads/	It is a commercially available standalone automated flexible docking algorithm.	Jain (2003)
iGEMDOCK	http://gemdock.life.nctu.edu.tw/dock/igemdock.php	It is an open-source standalone docking software for noncommercial researchers. It could be used for docking, virtual screening, and post-docking analysis.	Hsu et al. (2011)
SwissDock	http://www.swissdock.ch/	It is a freely available web service that predicts the molecular interactions between a small molecule inhibitor and a target protein.	Grosdidier et al. (2011)
PatchDock	https://bioinfo3d.cs.tau.ac.il/PatchDock/	It is a freely available web service that performs the molecular docking of protein–ligand or protein–protein or protein–nucleic acid.	Duhovny et al. (2005)
iScreen	http://iscreen.cmu.edu.tw/basic.php	It is a freely available world's first cloud computing–based web server for docking, virtual screening, and de novo drug design. It uses the Traditional Chinese Medicine (TCM) database for the virtual screening and docking process.	Tsai et al. (2011)
iDock	https://github.com/HongjianLi/idock	It is a freely available multithreaded virtual screening tool for flexible ligand docking which is derived from AutoDock Vina.	Li et al. (2012)
LibDock	https://www.3ds.com/products-services/biovia/	It is a commercially available flexible docking program that uses protein geometric features such as hotspots (i.e., polar and apolar regions) to make contacts with ligands polar and apolar regions. LibDock program is implemented in Discovery Studio (Accelrys, USA).	Rao et al. (2007)
DOT	https://www.sdsc.edu/CCMS/DOT/	It is an open-source standalone desktop software package for molecular docking of protein–ligand, protein–protein, and protein–nucleic acids.	Roberts et al. (2013)
BSP-SLIM	https://zhanglab.ccmb.med.umich.edu/BSP-SLIM/	It is a freely available docking program for predicting low-resolution protein–ligand structures.	Lee and Zhang (2012)
Lead finder	https://www.cresset-group.com/software/lead-finder/	Lead finder is a commercially available docking program for virtual screening and binding energy calculation of protein–ligand interactions.	Stroganov et al. (2008)
ParDOCK	http://www.scfbio-iitd.res.in/pardock/	It is a freely available web server for automated protein–ligand docking program.	Gupta. et al. (2007)

purpose, all the database compounds are docked and scored; the top-scoring compounds would be considered as hits and further evaluated for biological activity. It is a fast and cost-effective method and widely employed in the drug discovery process for a hit or lead identification. There are several VS tools/web server including AutoDock Vina (Trott & Olson, 2010), FlexX (Schellhammer & Rarey, 2004), Glide virtual screening workflow (Workflow, 2015), Gold (Verdonk et al., 2003), iScreen (Tsai et al., 2011), and Lead finder (Stroganov et al., 2008) that are widely employed in the drug discovery. AutoDock Vina is a new generation of docking software for performing docking/VS. It is faster and significantly improved the accuracy of binding mode prediction compared to AutoDock. AutoDock Vina implements the Monte Carlo–based search algorithm (Trott & Olson, 2010). iScreen is another open-source world's first cloud computing web server for docking, VS, and de novo drug design. It uses the TCM database for the VS and docking process. FlexX is a commercially available standalone docking tool that performs VS.

Many of the docking programs discussed above can be used for VS purposes as well. For example, Glide virtual screening workflow (Workflow, 2015) combines HTVS, SP docking, and XP docking in a sequential way to filter the given set of compounds. There is an option in each stage to automatically move a certain percentage of top hits from one step to the next step. For example, the top 10% scored compounds from HTVS could be subjected to docking in SP docking and the top scoring 10% compounds could be moved to the last stage of XP docking to give final hit compounds.

Such a sequential treatment of compounds could help to subject a large set of compounds to VS. Other useful programs for VS are listed in Table 9.4.

3.3.2 Reverse docking tools

The docking described above can be categorized into two methods: one is conventional docking and the other is reverse docking. In conventional docking, one or more ligand molecules are docked to a single-target protein to identify the ligand molecule that strongly binds to the protein. Thus, the conventional docking method is generally used to screen a large number of chemical molecules to identify potential hit/lead molecules (VS). In reverse docking or inverse docking (Jenkins et al., 2006; Kharkar et al., 2014), one ligand molecule is docked to multiple-target proteins to identify the protein that strongly interacts with the query ligand molecule. Thus, reverse docking can be used to identify new targets for existing ligand molecules to avoid off-target/side effects and drug toxicity. Additionally, reverse docking can be effectively used in drug repurposing where the binding affinity of existing/approved drug molecules can be evaluated for other target proteins. Furthermore, the predicted ligand pose and binding site knowledge derived from reverse docking can be used in lead optimization and development. However, reverse docking has limitations including high computational time, shortage of available target structures, and interprotein noises of docking scores. There are several reverse docking tools/servers including idTarget, TarFisDock, and INVDOCK reported with various target databases and scoring schemes (Table 9.5).

TABLE 9.5
List of Available Inverse Docking Tools/Servers.

Tool/Server	Web link	Description	References
ACID	http://chemyang.ccnu.edu.cn/ccb/server/ACID	It is a free tool that uses consensus inverse docking strategy and clinically relevant macromolecular targets for drug repurposing.	Wang et al. (2019)
idTarget	http://idtarget.rcas.sinica.edu.tw	idTarget is a freely available web-based server that predicts possible targets of a ligand molecule. It uses a divide-and-conquer docking approach and Protein Data Bank (PDB) deposited proteins as possible targets for inverse docking.	Wang et al. (2012)

Continued

TABLE 9.5
List of Available Inverse Docking Tools/Servers.—cont'd

Tool/Server	Web link	Description	References
TarFisDock	http://www.dddc.ac.cn/tarfisdock/	TarFisDock is a web-based freely accessible tool that automatically searches for small molecule–protein interactions. It utilizes PDTD (potential drug target database) with 698 protein structures from 15 different therapeutic areas to make the prediction using reverse ligand–protein docking. TarFisDock can be used to target identification and also to study the drug mechanism of action.	Li et al. (2006)
Inverse-docking (INVDOCK)	http://bidd.nus.edu.sg/group/softwares/invdock.htm	INVDOCK is a reverse docking method that automatically identifies possible protein and nucleic acid (RNA or DNA) targets for a reported and newly designed drug molecule.	Chen and Ung (2001), Chen and Zhi (2001)

idTarget (Wang et al., 2012) is a freely available web-based server that predicts possible binding targets of a ligand molecule through a divide-and-conquer docking approach. TarFisDock (Li et al., 2006) is another web-based tool that automatically searches for small molecule and protein interactions over a large range of protein structures. Thus, TarFisDock can be used for target identification and to study the mechanism of existing drugs. INVDOCK (Chen & Ung, 2001; Chen & Zhi, 2001) is another reverse docking tool that automatically identifies potential protein and nucleic acid (RNA or DNA) targets for reported and newly designed drug molecules.

Due to limited space, only selected tools are listed/described in this chapter. Interested readers may refer to other recent publications that describe tools/resources related to CADD techniques (Agrawal et al., 2018; Bezhentsev et al., 2017; Krüger et al., 2016; Pagadala et al., 2017; Pirhadi et al., 2016; Singla et al., 2013).

3.4 Computational Facilities

DBVS demand computational facilities depending upon the size of the database used for screening and the quantum of calculations involved. The range of computing facilities used in drug discovery for VS is desktop computers, distributed computing/grid computing, cloud computing, hybrid multicore CPU/graphics processing unit (GPU) accelerator-based HPC, and supercomputers (Fig. 9.3).

Distributed computing is the computing model where software components are shared by several computers. This is specific to the programs with components that are shared among different computers located in a confined geographical location. An example of distributed computing is the three-tier model architecture. In grid computing, a large number of computers target a complex problem. In this model, servers, computers connected by the internet perform tasks independently. Individual computers can offer their processing power and time to solve complex problems. Volunteer computing is one component of grid computing. Grid computing is actually computers on a network that work together to solve complex tasks and may perform as supercomputers (Hwang et al., 2012).

As the efficiency and speed of the CPU-based computing are not sufficient for the computation-intensive tasks, GPUs with efficient hardware parallelization and high throughput are preferred nowadays. The tasks are performed in the GPU which has smaller parallel cores to perform massive HPC. NVIDIA invented a parallel programming model known as CUDA (Compute Unified Device Architecture). CUDA or OpenCL software platforms help in the implementation of several significantly faster applications. Tesla, GEForce, and Quadro support this CUDA. In fact, in the case of drug discovery, computational tasks need a balance between CPU and GPU-based computing approaches to handle both the inexpensive and costly computations. Hence, a hybrid CPU and GPU-based

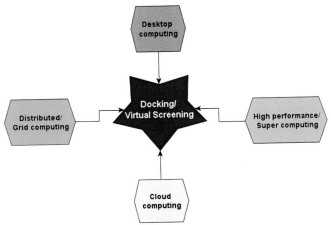

FIG. 9.3 Available computational facilities for carrying out virtual screening.

computing with multicore CPUs and multicore GPUs is preferred in VS experiments. This hybrid computing may involve GPUs, in-house clusters of computers, remote supercomputers, and distributed computing infrastructures (Hwang et al., 2012).

Computers that perform with the highest operational rate in terms of floating point of operations per second (FLOPS) compared to the other computers whose rate is measured in million instructions per second (MIPS) are called supercomputers. They consist of multiple computers performing parallel processing. Supercomputers have superior hardware architecture, huge data storage, and sophisticated and specialized network connectivity (Hwang et al., 2012). As of now, Fujitsu Fugaku (https://www.fujitsu.com/global/about/innovation/fugaku/) from Japan with 415 PFLOPS of performance is ranked as the top among supercomputers. Several countries have established their own supercomputers for research/academic activities.

Currently, the use of cloud computing, which utilizes the available information technology capabilities, is the trend. "A Cloud is a type of parallel and distributed system consisting of a collection of interconnected and virtualized computers that are dynamically provisioned and presented as one or more unified computing resources based on service-level agreements established through negotiation between the service provider and consumers" (Buyya, 2009). John McCarthy, in 1961, predicted that "computation may someday be organized as a public utility." Cloud computing is divided into two sections such as a front end and back end connected with the help of the network, i.e., through the internet. The front end is the end-user with desktop, GUI (graphical user interface), etc. The back end uses central server functioning with the help of protocols

and with the software called Middleware. Both CPU-based distributed computing and parallel computations using GPU are utilized in cloud-based computing environments.

Clouds are categorized into private, public, and hybrid clouds (Chauhan, 2012). Access to the cloud environment is obtained on a minimal pay-per-use basis. Public cloud computing environments, such as Amazon AWS (https://aws.amazon.com/), Microsoft Azure (https://azure.microsoft.com/en-in/), Google Cloud Platform (https://cloud.google.com/), IBM cloud services (https://www.ibm.com/in-en/cloud), VMWare Cloud (https://www.vmware.com/in/cloud-solutions.html), Oracle Cloud (https://www.oracle.com/cloud/), etc., have achieved remarkable improvements in computational performance for large-scale (bioinformatics) analysis in recent years (Ohue et al., 2020). Typical cloud architecture is illustrated in Fig. 9.4.

Normally, for large-scale VS, HPC environments such as computing clusters, grid computers, and cloud computing can be utilized. The next two subsections present some of the recently reported uses of such facilities for VS experiments.

3.4.1 Virtual screening in cloud, grid, and distributed computing environments

Raccoon2 (Temelkovski et al., 2019) is the GUI available in a desktop application that helps in pre- and post-docking analysis. VS using AutoDock and AutoDock Vina with cloud computing support such as cloud access service is utilized in Mell & Grance (2011). The WS-PGRADE/gUSE, a workflow framework with a remote application programming interface (API), helps extend Raccoon2 in the cloud through the CloudBroker

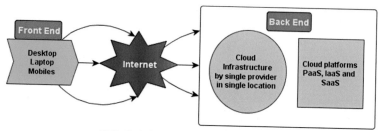

FIG. 9.4 A general cloud architecture.

platform. This CloudBroker platform interacts with various clouds. In Kiss et al. (2013), a desktop application developed in .NET allows submitting the docking jobs from AutoDock and AutoDock Vina to the Azure-based cloud computing service called VENUS-C. Cloud-Broker platform and WS-PGRADE/gUSE technology along with the docking software AutoDock4 or Auto-Dock Vina were utilized in Temelkovski et al. (2019). A web-based WS-PGRADE portal is used as the user interface, with a separate environment for pre- and post-docking processing.

In the case of rigid receptor docking experiments (Ellingson & Baudry, 2012), AutoDockCloud performs HTs of parallel tasks with the help of open-source Hadoop framework implementing the MapReduce paradigm on a cloud platform using AutoDock4.2. In this, ligands from the DUD (directory of useful decoys) database were utilized in the docking process. A novel method (Paris et al., 2018) to optimize ensemble docking-based experiments by reducing the size of an isoniazid-resistant enoyl reductase of *Mycobacterium tuberculosis* (InhA) fully flexible receptor (FFR) model for docking runtime and scaling docking workflow on cloud virtual machines was reported. In this work, an efficient cost-benefit pool of virtual machines was identified and the performance of docking in different configurations of Azure instances is evaluated. InhA FFR model was deployed on the cloud-based scientific workflow e-FReDock (uses e-SC API Java client to control the invocation of both the subworkflows) to perform in-depth molecular docking simulations of FFR models and several ligands. The cloud platforms used in this work include Microsoft Azure public cloud and Cloud Innovation Center private cloud.

Spark-VS is a parallel structure-based VS method on public cloud resources or commodity-based computer clusters. When the DBVS involves large molecular libraries, it becomes highly expensive as it involves very little parallelizable tasks as in Capuccini et al. (2017). Google's MapReduce transformed the large-scale analysis with the help of commodity hardware and cloud resources for processing massive datasets. It also provides transparent scalability and fault tolerance at the software level. Apache Hadoop and Apache Spark are the open-source implementations of MapReduce. Jaghoori et al. (Jaghoori et al., 2016) used the AutoDock Vina tool with features such as an improved level of parallelization and proper HPC infrastructure to enable screening of large ligand libraries.

Cluster and grid infrastructures used WS-PGRADE workflow management system in three steps such as preparing by splitting the input library into groups to be processed in parallel, running AutoDock Vina on these groups in parallel, and merging the outputs. Paris et al. (Paris et al., 2016) used a cloud-based workflow to dock between FFR model and the number of ligands by first discarding unpromising MD conformations and exploiting the Microsoft Azure on-demand cloud platform resources. This environment is built on a cloud-based workflow enactment system specialized to hold high-throughput tasks with better quality experiments in lesser time. A novel algorithm for docking prediction known as fast cloud-based protein–ligand docking prediction algorithm (FCPLDPA) to augment the performance of GA is reported. The simulation results show that the computation cost of GA is reduced by FCPLDPA with the help of cloud computing technologies and also the pattern reduction method improves the caliber of the end results (Chen et al., 2013).

3.4.2 SARS-CoV-2 research using supercomputing facilities

Using MOGONII supercomputer at JGU, Germany, molecular docking has arrived at several approved hepatitis C drugs simeprevir, paritaprevir, grazoprevir, and velpatasvir as promising candidates for the treatment of SARS-CoV-2. The study authors also found the natural substance from the Japanese honeysuckle (*Lonicera japonica*) as a strong candidate against SARS-CoV-2 (Kadioglu et al., 2020). In Grand et al. (2020), with the help of the world's most powerful GPU-accelerated supercomputer, ORNL SUMMIT with

CUDA API performed high-throughput docking to identify antiviral drugs for SARS-CoV-2. SWEETLEAD molecular library (Novick et al., 2013) and AutoDock Vina tools were used in this research. Smith & Smith (2020) performed ensemble docking to identify small molecules that bind either to the viral S-Protein at the host receptor site or to the S-Protein—human ACE2 interface using powerful supercomputer SUMMIT.

4 CONCLUSION

Searching for a new drug for a particular disease is an essential and challenging task. Developing a new drug molecule typically takes 10—15 years or in some cases even more. To reduce the time and also the cost involved, researchers have continuously strived to implement new strategies. CADD techniques, in general, have emerged as a useful strategy that enables cutting cost and time for bringing drugs from bench to bedside. Among the CADD techniques, molecular docking has a special place and is the most often used technique both in the academia and industry alike. Docking helps to understand drug and target interaction at the molecular level; this in turn aids in the identification of new leads and also in the optimization to bring out better drugs. In the ongoing COVID-19 pandemic, the scientific community world over has utilized various CADD techniques to speed the drug discovery effort (Amin & Jha, 2020; Mishra et al., 2020; Wang & Guan, 2020; Yadav et al., 2020; Zehra et al., 2020).

For the convenience of students and researchers, this chapter provides a bird's view of lead identification and optimization utilizing an integrated computational and experimental work. Also, the chapter provides the basics of docking and its various applications. Moreover, the chapter provides various resources that are essential for carrying out docking experiments for both lead identification and optimization purpose. Tools/software/ servers, both commercial and open access, that are available for a drug target database, drug/ligand/chemical databases, binding site prediction tools/servers, and docking tools/servers are listed and are also discussed. Also, various cloud and HPC facilities that can be used for computational work are highlighted. We believe that the resources provided in this chapter can aid the scientific community in their quest for new drugs.

REFERENCES

Acharya, C., Coop, A., Polli, J. E., & Alexander, D. M. (2011). Recent advances in ligand-based drug design: Relevance and utility. *Current Computer-Aided Drug Design*, 7(1), 10—22. https://doi.org/10.1038/jid.2014.371.

Agrawal, P., Raghav, P. K., Bhalla, S., Sharma, N., & Raghava, G. P. S. (2018). Overview of free software developed for designing drugs based on protein-small molecules interaction. *Current Topics in Medicinal Chemistry*, 18(13), 1146—1167. https://doi.org/10.2174/1568026618666180816155131.

Allen, W. J., Balius, T. E., Mukherjee, S., Brozell, S. R., Moustakas, D. T., Lang, P. T., Case, D. A., Kuntz, I. D., & Rizzo, R. C. (2015). DOCK 6: Impact of new features and current docking performance. *Journal Computer Chemistry*, 36(15), 1132—1156. https://doi.org/10.1002/jcc.23905.

Amin, S. A., & Jha, T. (2020). Fight against novel coronavirus: A perspective of medicinal chemists. *European Journal of Medicinal Chemistry*, 201(112559), 1—12. https://doi.org/10.1016/j.ejmech.2020.112559.

Baig, M. H., Ahmad, K., Roy, S., Ashraf, J. M., Adil, M., Siddiqui, M. H., Khan, S., Kamal, M. A., Provazník, I., & Choi, I. (2016). Computer aided drug design: Success and limitations. *Current Pharmaceutical Design*, 22(5), 572—581. https://doi.org/10.2174/1381612822666151125000550.

Banegas-Luna, A. J., Cerón-Carrasco, J. P., & Pérez-Sánchez, H. (2018). A review of ligand-based virtual screening web tools and screening algorithms in large molecular databases in the age of big data. *Future Medicinal Chemistry*, 10(22), 2641—2658. https://doi.org/10.4155/fmc-2018-0076.

Bateman, A., Martin, M. J., ODonovan, C., Magrane, M., Alpi, E., Antunes, R., Bely, B., Bingley, M., Bonilla, C., Britto, R., Bursteinas, B., Bye-AJee, H., Cowley, A., Da Silva, A., De Giorgi, M., Dogan, T., Fazzini, F., Castro, L. G., Figueira, L., ... (2017). UniProt: The universal protein knowledgebase. *Nucleic Acids Research*, 45, 158—169. https://doi.org/10.1093/nar/gkw1099.

Benson, D. A., Cavanaugh, M., Clark, K., Karsch-Mizrachi, I., Lipman, D. J., Ostell, J., & Sayers, E. W. (2013). GenBank. *Nucleic Acids Research*, 41(D1), D36—D42. https://doi.org/10.1093/nar/gks1195.

Berry, M., Fielding, B., & Gamieldien, J. (2015). Practical considerations in virtual screening and molecular docking. *Emerging Trends in Computational Biology, Bioinformatics, and Systems Biology*, 21(1), 487—502. https://doi.org/10.1155/2010/706872.

Bezhentsev, V. M., Druzhilovskii, D. S., Ivanov, S. M., Filimonov, D. A., Sastry, G. N., & Poroikov, V. V. (2017). Web resources for discovery and development of new medicines. *Pharmaceutical Chemistry Journal*, 51(2), 91—99. https://doi.org/10.1007/s11094-017-1563-x.

Binkowski, T. A., Jiang, W., Roux, B., Anderson, W. F., & Joachimiak, A. (2014). Virtual high-throughput ligand screening. *Methods*, 1140, 251—261. https://doi.org/10.1007/978-1-4939-0354-2.

Binkowski, T. A., Naghibzadeh, S., & Liang, J. (2003). CASTp: Computed atlas of surface topography of proteins. *Nucleic Acids Research*, 31(13), 3352—3355. https://doi.org/10.1093/nar/gkg512.

Bleicher, K. H., Böhm, H. J., Müller, K., & Alanine, A. I. (2003). Hit and lead generation: Beyond high-throughput screening. *Nature Reviews Drug Discovery*, 2(5), 369—378. https://doi.org/10.1038/nrd1086.

Brylinski, M., & Feinstein, W. P. (2013). EFindSite: Improved prediction of ligand binding sites in protein models using meta-threading, machine learning and auxiliary ligands. *Journal of Computer-Aided Molecular Design, 27*(6), 551–567. https://doi.org/10.1007/s10822-013-9663-5.

Brylinski, M., & Skolnick, J. (2008). A threading-based method (FINDSITE) for ligand-binding site prediction and functional annotation. *Proceedings of the National Academy of Sciences of the United States of America, 105*(1), 129–134. https://doi.org/10.1073/pnas.0707684105.

Burley, S. K., Berman, H. M., Kleywegt, G. J., Markley, J. L., Nakamura, H., & Velankar, S. (2017). Protein Data Bank (PDB): The single global macromolecular structure structure archive. *Methods in Molecular Biology, 1607*, 627–641. https://doi.org/10.1007/978-1-4939-7000-1.

Buvaneswari, C. N., Westbrook, J., Ghosh, S., Petrov, A. I., Sweeney, B., Zirbel, C. L., Leontis, N. B., & Berman, H. M. (2014). The nucleic acid database: New features and capabilities. *Nucleic Acids Research, 42*(D1), D114–D122. https://doi.org/10.1093/nar/gkt980.

Buyya, R. (2009). Market-oriented cloud computing: Vision, hype, and reality of delivering computing as the 5th utility. In *2009 9th IEEE/ACM international symposium on cluster computing and the grid* (Vol. 1). https://doi.org/10.1109/CCGRID.2009.97.

Capuccini, M., Ahmed, L., Schaal, W., Laure, E., & Spjuth, O. (2017). Large-scale virtual screening on public cloud resources with Apache spark. *Journal of Cheminformatics, 9*(15), 1–6. https://doi.org/10.1186/s13321-017-0204-4.

Chauhan, P. V. (2012). Cloud computing in distributed system. *International Journal of Engineering Research and Technology, 1*(10), 1–8.

Chen, L., Calin, G. A., & Zhang, S. (2012). Novel insights of structure-based modeling for RNA-targeted drug discovery. *Journal of Chemical Information and Modeling, 52*(10), 2741–2753. https://doi.org/10.1021/ci300320t.

Chen, X., Ji, Z. L., & Chen, Y. Z. (2002). TTD: Therapeutic target database. *Nucleic Acids Research, 30*(1), 412–415. https://doi.org/10.1093/nar/30.1.412.

Chen, J. H., Linstead, E., Swamidass, S. J., Wang, D., & Baldi, P. (2007). ChemDB update — full-text search and virtual chemical space. *Bioinformatics, 23*(17), 2348–2351. https://doi.org/10.1093/bioinformatics/btm341.

Chen, J. Le, Tsai, C. W., Chiang, M. C., & Yang, C. S. (2013). A high performance cloud-based protein-ligand docking prediction algorithm. *BioMed Research International*, 1–8. https://doi.org/10.1155/2013/909717.

Chen, Y. Z., & Ung, C. Y. (2001). Prediction of potential toxicity and side effect protein targets of a small molecule by a ligand-protein inverse docking approach. *Journal of Molecular Graphics and Modelling, 20*(3), 199–218. https://doi.org/10.1016/S1093-3263(01)00109-7.

Chen, Y. Z., & Zhi, D. G. (2001). Ligand - protein inverse docking and its potential use in the computer search of protein targets of a small molecule. *Proteins: Structure, Function and Genetics, 43*(2), 217–226. https://doi.org/10.1002/1097-0134(20010501)43:2.

Davies, M., Nowotka, M., Papadatos, G., Dedman, N., Gaulton, A., Atkinson, F., Bellis, L., & Overington, J. P. (2015). ChEMBL web services: Streamlining access to drug discovery data and utilities. *Nucleic Acids Research, 43*(W1), 612–620. https://doi.org/10.1093/nar/gkv352.

Devi, R. V., Sathya, S. S., & Coumar, M. S. (2015). Evolutionary algorithms for de novo drug design — A survey. *Applied Soft Computing Journal, 27*, 543–552. https://doi.org/10.1016/j.asoc.2014.09.042.

Dimitropoulos, D., Ionides, J., & K, H. (2006). Using PDBeChem to search the PDB ligand dictionary. In G. D. S. A. D. Baxevanis, R. D. M. Page, G. A. Petsko, & L. D. Stein (Eds.), *Current protocols in bioinformatics* (pp. 1–3). Hoboken, N. J.: John Wiley & Sons. https://doi.org/10.1002/0471250953.bi1403s15.

Duhovny, D. S., Inbar, Y., Nussinov, R., & Wolfson, H. J. (2005). PatchDock and SymmDock: Servers for rigid and symmetric docking. *Nucleic Acids Research, 33*(Web Server issue), 363–367. https://doi.org/10.1093/nar/gki481.

Ellingson, S. R., & Baudry, J. (2012). High-throughput virtual molecular docking with AutoDockCloud. *Concurrency and Computation: Practice and Experience*, 685–701. https://doi.org/10.1002/cpe.2926.

Erlanson, D. A., De Esch, I. J. P., Jahnke, W., Johnson, C. N., & Mortenson, P. N. (2020). Fragment-to-lead medicinal chemistry publications in 2018. *Journal of Medicinal Chemistry, 63*(9), 4430–4444. https://doi.org/10.1021/acs.jmedchem.9b01581.

Erlanson, D. A., McDowell, R. S., & O Brien, T. (2004). Fragment-based drug discovery. *Journal of Medicinal Chemistry, 47*(14), 3463–3482. https://doi.org/10.1021/jm040031v.

Ferreira, L. G., Dos Santos, R. N., Oliva, G., & Andricopulo, A. D. (2015). Molecular docking and structure-based drug design strategies. *Molecules, 20*. https://doi.org/10.3390/molecules200713384.

Forli, S. (2015). Charting a path to success in virtual screening. *Molecules, 20*, 18732–18758. https://doi.org/10.3390/molecules201018732.

Forli, S., Huey, R., Pique, M. E., Sanner, M. F., Goodsell, D. S., & Olson, A. J. (2016). Computational protein–ligand docking and virtual drug screening with the AutoDock suite. *Nature Protocols, 11*(5), 905–918. https://doi.org/10.1038/nprot.2016.051.

Fradera, X., & Babaoglu, K. (2017). Overview of methods and strategies for conducting virtual small molecule screening. *Current Protocols in Chemical Biology, 9*(3), 196–212. https://doi.org/10.1002/cpch.27.

Friesner, R. A., Banks, J. L., Murphy, R. B., Halgren, T. A., Klicic, J. J., Mainz, D. T., Repasky, M. P., Knoll, E. H., Shelley, M., Perry, J. K., Shaw, D. E., Francis, P., & Shenkin, P. S. (2004). Glide: A new approach for rapid, accurate docking and scoring. 1. Method and assessment of docking accuracy. *Journal of Medicinal Chemistry, 47*(7), 1739–1749. https://doi.org/10.1021/jm0306430.

Friesner, R. A., Murphy, R. B., Repasky, M. P., Frye, L. L., Greenwood, J. R., Halgren, T. A., Sanschagrin, P. C., & Mainz, D. T. (2006). Extra precision glide: Docking and scoring incorporating a model of hydrophobic enclosure for protein-ligand complexes. *Journal of Medicinal Chemistry*, *49*(21), 6177–6196. https://doi.org/10.1021/jm051256o.

Frolkis, A., Knox, C., Lim, E., Jewison, T., Law, V., Hau, D. D., Liu, P., Gautam, B., Ly, S., Guo, A. C., Xia, J., Liang, Y., Shrivastava, S., & Wishart, D. S. (2010). SMPDB: The small molecule pathway database. *Nucleic Acids Research, 38*(DI), 480–487. https://doi.org/10.1093/nar/gkp1002.

Gilson, M. K., Liu, T., Baitaluk, M., Nicola, G., Hwang, L., & Chong, J. (2016). BindingDB in 2015: A public database for medicinal chemistry, computational chemistry and systems pharmacology. *Nucleic Acids Research, 44*(D1), 1045–1053. https://doi.org/10.1093/nar/gkv1072.

Goodsell, D. S., & Olson, A. J. (1990). Automated docking of substrates to proteins by simulated annealing. *Proteins: Structure, Function, and Bioinformatics, 8*(3), 195–202. https://doi.org/10.1002/prot.340080302.

LeGrand, S., Scheinberg, A., Tillack, A. F., Thavappiragasam, M., Vermaas, J. V., Agarwal, R., Larkin, J., Poole, D., Santos-Martins, D., Solis-Vasquez, L., Koch, A., Forli, S., Hernandez, O., Smith, J. C., & Sedova, A. (2020). GPU-accelerated drug discovery with docking on the summit supercomputer: Porting, optimization, and application to COVID-19 research. *ArXiv*. Retrieved from https://pubmed.ncbi.nlm.nih.gov/32676519.

Gromiha, M. M., Yugandhar, K., & Jemimah, S. (2017). Protein-protein interactions: Scoring schemes and binding affinity. *Current Opinion in Structural Biology, 44*, 31–38. https://doi.org/10.1016/j.sbi.2016.10.016.

Grosdidier, A., Zoete, V., & Michielin, O. (2011). SwissDock, a protein-small molecule docking web service based on EADock DSS. *Nucleic Acids Research, 39*(Web Server issue), 270–277. https://doi.org/10.1093/nar/gkr366.

Guedes, I. A., Pereira, F. S. S., & Dardenne, L. E. (2018). Empirical scoring functions for structure-based virtual screening: Applications, critical aspects, and challenges. *Frontiers in Pharmacology, 9*, 1–18. https://doi.org/10.3389/fphar.2018.01089.

Guido, R. V. C., Oliva, G., & Andricopulo, D. (2011). Modern drug discovery technologies: Opportunities and challenges in lead discovery. *Combinatorial Chemistry & High Throughput Screening, 14*(10), 830–839. https://doi.org/10.2174/138620711797537067.

Guilloux, V. Le, Schmidtke, P., & Tuffery, P. (2009). Fpocket: An open source platform for ligand pocket detection. *BMC Bioinformatics, 10*, 1. https://doi.org/10.1186/1471-2105-10-168.

Gupta, A., Sharma, P., & Jayaram, B. (2007). ParDOCK: An all atom energy based Monte Carlo docking protocol for protein-ligand complexes. *Protein and Peptide Letters, 14*(7), 632–646. https://doi.org/10.2174/092986607781483831.

Halgren, T. A., Murphy, R. B., Friesner, R. A., Beard, H. S., Frye, L. L., Pollard, W. T., & Banks, J. L. (2004). Glide: A new approach for rapid, accurate docking and scoring. 2. Enrichment factors in database screening. *Journal of Medicinal Chemistry, 47*(7), 1750–1759. https://doi.org/10.1021/jm030644s.

Halperin, I., Ma, B., Wolfson, H., & Nussinov, R. (2002). Principles of docking: An overview of search algorithms and a guide to scoring functions. *Proteins: Structure, Function and Genetics, 47*(4), 409–443. https://doi.org/10.1002/prot.10115.

Hazai, E., Kovács, S., Demkó, L., & Bikádi, Z. (2009). Docking-Server: Molecular docking calculations online. *Acta Pharmaceutica Hungarica, 79*(1), 17–21.

Hecker, N., Ahmed, J., Von Eichborn, J., Dunkel, M., Macha, K., Eckert, A., Gilson, M. K., Bourne, P. E., & Preissner, R. (2012). SuperTarget goes quantitative: Update on drug-target interactions. *Nucleic Acids Research, 40*, 1113–1117. https://doi.org/10.1093/nar/gkr912.

Hoffmann, T., & Gastreich, M. (2019). The next level in chemical space navigation: Going far beyond enumerable compound libraries. *Drug Discovery Today, 24*(5), 1148–1156. https://doi.org/10.1016/j.drudis.2019.02.013.

Holenz, Jörg, Mannhold, R., Kubinyi, H., & Folkers, G. (2016). Lead generation: Methods and strategies, 2 volume set. *Pharmaceutical & Medicinal Chemistry, 2*, 1–824. https://doi.org/10.1017/CBO9781107415324.004.

Holenz, J., & Stoy, P. (2019). Advances in lead generation. *Bioorganic and Medicinal Chemistry Letters, 29*(4), 517–524. https://doi.org/10.1016/j.bmcl.2018.12.001.

Hsu, K.-C., Chen, Y.-F., Lin, S.-R., & Yang, J.-M. (2011). iGEM-DOCK: a graphical environment of enhancing GEMDOCK using pharmacological interactions and post-screening analysis. *Journal of Biological Chemistry, 12*(S33), 1–11. https://doi.org/10.1074/jbc.m010223200.

Hughes, J. P., Rees, S. S., Kalindjian, S. B., & Philpott, K. L. (2011). Principles of early drug discovery. *British Journal of Pharmacology, 162*(6), 1239–1249. https://doi.org/10.1111/j.1476-5381.2010.01127.x.

Hu, J., Li, Y., Zhang, Y., & Yu, D.-J. (2018). ATPbind: Accurate protein–ATP binding site prediction by combining sequence-profiling and structure-based comparisons. *Journal of Chemical Information and Modeling, 58*(2), 501–510. https://doi.org/10.1021/acs.jcim.7b00397.

Hwang, K., Fox, G. C., & Dongarra, J. J. (2012). *Distributed and cloud computing: From parallel processing to the internet of things*. Morgan Kaufmann.

Irwin, J. J., & Shoichet, B. K. (2005). ZINC – a free database of commercially available compounds for virtual screening. *Journal of Chemical Information and Modeling, 45*(1), 177–182. https://doi.org/10.1021/ci049714%2B.

Irwin, J. J., Sterling, T., Mysinger, M. M., Bolstad, E. S., & Coleman, R. G. (2012). ZINC: A free tool to discover chemistry for biology. *Journal of Chemical Information and Modeling, 52*(7), 1757–1768. https://doi.org/10.1021/ci3001277.

Jaghoori, M. M., Bleijlevens, B., & Olabarriaga, S. D. (2016). 1001 ways to run AutoDock Vina for virtual screening. *Journal of Computer-Aided Molecular Design, 30*(3), 237–249. https://doi.org/10.1007/s10822-016-9900-9.

Jain, A. N. (2003). Surflex: Fully automatic flexible molecular docking using a molecular similarity-based search engine. *Journal of Medicinal Chemistry, 46*(4), 499–511. https://doi.org/10.1021/jm020406h.

Jenkins, J. L., Bender, A., & Davies, J. W. (2006). In silico target fishing: Predicting biological targets from chemical structure. *Drug Discovery Today: Technologies*, *3*(4), 413–421. https://doi.org/10.1016/j.ddtec.2006.12.008.

Jewison, T., Su, Y., Disfany, F. M., Liang, Y., Knox, C., Maciejewski, A., Poelzer, J., Huynh, J., Zhou, Y., Arndt, D., Djoumbou, Y., Liu, Y., Deng, L., Guo, A. C., Han, B., Pon, A., Wilson, M., Rafatnia, S., Liu, P., & Wishart, D. S. (2014). SMPDB 2.0: Big improvements to the small molecule pathway database. *Nucleic Acids Research*, *42*(D1), 478–484. https://doi.org/10.1093/nar/gkt1067.

Jiménez, J., Doerr, S., Martínez-Rosell, G., Rose, A. S., & De Fabritiis, G. (2017). DeepSite: Protein-binding site predictor using 3D-convolutional neural networks. *Bioinformatics*, *33*(19), 3036–3042. https://doi.org/10.1093/bioinformatics/btx350.

Kadioglu, O., Saeed, M., Greten, H. J., & Efferth, T. (2020). Identification of novel compounds against three targets of SARS CoV2 coronavirus by combined virtual screening and supervised machine learning. *Bulletin of the World Health Organization*, 1–29. https://doi.org/10.2471/BLT.20.251561.

Kalidas, Y., & Chandra, N. (2008). PocketDepth: A new depth based algorithm for identification of ligand binding sites in proteins. *Journal of Structural Biology*, *161*(1), 31–42. https://doi.org/10.1016/j.jsb.2007.09.005.

Katsila, T., Spyroulias, G. A., Patrinos, G. P., & Matsoukas, M. T. (2016). Computational approaches in target identification and drug discovery. *Computational and Structural Biotechnology Journal*, *14*, 177–184. https://doi.org/10.1016/j.csbj.2016.04.004.

Kharkar, P. S., Warrier, S., & Gaud, R. S. (2014). Reverse docking: A powerful tool for drug repositioning and drug rescue. *Future Medicinal Chemistry*, *6*(3), 333–342. https://doi.org/10.4155/fmc.13.207.

Kim, S., Thiessen, P. A., Bolton, E. E., Chen, J., Fu, G., Gindulyte, A., Han, L., He, J., He, S., Shoemaker, B. A., Wang, J., Yu, B., Zhang, J., & Bryant, S. H. (2016). PubChem substance and compound databases. *Nucleic Acids Research*, *44*(D1), 1202–1213. https://doi.org/10.1093/nar/gkv951.

Kiss, T., Borsody, P., Terstyanszky, G., Winter, S., Greenwell, P., McEldowney, S., & Heindl, H. (2013). Large-scale virtual screening experiments on windows Azure-based cloud resources. *Concurrency and Computation: Practice and Experience*, *26*(10), 1760–1770. https://doi.org/10.1002/cpe.3113.

Krüger, J., Thiel, P., Merelli, I., Grunzke, R., & Gesing, S. (2016). Portals and web-based resources for virtual screening. *Current Drug Targets*, *17*(14), 1649–1660. https://doi.org/10.2174/1389450117666160201105806.

Kufareva, I., Ilatovskiy, A. V., & Abagyan, R. (2012). Pocketome: An encyclopedia of small-molecule binding sites in 4D. *Nucleic Acids Research*, *40*(D1), 535–540. https://doi.org/10.1093/nar/gkr825.

Kuhn, M., von Mering, C., Campillos, M., Jensen, L. J., & Bork, P. (2008). STITCH: Interaction networks of chemicals and proteins. *Nucleic Acids Research*, *36*, 684–688. https://doi.org/10.1093/nar/gkm795.

Kumar, A., & Zhang, K. Y. J. (2015). Hierarchical virtual screening approaches in small molecule drug discovery. *Methods*, *71*, 26–37. https://doi.org/10.1016/j.ymeth.2014.07.007.

Leelananda, S. P., & Lindert, S. (2016). Computational methods in drug discovery. *Beilstein Journal of Organic Chemistry*, *12*, 2694–2718. https://doi.org/10.3762/bjoc.12.267.

Lee, A., Lee, K., & Kim, D. (2016). Using reverse docking for target identification and its applications for drug discovery. *Expert Opinion on Drug Discovery*, *11*(7), 707–715. https://doi.org/10.1080/17460441.2016.1190706.

Lee, H. S., & Zhang, Y. (2012). BSP-SLIM: A blind low-resolution ligand-protein docking approach using predicted protein structures. *Proteins*, *80*(1), 93–110. https://doi.org/10.1038/jid.2014.371.

Li, J., Fu, A., & Zhang, L. (2019). An overview of scoring functions used for protein–ligand interactions in molecular docking. *Interdisciplinary Sciences: Computational Life Sciences*, *11*(2), 320–328. https://doi.org/10.1007/s12539-019-00327-w.

Li, H., Gao, Z., Kang, L., Zhang, H., Yang, K., Yu, K., Luo, X., Zhu, W., Chen, K., Shen, J., Wang, X., & Jiang, H. (2006). TarFisDock: A web server for identifying drug targets with docking approach. *Nucleic Acids Research*, *34*(W), 219–224. https://doi.org/10.1093/nar/gkl114.

Li, H., Leung, K. S., & Wong, M. H. (2012). Idock: A multi-threaded virtual screening tool for flexible ligand docking. In *2012 IEEE symposium on computational intelligence and computational biology, CIBCB 2012* (pp. 77–84). https://doi.org/10.1109/CIBCB.2012.6217214.

Lionta, E., Spyrou, G., Vassilatis, D., & Cournia, Z. (2014). Structure-based virtual screening for drug discovery: Principles, applications and recent advances. *Current Topics in Medicinal Chemistry*, *14*(16), 1923–1938. https://doi.org/10.2174/1568026614666140929124445.

Little, J. L., Williams, A. J., Pshenichnov, A., & Tkachenko, V. (2012). Identification of "known unknowns" utilizing accurate mass data and chemspider. *Journal of the American Society for Mass Spectrometry*, *23*, 179–185. https://doi.org/10.1007/s13361-011-0265-y.

Liu, R., Li, X., & Lam, K. S. (2017). Combinatorial chemistry in drug discovery. *Current Opinion in Chemical Biology*, *176*(3), 117–126. https://doi.org/10.1016/j.cbpa.2017.03.017.

Liu, T., Lin, Y., Wen, X., Jorissen, R. N., & Gilson, M. K. (2007). BindingDB: A web-accessible database of experimentally determined protein-ligand binding affinities. *Nucleic Acids Research*, *35*, 198–201. https://doi.org/10.1093/nar/gkl999.

Liu, M., & Wang, S. (1999). MCDOCK: A Monte Carlo simulation approach to the molecular docking problem. *Journal of Computer-Aided Molecular Design*, *13*(5), 435–451. https://doi.org/10.1023/A:1008005918983.

Luo, J., Wei, W., Waldispühl, J., & Moitessier, N. (2019). Challenges and current status of computational methods for docking small molecules to nucleic acids. *European Journal of Medicinal Chemistry*, *168*, 414–425. https://doi.org/10.1016/j.ejmech.2019.02.046.

Maia, E. H. B., Assis, L. C., de Oliveira, T. A., da Silva, A. M., & Taranto, A. G. (2020). Structure-based virtual screening: From classical to artificial intelligence. *Frontiers in Chemistry*, *8*, 1–18. https://doi.org/10.3389/fchem.2020.00343.

Maruca, A., Ambrosio, F. A., Lupia, A., Romeo, I., Rocca, R., Moraca, F., Talarico, C., Bagetta, D., Catalano, R., Costa, G., Artese, A., & Alcaro, S. (2019). Computer-based techniques for lead identification and optimization i: Basics. *Physical Sciences Reviews*, *4*(6), 1–14. https://doi.org/10.1515/psr-2018-0113.

Maveyraud, L., & Mourey, L. (2020). Protein X-ray crystallography and drug discovery. *Molecules*, *25*(5), 1–18. https://doi.org/10.3390/molecules25051030.

Mell, P., & Grance, T. (2011). *The NIST definition of cloud computing*. Special Publication 800-145 http://csrc.nist.gov/publications/nistpubs/800-145/sp800-145.pdf.

Mendez, D., Gaulton, A., Bento, A. P., Chambers, J., De Veij, M., Félix, E., Magariños, M. P., Mosquera, J. F., Mutowo, P., Nowotka, M., Gordillo-Marañón, M., Hunter, F., Junco, L., Mugumbate, G., Rodriguez-Lopez, M., Atkinson, F., Bosc, N., Radoux, C. J., Segura-Cabrera, A., ... (2019). ChEMBL: Towards direct deposition of bioassay data. *Nucleic Acids Research*, *47*(D1), 930–940. https://doi.org/10.1093/nar/gky1075.

Meng, X.-Y., Zhang1, H.-X., Mezei, M., & Cui2, M. (2011). Molecular docking: A powerful approach for structure-based drug discovery. *Current Computer-Aided Drug Design*, *7*(2), 146–157. https://doi.org/10.1038/jid.2014.371.

Mishra, D., Mishra, A., Chaturvedi, V. K., & Singh, M. P. (2020). An overview of COVID-19 with an emphasis on computational approach for its preventive intervention. *3 Biotech*, *10*(435), 1–13. https://doi.org/10.1007/s13205-020-02425-9.

Mohsa, R. C., & Greig, N. H. (2017). Drug discovery and development: Role of basic biological research. *Alzheimer's and Dementia: Translational Research and Clinical Interventions*, *3*(4), 651–657. https://doi.org/10.1016/j.trci.2017.10.005.

NCBI. (1988). *National Center for Biotechnology Information (NCBI)*. Bethesda (MD): National Library of Medicine (US), National Center for Biotechnology Information [1988] – [cited 2020 Oct 06] https://www.ncbi.nlm.nih.gov/.

de Souza Neto, L. R., Moreira-Filho, J. T., Neves, B. J., Maidana, R., Guimarães, A., Furnham, N., Andrade, C. H., & Silva, F. P., Jr. (2020). In silico strategies to support fragment-to-lead optimization in drug discovery. *Frontiers in Chemistry*, *8*(93), 1–18. https://doi.org/10.3389/fchem.2020.00093.

Neves, M. A. C. (2012). Docking and scoring with ICM: The benchmarking results and strategies for improvement marco. *The Journal of Computer-Aided Molecular*, *26*(6), 675–686. https://doi.org/10.1007/s10822-012-9547-0.Docking.

Ng, R. (2009). *Drugs from discovery to approval* (Vol. 2). Wiley-BlackWell.

Nicolaou, C., Kannas, C., & Loizidou, E. (2012). Multi-objective optimization methods in de novo drug design. *Mini Reviews in Medicinal Chemistry*, *12*(10), 979–987. https://doi.org/10.2174/138955712802762284.

Novick, P. A., Ortiz, O. F., Poelman, J., Abdulhay, A. Y., & Pande, V. S. (2013). SWEETLEAD: An in silico database of approved drugs, regulated chemicals, and herbal isolates for computer-aided drug discovery. *PLoS One*, *8*(11), 1–9. https://doi.org/10.1371/journal.pone.0079568.

Ohue, M., Aoyama, K., & Akiyama, Y. (2020). High-performance cloud computing for exhaustive protein-protein docking. *ArXiv:2006.08905 [Cs.DC]*, 1–11. Retrieved from http://arxiv.org/abs/2006.08905.

Pagadala, N. S., Syed, K., & Tuszynski, J. (2017). Software for molecular docking: A review. *Biophysical Reviews*, *9*(2), 91–102. https://doi.org/10.1007/s12551-016-0247-1.

Paris, R. De, Ruiz, D. A. D., & De Souza, O. N. (2016). A cloud-based workflow approach for optimizing molecular docking simulations of fully-flexible receptor models and multiple ligands. In *Proceedings – IEEE 7th international conference on cloud computing technology and science, CloudCom 2015* (pp. 495–498). https://doi.org/10.1109/CloudCom.2015.43.

Paris, R. De, Vahl Quevedo, C., Ruiz, D. D., Gargano, F., & de Souza, O. N. (2018). A selective method for optimizing ensemble docking-based experiments on an InhA fully-flexible receptor model. *BMC Bioinformatics*, *19*(1), 1–16. https://doi.org/10.1186/s12859-018-2222-2.

Pinzi, L., & Rastelli, G. (2019). Molecular docking: Shifting paradigms in drug discovery. *International Journal of Molecular Sciences*, *20*(18), 1–23. https://doi.org/10.3390/ijms20184331.

Pirhadi, S., Shiri, F., & Ghasemi, J. B. (2013). Methods and applications of structure based pharmacophores in drug discovery. *Current Topics in Medicinal Chemistry*, *13*(9), 1036–1047. https://doi.org/10.2174/1568026611313090006.

Pirhadi, S., Sunseri, J., & Koes, D. R. (2016). Open source molecular modeling. *Journal of Molecular Graphics and Modelling*, *69*, 127–143. https://doi.org/10.1016/j.jmgm.2016.07.008.

Pruitt, K. D., Tatusova, T., & Maglott, D. R. (2007). NCBI reference sequences (RefSeq): A curated non-redundant sequence database of genomes, transcripts and proteins. *Nucleic Acids Research*, *35*, 61–65. https://doi.org/10.1093/nar/gkl842.

Rabbani, G., Baig, M. H., Ahmad, K., & Choi, I. (2018). Protein-protein interactions and their role in various diseases and their prediction techniques. *Current Protein & Peptide Science*, *19*(10), 948–957. https://doi.org/10.2174/1389203718666170828122927.

Rankovic, Z., & Morphy, R. (2010). Lead generation approaches in drug discovery. *Lead Generation Approaches in Drug Discovery*, 1–290. https://doi.org/10.1002/9780470584170.

Rao, S. N., Head, M. S., Kulkarni, A., & LaLonde, J. M. (2007). Validation studies of the site-directed docking program LibDock. *Journal of Chemical Information and Modeling*, *47*(6), 2159–2171. https://doi.org/10.1021/ci6004299.

Renaud, J. P., Chari, A., Ciferri, C., Liu, W. T., Rémigy, H. W., Stark, H., & Wiesmann, C. (2018). Cryo-EM in drug discovery: Achievements, limitations and prospects. *Nature Reviews Drug Discovery*, *17*(7), 471–492. https://doi.org/10.1038/nrd.2018.77.

Roberts, V. A., Thompson, E. E., Pique, M. E., Perez, M. S., & Eyck, L. Ten (2013). DOT2: Macromolecular docking with improved biophysical models. *Journal of Computational Chemistry*, *34*(20), 1743–1758. https://doi.org/10.1002/jcc.23304.

Ruddigkeit, L., Blum, L. C., & Reymond, J. L. (2013). Visualization and virtual screening of the chemical universe database

GDB-17. *Journal of Chemical Information and Modeling,* 53(1), 56–65. https://doi.org/10.1021/ci300535x.

Ruyck de, J., Brysbaert, G., Blossey, R., & Lensink, M. F. (2016). Molecular docking as a popular tool in drug design, an in silico travel. *Advances and Applications in Bioinformatics and Chemistry,* 9(1), 1–11. https://doi.org/10.2147/AABC.S105289.

Saikia, S., & Bordoloi, M. (2019). Molecular docking: Challenges, advances and its use in drug discovery perspective. *Current Drug Targets,* 20(5), 501–521. https://doi.org/10.2174/1389450119666181022153016.

Sander, T., Freyss, J., Von Korff, M., & Rufener, C. (2015). DataWarrior: An open-source program for chemistry aware data visualization and analysis. *Journal of Chemical Information and Modeling,* 55(2), 460–473. https://doi.org/10.1021/ci500588j.

Santos, R. N., Ferreira, L. G., & Andricopulo, A. D. (2018). Practices in molecular docking and structure-based virtual screening. In *Computational drug discovery and design* (Vol. 1762, pp. 31–50). https://doi.org/10.1007/978-1-4939-7756-7_3.

Santos, R., Ursu, O., Gaulton, A., Bento, A. P., Donadi, R. S., Bologa, C. G., Karlsson, A., Al Lazikani, B., Hersey, A., Oprea, T. I., & Overington, J. P. (2017). A comprehensive map of molecular drug targets. *Nature Reviews Drug Discovery,* 16(1), 19–34. https://doi.org/10.1038/nrd.2016.230.A.

Sarvagalla, S., Basha, S. S., & Coumar, M. S. (2019). An overview of computational methods, tools, servers, and databases for drug repurposing. In *In silico drug design repurposing techniques and methodologies* (pp. 743–780). Academic Press. https://doi.org/10.1016/B978-0-12-816125-8.00025-0.

Schellhammer, I., & Rarey, M. (2004). FlexX-Scan: Fast, structure-based virtual screening. *Proteins: Structure, Function and Genetics,* 57(3), 504–517. https://doi.org/10.1002/prot.20217.

Schneider, G., & Clark, D. E. (2019). Automated de novo drug design: Are we nearly there yet? *Angewandte Chemie International Edition,* 58(32), 10792–10803. https://doi.org/10.1002/anie.201814681.

Scior, T., Bender, A., Tresadern, G., Medina-Franco, J. L., Martínez-Mayorga, K., Langer, T., Cuanalo-Contreras, K., & Agrafiotis, D. K. (2012). Recognizing pitfalls in virtual screening: A critical review. *Journal of Chemical Information and Modeling,* 52(4), 867–881. https://doi.org/10.1021/ci200528d.

Seidel, T., Schuetz, D. A., Garon, A., & Langer, T. (2019). The pharmacophore concept and its applications in computer-aided drug design. *Progress in the Chemistry of Organic Natural Products,* 110, 99–141. https://doi.org/10.1007/978-3-030-14632-0_4.

Selvaraj, C., & Singh, S. K. (2018). Computational and experimental binding mechanism of DNA-drug interactions. *Current Pharmaceutical Design,* 24(32), 3739–3757. https://doi.org/10.2174/1381612824666181106101448.

Shin, W. H., & Seok, C. (2012). GalaxyDock: Protein-ligand docking with flexible protein side-chains. *Journal of Chemical Information and Modeling,* 52(12), 3225–3232. https://doi.org/10.1021/ci300342z.

Shoichet, B. K. (2006). Virtual screening of chemical libraries. *Nature,* 432(7019), 862–865. https://doi.org/10.1038/nature03197.

Silverman, R. B. (2004). *The organic chemistry of drug design and drug action* (2nd ed.). Academic Press. https://doi.org/10.1016/B978-0-08-051337-9.50007-9.

Singla, D., Dhanda, S. K., Chauhan, J. S., Bhardwaj, A., Brahmachari, S. K., Consortium, O. S. D. D., & Raghava, G. P. S. (2013). Open source software and web services for designing therapeutic molecules. *Current Topics in Medicinal Chemistry,* 13(10), 1172–1191. https://doi.org/10.2174/1568026611313100005.

Siramshetty, V. B., Eckert, O. A., Gohlke, B. O., Goede, A., Chen, Q., Devarakonda, P., Preissner, S., & Preissner, R. (2018). SuperDRUG2: A one stop resource for approved/marketed drugs. *Nucleic Acids Research,* 46(D1), 1137–1143. https://doi.org/10.1093/nar/gkx1088.

Sliwoski, G., Kothiwale, S., Meiler, J., & Lowe, E. W. (2014). Computational methods in drug discovery. *Pharmacological Reviews,* 66(1), 334–395. https://doi.org/10.1124/pr.112.007336.

Smith, M. D., & Smith, J. C. (2020). Repurposing therapeutics for COVID-19: Supercomputer-based docking to the SARS-CoV-2 viral spike protein and viral spike protein-human ACE2 interface. *ChemRxiv.* https://doi.org/10.26434/chemrxiv.11871402.

Soni, N., & Madhusudhan, M. S. (2017). Computational modeling of protein assemblies. *Current Opinion in Structural Biology,* 44, 179–189. https://doi.org/10.1016/j.sbi.2017.04.006.

Stroganov, O. V., Novikov, F. N., Stroylov, V. S., Kulkov, V., & Chilov, G. G. (2008). Lead finder: An approach to improve accuracy of protein-ligand docking, binding energy estimation, and virtual screening. *Journal of Chemical Information and Modeling,* 48(12), 2371–2385. https://doi.org/10.1021/ci800166p.

Sulimov, V. B., Kutov, D. C., & Sulimov, A. V. (2019). Advances in docking. *Current Medicinal Chemistry,* 26(42), 7555–7580. https://doi.org/10.2174/0929867326566618 0904115000.

Takeuchi, K., Baskaran, K., & Arthanari, H. (2019). Structure determination using solution NMR: Is it worth the effort? *Journal of Magnetic Resonance,* 306, 195–201. https://doi.org/10.1016/j.jmr.2019.07.045.

Temelkovski, D., Kiss, T., Terstyanszky, G., & Greenwell, P. (2019). Extending molecular docking desktop applications with cloud computing support and analysis of results. *Future Generation Computer Systems,* 97, 814–824. https://doi.org/10.1016/j.future.2019.03.017.

Tian, W., Chen, C., Lei, X., Zhao, J., & Liang, J. (2018). CASTp 3.0: Computed atlas of surface topography of proteins. *Nucleic Acids Research,* 46(W), 363–367. https://doi.org/10.1093/nar/gky473.

Totrov, M., & Abagyan, R. (2008). Flexible ligand docking to multiple receptor conformations: A practical alternative. *Current Opinion in Structural Biology,* 18(2), 178–184. https://doi.org/10.1016/j.sbi.2008.01.004.

Trott, O., & Olson, A. J. (2010). Autodock vina: Improving the speed and accuracy of docking. *Journal of Computational Chemistry*, *31*(2), 455–461. https://doi.org/10.1002/jcc.21334.AutoDock.

Tsai, T. Y., Chang, K. W., & Chen, C. Y. C. (2011). IScreen: World's first cloud-computing web server for virtual screening and de novo drug design based on TCM database@Taiwan. *Journal of Computer-Aided Molecular Design*, *25*(6), 525–531. https://doi.org/10.1007/s10822-011-9438-9.

Tuszynska, I., Magnus, M., Jonak, K., Dawson, W., & Bujnicki, J. M. (2015). NPDock: A web server for protein-nucleic acid docking. *Nucleic Acids Research*, *43*(W), 425–430. https://doi.org/10.1093/nar/gkv493.

Ulrich, E. L., Akutsu, H., Doreleijers, J. F., Harano, Y., Ioannidis, Y. E., Lin, J., Livny, M., Mading, S., Maziuk, D., Miller, Z., Nakatani, E., Schulte, C. F., Tolmie, D. E., Kent Wenger, R., Yao, H., & Markley, J. L. (2008). BioMagResBank. *Nucleic Acids Research*, *36*(D1), 402–408. https://doi.org/10.1093/nar/gkm957.

Venkatachalam, C. M., Jiang, X., Oldfield, T., & Waldman, M. (2003). LigandFit: A novel method for the shape-directed rapid docking of ligands to protein active sites. *Journal of Molecular Graphics and Modelling*, *21*(4), 289–307. https://doi.org/10.1016/S1093-3263(02)00164-X.

Verdonk, M. L., Cole, J. C., Hartshorn, M. J., Murray, C. W., & Taylor, R. D. (2003). Improved protein–ligand docking using GOLD. *Proteins*, *52*(4), 609–623. https://doi.org/10.1002/prot.10465.

Voigt, J. H., Bienfait, B., Wang, S., & Nicklaus, M. C. (2001). Comparison of the NCI open database with seven large chemical structural databases. *Journal of Chemical Information and Computer Sciences*, *41*, 702–712. https://doi.org/10.1021/ci000150t.

Wang, J. C., Chu, P. Y., Chen, C. M., & Lin, J. H. (2012). idTarget: A web server for identifying protein targets of small chemical molecules with robust scoring functions and a divide-and-conquer docking approach. *Nucleic Acids Research*, *40*(W), 393–399. https://doi.org/10.1093/nar/gks496.

Wang, X., & Guan, Y. (2020). COVID-19 drug repurposing: A review of computational screening methods, clinical trials, and protein interaction assays. *Medicinal Research Reviews*, (July), 1–24. https://doi.org/10.1002/med.21728.

Wang, Z., Sun, H., Shen, C., Hu, X., Gao, J., Li, D., Cao, D., & Hou, T. (2020). Combined strategies in structure-based virtual screening. *Physical Chemistry Chemical Physics*, *22*(6), 3149–3159. https://doi.org/10.1039/c9cp06303j.

Wang, F., Wu, F. X., Li, C. Z., Jia, C. Y., Su, S. W., Hao, G. F., & Yang, G. F. (2019). ACID: A free tool for drug repurposing using consensus inverse docking strategy. *Journal of Cheminformatics*, *11*(73), 1–11. https://doi.org/10.1186/s13321-019-0394-z.

Wass, M. N., Kelley, L. A., & Sternberg, M. J. E. (2010). 3DLigandSite: Predicting ligand-binding sites using similar structures. *Nucleic Acids Research*, *38*(W), 469–473. https://doi.org/10.1093/nar/gkq406.

Wermuth, C. G. (2008). *The practice of medicinal chemistry* (3rd ed.). Academic Press.

Wishart, D. S., Feunang, Y. D., Guo, A. C., Lo, E. J., Marcu, A., Grant, J. R., Sajed, T., Johnson, D., Li, C., Sayeeda, Z., Assempour, N., Iynkkaran, I., Liu, Y., Maciejewski, A., Gale, N., Wilson, A., Chin, L., Cummings, R., Le, D., … (2018). DrugBank 5.0: A major update to the DrugBank database for 2018. *Nucleic Acids Research*, *46*(D1), 1074–1082. https://doi.org/10.1093/nar/gkx1037.

Wishart, D. S., Tzur, D., Knox, C., Eisner, R., Guo, A. C., Young, N., Cheng, D., Jewell, K., Arndt, D., Sawhney, S., Fung, C., Nikolai, L., Lewis, M., Coutouly, M. A., Forsythe, I., Tang, P., Shrivastava, S., Jeroncic, K., Stothard, P., … (2007). HMDB: The human metabolome database. *Nucleic Acids Research*, *35*(Suppl. 1), 521–526. https://doi.org/10.1093/nar/gkl923.

Workflow, V. screening. (2015). *Virtual screening workflow 2015-2, Glide version 6.6, LigPrep version 3.3, QikProp version 4.3*. New York, NY: Schrödinger, LLC.

Wouters, O. J., McKee, M., & Luyten, J. (2020). Estimated research and development investment needed to bring a new medicine to market, 2009–2018. *JAMA - Journal of the American Medical Association*, *323*(9), 844–853. https://doi.org/10.1001/jama.2020.1166.

Yadav, M., Dhagat, S., & Eswari, J. S. (2020). Emerging strategies on in silico drug development against COVID-19: Challenges and opportunities. *European Journal of Pharmaceutical Sciences*, *155*, 105522. https://doi.org/10.1016/j.ejps.2020.105522.

Yang, J., Roy, A., & Zhang, Y. (2013). Protein-ligand binding site recognition using complementary binding-specific substructure comparison and sequence profile alignment. *Bioinformatics*, *29*(20), 2588–2595. https://doi.org/10.1093/bioinformatics/btt447.

Yan, Y., & Huang, S. Y. (2019). Pushing the accuracy limit of shape complementarity for protein-protein docking. *BMC Bioinformatics*, *20*(25), 1–10. https://doi.org/10.1186/s12859-019-3270-y.

Yan, X., Liao, C., Liu, Z., Hagler, A. T., Gu, Q., & Xu, J. (2016). Chemical structure similarity search for ligand-based virtual screening: Methods and computational resources. *Current Drug Targets*, *17*(14), 1580–1585. https://doi.org/10.2174/1389450116666151102095555.

Yu, W., & Jr, A. D. M. (2017). Chapter 5 computer-aided drug design methods. *Methods in Molecular Biology*, *1520*, 85–106. https://doi.org/10.1007/978-1-4939-6634-9.

Zehra, Z., Luthra, M., Siddiqui, S. M., Shamsi, A., Gaur, N. A., & Islam, A. (2020). Corona virus versus existence of human on the earth: A computational and biophysical approach. *International Journal of Biological Macromolecules*, *161*, 271–281. https://doi.org/10.1016/j.ijbiomac.2020.06.007.

Zhang, W., Ji, L., Chen, Y., Tang, K., Wang, H., Zhu, R., Jia, W., Cao, Z., & Liu, Q. (2015). When drug discovery meets web search: Learning to Rank for ligand-based virtual screening. *Journal of Cheminformatics*, *7*(5), 1–13. https://doi.org/10.1186/s13321-015-0052-z.

Zhang, Q., Sanner, M., & Olson, A. J. (2009). Shape complementarity of protein-protein complexes at multiple resolutions. *Proteins*, *75*(2), 453–467. https://doi.org/10.1002/prot.22256.

CHAPTER 10

Do It Yourself—Dock It Yourself: General Concepts and Practical Considerations for Beginners to Start Molecular Ligand—Target Docking Simulations

THOMAS SCIOR

1 INTRODUCTION

In a stepwise manner, this user guide introduces molecular docking for beginners with pharmaceutical or medicinal chemistry background knowledge. The focus lies on practical aspects while some scientific aspects are described as and when required. The book chapter provides knowledge to practice simple molecular ligand–target docking simulations on a desktop computer.

Two seminal articles report about the docking software features and taken together propose a vast list of choices for docking and general methods with different type of hardware and operating systems (Pagadala et al., 2017; Pirhadi et al., 2016). Free to use programs, sometimes labeled as "open source software," can be recommended to gain practical experience with small organic molecule (SOM) docking into protein targets. To further understanding of the basics and the wider range of implications for docking as a computer tool, we present a short list of recommended literature which by no means could be exhaustive, it is just a sort of awareness training for the beginners to start with (Table 10.1).

2 SELECTING THE DOCKING PROGRAM AND ASSOCIATED TOOLS

It lies beyond the scope of this book chapter to provide an exhaustive synopsis about computational tools that are available for free academic or private use, for instance, HADDOCK (Dominguez et al., 2003), Chimera X (Goddard et al., 2018), or web-based SwissDock (Grosdidier et al., 2011), just to name three of them. Hence, a personal choice had to be made by combining popular programs for beginners to prepare molecular input data, run docking, and analyze the output. Four programs are proposed because they all can be installed together under the same operating system on a computer, namely MS-Windows. Together they carry out the tasks to input data, process docking, and analyze the output data: (1) **Autodock Vina (ADV)** is a popular docking program. (2) Input and output data of ADV can be generated and visualized by **Autodock Tools (ADT)** from the same academic site (AKA MGL-Tools). Two more programs can be helpful to prepare targets and ligands, namely proteins and small organic compounds: (3) a protein-specific tool box named **Swiss PDB Viewer (SPDBV)** and (4) **Vega ZZ**, which is a versatile molecular modeling program.

Indeed, **ADV** is a popular offline tool. So, it has to be downloaded first from the official web page. It can be installed on any desktop or laptop computer. Versions for three operating systems exist: MS-Windows, Linux, or MAC OS X. MGL tools—**ADT**—offer the possibility for ADV users to manipulate and visualize input and output data. ADV and ADT were developed at the same academic group where Autodock4 (**AD4**) was created, what may explain the word creation of **VINA** as an acronym for "Vina Is Not Autodock." Other authors cite it as VINA. Here, we use the abbreviation ADV, next to ADT. Around 2008, ADV was introduced by its program developer Oleg Trott to simplify

Molecular Docking for Computer-Aided Drug Design. https://doi.org/10.1016/B978-0-12-822312-3.00003-5

TABLE 10.1

Recommended Reading to Further the Understanding About Ligand Docking to Target Proteins.

Topic	Recommendation	Reference
General overview—Introduction	In a way, the history of structure-based virtual screening (SBVS) is summarized with succinct descriptions of early developments of docking tools, different search and scoring methods, homology models, data sources, ROC diagrams, etc. Consensus virtual screening (CVS) is a response to deal with the fact that different software often do not yield identical results.	Maia, Assis, et al. (2020)
Classes or types of docking techniques	Synopsis of different ligand–target molecular docking approaches and their applicability range, i.e., different program types, algorithms, and scoring functions. Introduction to benchmarking, i.e., comparing performance and success of software (hit rates). Rigid body versus protein flexibility is discussed.	Meng et al. (2011)
Docking workflow	Unattended by the researcher, an SBVS is performed by ADV or DOCK 6 or CVS. The automatization comprises three-dimensional (3D) model preparations of targets, homology modeling of unavailable 3D structures, protonation states, etc.	Maia, Medaglia, et al. (2020)
General workflow and binding site (BS) detection	An alternative scheme is presented for unsupervised ligand and target protein preparations for docking input. An automatic BS identification is also proposed.	Zhang et al. (2014)
Benchmarking	Testing the performance and verifying the outcome of docking simulations depend also on the data source. Concepts and their challenges are discussed, e.g., enrichment factor optimization, VSC, or the directory of useful decoys datasets (DUD).	Pedretti et al. (2019)
Validation of docked poses	Because different software differ in the final ligand poses, concepts and tools (Clusterizer and DockAccessor) for validation are introduced: re-/back-/self-docking to verify if the crystal binding pose (from PDB entry) can be reproduced correctly, docking accuracy, best-docked, best-cluster, or best-fit poses.	Ballante and Marshall (2016)

TABLE 10.1
Recommended Reading to Further the Understanding About Ligand Docking to Target Proteins.—cont'd

Topic	Recommendation	Reference
Consensus docking and CVS	Various docking methods can be combined aiming at better success or hit rates. They are called consensus docking or CVS. The combination is as reliable as the best available methods from literature. Comment: The hit rate of CVS objectively depends on the hand-selected combination of case-suited docking software and the input data (the studied cases, here: DUD).	Poli et al. (2016)
Rigid versus flexible geometry	A simplified docking simulation accounts for the conformational flexibility of the ligand, but not of the biomolecular target due to the complicated concert of side chain and backbone flexibilities and computational challenges thereof. In molecular dynamics (MD) studies, conformational changes (geometries) over time are simulated (torsion angles with rotatable bonds). A workflow is proposed for automated docking with flexible targets; also certain conformations can be selected in the MD play back (snapshots of MD frames).	Machado et al. (2011)
Unusual binding	Due to sigma hole interactions of halogenated ligands with the target (especially -F or $-CF_3$), predicting the noncovalent halogen bond correctly requires either QM calculations or special adaptations to force field under molecular mechanics.	Costa et al. (2019)
Structure–activity relationships	The "credo" or paradigm: Similar structures will behave in a similar—and therefore predictable—way for SBVS or QSAR. The underlying problem is how similar is similar, actually in chemistry?	Bender and Glen (2004), Scior et al. (2012, 2009)
Docking problems	Presentation of general aspects about underpinnings, shortcomings, and implications which are independent from the specific docking program.	Forli (2015)

docking, claiming it be even faster and more accurate than its predecessor AD4. Hence, the four selected programs are combined for docking:

1. ADV (Trott & Olson, 2010) (http://vina.scripps.edu/download.html)
2. ADT (Morris et al., 2009) (http://autodock.scripps.edu/resources/adt)
3. SPDBV (Guex & Peitsch, 1997)
4. Vega ZZ (Pedretti et al., 2020) (http://ww.ddl.unimi.it/VegaZZ).

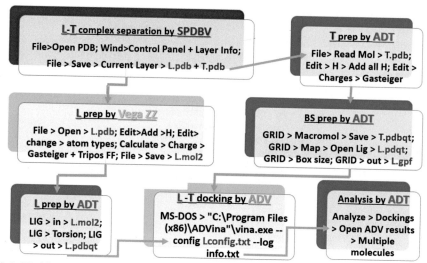

FIG. 10.1 Workflow chart for docking. With Swiss PDB Viewer (SPDBV), a ligand—target complex (L—T) is read in (top left box). And after separating L from T, both follow different preparation steps by Vega ZZ, SPDBV and ADT following the light *blue arrows* from top left to bottom right. The fundamental steps are indicated by the corresponding commands from the respective main menus. All steps are explained in details in the text (mainly Sections 5—8). The four background colors reflect the four programs. *BS*, binding site; *FF*, force field; *H*, hydrogen atoms; *L*, ligand; *T*, target molecule. Of note, the MS-DOS path to *vina.exe* is an example.

It constitutes a general molecular modeling program with options of reading and saving three-dimensional (3D) models of molecules in many file formats.

The following two sections explain how to install the selected four programs and integrate them into a virtual docking laboratory on the computer. A graphical work-flow chart displays the input and output data flow (Fig. 10.1). It is explained in details in Sections 5—8 and summarized in Section 10.

3 INSTALLING AND LAUNCHING ADV, ADT, SPDBV, AND VEGA ZZ

3.1 Installing the Four Programs on the Computer

The simplest way to use a combination of molecular modeling and docking programs is to install them together on a computer under the same operating system, here under MS-Windows. After downloading the installation for ADT and Vega ZZ, proceed by simply mouse-clicking on the setup executable file (file extension ".exe") with the exception of SPDBV and ADV. SPDBV is sent as a compressed zip file which can be opened with 7-zip file manager, WinRAR archive, or pk-unzip, among many other free tools. ADV is packed into a file with a "dot" msi extension. MSI stands for "MicroSoft-Installer" and is furnished by the program distributor. Like the other two files with a "dot" exe

extension for MS-Windows installation, MSI automatically handles the procedure to create a program command executable file on the computer, here vina.exe. The end user may accept the proposed location (**directory path**) for the *vina.exe* file which constitutes the command to run a docking simulation later (Fig. 10.1).

Under MS-Windows, ADV and ADT are both installed in a system folder called "Program Files (*x86*)" to reflect that they were designed to run-in 32-Bit mode.

3.2 Starting the Four Programs

With **ADT**, **ADV**, **SPDBV**, and **Vega ZZ** properly installed on a computer running under the operating system MS-Windows, many options to run ADV exist because a variety of molecular modeling packages offer special user interfaces to run ADV.

It is a trivial task to start **SPDBV** or **Vega ZZ** by a mouse click on their respective icons which appear on the desktop after installation. Alternatively, both can be launched through the main menu from MS-Windows under "all programs" at the bottom-left corner of the computer screen. This main menu of the MS-Windows operating system can be opened by a direct mouse click or after pressing the special Windows key. In its search text field for programs or files, the program name is typed in.

Starting ADT requires the **command line inter-preter**. It can be evoked in the search text field from aforementioned main menu window with the word "cmd." This seemingly old-fashioned software interface requires the user to get familiar with a few commands (*cd, dir, help, help cd, help dir*). Internet pages provide more details for beginners who wish to understand the complete syntax of MS-DOS scripting language (MiscroSoft Disk Operating System). Since ADV was evoked from the location where all the model data are stored on disk, all input and output data (streams) are now directed to this current working directory.

To avoid errors, the end user opens a file explorer (browser) window and clicks into its file location text field to generate a complete path in MS-DOS language syntax (Fig. 10.2). Now, the MS-DOS path can be copied and pasted into the black MS-DOS window (Fig. 10.3). Copy and paste is a common operation for MS-Windows operating systems by pressing the "cntrl" key with "c," followed by the combination "cntrl" with "v." Under MS-DOS, the copy part is the same and expects again the user to hold down the key "cntrl" while pressing the "c" key on the keyboard.

However, the second part is different. Here, for pasting, the user must click on the rightmost mouse bottom (see Subsection 3.3 for more details).

3.3 Starting ADT and ADV for Newcomers to MS-DOS

Newcomers to MS-DOS are advised to read the following segment, and all other users can skip it to proceed to the next Section 4.

After localizing the working directory with the docking data by an MS-Windows file explorer, the complete file path can be typed in after the command word *cd*. In Fig. 10.3 the example shows a complete path: *cd C:\20_Data_work_HPlapCasa4covid\   20_CADD\20_-CADD_Projects\20_dock_pitfalls_ElsevierBookChap*. The first part of "*C:\User\tscior>*" symbolizes the physical disk on the computer (*C:*), and the backslash (...\...\...) separates directories and subdirectories and is ended by the prompt symbol (>) where the user types in commands such as cd or dir.

The MS-DOS window is a nongraphical (rudimentary) window and is required to launch ADT and ADV when used as a standalone program. Instead of typing

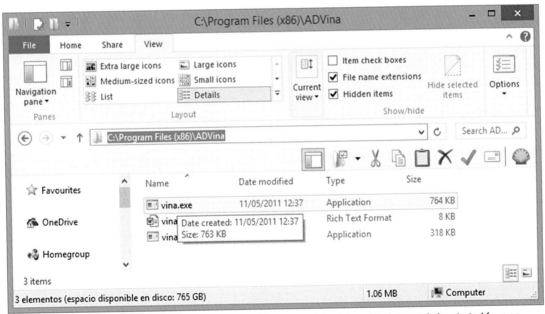

FIG. 10.2 Screenshot of a typical MS-Windows file explorer. Standalone VINA can only be started from an MS-DOS window, so user must know the location where it is installed (MS-DOS path). Upon clicking into the text field which indicates the directories (and subdirectories or folders) of the file explorer, a complete directory (or folder) path is generated in MS-DOS command language. It is highlighted in blue background color in this text field (which is preceded by a *black upward arrow*). The option to toggle on and off the "File name extensions" to all files is located in the upper right-hand corner. The highlighted text field displays the path, here for *C:\"Program Files (x86)"\ADVina*. This is the path to the executable file for VINA (*vina.exe*).

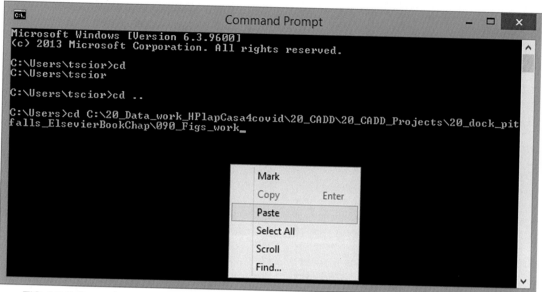

FIG. 10.3 Screenshot of the MS-DOS window. It gives access to the command line interpreter for MS-DOS language. Here, the current working directory was set to *C:\User\tscior* followed by the prompt symbol >. After the prompt symbol (>), the user types in commands; here, *cd* to change directory. Here a very long path was not typed in but pasted in.

in long path instructions, it is much easier to open the Microsoft file explorer window in the folder where the docking study data files are kept. By a mouse click into the text field of the file explorer window, the complete path is displayed in MS-DOS scripting language. For instance, the MS-Windows label *"This PC: Windows (C:) 20_MyData 20_CADD"* expands to *"C:\20_MyData\20_CADD,"* which is the path in MS-DOS language. Such MS-DOS notations for full paths are needed to access the docking data and to start the ADT batch file (adt.bat) as well as the ADV launch file (*vina.exe*).

In the file explorer window of MS-Windows, the highlighted path appears in blue and is copied by a (double) key stroke "cntrl" with "c." Unluckily, the key combination "cntrl" with "v" for the commonly used paste operation under MS-Windows does not work at all under MS-DOS. This is so (complicated) because MS-DOS was invented prior to the MS-Windows operating systems with graphical user interface (GUI). Of note, GUI is another term to say that the software are mouse-clickable window programs which run under Microsoft Windows, Linux, or Mac OS. When the mouse pointer is located on said MS-DOS window, the so-called context menu, which is a click on the right bottom of the computer mouse (or laptop touch pad), yields the wanted paste operation. Without the need of error-prone typing, the complete

path is known to the command line interpreter. The computer user clicks the return or enter key. If she or he has forgotten to put the change directory command *cd* in front of the path, then the path is recalled by striking the "upward arrow" key, which is one of the four keys with the arrow symbols next to the number pad on the bottom right side of many keyboard. The docking study data are normally stored on a hard disk drive (HDD, also referred to as "disk" under the MS-DOS name C:) or any existing subdivision of it called partition (e.g., D: or E:, etc.). To access the data under MS-DOS, it is necessary to change to that location on the disk or partition (e.g., C: or D: or E:, etc.). After the MS-DOS prompt (C:|>), the letter (C or D or E, etc.) followed by a colon (:) has to be typed in (D: or E:, etc.). With the command *dir*, all files in this current working directory or folder can be listed. For users who are unfamiliar with MS-DOS commands, the word *help* will list common **MS-DOS commands**. Moreover, *help cd* informs about the *cd* command syntax, i.e., how to use it properly.

One MS-DOS window is needed for ADT and another one for ADV (as a standalone command, albeit it could also be invoked through special program interfaces). Nevertheless, it is a good idea to repeat the steps to open more than two MS-DOS windows in case a window is accidently closed or the program stops. In the current working directory, the instruction to launch ADT can be copied from a protocol file of the CADD folder (see Section 4.1) into the MS-DOS prompt to launch ADT, for instance, here, *C:\"Program Files (x86)\MGLTools-1.5.6"\adt.bat*. Here, the quotation marks are part of the command language (syntax). They are necessary to indicate that spaces do form part of a command argument (one word includes spaces) because spaces separate the arguments of commands— just like spaces separate words in normal text.

At this stage the *vina.exe* command with its full path can be tested in the second MS-DOS prompt window, for instance, *"C:\Program Files (x86)\ADVina"\vina.exe*. The given path is an example and the valid path can be seen in the MS-Windows main menu (see above). As both MS-DOS prompt windows had been set to the current working directory, ADT readily finds the input data files and stores the output data files in the same place.

4 CREATING THE VIRTUAL DOCKING LABORATORY

4.1 Creating a Protocol File for the ADT and ADV Start Procedure

Prior to set up and run ADV, however, it is best to create a small document to always remember the unusual procedure to launch ADT or ADV. Said memo file could read the following:

1. Open the default command line interpreter for MS-DOS via the MS-Windows main menu (bottom-left corner of the MS-Windows screen) and type *cmd* into the search text field.

2. An MS-DOS window opens with a black background (Fig. 10.3).

3. In this new window, it is necessary to type in the complete path toward the current working directory with the data files. For instance, the following spaceless combination D:\ (letter D Colon Backslash) is the first command followed by an "enter" or "return" key stroke. It is followed by a second command with something like *cd D:\"my documents"\firstleveldirectory\subdirectory*.

4. Finally, ADV can be started by typing in the path to its executable file, e.g., *C:\"Program Files (x86)\ADVina"\vina.exe*.

The example here encloses the expression with blank spaces by "…" to tell the command interpreter that this is "one word only," say one location in the directory path. Conversely, *vina.exe* does not need to be put into quotation marks (AKA double quotes).

4.2 Organizing the Virtual Docking Laboratory on the Computer Desktop

The protocol file with the documentation of the MS-DOS commands and path information to launch ADV could be linked to a folder on the desktop (Fig. 10.4). In this example, the folder is called CADD as acronym for "computer-aided drug design." This virtual molecular modeling and docking laboratory can be created as a simple desktop folder after opening the context menu by clicking the right bottom of mouse or touch pad. Into this desktop folder CADD, the user can drag and drop the icons of all modeling applications and related tools. The user can also create the links if icons are missing for all other docking or general-purpose molecular modeling software, tutorials, help manuals (PDF files), and protocol files with information about starting non-GUI applications, in addition to the www addresses for web-based tools from the Internet. In this context, the beginner is advised not to create or store user data-containing files in any desktop folder, only their links to the physical information on the HDD of the computer, as desktop items are not necessarily saved in backup operations (on a routine basis).

Concerning **Vega ZZ**, the user must register with an email account which has to be accessed probably by several persons over time. Vega ZZ expects neither hardware nor email to be changed for renewing monthly licenses. A work around is to create a live system on CD or USB stick. During the installation process, the option for Live CD creations must have been activated; otherwise no **Vega ZZ Life-system** can be created after Vega ZZ installation on the computer. A direct link to the executable on the CD (e.g., a DVD/CD drive labeled "E:" the path has been set to "E:\Vega ZZ\Bin\Win32\-VegaZZ.exe") can be stored into a folder and listed in the virtual laboratory, i.e., the CADD folder on the desktop. To avoid problems among changing users on a given computer, a simple text file could inform about the required email to be shared with all work team members (cf. file VegaZZ_licence_run.txt **in** Fig. 10.4). A shortcut to Vega ZZ Live system could also lead directly to the executable *VegaZZ.exe* file, provided the "Live CD" had been put into the CD-ROM/DVD drive (see more details below in Subsection 4.4).

4.3 Using Molecular File Formats in the Virtual Docking Laboratory

At all times it is extremely useful to recognize the file extensions of all input and output data saved on the storage device(s) of the computer—in particular, those files which hold structural information of molecules, mostly

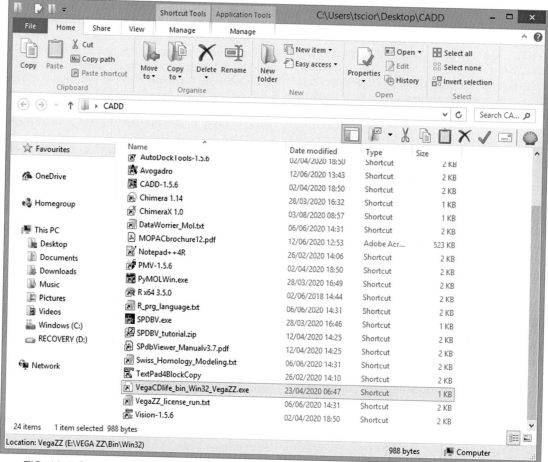

FIG. 10.4 Screenshot of a virtual docking laboratory. The desktop folder collects the links (shortcuts) to the corresponding files on the computer. It provides an overview of the installed molecular modeling programs, help manuals or tutorials, and special editors (Textpad, Notepad++). Links to descriptions on how to start a complicated tool or scripts to run non−Windows-based applications can also be stored in alphabetic order. Here, an EXE file is highlighted. Below it a TXT file reports how to renew Vega ZZ license on a monthly basis or launch a certain (perpetual) Vega ZZ version as a "Life CD" without need to renew the license.

3D models or molecular property descriptions. They are of utmost importance to properly connect a larger number of modeling tools or programs into a data stream. Each program can be seen as an independent unit of input/processing/output (IPO) operations (IPO → IPO → IPO → IPO, etc.) according to the workflow of the docking study (Fig. 10.1).

Under MS-Windows, the modeler opens the file explorer (tool to browse between directories or folders and files) and looks for the VIEW option in its top-level menu or icon (pictogram or symbol) which is somewhere in the upper part or at the bottom of the explorer window. Here, the user activates the display of

file extensions: VIEW > File name extensions ON/OFF (Fig. 10.2).

Not all programs can read or write the same file formats, and if an output format in one step is not readable, the following step—a so-called <u>file format conversion</u>—will be necessary to bridge the missing data format compatibility at this particular stage in the workflow of the study (IPO → IPO → conversion tool → IPO, etc.).

To achieve this goal, **special file conversion tools** have been created. Offline **BABEL** was developed more than a decade ago to bridge the data flow between output and input between CADD software. A modern solution has been developed with its online counterpart **Open-BABEL** (visit at https://openbabel.org). Both have become very popular in the modeling community. Here, the general modeling package **Vega ZZ** (visit at https://www.ddl.unimi.it/) with its huge capacity of reading and writing tens of formats can be useful in this context. In more general terms, file extensions reflect the file formats which are associated to certain modeling programs. They constitute the terminal part of the file name separated by (last occurrence of) a dot or period and commonly followed by a three- or four-letter code; for instance, the rather universal format for biomolecules is "dot pdb" (**Protein Data Bank** at http://www.rcsb.org/). Although this **PDB format** is commonly used to store 3D models of proteins, RNA, or DNA and their combinations, e.g., a polymerase with DNA, it can also save SOMs, especially the ligands for target protein docking. Ligands and their targets can also be stored in other formats, such as **MOL** (from program of MDL) or **MOL2/ML2** (from program of Tripos) or **PDBQT** (from program ADT). Of note, PDBQT is based on the PDB format and enriched with additional information about partial atomic charges (Q) and the assignment of rotatable bonds around dihedral angles or torsions (T) to produce conformational flexibility (of the ligand).

A useful program to add H atoms manually and save them in many file formats is **Vega ZZ** as it can read in and out an impressively large(r) set of format types for 3D models (than it is seen with other programs in the field). It is noteworthy to mention that MOL2 as well as PDBQT formats can contain more than one molecule, one after the other, in numbers up to the tens or even sheer numbers to the hundreds or thousands for virtual screening (Scior et al., 2012). Nevertheless, they are readily distinguishable. In PDBQT, multiple compounds are clearly separated by keywords, e.g., the molecular model 341 ends by "ENDMDL," while "MODEL 342" marks the beginning of the following model 342. Mostly, all (free) model formats are not binary files, i.e., they are readable and editable (so-called ASCII) files.

4.4 Combining Swiss PDB Viewer and Vega ZZ

SPDBV is a specialized tool for proteins. This is because its selection levels (or hierarchy) can handle proteins on a chain or residue level: chains are labeled by letters such as A, B, C, etc., while amino acids or other residues have a three-letter code, all of which can be inspected by clicking through SPDBV menu > WIND > CONTROL PANEL. But it does not descend to the atomic level in general (only in menu options for mending PDB input files with missing data). Hence, ligands and other compounds are only treated as residues through SPDBV

menu > WIND > LAYERS INFO. As a direct consequence of these built-in molecular concepts, no general manipulation at the atomic scale was foreseen by the program developers. As proteins, DNA, and RNA are biopolymers, they are composed of a priori known building blocks. These built-in monomers do not need to show their hydrogen atoms, but they can be shown by ADT or Vega ZZ without ambiguity. Not so, SOMs, where a five- or six-membered ring could be aromatic or not, etc.

In contrast, atom (level) manipulations can be carried out by **Vega ZZ** (main menu > EDIT) and atom manipulation options in addition to two icons ("ADD ATOMS" and "ADD FRAGMENTS") which both also allow removing atoms or fragments. Vega ZZ even has options for electron treatments by quantum chemical tools (main menu > CALCULATE). It has options not only to deal with SOMs but also large biomolecules in a user-friendly **bioinformatic tool box** (main menu > BIOINFORMATICS).

In a formal way, speaking of applicability range of software, the following short notion reflects the proposed combination here:

$$SPDBV + Vega\ ZZ = target + ligand.$$

Vega ZZ is recommended to generate 3D ligand models of drugs or SOMs in general because it can become a daunting task to create 3D models of ligand molecules with unusual or extreme geometries and bonding, such as antituberculotic isoniazid (Scior et al., 2002) or antiprotozoal nitazoxanide (Scior et al., 2015). Under ADT, hydrogens and charges can be added to the ligands to prepare them for input, but chemically complicated molecules are better treated under Vega ZZ as outlined in Section 6. In addition, Vega ZZ also has options to manipulate atom names, numbers or types, bonds, and rotate bonds (set torsion angles) under its main menu > EDIT > ADD/REMOVE/BUILD or CHANGE. Similar small molecules can be superpositioned, i.e., fitted or aligned in space under Vega ZZ main menu > EDIT > SIMILARITY. But proteins which are related by a common ancestral structure or function should be 2D or 3D aligned by clicking on SPDBV main menu > FIT > ALIGN or MAGIC FIT.

5 PREPARING THE INPUT STRUCTURES OF THE TARGET PROTEINS

5.1 Downloading the 3D Models of the Target Proteins

After installing the programs and linking them into the virtual laboratory (Fig. 10.4), the protein to be studied has to enter the SPDBV canvas (the black molecular working area) in two ways. Either it is retrieved from

FIG. 10.5 Display of the main menu of SPDBV. It is organized in one row of command options and another row of icons. The user mouse-clicks on the first option "FILE" to import a protein structure from PDB or open a PDB file from the computer. A large black canvas with the molecule will start automatically. All options are started by mouse clicks whose navigation symbol will be in this book chapter ">," e.g., SPDBV main menu > FILE > IMPORT; or SPDBV main menu > FILE > OPEN PDB FILE.

the PDB data bank (https://www.rcsb.org/) and saved to disk or it is read from local sources (HDD, CD, USB stick) on the computer (Fig. 10.5). The target is either a liganded complex (ligand–target complex) or an unliganded protein, i.e., without a bound ligand. SPDBV appears under a so-called "Graphical User Interface" (GUI for short). This means it is MS-Windows–operated and has a message for the user that the SPDBV canvas is in a mode to "Move All" molecules (text in red), not just the visible ones but also the invisible molecules (Fig. 10.5). The CONTROL PANEL under WIND enables showing and hiding of residues, chains, and entire proteins. Upon clicking on the Earth icon (third row), the red message to the right changes from "Move All" to "Move Selection." Now a residue, chain, or molecule which was selected by WIND -> CONTROL PANEL can be moved around manually in the canvas which is an important feature for Section 7 where a ligand is manually placed ("docked") near the binding site (BS).

For 3D models of target proteins on the computer, the practitioner selects the directory path to the folder with the PDB entry; for example, on an MS-Windows operating system, this would be something like $C:\SPDBV\_4.10\_PC\1CRN.pdb$. Here, 1CRN.pdb is the PDB code of a sample structure which is located in the same directory (folder) where the SPDBV executable file is located. Both *1CRN.pdb* and *SPDBV.exe* can be found by the MS-Windows file explorer.

One out of many sources for original PDF files of proteins is the **RCSB Protein Data Bank** at https://www.rcsb.org/ (Berman et al., 2000). Upon displaying the 3D model, saving a PDB file with a general molecular modeling program often implies that it is not any more stored in a completely identical format to the original PDB file from RCSB. Probably it is still compatible with a given program, but remembering this fact might explain unexpected behavior in other programs. Incompatibilities may also arise after file format conventions of molecular models for file transfer between programs, even for the very same file extension, as just mentioned

for PDB. The more simplistic the program (so-called parser code) was written, the higher the likelihood that such incompatibility problems become relevant. An original PDB file from PDB server (or 1CRN.pdb included in SPDBV) and its versions saved to disk by SPDBV or Vega ZZ can be opened with a text editor to see such format differences.

5.2 Inspecting the Target 3D Models Under SPDBV

SPDBV offers an option to isolate a particular molecule from all others and execute operations on it without affecting the others. This is useful to manually dock a ligand into a cavity, save to disk the ligand with its new position for comparison, superposition, or as a new start position for docking itself. To get assistance in this context, SPDBV offers online help (at https://spdbv.vital-it.ch/moving_guide.html)—but not in the (offline) PDF manual called "DeepView: The Swiss-PdbViewer User Guide version 3.7."

But first, the beginner has to learn about some pivotal menu operations. Hence, we will have a closer look at the SPDBV main menu > WIND > CONTROL PANEL. Top down the window tells "Control Panel" in blue and then follows a bar (say a line) stating the molecule or PDB entry's file name in gray color. That now is extremely tricky, not even experienced computer scientist would guess that the user must click on this bar, and all other molecules will now be listed in a new pull down menu. Those other molecules had been called into the canvas by mouse clicking along SPDBV main menu > FILE > OPEN PDB FILE or by FILE > OPEN MOL FILE (or SDF). The modeler can toggle between the molecules to list their residues in the CONTROL PANEL (Fig. 10.6). For ligands, no amino acids are listed, just unusual residue names for non–amino acid monomers.

To reproduce Fig. 10.6, the following steps and display settings are necessary: (1) the main menu option "DISPLAY" was set to "Render in solid 3D"; (2) in the main menu option "WIND," three additional

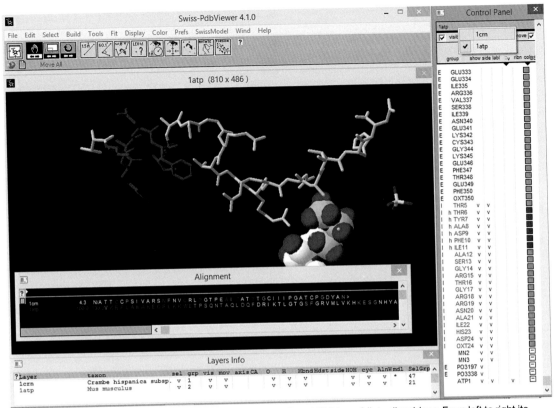

FIG. 10.6 Display of the SPDBV in action. The Control Panel (rightmost) lists all residues. From left to right its columns host the chains (here A); the secondary structure (here h and s for helices and strands or beta sheets); "+ - **show**" to display or undisplay the residues (hide); "+ - **side**" to show or hide the side chains of amino acids; "+ - **labl**" to show or hide the residue labels (names and numbers to identify the residues); "+ - **V**" to contour the atom with its dotted volume (useful to localize the residue, e.g., the ligand); "+ - **rib**" to show or hide the backbone line or ribbon (connecting peptide bonds of the amino acid chain); and finally the color option for each residue (useful to localize the residue, e.g., the ligand).

windows "CONTROL PANEL," "ALIGNMENT," and "LAYER INFO" were opened; (3) in the main menu option "COLOR" by "ALIGNMENT DIVERSITY" was activated for the multicolored 1CRN sequence in the "Alignment" window; (4) after selecting Chain I in the "CONTROL PANEL" for display, the "SECONDARY STRUCTURE" was activated in the main menu option "COLOR," which takes effect in all three windows (control panel, canvas, and sequence) with a simple color code: gray for neither helices (red) nor beta strands (yellow); (5) the ligand (ATP1 in 1ATP) is the last residue entry in the listing of "CONTROL PANEL" and displayed in atom colors ("CPK" option under "Color") with its volume activated in the "CONTROL PANEL." Of note, 1CRN and 1ATP are the PDB codes of two proteins from SPDBV (here, *C:\SPDBV_4.10_PC\download*).

With the **3D protein model** in the canvas of SPDBV, its overall structure with domain folds, chains, and individual amino acids of interest—especially those at the BS with all molecular components around the protein—can be inspected and identified by mouse clicking through SPDBV main menu > WIND > CONTROL PANEL (Fig. 10.6). The residues are listed, chainwise, e.g., chain A, or a second one B or even C, and so forth. First, the amino acids are listed for each chain with a sequence number, e.g., "A ILE16" stands for isoleucine number 16 in chain A. At the bottom of the list, the **non—amino acid components** associated to the protein are recorded: counterions, crystal water moieties, enzyme cofactors, heme groups, cocrystallized drugs or ligands can be found here. Their names are remarkably different from the three-letter codes of the 20 **standard amino acids**, which are also called residues or monomers of a biopolymeric molecule (or polymeric biomolecule such as proteins or DNA). In the "CONTROL PANEL," the user clicks on the first "A" label in the first column, which designates the chains with A, B, C, and so forth. The entire "A" labels of the first chain A will get highlighted in red. After clicking on the "-" sign in the (third) SHOW column, the molecule is undisplayed in the canvas; this means it is not removed, it just cannot be seen. At that stage, to unselect the chain "A" in red, the user clicks again but this time in combination with the SHIFT key of the keyboard. The red color of label "A" will turn to black. To continue the inspection tour, the practitioner scrolls down to the bottom of the list (see rightmost vertical window in Fig. 10.6), where the non−amino acid components (or residues) are listed. Next, the user can select all non−amino acid components by clicking on those GROUP labels in combination with the SHIFT key. Then the user clicks SHOW "+." Now, it is possible to identify the ligand for docking and with it all unnecessary moieties to be deleted (cf. residue entries at the bottom of the rightmost vertical window in Fig. 10.6). If the latter are lying far away from the ligand which occupies the BS, they can be left because they do not interfere with docking at the BS. A safe separation distance yields approximately twice the largest possible diameter of that ligand. To delete items, select their respective GROUP labels again in the CONTROL PANEL to highlight them in red, followed by SPDBV main menu > BUILD > REMOVE SELECTED RESIDUES. Of note, the beginner has to get used to naming inconsistencies between software and also within the same program. Here, in the delete command REMOVE, the object is called RESIDUE, while in the CONTROL PANEL, it is called a GROUP, and so forth, all of which make maneuvering sometimes a confusing endeavor for novices because they just do not understand the inconsistencies in terminology and naming conventions.

At the end of this operation, the ligand was successfully identified and isolated by eliminating all other groups. Finally, it was saved to disk: SPDBV main menu > FILE > SAVE > CURRENT LAYER by typing in a new file name. The molecule can only be saved in PDB format, and no other format is available. If the molecule is a small ligand, its PDB file can be read into **Vega ZZ** to generate a 3D model and save it in a MOL2 file with all atom and bond types including all hydrogen atoms correctly added.

5.3 Separating Ligand and Target Molecules Under SPDBV

For docking with ADV, two input files in PDBQT format are needed, one with the target molecule and the other with the ligand model. If both molecules are united in a ligand−target complex, they have to be separated before they can be redocked (back docking) under ADV. They can be extracted from their original complex structures under SPDBV. The ligand can even be moved away a few Angstroms (Å) and rotated a little bit from its true (observed and correct) binding position. As a direct result, it stays in a new position at the BS and is saved to disk as the new start position for ADV. A docking study could be biased when the docking program is instructed to test the start position (of the input data file) as a possible good solution in the correct assumption that the ligand's input data (Cartesian coordinates) might be already the ligand's true position from a crystal complex with its target (PDB entry). Conversely, when the ligand cannot be found in complex with the target protein, albeit it was built from scratch or published as the ligand of another crystallized protein complex, then the created or extracted ligand must be placed in proximity to the BS of the target. Only specialized programs are designed to discover BSs of a target protein (Table 10.1). With the same idea of unbiased docking, certain programs only start with random ligand positions or explicitly offer a menu option for random start positions (e.g., Autodock4).

This separation step will be carried out under SPDBV. To this end, the modeler clicks SPDBV > WIND > CONTROL PANEL to select the residue of the ligand at the bottom of the residue list, and its label becomes highlighted in red. Now, it follows the mouse operation SPDBV main menu > BUILD > REMOVE SELECTED RESIDUES. Upon removing the ligand, the unliganded target model can be saved as T.pdb file. The same procedure is used to create the ligand file without its target molecule.

5.4 Cleaning the 3D Models of Target Proteins

The crystal water molecules in the binding cavity have to be removed prior to start a docking simulation by ADV to guarantee a complete search and sampling of ligand conformations in space—unless a water molecule has been reported to contribute to the binding mechanism. To remove the ligand and any unnecessary moiety (e.g., a phosphate ion in the canvas of Fig. 10.6) from the input data, either the file can be opened and edited by a simple text editor or removing can be performed under visual control in the canvas under SPDBV (rightmost vertical window in Fig. 10.6). The new version of the protein can be kept in the canvas, but it is better to save it for later reuse, as it constitutes the cleaned, unliganded protein, sometimes called the "**apo-form.**" Sometimes, components are not impurities or cofactors, but they had been intentionally added during **crystallogenesis**. In trial and error experiments to grow the crystals, chemicals were added such as counterions, metal ions, glycine, or other organic compounds. When more than one ligand is present in the crystal, it could mean that the ligand had poor affinity and the ligand had to be added in larger quantities to "soak" the crystal. To know the biologically meaningful ligand, the publication of this PDB entry has to be consulted.

All biomolecular target models need to be prepared for input. The basic Arg, Lys, and His have cationic side chains, while the acidic ASP and GLU have anionic terminal groups under normal conditions in solution. Histidine bears two nitrogen atoms in its side chain ring. Hence, two potential protonation sites exist, called H1 (the distal site) and H3 (in ortho position); their occurrence is approximately 4:1 (H1:H3). The protonation site on histidine can be readily changed in ADT main menu > EDIT > HYDROGENS > EDIT HISTIDINE HYDROGENS. Also, missing atoms or residues in the crystal structures (PDB files) should be mended in the main menu > EDIT > MISC(ellaneous) > CHECK FOR MISSING ATOMS. If there are some atoms which could not be located in the X-ray image, then they can be modeled with > REPAIR MISSING ATOMS. Entire residues (not atoms) can be easily fixed by SPDBV by clicking through its main menu for five options: (1) > BUILD > BREAK BACKBONE; (2) > BUILD > LIGATE BACKBONE; (3) > BUILD > ADD RESIDUE; (4) > BUILD > REMOVE BOND; or (5) > BUILD > REMOVE SELECTED RESIDUES. Also, point mutations can be generated to build mutant types from wild types or vice versa. This task changes an existing amino acid against another one at a certain position, e.g., A24C (Ala24Cys) means alanine in position 24 has been changed to cysteine. Under certain conditions (lipophilic, water-free pockets), the side chain -COOH groups will not act as acids, i.e., some Asp or Glu side chains will remain neutral. The same applies to basic amine groups of Arg, Lys, or His. In a typical **hydrophobic pocket**, bulk water can be excluded. Hence, a polar solvent to assist dissociation is missing. In nonpolar organic solvent, it is about 10^{15} times as

difficult for an organic acid to dissociate into a carboxylate anion. In more general terms, for proteins, the ionization of acidic or basic head groups on the side chains of proteins or on the ligand itself greatly depends on the surrounding water solvent or nonpolarity. In addition, aforementioned residues may belong to the ligand's BS or play a role in the molecular mechanism of binding. Their hydrogen bond donor faculty could be critical for the protein–ligand interactions. The H atoms of the weak basic amino acid histidine can be rearranged under ADT by clicking along its main menu > EDIT > HYDROGENS > HISTIDINE.

Certain reports (Table 10.1) propose unsupervised (fully automated) preparations of the biomolecules, mostly proteins either as scripts or program implementation. Under SPDBV, much of this can be achieved by the users themselves.

The model quality can be inspected by coloring the **B-factors** or displaying the allowed and unfavorable (unstable and therefore less likely torsion angles) in Ramachandran plots by SPDBV. The B-factor ("B" may hint at "to blur"), sometimes called temperature factor, constitutes a descriptor for thermal displacement of electron densities (of atoms or groups). It reflects which segments of the sequence moved more or less during crystallographic image taking, i.e., X-ray scattering. Its acceptable range depends on the image resolution. A typical protein resolution ranges between 2 and 3 [Å] and has an acceptable B-factor around 100 up to 200. Higher resolutions need a smaller B-factor below 100; otherwise the electron density becomes a smeared spot or stain, i.e., it spreads out and the contours are no longer sharp, all of which hampers the identification and/or location in space of the underlying side chain atoms or groups like the aromatic ring systems in His, Phe, Trp, or Tyr. This is also the main reason why some PDB entries do not show complete structures or chimeric proteins are created to reduce side chain and backbone mobility. For the same reasons, the crystals are cooled down to extremely low temperatures. Of note, the B-factor is not always included in the PDB entry or publication, but the image resolution is always indicated in the PDB entries.

The **R-factor** (R = resolution) tells how difficult it was to convert the (2D) imaging into a 3D model or structure and therefore how much computer-aided corrections (guessing) had to be applied. For our example (1A9U.pdb) (**Protein Data Bank** at http://www.rcsb.org/), the R-factor should be not more than 0.3. This threshold is estimated as follows: a tenth of its resolution value (here 2.5 [Å] yields 2.5:10 = 0.25 and with an extra tolerance of 0.05 gives 0.3).

6 PREPARING THE SMALL ORGANIC COMPOUND LIGANDS

6.1 Generating New or Modified 3D Ligand Models Under Vega ZZ

In general, the use of Vega ZZ—running from either HDD under a monthly license or an external CD—offers four options to create new molecules: (1) either from scratch (from zero) or (2) from built-in fragments. As third option (3) Vega ZZ has also a small library of entire basic organic molecules and even drugs. Finally, (4) any ligand can be extracted from PDB entries under SPDBV (main menu > WIND > CONTROL PANEL > click on ligand residue; main menu > SELECT > INVERSE SELECTION; main menu > BUILD > REMOVE SELECTED RESIDUES) and modified thereafter under Vega ZZ (Fig. 10.1).

The building tool box in Vega ZZ allows to remove or add atoms or fragments handily. For data safety and human error concerns, each time-consuming step should be saved down to disk (recommended file format is ML2/MOL2), as there is no UNDO/REDO option for the last editing operations (Fig. 10.7).

The molecular structures can be <u>constructed from scratch</u> or from proposed built-in compounds or <u>fragments</u>. For an atom to be added, the required geometry is selected with the **hybridization** types: "sp3" is tetrahedral, "sp2" is trigonal planar, and "sp" lies on a straight line. The new atom is just characterized by its element label (H, C, O, N, P, etc.). The "Add/Remove atom" window is opened by the "pencil" icon to the left side of the main window (Fig. 10.7). Like in SPDBV in Vega ZZ options by clicking icons complement the

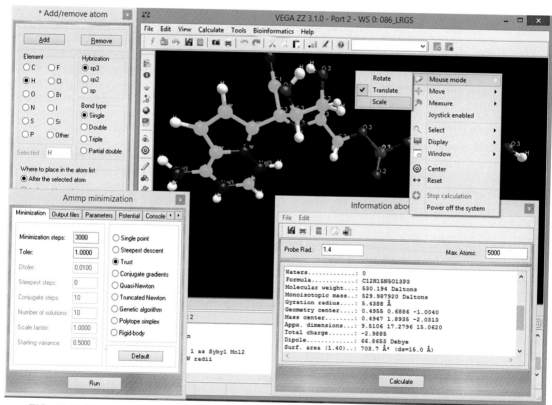

FIG. 10.7 Display of a 3D model in Vega ZZ. An ATP-like ligand is presented as ball and stick model in the black canvas. Color code: green C, white H, blue N, red O, and magenta P atoms. The information sheet about molecular properties of the ligand (bottom-left panel) was created by mouse clicking through main menu option "View" > "Information," while main menu option "Calculate" > "AMMP" opens the MM geometry optimizer (bottom-left). A single-point calculation yields the potential energy of the present geometry without changes in bond distances, angles, and torsions (dihedral angles). The "Add/Remove atom" window allows structural modification of an existing model or building from scratch. It is opened by clicking on the "pencil" icon.

options in the main menu window, i.e., they are not identical or redundant. The "Add/Remove atom" window allows the attachment of an atom by clicking on an existing atom and specifying the "Element" type, "Hybridization" status, and "Bond type." It seems chemically unreasonable that for a hydrogen atom to be added, one must specify a hybridization status in view of its "s" orbital. However, this status reflects the geometry of the attachment point—and not for the new atom to be attached. For instance, to change from toluene to benzene, after removing the methyl side chain of the aryl ring, the ring C atom (C.ar) now is in need of a new hydrogen. Following the aforementioned piece of advice, now the modeler chooses the element "H," hybridization status "sp2," and "Single" bond type. Another illustration is the change of toluene into anilinium mono cation. First, the exocyclic methyl group is removed. The operation needs the following selections: Element "N," "Hybridization sp2," and "Bond type Single." Then, the modeler adds three times the element "H" with an sp3 hybridization and a single bond type onto the nitrogen atom. The choice of "sp3" for each "H" element gives a tetrahedral geometry around the nitrogen atom for the three hydrogen atoms ($-NH_3^+$).

The building of a neutral aniline is explained here to compare it with anilinium example above. The user chooses from the construction menu the "sp2" feature instead, followed by the element choice "N," Hybridization "sp2," and Bond type "Single," and fills the valences of N with only two hydrogens: Element "H," Hybridization "sp2," and Bond type "Single." It is highly recommended to add all hydrogen atoms manually, i.e., in a user-attended fashion. The loss in speed against fully automated valence filling with H atoms (Vega ZZ main menu > EDIT > ADD HYDROGENS) is compensated by a gain in less error-prone structures, as the algorithms (software code by humans) cannot foresee all cases of rare and weird structures, such as antituberculotic isoniazid (Scior et al., 2002, 2006), antiprotozoal nitazoxanide (Scior et al., 2015), or dapsone (Scior et al., 2020).

At this stage with anilinium mono cation and aniline saved to disk in the MOL2 (ML2) format, the charges can be calculated by **Gasteiger method** (main menu > CALCULATE > CHARGE & POT). A window opens to choose from the Force Field pull down menu > "TRIPOS," and from the "CHARGES" pull down menu > "GASTEIGER" and run the calculation "FIX." In the canvas the results can be visually verified (main menu > VIEW > LABEL ATOM > CHARGE or TYPE) before they are stored in a file with MOL2 format.

The total charge is either 0 or 1 for aniline and anilinium, respectively. Under VIEW, molecular properties can be (calculated and) viewed in a table (VEGA menu > VIEW > INFORMATION), e.g., lipophilicity, dipole moment, total charge, molecular weight and volume, total and polar surface area.

6.2 Optimizing the Geometry of the 3D Models

Two main approaches can be safely used by the newcomer to optimize molecular geometries: molecular mechanics (**MM**) or quantum mechanics (**QM**) under Vega ZZ main menu > CALCULATE > AMMP or main menu > CALCULATE > MOPAC > MNDO). Precisely, MNDO (modified neglect of diatomic overlap) is a so-called semiempirical approach (Fig. 10.7).

In Vega ZZ, the calculation option **MOPAC** (molecular orbital package) bundles semiempiric tools to optimize geometries of SOMs. MOPAC is implemented here into Vega ZZ and is also a standalone tool set for Windows, Linux, and Mac OS. **AMMP** (advanced molecular modeling program) is an empiric force field (FF) method with the same purpose. It is implemented in Vega ZZ and belongs to the MM approach. MM takes up simplistic ideas and equations of the mechanical world of physics with Newton's classical laws of macroscopic objects and its extensions over the centuries. A structure is described as a mathematical function of spherical atoms (balls and springs) in terms of its molecular internal energy, called **FF** or potential (energy). **Molecular geometry optimization** by AMMP or MOPAC implies that the potential energy is lowered to find less distorted (incorrect) geometries. Structural relaxation will find one of the many (hundreds of) **local potential energy minima** and depends on the starting geometry and the search algorithm (see below). If a first "valley" (energy depression) is met, surrounded by fairly higher "hills" (energy barriers), then only a local minimum is met, but the total, absolute, or global minimum could be hidden behind the valley. By the way, molecular dynamics (MD) is an extension to MM (cf. statistical mechanics) and simulates molecular movements in solution over time in a thermal bath (normally at room temperature) and thereby overcomes those energy barriers, albeit search algorithms for minima usually cannot climb uphill (too much).

To explore nearby potential energy landscapes, the **geometry minimization** can be repeated applying different search algorithms for potential energy minima: the first choice for badly distorted molecules constitutes so-called steepest descent, followed by conjugate gradients, BFGS (Broyden–Fletcher–Goldfarb–

Shanno), and/or Newtonian methods for further fine-tuning. An optical control is the out-of-plane (oop) bending of a side chain to an aromatic ring which should show coplanarity, e.g., the trans-amide bond, or the exocyclic neutral amino group (-NH$_2$) on the benzene ring for aniline (C$_6$H$_5$−NH$_2$), not to forget the aryl ring system itself. If the FF treatment cannot reach such well-known landmarks of correct geometries, then the modeler switches to old but trustworthy **MNDO** of MOPAC. It brings badly distorted geometries (e.g., oop) straight back into (co)planarity, i.e., "back in plane" (Dewar & Thiel, 1977).

What about **electrons or charges**? Any beginner may ask this question. A concise view of the matter is to state that mostly older FFs like the Tripos FF can "live" without them during geometry optimization operations, i.e., minimizing the potential energy (AKA strain energy). As a direct consequence, neither partial charges nor total charges have to be loaded onto a neutral molecule (+1, +2, −1, −2, etc.) for intramolecular energy minimization. In stark contrast, a far more elaborated FF with tens of atom types for certain elements or atom groups constitutes <u>MMFF94</u> (Merck molecular force field) (Halgren, 1996). Its parameters stem from precalculated parameters for atom combinations by QC, including charges (suggested reading: https:// openbabel. readthedocs. io/en/latest/Forcefields/ mmff94.html).

A very brief sidestep to illustrate the FF implementation with commercial software seems appropriate here: what Tripos FF means for Sybyl is MMFF94 for MOE or MM+ for Hyperchem, namely the program's default (standard) FF. Of note, MM+ represents another elaborated FF with tens of atom types per element, much like MMFF94. Intriguingly, all three programs implement user interfaces to semiempiric (MOPAC) and ab initio (from first principles) QM calculations, too.

7 GENERATING THE SEARCH SPACE AND RUNNING THE DOCKING SIMULATIONS

For docking, the user has to inform ADV about the location of the grid box, which is the search space where to dock the ligand into the target (see keyword "center_x" in Table 10.2). Prior to loading the ligand to be docked into the ADT work area (canvas) in presence of the unliganded protein, the ligand has to be manually moved into the proximity of the BS by SPDBV or Vega ZZ. Upon saving the ligand in any position near the BS, it can be read into ADT again. ADT has a built-in option to choose "center grid box around ligand." This is a very convenient way to inform about the location of

the search area which now lies at the BS, thanks to the manually moved ligand. The Cartesian coordinates of its center (of mass) are indicated in the ADT **grid box** window and needed in the ADV script (Table 10.2). Moreover, to manually move either all ligands—for ligands of variable size, shape, and length—or only the biggest ligand into a start position in the binding cavity is also a good practice to define the location and size of the grid box. Under Vega ZZ main menu option "EDIT," the submenu "SIMILARITY" allows the superposition of ligands (AKA 3D alignment, match, or fit, whereas related proteins should be 2D or 3D aligned with SPDBV main menu "FIT" > "ALIGN" or "MAGIC FIT"). In addition, the grid box (search space) should comprise the (known and putative) interacting amino acids and this operation can be elegantly controlled by eyesight under ADT which displays a transparent grid box (Fig. 10.8). A trade-off has to be made between a larger search area and the computational burden. The rule of thumb is to take an extra length of a third part of the longest ligand for one direction (x, y, or z axis) where the most interesting residues lie (basic, acidic, aromatic side chains).

ADV simulates the ligand docking to target and as a result proposes up to nine final binding positions (docked poses). A script can be set up for running

TABLE 10.2
Listing of the Commands to Launch ADV.
Receptor = 1A9U_T.pdbqt
Ligand = 1A9U_L.pdbqt
Out = 1A9U_10bestDockedPoses.pdbqt
center_x = −11.663
center_y = 11.606
center_z = 69.273
size_x = 34
size_y = 58
size_z = 36
Exhaustiveness = 8

The script file can be modified to the needs of the user to run ADV. Here a ligand - target complex (PDB code: 1A9U) is proposed as an example. ADV defines all target molecules as receptor. In the preparation step the ligand was removed from the ligand − target complex under SPDBV and the target protein was saved as 1A9U_T.pdb file. Conversely, 1A9U_L.pdbqt means that only the ligand is present after SPDBV deleted the biomolecule from the original complex and saved as PDB file. XYZ coordinates determine the box center; while the box size in Angstroms [Å] determines the 3D search area for the best poses.

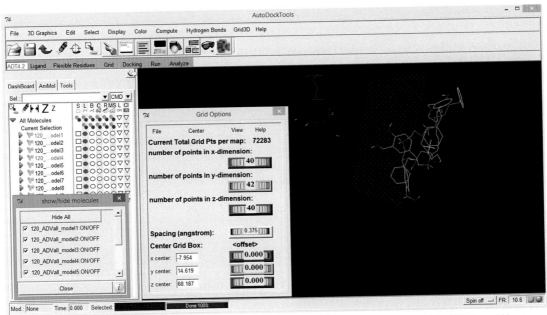

FIG. 10.8 ADT preparation of the binding site (BS). To define the search space around the BS, the grid box option of ADT is used and the Cartesian coordinates (x,y,z) of the box center and size are saved to a **grid parameter file** (file.gpf). The data stored in this file are needed to write the script to run ADV, albeit the proper gpf file itself is not used by ADV (but by AD4). To toggle on or off the display of molecules, the "show/hide molecules" panel was opened under DISPLAY in the main menu. The grid box was centered around the ligand while the target structure is not shown.

ADV automatically (Table 10.2). The script needs the following input data which were prepared in the preparation steps (Fig. 10.8): (1) two separate input files in PDBQT format under the keywords "receptor = " and "ligand = "—which means one for the target and the other for the ligand itself; (2) the location and dimensions of the grid box at the BS under the keywords "center = " and "size = "; (3) the definition of the keyword "out = " to store up to nine docked poses as 3D models in a PDBQT output file after docking (Table 10.2).

A second output file in TXT format will contain the numeric output data, i.e., energy results. It is evoked by an argument in the command line of the MS-DOS window ("–log" anyFileName.TXT). The instructions for ADV in the proposed CADD folder also could document the following two commands to launch ADV automatically and finally analyze the results under ADT (MGL tools) in an MS-DOS window:

"C:\Program Files (x86)\ADVina\vina.exe" –config ADVconfig.txt –log ADVrun_log.txt

"C:\"Program Files (x86)\MGLTools-1.5.6"\adt.bat"

Because calculation time span for ADV simulations is variable, running a job over night is a good idea for a personal computer otherwise used for daily work.

After finishing the docking simulation, the results can be analyzed with ADT any time later.

8 ANALYZING THE GRAPHICAL DOCKING RESULTS

The output with the computed 3D models can be analyzed under ADT. In case of a back or redocking study, a computed ligand–target complex can be created and compared to the experimentally observed complex, i.e., the original crystal complex of the PDB entry. First, the target in PDBQT format is read into the ADT canvas (3D display area) by clicking through its top-level main menu option "FILE" > "READ MOLECULE." Next the ligand with its computed positions is displayed clicking "ANALYZE" > "DOCKINGS" > "OPEN ADV RESULT." A new window will be opened to read in the output file in PDBQT format with the 3D models of the final docked poses of the studied ligand (cf. "out = 1A9U_10bestDockedPoses.pdbqt" in Table 10.2). The reading mode can be set to "Multiple Molecules" or "SINGLE MOLECULE" to read in all at once or one after the other. Alternatively, the main menu option "FILE" > "READ MOLECULE"

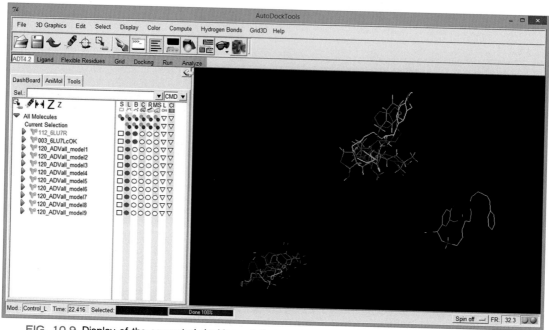

FIG. 10.9 Display of the computed docking solutions under ADT. The output file in PDBQT format of a successful ADV simulation is read into ADT. In the top left corner of the ADT window is the "File" option of the main menu. A small pull-down list can be opened with the "Read Molecule" option to select the 3D models in PDB, MOL2, or PDBQT file formats.

can be used to read in the PDBQT output file of ADV. Not only a file selection filter can be set to "ALL SUPPORTED FILE" but also "PDBQT" can be selected properly to read in the ADV file in pdbqt format (Fig. 10.9). The analysis is the same. All final docked poses can be read in as either discrete molecules or in conformation-like display mode (Fig. 10.10). When the modality "CONFORMATIONS" is chosen, only one conformation after the other can be called in for viewing. To see the other solutions one after the other, the user has to click on the "left" or "right" arrow keys of the keyboard (Fig. 10.11). At http://autodock.scripps.edu/faqs-help/tutorial/, the download of a tutorial file named "2012_ADTtut.pdf" by R. Huey, G. M. Morris, and S. Forli is highly recommended (Huey et al., 2012). Because ADV creates PDBQT files to store the output data with the final docked poses (3D models), the general information in the tutorial about PDBQT files for AD4 is still helpful for ADV, too. It also includes information about the analysis of results. For the hydrogen bonds, a top-level menu option called "Hydrogen Bonds" exist in ADT (Fig. 10.8). Under Vega ZZ, both PDBQT files of target and ligand poses can also be displayed and analyzed (Fig. 10.7). Under Vega ZZ main menu "FILE" > "OPEN" with file type "*.*," the user

chooses the file name of the multiple ligand file first and then the target file is read in. Now the hydrogen bond network can be elegantly displayed by right bottom clicking into the canvas of Vega ZZ. A context menu is opened (Fig. 10.7). Under "MEASURE" is an option of "HYDROGEN BONDS," which not only

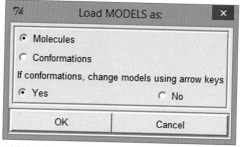

FIG. 10.10 Two modalities to read in the final poses. All final poses are read into the black canvas of ADT and either are displayed at once as independent molecules or can be viewed one after the other by activating the radio button labeled "Conformations." To scroll through the proposed docking solutions one by one, the user should activate the "Yes" option, which enables viewing by pressing the arrow keys on the keyboard.

FIG. 10.11 The output file in pdbqt format is read into ADT. In the black canvas of ADT, a single docked pose (magenta sticks) is displayed. To compare the predicted position, the ligand in its known position from the crystal structure was read in. It was isolated from its protein and then saved by SPDBV as a single molecule file in PDB format. Its carbon atoms are displayed in pink sticks. By convention, H atoms are white, O atoms red, and N atoms blue. The space fill model is also visible and was generated from the main menu option "Display" > "CPK" (not shown).

measures the length of H-bonds and salt bridges (Arg/His/Lys—Asp/Glu) but also toggles their display ON/OFF. This context menu has many more useful modeling options, among them one for the function of mouse movements for translation (X,Y plane) and resizing (depth along Z axis) or rotation around the molecule center (Fig. 10.7). Another useful option is found here with "SELECT" > "CUSTOM" > "PROXIMITY" where a radius around the ligand can be defined. Precisely, to focus on the BS, all entire "residues" (not all "atoms" selection mode) which fall within a certain radius (5—8 Angstroms [Å] is recommended) can be taken into account to only displaying them (and not showing the entire target protein). If the user detects a possible contact of interest, the context menu option "MEASURE" > "DISTANCE" can be evoked to draw a straight intermittent line with the value in Å (Fig. 10.7).

9 ANALYZING THE NUMERICAL DOCKING RESULTS

ADV users who also know AD4 will notice that the output data contain neither "**Estimated Free Energy of Binding**" in concentration units of Kcal/Mol nor

"**Estimated Inhibition Constant**" (Ki or K_i) expressed in molar concentrations. In social media, blogs, or online discussion forum, the newcomer may receive help, but answers may also be unqualified or wrong. In this context, certain comments in the Internet state that binding constants cannot be given for ADV. Actually, yes, Ki values can be calculated as it can be seen in Autodock4 which is the precursor for ADV. The underlying thermodynamic equation describes how the Gibbs free energy of binding, ΔG, is related to the equilibrium constant of the complex formation (association), **K**, with the unit [1/Mol], while **C** is the molar concentration. While "[K]" symbolizes the unit of absolute temperature, **T**, in Kelvin—here "K" is the symbol of the equilibrium constant in the following equation Eq. (10.1):

$$\Delta G = R \cdot T \cdot ln(K/C) \qquad (10.1)$$

with **R** = 8.315 [J/(K · Mol)]; room temperature **T** = 298 [K], and 1 [J] = 4.2 [cal].

The following outlines briefly the complex formation for two receptor inhibitors. They form ligand—target complexes in reversible reactions (subscripts _i and _ii). After the docking simulation, both estimated free energies of binding (ΔG_i and ΔG_ii) differ by 1.4 (Kcal/Mol). As a rule of thumb, this **energy difference between two ligands** corresponds approximately to a tenfold difference of their two inhibitory constants. Any multiple of this quantity (1.4 times 2, 3, or 4, etc.) yields a 2, 3, or 4 times ten-fold difference in their inhibition constants, Ki. But better is to carry out the calculus.

Here is a "nonmathematical" recipe to do this **binding energy to binding concentration conversion** with an ordinary calculator on a sheet of paper. First, the output log file of ADV ("ADVrun_log.txt"), which was defined as an argument in the command to start ADV ("–log ADVrun_log.txt"), is opened with a simple text editor to look for the energy values in [kcal/mol] of the final docked pose. The next step is to use any calculator for the **conversion of those energy values into equilibrium constants**. For example, for one inhibitor ADV calculated $\Delta G = -7.7$ [Kcal/Mol]. This value is multiplied by 1000 divided by 1.98 ($-7.7 \cdot$ **1000/1.98 = -3888.89**). The result is divided by 298 ($3888.89/298 = -13.05$). This value corresponds to the variable x of the exponential function "e^x" (value "E" raised to the power of variable "X"). On many devices where the "e^x" function is missing, the inverse natural logarithm does the same—pressing the "INV" key and the ln() key. The result here is 0.000 002 15 and corresponds to the inhibition constant in molarity unit. It can be converted into 2.15 micromolarity (2.15×10 raised to the power of -6). Another example "to do it yourself" is given: ADV estimates the free energy of binding of -9.52 [Kcal/Mol]. It yields 0.000 000 098, what corresponds to a nanomolar concentration range with Ki = 98 [nM]. The rule of thumb can be applied with caution. The difference between both affinities is calculated ($-9.52 - (-7.7) = 1.82$). The result is now taken as if it were the reference value of 1.4 [Kcal/Mol]. According to the aforementioned rule of thumb, this corresponds to a tenfold difference or more for both Ki values. Indeed, one is Ki = 2.15 [microM] and rounded to Ki = 1 [microM], while the other value is converted into micromolar concentration of Ki = 0.098 [microM], rounded to Ki = 0.1 [microM]. Hence, both Ki values approximately reflect the expected tenfold difference in good keeping with the aforementioned rule of thumb.

10 SYNOPSIS OF THE INPUT, PROCESSING, AND OUTPUT STEPS FOR ADV DOCKING

The proposed <u>workflow for the ligand</u> (Fig. 10.1) for ADV can be summarized starting from the top-left box

and following the two arrows to the last box at the bottom center:

1. separating the ligand from the target if a complex is provided and saving them in two PDB files under SPDBV;
2. → reading the L.pdb file into Vega ZZ and preparing the ligand with total and partial Gasteiger charges and saving it in an L.mol2 file;
3. → reading the MOL2 file of the ligand (L.mol2) into ADT and generating input data for ADV in PDBQT format, which imports the Gasteiger charges from MOL2 (Q) and includes the rotatable bonds (T).

The <u>workflow for the target</u> from the top-left to the top-right box is as follows:

4. → preparing the target under ADT following the advice given in the official ADT tutorial (and not the AD4 user guide) starting on page 10 (Huey et al., 2012) with "preparing a macromolecule";
5. → preparing the BS by defining a search box with its center and volume in space;
6. → adapting the ADV script called configuration file (config.txt) to suite the given docking study (Table 10.2);
7. → running fully automated ADV under MS-DOS with a command similar to "C:\Program Files (x86)\ADVina"\vina.exe;
8. → analyzing the final docked poses of ADV by reading the output file in PDBQT format into ADT and read the log file with the stored estimated binding energies of the final docked poses of the ligand.

11 SOLUTIONS FOR PROBLEMS WITH THE PROGRAMS (TROUBLE SHOOTING)

Of note, commercial or proprietary software are sometimes considered to be more <u>user-friendly</u> because they are distributed with a complete bibliographic documentation with tutorials, demos, user manuals, or scientific descriptions for experts. However, depending on the study design or workflow, any program may need some tinkering to fix problems during all three stages: IPO collection. It is possible to find solutions, but patience and experience is an asset here. Not only for beginners but it can be helpful to go back step by step in the procedure and compare the wanted case with the program's demo example to find and mend inconsistencies, file format incompatibilities program, or user errors (bugs). At times, it is recommended to close unnecessary applications (to free memory). Occasionally, it is even the best choice to sacrifice some unsaved information and restart the program or even the computer, in case of unexpected behavior. If a molecular item has to be mended, an option to work around is to make a file copy of the 3D model and open its copy by an editor like Textpad or Notepad++ and try to fix the issue according to a reference 3D model. After comparing the file contents, those differences may become evident which led to problems in the given 3D model. For instance, the total charge of −2 for a dianion can be manually redistributed on the partial charges of certain atoms, according to the outcome of an incompatible quantum chemistry data file. Text editing may also be used to restore the format requirements to avoid input reading problems for the next modeling step. To this end, it is noteworthy to mention that Textpad not only copies text blocks in a conventional linewise (horizontal) way but also has a vertical selection feature for column-wise copying.

12 APPLICABILITY OF THE PRESENTED DOCKING PROCEDURES

Instead of trying to cite examples of representative docking studies in the hundreds to the thousands (Hafeez et al., 2013), we conclude this book chapter referring to a few docking studies of our entourage to underscore the applicability range for docking of small drug-like ligands (Cortés-Percino et al., 2019; Hafeez et al., 2013; Kumar, Ahmad, Ahmad, Saifi, & Khan, 2012; Meloun & Bordovská, 2007; Pliego & Riveros, 2002; Scior et al., 2013, 2014, 2015; Scior, Abdallah, Salvador-Atonal, & Laufer, 2020). No special changes for the docking method is needed for allosteric BSs. They are special cases only for the biochemical function of that site, i.e., they do not possess the enzymatic or signaling function of "normal" BSs. Of note, the word allosteric has Greek language roots and could be translated literally as "[binding sites]" at other places (Marmolejo & Martínez, 2017; Méndez al. 2019). Docking can also be embedded in a screening strategy to filter large sets of potential active candidates (Allouche, 2012; Charles et al., 2020; Do et al., 2015; Macindoe et al., 2010; Singh & Coumar, 2017). Given a series of ligands for a target, QSAR studies can be carried out to complement the insight on the binding mode gained by docking studies (Scior et al., 2009); docking can be combined not only with QSAR but also with MD simulations (Arya et al., 2019; Cob et al., 2019; López et al., 2020).

13 CONCLUSION

Four docking programs (ADV, ADT, SPDBV, and Vega ZZ) were introduced with practical hints about how to combine them in a virtual laboratory for SOM docking into protein targets for newcomers with general knowledge in the field of medicinal chemistry. ADV is a popular and free docking program and its use under ADT is presented for beginners in the field. ADT is combined with SPDBV to prepare proteins as target molecules. Vega ZZ is proposed to generate or modify the ligand structure, especially the atom and bond types, along with valence and hybridization states. Particular attention was drawn on how to optimize the molecule geometry in connection with the addition of hydrogen atoms for small organic drug-like compounds as potential ligands for docking to target. Under Vega ZZ, the molecular geometry can be optimized by minimization of the potential energy. After docking with ADV, the computed ligand−target complexes can be displayed and results analyzed under ADT.

This nontheoretical guide invited the beginners to "do it yourself—dock it yourself."

REFERENCES

Allouche, A. (2012). Software news and updates gabedit — a graphical user interface for computational chemistry softwares. *Journal of Computational Chemistry*, 32, 174−182. https://doi.org/10.1002/jcc.

Arya, H., Yadav, C. S., Lin, S. Y., Syed, S. B., Charles, M. R. C., Kannadasan, S., Hsieh, H. P., Singh, S. S., Gajurel, P. R., & Coumar, M. S. (2019). Design of a potent anticancer lead inspired by natural products from traditional Indian medicine. *Journal of Biomolecular Structure and Dynamics*, 0(0), 000. https://doi.org/10.1080/07391102.2019.1664326.

Ballante, F., & Marshall, G. R. (2016). An automated strategy for binding-pose selection and docking assessment in structure-based drug design. *Journal of Chemical Information and Modeling*, 56(1), 54−72. https://doi.org/10.1021/acs.jcim.5b00603.

Bender, A., & Glen, R. C. (2004). Molecular similarity: A key technique in molecular informatics. *Organic and Biomolecular Chemistry*, 2(22), 3204−3218.

Berman, H. M., Westbrook, J., Feng, Z., Gilliland, G., Bhat, T. N., Weissig, H., Shindyalov, I. N., & Bourne, P. E. (2000). The Protein Data Bank. *Nucleic Acids Research*, 28, 235−242.

Charles, M. R. C., Mahesh, A., Lin, S. Y., Hsieh, H. P., Dhayalan, A., & Coumar, M. S. (2020). Identification of novel quinoline inhibitor for EHMT2/G9a through virtual screening. *Biochimie*, 168, 220−230. https://doi.org/10.1016/j.biochi.2019.11.006.

Cob-Calan, N. N., Chi-Uluac, L. A., Ortiz-Chi, F., Cerqueda-García, D., Navarrete-Vázquez, G., Ruiz-Sánchez, E., & Hernández-Núñez, E. (2019). Molecular docking and dynamics simulation of protein β-tubulin and antifungal cyclic lipopeptides. *Molecules*, 24(18), 1−10. https://doi.org/10.3390/molecules24183387.

Cortés-Percino, A., Vega-Báez, J. L., Romero-López, A., Puerta, A., Merino-Montiel, P., Meza-Reyes, S., Padrón, J. M., & Montiel-Smith, S. (2019). Synthesis and evaluation of pyrimidine steroids as antiproliferative agents. *Molecules, 24*(20), 1–10. https://doi.org/10.3390/molecules24203676.

Costa, P. J., Nunes, R., & Diogo Vila-Viçosa, D. (2019). Halogen bonding in halocarbon-protein complexes and computational tools for rational drug design. *Expert Opinion on Drug Discovery, 14*(8), 805–820. https://doi.org/10.1080/17460441.2019.1619692.

Dewar, M. J. S., & Thiel, W. (1977). Ground states of molecules. 38. The MNDO method. Approximations and parameters. *Journal of the American Chemical Society, 99*(15), 4899–4907. https://doi.org/10.1021/ja00457a004.

Do, Q. T., Medina-Franco, J. L., Scior, T., & Bernard, P. (2015). How to valorize biodiversity? Lets go hashing, extracting, filtering, mining, fishing. *Planta Medica, 81*(6), 436–449. https://doi.org/10.1055/s-0034-1396314.

Dominguez, C., Boelens, R., & Bonvin, A. M. (2003). HADDOCK: A protein-protein docking approach based on biochemical or biophysical information. *Journal of the American Chemical Society, 125*(7), 1731–1737. https://doi.org/10.1021/ja026939x.

Forli, S. (2015). Charting a path to success in virtual screening. *Molecules, 20*(10), 18732–18758. https://doi.org/10.3390/molecules201018732.

Goddard, T. D., Huang, C. C., Meng, E. C., Pettersen, E. F., Couch, G. S., Morris, J. H., & Ferrin, T. E. (2018). UCSF ChimeraX: Meeting modern challenges in visualization and analysis. *Protein Science, 27*(1), 14–25. https://doi.org/10.1002/pro.3235.

Grosdidier, A., Zoete, V., & Michielin, O. (2011). SwissDock, a protein-small molecule docking web service based on EADock DSS. *Nucleic Acids Research, 39*(Suppl. 2), 270–277. https://doi.org/10.1093/nar/gkr366.

Guex, N., & Peitsch, M. C. (1997). SWISS-MODEL and the Swiss-PdbViewer: An environment for comparative protein modeling. *Electrophoresis, 18*, 2714–2723.

Hafeez, A., Saify, Z. S., Naz, A., Yasmin, F., & Akhtar, N. (2013). Molecular docking study on the interaction of riboflavin (vitamin B2) and cyanocobalamin (vitamin B12) coenzymes. *Journal of Computational Medicine, 2013*, 1–5. https://doi.org/10.1155/2013/312183.

Halgren, T. A. (1996). Merck molecular force field. II. MMFF94 Van Der Waals and electrostatic parameters for intermolecular interactions. *Journal of Computational Chemistry, 17*, 520–552. https://doi.org/10.1002/(SICI)1096-987X(199604)17:5/6<520::AID-JCC2>3.0.CO;2-W.

Hauser, D. R. J., Scior, T., Domeyer, D. M., Kammerer, B., & Laufer, S. A. (2007). Synthesis, biological testing, and binding mode prediction of 6,9-diarylpurin-8-ones as p38 MAP kinase inhibitors. *Journal of Medicinal Chemistry, 50*(9), 2060–2066. https://doi.org/10.1021/jm061061w.

Huey, R., Morris, G. M., & Forli, S. (2012). *Using Autodock 4 and Autodock Vina with autodocktools : A tutorial.* Retrieved from http://autodock.scripps.edu/faqs-help/tutorial/using-autodock-4-with-autodocktools/2012_ADTtut.pdf.

Kumar, M., Ahmad, S., Ahmad, E., Saifi, M. A., & Khan, R. H. (2012). In silico prediction and analysis of caenorhabditis EF-hand containing proteins. *PloS One, 7*(5), 1–12. https://doi.org/10.1371/journal.pone.0036770.

López-López, E., Rabal, O., Oyarzabal, J., & Medina-Franco, J. L. (2020). Towards the understanding of the activity of G9a inhibitors: An activity landscape and molecular modeling approach. *Journal of Computer-Aided Molecular Design, 34*(6), 659–669. https://doi.org/10.1007/s10822-020-00298-x.

Machado, K. S., Schroeder, E. K., Ruiz, D. D., Cohen, E. M., & de Souza, O. N. (2011). FReDoWS: a method to automate molecular docking simulations with explicit receptor flexibility and snapshots selection. *BMC Genomics, 12*(Suppl. 4), S6. https://doi.org/10.1186/1471-2164-12-S4-S6.

Macindoe, G., Mavridis, L., Venkataraman, V., Devignes, M. D., & Ritchie, D. W. (2010). HexServer: An FFT-based protein docking server powered by graphics processors. *Nucleic Acids Research, 38*(Suppl. 2), 445–449. https://doi.org/10.1093/nar/gkq311.

Maia, E., Assis, L. C., de Oliveira, T. A., da Silva, A. M., & Taranto, A. G. (2020). Structure-based virtual screening: From classical to artificial intelligence. *Frontiers in chemistry, 8*, 343–361. https://doi.org/10.3389/fchem.2020.00343.

Maia, E. H. B., Medaglia, L. R., Da Silva, A. M., & Taranto, A. G. (2020). Molecular architect: A user-friendly workflow for virtual screening. *ACS Omega, 5*(12), 6628–6640. https://doi.org/10.1021/acsomega.9b04403.

Marmolejo-Valencia, A. F., & Martínez-Mayorga, K. (2017). Allosteric modulation model of the mu opioid receptor by herkinorin, a potent not alkaloidal agonist. *Journal of Computer-Aided Molecular Design, 31*(5), 467–482. https://doi.org/10.1007/s10822-017-0016-7.

Meloun, M., & Bordovská, S. (2007). Benchmarking and validating algorithms that estimate pK a values of drugs based on their molecular structures. *Analytical and Bioanalytical Chemistry, 389*(4), 1267–1281. https://doi.org/10.1007/s00216-007-1502-x.

Méndez, S. T., Castillo-Villanueva, A., Martínez-Mayorga, K., Reyes-Vivas, H., & Oria-Hernández, J. (2019). Structure-based identification of a potential non-catalytic binding site for rational drug design in the fructose 1,6-biphosphate aldolase from Giardia lamblia. *Scientific Reports, 9*(1), 11779. https://doi.org/10.1038/s41598-019-48192-3.

Meng, X. Y., Zhang, H. X., Mezei, M., & Cui, M. (2011). Molecular docking: A powerful approach for structure-based drug discovery. *Current Computer-Aided Drug Design, 7*(2), 146–157. https://doi.org/10.2174/157340911795677602.

Morris, G. M., Huey, R., Lindstrom, W., Sanner, M. F., Belew, R. K., Goodsell, D. S., & Olson, A. J. (2009). Autodock4 and AutoDockTools4: Automated docking with selective receptor flexibility. *Journal of Computational Chemistry, 16*, 2785–2791. https://doi.org/10.1002/jcc.21256.

Pagadala, N. S., Syed, K., & Tuszynski, J. (2017). Software for molecular docking: A review. *Biophysical Reviews, 9*(2), 91–102. https://doi.org/10.1007/s12551-016-0247-1.

Pedretti, A., Mazzolari, A., Gervasoni, S., Fumagalli, L., & Vistoli, G. (2020). The VEGA suite of programs: A versatile platform for cheminformatics and drug design projects. *Bioinformatics*, btaa774. https://doi.org/10.1093/bioinformatics/btaa774.

Pedretti, A., Mazzolari, A., Gervasoni, S., & Vistoli, G. (2019). Rescoring and linearly combining: A highly effective consensus strategy for virtual screening campaigns. *International Journal of Molecular Sciences*, 20(9), 2060. https://doi/10.3390/ijms20092060.

Pirhadi, S., Sunseri, J., & Koes, D. R. (2016). Open source molecular modeling. *Journal of Molecular Graphics and Modelling*, 69, 127—143. https://doi.org/10.1016/j.jmgm.2016.07.008.

Pliego, J. R., & Riveros, J. M. (2002). Theoretical calculation of pka using the cluster—continuum model. *The Journal of Physical Chemistry A*, 106(32), 7434—7439. https://doi.org/10.1021/jp025928n.

Poli, G., Martinelli, A., & Tuccinardi, T. (2016). Reliability analysis and optimization of the consensus docking approach for the development of virtual screening studies. *Journal of Enzyme Inhibition and Medicinal Chemistry*, 31, 167—173.

Scior, T., Abdallah, H. H., Salvador-Atonal, K., & Laufer, S. (2020). Dapsone is not a pharmacodynamic lead compound for its aryl derivatives. *Current Computer-Aided Drug Design*, 16(3), 327—339.

Scior, T., Bender, A., Tresadern, G., Medina-Franco, J. L., Martínez-Mayorga, K., Langer, T., Cuanalo-Contreras, K., & Agrafiotis, D. K. (2012). Recognizing pitfalls in virtual screening: A critical review. *Journal of Chemical Information and Modeling*, 52(4), 867—881. https://doi.org/10.1021/ci200528d.

Scior, T., & Garces-Eisele, S. J. (2006). Isoniazid is not a lead compound for its pyridyl ring derivatives, isonicotinoyl amides, hydrazides, and hydrazones: A critical review. *Current Medicinal Chemistry*, 13(18), 2205—2219. https://doi.org/10.2174/092986706777935249.

Scior, T., Lozano-Aponte, J., Figueroa-Vazquez, V., Yunes-Rojas, J. A., Zähringer, U., & Alexander, C. (2013). Three-dimensional mapping of differential amino acids of human, murine, canine and equine TLR4/MD-2 receptor complexes conferring endotoxic activation by lipid A, antagonism by Eritoran and species-dependent activities of Lipid IVA in the mammalian LPS sensor system. *Computational and Structural Biotechnology Journal*, 7, e201305003. https://doi.org/10.5936/csbj.201305003.

Scior, T., Lozano-Aponte, J., Ajmani, S., Hernández-Montero, E., Chávez-Silva, F., Hernández-Núñez, E., Moo-Puc, R., Fraguela-Collar, A., & Navarrete-Vázquez, G. (2015). Antiprotozoal nitazoxanide derivatives: Synthesis, bioassays and QSAR study combined with docking for mechanistic insight. *Current Computer-Aided Drug Design*, 11(1), 21—31. https://doi.org/10.2174/1573409911666150414145937.

Scior, T., Medina-Franco, J. L., Do, Q. T., Martínez-Mayorga, K., Yunes Rojas, J. A., & Bernard, P. (2009). How to recognize and workaround pitfalls in QSAR studies: A critical review. *Current Medicinal Chemistry*, 16(32), 4297—4313. https://doi.org/10.2174/092986709789578213.

Scior, T., Meneses Morales, I., Garcés Eisele, S. J., Domeyer, D., & Laufer, S. (2002). Antitubercular isoniazid and drug resistance of *Mycobacterium tuberculosis* — a review. *Archiv der Pharmazie*, 335(1112), 511—525. https://doi.org/10.1002/ardp.200290005.

Scior, T., Verhoff, M., Gutierrez-Aztatzi, I., Ammon, H. P. T., Laufer, S., & Werz, O. (2014). Interference of boswellic acids with the ligand binding domain of the glucocorticoid receptor. *Journal of Chemical Information and Modeling*, 54(3), 978—986. https://doi.org/10.1021/ci400666a.

Singh, V. K., & Coumar, M. S. (2017). Ensemble-based virtual screening: Identification of a potential allosteric inhibitor of Bcr-Abl. *Journal of Molecular Modeling*, 23(7), 218. https://doi.org/10.1007/s00894-017-3384-y.

Trott, O., & Olson, A. J. (2010). AutoDock Vina: Improving the speed and accuracy of docking with a new scoring function, efficient optimization and multithreading. *Journal of Computational Chemistry*, 31, 455—461.

Zhang, X., Wong, S. E., & Lightstone, F. C. (2014). Toward fully automated high-performance computing drug discovery: A massively parallel virtual screening pipeline for docking and molecular mechanics/generalized born surface area rescoring to improve enrichment. *Journal of Chemical Information and Modeling*, 54(1), 324—337. https://doi.org/10.1021/ci4005145.

CHAPTER 11

Use of Molecular Docking as a Decision-Making Tool in Drug Discovery

AZIZEH ABDOLMALEKI • FERESHTEH SHIRI • JAHAN B. GHASEMI

1 INTRODUCTION

Biomolecular recognition takes place based on the chemical interactions between a ligand and its specific receptor (target/protein) in a living organism. Proteins have advanced to use "high specificity" to bind ligands with tendencies to meet the precise necessities of the cell (Heifetz et al., 2016). Our understanding and knowledge of the chemical interactions are regularly improved by a crystal structure in atomic resolution level. In this regard, there are many biochemical databases and tools that presently provide powerful systems to gather, organize, store, and analyze vast volumes of different collected data. Besides, researchers can access various PDB structures to investigate receptors or targets engaged in the development of drugs. In this regard, data science presents many new algorithms, architecture, techniques, and analytics to handle the complexity and diversity of such data. Also, these tools can extract hidden knowledge and valuable information to facilitate decision making to identify and solve problems (Dagliati et al., 2018; Harper, 2014; Song & Zhu, 2016). Proteins' structural information makes a new way to use various tools such as virtual screening to find other hits and to progress off to an effective drug (Abdolmaleki et al., 2018; de Ávila et al., 2020).

In the early stages of drug discovery, researchers are testing thousands of natural products or plant extracts, small molecules, and looking for a potential molecule to develop as a drug. If they get a potential hit candidate, they move it to the next stage of hit-to-lead optimization. Therefore, the first stage of a significant drug finding project is "hit identification." In this practice, hits or small chemical compounds are identified, which bind to the protein and modify its task. So, hits ideally show some degree of specificity and potency against the target.

High-quality beginning hits gain quicker movement with lower attrition rates in drug finding plan (Goodnow, 2006; Kalyaanamoorthy & Chen, 2011). Fig. 11.1 describes a flowchart of hit identification phase and hit-to-lead phase in the structure-based drug design (SBDD) procedure.

The practice of screening for hit consists of high-throughput screening (HTS), a process using automated lab equipment which can fast examine the interaction between biological targets and drug candidates. The progress of modern computing has supported HTS facilities to screen daily so many drug candidates. In addition to HTS, current drug discovery programs rely on in silico computational process such as molecular docking. Thus, the ligand–receptor docking simulations are early methods to explore potential new hits in drug discovery stages. Molecular docking, as an SBDD method, may be used to predict a receptor–ligand complex structure. This is similar to lock and key concept of the enzyme, where the key is small molecule/ligand and lock is the receptor with a protein-binding pocket. In a computational opinion, molecular docking is an optimization problem for finding the best solution. It is equal to fit and/or adjust a ligand in the position of the pocket of the receptor. Thus, it is important as a decision support system for making the decision easier (Fischer, 1894a; de Ruyck et al., 2016; Santana Azevedo et al., 2012; Screen, 2001).

This chapter discusses how the ligand is docked with the receptor. In fact, it describes the adjustment of the ligand into the protein binding pocket and the interactions affecting the optimal positioning of the protein–ligand complexes. The aim is finding the exact position of a ligand from several potential positions. It also

Molecular Docking for Computer-Aided Drug Design. https://doi.org/10.1016/B978-0-12-822312-3.00010-2

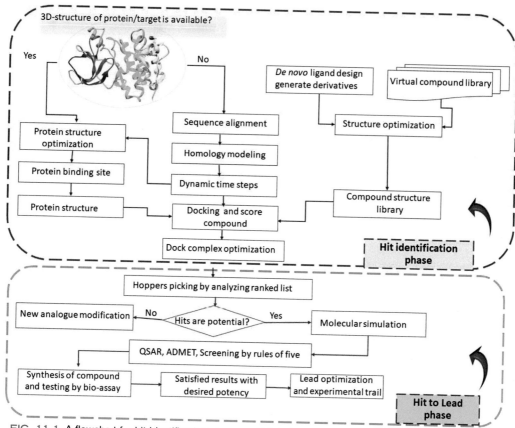

FIG. 11.1 A flowchart for hit identification phase and hit-to-lead phase in the structure-based drug design procedure.

points out what types of intermolecular interactions are well suited to define the strength of the binding affinity and its features. Also, the following sections describe integration techniques and the effective interactions in docking computations. The emphasis of this chapter is on the recent applications and developments in docking simulations for target proteins.

2 MOLECULAR DOCKING SIMULATION

Molecular/biochemical interaction naturally occurs between molecules in a cell within seconds. As indicated in the previous section, molecular docking process models these biochemical interactions to guess where and how two molecules would bind. The results of large-scale molecular docking simulations can offer valuable insight into the relationship between two molecules. This can be considered as a useful method for a biomedical researcher before performing in vivo or in vitro tests. In molecular modeling (molecular

dynamics [MD]) field, the molecular docking term discusses a computer-based technique which predicts the interaction energy between two molecules based on their three-dimensional (3D) structures. The process of molecular docking contains insertion molecules in suitable conformation to relate with a target. In fact, the docking process is the search to find the lowest energy conformation of a drug candidate in the active site of a macromolecule (proteins and sometimes DNA). It uses both information describing the conformation of ligand and information about the orientation of the drug candidate relative to the active site. This process mainly joins algorithms such as genetic algorithm, Monte Carlo simulation, MD, and fragment-based search systems which find the best fit for the candidate drug within protein's cavity. In summary, the docking process can identify the binding conformation and predict the binding affinity of the drug candidate in the protein active site.

In the next sections, major types of interactions in a protein–ligand complex are described.

2.1 Major Types of Interaction Between Ligand and a Protein (Target)

Proteins complex is a fundamental functional unit in biological processes almost in all cells. Proteins have many tasks in the biological system, including:

- a cell's messenger
- a regulatory molecule (inhibit or activate other proteins by binding them)
- an enzyme (accelerate biochemical reactions)
- an engine (conformational changes act as molecular motors)
- a signal detector (by binding small molecules)
- a selective channel (through cell's membrane)

To carry out various cellular tasks, proteins can bind to nucleic acids, other protein partners, and small molecules or therapeutic monoclonal antibody in a specific manner. Such cellular complexes lead to small effects on kinetics and or thermodynamics of biochemical reactions or switch cell systems from one state to another. Accordingly, a detailed description of protein interactions is important for a deep understanding of biological systems regarding drug potency and efficacy.

The made equilibrium between the bound and unbound states of the ligand–protein complex is a consequence of all or some of the interactions such as London dispersion interactions, van der Waals interactions, hydrogen bonds, dipole–dipole interactions, and ionic bonds. Besides these noncovalent interactions, the proteins can be modified covalently in the presence of specific chemical moieties that form a covalent bond with the appropriate amino acid in the protein (Aljoundi et al., 2020).

Noncovalent interactions are an important aspect of structure-based drug discovery that has been recognized in diverse fields associated with materials, chemical, physical, pharmacological, and biological sciences. In a typical interaction model between two molecules, the following energies are relevant:

- Energy (de)solvation
- Hydrophobic
- Electrostatics (nonbonding interactions)
- H-bonding (H-bond donating and accepting)
- Van der Waals (nonbonding interactions)
- Contacts (contact energies denote a desolvation process and thus significantly rely on the solvent. It can be calculated for a pairs of amino acids in the protein folding problem based on the potential of mean force theory and structural information of macromolecular complexes or membrane proteins) (Berrera et al., 2003; Miyazawa, 1996)
- Covalent bond (Aljoundi et al., 2020)

Several methods, including spectroscopy, biochemical/bulk techniques, single molecular studies, and structural analysis to monitor the interaction between proteins and ligands, were developed. Also, complimentary computational tools for searching new ligands for a certain protein binding site are an attractive choice in the pharmaceutical industry and academia research. MD simulations or protein docking can be used as a complementary way to predict macromolecular assemblies in protein–protein interaction (PPI) networks and the ensemble of possible interactions (Russel et al., 2012; de Ruyck et al., 2016). Of course, the flexibility of a small molecule in the binding process is the noticeably higher, and the computational cost is lower than the one required to assess PPI. Thus, small molecule docking has become an active research area in computational drug discovery (de Ruyck et al., 2016). Also, the molecular docking can be combined with other methods such as homology modeling and integrated PPI to increase the structural data (Naveed & Han, 2013; Wang et al., 2012). The process involving binding of two molecules is described by the terms as shown in (Fig. 11.2).

A dominant strategy for the new drug development is the study of chemical relations between a drug and its target. The researcher uses the information obtained from the interactions characterizing candidates' drugs and their targets to finding a new molecule for development. Computational techniques have played an important role in these investigations by contributing efficient, accurate, and comprehensive ways to analyze and interpret the massive data. These methods have assisted to compensate for any limitations in experimental methods, including the presence of biological and technical noise. These methods are

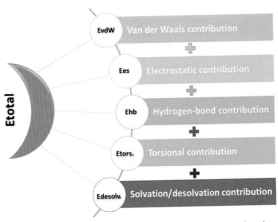

FIG. 11.2 Some typical interactions between two molecules.

potentially useful in helping to identify small molecules in biological studies. For example, ligand binding in enzymatic reactions and then for their inhibitory effects plays an important role in a wide range of diseases such as cardiovascular disease and cancer to design new drugs (de Ruyck et al., 2016).

Here, we describe effective interactions in high-throughput docking (HTD) starting with a typical target protein's structure. In this regard, a number of software and popular programs have been developed. In most cases, a useful search can be performed utilizing free programs. Thus, any resulting hits remain in a virtual form till they are tested experimentally.

2.1.1 Electrostatic energy in protein—ligand complexes

Electrostatic energy has an important task in drug design, virtual screening applications, and protein—protein docking. When a drug binds to its protein target, the frequent protonation state change is observed. In calculating the binding energy of small organic molecules, we have the opportunity to quickly look at the potential of a small molecule at the "binding site" of a protein target. Kukić and Nielsen have reviewed modeling of electrostatic effects in protein—ligand complexes and proteins. They have also discussed a few algorithms for computing electrostatic energies in proteins using experimental data (Kukić & Nielsen, 2010). Molegro Virtual Docker (Kusumaningrum et al., 2014), AutoDock4 (Morris et al., 2009), and AutoDock Vina (Seeliger & de Groot, 2010; Trott & Olson, 2010) are a few available docking programs for studying these interactions. Considering these programs and their scoring functions, it can be said that the scoring function equations add up different types of protein—ligand interaction to model the binding affinity.

Most of the scoring functions take into account electrostatic interactions between two molecules in the protein—ligand interactions. In this regard, an example in pairing studies of proteins in the field of PPI networks, ZDOCK (Pierce et al., 2014), a common docking algorithm, uses the fast Fourier transform for global searching to find docking sites considering shape complementarity, statistical potential, and electrostatic energy.

Generally, a typical scoring function uses several approximations to describe intermolecular and intramolecular interactions in an association of receptor and ligand to decide the intermolecular energy interactions. As a result, singular scoring functions are not perfect, and the researcher must sensibly check or combine other functions to increase the worth of docking results.

2.1.2 Van der Waals potential in protein—ligand complexes

Van der Waals force is a factor that contributes to the formation of protein—ligand complexes. This force can be computed by physical models considering the potential of Lennard-Jones which can accurately predict van der Waals interactions with a complex calculation, which enable us for its usage to simulations of molecular docking and virtual screening of large ligands' database. One of the most challenging topics in structural biology is predicting the molecular binding of protein—ligand or protein—protein pairs (Fahmy & Wagner, 2002). The main problems are the following:

- Reliable estimation of the binding free energies for docked states
- Take into account the possible docking orientations in a high-resolution situation
- Consider structural relocations and movement of the docking surfaces upon interaction

Fahmy and Wagner have proposed TreeDock algorithm by minimizing the van der Waals energies for protein docking. It focuses on the problems in a rigid body docking search by exploring enough number of docking orientations and indicating molecules as multidimensional binary search trees (Fahmy & Wagner, 2002). The traditional mechanism of action for a noncovalent inhibitor is commonly reliant on the small inhibitor that binds to a protein—target noncovalently and the consequent improvement of selectivity and potency. This mechanism is usually made by refining the shape and noncovalent interfaces (van der Waals interactions, hydrogen bonds, salt bridges, etc.) between the inhibitor and target binding site residues.

Some empirical scoring functions use approximate van der Waals interactions with the Lennard-Jones potential to estimate protein—ligand interactions for a specific biological system. Docking processes are different for the covalent bonding from the noncovalent bonding, even using the same software. Developing the mainstream docking techniques has been emphasized on the operational estimate of the binding modes of noncovalent inhibitors (Aljoundi et al., 2020).

2.1.3 Hydrogen bonds in protein—ligand complexes

The significance of H-bonds in the binding affinity of a target drug is an important subject. In fact, energy analysis and evaluation of hydrogen-bonding potential is of special importance in drug design and development due to its presence and possibility in protein—ligand interaction (Bitencourt-Ferreira et al., 2019; Heifetz et al., 2016; Pauling & Corey, 1951). H-bonding stabilizes proteins in the secondary structure of the protein, and it determines the alpha helix and beta sheets in protein structures. Thus, H-bonds are present everywhere in nature and play a different role in some natural processes such as

- protein folding
- protein—ligand interactions
- catalysis

In each H-bond interaction, we can imagine the dipole—dipole interaction as a result of contributing acceptor (A) and donor (D) atoms. H-bonds increase the potency of different functions of the cellular system by enhancing molecular interactions. Therefore, the magnitude and mechanism by which hydrogen bonds adjust the molecular interactions is of great concern in understanding biomolecular interaction due to continuous competition between H-bonding process and bulk water (Chen et al., 2016). Patil et al. calculated the relative impact of H-bonds and hydrophobic interaction of the docked molecules of 4-amino-substituted compounds at the binding sites of c-Abl and c-Src. They have shown how H-bonds optimize the hydrophobic interactions at the ligand—protein interface (Patil et al., 2010).

Hydrogen atoms' location in a functional group directs the capacity of H-bonds formation. Therefore, precise data of the places of H-atoms in proteins are essential to detect H-bond networks and their aspects indeed. The high mobility of H atoms leads to several degrees of freedom: protonation states, where the number of H atoms at a functional group can change, torsional changes, where the H atom site is rotated around the last heavy-atom bond in a residue, and tautomeric states, where an H atom changes its binding partner. Similarly, side chain flips in histidine, glutamine, and asparagine residues that are even seen in public crystallographic databases. Such information must be known before structure-based calculations (Lippert & Rarey, 2009).

LigPlot program is a powerful tool to make 2D plots to characterize interactions of protein and ligand. It enables us to specify structural status to determine intermolecular H-bonds for protein—ligand complexes for theoretical and experimental structures (Laskowski & Swindells, 2011; Wallace et al., 1995).

2.1.4 Hydrophobic interaction in protein—ligand complexes

The process of binding a ligand to a protein is a complex phenomenon impacted by solvation and desolvation where the associating elements become somewhat desolvated. This thermodynamically determined chain of occasions prompts the arrangement of ideal cooperation between the ligand and the protein where hydrophobic contacts are the main impetuses. Hydrophobic moieties associate partner together to decrease the connections with the encompassing. In fact, hydrophobic interactions are the major driving force for protein—ligand interaction and/or formation. It is well known that increased hydrophobic interactions are a key concern for lead optimization, as this often involves increasing ligand molecular weight, rotatable bonds, and lipophilicity, all disturbing the ADMET features of ligands. Designing ligands with high binding affinity and satisfactory ADMET properties is, therefore, a major and often cost-prohibiting challenge (Chen et al., 2016). In this regards, early studies showed a correlation in the change in heat capacity in protein folding or unfolding and the change in solvation of a polar or hydrophobic surface area with the heat capacity effects of the transfer of small hydrocarbon molecules into an aqueous solution (Ha et al., 1989; Livingstone et al., 1991). Characterization of the interaction between small hydrophobic ligands and proteins by direct mass spectrometry (MS) is usually problematic due to the instability of this complex in the gas phase. Xiao et al. have developed an indirect method of assessing small hydrophobic ligand binding to a typical protein by combining MS with electrospray ionization and hydrogen/deuterium exchange. It examines the effectiveness of ligand—protein interactions using monitoring of changes in protein flexibility (Xiao et al., 2003).

Patil et al. (2010) have investigated a multi-targeted small molecule binding with low affinity to its respective targets. They have studied the task of drug binding affinity at the hydrophobic pocket of c-Src and c-Abl kinase. They performed in silico docking studies using ZDOCK, CDOCKER, LigandFit, and Discovery Studio software modules. Also, they have compared H-bonding interactions of docked molecules through LigPlot program. They showed the binding affinity can be transformed by integrating the conformationally preferred functional groups at the active site of the ligand—target interface. Their results demonstrated that optimized hydrophobic interactions and hydrogen bonding both stabilize the ligands at the binding site.

2.1.5 Covalent bond formation in protein—ligand complexes

Another form of protein—ligand interaction in their complex is called covalent bonding, which is the basis for reassembling of a series of drugs in drug discovery. Covalent inhibitors retain significant advantages over

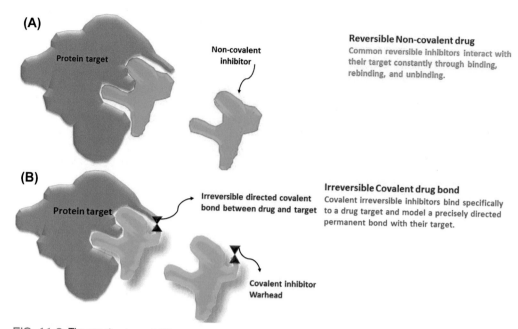

FIG. 11.3 The mechanism of **(A)** noncovalent binding to the drug target protein and **(B)** covalent binding through a functional group in the ligand.

noncovalent inhibitors and result in highly selective inhibitors (Aljoundi et al., 2020).

Covalent bonds are usually made by the interaction between a reactive functional group of the ligand such as epoxy, hydroxyl, or carbonyl and a nucleophilic cysteine, threonine, serine, or, rarely, lysine, thus leading to the formation of covalent adducts (Adeniyi et al., 2016). Irreversibility of this bond is inside the half-life of the protein and necessarily long-lived, causing a drug–protein complex, which is not subject to classical equilibrium (Fig. 11.3).

The improvement of selective irreversible covalent inhibitors has been the focus of thoughtful investigations. Covalent drugs have a very positive effect on human health and also help in the treatment of various symptoms of the disease. In the pharmaceutical industry, these helpful drugs have already demonstrated to be a very useful class. Nearly 30% of the total drugs in the market place that target enzymes are covalent drugs (Aljoundi et al., 2020).

Several in silico methods have been improved to simulate covalent interactions and irreversible inhibitions; however, this is still a challenging field to discover. Molecular docking software such as Gold and Autodock handled this concept by using "link atom," "grid-based method," and a "modification of the flexible side chain" tactic, respectively. Covalent-Dock is another program which has addressed this issue

by automatic preparation of ligand files but is limited in reactions between a nucleophilic receptor and electrophilic ligand. Kumalo et al. have reviewed the theory, applications, and several case studies of covalent docking in drug discovery (Kumalo et al., 2015).

2.2 Different Approaches for Molecular Docking

2.2.1 Rigid and flexible docking models

In rigid docking, both protein and ligand are considered to be fixed so that bond angles or lengths are not changeable during sampling. During a flexible docking, receptor and ligand are considered flexible, and it requires much more time and computation (Rosenfeld et al., 1995). Before docking is performed, protein and ligand structures are subjected to energy minimization. If the protein 3D structure is not available, it is modeled using the homology modeling of the available target protein sequence. Furthermore, the accuracy of the protein 3D structure obtained could be optimized by MD simulation.

To provide a conformer of the protein target that is different from the reported crystal structure, MD simulations have been applied. This conformer should be generated before the molecular docking process. Moreover, after the molecular docking, MD runs have been done to appraise the predicted binding modes of the best-ranking compounds as a final criterion for

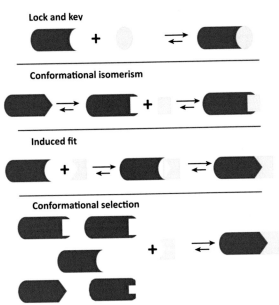

Lock and key

Conformational isomerism

Induced fit

Conformational selection

FIG. 11.4 Four types of docking mode.

in silico filtering that is useful for the synthesis of the hit compounds (Zhao & Caflisch, 2015).

Fig. 11.4 shows four models for docking mode including lock and key (Fischer, 1894b), conformation isomerism (Frauenfelder et al., 1991; Monod et al., 1965; Sandak et al., 1998), induced fit (Bosshard, 2001; Koshland, 1958), and conformation selection (Kumar et al., 2000; Weikl & Paul, 2014). In the lock and key model, the ligand fits accurately and perfectly binds into the active site due to their complementary shapes. The main concept of docking is developed from the concept of "lock and key" for rational drug design, but the precise algorithm to match the shape and size for the "key" (the ligand) into the "lock" (the target protein) is changeable across programs.

In the conformation isomerism model, flexibility is allowed in the ligand molecule to create a collection of molecular conformations. The basic prerequisite for conformational analysis is that the results from conformational sample, namely "ensemble," will be representative of the system generally. The conformer that has the lowest energy is the global minimum energy conformation. Bioactive conformation is a ligand geometry adopted when it binds to the target, not necessarily in the global energy minimum conformation. The analysis of the conformational set that was sampled and optimized is required to ensure the conformational characteristics of the molecule under study.

From Fig. 11.4, it sounds that just the induced fit model will vary the shape of the binding pocket. In induced fit binding, following ligand binding, the conformational change of the protein occurs. This change is a conformational relaxation of the intermediate state into the bound ground state that is seemingly "induced" by the ligand, as the shape of enzyme active site is not exactly complementary. However, the entry into the binding pocket will sometimes vary the shape of the binding site, as shown by simulation from dynamic molecular experiments. Hence, upon binding of the ligand to the active site, the binding causes the active site to become complementary to the ligand. Existence of such a variation in ligand binding mode will make the results from rigid docking unreliable. This probably suggests that there may be faults in docking programs that consider a rigid protein structure.

In the conformational selection mode, before binding a ligand, conformational excitation from the unbound ground state conformation of the protein occurs. In this mechanism, the ligand sounds to "select" and fix a higher-energy conformation for binding. To sum, the "induced fit" mechanism accompanies binding first and in the "conformational selection," conformational change happens initially (Chen, 2015; Paul & Weikl, 2016).

2.2.2 Fragment-based molecular docking

Fragment-based drug design (FBDD) has turned out to be an effective methodology for drug development and could yield fruitful results (Rosenfeld et al., 1995). Fragment binding can elucidate hot spots on proteins, which can lead to high-affinity receptor–ligand interactions. The fragment-based approach more especially is suitable for detecting interaction hot spots among the protein–ligand binding site because the fragments can interact with an important region of the target protein. Fig. 11.5 shows the methodology of fragment docking for inhibitor discovery. Screening of fragment libraries using molecular docking is done in order to select compounds that complement subsites in the target protein binding pocket.

Top ranking fragments are chosen for experimental testing. Recognized compounds with confirmed affinity to the target can be linked, merged, or grown into specific and higher affinity inhibitors (Fig. 11.6) (Chen & Pohlhaus, 2010). These three strategies are briefly described below: (1) Growing: After the recognition of a suitable fragment inside the binding pocket, substitutions can be attached to the recognized fragment. The growing process will enhance the lead

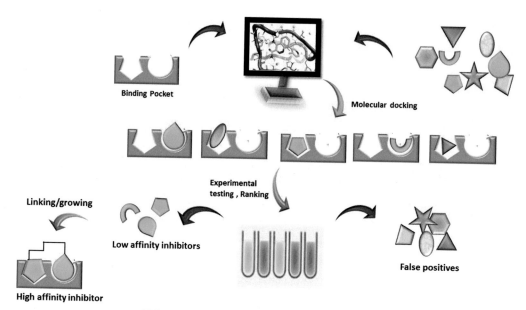

FIG. 11.5 Fragment docking in inhibitor discovery.

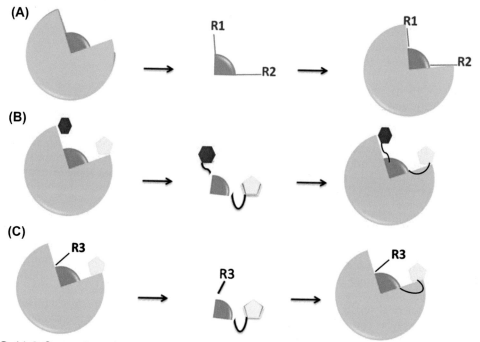

FIG. 11.6 Commonly used strategies for fragment-based drug design. **(A)** Growing; **(B)** linking; **(C)** merging.

likability of the original fragment to increase receptor–ligand interactions. (2) Linking: Multiple fragments that can bind simultaneously to different pockets are connected to enhance the lead likability and to create a new compound with the potential affinity toward

the pocket; linkers are used to connect the fragment. (3) Merging: This is carried out when known lead compounds for a specified target are able to some extent occupy different binding pockets. One or more fragment(s) can be found to be appropriate for the remaining space. Therefore, one or more linker(s) can be introduced to join the known lead and the fragment(s) to enhance the strength of receptor— ligand interactions. Fragment merging is not only used in novel compound design but also applied as a tool for chemical modification and derivative generation (Bian & Xie, 2018).

Caflisch research group has developed free, open-source fragment-based docking suite, namely DAIM—SEED—FFLD. The docking can be explained in three steps. Decomposition and identification of molecules or briefly DAIM decomposes the molecules into molecular fragments. Then, these fragments are docked using SEED in which the programs dock the libraries of fragments with solvation energy evaluation. In the final step, the molecules from the docked fragments are reconstructed by using the program for fragment-based flexible ligand docking, namely the FFLD program (Budin et al., 2001; Kolb & Caflisch, 2006; Majeux et al., 2001). Another approach is an MD-based protocol called SILCS (site identification by ligand competitive saturation). In the SILCS approach, different fragments simultaneously compete for binding of a flexible protein, using explicit solvent MD simulation. FragMap is a binding probability map for each fragment on the protein surface that is generated by the snapshots of the MD trajectories. In a case study on BCL-6 oncoprotein, 1 M propane and benzene were applied as aliphatic and aromatic probes, and water reflects both hydrogen bond donor and acceptor. Output results presented key binding hot spots as well as minor conformational changes of receptor like side chain rearrangements (Guvench & MacKerell, 2009).

3 INTEGRATED COMPUTATIONAL METHODS INVOLVING MOLECULAR DOCKING

3.1 Combining Molecular Docking With Molecular Dynamics Simulations

MD is a computer simulation technique that can provide valuable information on the changes in conformations of biomolecules such as protein by calculation the time-dependent behavior of molecular systems using numerical integration of Newton's equation of motion for a specific interatomic potential defined. MD simulations are mainly used in the prediction of atomistic-level detail of a protein structure and in the investigation of biomolecular interactions and mechanisms understanding. Because of their atomistic resolution, MD simulations have also become useful tools for structure-based drug design.

3.1.1 Protein—ligand binding and solvent treatment with molecular dynamic simulations

The experimental methods in contrast to computational techniques have limitations in their temporal and spatial resolution, while the latter suffers from lower accuracy, especially in the assessment of binding affinity. Therefore, MD is used to analyze the intermolecular interactions such as free energy surface and pathways of molecules binding to proteins. Ligand unbinding makes available the insight into the complex affinity, and it enables us to study the dynamic phenomena such as complex formation and dissociation and the calculation of energy and kinetic parameters for the processes (Huang & Caflisch, 2011a). Huang et al. using the X-ray crystallography studied the free energy surface and unbinding pathways of six ligands (4—11 nonhydrogen atoms) from the active site of FKBP—the FK506 binding protein that is also a peptidyl-prolyl cis-trans isomerase—by explicit solvent MD simulations. MD simulations of (un)binding at the atomic-level detail provide a complete image of the free energy surface and (un)binding pathways (Huang & Caflisch, 2011b, 2011c).

MD simulation with an explicit solvent is also used for the determination of water stability in the binding site. Another approach is based on the analysis of multiple crystal structures (Huang et al., 2014). MD simulations of two bromodomains to measure the structural stability of the six water molecules were carried out, which sound to be conserved at the bottom of the binding pocket in most crystal structures of bromodomains. To summarize, the following observations arise when using the MD simulation of molecular fragments and cosolvent binding to FKBP and bromodomains: (1) existence of metastable states according to multiple binding poses and (2) the prominent role of solvent molecules in molecular recognition. These features prove that molecular docking is greatly cheaper computationally than MD simulations (Zhao & Caflisch, 2015). Another example of the treatment of solvent reported a very high hit rate for high-throughput fragment docking to the BRPF1 bromodomain, which resulted in six chemotypes with many favorable ligand efficiencies which are validated by X-ray crystallography (Zhu & Caflisch, 2016). In another study, fragment hits were recognized by HTD into the CBP bromodomain and implicit solvent force field ranking. Then, they were validated using nuclear magnetic resonance spectroscopy and X-ray crystallography (Spiliotopoulos et al., 2017). In the last example, two chemotypes of CREBBP bromodomain ligands were recognized by fragment-based HTD and implicit solvent force field ranking. Information arising from the analysis of explicit solvent MD simulations resulted in the chemical synthesis of derivatives for hit optimization (Xu et al., 2016). To sum, there are two treatments for solvents during HTD methodology (Huang et al., 2014; Marchand et al., 2016): first, water molecules structurally preserved in the binding site, for example, the water molecules that act as bridges in intermolecular hydrogen bonds which are usually considered explicitly; in the second type, some water molecules in bulk can be approximated efficiently by implicit solvent models (Majeux et al., 2001).

3.1.2 Optimization of ligand conformation by Molecular dynamic simulations

MD simulations are sometimes used to obtain further optimization of the conformation of the active compound (hits) originating from in silico discovery campaigns. While the MD sampling incorporates more information present at the atomistic level, this method is more useful when the crystal structure of the complex is not available. Even if the structure has been solved, MD is an effective way to obtain knowledge about the stability of the interaction over time and to identify which ones contribute most to the binding. One example in ligand optimization by MD simulation was introduced as a new chemical class of inhibitors of the EphB4 tyrosine kinase group by HTD. After docking, to analyze the four top-ranking scaffolds, MD simulation was applied. It was observed in the protein active site that a stable hydrogen bond network was maintained; this validated the four compounds as verifiable inhibitors with nanomolar affinity (Zhao et al., 2012). Though the rigid molecular docking can be done on desktop computers, however, for the MD-based calculations depending on what we are trying to achieve, for example, to compute free energies and binding kinetics needs high-performance computing resources such as cluster computing and/or graphical processing units (Samsonov et al., 2014).

3.2 Combining Molecular Docking and Pharmacophore Models

In the early 1900s, Alder, Paul Ehrlich first coined the pharmacophore term (Ehrlich, 1909). Then between 1967 and 1971, the physical–chemical concept of pharmacophore was introduced by Monty Kier (Kier, 1971, 1967). With the advancement of computer science in modern drug discovery, pharmacophore has consequently become a useful interface between the medicinal chemistry and computational chemistry, not only in virtual screening but also in library design

for efficient hit discovery. Besides, pharmacophore is useful in the optimization of lead compounds to final drug candidates. According to Applied Chemistry (IUPAC) definition, "A pharmacophore is the ensemble of steric and electronic features that is necessary to ensure the optimal supramolecular interactions with a specific biological target structure and to trigger (or to block) its biological response" (Wermuth et al., 1998). Also, Guner defined pharmacophore as an ensemble of structural features that are indispensable for molecular identification. These features mostly include hydrogen bond acceptors and donors, acidic and basic groups, partial charge, and aliphatic and aromatic hydrophobic moieties (Abdolmaleki et al., 2018).

A pharmacophore model can be generated by either ligand-based or structure-based methods (Pirhadi et al., 2013, 2014). In ligand-based methods, the target protein structure is not available; pharmacophoric features are generated by the alignment of a set of bioactive conformations of molecules and subsequently extracting the best common features for binding activity. On the other hand, in structure-based pharmacophore techniques, interaction points within the protein binding site based on a target–ligand complex structure are used for model generation.

When a ligand-based or structure-based pharmacophore model is generated, the pharmacophoric feature landscapes (model) can be utilized in the recognition of hit molecules in a large database, which is named "pharmacophore-based virtual screening" (Balakumar et al., 2018). In order to identify the hits, a pharmacophore model encodes the correct 3D organization of the needed interaction pattern. Based on the available information about the specific protein target, different options are present to create such a query (Qing et al., 2014). As can be seen in Fig. 11.7, for virtual screening,

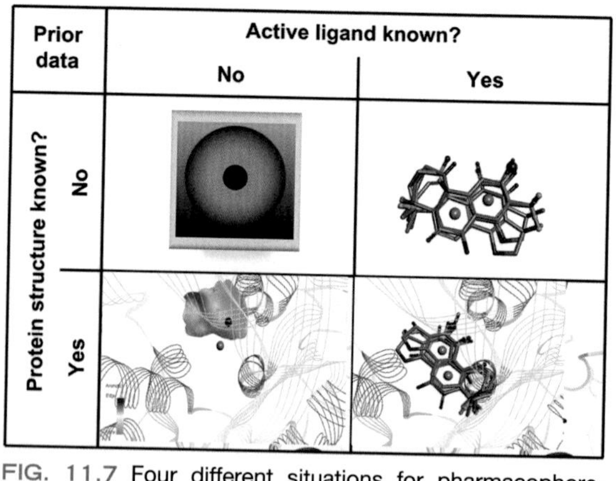

FIG. 11.7 Four different situations for pharmacophore-based searching.

there are four different conditions that may be encountered. When the ligand and protein structure information is not available, the option is experimental screening. In the second option, when the active ligands are existing, and the protein structure is unknown, then the pharmacophore ligand-based virtual screening strategy can be employed.

The best condition is when binding ligand and structural information is available, whereas the most challenging option is when only a protein structure is present. In the pharmacophore-based virtual screening method, the screening process involves two steps: management of the conformational flexibility of the molecules and pharmacophore pattern recognition. When the molecules considered are flexible, they have multiple conformations that each one of the conformations binds to the binding site of the protein with a special mode. Therefore, considering the flexibility of molecules throughout the pharmacophore modeling process is vital. In the conformational search step, a plethora of conformations is generated for each ligand. To achieve the recognition of a pharmacophore pattern named "substructure searching," the task checks the existent of a query pharmacophore in a given conformer of a molecule (Yang, 2010). Although pharmacophore-based virtual screening approach is useful, there is no reliable scoring metrics (Qing et al., 2014). Irrespective of the performance of the method, due to this limitation, the ligands that possess the necessary pharmacophoric features may fail to show in vitro/in vivo pharmacological activity. The most challenging problem of pharmacophore-based virtual screening indicates a higher "false positive" rate. This renders that only a low percentage of virtual hits are actually bioactive. In very simple terms, the hit molecules identified from virtual screening may not be pharmacologically "active" (Yang, 2010). The following factors are helpful to overcome this challenge.

1. Validation and optimization of the pharmacophore model
2. Integrating knowledge of target macromolecule, i.e., flexibility of target

It is known, although the active sites of the proteins are flexible, the structure-based pharmacophore models are taken from a single conformational state of the protein. Such a model is not acceptable for all the possible potential ligand—receptor interactions. In such a condition, the useful method consistently used to address the target flexibility issues in structure-based drug design is the MD simulation. To do this, the knowledge of protein—ligand interactions and the availability of the 3D structure are needed for the generation of a dynamic pharmacophore model. Dynamic pharmacophore

technique distinguishes compounds that tightly bind to the target considering the flexibility of their binding pockets, theoretically decreasing the entropic penalties due to the target—ligand binding. Trajectories obtained from MD simulation provide multiple conformations of a protein active site that they describe the targets' intrinsic flexibilities (Choudhury & Sastry, 2019). Some successful studies were reported (Giménez-Oya et al., 2009; Shiri et al., 2019) that used MD simulations in pharmacophore modeling. These results confirm that the pharmacophore model generated by MD simulation reveals a meaningful and better representation of the flexibility of pharmacophore.

3. Applying a combination of pharmacophore-based and docking-based virtual screening that ignored the steric restriction by the macromolecular target, albeit it is partially considered by excluded volume features. In many cases, the results show that the interactions between a ligand and a target are sensitive to distance, particularly the short-range interactions, such as the electrostatic interaction; thus, it is difficult for a pharmacophore model to account for this. As a result, the combination of pharmacophore-based and docking-based virtual screening (subtype of hierarchical virtual screening approaches) can be considered as an efficient approach for virtual screening. Several combined virtual screening strategies and their validity has been well reviewed (Kirchmair et al., 2008; Kumar & Zhang, 2015; Muegge, 2008; Talevi et al., 2009).

As described above, this consecutive combination of different methods is called hierarchical virtual screening approach. Applying the hierarchical combination of ligand-based virtual screening such as similarity search and pharmacophore screening and structure-based virtual screening such as molecular docking and MD simulation approaches resulted in noteworthy success in a plethora of drug discovery campaigns. Due to very large screening library size (Ruddigkeit et al., 2012) and computational cost related with some virtual screening approaches, it is indispensable to merge various virtual screening approaches to filter compounds. As can be seen in Fig. 11.8, the combination of ligand-based and structure-based approaches can be used in a sequential or parallel way. The major popular manner of combining these approaches is to use them in a sequential funnel-like way, typically known as a hierarchical virtual screening (Fig. 11.8A). By contrast, in the parallel virtual screening manner, several approaches are performed in parallel, and then the best hits ranked according to each approach are selected (Fig. 11.8B).

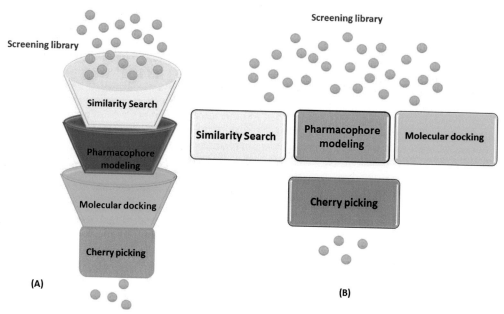

FIG. 11.8 Integration of ligand-based and structure-based approaches. **(A)** Hierarchical or sequentially virtual screening and **(B)** parallel virtual screening.

The best outcomes of both methods in terms of biological activity are reported (Kumar & Zhang, 2015). In a majority of hierarchical virtual screening, the visual selection of compounds by specialist researchers is essential and is usually known as "cherry-picking." Herein, ranking from virtual screening approaches is merged with chemical information and with literature-based knowledge for better outcome. Thus, molecular docking has become a central filtering component used in virtual screening. In hierarchical virtual screening, incorporating molecular docking with various levels of filtering such as similarity search, pharmacophore, and MD simulation will be faster and more effective.

Before a docking study, a pharmacophore model can be used either to filter hits from the virtual screen or to trim down the size of a ligand library. In this way; when a pharmacophore query was applied for filtering, molecular docking can be used as a postfiltered stage. The procedure can omit compounds that would result in the high ranking in a pharmacophore search, but that fail to bind due to incompatibility of the general ligand structure with the receptor site (Hindle et al., 2002). Using the pharmacophore alignment to guide the placement during the docking simulations is another option (Hu & Lill, 2012, 2014). In this case, the pharmacophore model is used to define the placement of the ligand, likewise the fitting of a molecule into the pharmacophore query, or to lead the placement via a constraint in the process of scoring the different docking poses (Goto et al., 2004).

4 CONCLUSION AND OUTLOOK

Nowadays, we have witnessed a massive growth of successful applications of computational strategies in various drug discovery campaigns. Molecular docking is often used to discover how a small molecule interacts with a protein target. Molecular docking anticipates the behavior of small molecules in the binding pockets of target proteins and recognizes correct poses of the small molecule in the binding pocket and also predicts the affinity between the ligand and the protein. Unequivocally, molecular docking plays promising roles in the hit identification and lead optimization in modern drug discovery. A significant development in molecular docking has happened, but it needs further improvements in this methodology. Target binding site information is a gold point and assists in developing this methodology. As discussed in the manuscript, integrating MD and pharmacophore modeling to molecular docking helps to use molecular docking simulation in virtual screening better. Hence, there are multiple new horizons to be explored through the application of molecular docking techniques in designing therapeutics.

REFERENCES

Abdolmaleki, A., Shiri, F., & Ghasemi, J. B. (2018). Computational multi-target drug design. In *Multi-target drug design using chem-bioinformatic approaches* (pp. 51–90). Springer.

Adeniyi, A. A., Muthusamy, R., & Soliman, M. E. (2016). New drug design with covalent modifiers. *Expert Opinion on Drug Discovery, 11*(1), 79–90.

Aljoundi, A., Bjij, I., El Rashedy, A., & Soliman, M. E. (2020). Covalent versus non-covalent enzyme inhibition: Which route should we take? A justification of the good and bad from molecular modelling perspective. *The Protein Journal,* 1–9.

Balakumar, C., Ramesh, M., Tham, C. L., Khathi, S. P., Kozielski, F., Srinivasulu, C., Hampannavar, G. A., Sayyad, N., Soliman, M. E., & Karpoormath, R. (2018). Ligand-and structure-based in silico studies to identify kinesin spindle protein (KSP) inhibitors as potential anti-cancer agents. *Journal of Biomolecular Structure and Dynamics, 36*(14), 3687–3704.

Berrera, M., Molinari, H., & Fogolari, F. (2003). Amino acid empirical contact energy definitions for fold recognition in the space of contact maps. *BMC Bioinformatics, 4*(1), 8.

Bian, Y., & Xie, X.-Q. S. (2018). Computational fragment-based drug design: Current trends, strategies, and applications. *The AAPS Journal, 20*(3), 59.

Bitencourt-Ferreira, G., Veit-Acosta, M., & de Azevedo, W. F. (2019). Hydrogen bonds in protein-ligand complexes. In *Docking screens for drug discovery* (pp. 93–107). Springer.

Bosshard, H. R. (2001). Molecular recognition by induced fit: How fit is the concept? *Physiology, 16*(4), 171–173.

Budin, N., Majeux, N., & Caflisch, A. (2001). Fragment-based flexible ligand docking by evolutionary optimization. *Biological Chemistry, 382*(9), 1365–1372.

Chen, Y.-C. (2015). Beware of docking! *Trends in Pharmacological Sciences, 36*(2), 78–95.

Chen, D., Oezguen, N., Urvil, P., Ferguson, C., Dann, S. M., & Savidge, T. C. (2016). Regulation of protein-ligand binding affinity by hydrogen bond pairing. *Science Advances, 2*(3), e1501240.

Chen, Y., & Pohlhaus, D. T. (2010). In silico docking and scoring of fragments. *Drug Discovery Today: Technologies, 7*(3), e149–e156.

Choudhury, C., & Sastry, G. N. (2019). Pharmacophore modelling and screening: Concepts, recent developments and applications in rational drug design. In *Structural bioinformatics: Applications in preclinical drug discovery process* (pp. 25–53). Springer.

Dagliati, A., Tibollo, V., Sacchi, L., Malovini, A., Limongelli, I., Gabetta, M., Napolitano, C., Mazzanti, A., De Cata, P., Chiovato, L., & Priori, S. (2018). Big data as a driver for clinical decision support systems: A learning health systems perspective. *Frontiers in Digital Humanities, 5,* 8.

Ehrlich, P. (1909). Über den jetzigen Stand der Chemotherapie. *Berichte der Deutschen Chemischen Gesellschaft, 42*(1), 17–47.

Fahmy, A., & Wagner, G. (2002). TreeDock: A tool for protein docking based on minimizing van der Waals energies. *Journal of the American Chemical Society, 124*(7), 1241–1250.

Fischer, E. (1894a). Einfluss der configuration auf die Wirkung der enzyme. *Berichte der Deutschen Chemischen Gesellschaft, 27*(3), 2985–2993.

Fischer, E. (1894b). Influence of configuration on the action of enzymes. *Berichte der Deutschen Chemischen Gesellschaft, 27,* 2985–2993.

Frauenfelder, H., Sligar, S. G., & Wolynes, P. G. (1991). The energy landscapes and motions of proteins. *Science, 254*(5038), 1598–1603.

Giménez-Oya, V., Villacañas, Ó., Fernàndez-Busquets, X., Rubio-Martinez, J., & Imperial, S. (2009). Mimicking direct protein–protein and solvent-mediated interactions in the CDP-methylerythritol kinase homodimer: A pharmacophore-directed virtual screening approach. *Journal of Molecular Modeling, 15*(8), 997–1007.

Goodnow, R. A., Jr. (2006). Hit and lead identification: Integrated technology-based approaches. *Drug Discovery Today: Technologies, 3*(4), 367–375.

Goto, J., Kataoka, R., & Hirayama, N. (2004). Ph4Dock: Pharmacophore-based protein– ligand docking. *Journal of Medicinal Chemistry, 47*(27), 6804–6811.

Guvench, O., & MacKerell, A. D., Jr. (2009). Computational fragment-based binding site identification by ligand competitive saturation. *PLoS Computational Biology, 5*(7).

Harper, E. (2014). Can big data transform electronic health records into learning health systems?. In *Paper presented at the nursing informatics.*

Ha, J.-H., Spolar, R. S., & Record, M. T., Jr. (1989). Role of the hydrophobic effect in stability of site-specific protein-DNA complexes. *Journal of Molecular Biology, 209*(4), 801–816.

Heifetz, A., Chudyk, E. I., Gleave, L., Aldeghi, M., Cherezov, V., Fedorov, D. G., Biggin, P. C., & Bodkin, M. J. (2016). The fragment molecular orbital method reveals new insight into the chemical nature of GPCR–ligand interactions. *Journal of Chemical Information and Modeling, 56*(1), 159–172.

Hindle, S. A., Rarey, M., Buning, C., & Lengauer, T. (2002). Flexible docking under pharmacophore type constraints. *Journal of Computer-Aided Molecular Design, 16*(2), 129–149.

Huang, D., & Caflisch, A. (2011a). The free energy landscape of small molecule unbinding. *PLoS Computational Biology, 7*(2).

Huang, D., & Caflisch, A. (2011b). Small molecule binding to proteins: Affinity and binding/unbinding dynamics from atomistic simulations. *ChemMedChem, 6*(9), 1578–1580.

Huang, D., & Caflisch, A. (2011c). *PLoS Computational Biology, 7,* e1002002.

Huang, D., Rossini, E., Steiner, S., & Caflisch, A. (2014). Structured water molecules in the binding site of bromodomains can be displaced by cosolvent. *ChemMedChem, 9*(3), 573–579.

Hu, B., & Lill, M. A. (2012). Protein pharmacophore selection using hydration-site analysis. *Journal of Chemical Information and Modeling, 52*(4), 1046–1060.

Hu, B., & Lill, M. A. (2014). PharmDock: A pharmacophore-based docking program. *Journal of Cheminformatics, 6*(1), 14.

Kalyaanamoorthy, S., & Chen, Y.-P. P. (2011). Structure-based drug design to augment hit discovery. *Drug Discovery Today, 16*(17–18), 831–839.

Kier, L. B. (1967). Molecular orbital calculation of preferred conformations of acetylcholine, muscarine, and muscarone. *Molecular Pharmacology, 3*(5), 487–494.

Kier, L. (1971). *Molecular orbital theory in drug research*. Academic Press.

Kirchmair, J., Distinto, S., Schuster, D., Spitzer, G., Langer, T., & Wolber, G. (2008). Enhancing drug discovery through in silico screening: Strategies to increase true positives retrieval rates. *Current Medicinal Chemistry*, *15*(20), 2040–2053.

Kolb, P., & Caflisch, A. (2006). Automatic and efficient decomposition of two-dimensional structures of small molecules for fragment-based high-throughput docking. *Journal of Medicinal Chemistry*, *49*(25), 7384–7392.

Koshland, D. E. (1958). Application of a theory of enzyme specificity to protein synthesis. *Proceedings of the National Academy of Sciences*, *44*(2), 98–104.

Kukić, P., & Nielsen, J. E. (2010). Electrostatics in proteins and protein–ligand complexes. *Future Medicinal Chemistry*, *2*(4), 647–666.

Kumalo, H. M., Bhakat, S., & Soliman, M. E. (2015). Theory and applications of covalent docking in drug discovery: Merits and pitfalls. *Molecules*, *20*(2), 1984–2000.

Kumar, S., Ma, B., Tsai, C.-J., Sinha, N., & Nussinov, R. (2000). Folding and binding cascades: Dynamic landscapes and population shifts. *Protein Science*, *9*(1), 10–19.

Kumar, A., & Zhang, K. Y. (2015). Hierarchical virtual screening approaches in small molecule drug discovery. *Methods*, *71*, 26–37.

Kusumaningrum, S., Budianto, E., Kosela, S., Sumaryono, W., & Juniarti, F. (2014). The molecular docking of 1, 4-naphthoquinone derivatives as inhibitors of Polo-like kinase 1 using Molegro virtual docker. *Journal of Applied Sciences*, *4*, 47–53.

Laskowski, R. A., & Swindells, M. B. (2011). *LigPlot+: Multiple ligand–protein interaction diagrams for drug discovery*. ACS Publications.

Lippert, T., & Rarey, M. (2009). Fast automated placement of polar hydrogen atoms in protein-ligand complexes. *Journal of Cheminformatics*, *1*(1), 13.

Livingstone, J. R., Spolar, R. S., & Record, M. T., Jr. (1991). Contribution to the thermodynamics of protein folding from the reduction in water-accessible nonpolar surface area. *Biochemistry*, *30*(17), 4237–4244.

Majeux, N., Scarsi, M., & Caflisch, A. (2001). Efficient electrostatic solvation model for protein-fragment docking. *Proteins: Structure, Function, and Bioinformatics*, *42*(2), 256–268.

Marchand, J.-R. m., Lolli, G., & Caflisch, A. (2016). Derivatives of 3-amino-2-methylpyridine as BAZ2B bromodomain ligands: In silico discovery and in crystallo validation. *Journal of Medicinal Chemistry*, *59*(21), 9919–9927.

Miyazawa, S. (1996). *Macromolecules* https://doi.org/10.1021/ma00145a039 Vol. 18, 534 (1985); Google Scholar Crossref S. Miyazawa and RL Jernigan. *Journal of Molecular Biology*. https://doi.org/10.1006/jmbi, 256(623), 1996.

Monod, J., Wyman, J., & Changeux, J.-P. (1965). On the nature of allosteric transitions: A plausible model. *Journal of Molecular Biology*, *12*(1), 88–118.

Morris, G. M., Huey, R., Lindstrom, W., Sanner, M. F., Belew, R. K., Goodsell, D. S., & Olson, A. J. (2009). Auto-Dock4 and AutoDockTools4: Automated docking with selective receptor flexibility. *Journal of Computational Chemistry*, *30*(16), 2785–2791.

Muegge, I. (2008). Synergies of virtual screening approaches. *Mini Reviews in Medicinal Chemistry*, *8*(9), 927–933.

Naveed, H., & Han, J. J. (2013). Structure-based protein-protein interaction networks and drug design. *Quantitative Biology*, *1*(3), 183–191.

Patil, R., Das, S., Stanley, A., Yadav, L., Sudhakar, A., & Varma, A. K. (2010). Optimized hydrophobic interactions and hydrogen bonding at the target-ligand interface leads the pathways of drug-designing. *PLoS One*, *5*(8).

Pauling, L., & Corey, R. B. (1951). Atomic coordinates and structure factors for two helical configurations of polypeptide chains. *Proceedings of the National Academy of Sciences of the United States of America*, *37*(5), 235.

Paul, F., & Weikl, T. R. (2016). How to distinguish conformational selection and induced fit based on chemical relaxation rates. *PLoS Computational Biology*, *12*(9).

Pierce, B. G., Wiehe, K., Hwang, H., Kim, B.-H., Vreven, T., & Weng, Z. (2014). ZDOCK server: Interactive docking prediction of protein–protein complexes and symmetric multimers. *Bioinformatics*, *30*(12), 1771–1773.

Pirhadi, S., Shiri, F., & Ghasemi, J. B. (2013). Methods and applications of structure based pharmacophores in drug discovery. *Current Topics in Medicinal Chemistry*, *13*(9), 1036–1047.

Pirhadi, S., Shiri, F., & Ghasemi, J. B. (2014). Pharmacophore elucidation and 3D-QSAR analysis of a new class of highly potent inhibitors of acid ceramidase based on maximum common substructure and field fit alignment methods. *Journal of the Iranian Chemical Society*, *11*(5), 1329–1336.

Qing, X., Lee, X. Y., De Raeymaecker, J., Tame, J. R., Zhang, K. Y., De Maeyer, M., & Voet, A. R. (2014). Pharmacophore modeling: Advances, limitations, and current utility in drug discovery. *Journal of Receptor, Ligand and Channel Research*, *7*, 81–92.

Rosenfeld, R., Vajda, S., & DeLisi, C. (1995). Flexible docking and design. *Annual Review of Biophysics and Biomolecular Structure*, *24*(1), 677–700.

Ruddigkeit, L., Van Deursen, R., Blum, L. C., & Reymond, J.-L. (2012). Enumeration of 166 billion organic small molecules in the chemical universe database GDB-17. *Journal of Chemical Information and Modeling*, *52*(11), 2864–2875.

Russel, D., Lasker, K., Webb, B., Velázquez-Muriel, J., Tjioe, E., Schneidman-Duhovny, D., Peterson, B., & Sali, A. (2012). Putting the pieces together: Integrative modeling platform software for structure determination of macromolecular assemblies. *PLoS Biology*.

de Ruyck, J., Brysbaert, G., Blossey, R., & Lensink, M. F. (2016). Molecular docking as a popular tool in drug design, an in silico travel. *Advances and Applications in Bioinformatics and Chemistry*, *9*, 1.

Samsonov, S. A., Gehrcke, J.-P., & Pisabarro, M. T. (2014). Flexibility and explicit solvent in molecular-dynamics-based docking of protein–glycosaminoglycan systems. *Journal of Chemical Information and Modeling, 54*(2), 582–592.

Sandak, B., Wolfson, H. J., & Nussinov, R. (1998). Flexible docking allowing induced fit in proteins: Insights from an open to closed conformational isomers. *Proteins: Structure, Function, and Bioinformatics, 32*(2), 159–174.

Santana Azevedo, L., Pretto Moraes, F., Morrone Xavier, M., Ozorio Pantoja, E., Villavicencio, B., Aline Finck, J., Menegaz Proenca, A., Beiestorf Rocha, K., & Filgueira de Azevedo, W. (2012). Recent progress of molecular docking simulations applied to development of drugs. *Current Bioinformatics, 7*(4), 352–365.

Screen, T. (2001). 1. Fischer E. (1894). Einfluss der Configuration auf die Wirkung der Enzyme. *Berichte der Deutschen Chemischen Gesellschaft, 27*, 2985-2993. 2. Weber E. (2000). Molecular recognition. In *Kirk-Othmer Encyclopedia of Chemical Technology*, John Wiley & Sons, Inc., online article. 3. Chen B., Piletsky S. & Turner A. P. F. Molecular recognition: Design of" keys". Comb. Chem. High. *Accounts of Chemical Research, 34*(12), 938–945.

Seeliger, D., & de Groot, B. L. (2010). Ligand docking and binding site analysis with PyMOL and Autodock/Vina. *Journal of Computer-Aided Molecular Design, 24*(5), 417–422.

Shiri, F., Pirhadi, S., & Ghasemi, J. B. (2019). Dynamic structure based pharmacophore modeling of the acetylcholinesterase reveals several potential inhibitors. *Journal of Biomolecular Structure and Dynamics, 37*(7), 1800–1812.

Song, I. Y., & Zhu, Y. (2016). Big data and data science: What should we teach? *Expert Systems, 33*(4), 364–373.

Spiliotopoulos, D., Zhu, J., Wamhoff, E. C., Deerain, N., Marchand, J. R., Aretz, J., Rademacher, C., & Caflisch, A. (2017). Virtual screen to NMR (VS2NMR): Discovery of fragment hits for the CBP bromodomain. *Bioorganic & Medicinal Chemistry Letters, 27*(11), 2472–2478.

Talevi, A., Gavernet, L., & Bruno-Blanch, L. E. (2009). Combined virtual screening strategies. *Current Computer-Aided Drug Design, 5*(1), 23–37.

Trott, O., & Olson, A. J. (2010). AutoDock Vina: Improving the speed and accuracy of docking with a new scoring function, efficient optimization, and multithreading. *Journal of Computational Chemistry, 31*(2), 455–461.

Wallace, A. C., Laskowski, R. A., & Thornton, J. M. (1995). LIGPLOT: A program to generate schematic diagrams of protein-ligand interactions. *Protein Engineering, Design and Selection, 8*(2), 127–134.

Wang, X., Wei, X., Thijssen, B., Das, J., Lipkin, S. M., & Yu, H. (2012). Three-dimensional reconstruction of protein networks provides insight into human genetic disease. *Nature Biotechnology, 30*(2), 159.

Weikl, T. R., & Paul, F. (2014). Conformational selection in protein binding and function. *Protein Science, 23*(11), 1508–1518.

Wermuth, C.-G., Ganellin, C., Lindberg, P., & Mitscher, L. (1998). Glossary of terms used in medicinal chemistry (IUPAC Recommendations 1998). *Pure and Applied Chemistry, 70*(5), 1129–1143.

Xiao, H., Kaltashov, I. A., & Eyles, S. J. (2003). Indirect assessment of small hydrophobic ligand binding to a model protein using a combination of ESI MS and HDX/ESI MS. *Journal of the American Society for Mass Spectrometry, 14*(5), 506–515.

Xu, M., Unzue, A., Dong, J., Spiliotopoulos, D., Nevado, C., & Caflisch, A. (2016). Discovery of CREBBP bromodomain inhibitors by high-throughput docking and hit optimization guided by molecular dynamics. *Journal of Medicinal Chemistry, 59*(4), 1340–1349.

Yang, S.-Y. (2010). Pharmacophore modeling and applications in drug discovery: Challenges and recent advances. *Drug Discovery Today, 15*(11–12), 444–450.

Zhao, H., & Caflisch, A. (2015). Molecular dynamics in drug design. *European Journal of Medicinal Chemistry, 91*, 4–14.

Zhao, H., Dong, J., Lafleur, K., Nevado, C., & Caflisch, A. (2012). Discovery of a novel chemotype of tyrosine kinase inhibitors by fragment-based docking and molecular dynamics. *ACS Medicinal Chemistry Letters, 3*(10), 834–838.

Zhu, J., & Caflisch, A. (2016). Twenty crystal structures of bromodomain and PHD finger containing protein 1 (BRPF1)/ligand complexes reveal conserved binding motifs and rare interactions. *Journal of Medicinal Chemistry, 59*(11), 5555–5561.

CHAPTER 12

Biomolecular Talks—Part 2: Applications and Challenges of Molecular Docking Approaches

AMUTHA RAMASWAMY • SANGEETHA BALASUBRAMANIAN •
MUTHUKUMARAN RAJAGOPALAN

1 INTRODUCTION

The emergence of novel theoretical techniques has enhanced the analytical space of molecular dynamics (MD) as well as docking approaches with added features to fine-tune as well as speed up the process of drug design. Traditional identification of therapeutic agents mainly lodged on the information about target structure, design and synthesis of lead compounds, and finally the estimation of druggability by looking at the effective interactions between target and drug molecules and hence was a team play between biologist, chemist, and pharmacologist, respectively, for a significantly lengthy and tedious process. Even though the current development of therapeutic compounds dwells upon their skills, the entry of computational biologists played a vital role in identifying possible lead compounds by screening several millions of candidate molecules in a relatively shorter time frame with the help of molecular docking approaches and hence the laborious experimental efforts toward drug discovery are rationalized significantly in a simplified manner.

For the last few decades, molecular docking has been instrumental in designing significant number of new drugs (Matter, 2011). The success rate of drug discovery process is greatly improved by docking-based virtual screening (Vs) approaches, and hence, docking has emerged as an irreplaceable technology in pharmaceutical research. Besides, the limitations in applying docking techniques to various complex problems stress upon the need of novel approaches in drug design. It is obvious that the extensive application of molecular docking and related technologies in both basic and industrial researches would bring out more cost-effective novel therapies even against chronic diseases.

This chapter forms the continuation of Chapter 2 "Biomolecular Talks: a theoretical revisit on docking principles," in which the structural and theoretical bases of majority of the molecular interactions responsible for biomolecular interactions were highlighted. In continuation to that chapter, the applications of molecular docking approaches in demonstrating the biomolecular interactions at in silico level and its existing as well as forthcoming challenges have been elaborated here.

2 APPLICATIONS OF MOLECULAR DOCKING

The process of drug development is a time-consuming process with a high rate of failure in clinical trials. Molecular docking has invariably revolutionized the pipeline of drug discovery. The following sections address some of the major applications (Fig. 12.1) of molecular docking approaches.

2.1 Protein—Drug Interactions

2.1.1 Hit identification by virtual screening

Docking methods play significant role in lead generation and optimization. The process of hunting lead molecules from a large database of lead-like small molecules is a lengthy and tedious process, and the application of docking methods has significantly shortened the process time (Hillisch et al., 2004; Joseph-McCarthy et al., 2007; Shoichet et al., 2002). The database of lead-like small molecules can be generated from literature, reported inhibitors of specific targets, fragment screening, drug-like small molecules, etc. With the advent of supercomputing facilities, screening of such large database using scoring schemes eventually takes less time for docking individual compounds (Macarron

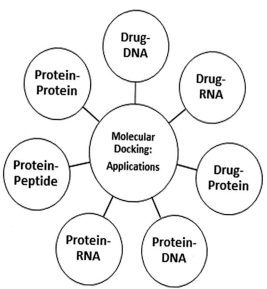

FIG. 12.1 Applications of molecular docking techniques in exploring various biomolecular interactions.

et al., 2011). The foremost application of molecular docking is VS or high-throughput virtual screening (HTVS) (Perez-Castillo et al., 2017). The VS process can be broadly classified into two types: (1) ligand-based virtual screening (LBVS) and (2) structure-based virtual screening (SBVS) (Damm-Ganamet et al., 2016) (Fig. 12.2).

In LBVS, the physicochemical properties of the ligands are used to create a pharmacophore model, which is further used to screen databases (Geppert et al., 2010). The pharmacophore model or feature or fingerprint (FP) is the ensemble of common electronic features (properties such as geometry, ring content, rotatable bonds, planar and nonplanar systems, partial charges, polarizabilities, solvation properties, etc.) that initiate essential molecular interactions with the biological target of interest either to trigger or block its biological response.

In SBVS, the target structure of therapeutic interest is known and the pharmacophore feature of active/binding site is generated. The small molecules deposited in publicly available databases such as ZINC (Irwin & Shoichet, 2005), PubChem (Kim et al., 2016), ChemSpider (Ayers, 2012), ChEMBL (Gaulton et al., 2012), ChemBank (Seiler et al., 2008), DrugBank (Wishart et al., 2006), and Binding DB are directly screened to ensure their binding efficiency with the pharmacophoric feature of the receptor (Armacost et al., 2020; Cournia et al., 2020). The general steps followed in

SBVS are (1) receptor preparation; (2) compound database selection; (3) high-throughput molecular docking; and (4) post-docking analysis (Ferreira et al., 2015).

There are numerous VS studies which were successful in identifying lead molecules against disease-related target proteins. A few success stories, in which the lead molecules identified by VS were further tested in experimental setup to give good inhibition, are discussed here.

InhA, the enoyl-ACP reductase which is responsible for biosynthesis of long-chain fatty acids in *Mycobacterium tuberculosis*, is an interesting molecular target for the development of novel drugs (Marrakchi et al., 2000). A multistep approach combining pharmacophore modeling and molecular docking was employed to screen a database of 999,853 compounds for the identification of new lead molecules. Several docking software such as GOLD, AutoDock, FlexX, and Surflex-Dock implementing different scoring functions were used to improve screening efficiency. By using computational screening methods alone, the study was able to identify three compounds with IC_{50} and K_i values in the micromolar range which could be tested further for their inhibition potency (Pauli et al., 2013).

Signal transducers and transcription activator proteins (STATs) are the transcription factors activated by phosphorylation and dimerization. They bind to specific motif in DNA and facilitate gene transcription (Mitchell & John, 2005). There are seven isoforms of STAT, among which STAT3 is an attractive target for cancer therapeutics. Matsuno, K et al. employed SBVS strategy to screen a database containing 360,000 molecules. Docking and visual inspection resulted in 136 compounds, of which STX-0119 is proposed as the most active STAT3 dimerization inhibitor from fluorescence resonance energy transfer assays and in vivo studies in mice models (Matsuno et al., 2010).

Similarly, docking and intermolecular interaction FPs (see Fig. 12.2) were used to find inhibitors against human aldose reductase (ALR2) protein in the treatment of diabetes (Brownlee, 2001). About 128 compounds, showing enhanced affinity for ALR2, were selected from 7200 compounds, when traditional molecular docking was further fine-tuned by implementing intermolecular interaction fingerprinting. Further experimental studies proposed 15 compounds, with IC_{50} value in the micromolar or nanomolar range, as novel ALR2 inhibitors (Wang et al., 2013). Ren et al. followed a hierarchical multistep approach of support vector machine (SVM) modeling, pharmacophore modeling, and docking to screen Pim-1 (a proto-oncogene that encodes a serine/threonine kinase) inhibitors. A database

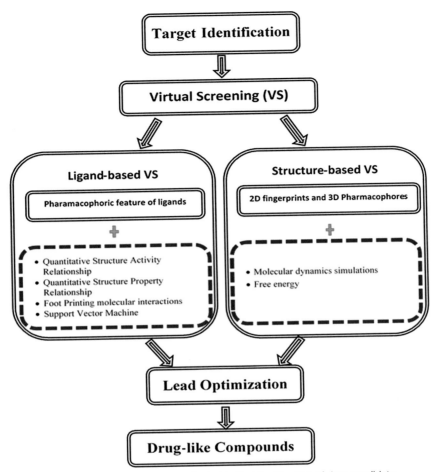

FIG. 12.2 General scheme adopted in the identification of drug candidate.

of ~20 million compounds was reduced to 56,583 and 10,631 compounds through SVM and pharmacophore modeling, respectively. While docking these compounds against Pim-1, about 47 lead compounds were selected for further experimental testing (Ren et al., 2011).

The inhibition of Feline McDonough Sarcoma (FMS)-like tyrosine kinase 3 (FLT3) is recognized as a therapeutic approach for acute myeloid leukemia. Being a tyrosine kinase, the active site loop (consisting of Asp-Phe-Gly or "DFG" motif) takes DFG-in and DFG-out conformations and hence both should be considered for designing inhibitors. The authors modeled both conformations of FLT3 and tested their accuracy by docking known in/out inhibitors. Furthermore, SBVS of a library of 125,000 compounds against FLT3 listed 97 compounds as hits, which were further tested for their inhibition potency. Two compounds with micromolar inhibition constant ($IC_{50} = 2.3\ \mu M$, $10.7\ \mu M$) identified from SBVS and MD studies showed that one of these compounds could be a promising lead molecule for FLT3 inhibition (Ke et al., 2015). Complexes formed by docking these two compounds with DFG-in conformation of FLT3 structure are shown in Fig. 12.3. This study is a clear example to show how computational and experimental methods mutually benefit in drug discovery.

In brief, it can be stated that HTVS is a reliable technique used to increase the probability of identifying potential leads in a shorter time frame when compared with the expensive and time-consuming experimental methods. Usage of reduced dataset for further experimental analysis finds merit in identifying new drug candidates more quickly and cost-efficiently.

Even though all VS studies may result in compounds with high inhibitory potency, screening of a wide

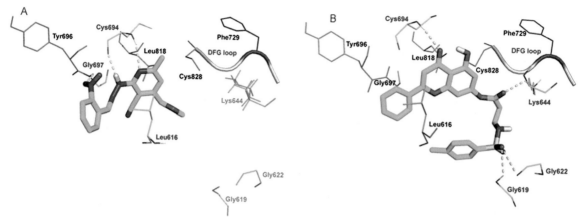

FIG. 12.3 Docked poses of two hits (A) Hit BPR056 and (B) Hit BPR080 while forming complexes with the DFG-in conformation of modeled FLT3 structure. (Figure reproduced with permission from Ke, Y. Y., Singh, V. K., Coumar, M. S., Hsu, Y. C., Wang, W. C., Song, J. S., ... Hsieh, H. P. (2015). Homology modeling of DFG-in FMS-like tyrosine kinase 3 (FLT3) and structure-based virtual screening for inhibitor identification. *Scientific Reports, 5*, 11702. doi: 10.1038/srep11702. Copyright (2015) Springer Nature.)

chemical space can give rise to several "lead" molecules, which can be further optimized to improve their binding efficiency. A few such examples where structure–activity relationship (SAR) studies on VS identified lead compounds resulting in inhibitors with increased binding efficiency are discussed (Slater & Kontoyianni, 2019).

A VS study of fatty acid binding protein 4 (FABP4) (Cai et al., 2015) against 80,000 compounds from Specs and Maybridge databases using Glide package (Friesner et al., 2004, 2006) led to the identification of 52 and 80 compounds using docking scores and interaction pattern, respectively, as selection criteria. When tested experimentally, 11 compounds showed better inhibition against FABP4. To further improve their potency, a similarity search was performed using these 11 compounds, and an additional 36 compounds were identified. One of these 36 compounds showed inhibition potency higher than the endogenous ligand, and concurrently, two more compounds from this class resulted in better inhibition as verified by crystallography methods (Cai et al., 2015).

Xue X et al. performed Vs of their in-house chemical library containing 10,000 compounds (Specs, ChemDiv, Lifechemicals, and ChemBridge) to target the first bromodomain of BDR4 (bromodomain-containing protein 4). Based on the docking score and interaction pattern, 220 compounds were selected and 15 representative compounds were further evaluated. Among these compounds, two compounds at concentrations of 6.93 and 4.70 μM exhibit inhibitory activity and further extensive SAR modifications improved the physicochemical properties (excellent bioavailability, cellular

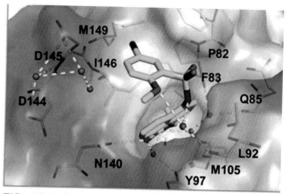

FIG. 12.4 Co-crystal structure of lead compound with BRD4 (PDB ID: 5DX4). (Figure reproduced with permission from Xue, X., Zhang, Y., Liu, Z., Song, M., Xing, Y., Xiang, Q., ... Xu, Y. (2016). Discovery of benzo[cd]indol-2(1H)-ones as potent and specific BET bromodomain inhibitors: Structure-based virtual screening, optimization, and biological evaluation. *Journal of Medicinal Chemistry, 59*(4), 1565–1579. doi: 10.1021/acs.jmedchem.5b01511. Copyright (2016) American Chemical Society.)

permeability). Both alphaScreen and ITC assays revealed an IC_{50} value up to 410 nM and a K_d value of 137 nM, respectively. Such lead compound could be considered as a new potent selective BDR4 inhibitor to treat cancer and inflammatory diseases (Xue et al., 2016) and its key interactions shared with the active site residues of BDR4 are shown in Fig. 12.4.

In an attempt to design autotaxin inhibitors, Pantsar et al. have identified a lead molecule with an IC_{50} value of 134 nM. The search began by screening 11,817,620 compounds, which were reduced to ~1000 compounds based on the scoring function evaluated using both standard precision (SP) and extra precision (XP) modules of Glide docking. Further refinement based on visual inspection and induced fit docking resulted about 26 compounds and finally 4 compounds revealed better inhibition activity at the range of nM when improved by SAR modifications (Pantsar et al., 2017).

Gandhimathi and Sowdhamini performed SBVS against human 5-hydroxytryptamine receptor (5-HT2A) protein on a dataset of 1,40,809 molecules from the clean lead-like category of ZINC database. The cascade of screening processes such as HTVS, SP, XP, interaction pattern and visual inspection, and the docking score could filter out 64 compounds. Final refinement using induced fit docking resulted 15 compounds, which were further investigated for specific agonist- and antagonist-type interactions with 5-HT2A (Gandhimathi & Sowdhamini, 2016).

These studies highlight that the activity of lead-like compounds could greatly be improved by complementing VS with SAR modifications, particularly when the core scaffold of lead molecules was obtained by VS. The advantages offered by docking toward drug discovery are well established (Jorgensen, 2004; Macalino et al., 2015; Shoichet et al., 2002). The docking protocol extends greater flexibility while integrating with different approaches such as ligand-based, MD, and binding affinity, etc., to enhance the selection efficiency of Vs approach. It is also important to mention that different approaches can be applied/implemented at different phases of screening process to improve docking prediction (Ferreira et al., 2015; Lavecchia, 2015). It is quite obvious that MD simulation is an interesting tool that allows the investigation of molecular interaction phenomenon in spatial and temporal scale, which is not feasible with experimental methods (Salmaso & Moro, 2018).

2.1.2 Docking integrated with advanced methods

Despite the strength of docking approach in identifying lead-like compounds, its filtering efficiency could greatly be enhanced multifold, when combined with other computational methods such as ligand-based approach, MD simulations, and binding free energy analysis, etc. (Pinzi & Rastelli, 2019). Integration of MD simulations facilitates the conformational sampling in a wide landscape of binding energy, while the incorporation of binding free energy analysis extends a better rescoring method. The ligand-based approaches

enhance the docking results by various methods. For example, Perryman et al. demonstrated an improved docking prediction by implementing pharmacophore-based rescoring scheme rather than using standard Vs method (Perryman et al., 2014).

Similar results were also obtained by Jiang and Rizzo who retrospectively evaluated the performance of the DOCK program against three clinical drug targets such as epidermal growth factor receptor (EGFR), insulin-like growth factor 1 receptor (IGF-1R), and human immunodeficiency virus glycoprotein 41 (HIVgp41). The authors have demonstrated that the combination of standard DOCK energy function and a specially devised pharmacophore-based scoring function outperformed other approaches in discriminating active from inactive compounds (Jiang & Rizzo, 2015; Moustakas et al., 2006).

While considering the current pandemic of SARS-CoV-2 virus, several attempts are being carried out across the globe to find out drugs against SARS-CoV-2 viral infection. Very recently, Gentile, D. et al. (Gentile et al., 2020) have proposed inhibitors against SARS-CoV-2 target protein, chymotrypsin-like protease (which is the main protease), by employing docking with pharmacophore modeling. In their study, a pharmacophore model of co-crystallized inhibitor N3 with this protease was generated through Pharmit server as shown in Fig. 12.5 (Sunseri & Koes, 2016). This pharmacophore model was used to screen a database of marine natural products containing 14,064 compounds, which generated 164,952 conformers. Initial screening filtered out 770 compounds and the subsequent screening based on the generated pharmacophore filter and a root mean square deviation cutoff of <2.0 Å (from the co-crystal ligand N3) resulted in 180 compounds. Docking was performed for these 180 compounds using Autodock Vina (Trott & Olson, 2010), and 17 compounds with best docking scores were subjected to MD simulations. These 17 compounds that belong to phlorotannins from *Sargassum spinuligerum* brown algae are extensively used in Traditional Chinese Medicine (Gentile et al., 2020).

Recently, Pant et al. (Pant et al., 2020) used structures of SARS-CoV-2 main protease co-crystallized with peptide-like and small molecular inhibitor and performed a similarity search against ZINC15 (Sterling & Irwin, 2015) and CHEMBL (Gaulton et al., 2012) database with approximately 5 million compounds. The initial search yielded 4000 compounds from docking using Glide module. The 50 compounds with highest docking score were selected through manual inspection and were subjected to molecular mechanics

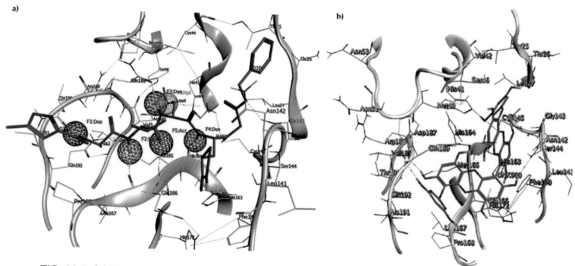

FIG. 12.5 **(A)** Pharmacophore model generated by Pharmit server. The hydrogen bond donors, hydrogen bond acceptors, and hydrophobic centers are indicated by blue, orange, and gray spheres, respectively. **(B)** Interaction profile of the best docked compound in complex with SARS-CoV-2 main protease. (Figure reproduced with permission from Gentile, D., Patamia, V., Scala, A., Sciortino, M.T., Piperno, A., & Rescifina, A. (2020). Putative inhibitors of SARS-CoV-2 main protease from a library of marine natural products: A virtual screening and molecular modeling study. Marine Drugs, 18(4). doi: 10.3390/md18040225. Copyright (2020) MDPI.)

Poisson—Boltzmann surface area (MM/PBSA) rescoring method. The compounds with better binding energy than co-crystallized inhibitor with SARS-CoV-2 main protease were further analyzed. The hit compounds formed strong H-bonding and salt bridge interactions with Gln-192, Glu-166, His-166, and His-41 residues and are in the pipeline for further experimental validation (Pant et al., 2020). Overall, these studies demonstrate the advantage of using ligand-based drug design together with structure-based docking protocol for a faster drug discovery process.

2.1.3 Docking integrated with molecular dynamic simulations and machine learning

Apart from classical MD simulations, advanced sampling techniques, such as umbrella sampling, metadynamics (Laio & Parrinello, 2002), and replica exchange MD (Sugita & Okamoto, 1999), could also be applied to identify both protein conformations and binding pockets that could be exploited further using docking and screening methods for the design of novel inhibitors. Nonetheless, these methods demand more computational power than classical MD simulations. Apart from the application of MD simulation prior to docking, the efficiency of VS process could be improved by other computational methods even at the post-docking stage. BEAR (binding estimation after refinement) is one such tool which performs MD-based refinement of the docked protein—ligand complexes followed by binding free energy analyses using MM/PBSA and molecular mechanics generalized Born surface area methods (Kollman et al., 2000; Rastelli et al., 2009).

Statistical and Artificial Intelligence approaches have gained enormous focus in the recent years due to the growth of information available in structural, chemical, and bioactivity databases (Lavecchia, 2015). These methods can increase the speed and accuracy of binding affinity predictions based on the available information. Machine learning (ML) approaches implemented using random forest algorithm or SVMs have particularly been successful in improving the binding affinity predictions and have also been used to improve the scoring functions (Breiman, 2001; Cortes & Vapnik, 1995; Wojcikowski et al., 2017). A computational workflow integrating SVM and docking was employed to identify novel inhibitors for c-Met tyrosine kinase. In this study, the SVM model which could discriminate between active and inactive c-Met inhibitor was developed. Combining this SVM with docking approach, a huge library of ligands was screened that eventually lead to the identification of eight active compounds (Xie et al., 2011).

The most incredible application of ML method is its ability to differentiate between active and inactive molecules. The ability of two different scoring functions such as Glide and neural network–based scoring function (NNScore) (Ragoza et al., 2017) to identify active estrogen receptor alpha (ERα) ligands from a set of decoys was demonstrated. About 1560 known ERα inactive ligands were included as decoys in the library of compounds chosen for screening (Durrant et al., 2015). The top scoring ligands identified by these two scoring functions were chemically diverse, and the ML algorithms were able to correctly predict the active ligands. A similar study for the identification of anaplastic lymphoma kinase inhibitors by screening Specs library (220,000 compounds) using molecular interaction energy components (MIEC)-SVM model predicted novel chemical scaffolds as potent inhibitors. Among these screened compounds, seven active compounds with IC_{50} values lesser than 10 μM and a success rate of 14% were identified by MIEC-SVM. The predictive ability of MIEC-SVM is significantly higher than other scoring functions such as Auto Dock, which render a success rate of 6% only (Sun et al., 2016).

2.1.4 Docking in drug repositioning

Drug repositioning or repurposing allows one to explore the application of already approved drugs beyond the scope of their initial purpose. As there is a wealth of information on (1) the activity and side effects of approved drugs and (2) their targets and disease conditions, numerous efforts have been made in the recent years to repurpose the known drugs toward novel therapeutic targets (Hurle et al., 2013; March-Vila et al., 2017). For example, the drugs entacapone and tolcapone approved for the treatment of Parkinson's disease have been explored against InhA of *M. tuberculosis* by Kinnings et al. (Kinnings et al., 2009). The primary targets of entacapone and tolcapone are catechol-O-methyltransferase (COMT). In an evolutionary study, Xie and Bourne (Xie & Bourne, 2008) developed an algorithm to detect common binding sites in proteins with diverse sequence and function. This revealed an evolutionary relationship between the NAD-binding Rossmann fold and the SAM-binding domain of the SAM-dependent methyltransferases. Interestingly, COMT is a SAM-dependent methyltransferase and might possess a ligand binding pocket similar to those found in protein domains belonging to the NAD-binding Rossmann fold, according to Xie and Bourne. Hence, Kinnings et al. docked entacapone and tolcapone into 215 NAD-binding proteins of different species using Surflex (Jain, 2007) and eHiTs (Zsoldos et al., 2007). Interestingly, InhA enzyme from *M. tuberculosis* also appeared as one of the top ranked targets. Further structural alignment of COMT and InhA revealed a similar binding pattern (Fig. 12.6). These

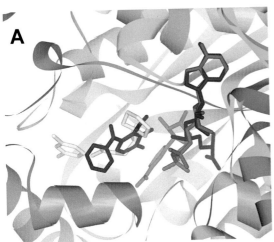 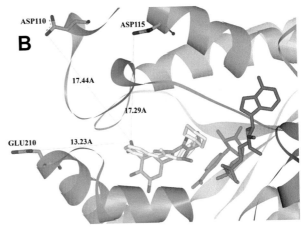

FIG. 12.6 The similarity in ligand binding sites of catechol-O-methyltransferase (COMT) (green) and InhA (blue). The cofactors SAM (purple) and NAD (orange) as well as their respective ligands (red and yellow) of COMT and InhA are shown in subset **(A)**. The binding pose of entacapone (colored in all atom type) in comparison with the native InhA ligand (yellow) is shown in subset **(B)**. (Figure reproduced with permission from Kinnings, S. L., Liu, N., Buchmeier, N., Tonge, P. J., Xie, L., & Bourne, P. E. (2009). Drug discovery using chemical systems biology: Repositioning the safe medicine comtan to treat multi-drug and extensively drug resistant tuberculosis. *PLoS Computational Biology, 5*(7), e1000423. doi: 10.1371/journal.pcbi.1000423; Copyright (2009) PLOS Computational biology.)

results ensured entacapone as a promising lead compound against drug-resistant strains of *M. tuberculosis* (Kinnings et al., 2009).

Dakshanamurthy et al. have demonstrated a novel rapid methodology called "train, match, fit, streamline" that uses shape, topology, and chemical features such as docking score and contact points to evaluate potential drug–target interactions with high accuracy. In their study, compounds from DrugBank, BindingDB (Liu et al., 2007), and FDA databases were screened against 2335 X-ray crystal structures of human proteins reported in Protein Data Bank. One of the successful outcomes of their methodology is the antiparasitic drug mebendazole, which is known to inhibit VEGFR2, showing a kinase inhibition activity of 3.6 μM and also blocks angiogenesis in a human umbilical vein endothelial cell–based angiogenesis functional assay. Surprisingly, the approved antiinflammatory cyclooxygenase-2 inhibitor (COX-2) celecoxib and its inactive analog dimethyl celecoxib found to be the top interacting/binding inhibitor candidates to cadherin-11, an important target of rheumatoid arthritis. The cell line assays further showed an inhibition of IC_{50} of 40 and 36 μM, respectively, and such compounds with weak affinity could be optimized by further SAR investigations (Dakshanamurthy et al., 2012).

Another interesting drug repurposing study is the identification of an EGFR inhibitor as a promising allosteric binder of Bcr-Abl kinase. This prediction was done by combining SBVS and LBVS against the DrugBank (Wishart et al., 2006) dataset to identify allosteric inhibitors of Bcr-Abl kinase as the long term usage of active site inhibitors of Bcr-Abl kinase shows increased resistance. The EGFR inhibitor, gefitinib, showed efficient binding to the allosteric site, while simultaneously enhancing the active site binding efficiency of imatinib at the active site, which was also experimentally tested by MTT assay (Singh, Chang et al., 2017a, Singh, Ganjiwale, et al., 2017b).

Drug repurposing has also been very useful for identifying promising drug candidates against the recent outbreak of SARS-CoV-2 virus. A dataset of (1) 2201 approved drugs from DrugBank and (2) compounds similar to lopinavir from PubChem was screened by employing docking and MD to identify inhibitors against SARS-CoV-2 main protease. A preliminary screening based on Glide docking score was further refined using MM/PBSA–weighted solvent-accessible surface area (WSAS) with a cutoff docking score of −8.5 kcal/mol and the resulted top five compounds were subjected to MD simulation studies. The study

identified (1) carfilzomib, an approved anticancer drug against proteasome, (2) eravacycline, a synthetic antibiotic, (3) a bioactive PubChem compound, and (4) valrubicin, a chemotherapy agent as the potential drug candidate against SARS-CoV main protease (Wang, 2020).

A similar screening study was performed on a dataset of 4384 drugs (all approved for human use) against SARS-CoV-2 main protease. Docking was carried out using iDOCK, and the complexes were selected based on both docking score and the ability to form interactions with active site residues His41 or Cys145. Finally, the MD simulations resulted in potential inhibitors such as daunorubicin, amrubicin, and valrubicin (N-trifluoroacetyladriamycin), which are already being used as chemotherapy agents. Interestingly, valrubicin appeared as one of the top compounds in both the studies discussed here (Jimenez-Alberto et al., 2020).

Also, docking methodology employed for drug repurposing was useful in identifying SARS-CoV-2 inhibitors against various targets such as (1) surface spike glycoprotein, S-protein (Choudhary et al., 2020), (2) nonstructural protein 16 (nsp-16), a viral RNA methyltransferase (MTase) (Tazikeh-Lemeski et al., 2020), (3) RNA-dependent RNA polymerases (Elfiky, 2020), (4) human ACE2 (Terali et al., 2020), and (v) chymotrypsin-like protease (Eleftheriou et al., 2020; Hage-Melim et al., 2020; Meyer-Almes, 2020).

2.2 Drug–DNA Interactions

Similar to protein–ligand interactions, nucleic acids (NAs) also interact with small organic molecules (SOMs). An understanding of interactions between NAs and SOMs is important for designing and optimizing gene regulatory ligands. The interaction between small molecules and NAs may be covalent or noncovalent. In general, NA–ligand interactions could be characterized into three different binding modes: (1) nonspecific, via electrostatic interactions, (2) groove binding, and (3) intercalation or π-stacking mode (Kruger, Zimbres, Kronenberger, & Wrenger, 2018; Tessaro & Scapozza, 2020).

Lee, M et al. performed HTVS of a database of ∼20,000 natural compounds using ICM and the resulted 5 compounds were promising enough for further biological testing. Among these compounds, fonsecin B emerged as a G-quadruplex stabilizer (Lee et al., 2010). Daldrop et al. carried out VS of 2500 compounds against adenine-binding riboswitch (AR) using DOCK 3.5. The top five compounds studied experimentally revealed four compounds with binding affinity to AR in mM range (Daldrop et al., 2011). An interesting application of NA—protein interactions can be observed with retinol binding protein 4 (RBP4). RBP4, which is secreted in adipocytes, is used as a biomarker for detecting type 2 diabetes. Increase in RBP4 secretion level indicates higher chances of obesity and insulin resistance (Rocha et al., 2013). The traditional RBP4 detection methods include enzyme-linked immunosorbent assay, which was replaced by aptamers (single-stranded NAs) in the recent years (Graham, Wason, Bluher, & Kahn, 2007). Accordingly, a 76-mer ssDNA aptamer (RBA) binds to RBP4 with high affinity (K_d: 0.2 ± 0.03 μM), but the mode of interaction was not understood. Torabi, R et al. analyzed the binding interaction of RBP4 with RBA through molecular docking and MD simulations. The polar, positively charged residues (arginine, asparagine, and lysine) and the thymine bases in the RBA were identified as the key residues stabilizing RBP4—RBA interactions (Torabi, Bagherzadeh, Ghourchian, & Amanlou, 2016). In addition to protein—small molecule interactions, the importance of docking procedures to study NA—biomolecular interactions cannot be neglected.

2.3 Protein—Nucleic Acid Interactions

In addition to ligand—NA docking, protein—NA forms a major part of biomolecular interactions. The biomolecular sensing mechanisms, especially observed between proteins and DNAs, follow either direct or indirect recognition/readout mechanisms. In the direct recognition mechanism, the amino acids of DNA-binding proteins interact directly (mainly mediated through the complementary H-bonding interactions) with the NA bases involved in the respective biological event (Garvie et al., 2001; Garvie & Wolberger, 2001). In the indirect recognition mechanism, the protein interacts with DNA by recognizing its structural properties and is predominantly observed when less number of direct base-specific interactions are found (Rohs et al., 2009).

Some examples of important protein—NA complexes studied using docking and MD simulations are discussed below.

L1 endonuclease (L1-EN), which is a catalytic enzyme involved in L1 retrotransposition in humans, recognizes specific DNA sequences and initiates structural changes to catalyze the insertion of L1 DNA. To understand the sequence-specific recognition of L1-EN, specific and nonspecific NA sequences such as 5′-CGTTTTAAAACG-3′ and 5′- GGCGCGCGCGCC-3′, respectively, were docked using HADDOCK server, and the structural dynamics of these L1 complexes were studied for a period of 50 ns. It is observed that L1 complex with specific DNA sequence forms H-bonding interactions with the residues such as Arg155, Asn118, Tyr115, Lys70, Lys71, His45, His198, and Ser202, whereas L1 complex with nonspecific DNA sequence forms interaction only with Ser202 and Ser195. This observation ensures the sequence specificity of L1-EN toward 5′-CGTTTTAAAACG-3′ sequence and is shown in Fig. 12.7 (Rajagopalan et al., 2017).

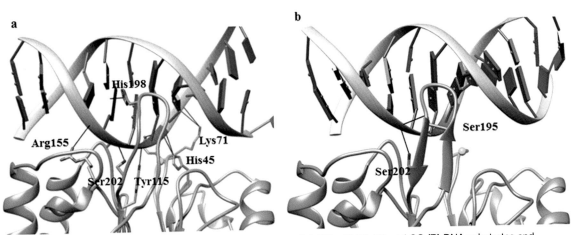

FIG. 12.7 The docked complexes of L1 endonuclease (L1-EN) with TA **(A)** and CG **(B)** DNA substrates and the H-bonds formed between L1-EN and DNA are shown in black lines. The TA junction in the substrate DNA is colored brown. (Figure reproduced with permission from Rajagopalan, M., Balasubramanian, S., & Ramaswamy, A. (2017). Structural dynamics of wild type and mutated forms of human L1 endonuclease and insights into its sequence specific nucleic acid binding mechanism: A molecular dynamics study. *Journal of Molecular Graphics and Modelling, 76,* 43–55. doi: 10.1016/j.jmgm.2017.07.002. Copyright (2017) Elsevier.)

Giri et al. used combined nuclear magnetic resonance (NMR) spectroscopy, isothermal titration calorimetry, and docking to characterize the interaction of ARID1A (AT-rich interactive domain-containing protein 1A), also known as BAF250a with DNA. The AT-rich interaction domain (ARID) containing BAF250a is a subunit of the BAF-A class of SWI/SNF chromatin remodeling complexes (Wang et al., 2004). However, the exact mechanism of BAF250a binding with DNA was not well understood. Docking results show that BAF250a interacts with AT-rich DNA signatures due to the structural flexibility exhibited by AT DNA (Giri et al., 2020). Thus, molecular docking tools prove useful in understanding NA—protein interactions as well.

2.4 Protein—Protein Interactions

In the recent years, protein—protein interactions (PPIs) have been considered as novel therapeutic targets (Sun & Zhao, 2010). Proteins are dynamic biomolecules which can interact with proteins, peptides, NAs, or lipids to perform various functions. As PPIs are regulatory by nature, they are implicated in cancers, metabolic, and neurological disorders. Targeting PPIs is a challenging task since proteins have very diverse interacting partners according to their functions. Identifying the mode of PPIs is necessary to understand the associated biological processes, mutational effects, and the rational design of PPI inhibitors. Experimental methods such as pull-down assay, yeast two-hybrid method, co-immunoprecipitation, and mass spectrometry methods provide evidences for PPIs.

Identification of PPI interaction modes using computational methods can be performed by molecular docking either using a template or by a template-free method. The template-based docking method is fast and reliable because a known dimer is used to guide the modeling of protein—protein complex. As there is no actual docking process, template-based method does not require any scoring functions (Macalino et al., 2018; Sable & Jois, 2015). On the other hand, template-free docking involves docking and scoring algorithms as in SOMs docking protocols. Here, we present a few case studies where VS and ligand docking protocols have jointly been used to identify small molecule PPI inhibitors.

2.5 Protein—Peptide and Small Molecule Interactions

HIV-1 Nef (HIV-1 negative factor) protein is known to interact with its own SH3 binding surface and involves in the infection, pathogenicity, and disease development. A combined VS and high-throughput docking of the National Cancer Institute diversity library (1420 compounds)

was performed using FlexX docking program to identify inhibitors for this PPI. The 335 lowest energy compounds were subsequently subjected to an SH3-based pharmacophoric filter and the resulting 33 potential hits were clustered. About 10 compounds were tested using in vitro cell-based assays and finally two potent drug-like hits that could inhibit HIV-1 Nef/SH3 interaction competitively have been identified (Betzi et al., 2007).

During thrombus formation, the von Willebrand factor (VWF) undergoes conformational changes due to stress, enabling the platelet glycoprotein (GP) Ib-V-IX to interact with VWF-A1 domain and antithrombotic agents target this complex formation. In a study to identify novel antithrombotic agents, MD simulations of VWF-GP-Ib-V-IX complex identified several hot spots for small molecule interaction. Using these hotspots as features, the databases Asinex and Enamine (1,500,000 compounds) were screened. The resulting compounds were again screened in MOE to bring down to 80,000 compounds. Docking was performed for all these compounds against VWF-A1 and GPIbα using GOLD program, which finally identified 24 compounds to be tested by in vitro assays for their potency (Broos et al., 2012).

A similar study was performed by Sarvagalla et al. to identify novel inhibitors against the anticancer target protein, survivin. Survivin belongs to the inhibitor of apoptosis family of proteins and forms PPIs with several other proteins including the chromosomal passenger complex (CPC: survivin/borealin/INCENP). The activity of survivin can be arrested by inhibiting its interaction with other proteins and hence forms a promising target for anticancer therapeutics. The study identified the hot spot residues in the survivin and CPC. Next, a pharmacophore model of the hot spot residues was generated and screened against the DrugBank database using VS, which revealed indinavir as a potent binder, supported by biochemical assays (Sarvagalla et al., 2016).

Abdulrahman et al. targeted the structurally known flaviviral virulent factor NS2B-NS3 serine protease of ZIKV with peptide inhibitors. They designed two short cationic peptides AYA2 and AYA9 using HyperChem 8 based on a template peptide (CP5-46-A) and were docked at the interacting surface of NS2B-NS3 using HADDOCK server. The docking energy of both peptides was lower than the well-known peptide inhibitor, aprotinin. Further biological testing showed that the IC_{50} values of 30.9 and 22.1 μM are low when compared to that of aprotinin (35.4 μM). It is also observed with less cytotoxic effects and this study proposed a cost-effective strategy for developing peptide inhibitors (Abdulrahman et al., 2019).

Due to the devastating current outbreak of SARS-CoV-2, researchers globally are racing to develop drugs for SARS-CoV-2 infection via various computational screening approaches. The entry of SARS-CoV-2 into the host cell is mediated by the attachment/interaction of viral glycoprotein spike (S) protein with angiotensin I converting enzyme 2 (ACE2) receptor in human cells. In particular, the receptor binding domain (RBD) interacts with the ACE2 receptor. Souza et al. targeted the S-protein with eight synthetic antimicrobial peptides to block its entry. The synthetic peptides are docked into SARS-CoV-2 spike in open and closed states using ClusPro 2.0 server. The best complex was selected based on the interface energy, interacting residues, and analyzed through MD simulations. Out of the eight peptides, six peptides interact with S1 domain and two interact with S2 domain of S-protein in open and closed states. The Mo-CBP$_3$-PepII that interacts with α-helices of S1 domain shows lowest binding energy with highest number of interactions shown in Fig. 12.8. Usually the RBD is targeted by peptides in order to inhibit/prevent the interaction of S-protein with ACE2 receptor. The Mo-CBP$_3$-PepII did not bind with RBD domain but induced conformational changes in S-protein which resulted in the incorrect binding mode with ACE2 receptor. Also the peptide shows more binding affinity toward S-protein than ACE2. These peptides act as good lead molecule but need further analysis and experimental validation (Souza et al., 2020).

This chapter deals with numerous instances where docking can be used as standalone or in combination with other techniques. Docking finds application in immunoinformatics and vaccine development. Mukherjee, S., et al. mined the proteome of SARS-CoV-2 to identify immunodominant epitopes which can induce both antibody and cell-mediated immunity toward current pandemic viral infection. They employed various ML approaches such as (1) BpiPred, (2) ABCpred, and (3) LBtope and identified 15 potential immunogenic regions from three proteins. It mapped 25 epitopes (that are identical to experimentally validated SARS-CoV epitopes) ranging from 8 to 28 amino acids. To understand the binding pattern, the peptides are docked with MHC-I and MHC-II using Autodock Vina. Docking analysis revealed higher affinity of (1) smaller peptides with HLA proteins from MHC-I and (2) longer peptides with MHC class II proteins.

Furthermore, seven epitopes exhibit nontoxicity and nonallergen and can be considered as a potential vaccine candidate (Mukherjee et al., 2020).

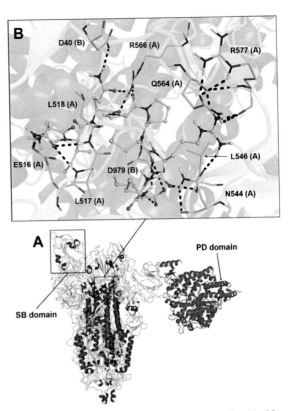

FIG. 12.8 Structure of S-protein complexed with Mo-CBP3-PepII and ACE2 **(A)** and the interactions at S1 domain shown as inset **(B)**. (Figure reproduced with permission from Souza, P. F. N., Lopes, F. E. S., Amaral, J. L., Freitas, C. D. T., & Oliveira, J. T. A. (2020). A molecular docking study revealed that synthetic peptides induced conformational changes in the structure of SARS-CoV-2 spike glycoprotein, disrupting the interaction with human ACE2 receptor. *International Journal of Biological Macromolecules, 164*, 66—76. doi: 10.1016/j.ijbiomac.2020.07.174. Copyright (2020) Elsevier.)

2.6 Protein–Protein Interactions

The PPIs play a vital role in cellular functions. PPIs facilitate majority of the biological processes, including cell-to-cell communication that forms complex network called interactome (Macalino et al., 2018). About 650,000 PPI networks have been identified in several organisms (Stumpf et al., 2008). Modulating these interactions would pave way to regulate these biological processes in disease conditions and hence their characterization gains importance. Here, few examples of computational methods assisting PPI studies are discussed.

In colorectal cancer, the Wnt and Frizzled (Fzd) family members play decisive roles (Akiyama, 2000). Kalhor, H et al. used molecular docking and MD simulation studies to explore their interaction mechanism using HADDOCK and GROMACS packages. Their study suggested a potential interaction mode of Wnt2 with the Fzd7 cysteine-rich domain. It is observed that N-terminal sites of Fzd7 interact with the C-terminal of Wnt2, which is stabilized by both electrostatic and hydrophobic interactions (Kalhor et al., 2018).

In another case, the CB1 cannabinoid receptors (CB1R) have been considered as drug targets for inflammatory, neuropathic pain, and cardiometabolic disorders. The function of CB1R is regulated by its C-terminal domain, mediating PPIs with several proteins. One among the PPI is with cannabinoid receptor interacting proteins 1a and 1b (CRIP1a/b). Singh, P

et al. (Singh et al., 2019) modeled the C-terminal region (417–472 a.a) of CBIR. Using Rho-GDI2 as template, the three-dimensional (3D) models of CRIP1a/b were generated by homology modeling approach implemented in Modeller package (Eswar et al., 2006). The CRIP1a/b was docked into the C-terminal domains of CB1R using HADDOCK. The best docked complexes were selected based on their energy values and previous results (Singh, Ganjiwale, et al., 2017). The CRB1-CRIP1a/b complexes were stabilized by hydrophobic and multiple hydrogen bond interactions, particularly the terminal nine residues of CB1R (S^{464}TDTSAEAL^{472}) form interaction with Asn61, Asp81, Arg82, Val83, Tyr85, Asn127, Tyr128, and Lys130 of CRIP1a/b as shown in Fig. 12.9.

Despite the advantages of molecular docking protocols in exploring the binding, interaction, and

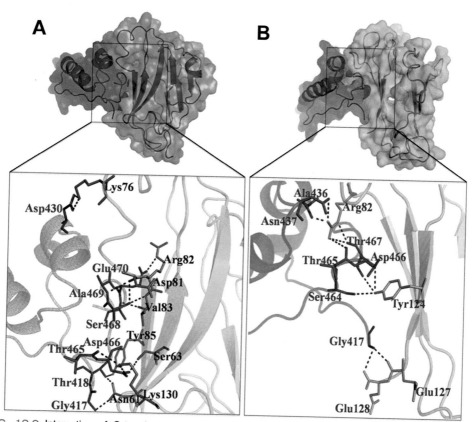

FIG. 12.9 Interaction of C-terminal CB1R (magenta) with **(A)** CRIP1a (blue) and **(B)** CRIP1b (green). (Figure reproduced with permission from Singh, P., Ganjiwale, A., Howlett, A. C., & Cowsik, S. M. (2019). Molecular interaction between distal C-terminal domain of the CB1 cannabinoid receptor and cannabinoid receptor interacting proteins (CRIP1a/CRIP1b). *Journal of Chemical Information and Modeling, 59*(12), 5294–5303. doi: 10.1021/acs.jcim.9b00948. Copyright (2019) American Chemical Society.)

inhibition efficiencies of most of the biomolecular complexes, these methods still have their own challenges while applying universally on all biomolecular complexes.

3 CHALLENGES IN MOLECULAR DOCKING STUDIES

Molecular docking protocols are mainly developed for the most prevalent therapeutic targets, namely proteins. As a rule of thumb, the results obtained from any study depend on the quality of input data, which, in case of docking studies, are the structures of ligands and receptors. Apart from the quality of inputs, several other factors such as the potential energy functions, search algorithms, and scoring functions also play major role in deciding the quality of results (Forli, 2015).

3.1 Ligand Structures

3.1.1 Optimal three-dimensional geometries

The starting structures of ligands are obtained from various sources either by manual sketching or even downloading from any of the compound databases. The available ligand structure data can also vary from 1D (SMILES), 2D (.mol), or 3D (.mol2, .sdf, .pdb, etc.) formats. A few freely available software for generating chemical structures (for academic purpose only) are OpenBabel (O'Boyle et al., 2011), ChemAxon, and Avogadro GUI (Hanwell et al., 2012), while databases such as ZincDB, ChemBank, and PubChem store a large number of chemical compounds. Obtaining a reliable 3D geometry representing proper bond lengths, angles, and enantiomeric information, among other properties is of prime importance (Uitdehaag et al., 1999). As an example, the conversion of structures from 1D SMILES format to 3D format might lead to improper conformations of aromatic rings in macrocyclic compounds, which has to be keenly inspected. To some extent, different docking tools circumvent these problems by (1) generating multiple low-energy conformations prior to docking (Poulsen et al., 2013), followed by docking all the conformations, or (2) allowing flexible degrees of freedom during docking (Forli & Botta, 2007). The former approach of docking all possible conformations is a computationally intensive task, while the latter approach is practically feasible to sample a majority of conformations. Though the extent of sampling is high by generating all possible conformations, the accuracy in a flexible docking approach has been vastly improved by developing rule-based search algorithms.

In addition to the ligand geometries, one has to carefully inspect the definition of atom types representing the chemical environment of each atom, as few docking algorithms replace unknown or unique atom types as dummy atoms leading to inaccurate scores and energies (Sivakumar et al., 2012).

3.1.2 Tautomeric equilibria and protonation states

The isomers of organic compounds differing in the arrangement of one or more hydrogen atoms are referred as tautomers. The occurrence of multiple tautomers is expected to increase the structural and chemical diversity of the molecule. More than 25% of drugs reveal tautomerism. Such ambiguity in H-atom position dictates the relative orientation of interacting complex partners and thus influences the mode of binding as well. Hence, the docking approaches emphasize precise placement of hydrogen atoms while accounting the strength of hydrogen bonding interaction.

Similarly, different protonation states can also have serious impact on the binding modes of ligands by modifying their chirality and hydrogen bonding pattern, as in case of the drug methotrexate. When protonated, the N1 nitrogen of methotrexate can establish two hydrogen bonds with the aspartic acid of the target (Cocco et al., 1983). Contrastingly, if the protonation state of N1 nitrogen is not defined correctly, the hydrogen bond network will be lost leading to unfavorable binding. The protonation state of a ligand also depends on the physiological environment such as pH, solvent, and temperature (Sitzmann et al., 2010). The available search and scoring algorithms compromise on accuracy to achieve speed and hence these algorithms might not sample all the functional tautomers of ligands. As the tautomeric state also depends on other factors such as pH, solvent, and active site of receptor, care must be taken in preparing the inputs for docking as well as while interpreting the docking results. The docking algorithms generally consider a pH range of 7.4 ± 2; however, the physiological pH of the active site might deviate considerably based on the solvent exposure, hydrogen bonding environment, and pKa of neighboring residues. Several examples of docking results depending on the correct pH can be seen in case of HIV-1 protease due to the presence of acidic residues in its catalytic site (Brik & Wong, 2003), in carbonic anhydrase due to the presence of basic metals such as zinc (Stams et al., 1998), and HDAC enzymes (Vannini et al., 2004).

3.1.3 Formal charges

The nonbonded interactions between ligand and receptor depend on the quality of the ligand and receptor description. The van der Waals and electrostatic interactions are accurately described using the ionic radii and charge of the individual atoms. The ionic radii for different chemical species are defined in standard force fields and are identified based on the atom type description in force fields. However, defining the charge of a ligand is a complicated process due to the vast chemical diversity of ligands. The formal charge of a ligand depends on its constituent chemical groups, while the partial charges on each atom depend on the polarization effect. In other words, the partial charges represent the distribution of electrons in a molecule across its individual atoms. The partial charges of ligand atoms can be assigned using several methods such as quantum mechanics (QM) (Jones et al., 1991), semiempirical (Gregory et al., 1998), force field–based (Jorgensen & Tirado-Rives, 1988), or fully empirical (Gasteiger & Marsili, 1980) methods.

The QM and semiempirical methods are the most accurate and computationally expensive. Force field–based or empirical methods are faster providing acceptable results for most drug-like molecules, but it is not suitable for inorganic compounds or metal-coordinating species. The use of incorrect charge sets could lead to mismatched charge pairing between the ligand and receptor leading to false negatives. For this reason, a majority of the scoring functions cannot handle highly charged systems.

3.1.4 Physicochemical properties

The ever-increasing database of chemical species has made available widely SOMs to be studied as ligands. To reduce the complexity, these databases are screened for ligands that match a set of required criteria. These criteria are a set of parameters based on the physicochemical properties of ligands and the most commonly used ones are the "rule of five" by Lipinski et al. (Lipinski et al., 2001) for drug-like molecules and a similar "rule of three" for fragment-like molecules. The common properties among these two rules are the number of hydrogen bonds, molecular weight, and octanol-water partition coefficient (logP) to define the oral bioavailability and drug likeness (Abad-Zapatero, 2007). Another available criterion to screen a library of compounds is the pan-assay interference compounds (PAINS). A simpler way of screening compound databases is the removal of compounds containing reactive

groups such as acrylamide and chloroacetamide, which can lead to undesirable interactions and thereby cause deleterious effects in high-throughput screening assays (Baell & Holloway, 2010). The screening criteria mentioned here, along with other criteria defined based on the target, can help to reduce the initial pool to remove undesirable ligands and thereby save time and resources during VS. It is also beneficial to use a fragment-based drug design approach to reduce the library size and for high coverage of chemical space (Chen & Shoichet, 2009).

3.2 Target Structure

Similar to the challenges faced in obtaining and preparing ligands for a successful docking study, there are several structure- and property-related challenges in preparing the target receptor.

3.2.1 Refinement of three-dimensional structure

The basic requisite of any docking study is the structure of target molecule, which can be obtained by experimental methods such as X-ray crystallography and NMR and computationally through homology modeling approaches. The overall quality of the target structure should be highly refined irrespective of the method used in structure prediction. The R-factor and resolution gives an idea of the overall quality of structures obtained by X-ray crystallography, while the B-factor and occupancy terms (lower the value, lesser the dynamics) measure the quality at the atomic level (Cooper, Porebski, Chruszcz, & Minor, 2011). High occupancy alternate conformations of amino acid side chains are also reported in X-ray crystallography structures, and the choice of conformation requires knowledge on the function of the target protein. Regions with very high B-factors, such as disordered regions of protein loops, could not be captured by X-ray crystallography and hence these missing regions should be handled with care. Also, there are few reports where human errors and issues with software have led to inaccuracies in interpretations (Miller, 2006).

In cases where the target structures are not available or only low-quality structures are available, homology modeling approaches can be used. However, these methods also pack their own challenges, especially, the quality of homology models depends on the quality of the sequence alignment and the template structure (Hillisch et al., 2004). Once a starting structure with reasonably good quality is available, the structure is further corrected for missing atoms, bond lengths, angles, and an overall conformational assessment of the backbone and side chains, etc. While solving the structure by experimental methods, the coordinates of salts and other molecules in solution are also reported along with the molecule of interest. Before docking, these molecules have to be removed from the target structure in order to prevent errors in docking. Also, docking tools are generally designed to recognize only the standard amino acids and NA bases as receptors and hence nonnatural amino acids such as selenomethionine and posttranslationally modified amino acids have to be modified to their native states before docking (Forli, 2015). The challenges in acquiring a satisfactory model for NA docking are stronger than that for proteins. The presence of various secondary structures such as bulges, hairpins, stem loops, triplexes, and G-quadruplexes poses additional complexity. All these factors must be carefully considered while choosing a target structure for molecular docking.

3.2.2 Protonation states

Similar to the tautomeric states of ligands, the protonation states of amino acids also play a key role to obtain a successful docking result. All the arguments and challenges in identifying the protonation states of ligands stand valid even in the case of target protein or NA due to their high variability in pH. The presence of multiple protonable residues in the active site poses an additional challenge because there can be more than one probable combination of protonation states. Identifying the right combination of protonated amino acids requires deep knowledge on the structure and function of the target molecule. The best example and a still ongoing debate on the combination of protonation states is the catalytic residue Asp25 of HIV protease (Harte & Beveridge, 1993). Similarly, in secretase (BACE1), the right combination of protonation state for the catalytic dyad Asp32 and Asp228 is essential for proper ligand binding (Barman & Prabhakar, 2012).

3.2.3 Presence of metal ions, cofactors, and waters

During enzymatic catalysis, metals and cofactors also play crucial roles, and hence, modeling such systems needs special attention. Modeling ligand–metal and protein–metal coordination in scoring functions is a challenging task due to the partial covalent nature of the interaction (Moitessier et al., 2008). As a common

practice, it is recommended to select docking programs with dedicated scoring functions to deal with metal complexes (Irwin et al., 2005). The presence of dominant charges in the metal ions poses major challenges in scoring ligand binding. The removal of bound cofactors such as NADP and heme would bias the docking results because these molecules might involve in direct interaction with the substrate or might be absolutely necessary for the function (Scior & Quiroga, 2019).

However, the decision on retaining or removal of the cofactors depends on whether the docking study follows competitive or allosteric inhibition. For instance, the antimalarial drug chloroquine displaces the NADH cofactor in the active site of lactate dehydrogenate of *Plasmodium falciparum* (Read et al., 1999). As this is a competitive inhibition, the NADH cofactor has to be removed before docking. The consideration of water molecules during docking requires extra care and knowledge on the role of waters at the active site. In several enzymes such as proteases, waters play key roles in enzyme catalysis and hence these "active" waters have to be treated properly. In such cases, ligands can be designed to replace the water molecule (hence removed before docking) or to interact strongly (hence retained during docking). Hence, the inclusion or removal of cofactors and water molecules for docking studies requires prior knowledge on the structural and functional role of these molecules in the target of interest.

3.2.4 Selection of binding site

Defining the binding site is another crucial step in a screening and docking process. When a known binding site is targeted, the search area should cover the regions occupied by known ligands but not limited to them because alternative binding mode is always possible. Contrarily, a large search boundary will reduce the accuracy and is computationally expensive. Hence, it is recommended to include some endogenous molecules in the screening library as a validation to docking (to ensure the scoring and search parameters are optimal) and also to be used as a reference while evaluating VS results (Forli, 2015). Similar to protein, binding sites in NAs can also be defined by few structural features, for example, the major and minor grooves of DNA. Structurally, the major groove is wider and hence accessible to proteins and peptides, whereas the minor groove is narrow and accommodates only small molecules. However, this trend is vice versa in the case of RNA, where the major groove being narrower interacts with small molecules and minor groove with proteins. The highly charged phosphate backbone strongly binds metal ions and waters. Because a small molecule has to disrupt this hydration shell to bind strongly with the backbone, another important requirement for a NA binding small molecule is a high polarity (Luo, Wei, Waldispuhl, & Moitessier, 2019). Apart from groove binding, the ligands can intercalate or stack over the terminal bases of NAs which are more challenging to observe by docking studies.

3.2.5 Scoring functions and search algorithms

Even with high-quality inputs, the key to success of any docking program is its search method and scoring function, as they always possess a fair amount of correlation between speed and accuracy. Docking programs are successfully applied in screening large databases, and hence, the scoring functions should be accurate enough to yield less false positive hits. The scoring functions have two roles: (1) predicting the possible pose (depends upon the sampling) and (2) ranking among the predicted poses (depend on the energy terms). The scoring functions are broadly divided into three types: (1) empirical, (2) knowledge-based, and (3) force field–based (Wang, Lu, & Wang, 2003) as discussed before in Chapter 2. The scoring functions mostly depend on the dataset used for their initial development. Although good amount of success is achieved, scoring functions dealing with strongly charged sites (Moitessier, Westhof, & Hanessian, 2006) or those with hydrophobic/hydrophilic dominant components show limitations (Perola, Walters, & Charifson, 2004). Hence, in such cases, the implementation and development of target-specific scoring functions are encouraged for better results (Ross, Morris, & Biggin, 2013). Also, scoring functions should be validated by the user with a known set of ligands specific for their target of interest, prior to a large-scale VS. Protein sampling also remains one of the problems for advanced scoring functions that include polarizability. Alternatively, some empirical scoring functions such as GlideScore and ChemScore fit the scoring function terms against experimental binding affinity data, thereby improving the accuracy (Cournia et al., 2020).

The two major pitfalls of scoring functions are (1) overestimating enthalpic components due to their additive nature; for example, a large ligand might possess higher scores probably due to more number of interactions rather than the quality of binding (Schulz-Gasch & Stahl, 2004); and (2) dependency on the search algorithm that samples the entire conformational landscape. The performance of scoring functions also depends on the complexity of molecules arising from the size, increased degrees of freedom by the flexibility of both ligand and receptor, etc. One of the best ways to limit the exhausting use of scoring function is to reduce the complexity in the degrees of freedom of the conformational landscape either by accounting all possible experimental data supporting sampling or by introducing a united scoring model that accounts every vital parameters influencing the conformational landscape.

4 COMPUTATIONAL TOOLS/SERVERS AVAILABLE FOR DOCKING

There are several standalone applications as well as web servers available to perform docking studies. A list of few such tools, in no particular order, is included here for the reader's reference (Table 12.1).

5 CONCLUSION

Overall, this chapter outlines the applications and challenges of molecular docking approaches in various fields' right from structural biology to pharmaceutical sciences. Despite the potential applications of docking methods in handling various molecular interactions, there is lot of scopes in improving the scoring function by considering the features addressed in the section describing the challenges in molecular docking. Due to the application of molecular docking approaches, the recent decades have witnessed higher success rate in the process of drug design. Besides the multi-facet role of molecular docking approaches, it is also mandate to fine-tune and accelerate the drug design processes through emerging novel approaches, techniques and tools to face the future challenges and the days are not far to solely confront personalized medicine through computational approaches.

TABLE 12.1
List of Various Docking Tools and Servers.

S.No	Tool	Scoring Function	URL	Online/ Standalone	References
1	ATTRACT	Normal mode analysis	www.attract.ph.tum.de	Online	de Vries et al. (2015)
2	AutoDock	Lamarckian genetic algorithm and empirical free energy scoring function	http://autodock. scripps.edu/	Standalone	Morris et al. (2009)
3	CDOCKER	simulated annealing—based algorithm	https://www.3dsbiovia. com/	Standalone	Gagnon et al. (2016)
4	ClusPro	Fast Fourier transform	https://cluspro.org	Online	Kozakov et al. (2017)
5	DOCK	Incremental construction and random conformation search, Coulombic and Lennard-Jones grid-based scoring function	http://dock.compbio. ucsf.edu/	Standalone	Allen et al. (2015)
6	DSX[ONLINE]	Knowledge-based scoring function	http://pc1664. pharmazie.uni-marburg.de/drugscore/	Online	Neudert and Klebe (2011)
7	EADock2	Hybrid sampling engine and multiobjective scoring function, tree-based DSS algorithm	http://www.swissdock. ch	Online	Grosdidier et al. (2011a, 2011b)
8	FDS	Graph theory and a recursive distance geometry algorithm	NA	Standalone	Taylor et al. (2003)
9	FLEXX	Incremental build-based Bohm's empirical scoring function	https://www.biosolveit. de/FlexX/	Standalone	Rarey et al. (1996)
10	FTDOCK	Genetic algorithm, Fourier transform rigid body docking	http://www.sbg.bio.ic. ac.uk/docking/ftdock. html	Standalone	Gabb et al. (1997)
11	GalaxyPepDock	Local peptide docking	http://galaxy.seoklab. org/cgi-bin/submit.cgi? type=PEPDOCK	Online	Lee et al. (2015)
12	GEMDOCK	Empirical scoring function, genetic algorithm	http://gemdock.life. nctu.edu.tw/dock/	Standalone	Yang and Chen (2004)
13	Glide	Empirical scoring function	https://www. schrodinger.com/glide	Standalone	Friesner et al. (2006)
14	GOLD	Force field—based scoring function and genetic algorithm	https://www.ccdc.cam. ac.uk/solutions/csd-discovery/components/ gold/	Standalone	Verdonk et al. (2003)
15	HADDOCK	Simulated annealing	http://www.bonvinlab. org/software/	Online or standalone	de Vries et al. (2010)
16	Hex	Spherical polar Fourier correlations	http://hexserver.loria.fr/	Online or standalone	Macindoe et al. (2010)

Continued

TABLE 12.1
List of Various Docking Tools and Servers.—cont'd

S.No	Tool	Scoring Function	URL	Online/ Standalone	References
17	ICM-Dock	Force field—based scoring function and Monte Carlo—based sampling	http://www.molsoft.com/docking.html		Neves et al. (2012)
18	MEGADOCK	Fast Fourier transform	http://www.bi.cs.titech.ac.jp/megadock/	Standalone	Ohue et al. (2014)
19	Molegro Virtual Docker	Force field—based scoring function	http://molexus.io/molegro-virtual-docker/		Bitencourt-Ferreira and de Azevedo (2019)
20	PatchDock	Geometric hashing	http://bioinfo3d.cs.tau.ac.il/PatchDock/	Online	Schneidman-Duhovny et al. (2005)
21	pyDock	Fast Fourier transform	https://life.bsc.es/pid/pydock/	Standalone	Cheng et al. (2007)
22	rDock	Force field—based scoring function and genetic algorithm	http://sourceforge.net/projects/rdock/		Ruiz-Carmona et al. (2014)
23	RosettaDock	Monte Carlo	http://rosie.rosettacommons.org/docking2/submit	Online	Lyskov and Gray (2008)
24	Surflex-Dock	Hammerhead fragment-based algorithm and genetic algorithm	https://omictools.com/surflex-dock-tool	Standalone	Spitzer and Jain (2012)
25	SwissDock	Hybrid sampling engine and multiobjective scoring function, tree-based DSS algorithm	http://www.swissdock.ch/docking	Online	Grosdidier et al. (2011a, 2011b)
26	ZDOCK	Fast Fourier transform	http://zdock.umassmed.edu/	Standalone Online	Pierce et al. (2014)

REFERENCES

Abad-Zapatero, C. (2007). A Sorcerer's apprentice and the rule of five: From rule-of-thumb to commandment and beyond. *Drug Discovery Today*, *12*(23—24), 995—997. https://doi.org/10.1016/j.drudis.2007.10.022.

Abdulrahman, A. Y., Khazali, A. S., Teoh, T. C., Rothan, H. A., & Yusof, R. (2019). Novel peptides inhibit Zika NS2B-NS3 serine protease and virus replication in human hepatic cell line. *International Journal of Peptide Research and Therapeutics*, *25*(4), 1659—1668. https://doi.org/10.1007/s10989-019-09808-4.

Akiyama, T. (2000). Wnt/beta-catenin signaling. *Cytokine & Growth Factor Reviews*, *11*(4), 273—282. https://doi.org/10.1016/s1359-6101(00)00011-3.

Allen, W. J., Balius, T. E., Mukherjee, S., Brozell, S. R., Moustakas, D. T., Lang, P. T., Case, D. A., Kuntz, I. D., & Rizzo, R. C. (2015). DOCK 6: Impact of new features and current docking performance. *Journal of Computational Chemistry*, *36*(15), 1132—1156. https://doi.org/10.1002/jcc.23905.

Armacost, K. A., Riniker, S., & Cournia, Z. (2020). Novel directions in free energy methods and applications. *Journal of Chemical Information and Modeling*, *60*(1), 1—5. https://doi.org/10.1021/acs.jcim.9b01174.

Ayers, M. (2012). ChemSpider: The free chemical database. *Reference Reviews*, *26*(7), 45—46. https://doi.org/10.1108/09504121211271059.

Baell, J. B., & Holloway, G. A. (2010). New substructure filters for removal of pan assay interference compounds (PAINS) from screening libraries and for their exclusion in bioassays. *Journal of Medicinal Chemistry*, *53*(7), 2719—2740. https://doi.org/10.1021/jm901137j.

Barman, A., & Prabhakar, R. (2012). Protonation states of the catalytic dyad of beta-secretase (BACE1) in the presence of chemically diverse inhibitors: A molecular docking study. *Journal of Chemical Information and Modeling, 52*(5), 1275–1287. https://doi.org/10.1021/ci200611t.

Betzi, S., Restouin, A., Opi, S., Arold, S. T., Parrot, I., Guerlesquin, F., Morelli, X., & Collette, Y. (2007). Protein protein interaction inhibition (2P2I) combining high throughput and virtual screening: Application to the HIV-1 Nef protein. *Proceedings of the National Academy of Sciences of the United States of America, 104*(49), 19256–19261. https://doi.org/10.1073/pnas.0707130104.

Bitencourt-Ferreira, G., & de Azevedo, W. F., Jr. (2019). Molegro virtual docker for docking. *Methods in Molecular Biology, 2053,* 149–167. https://doi.org/10.1007/978-1-4939-9752-7_10.

Breiman, L. (2001). Random forests. *Machine Learning, 45*(1), 5–32. https://doi.org/10.1023/A:1010933404324.

Brik, A., & Wong, C. H. (2003). HIV-1 protease: Mechanism and drug discovery. *Organic and Biomolecular Chemistry, 1*(1), 5–14. https://doi.org/10.1039/b208248a.

Broos, K., Trekels, M., Jose, R. A., Demeulemeester, J., Vandenbulcke, A., Vandeputte, N., Venken, T., Egle, B., De Borggraeve, W. M., Deckmyn, H., & De Maeyer, M. (2012). Identification of a small molecule that modulates platelet glycoprotein Ib-von Willebrand factor interaction. *Journal of Biological Chemistry, 287*(12), 9461–9472. https://doi.org/10.1074/jbc.M111.311431.

Brownlee, M. (2001). Biochemistry and molecular cell biology of diabetic complications. *Nature, 414*(6865), 813–820. https://doi.org/10.1038/414813a.

Cai, H., Liu, Q., Gao, D., Wang, T., Chen, T., Yan, G., Chen, K., Xu, Y., Wang, H., Li, Y., & Zhu, W. (2015). Novel fatty acid binding protein 4 (FABP4) inhibitors: Virtual screening, synthesis and crystal structure determination. *European Journal of Medicinal Chemistry, 90,* 241–250. https://doi.org/10.1016/j.ejmech.2014.11.020.

Cheng, T. M., Blundell, T. L., & Fernandez-Recio, J. (2007). pyDock: electrostatics and desolvation for effective scoring of rigid-body protein-protein docking. *Proteins, 68*(2), 503–515. https://doi.org/10.1002/prot.21419.

Chen, Y., & Shoichet, B. K. (2009). Molecular docking and ligand specificity in fragment-based inhibitor discovery. *Nature Chemical Biology, 5*(5), 358–364. https://doi.org/10.1038/nchembio.155.

Choudhary, S., Malik, Y. S., & Tomar, S. (2020). Identification of SARS-CoV-2 cell entry inhibitors by drug repurposing using in silico structure-based virtual screening approach. *Frontiers in Immunology, 11,* 1664. https://doi.org/10.3389/fimmu.2020.01664.

Cocco, L., Roth, B., Temple, C., Jr., Montgomery, J. A., London, R. E., & Blakley, R. L. (1983). Protonated state of methotrexate, trimethoprim, and pyrimethamine bound to dihydrofolate reductase. *Archives of Biochemistry and Biophysics, 226*(2), 567–577. https://doi.org/10.1016/0003-9861(83)90326-0.

Cooper, D. R., Porebski, P. J., Chruszcz, M., & Minor, W. (2011). X-ray crystallography: assessment and validation of protein–small molecule complexes for drug discovery.

Expert Opinion on Drug Discovery, 6(8), 771–782. https://doi.org/10.1517/17460441.2011.585154.

Cortes, C., & Vapnik, V. (1995). Support-vector networks. *Machine Learning, 20*(3), 273–297. https://doi.org/10.1007/BF00994018.

Cournia, Z., Allen, B. K., Beuming, T., Pearlman, D. A., Radak, B. K., & Sherman, W. (2020). Rigorous free energy simulations in virtual screening. *Journal of Chemical Information and Modeling.* https://doi.org/10.1021/acs.jcim.0c00116.

Dakshanamurthy, S., Issa, N. T., Assefnia, S., Seshasayee, A., Peters, O. J., Madhavan, S., Uren, A., Brown, M. L., & Byers, S. W. (2012). Predicting new indications for approved drugs using a proteochemometric method. *Journal of Medicinal Chemistry, 55*(15), 6832–6848. https://doi.org/10.1021/jm300576q.

Daldrop, P., Reyes, F. E., Robinson, D. A., Hammond, C. M., Lilley, D. M., Batey, R. T., & Brenk, R. (2011). Novel ligands for a purine riboswitch discovered by RNA-ligand docking. *Chemical Biology, 18*(3), 324–335. https://doi.org/10.1016/j.chembiol.2010.12.020.

Damm-Ganamet, K. L., Bembenek, S. D., Venable, J. W., Castro, G. G., Mangelschots, L., Peeters, D. C., McAllister, H. M., Edwards, J. P., Disepio, D., & Mirzadegan, T. (2016). A prospective virtual screening study: Enriching hit rates and designing focus libraries to find inhibitors of PI3Kdelta and PI3Kgamma. *Journal of Medicinal Chemistry, 59*(9), 4302–4313. https://doi.org/10.1021/acs.jmedchem.5b01974.

Durrant, J. D., Carlson, K. E., Martin, T. A., Offutt, T. L., Mayne, C. G., Katzenellenbogen, J. A., & Amaro, R. E. (2015). Neural-network scoring functions identify structurally novel estrogen-receptor ligands. *Journal of Chemical Information and Modeling, 55*(9), 1953–1961. https://doi.org/10.1021/acs.jcim.5b00241.

Eleftheriou, P., Amanatidou, D., Petrou, A., & Geronikaki, A. (2020). In silico evaluation of the effectivity of approved protease inhibitors against the main protease of the novel SARS-CoV-2 virus. *Molecules, 25*(11). https://doi.org/10.3390/molecules25112529.

Elfiky, A. A. (2020). Anti-HCV, nucleotide inhibitors, repurposing against COVID-19. *Life Sciences, 248,* 117477. https://doi.org/10.1016/j.lfs.2020.117477.

Eswar, N., Webb, B., Marti-Renom, M. A., Madhusudhan, M. S., Eramian, D., Shen, M. Y., Pieper, U., & Sali, A. (2006). Comparative protein structure modeling using modeller. *Current Protocols in Bioinformatics.* https://doi.org/10.1002/0471250953.bi0506s15 (Chapter 5), Unit-5 6.

Ferreira, L. G., Dos Santos, R. N., Oliva, G., & Andricopulo, A. D. (2015). Molecular docking and structure-based drug design strategies. *Molecules, 20*(7), 13384–13421. https://doi.org/10.3390/molecules200713384.

Forli, S. (2015). Charting a path to success in virtual screening. *Molecules, 20*(10), 18732–18758. https://doi.org/10.3390/molecules201018732.

Forli, S., & Botta, M. (2007). Lennard-Jones potential and dummy atom settings to overcome the AUTODOCK limitation in treating flexible ring systems. *Journal of Chemical Information and Modeling,* 47(4), 1481−1492. https://doi.org/10.1021/ci700036j.

Friesner, R. A., Banks, J. L., Murphy, R. B., Halgren, T. A., Klicic, J. J., Mainz, D. T., Repasky, M. P., Knoll, E. H., Shelley, M., Perry, J. K., Shaw, D. E., Francis, P., & Shenkin, P. S. (2004). Glide: A new approach for rapid, accurate docking and scoring. 1. Method and assessment of docking accuracy. *Journal of Medicinal Chemistry,* 47(7), 1739−1749. https://doi.org/10.1021/jm0306430.

Friesner, R. A., Murphy, R. B., Repasky, M. P., Frye, L. L., Greenwood, J. R., Halgren, T. A., Sanschagrin, P. C., & Mainz, D. T. (2006). Extra precision glide: Docking and scoring incorporating a model of hydrophobic enclosure for protein-ligand complexes. *Journal of Medicinal Chemistry,* 49(21), 6177−6196. https://doi.org/10.1021/jm051256o.

Gabb, H. A., Jackson, R. M., & Sternberg, M. J. (1997). Modelling protein docking using shape complementarity, electrostatics and biochemical information. *Journal of Molecular Biology,* 272(1), 106−120. https://doi.org/10.1006/jmbi.1997.1203.

Gagnon, J. K., Law, S. M., & Brooks, C. L., 3rd (2016). Flexible CDOCKER: Development and application of a pseudo-explicit structure-based docking method within CHARMM. *Journal of Computational Chemistry,* 37(8), 753−762. https://doi.org/10.1002/jcc.24259.

Gandhimathi, A., & Sowdhamini, R. (2016). Molecular modelling of human 5-hydroxytryptamine receptor (5-HT2A) and virtual screening studies towards the identification of agonist and antagonist molecules. *Journal of Biomolecular Structure and Dynamics,* 34(5), 952−970. https://doi.org/10.1080/07391102.2015.1062802.

Garvie, C. W., Hagman, J., & Wolberger, C. (2001). Structural studies of Ets-1/Pax5 complex formation on DNA. *Molecular Cell,* 8(6), 1267−1276. https://doi.org/10.1016/s1097-2765(01)00410-5.

Garvie, C. W., & Wolberger, C. (2001). Recognition of specific DNA sequences. *Molecular Cell,* 8(5), 937−946. https://doi.org/10.1016/s1097-2765(01)00392-6.

Gasteiger, J., & Marsili, M. (1980). Iterative partial equalization of orbital electronegativity—a rapid access to atomic charges. *Tetrahedron,* 36(22), 3219−3228. https://doi.org/10.1016/0040-4020(80)80168-2.

Gaulton, A., Bellis, L. J., Bento, A. P., Chambers, J., Davies, M., Hersey, A., Light, Y., McGlinchey, S., Michalovich, D., Al-Lazikani, B., & Overington, J. P. (2012). ChEMBL: A large-scale bioactivity database for drug discovery. *Nucleic Acids Research,* 40(Database issue), D1100−D1107. https://doi.org/10.1093/nar/gkr777.

Gentile, D., Patamia, V., Scala, A., Sciortino, M. T., Piperno, A., & Rescifina, A. (2020). Putative inhibitors of SARS-CoV-2 main protease from A library of marine natural products: A virtual screening and molecular modeling study. *Marine Drugs,* 18(4). https://doi.org/10.3390/md18040225.

Geppert, H., Vogt, M., & Bajorath, J. (2010). Current trends in ligand-based virtual screening: Molecular representations, data mining methods, new application areas, and performance evaluation. *Journal of Chemical Information and Modeling,* 50(2), 205−216. https://doi.org/10.1021/ci900419k.

Giri, M., Maulik, A., & Singh, M. (2020). Signatures of specific DNA binding by the AT-rich interaction domain of BAF250a. *Biochemistry,* 59(1), 100−113. https://doi.org/10.1021/acs.biochem.9b00852.

Graham, T. E., Wason, C. J., Bluher, M., & Kahn, B. B. (2007). Shortcomings in methodology complicate measurements of serum retinol binding protein (RBP4) in insulin-resistant human subjects. *Diabetologia,* 50(4), 814−823. https://doi.org/10.1007/s00125-006-0557-0.

Gregory, D.,H., Christopher, J.,C., & Donald, G.,T. (1998). *Universal quantum mechanical model for solvation free energies based on gas-phase geometries.*

Grosdidier, A., Zoete, V., & Michielin, O. (2011a). Fast docking using the CHARMM force field with EADock DSS. *Journal of Computational Chemistry,* 32(10), 2149−2159. https://doi.org/10.1002/jcc.21797.

Grosdidier, A., Zoete, V., & Michielin, O. (2011b). SwissDock, a protein-small molecule docking web service based on EADock DSS. *Nucleic Acids Research,* 39(Web Server issue), W270−W277. https://doi.org/10.1093/nar/gkr366.

Hage Melim, L., Federico, L. B., de Oliveira, N. K. S., Francisco, V. C. C., Correia, L. C., de Lima, H. B., Gomes, S. Q., Barcelos, M. P., Francischini, I. A. G., & da Silva, C. (2020). Virtual screening, ADME/Tox predictions and the drug repurposing concept for future use of old drugs against the COVID-19. *Life Sciences,* 256, 117963. https://doi.org/10.1016/j.lfs.2020.117963.

Hanwell, M. D., Curtis, D. E., Lonie, D. C., Vandermeersch, T., Zurek, E., & Hutchison, G. R. (2012). Avogadro: An advanced semantic chemical editor, visualization, and analysis platform. *Journal of Cheminformatics,* 4(1), 17. https://doi.org/10.1186/1758-2946-4-17.

Harte, W. E., & Beveridge, D. L. (1993). Prediction of the protonation state of the active site aspartyl residues in HIV-1 protease-inhibitor complexes via molecular dynamics simulation. *Journal of the American Chemical Society,* 115(10), 3883−3886. https://doi.org/10.1021/ja00063a005.

Hillisch, A., Pineda, L. F., & Hilgenfeld, R. (2004). Utility of homology models in the drug discovery process. *Drug Discovery Today,* 9(15), 659−669. https://doi.org/10.1016/S1359-6446(04)03196-4.

Hurle, M. R., Yang, L., Xie, Q., Rajpal, D. K., Sanseau, P., & Agarwal, P. (2013). Computational drug repositioning: From data to therapeutics. *Clinical Pharmacology & Therapeutics,* 93(4), 335−341. https://doi.org/10.1038/clpt.2013.1.

Irwin, J. J., Raushel, F. M., & Shoichet, B. K. (2005). Virtual screening against metalloenzymes for inhibitors and substrates. *Biochemistry,* 44(37), 12316−12328. https://doi.org/10.1021/bi050801k.

Irwin, J. J., & Shoichet, B. K. (2005). ZINC—a free database of commercially available compounds for virtual screening. *Journal of Chemical Information and Modeling,* 45(1), 177−182. https://doi.org/10.1021/ci049714+.

Jain, A. N. (2007). Surflex-dock 2.1: Robust performance from ligand energetic modeling, ring flexibility, and knowledge-based search. *Journal of Computer-Aided Molecular Design*, *21*(5), 281−306. https://doi.org/10.1007/s10822-007-9114-2.

Jiang, L., & Rizzo, R. C. (2015). Pharmacophore-based similarity scoring for DOCK. *The Journal of Physical Chemistry B*, *119*(3), 1083−1102. https://doi.org/10.1021/jp506555w.

Jimenez-Alberto, A., Ribas-Aparicio, R. M., Aparicio-Ozores, G., & Castelan-Vega, J. A. (2020). Virtual screening of approved drugs as potential SARS-CoV-2 main protease inhibitors. *Computational Biology and Chemistry*, *88*, 107325. https://doi.org/10.1016/j.compbiolchem.2020.107325.

Jones, T. A., Zou, J. Y., Cowan, S. W., & Kjeldgaard, M. (1991). Improved methods for building protein models in electron density maps and the location of errors in these models. *Acta Crystallographica Section A*, *47*(Pt 2), 110−119. https://doi.org/10.1107/s0108767390010224.

Jorgensen, W. L. (2004). The many roles of computation in drug discovery. *Science*, *303*(5665), 1813−1818. https://doi.org/10.1126/science.1096361.

Jorgensen, W. L., & Tirado-Rives, J. (1988). The OPLS [optimized potentials for liquid simulations] potential functions for proteins, energy minimizations for crystals of cyclic peptides and crambin. *Journal of the American Chemical Society*, *110*(6), 1657−1666. https://doi.org/10.1021/ja00214a001.

Joseph-McCarthy, D., Baber, J. C., Feyfant, E., Thompson, D. C., & Humblet, C. (2007). Lead optimization via high-throughput molecular docking. *Current Opinion in Drug Discovery & Development*, *10*(3), 264−274.

Kalhor, H., Poorebrahim, M., Rahimi, H., Shabani, A. A., Karimipoor, M., Akbari Eidgahi, M. R., & Teimoori-Toolabi, L. (2018). Structural and dynamic characterization of human Wnt2-Fzd7 complex using computational approaches. *Journal of Molecular Modeling*, *24*(10), 274. https://doi.org/10.1007/s00894-018-3788-3.

Ke, Y. Y., Singh, V. K., Coumar, M. S., Hsu, Y. C., Wang, W. C., Song, J. S., Chen, C. H., Lin, W. H., Wu, S. H., Hsu, J. T., Shih, C., & Hsieh, H. P. (2015). Homology modeling of DFG-in FMS-like tyrosine kinase 3 (FLT3) and structure-based virtual screening for inhibitor identification. *Scientific Reports*, *5*, 11702. https://doi.org/10.1038/srep11702.

Kim, S., Thiessen, P. A., Bolton, E. E., Chen, J., Fu, G., Gindulyte, A., Han, L., He, J., He, S., Shoemaker, B. A., Wang, J., Yu, B., Zhang, J., & Bryant, S. H. (2016). PubChem substance and compound databases. *Nucleic Acids Research*, *44*(D1), D1202−D1213. https://doi.org/10.1093/nar/gkv951.

Kinnings, S. L., Liu, N., Buchmeier, N., Tonge, P. J., Xie, L., & Bourne, P. E. (2009). Drug discovery using chemical systems biology: Repositioning the safe medicine comtan to treat multi-drug and extensively drug resistant tuberculosis. *PLoS Computational Biology*, *5*(7), e1000423. https://doi.org/10.1371/journal.pcbi.1000423.

Kollman, P. A., Massova, I., Reyes, C., Kuhn, B., Huo, S., Chong, L., Lee, M., Lee, T., Duan, Y., Wang, W., Donini, O., Cieplak, P., Srinivasan, J., Case, D. A., & Cheatham, T. E., 3rd (2000). Calculating structures and free energies of complex molecules: Combining molecular mechanics and continuum models. *Accounts of Chemical Research*, *33*(12), 889−897. https://doi.org/10.1021/ar000033j.

Kozakov, D., Hall, D. R., Xia, B., Porter, K. A., Padhorny, D., Yueh, C., Beglov, D., & Vajda, S. (2017). The ClusPro web server for protein-protein docking. *Nature Protocols*, *12*(2), 255−278. https://doi.org/10.1038/nprot.2016.169.

Kruger, A., Zimbres, F. M., Kronenberger, T., & Wrenger, C. (2018). Molecular modeling applied to nucleic acid-based molecule development. *Biomolecules*, *8*(3). https://doi.org/10.3390/biom8030083.

Laio, A., & Parrinello, M. (2002). Escaping free-energy minima. *Proceedings of the National Academy of Sciences of the United States of America*, *99*(20), 12562−12566. https://doi.org/10.1073/pnas.202427399.

Lavecchia, A. (2015). Machine-learning approaches in drug discovery: Methods and applications. *Drug Discovery Today*, *20*(3), 318−331. https://doi.org/10.1016/j.drudis.2014.10.012.

Lee, H. M., Chan, D. S., Yang, F., Lam, H. Y., Yan, S. C., Che, C. M., … Leung, C. H. (2010). Identification of natural product fonsecin B as a stabilizing ligand of c-myc G-quadruplex DNA by high-throughput virtual screening. *Chemical Communications (Cambridge)*, *46*(26), 4680−4682. https://doi.org/10.1039/b926359d.

Lee, H., Heo, L., Lee, M. S., & Seok, C. (2015). GalaxyPepDock: A protein-peptide docking tool based on interaction similarity and energy optimization. *Nucleic Acids Research*, *43*(W1), W431−W435. https://doi.org/10.1093/nar/gkv495.

Lipinski, C. A., Lombardo, F., Dominy, B. W., & Feeney, P. J. (2001). Experimental and computational approaches to estimate solubility and permeability in drug discovery and development settings (Reprinted from Advanced Drug Delivery Reviews, vol 23, pg 3−25, 1997) *Advanced Drug Delivery Reviews*, *46*(1−3), 3−26. https://doi.org/10.1016/S0169-409x(00)00129-0.

Liu, T., Lin, Y., Wen, X., Jorissen, R. N., & Gilson, M. K. (2007). BindingDB: A web-accessible database of experimentally determined protein-ligand binding affinities. *Nucleic Acids Research*, *35*(Database issue), D198−D201. https://doi.org/10.1093/nar/gkl999.

Luo, J., Wei, W., Waldispuhl, J., & Moitessier, N. (2019). Challenges and current status of computational methods for docking small molecules to nucleic acids. *European Journal of Medicinal Chemistry*, *168*, 414−425. https://doi.org/10.1016/j.ejmech.2019.02.046.

Lyskov, S., & Gray, J. J. (2008). The RosettaDock server for local protein-protein docking. *Nucleic Acids Research*, *36*(Web Server issue), W233−W238. https://doi.org/10.1093/nar/gkn216.

Macalino, S. J. Y., Basith, S., Clavio, N. A. B., Chang, H., Kang, S., & Choi, S. (2018). Evolution of in silico strategies for protein-protein interaction drug discovery. *Molecules*, *23*(8). https://doi.org/10.3390/molecules23081963.

Macalino, S. J., Gosu, V., Hong, S., & Choi, S. (2015). Role of computer-aided drug design in modern drug discovery. *Archives of Pharmacal Research*, *38*(9), 1686−1701. https://doi.org/10.1007/s12272-015-0640-5.

Macarron, R., Banks, M. N., Bojanic, D., Burns, D. J., Cirovic, D. A., Garyantes, T., Green, D. V., Hertzberg, R. P., Janzen, W. P., Paslay, J. W., Schopfer, U., & Sittampalam, G. S. (2011). Impact of high-throughput screening in biomedical research. *Nature Reviews Drug Discovery, 10*(3), 188−195. https://doi.org/10.1038/nrd3368.

Macindoe, G., Mavridis, L., Venkatraman, V., Devignes, M. D., & Ritchie, D. W. (2010). HexServer: An FFT-based protein docking server powered by graphics processors. *Nucleic Acids Research, 38*(Web Server issue), W445−W449. https://doi.org/10.1093/nar/gkq311.

March-Vila, E., Pinzi, L., Sturm, N., Tinivella, A., Engkvist, O., Chen, H., & Rastelli, G. (2017). On the integration of in silico drug design methods for drug repurposing. *Frontiers in Pharmacology, 8*, 298. https://doi.org/10.3389/fphar.2017.00298.

Marrakchi, H., Laneelle, G., & Quemard, A. K. (2000). InhA, a target of the antituberculous drug isoniazid, is involved in a mycobacterial fatty acid elongation system, FAS-II. *Microbiology, 146*(Pt 2), 289−296. https://doi.org/10.1099/00221287-146-2-289.

Matsuno, K., Masuda, Y., Uehara, Y., Sato, H., Muroya, A., Takahashi, O., Yokotagawa, T., Furuya, T., Okawara, T., Otsuka, M., Ogo, N., Ashizawa, T., Oshita, C., Tai, S., Ishii, H., Akiyama, Y., & Asai, A. (2010). Identification of a new series of STAT3 inhibitors by virtual screening. *ACS Medicinal Chemistry Letters, 1*(8), 371−375. https://doi.org/10.1021/ml1000273.

Matter, H. A. S.,C. (2011). Applications and success stories in virtual screening. In *Virtual screening* (pp. 319−358).

Meyer-Almes, F. J. (2020). Repurposing approved drugs as potential inhibitors of 3CL-protease of SARS-CoV-2: Virtual screening and structure based drug design. *Computational Biology and Chemistry, 88*, 107351. https://doi.org/10.1016/j.compbiolchem.2020.107351.

Miller, G. (2006). Scientific publishing - A scientist's nightmare: Software problem leads to five retractions. *Science, 314*(5807), 1856−1857. https://doi.org/10.1126/science.314.5807.1856.

Mitchell, T. J., & John, S. (2005). Signal transducer and activator of transcription (STAT) signalling and T-cell lymphomas. *Immunology, 114*(3), 301−312. https://doi.org/10.1111/j.1365-2567.2005.02091.x.

Moitessier, N., Westhof, E., & Hanessian, S. (2006). Docking of aminoglycosides to hydrated and flexible RNA. *Journal of Medicinal Chemistry, 49*(3), 1023−1033. https://doi.org/10.1021/jm0508437.

Moitessier, N., Englebienne, P., Lee, D., Lawandi, J., & Corbeil, C. R. (2008). Towards the development of universal, fast and highly accurate docking/scoring methods: A long way to go. *British Journal of Pharmacology, 153*, S7−S26. https://doi.org/10.1038/sj.bjp.0707515.

Morris, G. M., Huey, R., Lindstrom, W., Sanner, M. F., Belew, R. K., Goodsell, D. S., & Olson, A. J. (2009). AutoDock4 and AutoDockTools4: Automated docking with selective receptor flexibility. *Journal of Computational Chemistry, 30*(16), 2785−2791. https://doi.org/10.1002/jcc.21256.

Moustakas, D. T., Lang, P. T., Pegg, S., Pettersen, E., Kuntz, I. D., Brooijmans, N., & Rizzo, R. C. (2006). Development and validation of a modular, extensible docking program: DOCK 5. *Journal of Computer-Aided Molecular Design, 20*(10−11), 601−619. https://doi.org/10.1007/s10822-006-9060-4.

Mukherjee, S., Tworowski, D., Detroja, R., Mukherjee, S. B., & Frenkel-Morgenstern, M. (2020). Immunoinformatics and structural analysis for identification of immunodominant epitopes in SARS-CoV-2 as potential vaccine targets. *Vaccines (Basel), 8*(2). https://doi.org/10.3390/vaccines8020290.

Neudert, G., & Klebe, G. (2011). DSX: A knowledge-based scoring function for the assessment of protein-ligand complexes. *Journal of Chemical Information and Modeling, 51*(10), 2731−2745. https://doi.org/10.1021/ci200274q.

Neves, M. A., Totrov, M., & Abagyan, R. (2012). Docking and scoring with ICM: The benchmarking results and strategies for improvement. *Journal of Computer-Aided Molecular Design, 26*(6), 675−686. https://doi.org/10.1007/s10822-012-9547-0.

O'Boyle, N. M., Banck, M., James, C. A., Morley, C., Vandermeersch, T., & Hutchison, G. R. (2011). Open Babel: An open chemical toolbox. *Journal of Cheminformatics, 3*, 33. https://doi.org/10.1186/1758-2946-3-33.

Ohue, M., Shimoda, T., Suzuki, S., Matsuzaki, Y., Ishida, T., & Akiyama, Y. (2014). MEGADOCK 4.0: An ultra-high-performance protein-protein docking software for heterogeneous supercomputers. *Bioinformatics, 30*(22), 3281−3283. https://doi.org/10.1093/bioinformatics/btu532.

Pantsar, T., Singha, P., Nevalainen, T. J., Koshevoy, I., Leppanen, J., Poso, A., Niskanen, J. M. A., Pasonen-Seppanen, S., Savinainen, J. R., Laitinen, T., & Laitinen, J. T. (2017). Design, synthesis, and biological evaluation of 2,4-dihydropyrano[2,3-c]pyrazole derivatives as autotaxin inhibitors. *European Journal of Pharmaceutical Sciences, 107*, 97−111. https://doi.org/10.1016/j.ejps.2017.07.002.

Pant, S., Singh, M., Ravichandiran, V., Murty, U. S. N., & Srivastava, H. K. (2020). Peptide-like and small-molecule inhibitors against COVID-19. *Journal of Biomolecular Structure and Dynamics*, 1−10. https://doi.org/10.1080/07391102.2020.1757510.

Pauli, I., dos Santos, R. N., Rostirolla, D. C., Martinelli, L. K., Ducati, R. G., Timmers, L. F., Basso, L. A., Santos, D. S., Guido, R. V., Andricopulo, A. D., & Norberto de Souza, O. (2013). Discovery of new inhibitors of *Mycobacterium tuberculosis* InhA enzyme using virtual screening and a 3D-pharmacophore-based approach. *Journal of Chemical Information and Modeling, 53*(9), 2390−2401. https://doi.org/10.1021/ci400202t.

Perez-Castillo, Y., Helguera, A. M., Cordeiro, M., Tejera, E., Paz, Y. M. C., Sanchez-Rodriguez, A., Borges, M., & Cruz-Monteagudo, M. (2017). Fusing docking scoring functions improves the virtual screening performance for discovering Parkinson's disease dual target ligands. *Current Neuropharmacology, 15*(8), 1107−1116. https://doi.org/10.2174/1570159X15666170109143757.

Perola, E., Walters, W. P., & Charifson, P. S. (2004). A detailed comparison of current docking and scoring methods on systems of pharmaceutical relevance. *Proteins, 56*(2), 235–249. https://doi.org/10.1002/prot.20088.

Perryman, A. L., Santiago, D. N., Forli, S., Martins, D. S., & Olson, A. J. (2014). Virtual screening with AutoDock Vina and the common pharmacophore engine of a low diversity library of fragments and hits against the three allosteric sites of HIV integrase: Participation in the SAMPL4 protein-ligand binding challenge. *Journal of Computer-Aided Molecular Design, 28*(4), 429–441. https://doi.org/10.1007/s10822-014-9709-3.

Pierce, B. G., Wiehe, K., Hwang, H., Kim, B. H., Vreven, T., & Weng, Z. (2014). ZDOCK server: Interactive docking prediction of protein-protein complexes and symmetric multimers. *Bioinformatics, 30*(12), 1771–1773. https://doi.org/10.1093/bioinformatics/btu097.

Pinzi, L., & Rastelli, G. (2019). Molecular docking: Shifting paradigms in drug discovery. *International Journal of Molecular Sciences, 20*(18). https://doi.org/10.3390/ijms20184331.

Poulsen, A., William, A., Blanchard, S., Nagaraj, H., Williams, M., Wang, H., Lee, A., Sun, E., Teo, E. L., Tan, E., Goh, K. C., & Dymock, B. (2013). Structure-based design of nitrogen-linked macrocyclic kinase inhibitors leading to the clinical candidate SB1317/TG02, a potent inhibitor of cyclin dependant kinases (CDKs), Janus kinase 2 (JAK2), and Fms-like tyrosine kinase-3 (FLT3). *Journal of Molecular Modeling, 19*(1), 119–130. https://doi.org/10.1007/s00894-012-1528-7.

Ragoza, M., Hochuli, J., Idrobo, E., Sunseri, J., & Koes, D. R. (2017). Protein-ligand scoring with convolutional neural networks. *Journal of Chemical Information and Modeling, 57*(4), 942–957. https://doi.org/10.1021/acs.jcim.6b00740.

Rajagopalan, M., Balasubramanian, S., & Ramaswamy, A. (2017). Structural dynamics of wild type and mutated forms of human L1 endonuclease and insights into its sequence specific nucleic acid binding mechanism: A molecular dynamics study. *Journal of Molecular Graphics and Modelling, 76*, 43–55. https://doi.org/10.1016/j.jmgm.2017.07.002.

Rarey, M., Kramer, B., Lengauer, T., & Klebe, G. (1996). A fast flexible docking method using an incremental construction algorithm. *Journal of Molecular Biology, 261*(3), 470–489. https://doi.org/10.1006/jmbi.1996.0477.

Rastelli, G., Degliesposti, G., Del Rio, A., & Sgobba, M. (2009). Binding estimation after refinement, a new automated procedure for the refinement and rescoring of docked ligands in virtual screening. *Chemical Biology & Drug Design, 73*(3), 283–286. https://doi.org/10.1111/j.1747-0285.2009.00780.x.

Read, J. A., Wilkinson, K. W., Tranter, R., Sessions, R. B., & Brady, R. L. (1999). Chloroquine binds in the cofactor binding site of *Plasmodium falciparum* lactate dehydrogenase. *Journal of Biological Chemistry, 274*(15), 10213–10218. https://doi.org/10.1074/jbc.274.15.10213.

Ren, J. X., Li, L. L., Zheng, R. L., Xie, H. Z., Cao, Z. X., Feng, S., Pan, Y. L., Chen, X., Wei, Y. Q., & Yang, S. Y. (2011). Discovery of novel Pim-1 kinase inhibitors by a hierarchical multistage virtual screening approach based on SVM model, pharmacophore, and molecular docking. *Journal of Chemical Information and Modeling, 51*(6), 1364–1375. https://doi.org/10.1021/ci100464b.

Rocha, M., Banuls, C., Bellod, L., Rovira-Llopis, S., Morillas, C., Sola, E., … Hernandez-Mijares, A. (2013). Association of serum retinol binding protein 4 with atherogenic dyslipidemia in morbid obese patients. *PLoS One, 8*(11), e78670. https://doi.org/10.1371/journal.pone.0078670.

Rohs, R., West, S. M., Sosinsky, A., Liu, P., Mann, R. S., & Honig, B. (2009). The role of DNA shape in protein-DNA recognition. *Nature, 461*(7268), 1248–1253. https://doi.org/10.1038/nature08473.

Ross, G. A., Morris, G. M., & Biggin, P. C. (2013). One size does not fit all: The limits of structure-based models in drug discovery. *Journal of Chemical Theory and Computation, 9*(9), 4266–4274. https://doi.org/10.1021/ct4004228.

Ruiz-Carmona, S., Alvarez-Garcia, D., Foloppe, N., Garmendia-Doval, A. B., Juhos, S., Schmidtke, P., Barril, X., Hubbard, R. E., & Morley, S. D. (2014). rDock: a fast, versatile and open source program for docking ligands to proteins and nucleic acids. *PLoS Computational Biology, 10*(4), e1003571. https://doi.org/10.1371/journal.pcbi.1003571.

Sable, R., & Jois, S. (2015). Surfing the protein-protein interaction surface using docking methods: Application to the design of PPI inhibitors. *Molecules, 20*(6), 11569–11603. https://doi.org/10.3390/molecules200611569.

Salmaso, V., & Moro, S. (2018). Bridging molecular docking to molecular dynamics in exploring ligand-protein recognition process: An overview. *Frontiers in Pharmacology, 9*, 923. https://doi.org/10.3389/fphar.2018.00923.

Sarvagalla, S., Cheung, C. H. A., Tsai, J.-Y., Hsieh, H. P., & Coumar, M. S. (2016). Disruption of protein–protein interactions: Hot spot detection, structure-based virtual screening and in vitro testing for the anti-cancer drug target — survivin. *RSC Advances, 6*(38), 31947–31959. https://doi.org/10.1039/C5RA22927H.

Schneidman-Duhovny, D., Inbar, Y., Nussinov, R., & Wolfson, H. J. (2005). PatchDock and SymmDock: Servers for rigid and symmetric docking. *Nucleic Acids Research, 33*(Web Server issue), W363–W367. https://doi.org/10.1093/nar/gki481.

Schulz-Gasch, T., & Stahl, M. (2004). Scoring functions for protein-ligand interactions: A critical perspective. *Drug Discovery Today Technologies, 1*(3), 231–239. https://doi.org/10.1016/j.ddtec.2004.08.004.

Scior, T., & Quiroga, I. (2019). Induced fit for cytochrome P450 3A4 based on molecular dynamics. *ADMET and DMPK, 7*, 252. https://doi.org/10.5599/admet.729.

Seiler, K. P., George, G. A., Happ, M. P., Bodycombe, N. E., Carrinski, H. A., Norton, S., Brudz, S., Sullivan, J. P., Muhlich, J., Serrano, M., Ferraiolo, P., Tolliday, N. J., Schreiber, S. L., & Clemons, P. A. (2008). ChemBank: A small-molecule screening and cheminformatics resource database. *Nucleic Acids Research, 36*(Database issue), D351–D359. https://doi.org/10.1093/nar/gkm843.

Shoichet, B. K., McGovern, S. L., Wei, B., & Irwin, J. J. (2002). Lead discovery using molecular docking. *Current Opinion in Chemical Biology*, 6(4), 439–446. https://doi.org/10.1016/s1367-5931(02)00339-3.

Singh, V. K., Chang, H. H., Kuo, C. C., Shiao, H. Y., Hsieh, H. P., & Coumar, M. S. (2017). Drug repurposing for chronic myeloid leukemia: In silico and in vitro investigation of DrugBank database for allosteric Bcr-Abl inhibitors. *Journal of Biomolecular Structure and Dynamics*, 35(8), 1833–1848. https://doi.org/10.1080/07391102.2016.1196462.

Singh, P., Ganjiwale, A., Howlett, A. C., & Cowsik, S. M. (2017). In silico interaction analysis of cannabinoid receptor interacting protein 1b (CRIP1b) – CB1 cannabinoid receptor. *Journal of Molecular Graphics and Modelling*, 77, 311–321. https://doi.org/10.1016/j.jmgm.2017.09.006.

Singh, P., Ganjiwale, A., Howlett, A. C., & Cowsik, S. M. (2019). Molecular interaction between distal C-terminal domain of the CB1 cannabinoid receptor and cannabinoid receptor interacting proteins (CRIP1a/CRIP1b). *Journal of Chemical Information and Modeling*, 59(12), 5294–5303. https://doi.org/10.1021/acs.jcim.9b00948.

Sitzmann, M., Ihlenfeldt, W. D., & Nicklaus, M. C. (2010). Tautomerism in large databases. *Journal of Computer-Aided Molecular Design*, 24(6–7), 521–551. https://doi.org/10.1007/s10822-010-9346-4.

Sivakumar, P. M., Selvaraj, N. V., Ramesh, G., Mohanapriya, J., Prabhawathi, V., & Doble, M. (2012). Computational approaches to improve aggrecanase-1 inhibitory activity of (4-keto) phenoxy) methyl biphenyl-4-sulfonamide: Group based QSAR and docking studies. *Medicinal Chemistry*, 8(4), 673–682. https://doi.org/10.2174/157340612801216247.

Slater, O., & Kontoyianni, M. (2019). The compromise of virtual screening and its impact on drug discovery. *Expert Opinion on Drug Discovery*, 14(7), 619–637. https://doi.org/10.1080/17460441.2019.1604677.

Souza, P. F. N., Lopes, F. E. S., Amaral, J. L., Freitas, C. D. T., & Oliveira, J. T. A. (2020). A molecular docking study revealed that synthetic peptides induced conformational changes in the structure of SARS-CoV-2 spike glycoprotein, disrupting the interaction with human ACE2 receptor. *International Journal of Biological Macromolecules*, 164, 66–76. https://doi.org/10.1016/j.ijbiomac.2020.07.174.

Spitzer, R., & Jain, A. N. (2012). Surflex-Dock: Docking benchmarks and real-world application. *Journal of Computer-Aided Molecular Design*, 26(6), 687–699. https://doi.org/10.1007/s10822-011-9533-y.

Stams, T., Chen, Y., Boriack-Sjodin, P. A., Hurt, J. D., Liao, J., May, J. A., Dean, T., Laipis, P., Silverman, D. N., & Christianson, D. W. (1998). Structures of murine carbonic anhydrase IV and human carbonic anhydrase II complexed with brinzolamide: Molecular basis of isozyme-drug discrimination. *Protein Science*, 7(3), 556–563. https://doi.org/10.1002/pro.5560070303.

Sterling, T., & Irwin, J. J. (2015). ZINC 15–ligand discovery for everyone. *Journal of Chemical Information and Modeling*, 55(11), 2324–2337. https://doi.org/10.1021/acs.jcim.5b00559.

Stumpf, M. P., Thorne, T., de Silva, E., Stewart, R., An, H. J., Lappe, M., & Wiuf, C. (2008). Estimating the size of the human interactome. *Proceedings of the National Academy of Sciences of the United States of America*, 105(19), 6959–6964. https://doi.org/10.1073/pnas.0708078105.

Sugita, Y., & Okamoto, Y. (1999). Replica-exchange molecular dynamics method for protein folding. *Chemical Physics Letters*, 314(1), 141–151. https://doi.org/10.1016/S0009-2614(99)01123-9.

Sun, H., Pan, P., Tian, S., Xu, L., Kong, X., Li, Y., Dan, L., & Hou, T. (2016). Constructing and validating high-performance MIEC-SVM models in virtual screening for kinases: A better way for actives discovery. *Scientific Reports*, 6, 24817. https://doi.org/10.1038/srep24817.

Sunseri, J., & Koes, D. R. (2016). Pharmit: Interactive exploration of chemical space. *Nucleic Acids Research*, 44(W1), W442–W448. https://doi.org/10.1093/nar/gkw287.

Sun, J., & Zhao, Z. (2010). A comparative study of cancer proteins in the human protein-protein interaction network. *BMC Genomics*, 11(Suppl. 3), S5. https://doi.org/10.1186/1471-2164-11-S3-S5.

Taylor, R. D., Jewsbury, P. J., & Essex, J. W. (2003). FDS: Flexible ligand and receptor docking with a continuum solvent model and soft-core energy function. *Journal of Computational Chemistry*, 24(13), 1637–1656. https://doi.org/10.1002/jcc.10295.

Tazikeh-Lemeski, E., Moradi, S., Raoufi, R., Shahlaei, M., Janlou, M. A. M., & Zolghadri, S. (2020). Targeting SARS-COV-2 non-structural protein 16: A virtual drug repurposing study. *Journal of Biomolecular Structure and Dynamics*, 1–14. https://doi.org/10.1080/07391102.2020.1779133.

Terali, K., Baddal, B., & Gulcan, H. O. (2020). Prioritizing potential ACE2 inhibitors in the COVID-19 pandemic: Insights from a molecular mechanics-assisted structure-based virtual screening experiment. *Journal of Molecular Graphics and Modelling*, 100, 107697. https://doi.org/10.1016/j.jmgm.2020.107697.

Tessaro, F., & Scapozza, L. (2020). How "protein-docking" translates into the new emerging field of docking small molecules to nucleic acids? *Molecules*, 25(12). https://doi.org/10.3390/molecules25122749.

Torabi, R., Bagherzadeh, K., Ghourchian, H., & Amanlou, M. (2016). An investigation on the interaction modes of a single-strand DNA aptamer and RBP4 protein: a molecular dynamic simulations approach. *Organic and Biomolecular Chemistry*, 14(34), 8141–8153. https://doi.org/10.1039/c6ob01094f.

Trott, O., & Olson, A. J. (2010). AutoDock Vina: Improving the speed and accuracy of docking with a new scoring function, efficient optimization, and multithreading. *Journal of Computational Chemistry*, 31(2), 455–461. https://doi.org/10.1002/jcc.21334.

Uitdehaag, J. C., Mosi, R., Kalk, K. H., van der Veen, B. A., Dijkhuizen, L., Withers, S. G., & Dijkstra, B. W. (1999). X-ray structures along the reaction pathway of cyclodextrin glycosyltransferase elucidate catalysis in the alpha-amylase family. *Nature Structural Biology*, 6(5), 432–436. https://doi.org/10.1038/8235.

Vannini, A., Volpari, C., Filocamo, G., Casavola, E. C., Brunetti, M., Renzoni, D., Chakravarty, P., Paolini, C., De Francesco, R., Gallinari, P., Steinkuhler, C., & Di Marco, S.

(2004). Crystal structure of a eukaryotic zinc-dependent histone deacetylase, human HDAC8, complexed with a hydroxamic acid inhibitor. *Proceedings of the National Academy of Sciences of the United States of America, 101*(42), 15064–15069. https://doi.org/10.1073/pnas.0404603101.

Verdonk, M. L., Cole, J. C., Hartshorn, M. J., Murray, C. W., & Taylor, R. D. (2003). Improved protein-ligand docking using GOLD. *Proteins, 52*(4), 609–623. https://doi.org/10.1002/prot.10465.

de Vries, S. J., Schindler, C. E., Chauvot de Beauchene, I., & Zacharias, M. (2015). A web interface for easy flexible protein-protein docking with ATTRACT. *Biophysical Journal, 108*(3), 462–465. https://doi.org/10.1016/j.bpj.2014.12.015.

de Vries, S. J., van Dijk, M., & Bonvin, A. M. (2010). The HADDOCK web server for data-driven biomolecular docking. *Nature Protocols, 5*(5), 883–897. https://doi.org/10.1038/nprot.2010.32.

Wang, J. (2020). Fast identification of possible drug treatment of coronavirus disease-19 (COVID-19) through computational drug repurposing study. *Journal of Chemical Information and Modeling, 60*(6), 3277–3286. https://doi.org/10.1021/acs.jcim.0c00179.

Wang, L., Gu, Q., Zheng, X., Ye, J., Liu, Z., Li, J., Hu, X., Hagler, A., & Xu, J. (2013). Discovery of new selective human aldose reductase inhibitors through virtual screening multiple binding pocket conformations. *Journal of Chemical Information and Modeling, 53*(9), 2409–2422. https://doi.org/10.1021/ci400322j.

Wang, R., Lu, Y., & Wang, S. (2003). Comparative evaluation of 11 scoring functions for molecular docking. *Journal of Medicinal Chemistry, 46*(12), 2287–2303. https://doi.org/10.1021/jm0203783.

Wang, X., Nagl, N. G., Wilsker, D., Van Scoy, M., Pacchione, S., Yaciuk, P., Dallas, P. B., & Moran, E. (2004). Two related ARID family proteins are alternative subunits of human SWI/SNF complexes. *Biochemical Journal, 383*(Pt 2), 319–325. https://doi.org/10.1042/BJ20040524.

Wishart, D. S., Knox, C., Guo, A. C., Shrivastava, S., Hassanali, M., Stothard, P., Chang, Z., & Woolsey, J. (2006). DrugBank: A comprehensive resource for in silico drug discovery and exploration. *Nucleic Acids Research, 34*(Database issue), D668–D672. https://doi.org/10.1093/nar/gkj067.

Wojcikowski, M., Ballester, P. J., & Siedlecki, P. (2017). Performance of machine-learning scoring functions in structure-based virtual screening. *Scientific Reports, 7*, 46710. https://doi.org/10.1038/srep46710.

Xie, L., & Bourne, P. E. (2008). Detecting evolutionary relationships across existing fold space, using sequence order-independent profile-profile alignments. *Proceedings of the National Academy of Sciences of the United States of America, 105*(14), 5441–5446. https://doi.org/10.1073/pnas.0704422105.

Xie, Q. Q., Zhong, L., Pan, Y. L., Wang, X. Y., Zhou, J. P., Di-Wu, L., Huang, Q., Wang, Y. L., Yang, L. L., Xie, H. Z., & Yang, S. Y. (2011). Combined SVM-based and docking-based virtual screening for retrieving novel inhibitors of c-Met. *European Journal of Medicinal Chemistry, 46*(9), 3675–3680. https://doi.org/10.1016/j.ejmech.2011.05.031.

Xue, X., Zhang, Y., Liu, Z., Song, M., Xing, Y., Xiang, Q., Wang, Z., Tu, Z., Zhou, Y., Ding, K., & Xu, Y. (2016). Discovery of benzo[cd]indol-2(1H)-ones as potent and specific BET bromodomain inhibitors: Structure-based virtual screening, optimization, and biological evaluation. *Journal of Medicinal Chemistry, 59*(4), 1565–1579. https://doi.org/10.1021/acs.jmedchem.5b01511.

Yang, J. M., & Chen, C. C. (2004). GEMDOCK: A generic evolutionary method for molecular docking. *Proteins, 55*(2), 288–304. https://doi.org/10.1002/prot.20035.

Zsoldos, Z., Reid, D., Simon, A., Sadjad, S. B., & Johnson, A. P. (2007). eHiTS: a new fast, exhaustive flexible ligand docking system. *Journal of Molecular Graphics and Modelling, 26*(1), 198–212. https://doi.org/10.1016/j.jmgm.2006.06.002.

CHAPTER 13

Application of Docking for Lead Optimization

JEEVAN PATRA • DEEPANMOL SINGH • SAPNA JAIN • NEERAJ MAHINDROO

1 INTRODUCTION

Drug discovery is a costly affair with current estimates of cost for taking a molecule from discovery to market estimated at $2.6 billion and is increasing year on year (DiMasi et al., 2016). The hit-to-lead and lead-to-preclinical candidate optimization are iterative medicinal chemistry processes which require expertise, time, and patience (Fig. 13.1). The computational techniques have expanded the ability to screen large libraries of compounds to identify the potential hits at a fraction of cost and time required for high-throughput screening (HTS) of large libraries of compounds. Molecular docking has emerged as a reliable and cost-effective technique for lead identification and optimization. The orientation of the ligand in the binding pocket, its position, and interactions with the three-dimensional (3D) structure of the target protein are used to predict the relative activity of the molecules (Torres et al., 2019).

Despite all the hype, computational techniques have shown limited success in drug discovery efforts due to technical and scientific challenges associated with this very complex process. Until a decade ago, computational techniques were rarely considered for hit/lead optimization (Levoin et al., 2008). Applications of approaches such as relative binding free energy (RBFE) calculations using molecular simulations and statistical mechanics coupled with huge improvement in computational resources have significantly improved the prediction of relative ability of a small molecule to bind to the target protein accurately. Thus, these techniques are now contributing by improving accuracy of prediction and throughput and saving time in hit-to-lead and lead-to-candidate optimization by limiting the number of molecules synthesized and screened (Cournia et al., 2017). Furthermore, artificial intelligence, machine learning (ML), and deep learning (DL) are helping in designing and optimizing the lead compounds both in terms of activity and properties, thus improving the pace and efficiency of the process (Yang et al., 2020).

2 APPROACHES TO LEAD OPTIMIZATION

Lead optimization or modifications are chemical alterations of a known and previously characterized organic compound, a lead, or parent molecule, for enhancing its potency and drug-like properties. These enhancements can be accomplished by improving specificity for a particular host target site, increasing potency, improving rate and extent of absorption, minimizing toxicity, and certain alteration in physicochemical profiling for improving solubility, metabolism, crossing barriers, etc. Lead optimization is the multicomponent process performed using various standard medicinal chemistry techniques. The prime aim of lead optimization is to enhance the target binding affinity and ensure bioavailability at the site of action, thus ensuring in vitro and in vivo potency. The iterative process employs various strategies for optimizing lead compounds (Fig. 13.1). The standard medicinal chemistry techniques such as analog design, R-group enumerations (alkyl/aryl), ring variations (expansion/contraction, fusion), chain variations (extensions or contraction), structural alteration (simplification/rigidification), conformational changes, bioisosteric replacements, scaffold hopping, and dynamics docking include RBFE techniques (Fig. 13.2). It is well-documented that iterative process of lead optimization to improve affinity of the molecule to the target usually increases structural complexity, molecular weight (MW), and lipophilicity and flattens the structure and aromatic character of the molecule (Oprea et al., 2001). This negatively affects the physicochemical and pharmacokinetic properties of the molecule. Increase in MW, lipophilicity, aromatic rings, rotatable

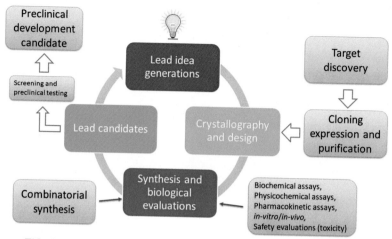

FIG. 13.1 Illustration depicting an iterative process for lead optimization.

FIG. 13.2 Standard medicinal chemistry lead optimization techniques.

bonds, hydrogen bond acceptors, and hydrogen bond donors increases the chances of failure of the drug candidate due to poor absorption, distribution, metabolism, excretion, and toxicity (ADMET) profile of the molecule (Hann, 2011; Hann & Keserü, 2012; Mignani et al., 2016). Analysis of 100 bestselling drugs showed that the simpler molecules turn out to be better drugs (Polanski et al., 2016). In order to save time, the pharmacological and pharmacokinetic properties are optimized simultaneously, thus ensuring drug-like characteristics and potency (Maynard & Roberts, 2016).

3 STRUCTURE-BASED LEAD OPTIMIZATION

In 1981, Jencks reported that small fragments can have high-quality interaction which can be optimized into complete molecules with high potency (Jencks, 1981). Later, this approach was studied and implemented to construct potent ligands from low molecular mass fragments. The fragments should obey the "Rule of Three" (Coyne et al., 2010) (Table 13.1). Though fragment hits are comparatively weak binding, they make high-quality interaction with the target to bind with sufficient affinity for detection as compared to HTS hits that interact with numerous suboptimal bindings but with limited optimized interactions (Kuntz et al., 1999).

In order to develop a virtual library, fragment-based lead optimization approach commences with the rule of three compliance followed by selection of functionalities that facilitate interactions, avoiding reactive, unstable, or toxic scaffolds. For structure-based lead optimization, the fragments of lead having weaker interactions with the binding pocket of the receptor are replaced with better fitting fragments, keeping the core of the ligand intact. The affinity of the lead is thus enhanced due to new favorable interactions. The strategy can include linking or merging of two nonoverlapping fragment hits directly or through a spacer while conserving the original orientation of the fragments in the binding pocket.

TABLE 13.1
The Table Summarizes Rule of Three for Fragments and Rule of Five for Drug-Like Molecules.

Category of Compound	Fragment-Like	Drug-Like
Rule	Rule of Three	Rule of Five
THRESHOLDS		
Molecular weight	<300	≤500
clogP	≤3	≤5
Hydrogen bond donors	≤3	≤5
Hydrogen bond acceptors	≤3	≤10
LIGAND EFFICIENCIES		
LE	0.38	0.29
Fit quality	0.55	0.81

This table is adapted with permission from Elsevier Schultes, S., De Graaf, C., Haaksma, E. E. J., De Esch, I. J. P., Leurs, R. & Krämer, O. (2010). Ligand efficiency as a guide in fragment hit selection and optimization. *Drug Discovery Today: Technologies, 7*(3) e157–e162.

Furthermore, to enhance druggability, the parameters such as rule of five (Lipinski et al., 2001), bRo5 (Doak et al., 2016), thermodynamics and kinetics profiling (Pan et al., 2013), and ligand efficiency (LE) metrics are used to optimize the leads (Abad-Zapatero & Metz, 2005). LE metrics ratio was first introduced in 1999 and burgeoned in the field of drug discovery processes (Kuntz et al., 1999). LE is defined as the ratio of the free energy of binding affinity of a ligand (ΔG) divided by its number of heavy atoms (nonhydrogen) (Hopkins, Groom, & Alex, 2004). The judicious application of LE metrics in the lead optimization, quantifies the molecular descriptors (molar mass, lipophilicity, etc.) required for obtaining the free energy calculation (binding affinity) for a macromolecular drug target for enhancing quality of lead drug candidates (Hopkins, Keserü, Leeson, Rees, & Reynolds, 2014).

4 MOLECULAR DOCKING FOR LEAD OPTIMIZATION

There are four broad strategies for lead optimization through docking:
- Improvement in van der Waals interactions during geometrical fit of the compound with the binding pocket,
- Incorporation of strong H-bonds,
- Enhancing hydrophobicity, and
- Introduction of restricted flexibility of the compound by curtailing conformational entropy.

Solvation and desolvation factors are also considered during lead optimization. The ligand–protein binding is predicted by estimating RBFE. For lead optimization using docking, the change in binding free energy with change in ligand structure is estimated (Fig. 13.3). Several factors influence the RBFE and are discussed in subsequent sections. The aim of the computational scientists is to enhance the accuracy and precision for the estimation of RBEF for protein–ligand interactions.

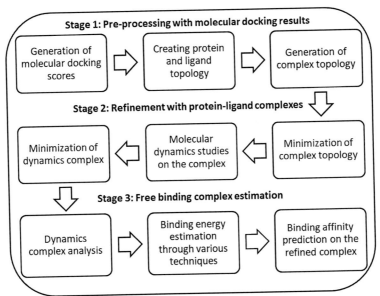

FIG. 13.3 Relative free binding energy estimation using molecular docking results.

The methods for calculating RBFE are based on molecular simulations and statistical mechanics (Cournia et al., 2017). In drug discovery and development, estimation of binding affinity helps to deduce a mechanism of action of the ligand, which is an important milestone for the study of drug action (De Vivo et al., 2016; Jorgensen, 2009; Mobley & Gilson, 2017; Wang et al., 2015; Williams-Noonan et al., 2018).

Advancements in computational technology, graphical processing units (GPUs), high-performance computers, and sampling algorithms (such as replica exchange with solute tempering method) have improved capability of accurate estimation of RBFE (Bergdorf et al., 2015; Harder et al., 2016; Harvey et al., 2009; Wang et al., 2004). The binding affinities are presented as docking scores which can be categorized into force field–based, knowledge-based, and empirical-based scoring functions, as discussed in Table 13.2.

5 METHODS FOR CALCULATING BINDING AFFINITIES FOR LEAD OPTIMIZATION

The binding affinities of library of compounds for lead optimization are more commonly calculated using following methods:

- Free energy perturbation (FEP)
- Linear interaction energy (LIE)
- Molecular mechanics Poisson–Boltzmann
- Quantum mechanics (QM)

5.1 Free Energy Perturbation

In 1954, Zwanzig gave a relationship for computing difference between binding free energy of bound phase (known as target phase) and unbound phases (known as initial or reference phase) (Irwin & Huggins, 2018; Jespers et al., 2019; Zwanzig, 1954). FEP-based simulations have been successfully employed by researchers for calculating binding free energy of the molecule with the target protein (Athanasiou et al., 2018; Cournia et al., 2017; Konze et al., 2019; Lenselink et al., 2016; Li et al., 2019; Matricon et al., 2017; Moraca et al., 2019; Reddy et al., 2020; Singh & Warshel, 2010; Stanton et al., 2016; Zeevaart et al., 2008). Unlike other techniques used for RBFE, FEP gives the most accurate results even when the difference in the free energy values is very small (Åqvist et al., 2002; Singh & Warshel, 2010). The FEP simulations can run efficiently, independently, and in parallel. For FEP simulation, the knowledge of binding mode is essential and constant motion of protein can be challenging. FEP calculations may vary from one ligand to another due to presence of solvent molecule (water) (Chen et al., 2018; Fratev et al., 2018; Jorgensen, 2009; Kaus et al., 2015; Lim et al., 2016; Sherborne et al., 2016; Wang et al., 2012). There are several reports in literature reporting use of FEP simulations for structural optimization (Athanasiou et al., 2018; Konze et al., 2019; Lenselink et al., 2016; Li et al., 2019; Matricon et al., 2017; Moraca et al., 2019; Reddy et al., 2020; Singh & Warshel, 2010; Stanton et al., 2016; Zeevaart et al., 2008). FEP has also shown good results in structure-based molecular target analysis for SARS-CoV-2 (Zhang & Zhou, 2020).

5.2 Linear Interaction Energy

The linear interaction energy (LIE) method involves simulating both dynamics simulations: ligand in

TABLE 13.2
Scoring Methods for Molecular Docking.

Scoring Method	Description	Tools	References
Force field scoring	Based on sum of all the energy terms including interaction energies, solvation energy, and entropy	DOCK, DockThor	Sun et al. (2012)
Knowledge-based	Based on knowledge-based force fields and includes weighted molecular features which are related to ligand–protein binding	DrugScore, PMF, and Astex statistical potential fitness function found in the GOLD docking program	Guedes et al. (2018)
Empirical	Based on sum of various empirical energy terms (known through experimental data) and to approximate binding energy	ChemScore, ChemPLP, and GlideScore	Eldridge et al. (1997), Korb et al. (2009)

solution and ligand in the target binding site (Aqvist et al., 1994; Åqvist et al., 2002). In 1994, Aqvist and co-workers introduced an end point method for predicting biomolecule—ligand binding affinities. LIE was used for optimization of proteases and dihydrofolate reductase inhibitors. This approach includes three important aspects: molecular mechanics force fields, solvation effects, and entropic effects. It shows good accuracy in energy calculations with a root mean square error of less than 1 kcal/mol compared with experimental determination (Gutiérrez-De-Terán & Åqvist, 2012; Hultén et al., 1997; Marelius et al., 1998).

The solvation energy calculation is important for lead optimization. Conventionally, it is calculated by creating a model molecular dynamics or Monte Carlo simulations followed by FEP or thermodynamic integration (TI) in the presence of water molecules. The drawbacks of this method include the limitation of size of mutations that can be studied and the time required to obtain a quality result (Jorgensen, 2009).

Molecular dynamics—based free energy calculation (using FEP and TI) is more accurate but is not a choice for HTS due to high cost involved. LIE provides an optimal trade-off between accuracy, efficiency, and speed. Recent studies have shown that improvement in LIE predictions can be done by incorporating entropy of (de)solvation into the calculation of free energy (Rifai, Ferrario, et al., 2020).

Linear response approximation measures free energy changes through interaction energies between solute and its environment. It is known as an end point method as simulations are done in initial and final states, i.e., when solute is free and when solute is bound to the biomolecule. Interactions are divided into two types, electrostatic and van der Waals interactions, which are related to free energy as shown in Eq. (13.1).

$$\Delta G_{bind} = \alpha \Delta(U_{vdw}) + \beta \Delta(U_{elec}) \qquad (13.1)$$

$\Delta(U_{vdw})$ is the difference between the van der Waals energies of solute and environment in free form and bound form. $\Delta(U_{elec})$ is the difference between the electrostatic energies of solute and environment in free form and bound form van Dijk et al. (2017); α and β are empirical parameters obtained from training set of compounds whose binding affinities are experimentally determined with linear regression analysis (Carlson & Jorgensen, 1995). The values of β depend on electrostatic properties and functional groups present in the compound of interest. A correlation between α value and binding site was reported by Wang et al. (Wang et al., 1999). Another parameter which is used optionally is denoted as γ and is known as fitting parameter which represents the hydrophobicity of the binding site (Almlöf et al., 2004).

In 2010, Stjernschantz and Oostenbrink introduced "iterative LIE," which is an extended version of previously existing LIE. It is also known as Boltzmann-weighting LIE scheme. Iterative LIE incorporates results of multiple molecular dynamics simulations from various orientation and conformations of proteins and combines them to calculate binding free energy (Stjernschantz & Oostenbrink, 2010). This helps to calculate BFE for flexible proteins and cytochromes, which can bind the ligands in different orientations. Boltzmann-like statistical weighing scheme given by Hritz and Oostenbrink was used to find the contribution of each of the individual simulation i from protein conformation and ligand binding (Eq. 13.2) (Hritz & Oostenbrink, 2009).

$$W_i = \frac{e^{-\Delta G_{bind_i t}/k_B T}}{\sum_i e^{-\Delta G_{bind_i t}/k_B T}} \qquad (13.2)$$

This is used in Eq. (13.1) to give a new correlation for finding ΔG_{bind} in Eq. (13.3).

$$\Delta G_{bind} = \alpha \sum_i^N W_i \Delta(U_{vdw}) + \sum_i^N W_i \beta W_i \Delta(U_{elec}) \qquad (13.3)$$

Iterative LIE has shown better accuracy when compared with single MD simulations per compound. This approach has been studied on different proteins including CYP1A4, CYP2D6, CYP19A1, and JAK2 kinase (Capoferri et al., 2015, 2017; van Dijk et al., 2017; Perić-Hassler et al., 2013). eTOX ALLIES and MDStudio are two automated platforms which are implementing iterative LIE (Capoferri et al., 2017). Fourier transform filtering strategy has been implemented in this approach where a predefined cutoff value is applied so as to remove fluctuations. This helps to study the interactions with lesser fluctuations resulting in better efficiency and accuracy (Rifai, Van Dijk, & Geerke, 2020; Vosmeer et al., 2016). In addition to the study of several binding poses, Rifai et al. have also predicted the probability of binding pose from the weighting values (Rifai et al., 2019).

5.3 Molecular Mechanics Poisson—Boltzmann

Molecular docking involves exhaustive conformational analysis in the binding space for the determination of binding mode. The accuracy of molecular docking method is determined by reliability of its scoring functions. All these energy-based scoring functions rely on pose ranking using hydrogen bonding, electrostatic interactions, π—π stacking, van der Waals interactions, desolvation effects, etc. (Raveh et al., 2011; Schindler et al., 2015; Spiliotopoulos et al., 2016). To achieve the desired result, a balance between accuracy, scoring functions, and computational costs is required. For the end point calculations, molecular mechanics

Poisson–Boltzmann surface area (MMPBSA) and molecular mechanics generalized Born surface area (MMGBSA) have been introduced (Massova & Kollman, 2000; Miller et al., 2012).

Although QM-based methods are more widely accepted due to their high accuracy and detailed description potential, still some classical approaches give better efficiency in case of complex biochemical systems and processes. The Poisson–Boltzmann solvent-accessible surface area (PBSA) method includes contribution of solvent in calculating free energy. The MMPBSA approach also includes enthalpy and entropy contributions as shown in Eqs. (13.4)–(13.7) (Srinivasan et al., 1998; Wang et al., 2018).

$$\Delta G_{bind,aq} = \Delta H - T\Delta S \approx \Delta E_{MM} + \Delta G_{bind,solv} - T\Delta S \quad (13.4)$$

$$\Delta E_{MM} = \Delta E_{covalent} + \Delta E_{electrostatic} + \Delta E_{vdw} \quad (13.5)$$

$$\Delta E_{covalent} = \Delta E_{bond} + \Delta E_{angle} + \Delta E_{torsion} \quad (13.6)$$

$$\Delta G_{bind,solv} = \Delta G_{polar} + \Delta G_{non-polar} \quad (13.7)$$

ΔE_{MM} represents gas-phase molecular mechanical energy change. It includes the covalent energy change (bond, angle, and torsional change), the electrostatic energy change, and van der Waals energy change. $\Delta G_{bind,solv}$ represents the solvation free energy change, which includes polar and nonpolar contributions. $T\Delta S$ represent conformational entropy.

Polar solvent contribution to ΔG_{bind} is calculated using generalized Born pairwise method and Poisson–Boltzmann distribution equation. Nonpolar solvent contributions are conventionally estimated through solvent-accessible surface area using Eq. (13.8) (Gallicchio & Levy, 2004; Pratt & Chandler, 1980).

$$\Delta G_{non-polar} = \gamma \times SASA + b \quad (13.8)$$

γ is the surface tension and b is a correction term.

A new equation (Eqs. 13.9 and 13.10) with dispersion free energy and cavity formation free energy (estimated through solvent-accessible volume) for calculation of $\Delta G_{non-polar}$ was proposed by Tan et al. (Tan et al., 2007):

$$\Delta G_{non-polar} = \Delta G_{dispersion} + \Delta G_{cavity} \quad (13.9)$$

$$\Delta G_{non-polar} = \Delta G_{dispersion} + \gamma SAV + b \quad (13.10)$$

Pearlman calculated free energies of a congeneric series of p38 MAP kinase inhibitors (16 ligands) and found that MMPBSA free energy method gave inferior results when compared with TI, OWFEG, ChemScore, PLPScore, and Dock energy Score (Pearlman, 2005). For assessing these molecular dynamics simulations, several tools have been developed. Over the last few years, many dynamics toolkits have been developed, for example, AMBER with both Python scripted (Miller et al., 2012) and Perl scripted (Homeyer & Gohlke, 2015), GROMACS (Kumari et al., 2014), GMXPBSA (Paissoni et al., 2015), and FEsetup (Loeffler et al., 2015).

Recent literature reports show the successful application of MMPBSA/MMGBSA rescoring functions in lead optimization process and more advanced simulations and free energy calculations (Guimarães & Cardozo, 2008; Homeyer et al., 2014; Knight et al., 2014). Employing this technique, novel inhibitors of *Pf*DHODH were optimized to identify a compound showing an IC$_{50}$ of 6 nM with 40% oral bioavailability (Fig. 13.4) (Xu et al., 2013). MMGBSA was also used for

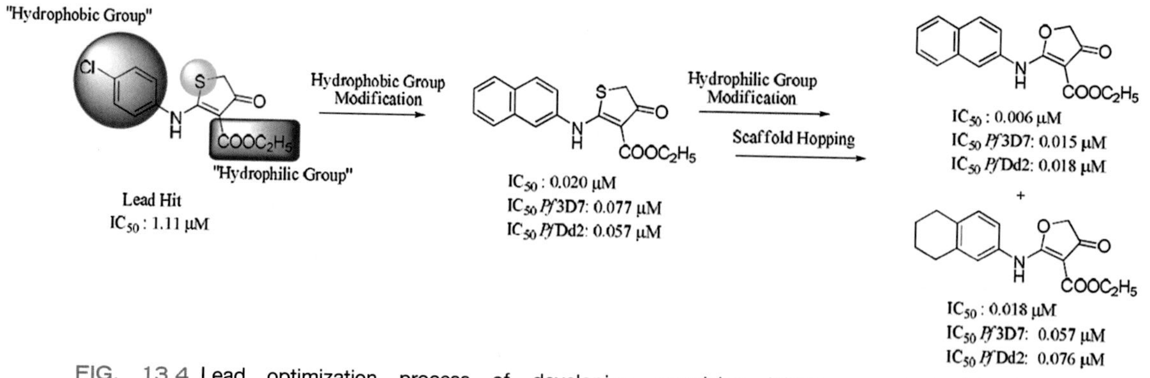

FIG. 13.4 Lead optimization process of developing promising inhibitors against *Pf*DHODH. (Figure reproduced with permission from Xu, M., Zhu, J., Diao, Y., Zhou, H., Ren, X., Sun, D., Huang, J., Han, D., Zhao, Z., Zhu, L., Xu, Y. & Li, H. (2013). Novel selective and potent inhibitors of malaria parasite dihydroorotate dehydrogenase: Discovery and optimization of dihydrothiophenone derivatives. *Journal of Medicinal Chemistry*, 56(20):7911–7924; copyright 2013 American Chemical Society.)

designing and optimizing a series 1,4,5-trisubstituted and 1,2,3-triazole scaffold inhibitors targeting Hsp90 in which compound, SST0287CL1, was shown to have activity comparable with clinically candidate NVP-AUY922 (Taddei et al., 2014). Pan et al. reported design and optimization of ALK inhibitor with IC_{50} of 0.27 nM and binding selectivity over 35 kinases using MMGBSA methodology (Pan et al., 2017).

5.4 Quantum Mechanics

Over last decade, there is an increase in interest in QM-based methods for lead optimization. These methods include all the components contributing to the energy of a system by thoroughly capturing the physics of the system (Cavasotto, 2020). QM explains contributions to the energy through electronic polarization, charge transfer, halogen bonding, and covalent-bond formation. It is applicable across the chemical space, is not system dependent, and is thus theoretically exact. With improvement in QM approaches, better accuracy for protein-binding interactions and improved scoring functions predicting binding affinities have been achieved (Cavasotto et al., 2018). Raha and Merz were among the pioneers in calculating binding free energy using QM approach. They developed a scoring function called QM score which studies the entire molecular structure at QM level to calculate the free energies (Raha & Merz, 2004).

The comparative performances of docking scoring functions including FEP, MMGBSA, and QM/MMGBSA were reported in 2017 by Pu et al. Although FEP gave the better r_s of 0.854 as compared to MMGBSA (r_s 0.76), computational cost was significantly lower with the latter approach. Hence, it was proposed that ligand treated by QM/MMGBSA can increase the accuracy of ranking (Pu et al., 2017). FEP results are better than other approaches but the accuracy is not to the expected level. Reddy and Erion found that QM/MM-based FEP method can promote FEP-based calculation by reducing time and increasing the accuracy level (Reddy & Erion, 2007). A series of studies using FEP-based QM/MM approach have been reported. Binding free energy of acetylcholinesterase inhibitors was calculated by FEP based on QM/MM potentials. The results showed interesting observations related to the role of van der Waals interactions, cation–π interactions, and involvement of water acting as a bridge between ligands and protein site (Nascimento et al., 2017).

In the last ten years, QM has been widely used for estimating protein–ligand affinities. Recently, research studies based on free energy calculations and application of advanced semi-empirical QM have been undefined (Chinnasamy, Saravanan, & Poomani, 2020; Peng, Wang, Zhijian, Cai, & Zhu, 2020; Souza, Thallmair, & Marrink, 2019).

6 MACHINE LEARNING–BASED LEAD OPTIMIZATION

Recently, research papers have reported estimation of docking scores using ML. This approach uses knowledge-based or empirical scoring method. Due to availability of good quality training data and improved methods, ML-based scoring functions (surrogate of ligand bioactivity) have better performance as compared with contemporary supervised learning algorithms (Li et al., 2020). Researchers have made models using neural networks for development of scoring functions for binding affinity prediction. Some of the most commonly used ML datasets are CASF, MoleculeNet, Binding DB, CSAR, and BindingDOAB. Most of ML datasets are present in an open-source package, Deep-Chem (Hughes et al., 2015; Wu et al., 2018). Different score functions, ML approaches, and descriptions are given in Table 13.3.

Due to the availability of large amount of genomic data, DL algorithms have been used for the drug repurposing using genomic modeling, transcription response–based prediction of therapeutic drugs, and sequence specification prediction. Ensemble learning methods were used to predict the chemotherapeutic response of potential anticancer drugs using drug-induced gene expression data (Tan et al., 2019).

ML has been used to explore the interactions between proteins and DNA (Hassanzadeh & Wang, 2016). From the 3D electron density and electrostatic potential fields, DL algorithms have been used to predict biological functions of the proteins (Golkov et al., 2017). ML algorithms have also been used for in silico prediction of ADMET properties and for drug protein interactions (Lipinski et al., 2019; Wen et al., 2017). Wen et al. (2017) worked on 2,146,240 drug–protein approved interaction pairs using extended Connectivity Fingerprints and Protein Sequence Composition Descriptors and DrugBank for dataset. The aim was to identify new drug and protein interaction pair using ML with a better accuracy as compared to quantitative structure–activity relationship approaches (Wen et al., 2017).

Rosalind, a DL algorithm, was used for in silico drug design and optimization acting as SARS-CoV-2 replication inhibitors targeting main protease of virus, M^{Pro} (Shaker et al., 2020). Another similar algorithm, named Deep Docking, was developed for fast prediction of docking scores. Deep Docking was able to identify 1000 potential ligands active against M^{Pro} (Ton et al., 2020). MT-DTI DL model was developed and used for repurposing of drugs for SARS-CoV-2 3C like proteinase.

TABLE 13.3
The Machine Learning Approaches for Lead Optimization.

Models for Predicting Score Function	Machine Learning Approaches	Datasets and Description	References
DLSCORE	Fully connected DNN	*PDBBind v2016 dataset* - Uses 348 binding analyzer (BINANA) descriptors	Hassan et al. (2018)
TopologyNet	Element-specific persistent homology method and CNN	*CASF-2007 and PDBbind v2016* Model predicts protein—ligand binding affinities and mutation-induced protein stability changes Helps in reducing dimensionality of three-dimensional (3D) biomolecular data	Cang and Wei (2017)
K_{DEEP}	3D CNN	*PDBbind v2016* - CSAR datasets, congeneric series sets	Li et al. (2015)
Pafnucy	4D DNN	*PDBbind v2016* - CSAR datasets 3D grid for the DNN model to utilize a 3D convolution to produce a feature map	Stepniewska-Dziubinska et al. (2018)
PotentialNet	Graph CNN (GCN)	*CASF-2007* - Steps include covalent-only propagation, dual noncovalent and covalent propagation, and ligand-based graph	Feinberg et al. (2018)
BgN-Score and BsN-Score	Multiple linear regression, multivariate adaptive regression splines, k nearest neighbors (kNN), SVM, RF, and boosted regression trees	*PDBbind v2007* - Combinations of the terms from X Score, AffiScore, GOLD, and RFScore v1	Ashtawy and Mahapatra (2015)
DeepMindRG	ResNet CNN model	*PDBbind v2018 dataset and DUD-E dataset* - Performance comparable to Pafnucy (4D DNN)	Zhang et al. (2019)
CSM-Lig	Gaussian process	*PDBbind v2007* - A web server for protein—ligand binding affinity prediction	Pires and Ascher (2016)
Feature functional theory -Bbinding predictor	Multiple additive regression tree (MART) also known as GBDT (gradient boosting decision tree)	*PDBbind v2007* - Uses six categories of microscopic features including Poisson—Boltzmann theory, nonpolar solvation models, and components in molecular mechanics Poisson—Boltzmann surface area and quantum models	Wang et al. (2017)
BT-Score	GBDT, gradient boosting decision tree	*PDBbind v2014* - For estimating binding affinity of out-of-sample test complexes	Ashtawy and Mahapatra (2018), Wang and Zhang (2017)

Atazanavir, remdesivir, and Kaletra were predicted to inhibit SARS-CoV-2. Remdesivir was approved by the FDA on May 1, 2020 for COVID-19 (Beck et al., 2020).

Generative tensorial reinforcement learning is a DL model for de novo small molecule design. This model helps in optimizing synthetic feasibility, novelty, and biological activity. The model can also identify how distinct the study molecule is from other existing molecules. It was applied by researchers for discovery of potent inhibitors of discoidin domain receptor 1 (DDR1), a kinase target implicated in fibrosis in just 21 days (Zhavoronkov et al., 2019).

Albeit ML has been potentially speeding up the in silico–based drug discovery and development process and reduces the chances of failure, it has some limitations. The ML process requires availability of large amount of high-quality data, absence of which can lead to compromise in the performance of the model. Pharma companies generate a lot of data but do not make it public, rather keep it as a commercial asset with themselves. Another limitation is poor rational interpretation of closely associated biological mechanisms (Zhang et al., 2017).

7 LIMITATIONS OF MOLECULAR DOCKING FOR LEAD OPTIMIZATION

Despite having remarkable benefits, lead optimization with molecular docking has certain limitations. The inaccuracy of the sampling from the libraries can lead to poor correlation of scoring functions with the experimental ligand free binding energies, thus leading to weak predictions of the ligand binding affinities. Furthermore, target models might not be accurate in terms of flexibility, solvent effects, and the conformational entropy (thermodynamics attribution) for both the drug target and the molecules (Salmaso & Moro, 2018).

Protein rigidity is one of the often neglected factors (Limongelli, 2020). Due to the use of rigid protein models, researchers underestimate ligand flexibility that can efficiently penetrate into binding pocket for proper interactions (Sulimov et al., 2019). Docking using rigid protein model leads to the formation of only one conformation or may display false negatives. This happens due to existence of constant motion having similar energies. Solvation effects lead to inaccurate binding modes in the canonical pockets during docking process, which is still an unsolved and neglected issue with the water molecules (Spyrakis & Cavasotto, 2015). The major reasons behind solvation effects are the following:

- Paucity of coordinates of crystal structures due to the diversified presence of smaller heteroatoms.
- Absence of hydrogen coordinates leads to unpredictable water molecules and acts as a linker in between ligand and receptor.
- Insufficiency of knowledge-based approach for accurate predictions for desolvation effects and its strength. The presence of many confounding factors leads to poor results in RBFE simulations.

RBFE works efficiently for homology protein models having high resolution with all binding residues (Spyrakis & Cavasotto, 2015). This strategy is compromised if the crystal structure has low resolution. Advancement in GPU hardware shows a tremendous change in these MD simulations. Nevertheless, it could not reduce the time of MD analysis. Molecular mechanics force fields seem to be of less accuracy with more time span during analysis. This timeline must be shortened to around 2×10^{-5} s. Covalent bond cannot break or be formed during standard simulations. Presence of disulfide cysteine residues and presence of acidic or basic amino acid residues can frequently loss or gain protons.

8 PREDICTIVE TOXICITY USING DOCKING AND MACHINE LEARNING

One of the major reasons for attrition of clinical candidates is poor efficacy or toxicity. Early prediction of selectivity and toxicity can help in saving the time and money. Toxicity and selectivity prediction through molecular docking studies is becoming part of the lead optimization process. Toxicity is conventionally determined through in vitro (HTS, cellular-based assay, etc.) and in vivo (preclinical and clinical) studies. These studies are expensive and time-consuming. Predicting an accurate potential toxicity profile of a compound in early development phases can reduce burden, time, and expenses.

Molecular docking is used for toxicity predictions by studying interactions between the antitargets and the lead molecules to predict the selectivity of the molecule and potential toxicity. Due to ethical concerns and practical limitations to animal testing specially for potential hazardous substances, there is a strong recommendation for use of in silico methods. Regulatory authorities also promote the use of in silico approaches as an alternative to animal studies. This not only reduces the sacrifice of animals but also is cost-effective and time-saving approach. There are many software, web-based servers, and online databases that are used for the toxicity predictions based on scoring functions, ML, and DL, which are enlisted in Table 13.4.

TABLE 13.4
Tools for Toxicity and Pharmacokinetic Prediction.

Sr. No.	Software/Web-Based Database	References
1	admetSAR	Cheng et al. (2012)
2	TOXNET (ToxLine and ChemIDplus)	Fowler and Schnall (2014)
3	ACToR	Judson et al. (2008)
4	DSSTox (ToxCast and Tox21)	Williams-DeVane et al. (2009)
5	ToxiM	Sharma et al. (2017)
6	ToxiPred	Mishra et al. (2014)
7	Tox21	Tice et al. (2013)
8	DeepTox	Mayr et al. (2016)
9	ProTox-II	Banerjee et al. (2018)
10	ToxCast	Kavlock et al. (2012)
11	pkCSM	Pires et al. (2015)
12	Lazar	Maunz et al. (2013)

Advancements of ML have helped in predicting toxicity, toxicological end points, and mechanism of actions. ML platforms have been reported for predictive toxicity, for example, WEKA (Hall et al., 2009; Markov & Russell, 2006), KNIME (Berthold et al., 2009; Fillbrunn et al., 2017), Scikit-learn (Garreta & Moncecchi, 2013; Pedregosa et al., 2011), and RapidMiner (Hofmann & Klinkenberg, 2014). The widely used tools/web servers/databases for toxicity and pharmacokinetic prediction are listed in Tables 13.4 and 13.5 lists applications for predictive toxicity.

9 CASE STUDIES

9.1 Case Study 1: Syk Inhibitors

The spleen tyrosine kinase (Syk) is a nonreceptor protein tyrosine kinase involved in diverse cellular processes. Syk is responsible for signal transduction of hematopoietic cells, platelet aggregation, neutrophils, and macrophages. Overactivation leads to hypersecretion of B cells in autoimmune diseases, hypersensitivity, and proliferation in chronic leukemia (Navara, 2004). Lovering et al. identified a type I inhibitor of Syk with imidazopyrazine hinge binding core, 3-(8-((3,4-dimethoxyphenyl)amino)imidazo[1,2-a]pyrazine-6-yl)-benzamide (**1**) showing IC_{50}

TABLE 13.5
Applications for Predictive Toxicity.

S. No.	Target	Inhibitors	Software	References
1	EGFR	EA1045 analogs	ADMET Schrodinger	Karnik et al. (2020)
2	β-Hematin	4-aminoquinoline hybrids	ADMET predictor	Fayyazi et al. (2019)
3	HER2	Isoxazole substituted 9-anilinoacridines	QikProp (Schrodinger suite)	Kalirajan et al. (2019)
4	*Helicobacter pylori* urease	Morin analogs	QikProp	Kataria and Khatkar (2019)
5	dDAT	Triazolobenzodiazepines	PreADMET server	Belhassan et al. (2019)
6	PPARα/γ	Scaffold hopping of saroglitazar	Discovery Studio v3.5	Jia et al. (2018)
7	FAAH	N-(2,4-dichlorobenzoyl) isatin schiff base hybrids	Molinspiration and PreADMET	Jaiswal et al. (2018)
8	MAPKAPK-2	Pyrrolopyrimidine analogs	admetSAR	Konidala et al. (2018)
9	Deformylase	Oxathiadiazole hybrids	TOPKAT, Discovery Studio v2.5	Yadav et al. (2018)
10	COX-1/2 and 5-LOX	Benzothiophene/benzofuran with rhodanine hybrids	PreADMET, Osiris Property Explorer	El-Miligy et al. (2017)

of 0.17 µM and LE of 0.32 (Lovering et al., 2016). The co-crystal structure of the (1) with Syk (PDB 5C26) showed favorable binding interaction of aniline nitrogen atom at position 8 with carbonyl oxygen atom of Ala451 residue and the interaction of ring nitrogen with backbone nitrogen of Ala451. The 3,4-dimethoxybenzylamino moiety of compound 1 interacts with charged amine of Lys458 in the solvent pocket. In an effort to improve the affinity by optimizing the ligand efficiency of the hinge binding region, a series of compounds were prepared and subjected to interactions and enzymatic studies. The synthesized compounds showed decrease in potency as compared to compound 1; for example, compound 2 showed IC_{50} of 20 µM, presumably due to conformational strain (6.1 kcal/mol) caused by clash between o-aniline proton and proton at position 7 of the core. Replacement of aryl group with pyridine in compound 3 decreased the conformational strain to 2 kcal/mol and improved IC_{50} to 0.19 µM (Fig. 13.5). This also resulted in replacement of a known toxicophore, aniline, with aminopyridine.

Authors prioritized the cores using Monte Carlo simulations with FEP for further synthesis of compounds, thus saving the resources and time. Difference in binding energies ($\Delta\Delta G$) of ligands with the protein was calculated with FEP, which indicated that the relative change in affinity is due to different head groups, for example, compounds 2 and 3. The FEP calculations using MCPRO+ could correctly predict the correlation of change in structure with potency in a series of compounds. Based on this result, a library of 17 compounds was designed with 5/6 ring system maintaining the hinge interactions. Two of the designed compounds were predicted by MCPRO+ to have better potency than the reference compound and were prioritized for synthesis. Compound 4 showed IC_{50} of 16 nM and LE of 0.37 and thus a substantial improvement over lead compound 1. However, compound 4 showed comparatively poor IC_{50} in cellular assay and was further optimized to compounds 5 and 6 showing LE of 0.38 and the cellular IC_{50} of 60 and 87 nM, respectively (Fig. 13.5). This case

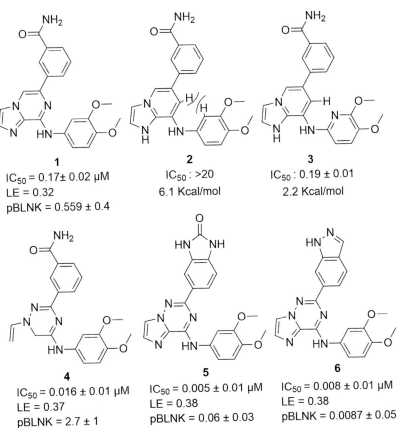

1
IC_{50} = 0.17± 0.02 µM
LE = 0.32
pBLNK = 0.559 ± 0.4

2
IC_{50} : >20
6.1 Kcal/mol

3
IC_{50} : 0.19 ± 0.01
2.2 Kcal/mol

4
IC_{50} = 0.016 ± 0.01 µM
LE = 0.37
pBLNK = 2.7 ± 1

5
IC_{50} = 0.005 ± 0.01 µM
LE = 0.38
pBLNK = 0.06 ± 0.03

6
IC_{50} = 0.008 ± 0.01 µM
LE = 0.38
pBLNK = 0.0087 ± 0.05

FIG. 13.5 Structures representing overlay of lead optimization using MC/FEP. Syk-1 inhibitor of PDB 5C26 **(1)**, conformational strains clashing **(2 and 3)**, and potential optimized compounds **(4, 5, and 6)**.

study shows successful application of FEP for lead optimization by calculating $\Delta\Delta G$ with MCPRO+.

9.2 Case Study 2: c-Src/Abl Inhibitor

Neuroblastoma is a rare type of cancer in the sympathetic nervous system. In 2015, Tintori and his co-workers studied series of compounds with pyrazolo[3,4-*d*]pyrimidines scaffold targeting c-Src/Abl for neuroblastoma (Tintori et al., 2015). Dasatinib (compound 7) is a well-known c-Src/Abl inhibitor with proven neuroblastoma growth reducing activity (Vitali et al., 2009). Previously reported compounds, 8, 9, and 10, containing pyrazolo[3,4-*d*]pyrimidine displayed promising antiproliferative activity (Fig. 13.6) (Matsunaga et al., 1993, 1994; Navarra et al., 2010; Radi et al., 2011).

Compound 8 showed appreciable selectivity for Src and this led to the development of a series of closely related compounds with same scaffold having presence of 6-methylthio group and different functional groups on C4 and N1 position. Although compound 9 showed potential inhibitory activity against *c*-Src (Ki = 90 nM), it had low solubility (0.12 µg/mL). Furthermore, development of more soluble compounds with similar scaffold was designed and compound 10 showed promising ADME profile with water solubility of 1.7 µg/mL and significant inhibitory activity against c-Src ($K_i = 0.21$ µM) (Fig. 13.7) (Matsunaga et al., 1993, 1994).

Tintori and his team worked to develop second generation of compounds on the basis of compound 10 with improved affinity and ADME profiles. A multidisciplinary approach was used combining X-ray crystallography, structure-based drug design, synthesis, in vitro ADME profiling, and biological evaluation (in vivo/in vitro).

Crystallographic complex of compound 10 and Src was taken as the starting point, and FEP calculations were used for optimizing the series of compounds.

Docking studies were used to understand the binding of the compound with c-Src (Fig. 13.8). The anilino group on C4 and side chain on N1 are present in the hydrophobic region of the c-Src. Src has two chains, chain A and chain B.

Optimization of compound 10 was done using Monte Carlo (MC) FEP with MCPRO program. C4 anilino ring was focused for increasing the binding affinity with c-Src. MC/FEP halogen (chlorine, bromine, and fluorine) and hydroxyl scans were used for the identification of the most promising sites and groups that can substitute C4 anilino hydrogens. The substitution of hydroxyl group gave a favorable response (positive $\Delta\Delta G_b$) at C2, C3, C4 ($\Delta\Delta G_b$ equal to 5.11, 4,89, and 1.49 kcal/mol, respectively) but is unfavorable at C5 and C6 ($\Delta\Delta G_b$ of −6.68 and −5.08 kcal/mol, respectively). Introduction of halogens such as chlorine, bromine, and fluorine at C2 also resulted in positive $\Delta\Delta G_b$ of 4.8, 1.28, and 1.96 kcal/mol, respectively.

Using FEP results, library of pyrazolo[3,4-*d*]pyrimidine analogs (29 compounds) were synthesized with m-OH and halogen present at the C4 anilino ring aiming for a better water solubility and c-Src binding affinities. Metasubstituted chlorine, bromine, and fluorine analogs were also synthesized and tested. The hydrogen bonding between 3-OH and the Glu310 side chain (involving formation of salt bridge with Lys295) contributed significantly to binding affinity. This interaction was not possible for chain A due to different orientations of Glu310 side chain. Halogen bonding was also studied using enhanced force field OPLS/CM1Ax. The effect of halogen bond interaction for chlorine and bromine was very marginal when substituted at C5 position in chain B simulations (1.14 and 0.35 kcal/mol, respectively) and negative results were obtained in case of simulation with chain A.

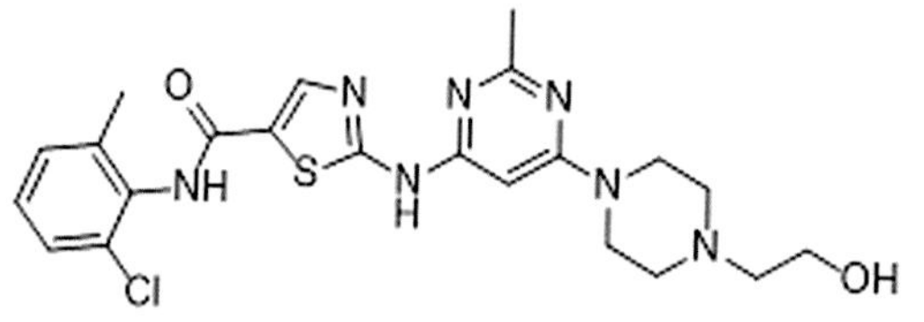

7 (Dasatinib)

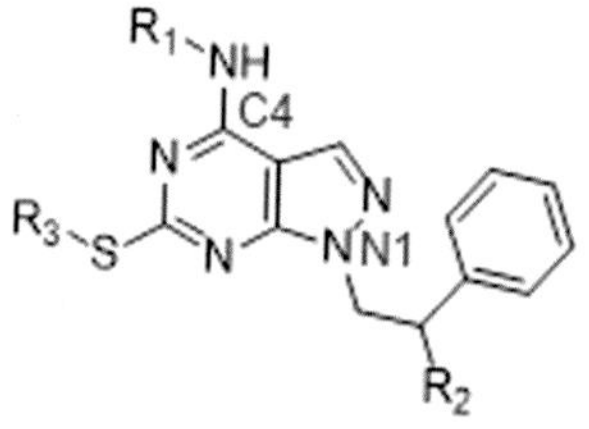

8; R_1 = -$CH_2C_6H_5$ R_2 = Cl R_3 = CH_3

9; R_1 = -$CH_2C_6H_4$-*p*Cl R_2 = H R_3 = CH_3

10; R_1 = -C_6H_5 R_2 = Cl R_3 = -CH_2CH_2-4-morpholino

FIG. 13.6 Previously reported analogues of pyrazolo[3,4-d]pyrimidines scaffold.

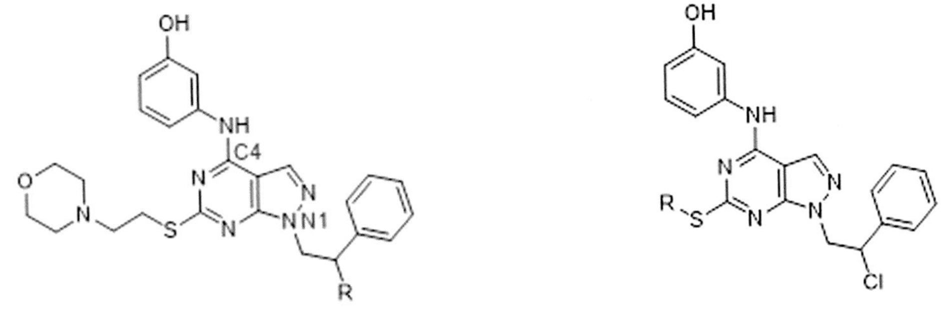

11 R = H
c-Src; Ki = 0.04 μM
Abl; Ki = 0.13 μM
IC_{50} = 0.9 μM

12 R = CH_3
c-Src; Ki = 0.07 μM
Abl; Ki = 0.43 μM
IC_{50} = 0.54 μM

13 R = CH_3
c-Src; Ki = 0.03 μM
Abl; Ki = 0.15 μM
IC_{50} = 0.12 μM

14 R = $CH(CH_3)_2$
c-Src; Ki = 0.01 μM
Abl; Ki = 0.12 μM
IC_{50} = 1.53 μM

15 R = Cyclopentyl
c-Src; Ki = 0.007 μM
Abl; Ki = 0.15 μM
IC_{50} = 0.62 μM

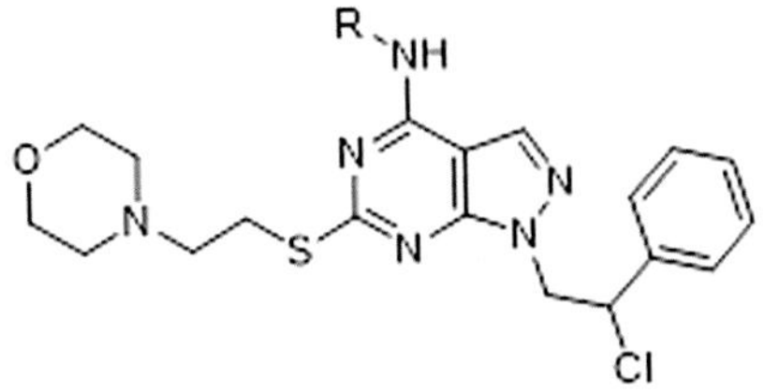

10; R = $-C_6H_5$
c-Src; Ki = 0.24 μM
Abl; Ki = 0.22 μM
Solubility = 1.7 μg/mL

16; R = $-C_6H_5$-*m*Br
c-Src; Ki = 0.13 μM
Abl; Ki = 0.12 μM
IC_{50} = 0.34 μM
Solubility = 3.7 μg/mL

FIG. 13.7 Optimized leads of c-Src/Ab1 inhibitor.

Introduction of halogen (**16**) or hydroxyl (**12**) substituents at position 3 was found to be favorable for c-Src inhibitory action. All 29 inhibitors were synthesized and subjected to biological evaluations against SH-SY5Y cells and ADME studies using parallel artificial membrane permeability and huma liver microsomes. In vitro action shows that addition of m-OH substituent on the C4 anilino ring results in potent agent with 2–30-fold increase in activity against c-Src. Most potent c-Src inhibitors were then evaluated for their inhibitory action (IC_{50}) against the proliferation of neuroblastoma SH-SY5Y cells. This case study shows successful application of FEP for lead optimization by calculating ΔΔG with MC/FEP, which resulted in optimized compound **12** and **16** from the parent compound **10** (Fig. 13.9).

9.3 Case Study 3: Replication Protein A Inhibitor

Replication protein A (RPA) interacts with the DNA and plays an important role in the DNA replication, repair, and recombination and DNA damage response. Inhibitors of RPA–DNA interactions are potential candidates for anticancer activity. In 2015, Gavande et al. discovered four compounds which were effective in blocking RPA–DNA interactions (Andrews & Turchi, 2004; Mishra et al., 2015; Neher et al., 2011; Shuck & Turchi, 2010). Out of these, compounds **17** (TDLR-505) and **18** (TDLR-551) were revealed to have reversible inhibition interaction with the central DNA binding domains A and B of RPA70. These compounds are active against the lung and ovarian cancer cell lines and showed synergism with cisplatin and etoposide (Mishra et al., 2015; Shuck & Turchi, 2010). Though the compounds had good potency, they had low solubility and cellular permeability. Gavande and team optimized compound **18** to improve these properties (Gavande et al., 2020).

Molecular docking studies were used to confirm high binding affinity of TDRL-551 with DNA binding domain B of RPA70. Hydrophobic and π–π

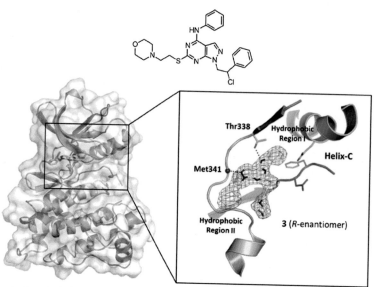

FIG. 13.8 Illustration depicting inhibitor **10** (reported as **3** in Tintori et al. (2015)) in complex with wild-type c-Src (PDB code: 4O2P). The kinase domain is in the active DFG-in conformation and hydrogen bond interactions of the inhibitor with Thr338 (gatekeeper) and the backbone amide of Met341 are illustrated as *red dotted lines*. Hinge region (orange), helix C (turquoise), DFG-motif (pink), and inhibitor **10** (yellow sticks). (Figure reproduced with permission from American Chemical Society Tintori, C., Fallacara, A. L., Radi, M., Zamperini, C., Dreassi, E., Crespan, E., Maga, G., Schenone, S., Musumeci, F., Brullo, C., Richters, A., Gasparrini, F., Angelucci, A., Festuccia, C., Delle Monache, S., Rauh, D. & Botta, M. (2015). Combining X-ray crystallography and molecular modeling toward the optimization of pyrazolo[3,4-d]pyrimidines as potent c-Src inhibitors active in vivo against neuroblastoma. *Journal of Medicinal Chemistry, 58*(1):347−361.)

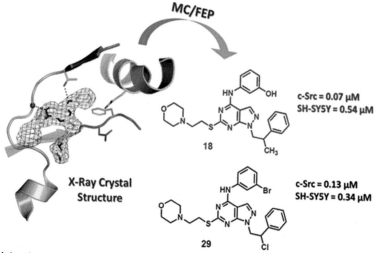

FIG. 13.9 Molecular overlay of optimized leads **12** (reported as **18** in Tintori et al. (2015)) and **16** (reported as **29** in Tintori et al. (2015)) using dynamics docking (MC/FEP) technique. (Figure reproduced with permission from American Chemical Society Tintori, C., Fallacara, A. L., Radi, M., Zamperini, C., Dreassi, E., Crespan, E., Maga, G., Schenone, S., Musumeci, F., Brullo, C., Richters, A., Gasparrini, F., Angelucci, A., Festuccia, C., Delle Monache, S., Rauh, D. & Botta, M. (2015). Combining X-ray crystallography and molecular modeling toward the optimization of pyrazolo[3,4-d]pyrimidines as potent c-Src inhibitors active in vivo against neuroblastoma. *Journal of Medicinal Chemistry, 58*(1) 347−361.)

interactions and the interaction between conserved carboxylic acid side chain and basic amino acid residues contributed to binding affinity. The docking studies also revealed that the extended terminal carboxylic acid side chain is toward the solvent-exposed region of the protein. This side chain is important for modulating the physiochemical properties of the compound.

Structure–activity relationship (SAR) studies coupled with docking studies were used to optimize the lead compound. Initially, aromatic ring A was targeted for optimization. A library of more than 20 compounds was prepared for SAR studies. RPA inhibitory activity was assessed using quantitative electrophoretic mobility shift assay. Docking studies helped to position an additional solubilizing group for improving the solubility. Potency and cellular uptake were further improved through SAR studies. Substituting terminal carboxylic group with morpholinopropane in compound **19** retained inhibition property and increased the potency upto threefolds as compared with parent lead compound **18** ($IC_{50} = 15.3$ μM). The morpholino derivatives, **19**, **20**, and **21**, showed superior cellular uptake, potency, and solubility as compared with carboxylic acid counterparts (Figs. 13.10 and 13.11). The inhibition mechanism of compound binding to the target protein was confirmed by fluorescent intercalator displacement assay with doxorubicin as positive control.

9.4 Case Study 4: Neuraminidase Inhibitor

Neuraminidase (NA), a surface glycoprotein in influenza virus, plays a significant role in viral replication and infection. It is a validated therapeutic target for designing of antiinfluenza drugs. Yu et al. (2019) discovered a new lead NA inhibitor **22** (AN-329/10738021) using structure-based virtual screening, molecular dynamics simulations, and bioassay validation (Yu et al., 2019). Optimization of lead compound **22** was done by targeting pocket named as 430-cavity which is adjacent to the active site of NA target (Fig. 13.12). Many previous reports have claimed 430-cavity as an attractive cavity/pocket in the target. This pocket has a large molecular volume and is directly connected to the active site and is present in different subtypes of influenza virus. Thus, the design of NA inhibitors based on the 430-activity is a reliable approach.

Lead generation: Ten NA inhibitors were used to generate a pharmacophore model with GALAHAD module of the SYBYL-X 2.1 package (Tripos Inc., St. Louis, USA). Out of 20 pharmacophore models, the best model was selected on the basis of Günner–Henry test score which was used to screen compounds from large SPECS database consisting of 212,713 compounds. Lipinski principles were used to filter the compounds with drug-like properties and finally a set of 774 compounds was obtained. Subsequent docking studies were done to study binding modes between inhibitors and NA crystal structures (PDB ID: 2HU0). After docking studies, the number of compounds to be further studied was reduced to 100. Scrutinizing these 100 compounds (using MD simulations) with respect to the spatial matching and higher binding score reduced the number to 10 compounds. MD simulations were again performed on these 10 compounds using Amber 12 and a lead was generated which was named as AN-329/10738021.

Lead optimization: N′-Benzylidene benzohydrazone was identified as a novel scaffold in the lead compound

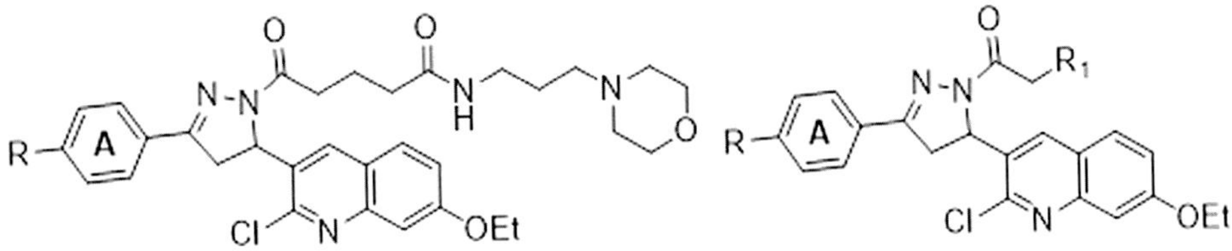

19 R = I
$IC_{50} = 4.9 \pm 0.28$ μM
20 R = chloro-pyridinyl
$IC_{50} = 4.0 \pm 0.56$ μM
21 R = fluoro-pyridinyl
$IC_{50} = 2.3 \pm 0.31$ μM

17 (TDRL-505)
R = Br
R_1 = -COOH

18 (TDRL-551)
R = I
$R_1 = -CH_2CH_2COOH$
$IC_{50} = 15.3 \pm 1.42$ μM

FIG. 13.10 Structure depicting previously reported (TDRL-505 and TDRL-551) and optimized compounds 19, 20, 21 as RPA inhibitors.

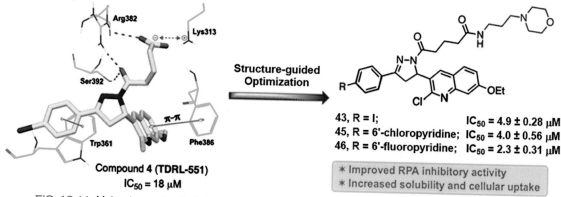

FIG. 13.11 Molecular overlay depicting optimized compounds **19** (reported as 43 (Gavande et al., 2020)), **20** (reported as 45 (Gavande et al., 2020)), and **21** (reported as 46 (Gavande et al., 2020)). (Figure reproduced with permission from American Chemical Society Gavande, N. S., Vandervere-Carozza, P. S., Pawelczak, K. S., Vernon, T. L., Jordan, M. R. & Turchi, J. J. (2020). Structure-guided optimization of replication protein A (RPA)—DNA interaction inhibitors. *Medicinal Chemistry Letters, 11*(6):1118—1124.)

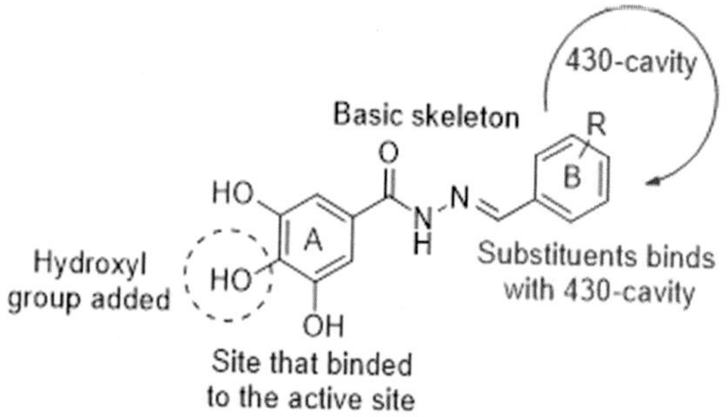

FIG. 13.12 Structure representing design of promising inhibitors based on parent lead compound **22**.

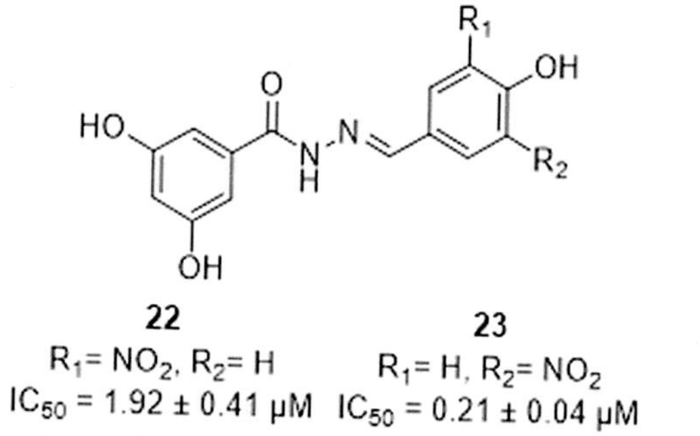

FIG. 13.13 Optimization of compound **22** to improve neuraminidase inhibition by 10 times by compound **23**.

(Yu et al., 2019). A series of inhibitors were generated (11 compounds) by optimization of lead compound. As 430-cavity is mainly composed of hydrophobic amino acids such as Trp403, Lys432, Ile427, Thr439, etc., the hydrophobic groups were introduced in the ring B (Fig. 13.12). The ring B was optimized by introducing electronegative groups such as nitro, halogen, hydroxyl, and methoxy.

In the series, compounds containing two electronegative groups showed activity comparable to oseltamivir carboxylate (OSC) while compound **23** showing better inhibition ($IC_{50} = 0.21$ μM) against NA than OSC ($IC_{50} = 3.04$ μM) and lead compound **22** (Figs. 13.13 and 13.14). The molecular docking indicated that the better activity of compound **23** can be attributed to the elongation of benzylidine molecule to the 430-cavity unlike OSC (a reference compound) that could not reach 430-cavity. The study showed that

presence of highly electronegative groups (nitro or halogens) in ring B enhances the activity as compared to methoxy or amide groups. Furthermore, hydroxyl group at para position to hydrazine decreases the activity. This was attributed to the strong hydrophilicity of hydroxyl group. Molecular docking studies revealed that the para position of hydrazine ring was close to the hydrophobic region of NA active site; thus, hydroxyl at this position decreases the activity. Comparison of binding poses of lead compound (AN-329/10738021), compound **23**, and OSC showed that all three were able to get implanted into the NA active site but the aryl B ring of the first two is able to extend into the 430-cavity due to which formation of hydrogen bonding and hydrophobic interactions were possible. These interactions were responsible for the good inhibition of lead compound **22** and compound **23**.

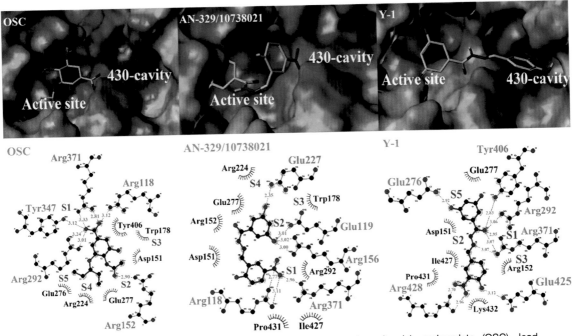

FIG. 13.14 Molecular overlay illustrating reference compound oseltamivir carboxylate (OSC), lead compound **22** (AN-329/10738021), and putative optimized compound **23** (reported as Y-1). (Figure reproduced with permission from American Chemical Society Yu, R., Cheng, L. P., Li, M. & Pang, W. (2019). Discovery of novel neuraminidase inhibitors by structure-based virtual screening, structural optimization, and bioassay. *ACS Medicinal Chemistry Letters, 10*(12):1667−1673.)

Furthermore, MD simulations were carried out for all the synthesized compounds to calculate the binding free energy (ΔG bind) using MMPBSA and MMGBSA between inhibitors and NA. The study showed that compound **23** has the lowest ΔG bind (-31.26 kcal mol^{-1}) and it exhibits the best inhibitory action.

10 CONCLUSION

A hit or lead is a starting point in the drug discovery; potency, specificity, and physicochemical properties are optimized concurrently to ensure potency and minimize side effects and toxicity. The efficiency of hit-to-lead-to-candidate optimization is essential for successful drug development. Advances in molecular structure determination actuated design of computational tools for optimizing lead compounds. Molecular docking has increasingly occupied a prominent role for hit finding and hit-to-lead optimization when structural information about the targeted molecule is available. Molecular docking predicts an optimized orientation of ligand in the binding pocket in its target. The iterations in binding mode in the active site of the target molecule of the ligand by incremental changes in the structure are used for developing more potent, selective, and efficient drug candidates. This is achieved by using several algorithms and scoring methods. The accuracy of these tools depends upon the data used for training sets and quality of the crystal structures of the target molecules. It is important to take into consideration the flexibility, solvent effects, and the conformational entropy (thermodynamics attribution) for both the drug target and the ligand (Salmaso & Moro, 2018). Protein rigidity is another important factor for these studies (Limongelli, 2020). This chapter discussed molecular docking tools for optimizing lead compounds in drug discovery process. The field is developing and is becoming efficient with introduction of ML and DL methods for lead optimization. The improvement in quality of data and computational tools can bring down the cost and time required for this important phase of drug discovery.

REFERENCES

Abad-Zapatero, C., & Metz, J. T. (2005). Ligand efficiency indices as guideposts for drug discovery. *Drug Discovery Today*, 10(7), 464–469.

Almlöf, M., Brandsdal, B. O., & Åqvist, J. (2004). Binding affinity prediction with different force fields: Examination of the linear interaction energy method. *Journal of Computational Chemistry*, 25(10), 1242–1254.

Andrews, B. J., & Turchi, J. J. (2004). Development of a high-throughput screen for inhibitors of replication protein A and its role in nucleotide excision repair. *Molecular Cancer Therapeutics*, 3(4), 385–391.

Åqvist, J., Luzhkov, V. B., & Brandsdal, B. O. (2002). Ligand binding affinities from MD simulations. *Accounts of Chemical Research*, 35(6), 358–365.

Aqvist, J., Medina, C., & Samuelsson, J. E. (1994). A new method for predicting binding affinity in computer-aided drug design. *Protein Engineering Design and Selection*, 7(3), 385–391.

Ashtawy, H. M., & Mahapatra, N. R. (2015). BgN-score and BsN-score: Bagging and boosting based ensemble neural networks scoring functions for accurate binding affinity prediction of protein-ligand complexes. *BMC Bioinformatics*, 16(S4), S8.

Ashtawy, H. M., & Mahapatra, N. R. (2018). Task-specific scoring functions for predicting ligand binding poses and affinity and for screening enrichment. *Journal of Chemical Information and Modeling*, 58(1), 119–133.

Athanasiou, C., Vasilakaki, S., Dellis, D., & Cournia, Z. (2018). Using physics-based pose predictions and free energy perturbation calculations to predict binding poses and relative binding affinities for FXR ligands in the D3R grand challenge 2. *Journal of Computer-Aided Molecular Design*, 32(1), 21–44.

Banerjee, P., Eckert, A. O., Schrey, A. K., & Preissner, R. (2018). ProTox-II: A webserver for the prediction of toxicity of chemicals. *Nucleic Acids Research*, 46(W1), W257–W263.

Beck, B. R., Shin, B., Choi, Y., Park, S., & Kang, K. (2020). Predicting commercially available antiviral drugs that may act on the novel coronavirus (2019-nCoV), Wuhan, China through a drug-target interaction deep learning model. *Computational and Structural Biotechnology Journal*, 18784–790.

Belhassan, A., Zaki, H., Benlyas, M., Lakhlifi, T., & Bouachrine, M. (2019). Study of novel triazolo-benzodiazepine analogues as antidepressants targeting by molecular docking and ADMET properties prediction. *Heliyon*, 5(9), e02446.

Bergdorf, M., Baxter, S., Rendleman, C. A., & Shaw, D. E. (2015). *Desmond/GPU performance as of October 2015*. DE Shaw Research Technical Report DESRES/TR–2015, 1.

Berthold, M. R., Cebron, N., Dill, F., Gabriel, T. R., Kötter, T., Meinl, T., Ohl, P., Thiel, K., & Wiswedel, B. (2009). KNIME-the Konstanz information miner: Version 2.0 and beyond. *ACM SIGKDD Explorations Newsletter*, 11(1), 26–31.

Cang, Z., & Wei, G. (2017). TopologyNet: Topology based deep convolutional and multi-task neural networks for biomolecular property predictions. *PLoS Computational Biology*, 13(7), e1005690.

Capoferri, L., Van Dijk, M., Rustenburg, A. S., Wassenaar, T. A., Kooi, D. P., Rifai, E. A., Vermeulen, N. P., & Geerke, D. P. (2017). eTOX ALLIES: An automated pipeLine for linear interaction energy-based simulations. *Journal of Cheminformatics*, 9(1), 1–13.

Capoferri, L., Verkade-Vreeker, M. C., Buitenhuis, D., Commandeur, J. N., Pastor, M., Vermeulen, N. P., & Geerke, D. P. (2015). Linear interaction energy based prediction of cytochrome P450 1A2 binding affinities with reliability estimation. *PLoS One*, 10(11), e0142232.

Carlson, H. A., & Jorgensen, W. L. (1995). An extended linear response method for determining free energies of hydration. *The Journal of Physical Chemistry*, 99(26), 10667–10673.

Cavasotto, C. N. (2020). Binding free energy calculation using quantum mechanics aimed for drug lead optimization. In A. Heifetz (Ed.), *Quantum mechanics in drug discovery*. New York: Springer.

Cavasotto, C. N., Adler, N. S., & Aucar, M. G. (2018). Quantum chemical approaches in structure-based virtual screening and lead optimization. *Frontiers in Chemistry*, 6188.

Chen, W., Deng, Y., Russell, E., Wu, Y., Abel, R., & Wang, L. (2018). Accurate calculation of relative binding free energies between ligands with different net charges. *Journal of Chemical Theory and Computation*, 14(12), 6346–6358.

Cheng, F., Li, W., Zhou, Y., Shen, J., Wu, Z., Liu, G., Lee, P. W., & Tang, Y. (2012). admetSAR: A comprehensive source and free tool for assessment of chemical ADMET properties. *Journal of Chemical Information and Modeling*, 52(11), 3099–3105.

Chinnasamy, K., Saravanan, M., & Poomani, K. (2020). Evaluation of binding and antagonism/downregulation of brila-nestrant molecule in estrogen receptor-α via quantum mechanics/molecular mechanics, molecular dynamics and binding free energy calculations. *Journal of Biomolecular Structure and Dynamics*, 38(1), 219–235.

Cournia, Z., Allen, B., & Sherman, W. (2017). Relative binding free energy calculations in drug discovery: Recent advances and practical considerations. *Journal of Chemical Information and Modeling*, 57(12), 2911–2937.

Coyne, A. G., Scott, D. E., & Abell, C. (2010). Drugging challenging targets using fragment-based approaches. *Current Opinion in Chemical Biology*, 14(3), 299–307.

De Vivo, M., Masetti, M., Bottegoni, G., & Cavalli, A. (2016). Role of molecular dynamics and related methods in drug discovery. *Journal of Medicinal Chemistry*, 59(9), 4035–4061.

Dimasi, J. A., Grabowski, H. G., & Hansen, R. W. (2016). Innovation in the pharmaceutical industry: New estimates of R&D costs. *Journal of Health Economics*, 47, 20–33.

Doak, B. C., Zheng, J., Dobritzsch, D., & Kihlberg, J. (2016). How beyond rule of 5 drugs and clinical candidates bind to their targets. *Journal of Medicinal Chemistry*, 59(6), 2312–2327.

El-Miligy, M. M. M., Hazzaa, A. A., El-Messmary, H., Nassra, R. A., & El-Hawash, S. A. M. (2017). New hybrid molecules combining benzothiophene or benzofuran with rhodanine as dual COX-1/2 and 5-LOX inhibitors: Synthesis, biological evaluation and docking study. *Bioorganic Chemistry, 72*102−72115.

Eldridge, M. D., Murray, C. W., Auton, T. R., Paolini, G. V., & Mee, R. P. (1997). Empirical scoring functions: I. The development of a fast empirical scoring function to estimate the binding affinity of ligands in receptor complexes. *Journal of Computer-Aided Molecular Design, 11*(5), 425−445.

Fayyazi, N., Esmaeili, S., Taheri, S., Ribeiro, F. F., Scotti, M. T., Scotti, L., Ghasemi, J. B., Saghaei, L., & Fassihi, A. (2019). Pharmacophore modeling, synthesis, scaffold hopping and biological β- hematin inhibition interaction studies for anti-malaria compounds. *Current Topics in Medicinal Chemistry, 19*(30), 2743−2765.

Feinberg, E. N., Sur, D., Wu, Z., Husic, B. E., Mai, H., Li, Y., Sun, S., Yang, J., Ramsundar, B., & Pande, V. S. (2018). PotentialNet for molecular property prediction. *ACS Central Science, 4*(11), 1520−1530.

Fillbrunn, A., Dietz, C., Pfeuffer, J., Rahn, R., Landrum, G. A., & Berthold, M. R. (2017). KNIME for reproducible cross-domain analysis of life science data. *Journal of Biotechnology, 261*149−261156.

Fowler, S., & Schnall, J. G. (2014). TOXNET: Information on toxicology and environmental health. *American Journal of Nursing, 114*(2), 61−63.

Fratev, F., Steinbrecher, T., & Jónsdóttir, S.Ó. (2018). Prediction of accurate binding modes using combination of classical and accelerated molecular dynamics and free-energy perturbation calculations: An application to toxicity studies. *ACS Omega, 3*(4), 4357−4371.

Gallicchio, E., & Levy, R. M. (2004). AGBNP: An analytic implicit solvent model suitable for molecular dynamics simulations and high-resolution modeling. *Journal of Computational Chemistry, 25*(4), 479−499.

Garreta, R., & Moncecchi, G. (2013). *Learning scikit-learn: Machine learning in python.* Packt Publishing Ltd.

Gavande, N. S., Vandervere-Carozza, P. S., Pawelczak, K. S., Vernon, T. L., Jordan, M. R., & Turchi, J. J. (2020). Structure-guided optimization of replication protein A (RPA)−DNA interaction inhibitors. *Medicinal Chemistry Letters, 11*(6), 1118−1124.

Golkov, V., Skwark, M. J., Mirchev, A., Dikov, G., Geanes, A. R., Mendenhall, J., & Meiler, J. C.,D. (2017). 3D deep learning for biological function prediction from physical fields. *arXiv.* preprint arXiv:1704.04039.

Guedes, I. A., Pereira, F. S., & Dardenne, L. E. (2018). Empirical scoring functions for structure-based virtual screening: Applications, critical aspects, and challenges. *Frontiers in Pharmacology, 9*, 1089.

Guimarães, C. R., & Cardozo, M. (2008). MM-GB/SA rescoring of docking poses in structure-based lead optimization. *Journal of Chemical Information and Modeling, 48*(5), 958−970.

Gutiérrez-De-Terán, H., & Åqvist, J. (2012). Linear interaction energy: Method and applications in drug design. In

R. Baron (Ed.), *Computational drug discovery and design.* (pp. 305−323). New York: Humana Press.

Hall, M., Frank, E., Holmes, G., Pfahringer, B., Reutemann, P., & Witten, I. H. (2009). The WEKA data mining software: An update. *ACM SIGKDD Explorations Newsletter, 11*(1), 10−18.

Hann, M. M. (2011). Molecular obesity, potency and other addictions in drug discovery. *MedChemComm, 2*(5), 349−355.

Hann, M. M., & Keserü, G. M. (2012). Finding the sweet spot: The role of nature and nurture in medicinal chemistry. *Nature Reviews Drug Discovery, 11*(5), 355−365.

Harder, E., Damm, W., Maple, J., Wu, C., Reboul, M., Xiang, J. Y., Wang, L., Lupyan, D., Dahlgren, M. K., & Knight, J. L. (2016). OPLS3: A force field providing broad coverage of drug-like small molecules and proteins. *Journal of Chemical Theory and Computation, 12*(1), 281−296.

Harvey, M. J., Giupponi, G., & Fabritiis, G. D. (2009). ACEMD: Accelerating biomolecular dynamics in the microsecond time scale. *Journal of Chemical Theory and Computation, 5*(6), 1632−1639.

Hassan, M. M., Mogollón, D. C., Fuentes, O., & Sirimulla, S. (2018). DLSCORE: A deep learning model for predicting protein-ligand binding affinities. *ChemRxiv, 1353.*

Hassanzadeh, & Wang, H. R. (2016). *DeeperBind: Enhancing prediction of sequence specificities of DNA binding proteins* (pp. 178−183).

Hofmann, M., & Klinkenberg, R. (2014). *RapidMiner: Data mining use cases and business analytics applications.* Boca Raton: CRC Press.

Homeyer, N., & Gohlke, H. (2015). Extension of the free energy workflow FEW towards implicit solvent/implicit membrane MM−PBSA calculations. *Biochimica et Biophysica Acta, 1850*(5), 972−982.

Homeyer, N., Stoll, F., Hillisch, A., & Gohlke, H. (2014). Binding free energy calculations for lead optimization: Assessment of their accuracy in an industrial drug design context. *Journal of Chemical Theory and Computation, 10*(8), 3331−3344.

Hopkins, A. L., Groom, C. R., & Alex, A. (2004). Ligand efficiency: a useful metric for lead selection. *Drug Discovery Today, 9*(10), 430−431.

Hopkins, A. L., Keserü, G. M., Leeson, P. D., Rees, D. C., & Reynolds, C. H. (2014). The role of ligand efficiency metrics in drug discovery. *Nature Reviews Drug Discovery, 13*(2), 105−121.

Hritz, J., & Oostenbrink, C. (2009). Efficient free energy calculations for compounds with multiple stable conformations separated by high energy barriers. *The Journal of Physical Chemistry B, 113*(38), 12711−12720.

Hughes, T. B., Miller, G. P., & Swamidass, S. J. (2015). Modeling epoxidation of drug-like molecules with a deep machine learning network. *ACS Central Science, 1*(4), 168−180.

Hultén, J., Bonham, N. M., Nillroth, U., Hansson, T., Zuccarello, G., Bouzide, A., Åqvist, J., Classon, B., Danielson, U. H., & Karlén, A. (1997). Cyclic HIV-1 protease inhibitors derived from mannitol: Synthesis, inhibitory potencies, and computational predictions of binding affinities. *Journal of Medicinal Chemistry, 40*(6), 885–897.

Irwin, B. W., & Huggins, D. J. (2018). Estimating atomic contributions to hydration and binding using free energy perturbation. *Journal of Chemical Theory and Computation, 14*(6), 3218–3227.

Jaiswal, S., Tripathi, R. K. P., & Ayyannan, S. R. (2018). Scaffold hopping-guided design of some isatin based rigid analogs as fatty acid amide hydrolase inhibitors: Synthesis and evaluation. *Biomedicine and Pharmacotherapy,* 1071611–1071623.

Jencks, W. P. (1981). On the attribution and additivity of binding energies. *Proceedings of the National Academy of Sciences of the United States of America, 78*(7), 4046–4050.

Jespers, W., Isaksen, G. V., Andberg, T. A. H., Vasile, S., Van Veen, A., Åqvist, J., Brandsdal, B. O., & Gutiérrez-De-Terán, H. (2019). QresFEP: An automated protocol for free energy calculations of protein mutations in Q. *Journal of Chemical Theory and Computation, 15*(10), 5461–5473.

Jia, W. Q., Jing, Z., Liu, X., Feng, X. Y., Liu, Y. Y., Wang, S. Q., Xu, W. R., Liu, J. W., & Cheng, X. C. (2018). Virtual identification of novel PPARα/γ dual agonists by scaffold hopping of saroglitazar. *Journal of Biomolecular Structure and Dynamics, 36*(13), 3496–3512.

Jorgensen, W. L. (2009). Efficient drug lead discovery and optimization. *Accounts of Chemical Research, 42*(6), 724–733.

Judson, R., Richard, A., Dix, D., Houck, K., Elloumi, F., Martin, M., Cathey, T., Transue, T. R., Spencer, R., & Wolf, M. (2008). ACToR–aggregated computational toxicology resource. *Toxicology and Applied Pharmacology, 233*(1), 7–13.

Kalirajan, R., Pandiselvi, A., Gowramma, B., & Balachandran, P. (2019). In-silico design, ADMET screening, MM-GBSA binding free energy of some novel isoxazole substituted 9-anilinoacridines as HER2 inhibitors targeting breast cancer. *Current Drug Research Reviews, 11*(2), 118–128.

Karnik, K. S., Sarkate, A. P., Lokwani, D. K., Narula, I. S., Burra, P. V. L. S., & Wakte, P. S. (2020). Development of triple mutant T790M/C797S allosteric EGFR inhibitors: A computational approach. *Journal of Biomolecular Structure and Dynamics,* 1–23.

Kataria, R., & Khatkar, A. (2019). Molecular docking, synthesis, kinetics study, structure-activity relationship and ADMET analysis of morin analogous as helicobacter pylori urease inhibitors. *BMC Chemistry, 13*(1).

Kaus, J. W., Harder, E., Lin, T., Abel, R., Mccammon, J. A., & Wang, L. (2015). How to deal with multiple binding poses in alchemical relative protein–ligand binding free energy calculations. *Journal of Chemical Theory and Computation, 11*(6), 2670–2679.

Kavlock, R., Chandler, K., Houck, K., Hunter, S., Judson, R., Kleinstreuer, N., Knudsen, T., Martin, M., Padilla, S., Reif, D., Richard, A., Rotroff, D., Sipes, N., & Dix, D. (2012). Update

on EPA's ToxCast program: Providing high throughput decision support tools for chemical risk management. *Chemical Research in Toxicology, 25*(7), 1287–1302.

Knight, J. L., Krilov, G., Borrelli, K. W., Williams, J., Gunn, J. R., Clowes, A., Cheng, L., Friesner, R. A., & Abel, R. (2014). Leveraging data fusion strategies in multireceptor lead optimization MM/GBSA End-point methods. *Journal of Chemical Theory and Computation, 10*(8), 3207–3220.

Konidala, K. K., Bommu, U. D., Yeguvapalli, S., & Pabbaraju, N. (2018). In silico insights into prediction and analysis of potential novel pyrrolopyridine analogs against human MAPKAPK-2: A new SAR-based hierarchical clustering approach. *3 Biotech, 8*(9), 385.

Konze, K. D., Bos, P. H., Dahlgren, M. K., Leswing, K., Tubert-Brohman, I., Bortolato, A., Robbason, B., Abel, R., & Bhat, S. (2019). Reaction-based enumeration, active learning, and free energy calculations to rapidly explore synthetically tractable chemical space and optimize potency of cyclin-dependent kinase 2 inhibitors. *Journal of Chemical Information and Modeling, 59*(9), 3782–3793.

Korb, O., Stützle, T., & Exner, T. E. (2009). Empirical scoring functions for advanced protein-ligand docking with PLANTS. *Journal of Chemical Information and Modeling, 49*(1), 84–96.

Kumari, R., Kumar, R., & Lynn, A. (2014). g_mmpbsa a GROMACS tool for high-throughput MM-PBSA calculations. *Journal of Chemical Information and Modeling, 54*(7), 1951–1962.

Kuntz, I. D., Chen, K., Sharp, K. A., & Kollman, P. A. (1999). The maximal affinity of ligands. *Proceedings of the National Academy of Sciences, 96*(18), 9997.

Lenselink, E. B., Louvel, J., Forti, A. F., Van Veldhoven, J. P., De Vries, H., Mulder-Krieger, T., Mcrobb, F. M., Negri, A., Goose, J., & Abel, R. (2016). Predicting binding affinities for GPCR ligands using free-energy perturbation. *ACS Omega, 1*(2), 293–304.

Levoin, N., Calmels, T., Poupardin-Olivier, O., Labeeuw, O., Danvy, D., Robert, P., Berrebi-Bertrand, I., Ganellin, C. R., Schunack, W., & Stark, H. (2008). Refined docking as a valuable tool for lead optimization: Application to histamine H3 receptor antagonists. *Archiv der Pharmazie, 341*(10), 610–623.

Li, Z., Huang, Y., Wu, Y., Chen, J., Wu, D., Zhan, C. G., & Luo, H. B. (2019). Absolute binding free energy calculation and design of a subnanomolar inhibitor of phosphodiesterase-10. *Journal of Medicinal Chemistry, 62*(4), 2099–2111.

Li, H., Leung, K. S., Wong, M. H., & Ballester, P. J. (2015). Improving AutoDock vina using random forest: The growing accuracy of binding affinity prediction by the effective exploitation of larger data sets. *Molecular Informatics, 34*(2–3), 115–126.

Limongelli, V. (2020). Ligand binding free energy and kinetics calculation in 2020. *Wiley Interdisciplinary Reviews: Computational Molecular Science, 10*(4), e1455.

Lim, N. M., Wang, L., Abel, R., & Mobley, D. L. (2016). Sensitivity in binding free energies due to protein reorganization. *Journal of Chemical Theory and Computation, 12*(9), 4620–4631.

Lipinski, C. A., Lombardo, F., Dominy, B. W., & Feeney, P. J. (2001). Experimental and computational approaches to estimate solubility and permeability in drug discovery and development settings. *Advanced Drug Delivery Reviews, 46*(1–3), 3–26.

Lipinski, C. F., Maltarollo, V. G., Oliveira, P. R., Da Silva, A. B. F., & Honorio, K. M. (2019). Advances and perspectives in applying deep learning for drug design and discovery. *Frontiers in Robotics and AI, 6*(108).

Li, H., Sze, K., Lu, G., & Ballester, P. J. (2020). Machine-learning scoring functions for structure-based drug lead optimization. *Wiley Interdisciplinary Reviews: Computational Molecular Science, 10*(5), e1465.

Loeffler, H. H., Michel, J., & Woods, C. (2015). FESetup: Automating setup for alchemical free energy simulations. *Journal of Chemical Information and Modeling, 55*(12), 2485–2490.

Lovering, F., Aevazelis, C., Chang, J., Dehnhardt, C., Fitz, L., Han, S., Janz, K., Lee, J., Kaila, N., Mcdonald, J., Moore, W., Moretto, A., Papaioannou, N., Richard, D., Ryan, M. S., Wan, Z. K., & Thorarensen, A. (2016). Imidazotriazines: Spleen tyrosine kinase (Syk) inhibitors identified by free-energy perturbation (FEP). *ChemMedChem, 11*(2), 217–233.

Marelius, J., Graffner-Nordberg, M., Hansson, T., Hallberg, A., & Åqvist, J. (1998). Computation of affinity and selectivity: Binding of 2, 4-diaminopteridine and 2, 4-diaminoquinazoline inhibitors to dihydrofolate reductases. *Journal of Computer-Aided Molecular Design, 12*(2), 119–131.

Markov, Z., & Russell, I. (2006). An introduction to the WEKA data mining system. *ACM SIGCSE Bulletin, 38*(3), 367–368.

Massova, I., & Kollman, P. A. (2000). Combined molecular mechanical and continuum solvent approach (MM-PBSA/GBSA) to predict ligand binding. *Perspectives in Drug Discovery and Design, 18*(1), 113–135.

Matricon, P., Ranganathan, A., Warnick, E., Gao, Z.-G., Rudling, A., Lambertucci, C., Marucci, G., Ezzati, A., Jaiteh, M., & Dal Ben, D. (2017). Fragment optimization for GPCRs by molecular dynamics free energy calculations: Probing druggable subpockets of the A 2A adenosine receptor binding site. *Scientific Reports, 7*(1), 1–12.

Matsunaga, T., Shirasawa, H., Tanabe, M., Ohnuma, N., Kawamura, K., Etoh, T., Takahashi, H., & Simizu, B. (1994). Expression of neuronal src mRNA as a favorable marker and inverse correlation to N-myc gene amplification in human neuroblastomas. *International Journal of Cancer, 58*(6), 793–798.

Matsunaga, T., Shirasawa, H., Tanabe, M., Ohnuma, N., Takahashi, H., & Simizu, B. (1993). Expression of alternatively spliced src messenger RNAs related to neuronal differentiation in human neuroblastomas. *Cancer Research,* 533179–533185.

Maunz, A., Gütlein, M., Rautenberg, M., Vorgrimmler, D., Gebele, D., & Helma, C. (2013). Lazar: A modular predictive toxicology framework. *Frontiers in Pharmacology, 4*, 38.

Maynard, A. T., & Roberts, C. D. (2016). Quantifying, visualizing, and monitoring lead optimization. *Journal of Medicinal Chemistry, 59*(9), 4189–4201.

Mayr, A., Klambauer, G., Unterthiner, T., & Hochreiter, S. (2016). DeepTox: Toxicity prediction using deep learning. *Frontiers in Environmental Science, 3*(80).

Mignani, S., Huber, S., Tomas, H., Rodrigues, J., & Majoral, J.-P. (2016). Compound high-quality criteria: A new vision to guide the development of drugs, current situation. *Drug Discovery Today, 21*(4), 573–584.

Miller Iii, B. R., Mcgee, T. D., Jr., Swails, J. M., Homeyer, N., Gohlke, H., & Roitberg, A. E. (2012). MMPBSA. Py: An efficient program for end-state free energy calculations. *Journal of Chemical Theory and Computation, 8*(9), 3314–3321.

Mishra, A. K., Dormi, S. S., Turchi, A. M., Woods, D. S., & Turchi, J. J. (2015). Chemical inhibitor targeting the replication protein A-DNA interaction increases the efficacy of Pt-based chemotherapy in lung and ovarian cancer. *Biochemical Pharmacology, 93*(1), 25–33.

Mishra, N. K., Singla, D., Agarwal, S., & Raghava, G. P. (2014). ToxiPred: A server for prediction of aqueous toxicity of small chemical molecules in T. Pyriformis. *Journal of Translational Toxicology, 1*(1), 21–27.

Mobley, D. L., & Gilson, M. K. (2017). Predicting binding free energies: Frontiers and benchmarks. *Annual Review of Biophysics, 46*531–46558.

Moraca, F., Negri, A., De Oliveira, C., & Abel, R. (2019). Application of free energy perturbation (FEP+) to understanding ligand selectivity: A case study to assess selectivity between pairs of phosphodiesterases (PDE's). *Journal of Chemical Information and Modeling, 59*(6), 2729–2740.

Nascimento, E. R. C., Oliva, M. N., Świderek, K., Martins, J. B., & Andrés, J. (2017). Binding analysis of some classical acetylcholinesterase inhibitors: Insights for a rational design using free energy perturbation method calculations with QM/MM MD simulations. *Journal of Chemical Information and Modeling, 57*(4), 958–976.

Navara, C. S. (2004). The spleen tyrosine kinase (Syk) in human disease, implications for design of tyrosine kinase inhibitor based therapy. *Current Pharmaceutical Design, 10*(15), 1739–1744.

Navarra, M., Celano, M., Maiuolo, J., Schenone, S., Botta, M., Angelucci, A., Bramanti, P., & Russo, D. (2010). Antiproliferative and pro-apoptotic effects afforded by novel Src-kinase inhibitors in human neuroblastoma cells. *BMC Cancer, 10*(1), 602.

Neher, T. M., Bodenmiller, D., Fitch, R. W., Jalal, S. I., & Turchi, J. J. (2011). Novel irreversible small molecule inhibitors of replication protein A display single-agent activity and synergize with cisplatin. *Molecular Cancer Therapeutics, 10*(10), 1796–1806.

Oprea, T. I., Davis, A. M., Teague, S. J., & Leeson, P. D. (2001). Is there a difference between leads and drugs? A historical perspective. *Journal of Chemical Information and Computer Sciences, 41*(5), 1308–1315.

Paissoni, C., Spiliotopoulos, D., Musco, G., & Spitaleri, A. (2015). GMXPBSA 2.1: A GROMACS tool to perform MM/PBSA and computational alanine scanning. *Computer Physics Communications*, 186105–186107.

Pan, A. C., Borhani, D. W., Dror, R. O., & Shaw, D. E. (2013). Molecular determinants of drug–receptor binding kinetics. *Drug Discovery Today*, 18(13–14), 667–673.

Pan, P., Yu, H., Liu, Q., Kong, X., Chen, H., Chen, J., Liu, Q., Li, D., Kang, Y., Sun, H., Zhou, W., Tian, S., Cui, S., Zhu, F., Li, Y., Huang, Y., & Hou, T. (2017). Combating drug-resistant mutants of anaplastic lymphoma kinase with potent and selective type-I1/2 inhibitors by stabilizing unique DFG-shifted loop conformation. *ACS Central Science*, 3(11), 1208–1220.

Pearlman, D. A. (2005). Evaluating the molecular mechanics Poisson– Boltzmann surface area free energy method using a congeneric series of ligands to p38 MAP kinase. *Journal of Medicinal Chemistry*, 48(24), 7796–7807.

Pedregosa, F., Varoquaux, G., Gramfort, A., Michel, V., Thirion, B., Grisel, O., Blondel, M., Prettenhofer, P., Weiss, R., & Dubourg, V. (2011). Scikit-learn: Machine learning in python. *The Journal of Machine Learning Research*, 122825–122830.

Peng, C., Wang, J., Zhijian, X., Cai, T., & Zhu, W. (2020). Accurate prediction of relative binding affinities of a series of HIV-1 protease inhibitors using semi-empirical quantum mechanical charge. *Journal of Computational Chemistry*, 41(19), 1773–1780.

Perić-Hassler, L., Stjernschantz, E., Oostenbrink, C., & Geerke, D. P. (2013). CYP 2D6 binding affinity predictions using multiple ligand and protein conformations. *International Journal of Molecular Sciences*, 14(12), 24514–24530.

Pires, D. E., & Ascher, D. B. (2016). CSM-lig: A web server for assessing and comparing protein–small molecule affinities. *Nucleic Acids Research*, 44(W1), W557–W561.

Pires, D. E., Blundell, T. L., & Ascher, D. B. (2015). pkCSM: Predicting small-molecule pharmacokinetic and toxicity properties using graph-based signatures. *Journal of Medicinal Chemistry*, 58(9), 4066–4072.

Polanski, J., Bogocz, J., & Tkocz, A. (2016). The analysis of the market success of FDA approvals by probing top 100 best-selling drugs. *Journal of Computer-Aided Molecular Design*, 30(5), 381–389.

Pratt, L. R., & Chandler, D. (1980). Effects of solute–solvent attractive forces on hydrophobic correlations. *The Journal of Chemical Physics*, 73(7), 3434–3441.

Pu, C., Yan, G., Shi, J., & Li, R. (2017). Assessing the performance of docking scoring function, Fep, Mm-GBSA, and QM/MM-GBSA approaches on a series of PLK1 inhibitors. *MedChemComm*, 8(7), 1452–1458.

Radi, M., Brullo, C., Crespan, E., Tintori, C., Musumeci, F., Biava, M., Schenone, S., Dreassi, E., Zamperini, C., Maga, G., Pagano, D., Angelucci, A., Bologna, M., & Botta, M. (2011). Identification of potent c-Src inhibitors strongly affecting the proliferation of human neuroblastoma cells. *Bioorganic & Medicinal Chemistry Letters*, 21(19), 5928–5933.

Raha, K., & Merz, K. M. (2004). A quantum mechanics-based scoring function: study of zinc ion-mediated ligand binding. *Journal of the American Chemical Society*, 126(4), 1020–1021.

Raveh, B., London, N., Zimmerman, L., & Schueler-Furman, O. (2011). Rosetta FlexPepDock ab-initio: Simultaneous folding, docking and refinement of peptides onto their receptors. *PloS One*, 6(4), e18934.

Reddy, M. R., & Erion, M. D. (2007). Relative binding affinities of fructose-1,6-bisphosphatase inhibitors calculated using a quantum mechanics-based free energy perturbation method. *Journal of the American Chemical Society*, 129(30), 9296–9297.

Reddy, K. K., Rathore, R. S., Srujana, P., Burri, R. R., Reddy, C. R., Sumakanth, M., Reddanna, P., & Reddy, M. R. (2020). Performance evaluation of docking programs- glide, gold, AutoDock and, SurflexDock, using free energy perturbation reference data: A case study of fructose-1, 6-bisphosphatase-AMP analogs. *Mini Reviews in Medicinal Chemistry*, 20(12), 1179–1187.

Rifai, E. A., Ferrario, V., Pleiss, J., & Geerke, D. P. (2020). Combined linear interaction energy and alchemical solvation free-energy approach for protein-binding affinity computation. *Journal of Chemical Theory and Computation*, 16(2), 1300–1310.

Rifai, E. A., Van Dijk, M., & Geerke, D. P. (2020). Recent developments in linear interaction energy based binding free energy calculations. *Frontiers in Molecular Biosciences*, 7(114).

Rifai, E. A., Van Dijk, M., Vermeulen, N. P., Yanuar, A., & Geerke, D. P. (2019). A comparative linear interaction energy and MM/PBSA study on SIRT1–ligand binding free energy calculation. *Journal of Chemical Information and Modeling*, 59(9), 4018–4033.

Salmaso, V., & Moro, S. (2018). Bridging molecular docking to molecular dynamics in exploring ligand-protein recognition process: An overview. *Frontiers in Pharmacology*, 9923.

Schindler, C. E., De Vries, S. J., & Zacharias, M. (2015). Fully blind peptide-protein docking with pepATTRACT. *Structure*, 23(8), 1507–1515.

Schultes, S., De Graaf, C., Haaksma, E. E. J., De Esch, I. J. P., Leurs, R., & Krämer, O. (2010). Ligand efficiency as a guide in fragment hit selection and optimization. *Drug Discovery Today: Technologies*, 7(3), e157–e162.

Shaker, N., Abou-Zleikha, M., Alamri, M., & Mehellou, Y. (2020). A generative deep learning approach for the discovery of SARS CoV 2 protease inhibitors. *ChemRxiv*.

Sharma, A. K., Srivastava, G. N., Roy, A., & Sharma, V. K. (2017). ToxiM: A toxicity prediction tool for small molecules developed using machine learning and chemoinformatics approaches. *Frontiers in Pharmacology*, 8(880).

Sherborne, B., Shanmugasundaram, V., Cheng, A. C., Christ, C. D., Desjarlais, R. L., Duca, J. S., Lewis, R. A., Loughney, D. A., Manas, E. S., & Mcgaughey, G. B. (2016). Collaborating to improve the use of free-energy and other quantitative methods in drug discovery. *Journal of Computer-Aided Molecular Design*, 30(12), 1139–1141.

Shuck, S. C., & Turchi, J. J. (2010). Targeted inhibition of replication protein A reveals cytotoxic activity, synergy with chemotherapeutic DNA-damaging agents, and insight into cellular function. *Cancer Research, 70*(8), 3189–3198.

Singh, N., & Warshel, A. (2010). Absolute binding free energy calculations: On the accuracy of computational scoring of protein-ligand interactions. *Proteins: Structure, Function, and Bioinformatics, 78*(7), 1705–1723.

Souza, P. C. T., Thallmair, S., & Marrink, S. J. (2019). An allosteric pathway in copper, zinc superoxide dismutase unravels the molecular mechanism of the G93A amyotrophic lateral sclerosis-linked mutation. *The Journal of Physical Chemistry Letters, 10*(24), 7740–7744.

Spiliotopoulos, D., Kastritis, P. L., Melquiond, A. S., Bonvin, A. M., Musco, G., Rocchia, W., & Spitaleri, A. (2016). dMM-PBSA: a new HADDOCK scoring function for protein-peptide docking. *Frontiers in Molecular Biosciences, 346.*

Spyrakis, F., & Cavasotto, C. N. (2015). Open challenges in structure-based virtual screening: Receptor modeling, target flexibility consideration and active site water molecules description. *Archives of Biochemistry and Biophysics, 583105–583119.*

Srinivasan, J., Cheatham, T. E., Cieplak, P., Kollman, P. A., & Case, D. A. (1998). Continuum solvent studies of the stability of DNA, RNA, and phosphoramidate–DNA helices. *Journal of the American Chemical Society, 120*(37), 9401–9409.

Stanton, R. A., Lu, X., Detorio, M., Montero, C., Hammond, E. T., Ehteshami, M., Domaoal, R. A., Nettles, J. H., Feraud, M., & Schinazi, R. F. (2016). Discovery, characterization, and lead optimization of 7-azaindole non-nucleoside HIV-1 reverse transcriptase inhibitors. *Bioorganic & Medicinal Chemistry Letters, 26*(16), 4101–4105.

Stepniewska-Dziubinska, M. M., Zielenkiewicz, P., & Siedlecki, P. (2018). Development and evaluation of a deep learning model for protein–ligand binding affinity prediction. *Bioinformatics, 34*(21), 3666–3674.

Stjernschantz, E., & Oostenbrink, C. (2010). Improved ligand-protein binding affinity predictions using multiple binding modes. *Biophysical Journal, 98*(11), 2682–2691.

Sulimov, V. B., Kutov, D. C., & Sulimov, A. V. (2019). Advances in docking. *Current Medicinal Chemistry, 26*(42), 7555–7580.

Sun, H., Xia, M., Austin, C. P., & Huang, R. (2012). Paradigm shift in toxicity testing and modeling. *The AAPS Journal, 14*(3), 473–480.

Taddei, M., Ferrini, S., Giannotti, L., Corsi, M., Manetti, F., Giannini, G., Vesci, L., Milazzo, F. M., Alloatti, D., Guglielmi, M. B., Castorina, M., Cervoni, M. L., Barbarino, M., Foderà, R., Carollo, V., Pisano, C., Armaroli, S., & Cabri, W. (2014). Synthesis and evaluation of new Hsp90 inhibitors based on a 1,4,5-trisubstituted 1,2,3-triazole scaffold. *Journal of Medicinal Chemistry, 57*(6), 2258–2274.

Tan, M., Özgül, O. F., Bardak, B., Ekşioğlu, I., & Sabuncuoğlu, S. (2019). Drug response prediction by ensemble learning and drug-induced gene expression signatures. *Genomics, 111*(5), 1078–1088.

Tan, C., Tan, Y. H., & Luo, R. (2007). Implicit nonpolar solvent models. *The Journal of Physical Chemistry B, 111*(42), 12263–12274.

Tice, R. R., Austin, C. P., Kavlock, R. J., & Bucher, J. R. (2013). Improving the human hazard characterization of chemicals: A Tox21 update. *Environmental Health Perspectives, 121*(7), 756–765.

Tintori, C., Fallacara, A. L., Radi, M., Zamperini, C., Dreassi, E., Crespan, E., Maga, G., Schenone, S., Musumeci, F., Brullo, C., Richters, A., Gasparrini, F., Angelucci, A., Festuccia, C., Delle Monache, S., Rauh, D., & Botta, M. (2015). Combining X-ray crystallography and molecular modeling toward the optimization of pyrazolo[3,4-d]pyrimidines as potent c-Src inhibitors active in vivo against neuroblastoma. *Journal of Medicinal Chemistry, 58*(1), 347–361.

Ton, A. T., Gentile, F., Hsing, M., Ban, F., & Cherkasov, A. (2020). Rapid identification of potential inhibitors of SARS-CoV-2 main protease by deep docking of 1.3 billion compounds. *Molecular Informatics, 39*(8), e2000028.

Torres, P. H., Sodero, A. C., Jofily, P., & Silva-, F. P., Jr. (2019). Key topics in molecular docking for drug design. *International Journal of Molecular Sciences, 20*(18), 4574.

Van Dijk, M., Ter Laak, A. M., Wichard, J. R. D., Capoferri, L., Vermeulen, N. P., & Geerke, D. P. (2017). Comprehensive and automated linear interaction energy based binding-affinity prediction for multifarious cytochrome P450 aromatase inhibitors. *Journal of Chemical Information and Modeling, 57*(9), 2294–2308.

Vitali, R., Mancini, C., Cesi, V., Tanno, B., Piscitelli, M., Mancuso, M., Sesti, F., Pasquali, E., Calabretta, B., Dominici, C., & Raschellà, G. (2009). Activity of tyrosine kinase inhibitor dasatinib in neuroblastoma cells in vitro and in orthotopic mouse model. *International Journal of Cancer, 125*(11), 2547–2555.

Vosmeer, C. R., Kooi, D. P., Capoferri, L., Terpstra, M. M., Vermeulen, N. P., & Geerke, D. P. (2016). Improving the iterative linear interaction energy approach using automated recognition of configurational transitions. *Journal of Molecular Modeling, 22*(1), 31.

Wang, L., Berne, B., & Friesner, R. A. (2012). On achieving high accuracy and reliability in the calculation of relative protein–ligand binding affinities. *Proceedings of the National Academy of Sciences, 109*(6), 1937–1942.

Wang, C., Greene, D. A., Xiao, L., Qi, R., & Luo, R. (2018). Recent developments and applications of the MMPBSA method. *Frontiers in Molecular Biosciences, 4*(87).

Wang, W., Wang, J., & Kollman, P. A. (1999). What determines the van der Waals coefficient β in the LIE (linear interaction energy) method to estimate binding free energies using molecular dynamics simulations? *Proteins: Structure, Function, and Bioinformatics, 34*(3), 395–402.

Wang, J., Wolf, R. M., Caldwell, J. W., Kollman, P. A., & Case, D. A. (2004). Development and testing of a general amber force field. *Journal of Computational Chemistry, 25*(9), 1157–1174.

Wang, L., Wu, Y., Deng, Y., Kim, B., Pierce, L., Krilov, G., Lupyan, D., Robinson, S., Dahlgren, M. K., Greenwood, J., Romero, D. L., Masse, C., Knight, J. L., Steinbrecher, T., Beuming, T., Damm, W., Harder, E., Sherman, W., Brewer, M., … Abel, R. (2015). Accurate and reliable prediction of relative ligand binding potency in prospective drug discovery by way of a modern free-energy calculation protocol and force field. *Journal of the American Chemical Society*, 137(7), 2695−2703.

Wang, C., & Zhang, Y. (2017). Improving scoring-docking-screening powers of protein−ligand scoring functions using random forest. *Journal of Computational Chemistry*, 38(3), 169−177.

Wang, B., Zhao, Z., Nguyen, D. D., & Wei, G. W. (2017). Feature functional theory−binding predictor (FFT−BP) for the blind prediction of binding free energies. *Theoretical Chemistry Accounts*, 136(4), 55.

Wen, M., Zhang, Z., Niu, S., Sha, H., Yang, R., Yun, Y., & Lu, H. (2017). Deep-learning-based drug−target interaction prediction. *Journal of Proteome Research*, 16(4), 1401−1409.

Williams-Devane, C. R., Wolf, M. A., & Richard, A. M. (2009). DSSTox chemical-index files for exposure-related experiments in arrayexpress and gene expression omnibus: Enabling toxico-chemogenomics data linkages. *Bioinformatics*, 25(5), 692−694.

Williams-Noonan, B. J., Yuriev, E., & Chalmers, D. K. (2018). Free energy methods in drug design: Prospects of "alchemical perturbation" in medicinal chemistry: Miniperspective. *Journal of Medicinal Chemistry*, 61(3), 638−649.

Wu, Z., Ramsundar, B., Feinberg, E. N., Gomes, J., Geniesse, C., Pappu, A. S., Leswing, K., & Pande, V. (2018). MoleculeNet: A benchmark for molecular machine learning. *Chemical Science*, 9(2), 513−530.

Xu, M., Zhu, J., Diao, Y., Zhou, H., Ren, X., Sun, D., Huang, J., Han, D., Zhao, Z., Zhu, L., Xu, Y., & Li, H. (2013). Novel selective and potent inhibitors of malaria parasite dihydroorotate dehydrogenase: Discovery and optimization of dihydrothiophenone derivatives. *Journal of Medicinal Chemistry*, 56(20), 7911−7924.

Yadav, M., Srivastava, R., Naaz, F., & Singh, R. K. (2018). Synthesis, docking, ADMET prediction, cytotoxicity and antimicrobial activity of oxathiadiazole derivatives. *Computational Biology and Chemistry*, 77226−77239.

Yang, Y., Zhang, R., Li, Z., Mei, L., Wan, S., Ding, H., Chen, Z., Xing, J., Feng, H., & Han, J. (2020). Discovery of highly potent, selective, and orally efficacious p300/CBP histone acetyltransferases inhibitors. *Journal of Medicinal Chemistry*, 63(3), 1337−1360.

Yu, R., Cheng, L. P., Li, M., & Pang, W. (2019). Discovery of novel neuraminidase inhibitors by structure-based virtual screening, structural optimization, and bioassay. *ACS Medicinal Chemistry Letters*, 10(12), 1667−1673.

Zeevaart, J. G., Wang, L., Thakur, V. V., Leung, C. S., Tirado-Rives, J., Bailey, C. M., Domaoal, R. A., Anderson, K. S., & Jorgensen, W. L. (2008). Optimization of azoles as anti-human immunodeficiency virus agents guided by free-energy calculations. *Journal of the American Chemical Society*, 130(29), 9492−9499.

Zhang, H., Liao, L., Saravanan, K. M., Yin, P., & Wei, Y. (2019). DeepBindRG: A deep learning based method for estimating effective protein−ligand affinity. *PeerJ*, 7e7362.

Zhang, L., Tan, J., Han, D., & Zhu, H. (2017). From machine learning to deep learning: Progress in machine intelligence for rational drug discovery. *Drug Discovery Today*, 22(11), 1680−1685.

Zhang, L., & Zhou, R. (2020). Structural basis of potential binding mechanism of remdesivir to SARS-CoV-2 RNA dependent RNA polymerase. *The Journal of Physical Chemistry B*, 124(32), 6955−6962.

Zhavoronkov, A., Ivanenkov, Y. A., Aliper, A., Veselov, M. S., Aladinskiy, V. A., Aladinskaya, A. V., Terentiev, V. A., Polykovskiy, D. A., Kuznetsov, M. D., & Asadulaev, A. (2019). Deep learning enables rapid identification of potent DDR1 kinase inhibitors. *Nature Biotechnology*, 37(9), 1038−1040.

Zwanzig, R. W. (1954). High-temperature equation of state by a perturbation method. I. nonpolar gases. *The Journal of Chemical Physics*, 22(8), 1420−1426.

CHAPTER 14

Multi-Target Drugs as Master Keys to Complex Diseases: Inverse Docking Strategies and Opportunities

PATRICIA SAENZ-MÉNDEZ

1 INTRODUCTION

Until recent years, the paradigm of drug discovery has been the development of selective compounds. This approach is intimately related to the concept of a "magic bullet" developed by Paul Ehrlich, the founder of chemotherapy, more than 150 years ago (Ehrlich, 1878, 1897; Strebhardt & Ullrich, 2008). Ehrlich postulated the existence of specific receptors and the idea that drugs go straightaway to their intended receptor. That idea evolved into the "magic bullet" concept, which suggests that there is one predetermined target for each drug. However, this reductionist concept is a simplification of the actual phenomenon, being the idea of drugs acting on multiple targets more appropriate. Therefore, "magic shotguns" or "silver bullets" better represent reality (Fig. 14.1).

Polypharmacology or pharmacological promiscuity has been raised as a breakthrough concept in drug discovery, shifting from the one target—one drug model to the multiple-target approach (Medina-Franco et al., 2013). Several aspects of drug design are related to polypharmacology (Fig. 14.2) (Anighoro et al., 2014; Méndez-Lucio et al., 2016; Peters, 2013).

First, multi-target drugs can produce undesired side effects. Only a low percentage of drug candidates finally reach the market. Late-stage drug attrition (i.e., failure) is often a consequence of toxicity, and the huge associated cost must be reduced. It was described that, on average, each potential drug binds to six target proteins instead of one (Azzaoui et al., 2007; Mestres et al., 2008). The well-known pharma Adagio "fail fast, fail cheap" states that it is better to fail in early stages of drug development to minimize the economic impact of failure (Hay et al., 2014; Waring et al., 2015). Besides this negative aspect of polypharmacology, multi-target

drugs may represent a valuable opportunity in treatment of complex diseases, such as cancer, psychosis, and infectious diseases (Peters, 2013; Roth et al., 2004; Zimmermann et al., 2007). Antimicrobial resistance is a growing public health threat worldwide and it has been described for almost all types of known antibiotics. If an antimicrobial agent hits several bacterial targets, it is more difficult for the microorganism to develop resistance (Brotzoesterhelt & Brunner, 2008). Last, but not least, off-target interactions are commonly observed for most drugs, and thus drug repurposing or drug repositioning is a faster and more economic shortcut strategy to make a drug candidate to reach the market (Ye et al., 2014).

Drug discovery strategies, both experimental (such as high-throughput screening) and computational, have evolved focusing on the single-target approach, which is indeed neglecting many other biological processes that are connected in the organisms. Therefore, the growing evidence of polypharmacology has boosted the development of multi-target approaches (Csermely et al., 2005; Hopkins, 2008; Maggiora, 2011; Medina-Franco et al., 2011). In fact, several multi-target drugs are already available on the market, mostly targeting kinases. Some examples of multi-target drugs (targeting kinases) approved by the FDA are included in Table 14.1 (Li et al., 2016).

Ramsay et al. reported an analysis of the FDA-approved new molecular entities (NMEs) from 2015 to 2017 (Fig. 14.3) (Ramsay et al., 2018). These data clearly reflect the increasing pharma interest in biologics (such as PROTACs, peptides, aptamers, etc.) (Valeur et al., 2017; Valeur et al., 2019), representing 34% of total approved NMEs. Also, though still smaller than the number of single-target drugs, multi-target ones

Molecular Docking for Computer-Aided Drug Design. https://doi.org/10.1016/B978-0-12-822312-3.00005-9

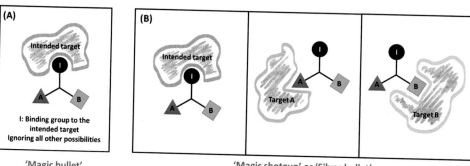

FIG. 14.1 **(A)** Single-target drug discovery. Only the pharmacophore binding to the intended target is considered. **(B)** Multi-target drug discovery. Instead of ignoring all other features of drug candidates, several targets are considered as interacting with different groups of the molecule. (Credit: Patricia Saenz-Méndez.)

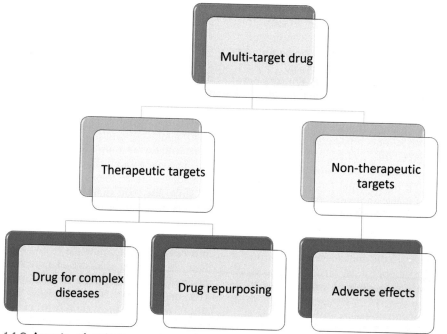

FIG. 14.2 Aspects of polypharmacology related to multi-target drugs. (Adapted from reference Méndez-Lucio, O., Naveja, J. J., Vite-Caritino, H., & Prieto-Martínez, F. D. (2016). Review. One drug for multiple targets: A computational perspective. *The Journal of the Mexican Chemical Society, 60*(3), 168–181.)

contribute up to 21% and continue growing. Undoubtedly, a change in drug discovery paradigm is ongoing.

Computer-aided multi-target drug design can be classified as ligand-based or structure-based (Koutsoukas et al., 2011; Kruger & Evers, 2010; Nantasenamat & Prachayasittikul, 2015; Westermaier et al., 2015). Ligand-based multi-target drug design does not require knowledge on the target receptor and thus can be based on descriptors (similarity search), based on 3D structure of the ligands (shape screening), or based on "understanding" the differences between known active and

inactive compounds (machine learning strategies) (Cereto-Massague et al., 2015; Liu et al., 2014; Medina-Franco et al., 2012; Saenz-Méndez, Genheden, et al., 2017). Structure-based drug design is based on the knowledge of the 3D structure of the target protein and it is frequently performed using high-throughput virtual screening, allowing to identify hits for single targets (Kalyaanamoorthy & Chen, 2011). This virtual screening can additionally rely on a pharmacophore-based search. However, this strategy is not sufficient when several receptors are being

TABLE 14.1
Examples of FDA-Approved Multi-Target Drugs (Li et al., 2016).

Drug Name	Indication	Targets
Afanitib	Metastatic non–small cell lung cancer	EGFR, HER2
Bosunitib	Ph + chronic myelogenous leukemia	ABL, Src
Cabozantinib	Metastatic medullary thyroid cancer	MET, VEGFR2
Dasatinib	Chronic myelogenous and acute lymphoblastic leukemia	ABL, Src
Imatinib	Ph + chronic myelogenous leukemia	ABL, PDGFR-β, c-Kit
Lapatinib	HER2-positive metastatic breast cancer	EGFR, HER2
Lenvatinib	RAI-refractory differentiated thyroid cancer	VEGFR2, c-Kit
Nintedanib	Idiopathic pulmonary fibrosis	VEGFR2, PDGFR-α, PDGFR-β
Palbociclib	ER-positive and HER2-negative breast cancer	CDK4, CDK6
Pazopanib	Renal cell carcinoma and soft tissue sarcoma	VEGFR2, c-Kit, PDGFR-β
Ponatinib	Chronic myeloid and acute lymphoblastic leukemia	ABL, c-Kit
Regorafenib	Metastatic colorectal cancer and advanced gastrointestinal stromal tumor	VEGFR2, c-Kit
Sorafenib	Renal cell carcinoma and unresectable hepatocellular carcinomas	RET, VEGFR2, PDGFR-β, B-Raf, c-Kit
Sunitinib	Renal cell carcinoma and imatinib-resistant gastrointestinal stromal tumor	VEGFR2, c-Kit, PDGFR-β, Raf, c-Kit
Vandetanib	Medullary thyroid cancer	EGFR, VEGFR2

CDK, cyclin-dependent kinase; *CLK*, cdc2-like kinase; *CML*, chronic myelogenous leukemia; *EGFR*, epidermal growth factor receptor; *FDA*, Food and Drug Administration; *GSK3*, glycogen synthase kinase 3; *HER2*, receptor tyrosine-protein kinase erbB-2; *ICD*, International Classification of Diseases; *JAK*, janus kinase; *MAPK*, mitogen-activated protein kinase; *MEK*, mitogen-activated protein kinase kinase; *PDGFR*, platelet-derived growth factor receptor; *PI3K*, phosphoinositide 3-kinase; *TK*, tyrosine kinase; *TKL*, tyrosine kinase like; *TNBC*, triple negative breast cancer; *TTD*, Therapeutic Target Database; *VEGFR2*, vascular endothelial growth factor receptor 2.

simultaneously considered. Finding a "master key" to target multiple receptors at once, **inverse docking** has emerged as enabling tool to address polypharmacology (Chen & Zhi, 2001). This strategy can be employed to find the best target (among a library of receptor structures) for a particular ligand (potential drug). Thus, the biological profile of a drug candidate can be anticipated.

In this chapter, I will review the computational strategies to address polypharmacology, particularly focusing on docking alternatives, while discussing case studies.

2 COMPUTATIONAL STRATEGIES FOR MULTI-TARGET DRUG DESIGN

Methodologies for multi-target drug design range from purely docking strategies to ligand profiling methodologies and combinations of these (Zhang et al., 2017). This chapter mainly focuses on docking-related strategies. However, it is worth to mention and briefly discuss other methodologies.

The great versatility of computational approaches in drug discovery is facilitated by the amazing amount of compound databases in public domain (Nicola et al.,

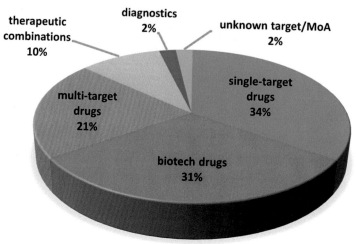

FIG. 14.3 Distribution of new molecular entities approved by the US FDA from 2015 to 2017. *MoA*, mechanism of action. (Adapted from reference Ramsay, R. R., Popovic-Nikolic, M. R., Nikolic, K., Uliassi, E., & Bolognesi, M. L. (2018). A perspective on multi-target drug discovery and design for complex diseases. *Clinical and Translational Medicine, 7*(1), 3. doi: 10.1186/s40169-017-0181-2.)

2012; Williams, 2008). These databases include natural products (Yongye et al., 2012), compounds commercially available (Barbosa & Del Rio, 2012), or known drugs, such as ZINC (https://zinc.docking.org/; Irwin et al., 2012), PubChem (https://pubchem.ncbi.nlm.nih.gov/; Kim et al., 2019), ChEMBL (https://www.ebi.ac.uk/chembl/; Mendez et al., 2019), Sweetlead (https://simtk.org/projects/sweetlead; Novick et al., 2013), GRAS (Burdock et al., 2006; https://www.accessdata.fda.gov/scripts/fdcc/?set=SCOGS), FDA-approved compounds (https://www.fda.gov/drugs/development-approval-process-drugs/drug-approvals-and-databases), the TCM database of herbal compounds used in traditional Chinese medicine (Chen, 2011), etc.

Chemogenomics or chemical genomics is a relatively new research field involving the systematic "screening" of libraries of small molecules against drug target families (such as G-protein–coupled receptors or kinases), aiming to identify all possible ligands of all possible targets (Andrusiak et al., 2012; Caron et al., 2001; Namchuk, 2002). Instead of focusing on particular targets, the chemogenomics approach considers target families and libraries of compounds with activity profiles (Bredel & Jacoby, 2004). To illustrate the approach, a library of approved drugs with annotated activities can be screened against a family of functionally related proteins outside the known activity profile. Through the systematic analysis of compound–target interactions, a known drug can be repurposed. Compounds active against a particular target frequently share several features, including

shape, hydrophobicity, ability to form hydrogen bonds, etc. For example, aiming to retrieve kinase inhibitors for the platelet-derived growth factor receptor (PDGFR) and the vascular endothelial growth factor receptor (VEGFR), a chemogenomics approach (e.g., similarity search) can be employed (Kim et al., 2018). It is possible to search for PDGFR inhibitors within a library of known VEGFR inhibitors, being both relevant receptors in anticancer drug development (Roskoski, 2007).

Computational chemogenomics integrates the chemical and biological data, being thus a merge of chemoinformatics and bioinformatics. Molecular information systems can be developed, including structure of the ligands, target sequence, mechanism of action, structure–activity relationship data, toxicological properties, etc. (Jacoby, 2011). The ligand and the target space are indeed connected. Several concepts reviewed in this chapter are represented in Fig. 14.4.

By exploring the target space, through target fishing methodologies, it is possible to predict the pharmacological profile for a ligand, detecting affinity toward several targets. This in turns allows for drug repurposing or detecting toxicological effects. When exploring the ligand space, using virtual screening techniques, it is possible to identify the best ligand for a distinct biological target, the conventional approach used in most cases (Bajorath, 2008; Rognan, 2007).

Molecular docking is a well-known tool to identify ligands for specific targets. The identification of the "best pose" involves sampling (i.e., searching the ligand conformational space) and scoring (i.e., ranking the

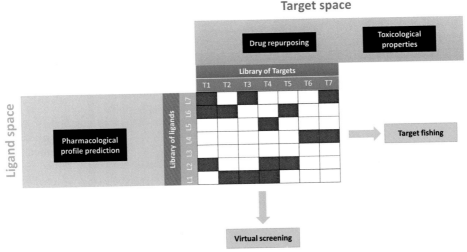

FIG. 14.4 Ligand—target space displaying all possible combinations. A ligand—target interaction is indicated in red. By exploring the target space (through target fishing), it is possible to achieve drug repurposing or advance toxicology effects. By exploring the ligand space (through virtual screening), the pharmacological profile prediction is advanced. (Credit: Patricia Saenz-Méndez.)

binding conformations). Several pitfalls of docking approaches are well known, such as accounting for receptor flexibility, solvation effects, or finding the best scoring function, which will not be further discussed in this chapter. Despite known pitfalls, docking has evolved and been developed, being currently a widespread methodology in drug discovery (Amaro et al., 2008; Lin et al., 2002; Saenz-Méndez, Genheden, et al., 2017; Trott & Olson, 2010; Yuriev et al., 2015; Yuriev & Ramsland, 2013).

Besides predicting ligands that bind to the active site of a receptor, molecular docking can be employed to identify putative targets for specific ligands. Instead of docking a database of ligands into a single target, a single ligand (or a ligand database) can be docked in a library of receptors (i.e., binding sites), an approach known as **inverse docking** (Chen & Zhi, 2001). Hits resulting from a round of inverse docking are targets that can be ranked (Fig. 14.5). Several inverse docking schemes are available and have been assessed and compared (Hui-Fang et al., 2010; Lee et al., 2016; Xu et al., 2018).

To implement an inverse docking scheme, a **target database** must be available. Inverse docking can take advantage of the exponential increase in protein coordinates deposited in the RSCB Protein Data Bank (PDB) (Berman et al., 2003). Since 1976 (13 structures available), the repository has been continuously growing, thanks to the analytical advancements in X-ray crystallography, protein NMR techniques, and lately the cryo-EM technology, containing today almost 155,000 protein structures (http://www.rcsb.org/pdb/home/home.do).

Therapeutic Target Database (TTD) (http://bidd.nus.edu.sg/group/cjttd/; Wang et al., 2020) focuses on known therapeutic protein and nucleic acid targets, having also information about the targeted disease, pathway, and the corresponding drugs for each of these targets. TTD contains 3419 targets, of which 461 are successful, 1191 clinical trial targets, 207 patented targets, and 1560 research targets. Several useful links are included, related to sequence, 3D structure, ligand binding properties, therapeutic class, EC number, drug structure, with the corresponding references.

Drug-target databases such as SuperTarget (Gunther et al., 2008; http://insilico.charite.de/supertarget/index.php?site=home), SIDER (Campillos et al., 2008; http://sideeffects.embl.de/), or DrugBank (https://www.drugbank.ca/; Wishart et al., 2006) are particularly useful. SuperTarget contains 6219 targets, 195,770 compounds, 282 drug-related pathways, and 332,828 drug-target interactions. SIDER contains information on marketed medicines and their known adverse drug reactions (ADRs), including 1430 drugs and 5868 side effects included. DrugBank combines chemo- and bioinformatics aiming to collect detailed drug data and drug target information. The latest version contains 13,579 drugs, and 5229 nonredundant protein sequences (drug targets, enzymes, transporters) are linked to these drugs.

In addition to publicly available databases, in-house databases have been created by several research groups.

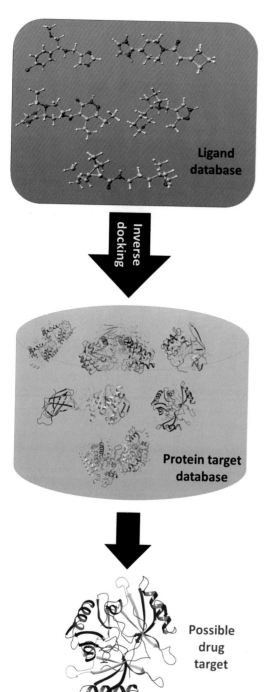

FIG. 14.5 Flowchart showing the inverse docking approach. (Credit: Patricia Saenz-Méndez.)

Lauro et al. built a database of 126 protein targets involved in tumor processes to perform an inverse docking study against a panel of ligands including 27 natural products, 3 semisynthetic compounds, and 13 synthetic compounds (Lauro et al., 2011).

It's worth mentioning that fragment-based docking has shown an increasing role in polypharmacology, which is particularly suited for multi-target drug design, as a consequence of the inherent promiscuous nature of fragments (Bottegoni et al., 2012; Chessari & Woodhead, 2009). Thus, docking experiments employing fragments databases can be performed to find "silver bullets." Fragments are usually defined as molecules with a molecular weight of <250−300 Da (Morphy & Rankovic, 2007), and as fragments are smaller and less functionalized, they usually display lower affinity for the receptor. However, fragment hits usually display high "ligand efficiency" (binding affinity per heavy atom) and thus are a convenient choice for hit optimization into leads. Fragment-based docking aims to find small molecules binding to a target. In a subsequent step, these fragments are turned into bigger molecules through "fragment growing," "fragment merging," and "fragment evolution" (Murray & Blundell, 2010; Murray & Rees, 2009; Schultes et al., 2010).

Normal docking programs can be employed for inverse docking studies. Most commonly used docking engines include UCSF DOCK (http://dock.compbio.ucsf.edu/; Lang et al., 2009), Autodock (Goodsell & Olson, 1990), Molecular Operating Environment (MOE) (Molecular Operating Environment (MOE), 2015), or Gold (Vendonk et al., 2003). The Schrödinger suite, through the Maestro interface, allows for performing docking on several targets simultaneously by simply including a search grid for each receptor. This approach certainly speeds up inverse docking experiments (Maestro, Version 11.9, Schrödinger, LLC, 2019).

To perform an inverse docking experiment with most commonly used docking engines, a given drug candidate (small molecule) should be independently docked, following the usual procedure, in the binding site of each protein in the target database. To finally rank the targets according to binding scores, a final "standardization" of docking scores must be performed. The prediction of the best target for a particular ligand represents an additional challenge. Scoring functions employed in all docking protocols (rigid, flexible, and induced fit) are designed to find the best ligand for a selected biological target (forward or normal docking), but they cannot be employed the other way around (inverse docking). Different receptors usually do not share identical binding pocket shape or internal energy (due to different protein sequence and structure). Therefore, to include this response, a normalization scheme must be implemented (Casey et al., 2009; Saenz-Méndez & Eriksson, 2018).

Lauro et al. suggested a normalization approach using Eq. (14.1) (Lauro et al., 2011; Lauro et al., 2012):

$$V = V_0 / [(M_L + M_R) / 2] \tag{14.1}$$

where V is the normalized score for a ligand on a target, V_0 is the predicted docking score obtained in the molecular docking calculation, M_L is the average binding energy of the ligand across different receptors, and M_R is the average binding energy of the target for all ligands studied. The higher the value of the normalized score V, the more promising is the interaction between ligand and receptor.

A different correction scheme to that used above is the so-called multiple active site correction (MASC), suggested by Vigers and Rizzi (Vigers & Rizzi, 2004), as shown in Eqs. (14.2)–(14.4).

$$\mu_i = \sum_j (S_{ij}) / N \quad j = 1, N \tag{14.2}$$

$$(\sigma_i)^2 = \sum_j (S_{ij} - \mu_i)^2 / (N - 1) \quad j = 1, N \tag{14.3}$$

$$S_{ij}' = (S_{ij} - \mu_i) / \sigma_i \tag{14.4}$$

where S_{ij} is the original calculated docking score for the ith compound and the jth pocket, and S_{ij}' is the modified score for compound i in active site j. μ_i and σ_i are the average and standard deviation of the scores for compound i across all pockets j, respectively. The MASC score is useful in that it includes information about how different from the average a value is and in which direction according to the sign. For example, a negative MASC score indicates that the ligand has a higher affinity for that particular receptor (more negative score) than the average among all receptors. This normalization protocol has been successfully employed in multi-target drug design (Saenz-Méndez, Eriksson, & Eriksson, 2017).

Web tools for inverse docking are available and became widely used. Wang et al. in 2012 launched idTarget (http://idtarget.rcas.sinica.edu.tw/index.php; Wang et al., 2012), a web-based tool for identifying biomolecular targets of small chemical molecules. Auto in silico Consensus Inverse Docking (ACID) is a web server mainly oriented to drug repurposing based on the consensus inverse docking method (http://chemyang.ccnu.edu.cn/ccb/server/ACID/; Wang et al., 2019).

3 CASE EXAMPLES

In this section, I will provide some interesting examples, though it does not pretend to be a comprehensive review. Many more examples of applying docking and inverse docking can be found in the literature (AbdulHameed et al., 2012; Calvaresi & Zerbetto, 2011; Chen, 2014; Chen & Ren, 2014; Chushak et al., 2015; Fan et al., 2012; Grinter et al., 2011; Ma et al., 2015; Park & Cho, 2017).

3.1 Polypharmacology in Drug Attrition and Side Effects

Besides the possibility to link ligands with new targets, thus exploring polypharmacology, chemogenomics allows to explore and map the target space. Cases and Mestres mined both 160 known protein targets and 44,032 small molecules associated with cardiovascular diseases. Some of these active compounds displayed affinity for another set of 421 proteins not linked before to cardiovascular diseases. This approach represents an indirect source to complete the cardiovascular target space or to prematurely detect off-target effects (Cases & Mestres, 2009).

An inverse docking study conducted by Chen and Ung (2001) emphasized the potential of the methodology to predict potential toxic effects of drugs. The study revealed that up to 83% of the experimentally known toxicity and side effects targets of eight clinically used agents (aspirin, gentamicin, vitamin C, penicillin G, ibuprofen, neomycin, indinavir, and 4H-tamoxifen; Table 14.2) were predicted by inverse docking (Chen & Ung, 2001).

Ji et al. (2006) applied inverse docking to search for ADR-related proteins. The authors tested 11 marketed anti-HIV drugs covering protease inhibitors, nucleoside reverse transcriptase inhibitors, and non−nucleoside reverse transcriptase inhibitors. They found that 86%−89% of the inverse docking predicted ADRs of tested drugs are consistent with the adverse reactions reported in the literature (Ji et al., 2006).

ADRs are induced when drugs bind to proteins other than their therapeutic targets, i.e., off-target effects are observed. Pan et al. reported a computational study to identify off-targets and thus to assess toxicity in the early stages of drug development for analgesics (Pan et al., 2014). To this end, the authors performed an inverse docking of analgesics and their active metabolites against human/mammal protein structures. Six marketed analgesics, with known serious side effects, were used, namely, oxycodone, fentanyl, morphine, acetaminophen, liquicet (acetaminophen−hydrocodone), and rofecoxib. Rofecoxib was withdraw from the market in 2004 due to observed risk of heart attack and stroke associated

TABLE 14.2
Chemical Structures of Some Clinically Used Agents and Representative Targets Associated to the Side Effects (Chen & Ung, 2001).

Name	Structure	Targets	Side Effects
Aspirin		Alcohol dehydrogenase Antithrombin L-3-hydroxyacyl-CoA dehydrogenase	Increased blood alcohol level Blood coagulation, thrombolysis Effect on Reye's syndrome patients
Gentamicin		Carbonic anhydrase Glutamate dehydrogenase	Nephrotoxicity Nephrotoxicity
Vitamin C		Alpha-amylase Phenylalanine hydroxylase	Interfere starch hydrolysis May cause phenylketonuria
Penicillin G		Glutathione S-transferase HLA-DR3 Catechol-O-methyltransferase (COMT)	Hepatotoxicity Allergic inflammatory response Competitive interference with normal COMT activity
Ibuprofen		Estrogen sulfotransferase	Sexual dysfunction

Neomycin — Carbonic anhydrase — Nephrotoxicity

Indinavir — Intestinal fatty acid–binding protein — Reduced triglyceride accumulation

4H-tamoxifen — Glutathione S-transferase, Alcohol dehydrogenase — Genotoxicity and carcinogenicity, Enhanced ethanol's sedative effect

TABLE 14.3
Number of Targets Predicted for the Tested Analgesics and Their Metabolites.

Drug Name	Active Metabolites	Number of Targets
Acetaminophen	Acetaminophen	642
	Acetimidoquinone	391
Fentanyl	Fentanyl	448
Morphine	Morphine	454
Oxycodone	Oxycodone	412
	Oxymorphone	487
Rofecoxib	Rofecoxib	842
Liquicet	Acetaminophen	642
	Acetimidoquinone	391
	Hydrocodone	257
	Hydromorphone	400

with its use. Results of inverse docking are shown in Table 14.3.

These results evidence the multi-target nature of the most commonly used analgesics and the toxicological risks associated to this nature. Moreover, in this study, three severe adverse drug reactions (SADRs) were chosen as case studies. For all six analgesics, 182 distinct off-targets associated with cardiac disorders were identified. For cardiac arrhythmias and lung disorders, 331 and 557 common off-targets were detected, respectively. This work clearly shows that the inverse docking approach may offer good starting points to explore sADRs.

3.2 Polypharmacology in Complex Diseases

Complex diseases, such as cancer, inflammatory conditions, infections, and central nervous system diseases, among others, represent a challenge in drug development as are known to possess complex pathways and involve several proteins. Ignoring such complexity usually results in the activation of alternative pathways, side effects, and increase in the cost and time of the development of an efficient drug (Gillespie & Singh, 2011). An effective strategy to effectively defeat those conditions should include a multi-target drug design protocol.

Availability of databases has been of utmost importance in multi-target drug design. For example, Cai et al. detected that the peptide deformylase of *Helicobacter pylori* is a new potential target for antibiotic development against the bacteria through rounds of inverse docking employing databases of natural products and

further validation by enzymatic assay and X-ray crystallography (Cai et al., 2006).

Lauro et al. in 2011 applied an inverse docking approach to natural bioactive compounds to screen their binding affinity against proteins involved in cancer processes. A library of 126 protein targets extracted from the PDB was inverse docked to analyze the binding modes with a library of 43 bioactive compounds. After normalizing the binding energies, some promising molecules with good agreement with reference compounds were identified (Lauro et al., 2011).

Kinase inhibitors are usually multi-target drugs, as the ATP site is frequently targeted by such inhibitors and the binding site of ATP is highly conserved in kinases. As a consequence, selectivity issues and undesired side effects are observed (Méndez-Lucio et al., 2016). However, in the treatment of complex diseases, such as cancer, promiscuous kinase inhibitors display good clinical efficiency when treating tumors, e.g., dasatinib, which binds about 159 kinases (Davis et al., 2011). Ideally, kinases inhibitors will be "selectively nonselective," which means that they should inhibit a few kinases (enabling treating complex diseases), but not all or many of them (avoiding undesired off-target effects) (Morphy, 2010). Zahler et al. used a virtual inverse docking approach to identify kinase targets for three derivatives of indirubin (indirubin-3'oxime, IOX), namely 5-bromo-indirubin-3'oxime (5BIO), 6-bromo-indirubin-3'oxime (6BIO), and 7-bromo-indirubin-3'oxime (7BIO) (Table 14.4) (Zahler et al., 2007).

Though the structure of the three derivatives is similar, their effects differ largely, which was assessed by inverse screening against a database of ca. 6000 protein binding sites. The screening showed significant enrichment of their known targets, and also several kinases not previously described as targets were identified. This work has led to the identification of the kinase PDK1 as a target for the indirubin derivative 6BIO.

Recently, Pandya et al. investigated the multi-target affinity of vinblastine against duplex DNA and human serum albumin (HSA). Vinblastine is a well-known anticancer agent effective against several types of cancer. Docking experiments were performed, and the interaction profile of vinblastine with duplex DNA and HSA was analyzed, providing the binding pattern of the anticancer drug (Pandya et al., 2014).

A structure-based virtual screening of 16 previously synthesized and tested aryl-aminopyridine and aryl-aminoquinoline derivatives was performed, against a library of potential targets. The 16 potential drugs were evaluated for cytotoxic activity against three

TABLE 14.4
Structure of the Derivatives Included in the Inverse Docking Approach Described in Zahler et al. (2007).

General Structure	Derivative	X	Y	Z
	indirubin-3'oxime	H	H	H
	5-bromo-indirubin-3'oxime	Br	H	H
	6-bromo-indirubin-3'oxime	H	Br	H
	7-bromo-indirubin-3'oxime	H	H	Br

cancer cell lines (human cervical cancer—HeLa, human chronic myeloid Leukemia—K562, and human melanoma—Fem-x) and two types of normal cells. Twelve of the compounds showed moderate cytotoxicity, with selectivity against K562, while the remaining four compounds were not active. An inverse virtual screening was thus performed to rationalize the experimental cytotoxicity, employing the PDTD database, which included 1207 entries of 841 known and potential drug targets from PDB, divided into 15 therapeutic areas and 13 biochemical criteria (Gao et al., 2008). Interestingly, the docking studies revealed that the active compounds interact with a number of kinases included in cell cycle regulation. This study shows the potential of inverse docking studies to detect potential targets for cytotoxic compounds and thus a starting point for further structure modification to improve activity (Eric et al., 2012).

Parkinson disease is a long-term degenerative disorder that mainly affects the motor system, the cause of which still remains unknown. The flavonoid baicalein has been proven to be effective in animal model, though the biological targets and mechanisms have not been clarified. Gao et al. reported a target fishing approach, including data mining, molecular docking, structure-based pharmacophore searching, and chemical similarity searching. The top two ranked proteins, i.e., catechol-O-methyltransferase and monoamine oxidase B, have been already reported as targets for baicalein. The third ranked receptor (N-methyl-D-aspartic acid receptor), not described before, was experimentally tested. The results showed that target fishing approaches are effective in drug target discovery and support the observed antiparkinsonian activity of baicalein (Gao et al., 2013).

In 2017, Saenz-Méndez et al. reported an inverse docking study aiming to assess the ligand selectivity between ADP-ribosylating virulence factors. These

toxins are encoded by human pathogens, such as *Pseudomonas aeruginosa* (exotoxin A), *Corynebacterium diphtheriae* (diphtheria toxin), and *Vibrio cholerae* (cholix toxin). This toxins ribosylate eukaryotic elongation factor 2, which catalyzes the translocation during protein synthesis, causing cell death. By targeting external toxins, it is possible to develop new antibiotics, while avoiding antimicrobial resistance. The authors analyzed and compared the three toxins and performed an inverse docking study with a small set of ligands. The normalized scores indicated that the developed methodology is suitable for identifying selective and promiscuous inhibitors and thus can be applied to find multi-target antibiotic design (Saenz-Méndez & Eriksson, 2018; Saenz-Méndez, Eriksson, & Eriksson, 2017).

3.3 Polypharmacology in Drug Repurposing

A straightforward application of polypharmacology is drug repurposing or drug repositioning, which has the potential to speed up the drug discovery process by identifying new uses for already approved drugs. This strategy allows to skip steps in the discovery process, thus saving time and money (Aube, 2012). This is particularly useful for those pathologies where a low amount of investment is observed, such as for rare diseases (Ekins et al., 2011), or for multifactorial diseases, such as cancer, where finding multi-target drugs is of particular relevance (Duenas-Gonzalez et al., 2008).

Chemogenomics has been applied for drug repurposing (Liu et al., 2014; Loging et al., 2007). Keiser et al. computed Tanimoto similarities of 65,000 biologically active compounds divided into sets of hundreds of targets, aiming to classify targets based on ligand features. By grouping ligands, links among unexpected clusters emerged. In particular, methadone, emetine, and loperamide may antagonize muscarinic M3, α2 adrenergic, and neurokinin NK2 receptors, respectively.

Thus, clustering receptors by ligand chemistry has the potential to reveal polypharmacology (Keiser et al., 2007).

In 2011, Li et al. developed a computational drug repurposing strategy, to perform molecular docking of small molecules against protein drug targets, aiming to find novel targets for existing drugs. About 252 human protein targets and 4621 approved and experimental small molecules from DrugBank were included in the study. In fact, the anticancer drug nilotinib targeted MAPK14, which was experimentally confirmed ($IC_{50} = 40$ nmol/L) (Li et al., 2011).

Herbs have been used to treat diseases since very early in history. However, the pharmacological mechanisms of many natural products remain unknown. An interesting study involving natural products from Cerrado (a typical Brazilian biome) has been described, aiming to find those biological targets involved in observed pharmacological effects. In this study, an inverse virtual screening experiment was performed using the sc-PDB database, which included 8166 protein—ligand complexes (http://bioinfo-pharma.u-strasbg.fr/scPDB/; Meslamani et al., 2011). Five natural compounds from Cerrado were used for the inverse virtual screening and six similar ligand compounds were obtained from six PDB deposited structures. The study revealed molecular targets for a set of natural compounds and described the interaction (Carregal et al., 2012).

Kinnings et al. carried out structural-based studies on nine different *Mycobacterium tuberculosis* InhA structures to evaluate the possible repurposing against tuberculosis, of entacapone and tolcapone (approved for the treatment of Parkinson's disease). InhA (enoyl-acyl carrier protein reductase) is essential for type II fatty acid biosynthesis and further synthesis of the bacterial cell wall, being thus an interesting target for antibiotic development. Their results showed that entacapone is a promising lead against resistant strains of *M. tuberculosis* (Kinnings et al., 2009).

Food components designed as "generally recognized as safe" provide molecules which are proved to be safe that can be easily repurposed.

Drug repurposing is a broad field and as such can find many applications. A study recently published aimed to identify a suitable food additive that interacts with transglutaminase 2 (TG2) binding motives in gluten-derived peptides (such as α2-gliadin) to prevent deamidation/transamidation, a process that triggers the initiation of celiac disease. This forward docking study screened a database of 1174 potential food grade ligands against a model of α2-gliadin. Among the five best ligands, ascorbyl palmitate decreased the TG2 transamination of gliadin by 82%. To completely stop transamidation, zinc chloride was added, reaching a 99% inhibition. Based on the docking studies, the authors proposed a combination of ascorbyl palmitate and zinc chloride as additives in celiac-safe foods (Engstrom et al., 2017).

4 CONCLUDING REMARKS AND FUTURE PERSPECTIVES

Several drugs interact with multiple targets, exerting complex effects. This well-known fact has shifted the paradigm of drug discovery and development from the one-target—centered model to the multiple-target approach. As the current goal is to identify "silver bullets" to treat complex diseases, for drug repurposing and for advance in toxicological issues, the computational approaches have evolved to follow this new paradigm.

To date, several successful applications of inverse docking have been described, evidencing the great potential for multi-target drug discovery. One clear limitation is the lack of structures for targets, with only a fraction of all druggable known proteins being available in the PDB database. This limitation can be sorted out by employing methodologies that do not depend on structure, such as ligand-based multi-target drug design. Despite unsolved challenges, docking and inverse docking approaches in polypharmacology are undoubtedly promising and will help in achieving the aims of the new paradigm in drug development.

ACKNOWLEDGMENTS

The author gratefully acknowledges Facultad de Química, UdelaR (Uruguay) and PEDECIBA for the permanent support, computer time, and funding.

REFERENCES

AbdulHameed, M. D. M., Chaudhury, S., Singh, N., Sun, H., Wallqvist, A., & Tawa, G. J. (2012). Exploring polypharmacology using a ROCS-based target fishing approach. *Journal of Chemical Information and Modeling, 52*(2), 492—505. https://doi.org/10.1021/ci2003544

Amaro, R. E., Baron, R., & McCammon, J. A. (2008). An improved relaxed complex scheme for receptor flexibility in computer-aided drug design. *Journal of Computer-Aided Molecular Design, 22*(9), 693—705. https://doi.org/10.1007/s10822-007-9159-2

Andrusiak, K., Piotrowski, J. S., & Boone, C. (2012). Chemical-genomic profiling: Systematic analysis of the cellular targets of bioactive molecules. *Bioorganic & Medicinal Chemistry,*

20(6), 1952—1960. https://doi.org/10.1016/j.bmc.2011.12.023

Anighoro, A., Bajorath, J., & Rastelli, G. (2014). Polypharmacology: Challenges and opportunities in drug discovery. *Journal of Medicinal Chemistry, 57*(19), 7874—7887. https://doi.org/10.1021/jm5006463

Aube, J. (2012). Drug repurposing and the medicinal chemist. *ACS Medicinal Chemistry Letters, 3*(6), 442—444. https://doi.org/10.1021/ml300114c

Azzaoui, K., Hamon, J., Faller, B., Whitebread, S., Jacoby, E., Bender, A., Jenkins, J. L., & Urban, L. (2007). Modeling promiscuity based on in vitro safety pharmacology profiling data. *ChemMedChem, 2*(6), 874—880. https://doi.org/10.1002/cmdc.200700036

Bajorath, J. (2008). Computational analysis of ligand relationships within target families. *Current Opinion in Chemical Biology, 12*(3), 352—358. https://doi.org/10.1016/j.cbpa.2008.01.044

Barbosa, A. J. M., & Del Rio, A. (2012). Freely accessible databases of commercial compounds for high- throughput virtual screenings. *Current Topics in Medicinal Chemistry, 12*(8), 866—877.

Berman, H., Henrick, K., & Nakamura, H. (2003). Announcing the worldwide protein Data Bank. *Nature Structural Biology, 10*(12), 980.

Bottegoni, G., Favia, A. D., Recanatini, M., & Cavalli, A. (2012). The role of fragment-based and computational methods in polypharmacology. *Drug Discovery Today, 17*(1—2), 23—34. https://doi.org/10.1016/j.drudis.2011.08.002

Bredel, M., & Jacoby, E. (2004). Chemogenomics: An emerging strategy for rapid target and drug discovery. *Nature Reviews Genetics, 5*(4), 262—275. https://doi.org/10.1038/nrg1317

Brotzoesterhelt, H., & Brunner, N. (2008). How many modes of action should an antibiotic have? *Current Opinion in Pharmacology, 8*(5), 564—573. https://doi.org/10.1016/j.coph.2008.06.008

Burdock, G. A., Carabin, I. G., & Griffiths, J. C. (2006). The importance of GRAS to the functional food and nutraceutical industries. *Toxicology, 221*(1), 17—27. https://doi.org/10.1016/j.tox.2006.01.012

Cai, J., Han, C., Hu, T., Zhang, J., Wu, D., Wang, F., Liu, Y., Ding, J., Chen, K., Yue, J., Shen, X., & Jiang, H. (2006). Peptide deformylase is a potential target for anti-helicobacter pylori drugs: Reverse docking, enzymatic assay, and X-ray crystallography validation. *Protein Science, 15*(9), 2071—2081. https://doi.org/10.1110/ps.062238406

Calvaresi, M., & Zerbetto, F. (2011). In silico carborane docking to proteins and potential drug targets. *Journal of Chemical Information and Modeling, 51*(8), 1882—1896. https://doi.org/10.1021/ci200216z

Campillos, M., Kuhn, M., Gavin, A.-C., Jensen, L. J., & Bork, P. (2008). Drug target identification using side-effect similarity. *Science, 321*, 263—266.

Caron, P. R., Mullican, M. D., Mashal, R. D., Wilson, K. P., Su, M. S., & Murcko, M. A. (2001). Chemogenomic approaches to drug discovery. *Current Opinion in Chemical Biology, 5*, 464—470.

Carregal, A. P., Comar, M., Alves, S. N., Siqueira, J. M.d., Lima, L. A., & Taranto, A. G. (2012). Inverse virtual screening studies of selected natural compounds from cerrado. *International Journal of Quantum Chemistry, 112*(20), 3333—3340. https://doi.org/10.1002/qua.24205

Cases, M., & Mestres, J. (2009). A chemogenomic approach to drug discovery: Focus on cardiovascular diseases. *Drug Discovery Today, 14*(9—10), 479—485. https://doi.org/10.1016/j.drudis.2009.02.010

Casey, F. P., Pihan, E., & Shields, D. C. (2009). Discovery of small molecule inhibitors of protein-protein interactions using combined ligand and target score normalization. *Journal of Chemical Information and Modeling, 49*, 2708—2717.

Cereto-Massague, A., Ojeda, M. J., Valls, C., Mulero, M., Pujadas, G., & Garcia-Vallve, S. (2015). Tools for in silico target fishing. *Methods, 71*, 98—103. https://doi.org/10.1016/j.ymeth.2014.09.006

Chen, C. Y. (2011). TCM database@Taiwan: The world's largest traditional Chinese medicine database for drug screening in silico. *PLoS One, 6*(1), e15939. https://doi.org/10.1371/journal.pone.0015939

Chen, S. J. (2014). A potential target of tanshinone IIA for acute promyelocytic leukemia revealed by inverse docking and drug repurposing. *Asian Pacific Journal of Cancer Prevention, 15*(10), 4301—4305. https://doi.org/10.7314/apjcp.2014.15.10.4301

Chen, S. J., & Ren, J. L. (2014). Identification of a potential anticancer target of danshensu by inverse docking. *Asian Pacific Journal of Cancer Prevention, 15*(1), 111—116. https://doi.org/10.7314/apjcp.2014.15.1.111

Chen, Y. Z., & Ung, C. Y. (2001). Prediction of potential toxicity and side effect protein targets of a small molecule by a ligand—protein inverse docking approach. *Journal of Molecular Graphics and Modelling, 20*, 199—218.

Chen, Y. Z., & Zhi, D. G. (2001). Ligand—protein inverse docking and its potential use in the computer search of protein targets of a small molecule. *Proteins, 43*, 217—226.

Chessari, G., & Woodhead, A. J. (2009). From fragment to clinical candidate—a historical perspective. *Drug Discovery Today, 14*(13—14), 668—675. https://doi.org/10.1016/j.drudis.2009.04.007

Chushak, Y. G., Chapleau, R. R., Frey, J. S., Mauzy, C. A., & Gearhart, J. M. (2015). Identifying potential protein targets for toluene using a molecular similarity search, in silico docking and in vitro validation. *Toxicology Research, 4*(2), 519—526. https://doi.org/10.1039/c5tx00009b

Csermely, P., Agoston, V., & Pongor, S. (2005). The efficiency of multi-target drugs: The network approach might help drug design. *Trends in Pharmacological Sciences, 26*(4), 178—182. https://doi.org/10.1016/j.tips.2005.02.007

Davis, M. I., Hunt, J. P., Herrgard, S., Ciceri, P., Wodicka, L. M., Pallares, G., Hocker, M., Treiber, D. K., & Zarrinkar, P. P. (2011). Comprehensive analysis of kinase inhibitor selectivity. *Nature Biotechnology, 29*(11), 1046—1051. https://doi.org/10.1038/nbt.1990

Duenas-Gonzalez, A., Garcia-Lopez, P., Herrera, L. A., Medina-Franco, J. L., Gonzalez-Fierro, A., & Candelaria, M. (2008).

The prince and the pauper. A tale of anticancer targeted agents. *Molecular Cancer*, 7, 82. https://doi.org/10.1186/1476-4598-7-82

Ehrlich, P. (1878). *Beiträge zur theorie und praxis der histologischen färbung.* Leipzig University.

Ehrlich, P. (1897). *Die wertbemessung des diphtherieheilserums und deren theoretischen grundlagen. Klinisches Jahrbuch,* 6, 299–326.

Ekins, S., Williams, A. J., Krasowski, M. D., & Freundlich, J. S. (2011). In silico repositioning of approved drugs for rare and neglected diseases. *Drug Discovery Today*, 16(7–8), 298–310. https://doi.org/10.1016/j.drudis.2011.02.016

Engstrom, N., Saenz-Méndez, P., Scheers, J., & Scheers, N. (2017). Towards celiac-safe foods: Decreasing the affinity of transglutaminase 2 for gliadin by addition of ascorbyl palmitate and $ZnCl_2$ as detoxifiers. *Scientific Reports*, 7(77), 1–8. https://doi.org/10.1038/s41598-017-00174-z

Eric, S., Ke, S., Barata, T., Solmajer, T., Antic Stankovic, J., Juranic, Z., Savic, V., & Zloh, M. (2012). Target fishing and docking studies of the novel derivatives of arylaminopyridines with potential anticancer activity. *Bioorganic & Medicinal Chemistry*, 20(17), 5220–5228. https://doi.org/10.1016/j.bmc.2012.06.051

Fan, S., Geng, Q., Pan, Z., Li, X., Tie, L., Pan, Y., & Li, X. (2012). Clarifying off-target effects for torcetrapib using network pharmacology and reverse docking approach. *BMC Systems Biology*, 6, 1–12.

Gao, L., Fang, J. S., Bai, X. Y., Zhou, D., Wang, Y. T., Liu, A. L., & Du, G. H. (2013). In silico target fishing for the potential targets and molecular mechanisms of baicalein as an antiparkinsonian agent: Discovery of the protective effects on NMDA receptor-mediated neurotoxicity. *Chemical Biology & Drug Design*, 81(6), 675–687. https://doi.org/10.1111/cbdd.12127

Gao, Z., Li, H., Zhang, H., Liu, X., Kang, L., Luo, X., Zhu, W., Chen, K., Wang, X., & Jiang, H. (2008). PDTD: A web-accessible protein database for drug target identification. *BMC Bioinformatics*, 9, 104. https://doi.org/10.1186/1471-2105-9-104

Gillespie, S. H., & Singh, K. (2011). XDR-TB, what is it; how is it Treated; and why is therapeutic failure so high? *Recent Patents on Antinfective Drug Discovery*, 6, 77–83.

Goodsell, D. S., & Olson, A. J. (1990). Automated docking of substrates to proteins by simulated annealing. *Proteins*, 8, 195–202.

Grinter, S. Z., Liang, Y., Huang, S. Y., Hyder, S. M., & Zou, X. (2011). An inverse docking approach for identifying new potential anti-cancer targets. *Journal of Molecular Graphics and Modelling*, 29(6), 795–799. https://doi.org/10.1016/j.jmgm.2011.01.002

Gunther, S., Kuhn, M., Dunkel, M., Campillos, M., Senger, C., Petsalaki, E., Ahmed, J., Urdiales, E. G., Gewiess, A., Jensen, L. J., Schneider, R., Skoblo, R., Russell, R. B., Bourne, P. E., Bork, P., & Preissner, R. (2008). SuperTarget and matador: Resources for exploring drug-target relationships. *Nucleic Acids Research*, 36(Database issue), D919–D922. https://doi.org/10.1093/nar/gkm862

Hay, M., Thomas, D. W., Craighead, J. L., Economides, C., & Rosenthal, J. (2014). Clinical development success rates for investigational drugs. *Nature Biotechnology*, 32, 40–51.

Hopkins, A. L. (2008). Network pharmacology: The next paradigm in drug discovery. *Nature Chemical Biology*, 4(11), 682–690. https://doi.org/10.1038/nchembio.118

http://bidd.nus.edu.sg/group/cjttd/. Therapeutic Target Database (TTD).

http://bioinfo-pharma.u-strasbg.fr/scPDB/. sc-PDB Database.

http://chemyang.ccnu.edu.cn/ccb/server/ACID/. ACID Tool.

http://dock.compbio.ucsf.edu/. UCSF DOCK.

http://idtarget.rcas.sinica.edu.tw/index.php. idTarget.

http://insilico.charite.de/supertarget/index.php?site=home. SuperTarget.

http://sideeffects.embl.de/. SIDER Database.

http://www.rcsb.org/pdb/home/home.do. Protein Data Bank.

https://pubchem.ncbi.nlm.nih.gov/. PubChem Database.

https://simtk.org/projects/sweetlead. SWEETLEAD Database.

https://www.accessdata.fda.gov/scripts/fdcc/?set=SCOGS. GRAS Database.

https://www.drugbank.ca/. DrugBank.

https://www.ebi.ac.uk/chembl/. ChEMBL Databse.

https://www.fda.gov/drugs/development-approval-process-drugs/drug-approvals-and-databases. FDA Databases.

https://zinc.docking.org/. ZINC Database.

Hui-Fang, L., Qing, S., Jian, Z., & Wei, F. (2010). Evaluation of various inverse docking schemes in multiple targets identification. *Journal of Molecular Graphics and Modelling*, 29(3), 326–330. https://doi.org/10.1016/j.jmgm.2010.09.004

Irwin, J. J., Sterling, T., Mysinger, M. M., Bolstad, E. S., & Coleman, R. G. (2012). ZINC: A free tool to discover chemistry for biology. *Journal of Chemical Information and Modeling*, 52(7), 1757–1768.

Jacoby, E. (2011). Computational chemogenomics. *WIREs Computational Molecular Science*, 1(1), 57–67. https://doi.org/10.1002/wcms.11

Ji, Z. L., Wang, Y., Yu, L., Han, L. Y., Zheng, C. J., & Chen, Y. Z. (2006). In silico search of putative adverse drug reaction related proteins as a potential tool for facilitating drug adverse effect prediction. *Toxicology Letters*, 164(2), 104–112. https://doi.org/10.1016/j.toxlet.2005.11.017

Kalyaanamoorthy, S., & Chen, Y. P. (2011). Structure-based drug design to augment hit discovery. *Drug Discovery Today*, 16(17–18), 831–839. https://doi.org/10.1016/j.drudis.2011.07.006

Keiser, M. J., Roth, B. L., Armbruster, B. N., Ernsberger, P., Irwin, J. J., & Shoichet, B. K. (2007). Relating protein pharmacology by ligand chemistry. *Nature Biotechnology*, 25(2), 197–206. https://doi.org/10.1038/nbt1284

Kim, S., Chen, J., Cheng, T., Gindulyte, A., He, J., He, S., Li, Q., Shoemaker, B. A., Thiessen, P. A., Yu, B., Zaslavsky, L., Zhang, J., & Bolton, E. E. (2019). PubChem 2019 update: Improved access to chemical data. *Nucleic Acids Research*, 47(D1), D1102–D1109. https://doi.org/10.1093/nar/gky1033

Kim, S., Shoemaker, B. A., Bolton, E. E., & Bryant, S. H. (2018). Finding potential multitarget ligands using PubChem. In J. B. Brown (Ed.), *Computational chemogenomics* (pp. 63–91). New York, NY: Humana Press.

Kinnings, S. L., Liu, N., Buchmeier, N., Tonge, P. J., Xie, L., & Bourne, P. E. (2009). Drug discovery using chemical systems biology: Repositioning the safe medicine comtan to treat multi-drug and extensively drug resistant tuberculosis. *PLoS Computational Biology, 5*(7), e1000423. https://doi.org/10.1371/journal.pcbi.1000423

Koutsoukas, A., Simms, B., Kirchmair, J., Bond, P. J., Whitmore, A. V., Zimmer, S., Young, M. P., Jenkins, J. L., Glick, M., Glen, R. C., & Bender, A. (2011). From in silico target prediction to multi-target drug design: Current databases, methods and applications. *Journal of Proteomics, 74*(12), 2554–2574. https://doi.org/10.1016/j.jprot.2011.05.011

Kruger, D. M., & Evers, A. (2010). Comparison of structure- and ligand-based virtual screening protocols considering hit list complementarity and enrichment factors. *ChemMedChem, 5*(1), 148–158. https://doi.org/10.1002/cmdc.200900314

Lang, P. T., Brozell, S. R., Mukherjee, S., Pettersen, E. F., Meng, E. C., Thomas, V., Rizzo, R. C., Case, D. A., James, T. L., & Kuntz, I. D. (2009). DOCK 6: Combining techniques to model RNA-small molecule complexes. *RNA, 15*(6), 1219–1230. https://doi.org/10.1261/rna.1563609

Lauro, G., Masullo, M., Piacente, S., Riccio, R., & Bifulco, G. (2012). Inverse Virtual Screening allows the discovery of the biological activity of natural compounds. *Bioorganic & Medicinal Chemistry, 20*(11), 3596–3602. https://doi.org/10.1016/j.bmc.2012.03.072

Lauro, G., Romano, A., Riccio, R., & Bifulco, G. (2011). Inverse virtual screening of antitumor targets: Pilot study on a small database of natural bioactive compounds. *Journal of Natural Products, 74*(6), 1401–1407. https://doi.org/10.1021/np100935s

Lee, A., Lee, K., & Kim, D. (2016). Using reverse docking for target identification and its applications for drug discovery. *Expert Opinion on Drug Discovery, 11*(7), 707–715. https://doi.org/10.1080/17460441.2016.1190706

Li, Y. Y., An, J., & Jones, S. J. (2011). A computational approach to finding novel targets for existing drugs. *PLoS Computational Biology, 7*(9), e1002139. https://doi.org/10.1371/journal.pcbi.1002139

Lin, J.-H., Perryman, A. L., Schames, J. R., & McCammon, J. A. (2002). Computational drug design accommodating receptor flexibility: The relaxed complex scheme. *Journal of the American Chemical Society, 124*, 5632–5633.

Liu, X., Xu, Y., Li, S., Wang, Y., Peng, J., Luo, C., Luo, X., Zheng, M., Chen, K., & Jiang, H. (2014). In silico target fishing: Addressing a "big data" problem by ligand-based similarity rankings with data fusion. *Journal of Cheminformatics, 6*, 33. https://doi.org/10.1186/1758-2946-6-33

Li, Y. H., Wang, P. P., Li, X. X., Yu, C. Y., Yang, H., Zhou, J., Xue, W. W., Tan, J., & Zhu, F. (2016). The human kinome targeted by FDA approved multi-target drugs and combination products: A comparative study from the drug-target interaction network perspective. *PLoS One, 11*(11), e0165737. https://doi.org/10.1371/journal.pone.0165737

Loging, W., Harland, L., & Williams-Jones, B. (2007). High-throughput electronic biology: Mining information for drug discovery. *Nature Reviews Drug Discovery, 6*(3), 220–230. https://doi.org/10.1038/nrd2265

Maestro, version 11.9. (2019). New York: Schrödinger, LLC.

Ma, S.-T., Feng, C.-T., Dai, G.-L., Song, Y., Zhou, G.-L., Zhang, X.-L., Miao, C. G., Yu, H., & Ju, W.-Z. (2015). In silico target fishing for the potential bioactive components contained in Huanglian Jiedu Tang (HLJDD) and elucidating molecular mechanisms for the treatment of sepsis. *Chinese Journal of Natural Medicines, 13*(1), 30–40. https://doi.org/10.1016/s1875-5364(15)60004-8

Maggiora, G. M. (2011). The reductionist paradox: Are the laws of chemistry and physics sufficient for the discovery of new drugs? *Journal of Computer-Aided Molecular Design, 25*(8), 699–708. https://doi.org/10.1007/s10822-011-9447-8

Medina-Franco, J. L., Giulianotti, M. A., Welmaker, G. S., & Houghten, R. A. (2013). Shifting from the single to the multitarget paradigm in drug discovery. *Drug Discovery Today, 18*(9–10), 495–501. https://doi.org/10.1016/j.drudis.2013.01.008

Medina-Franco, J. L., Martinez-Mayorga, K., Peppard, T. L., & Del Rio, A. (2012). Chemoinformatic analysis of GRAS (generally recognized as safe) flavor chemicals and natural products. *PLoS One, 7*(11), e50798. https://doi.org/10.1371/journal.pone.0050798

Medina-Franco, J. L., Yongye, A. B., Perez-Villanueva, J., Houghten, R. A., & Martinez-Mayorga, K. (2011). Multitarget structure-activity relationships characterized by activity-difference maps and consensus similarity measure. *Journal of Chemical Information and Modeling, 51*(9), 2427–2439. https://doi.org/10.1021/ci200281v

Méndez-Lucio, O., Naveja, J. J., Vite-Caritino, H., & Prieto-Martínez, F. D. (2016). Review. One drug for multiple targets: A computational perspective. *The Journal of the Mexican Chemical Society, 60*(3), 168–181.

Mendez, D., Gaulton, A., Bento, A. P., Chambers, J., De Veij, M., Felix, E., Magarinos, M. P., Mosquera, J. F., Mutowo, P., Nowotka, M., Gordillo-Maranon, M., Hunter, F., Junco, L., Mugumbate, G., Rodriguez-Lopez, M., Atkinson, F., Bosc, N., Radoux, C. J., Segura-Cabrera, A., ... Leach, A. R. (2019). ChEMBL: Towards direct deposition of bioassay data. *Nucleic Acids Research, 47*(D1), D930–D940. https://doi.org/10.1093/nar/gky1075

Meslamani, J., Rognan, D., & Kellenberger, E. (2011). sc-PDB: a database for identifying variations and multiplicity of 'druggable' binding sites in proteins. *Bioinformatics, 27*(9), 1324–1326. https://doi.org/10.1093/bioinformatics/btr120

Mestres, J., Gregori-Puigjane, E., Valverde, S., & Sole, R. V. (2008). Data completeness—the Achilles heel of drug—target networks. *Nature Biotechnology, 26*, 983–984.

Molecular Operating Environment (MOE). (2015). Montreal, Canada: Chemical Computing Group.

Morphy, R. (2010). Selectively nonselective kinase inhibition: Striking the right balance. *Journal of Medicinal Chemistry, 53*(4), 1413–1437. https://doi.org/10.1021/jm901132v

Morphy, R., & Rankovic, Z. (2007). Fragments, network biology and designing multiple ligands. *Drug Discovery Today, 12*(3–4), 156–160. https://doi.org/10.1016/j.drudis.2006.12.006

Murray, C. W., & Blundell, T. L. (2010). Structural biology in fragment-based drug design. *Current Opinion in Structural Biology, 20*(4), 497–507. https://doi.org/10.1016/j.sbi.2010.04.003

Murray, C. W., & Rees, D. C. (2009). The rise of fragment-based drug discovery. *Nature Chemistry, 1*(3), 187–192. https://doi.org/10.1038/nchem.217

Namchuk, M. N. (2002). Finding the molecules to fuel chemogenomics. *Targets, 1*(4), 125–129. https://doi.org/10.1016/s1477-3627(02)02206-7

Nantasenamat, C., & Prachayasittikul, V. (2015). Maximizing computational tools for successful drug discovery. *Expert Opinion on Drug Discovery, 10*(4), 321–329. https://doi.org/10.1517/17460441.2015.1016497

Nicola, G., Liu, T., & Gilson, M. K. (2012). Public domain databases for medicinal chemistry. *Journal of Medicinal Chemistry, 55*(16), 6987–7002. https://doi.org/10.1021/jm300501t

Novick, P. A., Ortiz, O. F., Poelman, J., Abdulhay, A. Y., & Pande, V. S. (2013). SWEETLEAD: An in silico database of approved drugs, regulated chemicals, and herbal isolates for computer-aided drug discovery. *PLoS One, 8*(11), e79568. https://doi.org/10.1371/journal.pone.0079568

Pandya, P., Agarwal, L. K., Gupta, N., & Pal, S. (2014). Molecular recognition pattern of cytotoxic alkaloid vinblastine with multiple targets. *Journal of Molecular Graphics and Modelling, 54,* 1–9. https://doi.org/10.1016/j.jmgm.2014.09.001

Pan, J. B., Ji, N., Pan, W., Hong, R., Wang, H., & Ji, Z. L. (2014). High-throughput identification of off-targets for the mechanistic study of severe adverse drug reactions induced by analgesics. *Toxicology and Applied Pharmacology, 274*(1), 24–34. https://doi.org/10.1016/j.taap.2013.10.017

Park, K., & Cho, A. E. (2017). Using reverse docking to identify potential targets for ginsenosides. *Journal of Ginseng Research, 41*(4), 534–539. https://doi.org/10.1016/j.jgr.2016.10.005

Peters, J.-U. (2013). Polypharmacology – foe or friend? *Journal of Medicinal Chemistry, 56*(22), 8955–8971. https://doi.org/10.1021/jm400856t

Ramsay, R. R., Popovic-Nikolic, M. R., Nikolic, K., Uliassi, E., & Bolognesi, M. L. (2018). A perspective on multi-target drug discovery and design for complex diseases. *Clinical and Translational Medicine, 7*(1), 3. https://doi.org/10.1186/s40169-017-0181-2

Rognan, D. (2007). Chemogenomic approaches to rational drug design. *British Journal of Pharmacology, 152*(1), 38–52. https://doi.org/10.1038/sj.bjp.0707307

Roskoski, R., Jr. (2007). Sunitinib: A VEGF and PDGF receptor protein kinase and angiogenesis inhibitor. *Biochemical and Biophysical Research Communications, 356*(2), 323–328. https://doi.org/10.1016/j.bbrc.2007.02.156

Roth, B. L., Sheffler, D. J., & Kroeze, W. K. (2004). Magic shotguns versus magic bullets: Selectively non-selective drugs for mood disorders and schizophrenia. *Nature Reviews Drug Discovery, 3,* 353–359.

Saenz-Méndez, P., & Eriksson, L. A. (2018). Exploring polypharmacology in drug design. In T. Mavromoustakos, & T. F. Kellici (Eds.), *Rational drug design: Methods and protocols* (pp. 229–243). Springer Science.

Saenz-Méndez, P., Eriksson, M., & Eriksson, L. A. (2017). Ligand selectivity between the ADP-ribosylating toxins: An inverse-docking study for multitarget drug discovery. *ACS Omega, 2*(4), 1710–1719. https://doi.org/10.1021/acsomega.7b00010

Saenz-Méndez, P., Genheden, S., Reymer, A., & Eriksson, L. A. (2017). Chapter 1. Computational chemistry and molecular modelling basics. In *Computational tools for chemical biology* (pp. 1–38).

Schultes, S., de Graaf, C., Haaksma, E. E. J., de Esch, I. J. P., Leurs, R., & Krämer, O. (2010). Ligand efficiency as a guide in fragment hit selection and optimization. *Drug Discovery Today: Technologies, 7*(3), e157–e162. https://doi.org/10.1016/j.ddtec.2010.11.003

Strebhardt, K., & Ullrich, A. (2008). Paul Ehrlich's magic bullet concept: 100 years of progress. *Nature Reviews Cancer, 8,* 473–480.

Trott, O., & Olson, A. J. (2010). AutoDock Vina: Improving the speed and accuracy of docking with a new scoring function, efficient optimization, and multithreading. *Journal of Computational Chemistry, 31*(2), 455–461. https://doi.org/10.1002/jcc.21334

Valeur, E., Gueret, S. M., Adihou, H., Gopalakrishnan, R., Lemurell, M., Waldmann, H., Grossmann, T. N., & Plowright, A. T. (2017). New modalities for challenging targets in drug discovery. *Angewandte Chemie International Edition in English, 56*(35), 10294–10323. https://doi.org/10.1002/anie.201611914

Valeur, E., Narjes, F., Ottmann, C., & Plowright, A. T. (2019). Emerging modes-of-action in drug discovery. *MedChemComm, 10*(9), 1550–1568. https://doi.org/10.1039/c9md00263d

Vendonk, M. L., Cole, J. C., Hartshorn, M. J., Murray, C. W., & Taylor, R. D. (2003). Improved protein–ligand docking using GOLD. *Proteins, 52*(4), 609–623.

Vigers, G. P. A., & Rizzi, J. P. (2004). Multiple active site corrections for docking and virtual screening. *Journal of Medicinal Chemistry, 47,* 80–89.

Wang, J. C., Chu, P. Y., Chen, C. M., & Lin, J. H. (2012). idTarget: a web server for identifying protein targets of small chemical molecules with robust scoring functions and a divide-and-conquer docking approach. *Nucleic Acids Research, 40*(Web Server issue), W393–W399. https://doi.org/10.1093/nar/gks496

Wang, F., Wu, F.-X., Li, C.-Z., Jia, C.-Y., Su, S.-W., Hao, G.-F., & Yang, G.-F. (2019). ACID: A free tool for drug repurposing using consensus inverse docking strategy. *Journal of Cheminformatics, 11*(1). https://doi.org/10.1186/s13321-019-0394-z

Wang, Y., Zhang, S., Li, F., Zhou, Y., Zhang, Y., Wang, Z., Zhang, R., Zhu, J., Ren, Y., Tan, Y., Qin, C., Li, Y., Li, X., Chen, Y., & Zhu, F. (2020). Therapeutic target database 2020: Enriched resource for facilitating research and early development of targeted therapeutics. *Nucleic Acids Research*, *48*(D1), D1031–D1041. https://doi.org/10.1093/nar/gkz981

Waring, M. J., Arrowsmith, J., Leach, A. R., Leeson, P. D., Mandrell, S., Owen, R. M., Pairaudeau, G., Pennie, W. D., Pickett, S. D., Wang, J., Wallace, O., & Weir, A. (2015). An analysis of the attrition of drug candidates from four major pharmaceutical companies. *Nature Reviews Drug Discovery*, *14*(7), 475–486. https://doi.org/10.1038/nrd4609

Westermaier, Y., Barril, X., & Scapozza, L. (2015). Virtual screening: An in silico tool for interlacing the chemical universe with the proteome. *Methods*, *71*, 44–57. https://doi.org/10.1016/j.ymeth.2014.08.001

Williams, A. J. (2008). Public chemical compound databases. *Current Opinion in Drug Discovery & Development*, *11*(3), 393–404.

Wishart, D. S., Knox, C., Guo, A. C., Shrivastava, S., Hassanali, M., Stothard, P., Chang, Z., & Woolsey, J. (2006). DrugBank: A comprehensive resource for in silico drug discovery and exploration. *Nucleic Acids Research*, *34*(Database issue), D668–D672. https://doi.org/10.1093/nar/gkj067

Xu, X., Huang, M., & Zou, X. (2018). Docking-based inverse virtual screening: Methods, applications, and challenges. *Biophysics Reports*, *4*(1), 1–16. https://doi.org/10.1007/s41048-017-0045-8

Ye, H., Liu, Q., & Wei, J. (2014). Construction of drug network based on side effects and its application for drug repositioning. *PLoS One*, *9*(2), e87864. https://doi.org/10.1371/journal.pone.0087864.t001

Yongye, A. B., Waddell, J., & Medina-Franco, J. L. (2012). Molecular scaffold analysis of natural products databases in the public domain. *Chemical Biology & Drug Design*, *80*(5), 717–724. https://doi.org/10.1111/cbdd.12011

Yuriev, E., Holien, J., & Ramsland, P. A. (2015). Improvements, trends, and new ideas in molecular docking: 2012–2013 in review. *Journal of Molecular Recognition*, *28*(10), 581–604. https://doi.org/10.1002/jmr.2471

Yuriev, E., & Ramsland, P. A. (2013). Latest developments in molecular docking: 2010–2011 in review. *Journal of Molecular Recognition*, *26*(5), 215–239. https://doi.org/10.1002/jmr.2266

Zahler, S., Tietze, S., Totzke, F., Kubbutat, M., Meijer, L., Vollmar, A. M., & Apostolakis, J. (2007). Inverse in silico screening for identification of kinase inhibitor targets. *Chemistry & Biology*, *14*(11), 1207–1214. https://doi.org/10.1016/j.chembiol.2007.10.010

Zhang, W., Pei, J., & Lai, L. (2017). Computational multitarget drug design. *Journal of Chemical Information and Modeling*, *57*(3), 403–412. https://doi.org/10.1021/acs.jcim.6b00491

Zimmermann, G. R., Lehar, J., & Keith, C. T. (2007). Multi-target therapeutics: When the whole is greater than the sum of the parts. *Drug Discovery Today*, *12*(1–2), 34–42. https://doi.org/10.1016/j.drudis.2006.11.008

Drug Repositioning: Principles, Resources, and Application of Structure-Based Virtual Screening for the Identification of Anticancer Agents

IMLIMAONG AIER • PRITISH KUMAR VARADWAJ

1 INTRODUCTION

Cancer is a disease responsible for the death of millions globally. With over 18.1 million new cancer cases and 9.6 million cancer deaths as of 2018, increasing population and lifestyle imposes a heavy burden upon cancer treatment (Bray et al., 2018). This blight is further made worrisome as cancer in advanced stages is often incurable, coupled with the fact that the efficacy of anticancer drugs was alarmingly low (25%) based on the study conducted by Spear et al. in 2001 and Bristow in 2012 (Mestroni et al., 2014; Spear et al., 2001) (Fig. 15.1A). Although the number of approved drugs for the treatment of oncology in the year 2019 was much higher than other ailments (Fig. 15.1B) (Mullard, 2020), the risk of failure in more advanced forms of cancer is still relatively high. Thus, the only way forward is to look for therapeutic options that could somehow address these issues. Drug discovery is the appropriate solution for finding novel therapeutics. However, the entire discovery process is a daunting task, made even more challenging by the complexity of human biology. At the heart of this matter lies two salient features: a drug molecule or a compound, also called a ligand, and a target protein for the drug, known as the receptor. With the ever increasing number of protein three-dimensional (3D) data, the discovery of ligands that can optimally bind to known proteins has increased progressively. Over the years, the cures for many terrible ailments have been identified with the help of traditional drug discovery. Although seemingly simple enough, the drug discovery process is a very high risk operation. Moreover, the number of novel drugs

approved from 2010 to 2019 is less than 50 on average per year (Fig. 15.1C) (Mullard, 2020).

The steps required for drug discovery consist of several complex procedures, each with a different set of equipment and timeline, spanning over 15–20 years (Fig. 15.2). The classical method of drug discovery relies heavily on the identification of small molecules that could alter the phenotype of an organism. Based on the desirability of the effect, these molecules are then refined, optimized, and tested in clinical trials. However, these traditional methods are often trial-and-error based, making it error prone and less systematic. On the other hand, rational drug discovery is based on the development of molecules with therapeutic value intended for known or unknown targets with a specific function. Thus, the identification of a target and a drug that could potentially bind to the target is the primary objective on which the whole operation of rational drug discovery stands. Once the target and a list of potential drug candidates have been obtained, a few selected candidates are profiled for safety. Compounds that pass this stage of the discovery phase are quickly subjected to preclinical trials in animal models, followed by clinical trials on humans. Finally, any drug that shows promising results is reviewed by the country or union's government agencies for approval, for example, the Food and Drug Administration (FDA) for the United States, the European Medicines Agency (EMA) for the European Union, or the Pharmaceuticals and Medical Devices Agency (PMDA) for Japan. However, the probability of success for a drug to move from Phase 1 of clinical trial to the approval stage is only

Molecular Docking for Computer-Aided Drug Design. https://doi.org/10.1016/B978-0-12-822312-3.00006-0

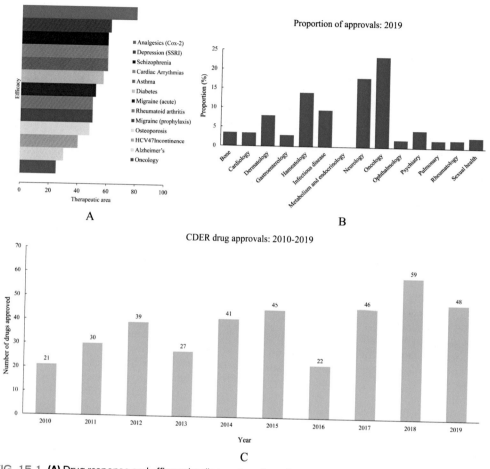

FIG. 15.1 **(A)** Drug response and efficacy by disease type from Spear et al., 2001, Bristow, 2012. **(B)** Portion of therapeutic drugs approved in 2019. **(C)** Number of drugs approved by the FDA CDER from 2010 to 2019.

13.8% (Wong et al., 2019). Thus, the total cost of drug discovery and development escalates to several millions of dollars, which is less than favorable for investors when time is also taken into account.

In rational drug discovery research, the first general step is to look for proteins that could serve as valid targets. The target in question should be associated with a certain disease and should be available for therapeutic intervention. Once the target has been identified, discerning chemical compounds that could act on the target is the next step. Some essential criteria are required for a chemical compound to be regarded as a drug compound or a lead molecule. The term used to describe the properties for a drug compound is drug-likeness. The general rule for evaluating drug-

likeliness is the "Lipinski's rule of five" (Table 15.1), which was formulated by Christopher A. Lipinski in 1997 (Lipinski et al., 2001), although other variations of the rule are also applied, for example, the "rule of three" (Congreve et al., 2003).

2 COMPUTER-AIDED VIRTUAL SCREENING

With the advent of computer-aided drug discovery (CADD), the entire operation of drug discovery has been made less troublesome. CADD compliments combinatorial chemistry, high-throughput screening (HTS), and quantum techniques in order to design and develop therapeutics. This technique falls under the development phase of drug designing, where the

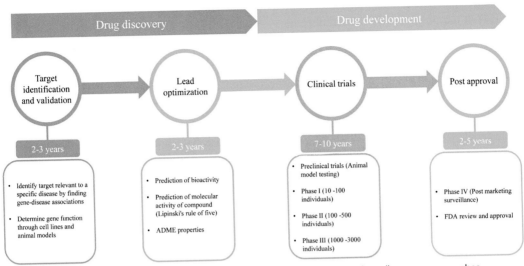

FIG. 15.2 Drug discovery and development process using modern drug discovery approaches.

TABLE 15.1
Drug-Likeliness Properties Based on Lipinski's Rule of Five.

Properties	Drug-Likeliness
Number of hydrogen bond donors (sum of NH and OH)	<5
Number of hydrogen bond acceptors (sum of N and O)	<10
Molecular mass	<500 Da
Octanol–water partition coefficient (logP)	<5

selection and filtering of compounds for testing are done. The systematic filtering of candidate compounds ensures that time and cost for the development phase are reduced. Due to the unprecedented amount of potential lead compounds available, several thousands of chemical compounds can be identified through in silico HTS, also known as virtual screening. This has reduced the time for lead identification and optimization by a drastic amount when compared with the traditional method of experimental techniques. Virtual screening is a computational method for scanning libraries of compounds to identify the binding of a suitable ligand to a target protein (Shoichet, 2004). This approach is used to enhance and accelerate lead discovery process.

The fundamental application of CADD and virtual screening is to serve two purposes: to identify potential

active compounds and to determine the absorption, distribution, metabolism, and excretion (ADME) profile of compounds. The identification of novel compounds requires stringent filtering, which is done to reduce the number of selected compounds. As mentioned earlier, specific rules for drug-likeliness have been proposed. Thus, filtering involves the application of certain rules such as Lipinski's rule of five. The second step is to determine the safety of the compounds. By filtering the compounds based on the ADME and toxicity properties, the side effect of compounds can be reduced. Increasing number of protein structures and the availability of vast libraries of lead molecules have paved the way for virtual screening techniques. Depending upon the information available with regard to proteins/ligands, virtual screening techniques are classified into two categories, ligand-based and structure-based, and are discussed further.

2.1 Ligand-Based Virtual Screening

One of the methods for screening is based on the availability of ligands, wherein a known ligand or a compound is used for identifying potential lead compounds based on common pharmacophore features. This technique is known as ligand-based virtual screening or LBVS. Pharmacophores are the essential features of a compound, such as hydrogen bond donors and acceptors, hydrophobic, and aromatic rings, based upon which interactions between the target and compound take place. This technique assumes the situation where all ligands bind to the same region of the target protein. Another popular technique used in LBVS is quantitative structure–activity relationship (QSAR) (Hansch, 1972). It is a powerful tool used to discover correlations between chemical structures and biological properties such as the half maximal inhibitory concentration (IC_{50}), half maximal effective concentration (EC_{50}), and inhibitor constant (Ki). The advantage of implementing this approach is that the computational complexity is low. On the other hand, variations in results are limited as chemical structures are derived from already known ligands.

Due to the advancements in computational chemistry, existing relationship between proteins and ligands can be used to establish meaningful links and extract knowledge. This form of relation is called structure–activity relationship (SAR), a concept introduced in 1865 by Crum-Brown and Fraser which states that the functional class of a protein shares similar properties with targets (Brown & Fraser, 1868). The most popular scheme of classification is the Enzyme Commission (EC), which divides enzymes into classes and assigns them four numbers based on different levels of classification. The first digit ranges from 1 to 6 and denotes the type of reaction catalyzed by the enzyme. The second digit indicates the subclass, the third indicates the sub-subclass, and the fourth denotes the serial number of the enzyme in its sub-subclass. Enzymes are broadly classified into six different classes based on the type of reaction, namely EC 1 for oxidoreductases, EC 2 for transferases, EC 3 for hydrolases, EC 4 for lyases, EC 5 for isomerases, and EC 6 for ligases (Tipton, 1994). By understanding the similarity in binding sites between different enzymes, the suitability of a ligand can be determined based on other similar known ligands for the same class, i.e., similar classes of enzymes will share ligands with similar properties while other classes will have different ligand properties. This knowledge can be reused among different enzymes belonging to the same class. A similar approach can be implemented in the case of ligands as well, where different subsets of the ligand are retained after the side chains are stripped off. Such structures are known as scaffolds and are used for chemical classification schemes. Using the information derived from all these steps, rules and models can be generated, which can be applied to virtual screening compounds. These rules and models are termed as molecular descriptors which take into account the number of atoms in a ligand, their distance, and the shortest path between them and are often used for lead optimization.

Machine learning methods, ranging from simple classifiers to deep learning, have also found their way into medicinal chemistry and have been proven to be quite effective when classifying compound libraries containing both active and inactive molecules. One of the best known implementations of SAR can be found in QSAR, where relation between the structure and biological effect is modeled mathematically through statistical means using machine learning principles. To determine the QSAR, >20 compounds with similar descriptors and interactions with the binding site are selected to build the prediction model. To deal with the abundance of molecular descriptors, principal component analysis is commonly used to identify relevant as well as independent features based on the variables present in the original descriptors. However, depending on the type of data, i.e., continuous or categorical, classification or regression tools can be used. Fig. 15.3 gives a visual depiction of QSAR generation. However, even this method is not without limitations, as some machine learning models can produce high false positive rates, which could be costly when it comes to drug discovery.

2.2 Structure-Based Virtual Screening

The other popular method for virtual screening technique is the structure-based virtual screening (SBVS), where the structural information of the protein is used (Lyne, 2002). The 3D structures in this method are generally derived using techniques such as X-ray crystallography and nuclear magnetic resonance, which reveal tiny details about the atoms and the interactions. This information is essential to decipher the mechanism of binding and inhibition or activation (Lounnas et al., 2013). It is worth noting that SBVS as a means of drug designing started trending only about 20 years ago due to the advancements in the field of computer science and technology, as the technique was relatively primitive in the early days for successful implementation and was thus considered nonessential. For sBVs, it is important for the small molecule to possess structural and chemical complementarity to the target receptor (Kuntz, 1992).

Proteins are the key components for regulating most cellular processes. The study of protein 3D structures and how protein–ligand complexes are formed can help understand their functional mechanisms and roles in the cell. Studying the structure of proteins helps provide information that could assist in drug design, providing researchers with the knowledge to discover and develop small molecules that could potentially inhibit or induce protein interactions. When information about the binding site is available, prior information about the ligand is not required. sBVs has gained attention due to the availability of protein structures that can serve as potential drug targets. The advantage of structure-based approach is that because the information

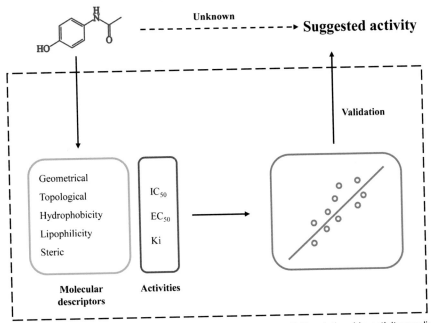

FIG. 15.3 Schematic representation of quantitative structure–activity relationship activity prediction.

of the receptor is generally already known, identification of ligand becomes relatively easier due to available details on the mechanism of action. However, some of the problems associated with this method are the computational complexities, such as assigning specific force field parameters, as well as finding optimal binding modes that could come close to experimental results.

3 DRUG REPOSITIONING

Discovering new indication for existing drugs, known as drug repositioning or drug repurposing, is a growing field in bioinformatics as it is used to identify new indications for existing drugs (Iorio et al., 2010; Gottlieb et al., 2011; Sirota et al., 2011). Traditional methods of drug discovery take billions of dollars and an average of 15 years to bring a new drug to the market (DiMasi et al., 2003), while an estimated 90% of the drugs fail in the early stage of drug development (Pantziarka et al., 2018). By repurposing of drugs, the cost associated with early stages can be reduced, while keeping the drugs safe for human trials (Ashburn & Thor, 2004) (Fig. 15.4). This technique of drug repurposing has been a success in both academia and pharmaceutical companies (Table 15.2).

Drug repurposing is generally carried out through computational means with the help of chemoinformatics tools that can identify structurally similar drugs

(Keiser et al., 2009) or with the help of machine learning methods that can identify drug–disease relations (Lamb et al., 2006; Sirota et al., 2011). Likewise, similarity of side effect (Campillos et al., 2008) and text mining of literature (Li et al., 2009) is often used. Network-based methods can also be applied to genes, which in the form of nodes can offer an opportunity to infer potential targets in important cell signaling pathways. While most of the approaches listed can only be applied when the targets or structures are known, gene expression profile–based approaches do not require prior knowledge of the drugs (Fig. 15.5).

3.1 Signature-Based Drug Repurposing

The application of signature-based drug repurposing involves the identification of uniquely expressed gene patterns across normal and pathological conditions. With the advent of large-scale genomics project beginning with the Human Genome Project (Lander et al., 2001), deciphering the biological and functional aspect of genes has been looked upon with great interest. The current availability of high-throughput genome sequence data has opened the gateway for developing new methods to study drug-disease associated response. Signature-based methods allow researcher to dwell into the molecular basis of the drug action and to uncover the interacting pathways linked to a disease. Methods involving gene expression profile involve comparing

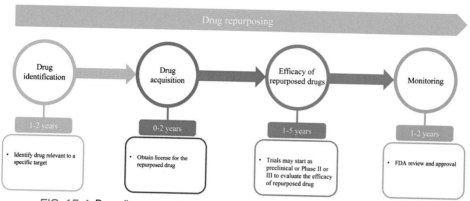

FIG. 15.4 Drug discovery and development process through repurposing strategy.

TABLE 15.2
List of Successfully Repurposed Drugs.

Sl. No.	Drug	Mechanism of Action	Original Indication	New Indication	References
1	Atomoxetine	Selective inhibition of presynaptic norepinephrine reuptake	Parkinson's disease	Attention deficient hyperactivity disorder	Bymaster et al. (2002)
2	Bupropion	Selectively inhibits dopamine and noradrenaline reuptake	Depression	Smoking cessation	Slemmer et al. (2000)
3	Celecoxib	Selective cyclooxygenase-2 (COX-2) inhibitor	Osteoarthritis and adult rheumatoid arthritis	Colon and breast cancer	Chen et al. (2015), Tseng et al. (2006)
4	Chlorpromazine	Dopamine receptor blockade	Antihistamine	Nonsedating tranquillizer	Lehmann and Hanrahan (1954)
5	Duloxetine	Serotonin and norepinephrine reuptake inhibitor	Depression	Chemotherapy-induced peripheral neuropathy	Hershman et al. (2014)
6	Finasteride	5-α-reductase inhibition	Benign prostatic hyperplasia	Hair loss	Salisbury and Tadi (2020)
7	Galantamine	Competitive inhibition of acetylcholinesterase	Polio, paralysis, and anesthesia	Alzheimer's disease	Scott and Goa (2000)
8	Lidocaine	Sodium channel blockade	Local anesthesia	Asthma	Ruddy (1971)
9	Mecamylamine	Nonselective, noncompetitive antagonist of the nicotinic acetylcholine receptors	Hypertension	Attention deficient hyperactivity disorder	Potter et al. (2009)
10	Mifepristone	Glucocorticoid receptor antagonism	Pregnancy termination	Psychotic major depression	Gallagher and Young (2006)
11	Milnacipran	Selective serotonin and norepinephrine reuptake inhibitor	Depression	Fibromyalgia syndrome	English et al. (2010)

TABLE 15.2
List of Successfully Repurposed Drugs.—cont'd

Sl. No.	Drug	Mechanism of Action	Original Indication	New Indication	References
12	Minoxidil	β-adrenoceptor blockade	Hypertension	Hair loss	Gottlieb et al. (1972)
13	Paclitaxel	Polymerization of tubulin by binding to the β subunit	Cancer	Restenosis	Gershlick et al. (2004)
14	Ropinirole	Non−ergoline dopamine agonist	Hypertension	Parkinson's disease and idiopathic restless leg syndrome	Kushida (2006)
15	Sibutramine	Non−selective serotonin reuptake inhibitor	Depression	Obesity	Stock (1997)
16	Thalidomide	TNF-α inhibition	Sedation, nausea, and insomnia	Leprosy and multiple myeloma	Crawford (1991), Rajkumar (2001)
17	Tofisopam	Isoenzyme-selective inhibitor of phosphodiesterases	Anxiety-related conditions	Irritable bowel syndrome	Leventer et al. (2008)
18	Topiramate	Na channel blockade, GABA stimulation	Epilepsy	Obesity	Kramer et al. (2011)
19	Zidovudine	Reverse transcriptase inhibition	Cancer	HIV/AIDS	De Clercq (1994)

the signature of a drug with that of a different drug or disease. This method is generally applied to transcriptome data, wherein samples from diseased cells before and after treatment with a particular drug are collected and compared with already analyzed disease data samples. Based on the correlation between the samples, the activity of the drug on the initial sample can be deciphered based on the negativity of the correlation. This technique has been proven to be effective in identifying chemotherapeutic compounds for otherwise drug-resistant cancers. Connectivity Map (cMap) (Lamb et al., 2006), a widely known database specifically

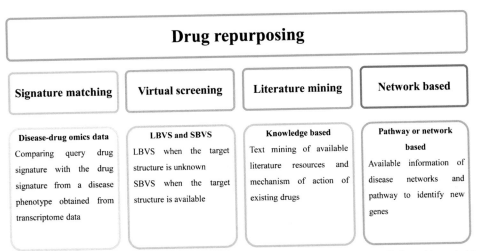

FIG. 15.5 An illustration of different computational drug repurposing strategies. *LBVS*, ligand-based virtual screening; *SBVS*, structure-based virtual screening.

developed for gene expression profiling, uses drug–disease and drug–drug signatures to predict the repositioning of compounds. An example for successful repurposing of drug using this method can be seen in the work presented by Dudley et al. (2011) where inflammatory bowel disease genomic data from NCBI were obtained and the gene signatures were matched using cMap, thus identifying topiramate, an anticonvulsant drug, as a suitable drug for the treatment of the disease. Although there are several advantages to this technique, the availability of the data and rigorousness of gene expression data analysis remains a limiting factor.

3.2 Network-Based Drug Repurposing

Another popular means of drug repurposing involves studying the biological aspects of the disease with respect to the drug. Network-based drug repurposing offers the means to reconstruct disease-linked pathways and protein interaction networks from genomics and proteomics data. This method uses interaction data and is represented in the form of nodes and edges, the nodes being the genes, proteins, disease phenotype, and the drug, while the edges represent the relation between the various components. By making use of this information, gene regulatory networks can be constructed to provide an understanding of the mechanism behind a drug's action. Similarly, by representing proteins as nodes, the function of interrelated proteins can be interpreted through the connecting edge via protein–protein interaction networks. In 2015, Zhang and his colleagues selected proteomics, metabolomics, and genome-wide associated studies to identify potential drug targets for the treatment of diabetes (Zhang et al., 2015). They found that cyclooxygenase target drugs, diflunisal, nabumetone, niflumic acid, and valdecoxib, could be used to treat type 1 diabetes, while alpha-2A adrenergic receptor (ADRA2A) target drugs phenoxybenzamine and idazoxan could be beneficial for type 2 diabetes. However, the pitfall for this technique is that biological data pertaining to disease–drug relation are often unavailable. Thus, the accuracy of prediction cannot be made certain.

3.3 Literature Mining and Knowledge-Based Drug Repurposing

Information from clinical trials, signaling pathways, and gene and protein interaction networks can be very useful when it comes to drug repurposing. Knowledge-based techniques, as the name suggests, can be a boon when it comes to collecting large numbers of information. In a way, this technique encompasses all the other methods discussed in this topic.

An example of this approach in research can be seen in a study conducted by Jin et al., in 2012 (Jin et al., 2012), in which literature for cancer-specific pathways and mechanisms, along with proteins and protein networks were analyzed using a statistical regression model and Bayesian factor regression model. After testing the model on breast and prostate cancer cells using repurposed drugs, they found that 90% of clinical responses were accurately predicted, thus illustrating the success behind implementing knowledge-based drug discovery. However, like any other approach, this technique faces certain limitations. The available data will vary across patients due to the difference in genetic traits and medical history and may not provide the best indication for a drug.

3.4 Target-Based Drug Repurposing Using Structure-Based Virtual Screening

3.4.1 Basic principles and resources for structure-based virtual screening

Due to the overwhelming need for potential drug candidates against a variety of diseases, the paradigm has shifted toward target-based drug repurposing. In contrast to signature matching, a specific protein can be targeted with a wide set of compounds, providing a fast and effective solution when dealing with large libraries of compounds. The compound libraries in question are FDA approved and often freely available for researchers, thus eliminating the need for purchasing expensive drugs (Table 15.3). For example, ChEMBL (Mendez et al., 2019) is a manually curated compound database maintained by the European Bioinformatics Institute (EBI), which has a vast collection of ∼2 million bioactive compound data. The database has over 13,000 FDA-approved drugs and over 13,000 targets mapped to various compounds. Another freely available database for drug repurposing is DrugBank (Wishart et al., 2008), which offers a variety of drugs and drug targets. The content of the database is limited to not only FDA-approved drugs, but is also home to over 6000 experimental compounds. Moreover, due to the fact that existing drugs have been deemed suitable in terms of bioavailability, coupled with the potential of identifying highly potent compounds that could be used for animal and human studies, interest to look for effective drugs has been renewed.

Despite all the shortcomings, virtual screening, particularly SBVS, becomes a powerful asset when combined with high-throughput docking of compound libraries. Docking is a technique in which molecules in a database are bound to the binding site of a target protein and is quite possibly the most popular tool in

TABLE 15.3
List of Compound Libraries Containing Small Molecules for Researchers.

Database	Number of Compounds	Website
ACD (Available Chemicals Directory)	3.9 million	http://accelrys. com/products/ databases/ sourcing/avaible chemicals directory.html
Asinex	550,000	http://www. asinex.com
ChemBank (Seiler et al., 2008)	1.2 million	http:// chembank. broadinstitute. org
ChEMBL (Mendez et al., 2019)	1,950,765 compounds including 13,308 drugs	https://www.ebi. ac.uk/chembl/
ChemBridge	700,000	http://www. chembridge.com
ChemDB (Chen et al., 2007)	5 million	http://cdb.ics. uci.edu
ChemDiv	1.5 million	http://www. chemdiv.com
ChemSpider	26 million	http://www. chemspider.com
Chimiothèque Nationale	48,370	http:// chimiotheque- nationale.enscm. fr/?lang=fr
CoCoCo (Del Rio et al., 2010)	7 million	http://cococo. unimore.it/tiki- index.php
Drug Discovery Center Collection	340,000	http://www. drugdiscovery. uc.edu/
DrugBank (Wishart et al., 2008)	13,570 drugs	https://www. drugbank.ca/
eMolecules	7 million	http://www. emolecules.com
Enamine	1.7 million	http://www. enammine.net
FDA-approved anticancer drugs	129 most current FDA-approved anticancer drugs	https://dtp.cancer. gov/organization/ dscb/obtaining/ available_plates. htm

Continued

TABLE 15.3
List of Compound Libraries Containing Small Molecules for Researchers.—cont'd

Database	Number of Compounds	Website
Maybridge	56,000	http://www. maybridge.com
MDDR (MDL Drug Data Report)	150,000	http://accelerys. com/products/ databases/ bioactivity/mddr. html
NCI open database	265,000	http://cactus.nci. nih.gov/ncidb2. 2/
PubChem (Kim et al., 2019)	30 million	http://pubchem. ncbi.nlm.nih.gov
Specs	240,000	http://www. specs.net
WOMBAT (Olah et al., 2007)	263,000	http://www. sunsetmolecular. com
ZINC (Irwin & Shoichet, 2005)	13 million	http://zinc. docking.org

computational drug discovery studies (Kitchen et al., 2004). Before the docking operation, the target structure or protein needs to undergo preprocessing in the form of target preparation. This constitutes the addition of hydrogen atoms and energy minimization of the target protein in order to avoid steric clashes. A problem commonly faced in molecular docking is the absence of a suitable target for the operation. The common solution is to model the 3D structure of the protein using techniques such as homology modeling, also known as comparative modeling. This refers to the technique of using the amino acid sequence of the target protein as a template, which is found in abundance. Similar known structures of proteins with greater than 20% identity are then identified through multiple sequence alignment and the 3D model of the target is generated using fragment assembly, segment matching, or through spatial restraints. Fragment-based methods rely on identifying conserved regions from template structures and then piecing together the complete model. On the other hand, segment matching is based on creating segments of the protein, which is subsequently matched to the template. In this case, the segments are selected by aligning similar regions rather than the entire protein. The last method involving spatial restraints is currently the most commonly used technique in homology modeling and it uses templates to generate probability density functions (PDFs) for estraints generated from a set of geometric data. These restraints are associated with structural aspects of the protein, such as the backbone and dihedral angles. By obtaining the PDFs from homologous structures, the final structure is obtained by selecting the model that shows the satisfactory spatial restraints. The most popular tool for homology modeling that uses spatial restraints is MODELLER (Eswar et al., 2006).

Molecular docking itself is a complex process that requires certain attention with regard to conformation. The flexibility of the ligand to be docked is crucial as it is directly correlated to the conformational space which it will occupy on the protein cavity. In cases where the target binding site or the cavity is unknown, certain pocket detection tools can be implemented, such as POCKET (Levitt & Banaszak, 1992), LIGSITE (Hendlich et al., 1997), and SURFNET (Laskowski, 1995). The importance of the binding site, besides being the region of the protein to which a ligand binds with specificity, is that cellular functions are regulated once the conformational landscape of the protein is changed. Over the years, several computational tools have been developed for the prediction of ligand binding site, mostly structure-based methods which are further classified into template-based methods and methods based upon buried pockets in the protein. Template-based methods rely on protein structures with close semblance to the target; thus, it is limited to only certain proteins with similar structures available. On the other hand, pocket-based methods rely on predictors to identify cavities. The predictors in question can be geometry, energy, or physicochemical-based or could even be a combination of different predictors. Each method uses its own scoring algorithm to predict the top binding site and residues.

After the identification of the ligand binding site on the protein surface, a grid is generated around the binding site to provide a search space for the ligand to find an optimal conformation. Once the grid has been generated, the actual docking process commences. The two main goals of a docking study are to accurately model the structure and to correctly predict the binding affinity. The general constituent of a docking algorithm is the search algorithm, which focuses on exploring available conformational space for ligand binding, and scoring function, method used to rank docked pose of a ligand to a receptor (Brooijmans & Kuntz,

2003; Kitchen et al., 2004). For search algorithms, the flexibility/rigidity of the ligand and receptor are two essential points to be taken into consideration. Thus, the entire docking operation is classified into rigid body docking and flexible docking. As the name suggests, rigid body docking does not consider the flexibility of both ligand and receptor, thus limiting the accuracy of the result. On the other hand, flexible docking provides a more optimized and accurate result but consumes high computational power, specifically on the CPU. A few of the search algorithms will be described in brief, namely systematic search algorithm and stochastic algorithm.

Systematic search is a flexible docking search algorithm, which is based on the degree of freedom of the ligand. This search method is further subdivided into three parts: exhaustive search, fragmentation, and conformational ensemble. Exhaustive search methods explore all possible rotatable bonds for the ligand, thus sampling all possible ligand conformations. However, computational time increases with an increase in the number of bonds. This method is applied to initial screening tools such as Glide. Fragmentation methods divide the ligand into rigid parts and each fragment is inserted into the binding site one after the other, gradually constructing the entire ligand. A tool that uses this method is FlexX (Rarey et al., 1996). The last type of systematic search, i.e., conformational ensemble, and tools such as OMEGA (OpenEye Scientific Inc, NM) (Hawkins et al., 2010) are used to generate an ensemble of ligand conformations, which is then ranked according to the binding energy. Glide (Friesner et al., 2006) uses ensemble docking for virtual screening workflow. The second search algorithm that will be discussed is the stochastic algorithm, a technique in which random changes in ligand conformation are made and are selected based on probability. Monte Carlo (MC) methods, evolutionary algorithms, Tabu search methods, and swarm optimization (SO) methods are the four types of stochastic search methods. MC method uses the Boltzmann probability function for calculating the random change, which is defined by Eq. (15.1):

$$P \sim exp\left[\frac{-(E_1 - E_0)}{k_B T}\right] \qquad (15.1)$$

where E_0 and E_1 represent energy scores of the ligand before and after the random change, respectively, k_B represents the Boltzmann constant, and T represents the absolute temperature of the system. AutoDock (Morris et al., 2009) has an option for MC-based docking.

Evolutionary algorithms are derived from evolutionary process in biology and it uses genetic operators, computational approximations, to select favorable features. An example of a popular evolutionary algorithm is the genetic algorithm (GA), where the best offspring from the combination of two genetic operators is selected for the subsequent iterations until the final offspring has all the best characteristics of all other generations. AutoDock is a tool that provides GA as an option for docking search process (Morris et al., 2009). Another stochastic approach is the Tabu search method, where ligand binding conformation is selected based on random exploration of all possible conformational space and discarding the pose if it fails to cross the cut-off from any previously recorded search space. The last stochastic search algorithm is SO, in which the search space for the ligand is guided by the information gathered from its neighbor with the best position, which is akin to the concept of swarm intelligence. AutoDock Vina (Ng et al., 2015) uses a framework of SO.

After virtual screening, the poses generated by the ligand in the cavity are scored using scoring functions, specifically physical-based or force field or empirical or knowledge-based functions. Force field–based scoring functions decomposes the ligand binding energy into bond stretching, torsion, bending, van der Waals (VDW) energies, and electrostatic energies. The parameters for the force fields are derived using tools such as AMBER (Case et al., 2005; Cornell et al., 1995) or CHARMM (Brooks et al., 2009) force fields. An example of force field scoring function can be represented using the tool DOCK in Eq. (15.2):

$$E = \sum_i \sum_j \left[\frac{A_{ij}}{r_{ij}^{12}} - \frac{B_{ij}}{r_{ij}^{6}} + \frac{q_i q_j}{\varepsilon(r_{ij}) r_{ij}} \right] \quad (15.2)$$

where A_{ij} and B_{ij} represent VDW parameters, q_i and q_j represent atomic charges, $\varepsilon(r_{ij})$ represents dielectric charge, and r_{ij} represents the distance between protein atom i and ligand atom j. Most force field–based methods treat water molecules as free energy perturbation and thermodynamic integration, which could be quite taxing on the computer. Thus, implicit solvent models such as Poisson–Boltzmann/surface area models and the generalized Born/surface area (GB/SA) models were developed to reduce the load while still considering water as a dielectric medium. In the case of empirical scoring method, all energy terms including VDW, electrostatic, bond, hydrophobic, and entropic energies are summed up to obtain the binding score and can be represented by Eq. (15.3):

$$\Delta G = \sum_i W_i \Delta G_i \quad (15.3)$$

where W_i is calculated using least square fitting for binding affinity, and ΔG_i represents individual empirical energy terms. This approach is computationally more efficient and simpler as compared to the force field method and is used in Glide for calculating the GlideScore.

Finally, knowledge-based technique is a class of scoring function which derives binding information from experimentally solved complexes using the potential of mean force, defined by the inverse Boltzmann relation equation given below (Eq. 15.4) (Thomas & Dill, 1996):

$$w(r) = -K_B T \ln \frac{\rho(r)}{\rho^*(r)} \quad (15.4)$$

where k_B represents the Boltzmann constant, T represents absolute temperature of the system, and $\rho(r)$ and $\rho^*(r)$ represent the protein–ligand atom pair number density at a distance r in the training set and the reference state pair density, respectively. GOLD (Verdonk et al., 2003) is a tool that makes use of this scoring function.

The docking operation is usually followed by molecular dynamics (MD) simulation to understand and verify the intricate characteristics of the protein–ligand complex. The dynamics, energy, and interaction of the system are essential aspects from therapeutic perspective. MD simulation is a widely used technique for monitoring the dynamic properties of microscopic systems, within a time scale, in full detail. MD simulations are suitable for the study of relatively small-scale movements, on time scale in terms of nanoseconds (Di Nola et al., 1994; Mangoni et al., 1999). MD simulations are calculated based on atomic position with respect to a function of time. To do this, the position of each atom is defined by a potential energy function. These functions are separated into covalent and noncovalent interactions, which is generally described as in Eq. (15.5):

$$V(R) = V_{bond} + V_{angle} + V_{dihedral} + V_{nonbond} \quad (15.5)$$

where

$$V_{bond} = \sum_{i=1}^{N_b} \frac{k_i}{2} (r_i - r_{0,i})^2 \quad (15.6)$$

$$V_{angle} = \sum_{i=1}^{N_\theta} \frac{k_i}{2} (\theta_i - \theta_{0,i})^2 \quad (15.7)$$

$$V_{dihedral} = \sum_{i=1}^{N_\varphi} \frac{k_i}{2} (1 + (n\varphi - \gamma)) \quad (15.8)$$

The force field parameters given in Eqs. (15.6)–(15.8) represent the bonded interactions between bond length r, bond angle θ, and the dihedral angle φ, and their respective force constant, n, represents the multiplicity and γ represents the phase of the dihedrals in Eq. (15.5).

For nonbonded interactions, the interactions between the particles are represented by Coulomb's law and VDW Eqs. (15.9) and (15.10), which can be represented as

$$V_{Coulomb} = \sum_{i>j} \frac{1}{4\pi\varepsilon_0\varepsilon_r} \frac{q_i q_j}{r_{i,j}} \tag{15.9}$$

$$V_{VDW} = \sum_{i>j} 4\varepsilon_{ij}\left[\left(\frac{\sigma_{ij}}{r_{i,j}}\right)^{12} - \left(\frac{\sigma_{ij}}{r_{i,j}}\right)^{6}\right] \tag{15.10}$$

The parameters for nonbonded interactions given in Eqs. (15.9) and (15.10) consist of partial charges for Coulomb's law, represented by q, and the depth and width of Lennard-Jones potential for VDW interactions by ε and σ, respectively. Force fields such as AMBER, GROMOS (Christen et al., 2005), OPLS (Jorgensen & Tirado-Rives, 1988), and CHARMM are used for defining the potential energy for all the interactions between atoms.

Another important factor for generating models during simulation is treatment of solvents. There are two general models for the solvent system: explicit and implicit solvent models. In the explicit model, atoms are represented as molecules, consisting of thousands of discrete solvent molecules, which is considered more accurate but requires more computational power. Such models make use of quantum mechanics and molecular mechanics, or even a combination of both. Solvent models calculated using explicit solvent range from simple point charge water models based on nonbonded interactions to interaction point–based methods such as the TIP2P, TIP3P, TIP4P, TIP5P, and even TIP6P models. On the other hand, implicit models consist of a continuous medium which represents the solvent. Models that describe implicit solvents include the solvent-accessible surface area (SASA) models, Poisson–Boltzmann equation, and GB/SA models, where the total solvation free energy is defined as the sum of cavity, VDW, and electrostatic polarization terms. Fig. 15.6 gives a visual depiction of the different steps involved in virtual screening, docking, and MD simulation.

A summary of tools commonly used for protein 3D structure modeling, binding site identification, docking and virtual screening, and MD simulation is provided in Table 15.4 with a brief description of each tool.

3.4.2 Application of structure-based virtual screening for drug repurposing in cancer

The use of sBVs has been widely reported over the past few years. In this section, various case studies of successful repositioning will be disseminated. The first example for application of SBVS discussed in this chapter was aimed at inhibiting the function of human androgen receptor (AR), which is highly essential for the development and progression of prostate cancer. The study reported by Bisson et al. (2007) involved the construction of the AR protein domains, which were screened against a library of ligands (Bisson et al., 2007). The ligands in question were divided into three databases: the first database consisting of 24 AR antagonists retrieved from literature; the second database consisting of 88 agonists and antagonists which specifically target androgen, estrogen, progesterone, and similar nuclear receptors; and the third database consisting of 5000 compounds retrieved from ChemBridge compound database. Molecular docking was performed using Molsoft ICM, and models with the highest docking score cutoff were retained (ICM docking scores from -60.0 to -32.0). Once the highest scoring compounds were obtained, the top 11 compounds were further subjected to in vitro assays. The resulting experiment yielded three compounds with high interaction: acetophenazine, fluphenazine, and pericyazine (Fig. 15.7A–C). These three ligands were originally marketed as antipsychotics.

The second case involved targeting protein–protein interaction, as opposed to the traditional use of enzymes and receptors as drug targets. In this case, tumor necrosis factor alpha (TNF-α), a well-known cytokine that is responsible for inflammatory response in various disorders, including cancer, was studied by Leung and his colleagues (Leung et al., 2011, 2012). Through high-throughput visual screening of more than 3000 FDA-approved drugs and using Molsoft ICM for evaluating the binding interaction between ligands and their target, and through enzyme assays, the team identified seven potent compounds. Out of these, darifenacin (Fig. 15.7D), used for treating overactive bladder syndrome, and ezetimibe (Fig. 15.7E), prescribed for regulating cholesterol levels, were found to disrupt TNF-α/TNFR interaction in vitro, thus downregulating TNF-α-driven gene expression in human cells.

Another study by Dakshanamurthy and his colleagues in 2012 discussed about performing docking of 3671 FDA-approved drugs across 2335 human protein crystal structures (Dakshanamurthy et al., 2012). In their work, the authors generated ligand-based descriptors using Schrodinger QikProp for calculating

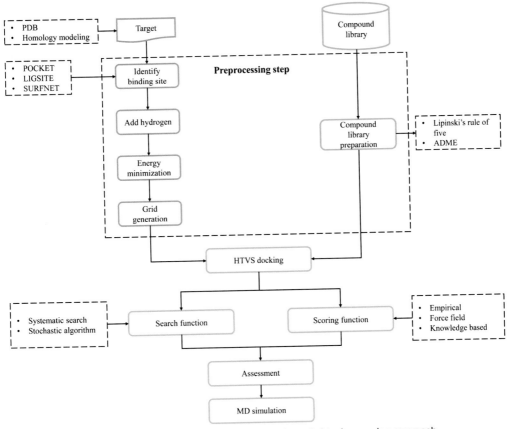

FIG. 15.6 A general workflow of structure-based virtual screening approach.

hydrogen bond acceptors, dipole moment, hydrogen bond donors, electron affinity, globularity, molecular weight, predicted log of the octanol/water partition coefficient (ClogP), SASA, and volume. Docking operation was performed using Glide with extra precision scoring function. They found that mebendazole (Fig. 15.7F), an antiparasitic drug, had an inhibitory effect on vascular endothelial growth factor receptor 2 (VEGFR2), which is a crucial component for angiogenesis and is routinely found in carcinoma cells. They further proved the inhibitory action of the drug experimentally.

Similar works in drug repurposing can be found where in silico, in vitro, and in vivo validations are carried out hand in hand. One such study was the validation of an antipsychotic drug, known as fluspirilene (Fig. 15.7G), as a potential candidate for the treatment of human hepatocellular carcinoma (HCC) through the inhibition of CDK2. This study (Shi et al., 2015) was carried out by screening for potential CDK2 inhibitors

from a list of 4311 FDA-approved small molecules using an open-source docking software called iDock (Li, Leung et al., 2014). Based on the docking score (−10.02 kcal/mol), fluspirilene among the top compounds was selected for in vitro studies. Furthermore, fluspirilene exhibited the highest inhibitory effect on HCC cell lines, along with inhibition of CDK2, decrease in cell viability, and inducing apoptosis. Additional in vivo tests via oral administration of the drug indicated that treatment with fluspirilene reduced the weight of tumor significantly.

Another study involving a combination of in silico, in vitro, and in vivo was reported by Xiao et al., 2017, where raloxifene (Fig. 15.7H) and bazedoxifene (Fig. 15.7I), a nonhormonal antiresorptive FDA-approved agent used for the treatment of postmenopausal osteoporosis, were found to possess properties useful for the treatment of rhabdomyosarcoma (Xiao et al., 2017). The researchers implemented multiple ligand simultaneous docking and repurposing

TABLE 15.4

Tools Used for Computer-Aided Drug Discovery.

Sl. No.	Tool	Description	Availability	Reference
HOMOLOGY MODELING				
1	MODELLER	A comparative or homology modeling tool that makes use of aligned sequences to generate spatial restraints. The tools also generate loop regions via ab initio prediction.	Free for academics	Eswar et al. (2006)
2	Prime	Prime is a part of the Schrodinger software suite which generates protein models based on physics-based energy functions.	Commercial	Jacobson et al. (2004)
3	SWISS-MODEL	A dedicated web server for homology modeling. It identifies structure templates for a given sequence and builds the model based on the alignment and through the implementation of fragment-based approaches.	Free for academics	Schwede et al. (2003)
4	YASARA	A molecular graphics and modeling program that includes docking and molecular dynamics (MD) simulation.	Free version available for academic use	Land and Humble (2018)
BINDING SITE PREDICTION				
5	3DLigandSite	A tool that predicts the binding site by superimposing ligand-bound structures with the target structure. This tool is available online as a web server.	Free for academics	Wass et al. (2010)
6	CASTp	An automatic cavity/binding site identification tool that uses Delaunay triangulation. This tool is available online as a web server or could be used as a plugin for PyMOL.	Free for academics	Binkowski et al. (2003)
7	LIGSITE	A binding site prediction tool that uses Connolly surface and degree of conservation.	Free for academics	Hendlich et al. (1997)
8	metaPocket	A ligand binding site detection metaserver which uses available tools such as LIGSITE, SURFNET, and various other methods.	Free for academics	Huang (2009)
9	POCKET	A tool where the protein structure is mapped onto a three-dimensional (3D) grid and a threshold is set. Grid points that exceed the threshold are retained.	Free for academics	Levitt and Banaszak (1992)
10	SURFNET	A ligand binding site finding tool similar to POCKET where the protein is mapped onto a 3D grid. However, SURFNET uses a sphere that defines the pocket. If any atoms fall under the sphere, its size is reduced until no more atoms are contained.	Free for academics	Laskowski (1995)

TABLE 15.4
Tools Used for Computer-Aided Drug Discovery.—cont'd

Sl. No.	Tool	Description	Availability	Reference
VIRTUAL SCREENING AND MOLECULAR DOCKING				
11	ArgusLab	A protein–ligand docking tool that uses a systematic search method for flexible ligands with X-score–based empirical scoring function.	Free for academics	Jones et al. (1997)
12	AutoDock	A docking software that uses genetic algorithm (Stochastic method) for flexible ligand and partially flexible protein with an empirical scoring function.	Free for academics	Morris et al. (2009)
13	DOCK	A docking software developed by UCSF that uses systematic search method for flexible ligands with a force field–based scoring function.	Free for academics	Ewing et al. (2001)
14	FlexX	A commercial docking software for virtual screening and molecular docking based on an incremental build–based approach.	Commercial	Rarey et al. (1996)
15	Glide	A docking software which is a part of the Schrodinger software suite. It can perform virtual screening with high accuracy and speed, using systematic searching with OPLS-AA force field for flexible ligands and an empirical scoring function.	Commercial	Friesner et al. (2006)
16	GOLD	A popular docking software that uses a stochastic search approach, genetic algorithm, on flexible ligands and partially flexible protein target, with a novel scoring function GoldScore, ChemScore (empirical), and ASP (knowledge based).	Commercial	Verdonk et al. (2003)
17	PyRx	An open-source virtual screening software used in combination with Vina and AutoDock.	Old version available for free; newer versions paid for academic and commercial use	Dallakyan and Olson (2015)
MD SIMULATION				
18	AMBER	An MD simulation tool that includes quantum mechanics/molecular mechanics.	Commercial	Case et al. (2005)
19	CHARMM	An MD simulation tool that includes Monte Carlo simulation method along with quantum mechanics/molecular mechanics.	Commercial	Brooks et al. (2009)
20	Desmond	A software package for MD simulation which is a part of the Schrodinger software suite. It supports replica exchange simulation methods.	Free for academics	Bowers et al. (2006)
21	GROMACS	A fast MD package developed for simulation that features Monte Carlo simulation method.	Free for academics	Van Der Spoel et al. (2005)
22	NAMD	A software for MD simulation which can be used with VMD to provide hybrid quantum mechanics/molecular mechanics simulations.	Free for academics	Phillips et al. (2005)

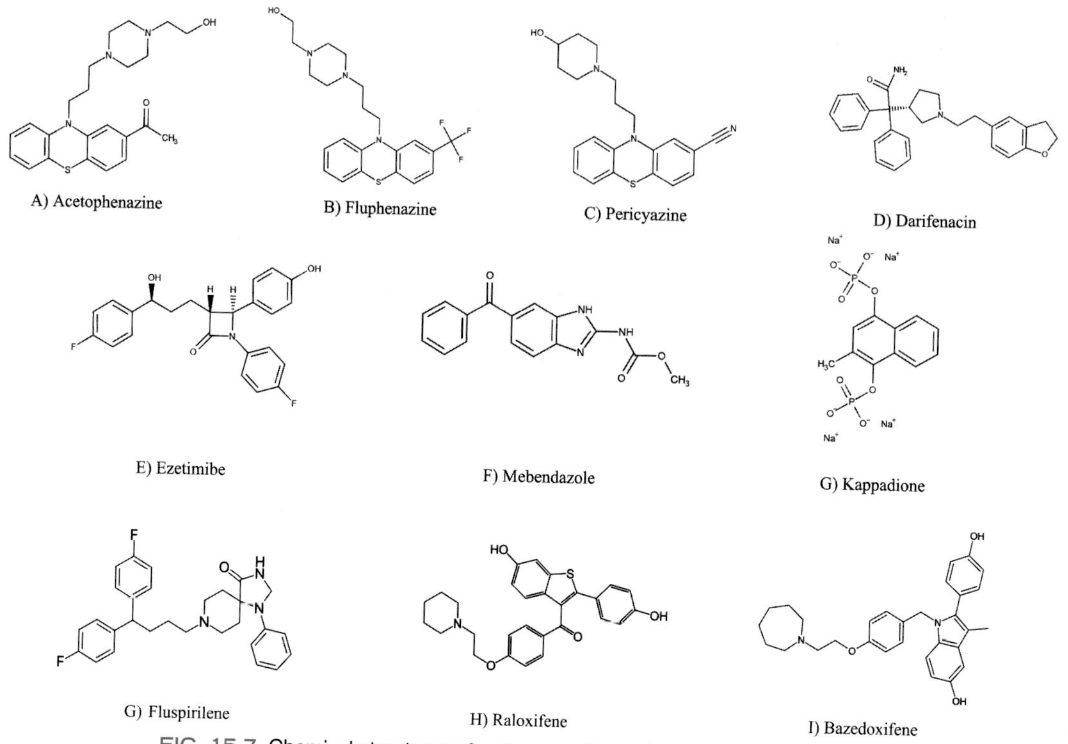

A) Acetophenazine

B) Fluphenazine

C) Pericyazine

D) Darifenacin

E) Ezetimibe

F) Mebendazole

G) Kappadione

G) Fluspirilene

H) Raloxifene

I) Bazedoxifene

FIG. 15.7 Chemical structures of anticancer drugs derived from case studies.

approaches using a drug scaffold database and Drug-Bank database. Raloxifene (−8.8 kcal/mol) and baze-doxifene (−8.2 kcal/mol), both of which selectively act as modulators of estrogen receptor for osteoporosis, were identified as the top compounds with the help of virtual screening (Li, Xiao, et al., 2014). These drugs disrupt the IL-6/GP130/STAT3 cancer signaling pathway, while bazedoxifene was reported to effectively block the activation of STAT3, inducing apoptosis and inhibiting human rhabdomyosarcoma cell growth in vitro or in vivo.

Other works reported in academics are often more hypothetical in nature; for example, Aier et al. (Aier et al., 2020) highlight the effort to look for repurposed drugs that could be used against pancreatic ductal adenocarcinoma by targeting a transcription factor, PAX2 (Aier et al., 2019). The target in question has been found to regulate the transcription of several antichemotherapeutic transport proteins belonging to the ABC family of transporter proteins. The authors used Schrodinger Glide and DrugBank database for the virtual screening process by selecting over 13,000

drug entries. It was found that kappadione (Fig. 15.7G), a derivative of vitamin K, formed a stable complex with the PAX2 at the DNA binding site (Glide score −9.819 kcal/mol) and prevented binding of DNA to the region via in vitro analysis.

Besides the case studies aforementioned, several recent studies have come up with potential repurposed drugs for the treatment of various tumors and cancers using a combination of repurposing approaches with in vitro and in vivo studies (Table 15.5).

Successful repositioning stories are not only limited to researchers and academics, as big pharma companies have ventured into the field of drug repurposing to undertake the challenges raised by various disorders. CRx-026, a novel syncretic anticancer agent, was developed by CombinatoRx Inc., a biotech company based in Boston, Massachusetts. Using a combination of HTS, pentamidine, an antiparasitic agent, and a phenothiazine called chlorpromazine, previously used as an antipsychotic, were found to work in synergistic combination. The approach for repositioning was based on HTS coupled with cell-based assays that could be used to

TABLE 15.5
Examples of Virtual Screening and Molecular Docking—Based Drug Repurposing Approaches With In Vitro or In Vivo Analysis.

Sl No.	Name of Drug	Original Indication	Proposed Indication	Repurposing Approach	Reference
1	Linagliptin	Treatment of type 2 diabetes mellitus	Multiple tumors, especially for colorectal cancer	Virtual screening and molecular docking, gene regulatory network analysis, in vitro, and in vivo experiments	Li et al. (2020)
2	Carvedilol	Treatment of high blood pressure and other heart-related issues	Target multiple cancers via inhibition of CYP1B	Virtual screening and molecular docking approaches and in vitro analysis	Wang et al. (2020)
3	Nomifensine	Antidepressant	Treatment of breast cancer by targeting NUDT5	Gene expression profile drug—disease relation, molecular docking, and in vitro analysis	Tong et al. (2020)
4	Isoconazole	Antifungal drug	Treatment of breast cancer by targeting NUDT5	Gene expression profile drug—disease relation, molecular docking, and in vitro analysis	Tong et al. (2020)
5	Argatroban	Treatment of thrombosis	Treatment of metastatic cancers	Virtual screening and molecular docking, gene regulatory network analysis, in vitro, and in vivo experiments	Goody et al. (2019)

target multiple disease pathways. The mechanism of action for this drug indicated that it acts in two distinct ways in mitosis: (1) by inhibiting the mitotic kinesin, hsEG5/KSP, which plays a critical role in centromere separation, and (2) by inhibiting PRL phosphatase, which regulates the progression of mitosis and ensures that the separation of chromosomes operates smoothly (Fig. 15.8). This new compound is well under Phase I/II of clinical trials where it is being tested for treatment of solid tumors (von Hoff et al., 2005; Lee et al., 2007). Besides being anticancerous, indications of the drug being applicable for respiratory and inflammatory ailments, as well as infectious diseases can be found.

Another example of drug repurposing by biopharmaceuticals is the search for therapeutics by ChemGenex Therapeutics. The company gained much attention when they unveiled a number of drugs which successfully completed Phase I/II of clinical studies (Targeted Therapies from 'Down Under', 2006). One

such drug is Ceflatonin, previously known as homoharringtonine, which was obtained from the plant extracts of a *Cephalotaxus* species. This drug was subcutaneously administered to induce programmed cell death in chronic myeloid leukemia (Quintas-Cardama & Cortes, 2008). Another compound from the same company that successfully completed clinical trial was a drug named Quinamed, also known as amonafide dihydrochloride. This drug is specifically administered to combat solid tumors and it acts through the inhibition of topoisomerase II, while also inhibiting the epidermal growth factor receptor (EGFR) which is highly essential for promoting growth in many types of cancer (Targeted Therapies from 'Down Under', 2006). The method by which the ChemGenex develops and repurposes lead candidates is through a combination of genetics-based screening and computational analysis, which is then tested on animal models to find relations between disease and genes (Fig. 15.9).

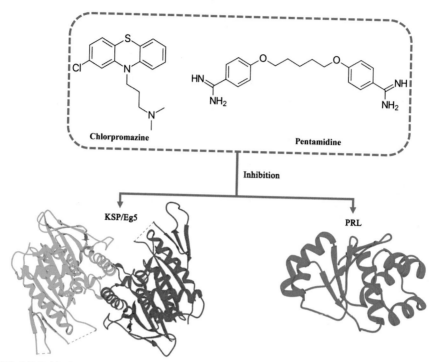

FIG. 15.8 Mechanism of action for synergistic combination of chlorpromazine and pentamidine.

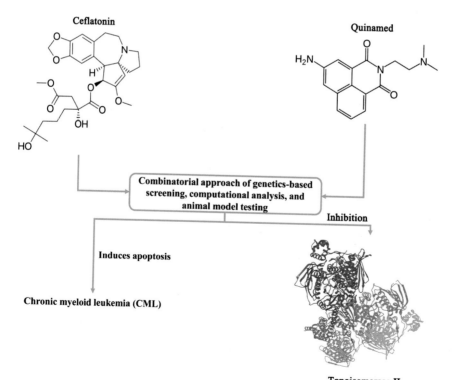

FIG. 15.9 Mechanism of action for Ceflatonin and Quinamed.

4 CONCLUSION

As described in this chapter, due to the rising cost as well as the risk of failure in drug development, pharmaceutical companies and academicians alike have sought out for an alternative in the form of drug repurposing, which not only solves the problem of discovery time but also reduces the cost drastically. The repurposed drugs discussed in this chapter have been used to target at least one form of cancer, besides being used for other indications. Given the amount of resource available in the form of compound libraries, drug repurposing by sBVs could be a great asset. Recent advancements in technology have made high-speed virtual screening of compounds accessible to researchers, while powerful graphical processing units have reduced the time for MD simulations drastically. Although designed to work as a standalone operation, molecular docking can quickly be integrated with workflows such as machine learning and ligand-based approaches to exceed the innovations made toward repurposing. Even with all the prospects of virtual screening and molecular docking, there are certain shortcomings when it comes to repurposing. An old drug for a new disease needs to take the means of administration and dosage into consideration. This could take a long time to achieve the desired outcome depending on the selection of patients for the drug administration. Moreover, incorrect dosage to the wrong patient could lead to terrible outcomes, which so often happens during Phase III of clinical trials. Other reasons that provide major obstacles for repurposing can be attributed to issues nonrelated to the discovery and development of the drug, such as issues with patents, where obtaining a second patent could prove to be troublesome in certain cases. Moreover, preventing access to data due to an existing patent hinders the process of repurposing. Another hurdle faced is that the efficacy of the repurposed drug has to be significantly better than the existing treatment. This is coupled with the fact that the repurposed drug may vary in dosage when compared with its generic counterpart, which would not only affect its success rate during clinical trials but also may prove to be more expensive if changes in the formulation or dosage are required. Any setback during these steps could lead to abandonment of the entire operation. Nonetheless, in silico approaches have provided new possibilities for drug discovery, which could be further improved upon in the future.

REFERENCES

Aier, I., Semwal, R., Dhara, A., Sen, N., & Varadwaj, P. K. (2019). An integrated epigenome and transcriptome analysis identifies PAX2 as a master regulator of drug resistance in high grade pancreatic ductal adenocarcinoma. *PLoS One, 14*(10), e0223554. https://doi.org/10.1371/journal.pone.0223554

Aier, I., Semwal, R., Raj, U., & Varadwaj, P. K. (2020). Comparative modeling and structure based drug repurposing of PAX2 transcription factor for targeting acquired chemoresistance in pancreatic ductal adenocarcinoma. *Journal of Biomolecular Structure and Dynamics,* 1–8. https://doi.org/10.1080/07391102.2020.1742793

Ashburn, T. T., & Thor, K. B. (2004). Drug repositioning: Identifying and developing new uses for existing drugs. *Nature Reviews Drug Discovery, 3*(8), 673–683. https://doi.org/10.1038/nrd1468

Binkowski, T. A., Naghibzadeh, S., & Liang, J. (2003). CASTp: Computed atlas of surface topography of proteins. *Nucleic Acids Research, 31*(13), 3352–3355. https://doi.org/10.1093/nar/gkg512

Bisson, W. H., Cheltsov, A. V., Bruey-Sedano, N., Lin, B., Chen, J., Goldberger, N., May, L. T., Christopoulos, A., Dalton, J. T., Sexton, P. M., Zhang, X. K., & Abagyan, R. (2007). Discovery of antiandrogen activity of nonsteroidal scaffolds of marketed drugs. *Proceedings of the National Academy of Sciences of the United States of America, 104*(29), 11927–11932. https://doi.org/10.1073/pnas.0609752104

Bowers, K. J., Chow, D. E., Xu, H., Dror, R. O., Eastwood, M. P., Gregersen, B. A., Klepeis, J. L., Kolossvary, I., Moraes, M. A., & Sacerdoti, F. D. (2006). Scalable algorithms for molecular dynamics simulations on commodity clusters. In *Paper presented at the SC'06: Proceedings of the 2006 ACM/IEEE conference on supercomputing.*

Bray, F., Ferlay, J., Soerjomataram, I., Siegel, R. L., Torre, L. A., & Jemal, A. (2018). Global cancer statistics 2018: GLOBOCAN estimates of incidence and mortality worldwide for 36 cancers in 185 countries. *CA: A Cancer Journal for Clinicians, 68*(6), 394–424. https://doi.org/10.3322/caac.21492

Bristow, M. R. (2012). Pharmacogenetic targeting of drugs for heart failure. *Pharmacology & Therapeutics, 134*(1), 107–115. https://doi.org/10.1016/j.pharmthera.2012.01.002

Brooijmans, N., & Kuntz, I. D. (2003). Molecular recognition and docking algorithms. *Annual Review of Biophysics and Biomolecular Structure, 32,* 335–373. https://doi.org/10.1146/annurev.biophys.32.110601.142532

Brooks, B. R., Brooks, C. L., 3rd, Mackerell, A. D., Jr., Nilsson, L., Petrella, R. J., Roux, B., Won, Y., Archontis, G., Bartels, C., Boresch, S., Caflisch, A., Caves, L., Cui, Q., Dinner, A. R., Feig, M., Fischer, S., Gao, J., Hodoscek, M., Im, W., ... Karplus, M. (2009). CHARMM: The biomolecular simulation program. *Journal of Computational Chemistry, 30*(10), 1545–1614. https://doi.org/10.1002/jcc.21287

Brown, A. C., & Fraser, T. R. (1868). On the connection between chemical constitution and physiological action; with special reference to the physiological action of the salts of the ammonium bases derived from Strychnia, Brucia, Thebaia, Codeia, Morphia, and Nicotia. *Journal of Anatomy and Physiology, 2*(2), 224–242.

Bymaster, F. P., Katner, J. S., Nelson, D. L., Hemrick-Luecke, S. K., Threlkeld, P. G., Heiligenstein, J. H., Morin, S. M., Gehlert, D. R., & Perry, K. W. (2002). Atomoxetine increases extracellular levels of norepinephrine and dopamine in prefrontal cortex of rat: A potential mechanism for efficacy in attention deficit/hyperactivity disorder. *Neuropsychopharmacology*, 27(5), 699–711. https://doi.org/10.1016/S0893-133X(02)00346-9

Campillos, M., Kuhn, M., Gavin, A. C., Jensen, L. J., & Bork, P. (2008). Drug target identification using side-effect similarity. *Science*, 321(5886), 263–266. https://doi.org/10.1126/science.1158140

Case, D. A., Cheatham, T. E., 3rd, Darden, T., Gohlke, H., Luo, R., Merz, K. M., Jr., Onufriev, A., Simmerling, C., Wang, B., & Woods, R. J. (2005). The amber biomolecular simulation programs. *Journal of Computational Chemistry*, 26(16), 1668–1688. https://doi.org/10.1002/jcc.20290

Chen, J. H., Linstead, E., Swamidass, S. J., Wang, D., & Baldi, P. (2007). ChemDB update—full-text search and virtual chemical space. *Bioinformatics*, 23(17), 2348–2351. https://doi.org/10.1093/bioinformatics/btm341

Chen, Q. W., Zhang, X. M., Zhou, J. N., Zhou, X., Ma, G. J., Zhu, M., Zhang, Y. Y., Yu, J., Feng, J. F., & Chen, S. Q. (2015). Analysis of small fragment deletions of the APC gene in Chinese patients with familial adenomatous polyposis, a precancerous condition. *Asian Pacific Journal of Cancer Prevention*, 16(12), 4915–4920. https://doi.org/10.7314/apjcp.2015.16.12.4915

Christen, M., Hunenberger, P. H., Bakowies, D., Baron, R., Burgi, R., Geerke, D. P., Heinz, T. N., Kastenholz, M. A., Krautler, V., Oostenbrink, C., Peter, C., Trzesniak, D., & van Gunsteren, W. F. (2005). The GROMOS software for biomolecular simulation: GROMOS05. *Journal of Computational Chemistry*, 26(16), 1719–1751. https://doi.org/10.1002/jcc.20303

Congreve, M., Carr, R., Murray, C., & Jhoti, H. (2003). A 'rule of three' for fragment-based lead discovery? *Drug Discovery Today*, 8(19), 876–877. https://doi.org/10.1016/s1359-6446(03)02831-9

Cornell, W. D., Cieplak, P., Bayly, C. I., Gould, I. R., Merz, K. M., Ferguson, D. M., Spellmeyer, D. C., Fox, T., Caldwell, J. W., & Kollman, P. A. (1995). A second generation force field for the simulation of proteins, nucleic acids, and organic molecules. *Journal of the American Chemical Society*, 117(19), 5179–5197. https://doi.org/10.1021/ja00124a002

Crawford, C. L. (1991). Use of thalidomide in leprosy. *British Medical Journal*, 303(6809), 1062–1063. https://doi.org/10.1136/bmj.303.6809.1062-c

Dakshanamurthy, S., Issa, N. T., Assefnia, S., Seshasayee, A., Peters, O. J., Madhavan, S., Uren, A., Brown, M. L., & Byers, S. W. (2012). Predicting new indications for approved drugs using a proteochemometric method. *Journal of Medicinal Chemistry*, 55(15), 6832–6848. https://doi.org/10.1021/jm300576q

Dallakyan, S., & Olson, A. J. (2015). Small-molecule library screening by docking with PyRx. *Methods in Molecular Biology*, 1263, 243–250. https://doi.org/10.1007/978-1-4939-2269-7_19

De Clercq, E. (1994). HIV resistance to reverse transcriptase inhibitors. *Biochemical Pharmacology*, 47(2), 155–169. https://doi.org/10.1016/0006-2952(94)90001-9

Del Rio, A., Barbosa, A. J., Caporuscio, F., & Mangiatordi, G. F. (2010). CoCoCo: A free suite of multiconformational chemical databases for high-throughput virtual screening purposes. *Molecular BioSystems*, 6(11), 2122–2128. https://doi.org/10.1039/c0mb00039f

Di Nola, A., Roccatano, D., & Berendsen, H. J. (1994). Molecular dynamics simulation of the docking of substrates to proteins. *Proteins*, 19(3), 174–182. https://doi.org/10.1002/prot.340190303

DiMasi, J. A., Hansen, R. W., & Grabowski, H. G. (2003). The price of innovation: New estimates of drug development costs. *Journal of Health Economics*, 22(2), 151–185. https://doi.org/10.1016/S0167-6296(02)00126-1

Dudley, J. T., Sirota, M., Shenoy, M., Pai, R. K., Roedder, S., Chiang, A. P., Morgan, A. A., Sarwal, M. M., Pasricha, P. J., & Butte, A. J. (2011). Computational repositioning of the anticonvulsant topiramate for inflammatory bowel disease. *Science Translational Medicine*, 3(96), 96ra76. https://doi.org/10.1126/scitranslmed.3002648

English, C., Rey, J. A., & Rufin, C. (2010). Milnacipran (Savella), a treatment option for fibromyalgia. *Pharmacy and Therapeutics*, 35(5), 261–266.

Eswar, N., Webb, B., Marti Renom, M. A., Madhusudhan, M. S., Eramian, D., Shen, M. Y., Pieper, U., & Sali, A. (2006). Comparative protein structure modeling using modeller. *Current Protocols in Bioinformatics*. https://doi.org/10.1002/0471250953.bi0506s15 (Chapter 5), Unit-5 6.

Ewing, T. J., Makino, S., Skillman, A. G., & Kuntz, I. D. (2001). DOCK 4.0: Search strategies for automated molecular docking of flexible molecule databases. *Journal of Computer-Aided Molecular Design*, 15(5), 411–428. https://doi.org/10.1023/a:1011115820450

Friesner, R. A., Murphy, R. B., Repasky, M. P., Frye, L. L., Greenwood, J. R., Halgren, T. A., Sanschagrin, P. C., & Mainz, D. T. (2006). Extra precision glide: Docking and scoring incorporating a model of hydrophobic enclosure for protein-ligand complexes. *Journal of Medicinal Chemistry*, 49(21), 6177–6196. https://doi.org/10.1021/jm051256o

Gallagher, P., & Young, A. H. (2006). Mifepristone (RU-486) treatment for depression and psychosis: A review of the therapeutic implications. *Neuropsychiatric Disease and Treatment*, 2(1), 33–42.

Gershlick, A., De Scheerder, I., Chevalier, B., Stephens-Lloyd, A., Camenzind, E., Vrints, C., Reifart, N., Missault, L., Goy, J. J., Brinker, J. A., Raizner, A. E., Urban, P., & Heldman, A. W. (2004). Inhibition of restenosis with a paclitaxel-eluting, polymer-free coronary stent: The European evaLUation of pacliTaxel Eluting Stent (ELUTES) trial. *Circulation*, 109(4), 487–493. https://doi.org/10.1161/01.CIR.0000109694.58299.A0

Goody, D., Gupta, S. K., Engelmann, D., Spitschak, A., Marquardt, S., Mikkat, S., Meier, C., Hauser, C., Gundlach, J. P., Egberts, J. H., Martin, H., Schumacher, T., Trauzold, A., Wolkenhauer, O., Logotheti, S., & Putzer, B. M. (2019). Drug repositioning inferred from E2F1-coregulator interactions studies for the prevention and treatment of metastatic cancers. *Theranostics, 9*(5), 1490−1509. https://doi.org/10.7150/thno.29546

Gottlieb, T. B., Katz, F. H., & Chidsey, C. A., 3rd (1972). Combined therapy with vasodilator drugs and beta-adrenergic blockade in hypertension. A comparative study of minoxidil and hydralazine. *Circulation, 45*(3), 571−582. https://doi.org/10.1161/01.cir.45.3.571

Gottlieb, A., Stein, G. Y., Ruppin, E., & Sharan, R. (2011). PREDICT: A method for inferring novel drug indications with application to personalized medicine. *Molecular Systems Biology, 7*, 496. https://doi.org/10.1038/msb.2011.26

Hansch, C. (1972). Quantitative relationships between lipophilic character and drug metabolism. *Drug Metabolism Reviews, 1*(1), 1−13. https://doi.org/10.3109/03602537208993906

Hawkins, P. C., Skillman, A. G., Warren, G. L., Ellingson, B. A., & Stahl, M. T. (2010). Conformer generation with OMEGA: Algorithm and validation using high quality structures from the Protein Databank and Cambridge structural database. *Journal of Chemical Information and Modeling, 50*(4), 572−584. https://doi.org/10.1021/ci100031x

Hendlich, M., Rippmann, F., & Barnickel, G. (1997). LIGSITE: Automatic and efficient detection of potential small molecule-binding sites in proteins. *Journal of Molecular Graphics and Modelling, 15*(6). https://doi.org/10.1016/s1093-3263(98)00002-3, 359−363, 389.

Hershman, D. L., Lacchetti, C., & Loprinzi, C. L. (2014). Prevention and management of chemotherapy-induced peripheral neuropathy in survivors of adult cancers: American Society of Clinical Oncology Clinical Practice Guideline Summary. *Journal of Oncology Practice, 10*(6), e421−e424. https://doi.org/10.1200/JOP.2014.001776

von Hoff, D. D., Gordon, M., Turner, J., White, E., Nichols, M. J., Elliott, P. J., & Mendelson, D. (2005). A phase I study with CRx-026, a novel dual action agent, in patients (pts) with advanced solid tumors. *Journal of Clinical Oncology, 23*(16_Suppl. l), 3073. https://doi.org/10.1200/jco.2005.23.16_suppl.3073

Huang, B. (2009). MetaPocket: A meta approach to improve protein ligand binding site prediction. *OMICS, 13*(4), 325−330. https://doi.org/10.1089/omi.2009.0045

Iorio, F., Bosotti, R., Scacheri, E., Belcastro, V., Mithbaokar, P., Ferriero, R., Murino, L., Tagliaferri, R., Brunetti-Pierri, N., Isacchi, A., & di Bernardo, D. (2010). Discovery of drug mode of action and drug repositioning from transcriptional responses. *Proceedings of the National Academy of Sciences of the United States of America, 107*(33), 14621−14626. https://doi.org/10.1073/pnas.1000138107

Irwin, J. J., & Shoichet, B. K. (2005). ZINC−a free database of commercially available compounds for virtual screening. *Journal of Chemical Information and Modeling, 45*(1), 177−182. https://doi.org/10.1021/ci049714+

Jacobson, M. P., Pincus, D. L., Rapp, C. S., Day, T. J., Honig, B., Shaw, D. E., & Friesner, R. A. (2004). A hierarchical approach to all-atom protein loop prediction. *Proteins, 55*(2), 351−367. https://doi.org/10.1002/prot.10613

Jin, G., Fu, C., Zhao, H., Cui, K., Chang, J., & Wong, S. T. (2012). A novel method of transcriptional response analysis to facilitate drug repositioning for cancer therapy. *Cancer Research,*

72(1), 33−44. https://doi.org/10.1158/0008-5472.CAN-11-2333

Jones, G., Willett, P., Glen, R. C., Leach, A. R., & Taylor, R. (1997). Development and validation of a genetic algorithm for flexible docking. *Journal of Molecular Biology, 267*(3), 727−748. https://doi.org/10.1006/jmbi.1996.0897

Jorgensen, W. L., & Tirado-Rives, J. (1988). The OPLS [optimized potentials for liquid simulations] potential functions for proteins, energy minimizations for crystals of cyclic peptides and crambin. *Journal of the American Chemical Society, 110*(6), 1657−1666. https://doi.org/10.1021/ja00214a001

Keiser, M. J., Setola, V., Irwin, J. J., Laggner, C., Abbas, A. I., Hufeisen, S. J., Jensen, N. H., Kuijer, M. B., Matos, R. C., Tran, T. B., Whaley, R., Glennon, R. A., Hert, J., Thomas, K. L., Edwards, D. D., Shoichet, B. K., & Roth, B. L. (2009). Predicting new molecular targets for known drugs. *Nature, 462*(7270), 175−181. https://doi.org/10.1038/nature08506

Kim, S., Chen, J., Cheng, T., Gindulyte, A., He, J., He, S., Li, Q., Shoemaker, B. A., Thiessen, P. A., Yu, B., Zaslavsky, L., Zhang, J., & Bolton, E. E. (2019). PubChem 2019 update: Improved access to chemical data. *Nucleic Acids Research, 47*(D1), D1102−D1109. https://doi.org/10.1093/nar/gky1033

Kitchen, D. B., Decornez, H., Furr, J. R., & Bajorath, J. (2004). Docking and scoring in virtual screening for drug discovery: Methods and applications. *Nature Reviews Drug Discovery, 3*(11), 935−949. https://doi.org/10.1038/nrd1549

Kramer, C. K., Leitao, C. B., Pinto, L. C., Canani, L. H., Azevedo, M. J., & Gross, J. L. (2011). Efficacy and safety of topiramate on weight loss: A meta-analysis of randomized controlled trials. *Obesity Reviews, 12*(5), e338−347. https://doi.org/10.1111/j.1467-789X.2010.00846.x

Kuntz, I. D. (1992). Structure-based strategies for drug design and discovery. *Science, 257*(5073), 1078−1082. https://doi.org/10.1126/science.257.5073.1078

Kushida, C. A. (2006). Ropinirole for the treatment of restless legs syndrome. *Neuropsychiatric Disease and Treatment, 2*(4), 407−419. https://doi.org/10.2147/nedt.2006.2.4.407

Lamb, J., Crawford, E. D., Peck, D., Modell, J. W., Blat, I. C., Wrobel, M. J., Lerner, J., Brunet, J. P., Subramanian, A., Ross, K. N., Reich, M., Hieronymus, H., Wei, G., Armstrong, S. A., Haggarty, S. J., Clemons, P. A., Wei, R., Carr, S. A., Lander, E. S., & Golub, T. R. (2006). The connectivity map: Using gene-expression signatures to connect small molecules, genes, and disease. *Science, 313*(5795), 1929−1935. https://doi.org/10.1126/science.1132939

Lander, E. S., Linton, L. M., Birren, B., Nusbaum, C., Zody, M. C., Baldwin, J., Devon, K., Dewar, K., Doyle, M., FitzHugh, W., Funke, R., Gage, D., Harris, K., Heaford, A., Howland, J., Kann, L., Lehoczky, J., LeVine, R., McEwan, P., … International Human Genome Sequencing, C. (2001). Initial sequencing and analysis of the human genome. *Nature, 409*(6822), 860−921. https://doi.org/10.1038/35057062

Land, H., & Humble, M. S. (2018). YASARA: A tool to obtain structural guidance in biocatalytic investigations. *Methods in Molecular Biology, 1685*, 43−67. https://doi.org/10.1007/978-1-4939-7366-8_4

Laskowski, R. A. (1995). SURFNET: A program for visualizing molecular surfaces, cavities, and intermolecular interactions. *Journal of Molecular Graphics, 13*(5). https://doi.org/10.1016/0263-7855(95)00073-9, 323−330, 307−328.

Lee, M. S., Johansen, L., Zhang, Y., Wilson, A., Keegan, M., Avery, W., Elliott, P., Borisy, A. A., & Keith, C. T. (2007). The novel combination of chlorpromazine and pentamidine exerts synergistic antiproliferative effects through dual mitotic action. *Cancer Research, 67*(23), 11359−11367. https://doi.org/10.1158/0008-5472.CAN-07-2235

Lehmann, H. E., & Hanrahan, G. E. (1954). Chlorpromazine; new inhibiting agent for psychomotor excitement and manic states. *AMA Archives of Neurology & Psychiatry, 71*(2), 227−237.

Leung, C. H., Chan, D. S., Kwan, M. H., Cheng, Z., Wong, C. Y., Zhu, G. Y., Fong, W. F., & Ma, D. L. (2011). Structure-based repurposing of FDA-approved drugs as TNF-alpha inhibitors. *ChemMedChem, 6*(5), 765−768. https://doi.org/10.1002/cmdc.201100016

Leung, C. H., Zhong, H. J., Yang, H., Cheng, Z., Chan, D. S., Ma, V. P., Abagyan, R., Wong, C. Y., & Ma, D. L. (2012). A metal-based inhibitor of tumor necrosis factor-alpha. *Angewandte Chemie International Edition in English, 51*(36), 9010−9014. https://doi.org/10.1002/anie.201202937

Leventer, S. M., Raudibaugh, K., Frissora, C. L., Kassem, N., Keogh, J. C., Phillips, J., & Mangel, A. W. (2008). Clinical trial: Dextofisopam in the treatment of patients with diarrhoea-predominant or alternating irritable bowel syndrome. *Alimentary Pharmacology & Therapeutics, 27*(2), 197−206. https://doi.org/10.1111/j.1365-2036.2007.03566.x

Levitt, D. G., & Banaszak, L. J. (1992). POCKET: A computer graphics method for identifying and displaying protein cavities and their surrounding amino acids. *Journal of Molecular Graphics, 10*(4), 229−234. https://doi.org/10.1016/0263-7855(92)80074-n

Li, H., Leung, K. S., Ballester, P. J., & Wong, M. H. (2014). istar: a web platform for large-scale protein-ligand docking. *PLoS One, 9*(1), e85678. https://doi.org/10.1371/journal.pone.0085678

Li, Y., Li, Y., Li, D., Li, K., Quan, Z., Wang, Z., & Sun, Z. (2020). Repositioning of hypoglycemic drug linagliptin for cancer treatment. *Frontiers in Pharmacology, 11*, 187. https://doi.org/10.3389/fphar.2020.00187

Li, H., Xiao, H., Lin, L., Jou, D., Kumari, V., Lin, J., & Li, C. (2014). Drug design targeting protein-protein interactions (PPIs) using multiple ligand simultaneous docking (MLSD) and drug repositioning: Discovery of raloxifene and bazedoxifene as novel inhibitors of IL-6/GP130 interface. *Journal of Medicinal Chemistry, 57*(3), 632−641. https://doi.org/10.1021/jm401144z

Li, J., Zhu, X., & Chen, J. Y. (2009). Building disease-specific drug-protein connectivity maps from molecular interaction networks and PubMed abstracts. *PLoS Computational Biology, 5*(7), e1000450. https://doi.org/10.1371/journal.pcbi.1000450

Lipinski, C. A., Lombardo, F., Dominy, B. W., & Feeney, P. J. (2001). Experimental and computational approaches to estimate solubility and permeability in drug discovery and development settings. *Advanced Drug Delivery Reviews, 46*(1−3), 3−26. https://doi.org/10.1016/s0169-409x(00)00129-0

Lounnas, V., Ritschel, T., Kelder, J., McGuire, R., Bywater, R. P., & Foloppe, N. (2013). Current progress in structure-based rational drug design marks a new mindset in drug discovery. *Computational and Structural Biotechnology Journal, 5*, e201302011. https://doi.org/10.5936/csbj.201302011

Lyne, P. D. (2002). Structure-based virtual screening: An overview. *Drug Discovery Today, 7*(20), 1047−1055. https://doi.org/10.1016/s1359-6446(02)02483-2

Mangoni, M., Roccatano, D., & Di Nola, A. (1999). Docking of flexible ligands to flexible receptors in solution by molecular dynamics simulation. *Proteins, 35*(2), 153−162. https://doi.org/10.1002/(sici)1097-0134(19990501)35:2<153::aid-prot2>3.0.co;2-e

Mendez, D., Gaulton, A., Bento, A. P., Chambers, J., De Veij, M., Felix, E., Magarinos, M. P., Mosquera, J. F., Mutowo, P., Nowotka, M., Gordillo-Maranon, M., Hunter, F., Junco, L., Mugumbate, G., Rodriguez-Lopez, M., Atkinson, F., Bosc, N., Radoux, C. J., Segura-Cabrera, A., Hersey, A., & Leach, A. R. (2019). ChEMBL: Towards direct deposition of bioassay data. *Nucleic Acids Research, 47*(D1), D930−D940. https://doi.org/10.1093/nar/gky1075

Mestroni, L., Begay, R. L., Graw, S. L., & Taylor, M. R. (2014). Pharmacogenetics of heart failure. *Current Opinion in Cardiology, 29*(3), 227−234. https://doi.org/10.1097/HCO.0000000000000056

Morris, G. M., Huey, R., Lindstrom, W., Sanner, M. F., Belew, R. K., Goodsell, D. S., & Olson, A. J. (2009). AutoDock4 and AutoDockTools4: Automated docking with selective receptor flexibility. *Journal of Computational Chemistry, 30*(16), 2785−2791. https://doi.org/10.1002/jcc.21256

Mullard, A. (2020). 2019 FDA drug approvals. *Nature Reviews Drug Discovery, 19*(2), 79−84. https://doi.org/10.1038/d41573-020-00001-7

Ng, M. C., Fong, S., & Siu, S. W. (2015). PSOVina: The hybrid particle swarm optimization algorithm for protein-ligand docking. *Journal of Bioinformatics and Computational Biology, 13*(3), 1541007. https://doi.org/10.1142/S0219720015410073

Olah, M., Rad, R., Ostopovici, L., Bora, A., Hadaruga, N., Hadaruga, D., Moldovan, R., Fulias, A., Mractc, M., & Oprea, T. I. (2007). WOMBAT and WOMBAT-PK: Bioactivity databases for lead and drug discovery. *Chemical Biology: From Small Molecules to Systems Biology and Drug Design, 1*, 760−786.

Pantziarka, P., Pirmohamed, M., & Mirza, N. (2018). New uses for old drugs. *British Medical Journal, 361*, k2701. https://doi.org/10.1136/bmj.k2701

Phillips, J. C., Braun, R., Wang, W., Gumbart, J., Tajkhorshid, E., Villa, E., Chipot, C., Skeel, R. D., Kale, L.,

& Schulten, K. (2005). Scalable molecular dynamics with NAMD. *Journal of Computational Chemistry*, 26(16), 1781–1802. https://doi.org/10.1002/jcc.20289

Potter, A. S., Ryan, K. K., & Newhouse, P. A. (2009). Effects of acute ultra-low dose mecamylamine on cognition in adult attention-deficit/hyperactivity disorder (ADHD). *Human Psychopharmacology*, 24(4), 309–317. https://doi.org/10.1002/hup.1026

Quintas-Cardama, A., & Cortes, J. (2008). Homoharringtonine for the treatment of chronic myelogenous leukemia. *Expert Opinion on Pharmacotherapy*, 9(6), 1029–1037. https://doi.org/10.1517/14656566.9.6.1029

Rajkumar, S. V. (2001). Thalidomide in the treatment of multiple myeloma. *Expert Rev Anticancer Ther*, 1(1), 20–28. https://doi.org/10.1586/14737140.1.1.20

Rarey, M., Kramer, B., Lengauer, T., & Klebe, G. (1996). A fast flexible docking method using an incremental construction algorithm. *Journal of Molecular Biology*, 261(3), 470–489. https://doi.org/10.1006/jmbi.1996.0477

Ruddy, J. M. (1971). Antidepressant overdosage in children—a new menace. *Medical Journal of Australia*, 2(22), 1148.

Salisbury, B. H., & Tadi, P. (2020). *5 Alpha reductase inhibitors StatPearls*. Treasure Island (FL).

Schwede, T., Kopp, J., Guex, N., & Peitsch, M. C. (2003). SWISS-MODEL: An automated protein homology-modeling server. *Nucleic Acids Research*, 31(13), 3381–3385. https://doi.org/10.1093/nar/gkg520

Scott, L. J., & Goa, K. L. (2000). Galantamine: A review of its use in Alzheimer's disease. *Drugs*, 60(5), 1095–1122. https://doi.org/10.2165/00003495-200060050-00008

Seiler, K. P., George, G. A., Happ, M. P., Bodycombe, N. E., Carrinski, H. A., Norton, S., Brudz, S., Sullivan, J. P., Muhlich, J., Serrano, M., Ferraiolo, P., Tolliday, N. J., Schreiber, S. L., & Clemons, P. A. (2008). ChemBank: A small-molecule screening and cheminformatics resource database. *Nucleic Acids Research*, 36(Database issue), D351–D359. https://doi.org/10.1093/nar/gkm843

Shi, X. N., Li, H., Yao, H., Liu, X., Li, L., Leung, K. S., Kung, H. F., Lu, D., Wong, M. H., & Lin, M. C. (2015). In silico identification and in vitro and in vivo validation of anti-psychotic drug fluspirilene as a potential CDK2 inhibitor and a candidate anti-cancer drug. *PLoS One*, 10(7), e0132072. https://doi.org/10.1371/journal.pone.0132072

Shoichet, B. K. (2004). Virtual screening of chemical libraries. *Nature*, 432(7019), 862–865. https://doi.org/10.1038/nature03197

Sirota, M., Dudley, J. T., Kim, J., Chiang, A. P., Morgan, A. A., Sweet-Cordero, A., Sage, J., & Butte, A. J. (2011). Discovery and preclinical validation of drug indications using compendia of public gene expression data. *Science Translational Medicine*, 3(96), 96ra77. https://doi.org/10.1126/scitranslmed.3001318

Slemmer, J. E., Martin, B. R., & Damaj, M. I. (2000). Bupropion is a nicotinic antagonist. *Journal of Pharmacology and Experimental Therapeutics*, 295(1), 321–327.

Spear, B. B., Heath-Chiozzi, M., & Huff, J. (2001). Clinical application of pharmacogenetics. *Trends in Molecular Medicine*, 7(5), 201–204. https://doi.org/10.1016/s1471-4914(01)01986-4

Stock, M. J. (1997). Sibutramine: A review of the pharmacology of a novel anti-obesity agent. *International Journal of Obesity and Related Metabolic Disorders*, 21(Suppl. 1), S25–S29.

Targeted Therapies from 'Down Under'. (2006). *Pharmaceutical & Diagnostic Innovation*, 4(2), 10–12. https://doi.org/10.1007/BF03257042

Thomas, P. D., & Dill, K. A. (1996). An iterative method for extracting energy-like quantities from protein structures. *Proceedings of the National Academy of Sciences of the United States of America*, 93(21), 11628–11633. https://doi.org/10.1073/pnas.93.21.11628

Tipton, K. F. (1994). Nomenclature Committee of the International Union of Biochemistry and Molecular Biology (NC-IUBMB). Enzyme nomenclature. Recommendations 1992. Supplement: Corrections and additions. *European Journal of Biochemistry*, 223(1), 1–5. https://doi.org/10.1111/j.1432-1033.1994.tb18960.x

Tong, X. Y., Liao, X., Gao, M., Lv, B. M., Chen, X. H., Chu, X. Y., Zhang, Q. Y., & Zhang, H. Y. (2020). Identification of NUDT5 inhibitors from approved drugs. *Frontiers in Molecular Biosciences*, 7, 44. https://doi.org/10.3389/fmolb.2020.00044

Tseng, P. H., Wang, Y. C., Weng, S. C., Weng, J. R., Chen, C. S., Brueggemeier, R. W., Shapiro, C. L., Chen, C. Y., Dunn, S. E., Pollak, M., & Chen, C. S. (2006). Overcoming trastuzumab resistance in HER2-overexpressing breast cancer cells by using a novel celecoxib-derived phosphoinositide-dependent kinase-1 inhibitor. *Molecular Pharmacology*, 70(5), 1534–1541. https://doi.org/10.1124/mol.106.023911

Van Der Spoel, D., Lindahl, E., Hess, B., Groenhof, G., Mark, A. E., & Berendsen, H. J. (2005). GROMACS: Fast, flexible, and free. *Journal of Computational Chemistry*, 26(16), 1701–1718. https://doi.org/10.1002/jcc.20291

Verdonk, M. L., Cole, J. C., Hartshorn, M. J., Murray, C. W., & Taylor, R. D. (2003). Improved protein-ligand docking using GOLD. *Proteins*, 52(4), 609–623. https://doi.org/10.1002/prot.10465

Wang, Y., He, X., Li, C., Ma, Y., Xue, W., Hu, B., Wang, J., Zhang, T., & Zhang, F. (2020). Carvedilol serves as a novel CYP1B1 inhibitor, a systematic drug repurposing approach through structure-based virtual screening and experimental verification. *European Journal of Medicinal Chemistry*, 193, 112235. https://doi.org/10.1016/j.ejmech.2020.112235

Wass, M. N., Kelley, L. A., & Sternberg, M. J. (2010). 3DLigandSite: Predicting ligand-binding sites using similar structures. *Nucleic Acids Research*, 38(Web Server issue), W469–W473. https://doi.org/10.1093/nar/gkq406

Wishart, D. S., Knox, C., Guo, A. C., Cheng, D., Shrivastava, S., Tzur, D., Gautam, B., & Hassanali, M. (2008). DrugBank: A knowledgebase for drugs, drug actions and drug targets.

Nucleic Acids Research, 36(Database issue), D901–D906. https://doi.org/10.1093/nar/gkm958

Wong, C. H., Siah, K. W., & Lo, A. W. (2019). Estimation of clinical trial success rates and related parameters. *Biostatistics, 20*(2), 273–286. https://doi.org/10.1093/biostatistics/kxx069

Xiao, H., Bid, H. K., Chen, X., Wu, X., Wei, J., Bian, Y., Zhao, C., Li, H., Li, C., & Lin, J. (2017). Repositioning Bazedoxifene as a novel IL-6/GP130 signaling antagonist for human rhabdomyosarcoma therapy. *PLoS One, 12*(7), e0180297. https://doi.org/10.1371/journal.pone.0180297

Zhang, M., Luo, H., Xi, Z., & Rogaeva, E. (2015). Drug repositioning for diabetes based on 'omics' data mining. *PLoS One, 10*(5), e0126082. https://doi.org/10.1371/journal.pone.0126082

CHAPTER 16

Design and Discovery of Kinase Inhibitors Using Docking Studies

TEODORA DJIKIC • ZARKO GAGIC • KATARINA NIKOLIC

1 INTRODUCTION

Protein kinases catalyze transfer of the γ-phosphate of ATP (adenosine triphosphate) to a protein substrate (alcoholic group of serine or threonine, or the phenolic group of tyrosine), through the following reaction:

$$MgATP^- + protein{-}O{:}H \rightarrow protein{-}O{:}PO_3^{2-} + MgADP + H^+$$

By phosphorylating the protein substrate, they mediate the signal transductions and regulate different cell activities, such as proliferation, survival, apoptosis, metabolism, transcription, differentiation, etc. Some phosphorylation sites on a substrate protein can be stimulatory while others can be inhibitory. The phosphorylation can increase or decrease activity of enzymes. Furthermore, it can affect association of substrate proteins with other proteins or their localization within the cell. Moreover, protein kinases can also modify other biological activities of proteins including transcription and translation (Cormier & Woodgett, 2016; Roskoski, 2015; Wu et al., 2015).

The human genome encodes at least 518 protein kinases (478 typical and 40 atypical). Accounting for around 2% of the human genome, they represent one of the most populated protein families (Manning, 2002). As they play vital role in almost every aspect of the cell activity, any alterations in their structure, function, and/or dynamics (mutations, overexpression, translocations, dysregulation, etc.) can lead to numerous diseases (Fabbro, 2015). Accordingly, they have become one of the central drug targets.

2 PROTEIN KINASES AS DRUG TARGETS

Until now, 52 kinase inhibitors have been approved by the FDA (Food and Drug Administration) for clinical use (Table 16.1), and many more are on clinical trials http://www.icoa.fr/pkidb/(Carles et al., 2018; Roskoski, 2019a). Majority of these kinase inhibitors have use in oncology, mainly in solid tumors including breast and lung cancers, followed by different types of leukemia. However, most of them are active against several types of cancer. Only eight kinase inhibitors are used in the treatment of noncancer diseases, such as myelofibrosis, rheumatoid arthritis, idiopathic pulmonary fibrosis, glaucoma, and against organ rejection (Fabbro, 2015; Roskoski, 2019a, 2019b). Besides anticancer effects, novel kinase inhibitors are now explored for therapy in other diseases such as inflammatory and autoimmune diseases, cardiovascular (hypertension), neurological (Parkinson's disease), etc. (Cohen & Alessi, 2013).

All of the approved and near-approval kinase inhibitors cover around 10% of the whole kinome (Fabbro, 2015). The most common targets of currently approved drugs are BCR-Abl tyrosine kinase, B-Raf serine-threonine kinase, vascular endothelial growth factor receptors (VEGFR), epidermal growth factor receptors (EGFR), and anaplastic lymphoma kinase ALK (Roskoski, 2019a). Moreover, most of them are multikinase inhibitors, and their therapeutic effectiveness might be the result of inhibition of several enzymes (Roskoski, 2019a). For example, imatinib was initially developed as an inhibitor of BCR-Abl tyrosine kinase fusion protein, specifically for the treatment of chronic myelogenous leukemia (CML). Afterward, it was discovered that it can be also used in gastrointestinal tumors because it inhibits the c-Kit receptor tyrosine kinase and chronic myelomonocytic leukemia because it inhibits the platelet-derived growth factor receptor (PDGFR) (Cohen & Alessi, 2013; Sawyers, 2003) Today, imatinib is used for the treatment of eight different diseases (Roskoski, 2019a).

Despite the great success in protein kinase drug discovery, two major hurdles still need to be overcome:
1. Low selectivity—The majority of approved kinase inhibitors bind the ATP kinase active site.

Molecular Docking for Computer-Aided Drug Design. https://doi.org/10.1016/B978-0-12-822312-3.00009-6

TABLE 16.1
Some of the FDA-Approved Protein Kinase Inhibitors With Their Primary Targets and Approved Indications.

FDA-Approved Protein Kinase Inhibitors	Chemical Structure	Primary Target	Indication
Dasatinib		BCR-Abl, Src	Chronic myelogenous leukemia
Erlotinib		Epidermal growth factor receptor (EGFR)	Non–small cell lung cancer (NSCLC), pancreatic cancer
Gefitinib		EGFR	NSCLC
Imatinib		BCR-Abl	Diverse types of leukemia and gastrointestinal stromal tumor
Nintedanib		FGFR1/2/3	Idiopathic pulmonary fibrosis

Osimertinib	EGFR, T970M	NSCLC
Ponatinib	BCR-Abl	Chronic myeloid leukemia, Philadelphia chromosome positive acute lymphoblastic leukemia
Sunitinib	VEGFR2	Renal cell carcinoma, gastrointestinal stromal tumor
Vemurafenib	B-Raf	BRAF V600-mutation positive melanoma

Accordingly, many of them have common core scaffolds and form similar interactions with the active site, which results in low selectivity (Maddox et al., 2016; Smyth & Collins, 2009). Sometimes lack of selectivity might lead to wanted therapeutic response, while in other cases, it can cause adverse side effects (Roskoski, 2019a).

2. Drug resistance—Mechanisms that lead to the resistance are diverse and specific; however, they all result in continued aberrant signaling, despite the application of appropriate inhibitor (Lovly & Shaw, 2014).

Considering these selectivity and resistance issues and the fact that only a small percentage of protein kinases have been explored as potential drug targets, this field offers a lot of potential for new drug discovery.

3 STRUCTURE OF PROTEIN KINASES

In 1988, Hanks et al. analyzed the sequences of 65 catalytic domains of protein kinases and identified conserved features between them (Hanks et al., 1988). From the alignment, they observed that the catalytic domains are not conserved homogeneously and that there are regions with high and low similarity. Based on this, they divided the catalytic domain into 11 conserved subdomains separated by regions of lower conservation. The conserved subdomains are important for catalytic function of enzymes. Amino acids of these regions either belong to the active site or contribute to the formation of the active site. Highly conserved amino acids play important role in catalysis. On the other hand, nonconserved subdomains, most probably, form flexible loops (Hanks et al., 1988). In conserved domain I, near the amino terminus, the GxGxxG (Gly-X-Gly-X-X-Gly) signature is found. Conserved domain II contains conserved lysine residue that appears to be involved in the phosphoryl transfer reaction. In 62 out of 65 aligned sequences, alanine was found two positions from lysine Ala-X-Lys. This conserved lysine residue also forms a salt bridge with the conserved glutamate residue that belongs to the domain III. In variable sequence of domain IV, three conserved hydrophobic residues (Leu, Ile, or Val) are found. Domain V contains another variable sequence.

The most conserved part of the catalytic domain is the central that contains subdomains VI to IX. Amino acid residues Asp and Asn in domain VI, and Asp, Phe, and Gly (DFG signature) of subdomain VII are involved in ATP binding. Additionally, domain VI contains an HRD (His-Arg-Asp) sequence, which forms part of the catalytic loop. In some cases, tyrosine residue replaces the histidine. Domain VIII contains a conserved APE motif (Ala-Pro-Glu). The DFG represents the beginning, while the APE motif represents the end of the protein kinase activation segment. Moreover, subdomains VI and VIII indicate substrate specificity. Serine/threonine protein kinases contain Asp-Leu-Lys-Pro-Glu-Asn, while the tyrosine kinases contain either Asp-Leu-Arg-Ala-Ala-Asn or Asp-Leu-Ala-Ala-Arg-Asn sequence. Domain IX contains a conserved aspartate and glycine residues (Hanks et al., 1988; Roskoski, 2015). Domain X contains variable sequence. Domain XI contains a conserved hydrophobic residue which forms a salt bridge with the glutamic acid from APE motif (Roskoski, 2015).

Few years after Hanks et al. analyzed the sequence, in 1991, Knighton et al. solved the crystal structure of cyclic AMP-dependent protein kinase in complex with a 20-amino acid inhibitor substrate analog (Protein Data Bank [PDB] accession code: 2CPK). Due to high sequence similarity and structure conservation, this crystal structure can be used to describe secondary and tertiary structure of all currently known protein kinases (Fig. 16.1). They described the typical architecture of protein kinase as bilobal—containing a small amino-terminal N-lobe and a large carboxy-terminal C-lobe, with a deep cleft between them filled by MgATP (Knighton et al., 1991). N-terminal lobe consists of five β-strands ($\beta1$–$\beta5$) and conserved α-helix (αC). The C-lobe is composed of eight α-helices (αD-I and αEF) and four short conserved β-strands ($\beta6$–$\beta9$) (Knighton et al., 1991; Roskoski, 2019b). ATP binding occurs in the cleft between the N- and C-terminal lobes, where hydrophobic residues form a pocket for adenine, while the charged residues bind the γ-phosphate (Knight et al., 2007). Both lobes contribute to its binding, but most of the interactions involve the N-lobe. Although C-lobe is mainly involved in protein–substrate binding, it also participates in ATP binding. Deep in the ATP pocket, there is "gatekeeper" residue which controls the entrance to the hydrophobic back cleft of the kinase (Fabbro, 2015; Kornev et al., 2006; Roskoski, 2015; Taylor & Kornev, 2011). As suggested by Hanks et al., in 1988, and later on confirmed by Knighton et al., in 1991, conserved residues play crucial roles in binding and positioning of ATP and catalytic mechanisms. They are mostly found in a vicinity of the active site, but can also reside in other parts of the protein kinase domain (Knight et al., 2007).

Glycine-rich GxGxxG motif, from domain I, belongs to the N-lobe and forms a conserved flexible loop between $\beta1$ and $\beta2$ (also called P-loop). It folds over the

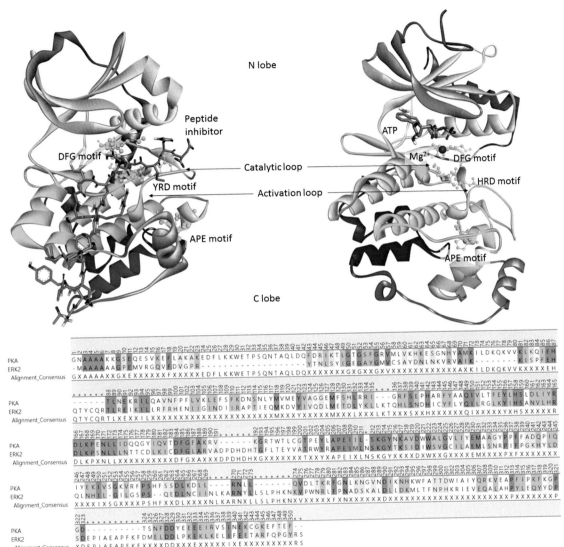

FIG. 16.1 The crystal structures of protein kinases (PKAs). cAMP-dependent PKA in complex with a peptide substrate mimicking inhibitor (PDB: 2CPK) (Knighton et al., 1991) (left) and ATP-bound form of the ERK2 kinase (PDB: 4GT3) with ATP and Mg^{2+} ion (right), colored as rainbow from N-lobe (red) to C-lobe (blue), and their sequence alignment (down) colored by alignment sequence similarity (dark green—identical, green—strong similarity, light green—weak similarity, white—nonmatching residues). Peptide and ATP are represented as stick. Residues of DGF, HRD, and APE motifs are represented as ball and stick.

ATP and positions γ-phosphate during the catalysis (Taylor & Kornev, 2011).

Lysine (from domain II, β3 strand) and glutamic acid (from domain III, αC helix) that belong to the N-lobe and two aspartic acids (from domain VI and VII) which are found in the C-lobe form conserved K/E/D/D (Lys/Glu/Asp/Asp) signature crucial for the catalysis. Lysine binds to the α- and β-phosphates of ATP.

Additionally, amino group of lysine residue forms a salt bridge with carboxylate group of aspartic acid. This salt bridge consolidates lysine's interactions with α- and β-phosphates of ATP and it is needed for kinase activation. The "first" aspartate positions the hydroxyl group of protein substrate and helps with the nucleophilic attack on the γ-phosphate of ATP. "Second" aspartate is the first residue of the DFG motif—the beginning of the activation loop (Roskoski, 2015, 2019a, 2019b).

H(Y) RD motif from domain VI belongs to the C-lobe and it is necessary for the orientation of the protein-substrate and transfer of the phosphoryl group. Aspartate residue interacts with hydroxyl group of the protein-substrate, while arginine residue interacts with the phosphorylated activation segment, thereby contributing to their proper orientation. It is thought that histidine (or tyrosine) residue is responsible for the upkeep of the structural position of the catalytic core (La Sala et al., 2016).

C-lobe also contains the activation loop—a flexible loop of 20–30 residues which can take open or closed conformation. As aforementioned, the activation loop begins with the DFG motif (domain VII) and extends up to an APE motif (domain VIII) (Hanks et al., 1988; Modi & Dunbrack, 2019). In the active conformation, aspartic acid residue from DFG motif interacts with Mg^{2+} (or Mn^{2+}) ion. It coordinates ATP binding and facilitates the phosphorylation (Adams, 2001). This segment, DFG motif and additional two residues after it, is also known as magnesium-binding loop (Kornev et al., 2006). At the other end of the activation loop, glutamic acid from APE motif is frozen by interaction with arginine from C-lobe and the formation of salt bridge (Roskoski, 2015, 2019b).

4 DYNAMICS OF PROTEIN KINASES

Thousands of crystal structures of protein kinases with various ligands, in both active and inactive conformations, have been deposited in the PDB database (https://www.rcsb.org/). This provided a deeper insight into their structure and function and a better understanding of their dynamics.

Typically, all protein kinases reside in a basal state and they are activated by a specific stimuli (Taylor et al., 2012). The exact mechanisms behind protein kinase activation and deactivation are specific for each kinase (Roskoski, 2015). Activation of protein kinases normally involves phosphorylation of a residue in the activation loop, which leads to its conformational rearrangement and increase in enzymatic activity (Kornev et al., 2006). In active conformation, the activation loop, most flexible part of the activation segment, is "opened." This conformation is also known as active, DFG-in conformation (Modi & Dunbrack, 2019).

Binding of a substrate peptide causes significant changes in the orientation of the αC-helix (from N-lobe) and the activation segment. When a substrate peptide binds, it interacts with the H(Y) RD motif.

Aspartate residue from this motif then orientates the hydroxyl group of the peptide substrate, while arginine supports the conformation of the activation segment (by linking the catalytic loop, phosphorylation site, and the magnesium-binding loop). Histidine/tyrosine binds to the aspartate from DFG motif which forms polar interactions with ATP phosphates, directly or through coordination of magnesium atom. Mg^{2+} ion interacts directly with an oxygen atom of the ATP β-phosphate. Furthermore, phenylalanine forms hydrophobic interactions with the HRD motif and αC-helix (Kornev et al., 2006; Taylor & Kornev, 2011). The phenylalanine is now positioned against or under the αC-helix to allow proper positioning of the DFG aspartate and αC-helix and to facilitate contact between lysine (from β3) and glutamic acid (from αC-helix). Formation of the salt bridge between lysine and glutamic acid is followed by formation of hydrogen bonds between lysine and oxygen atoms of α- and β-phosphates of ATP. This way, a hydrophobic pocket that engulfs the ATP's adenine ring is formed by the N-lobe. Additionally, glycine-rich loop covers ATP and positions its γ-phosphate for the phosphoryl transfer (Kornev et al., 2006; Modi & Dunbrack, 2019; Taylor & Kornev, 2011).

Protein kinases have numerous inactive conformations. Based on the spatial position of the phenylalanine side chain from DFG motif, inactive conformations can be DFG-in, DFG-out, and DFG-inter (Fig. 16.2). In majority of inactive conformations, the activation loop is "closed" and it is blocking the substrate binding. The DFG motif is now inconsistent with binding ATP and Mg^{2+} ion. In DFG-out conformation, phenylalanine is moved into the ATP binding pocket, therefore blocking its binding. DFG-inter represents the conformations in which the DFG-Phe side chain is out of the C-helix dividing the active site into two halves (Modi & Dunbrack, 2019).

Furthermore, protein kinases are also regulated in other ways, such as autoinhibition, interaction with regulatory subunits, binding of an allosteric modulators, etc. Binding of a modulator to an allosteric site leads to conformational changes of large group of residues resulting in the rearrangement of the activation loop (Kornev & Taylor, 2015; Shi et al., 2006). These modulators can act as effectors or inhibitors and can bind different protein kinase domains (Jahnke et al., 2010; Ohren et al., 2004; Rettenmaier et al., 2015; Shi et al., 2006; Vanderpool et al., 2009). Consequently, mechanisms underlying allosteric communications are very diverse and have to be analyzed on an individual basis (Kornev & Taylor, 2015).

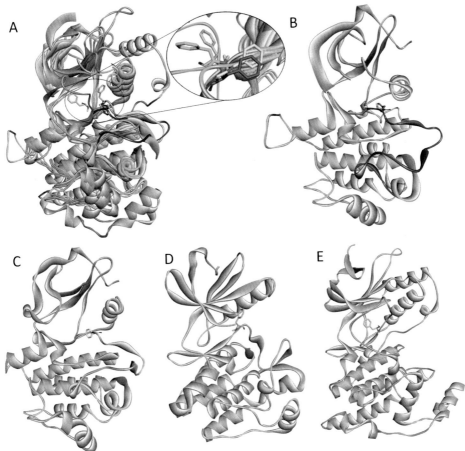

FIG. 16.2 Dynamics of protein kinases. **(A)** Alignment of one active (DFG-in) and different inactive (DFG-in, DFG-inter, and DFG-out) conformations and spatial position of phenylalanine residue from DFG motif; phenylalanine is represented as stick. **(B)** Insulin receptor kinase in complex in active DFG-in conformation (PDB: 3BU5) (Wu et al., 2008); **(C)** ABL1 kinase in inactive DFG-in conformation (PDB: 4TWP) (Pemovska et al., 2015); **(D)** BMX nonreceptor tyrosine kinase in DFG-inter inactive conformation (PDB: 3SXR) (Muckelbauer et al., 2011); **(E)** JNK2 kinase in inactive DFG-out conformation (PDB: 3NPC) (Kuglstatter et al., 2010).

5 TYPES OF INHIBITORS

Small molecule kinase inhibitors could bind the target in different ways. Basically, kinase inhibitors can be classified into noncovalent (types I–V) and covalent inhibitors (type VI). Noncovalent inhibitors can then be separated into competitive and noncompetitive, with respect to ATP. ATP competitive inhibitors can be further divided based on conformation of protein kinase they bind. Classical examples of different types of inhibitors are given in Table 16.2 and Fig. 16.3.

Type I inhibitors act as ATP mimetics. They bind in the ATP-binding pocket of the active DFG-in conformation (Fig. 16.3A). Due to ATP-binding site being conserved in all protein kinases, it may seem that these inhibitors have low selectivity and cause numerous side effects. Regardless, the majority of approved protein kinase inhibitors fall into this group. The differences within the ATP-binding pocket, such as size, shape, and polarity of side chain in the "gatekeeper" residue, can be used to improve selectivity of these inhibitors (Fabbro, 2015; Roskoski, 2016; Zuccotto et al., 2010).

Type I1/2 inhibitors also bind in the ATP-binding pocket, but to an inactive, DFG-in conformation.

Based on the part of ATP-binding pocket they occupy, type I1/2 inhibitors are further divided into two subgroups I1/2A and B. Inhibitors from subgroup

TABLE 16.2

Classification of the FDA-Approved Kinase Inhibitors. Some of the Inhibitors Could act as Cross-type Depending on the Kinase for Which They are Binding.

Type I	Alvocidib, bosutinib, brigatinib, cabozantinib, ceritinib, dasatinib, erlotinib, fedratinib, gefitinib, palbociclib, pazopanib, ponatinib, ruxolitinib, tofacitinib, vandetanib	
Type I1/2	I1/2A	Dabrafenib, dasatinib, erdafitinib, lapatinib, lenvatinib, palbociclib, vemurafenib
	I1/2B	Abemaciclib, alectinib, bosutinib, ceritinib, crizotinib, entrectinib, erlotinib, ribociclib
Type II	IIA	Axitinib, imatinib, nilotinib, Pexidartinib, ponatinib, regorafenib, sorafenib
	IIB	Bosutinib, nintedanib, sunitinib
Type III	Cobimetinib, trametinib	
Type IV	Everolimus, sirolimus, temsirolimus	
Type V	—	
Type VI	Afatinib, ibrutinib	

I1/2A bind to the front cleft gate area, as well as the back cleft (Fig. 16.3B), while subgroup I1/2B inhibitors occupy only the gate area and front cleft of the ATP-binding pocket (Fig. 16.3C). Hence, I1/2A are more selective because they bind back cleft hydrophobic pocket whose entrance is controlled by the "gatekeeper" residue (Fabbro, 2015; Roskoski, 2016).

Type II kinase inhibitors bind to the ATP-binding pocket, as well, but to an inactive, DFG-out conformation. In DFG-out state, an additional hydrophobic pocket in close proximity to the ATP site is exposed. Type II inhibitors occupy this hydrophobic pocket and lock the kinase in the inactive conformation. Even though they occupy additional pocket, their selectivity is similar to the selectivity of type I kinase inhibitors. In the same manner as with type I1/2, they are subsequently divided into subgroups IIA and IIB. Type IIA inhibitors bind the front cleft, gate area, and the back cleft (Fig. 16.3D), while type IIB bind only the front cleft and gate area (Fig. 16.3E) (Fabbro, 2015; Roskoski, 2016).

Type III and IV kinase inhibitors are allosteric, ATP-noncompetitive kinase inhibitors Type III inhibitors dock to a site near by the ATP-binding pocket (Fig. 16.3F), while type IV inhibitors dock to sites distant from the substrate binding sites (Fig. 16.3G) (Fabbro, 2015; Roskoski, 2016). Unlike previous types, they do not form conserved interactions with residues of ATP-binding pocket and therefore show highest selectivity (Roskoski, 2016). Additionally, drug resistance caused by mutations in the ATP-binding site was reported in almost all of ATP-competitive inhibitors. Therefore, overcoming these mutations could be another advantage of developing allosteric kinase inhibitors (Gibbons et al., 2012).

Type V inhibitors are bivalent/bisubstrate inhibitors that occupy two different regions of the kinase (Fig. 16.3H) (Roskoski, 2016). Designing the type V inhibitors, in which an ATP mimicking scaffold is linked to a scaffold, targeting a site outside of the ATP-binding pocket, became very popular in the past decade, as a potential solution for the selectivity problems (Gower et al., 2014; Hill et al., 2012).

Type VI inhibitors bind covalently to their protein kinase target. They may be both reversible and irreversible inhibitors (Fig. 16.3I). Majority of these inhibitors bind to the ATP site to form a covalent bond to a cysteine residue near the active site, disabling the binding of ATP to its pocket (Fabbro, 2015; Roskoski, 2016).

Besides small molecule kinase inhibitors, there are also peptide inhibitors. Peptide inhibitors are designed to disrupt protein—protein interactions and in that way block the phosphorylation activity. Commonly, they are synthetic peptides that mimic the substrate, but they can be also derived from other interacting proteins (endogenous inhibitory proteins, anchoring, or scaffold proteins) or even from protein kinase sequence itself (Bogoyevitch et al., 2005; Eldar-Finkelman & Eisenstein, 2009). Even though these inhibitors have some advantages comparing to

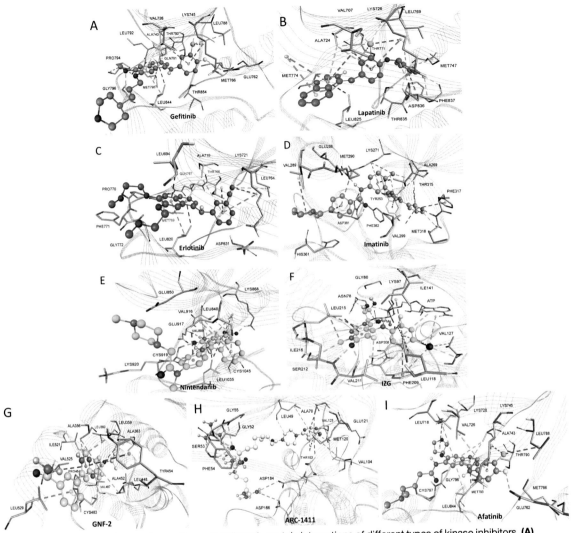

FIG. 16.3 Binding modes and important ligand–protein interactions of different types of kinase inhibitors. **(A)** Type I inhibitor: epidermal growth factor receptor (EGFR) in complex with type I inhibitor gefitinib (PDB:4WKQ) (Yosaatmadja et al., To be Published); **(B)** type I1/2A: ErbB4 kinase in complex with lapatinib (PDB:3BBT) (Qiu et al., 2008); **(C)** type I1/2B: EGFR tyrosine in complex with erlotinib (PDB: 4HJO) (Park et al., 2012); **(D)** type 2A: Abl kinase in complex with imatinib (PDB: 2HYY) (Cowan-Jacob et al., 2007); **(E)** type 2B: VEGFR2 in complex with nintedanib (PDB: 3C7Q) (Hilberg et al., 2008); **(F)** type III: MEK 1 in complex with ligand IZG and MgATP (PDB: 3PP1) (Dong et al., 2011); **(G)** type IV: Abl kinase in complex with allosteric inhibitor GNF-2 (PDB:3K5V) (Zhang et al., 2010); **(H)** type V: PKA with the bisubstrate inhibitor ARC-1411 (PDB: 5IZJ) (Ivan et al., 2016); (I) type VI: EGFR kinase in complex with irreversible inhibitor afatinib (PDB:4G5J) (Solca et al., 2012).

small molecule kinase inhibitors (such as specificity and safety) due to their pharmacokinetic properties (metabolic instability, difficulty to cross cell membranes, potential to provoke immune response, low oral bioavailability), they have been generally disregarded in drug design. However, with major advancements in pharmaceutical technology, it is now possible to improve the bioavailability and stability of peptide inhibitors through chemical modifications and/or by using different formulation techniques (Eldar-Finkelman & Eisenstein, 2009; Renukuntla et al., 2013; Wójcik & Berlicki, 2016).

6 MOLECULAR DOCKING IN DESIGN AND DISCOVERY OF KINASE INHIBITORS

In recent years, structure-based drug design methods and among them molecular docking have been extensively used in developing kinase inhibitors (Gagic et al., 2020). Since the approval of imatinib and recognition of protein kinases as important targets for the development of anticancer drugs, a great number of high-resolution X-ray crystal structures of protein kinases have been reported in PDB and became accessible for use in structure-based drug design, especially in docking studies. Many of them are available in complex with known inhibitors, and investigation of their binding modes represents valuable resource for assessing structural determinants that ligand should poses in order to achieve key interactions with active site residues. This information is further exploited for lead optimization and design of new, more potent, and selective kinase inhibitors.

Most of the approved kinase inhibitors are ATP-competitive, which means they bind in the ATP pocket of kinase active DFG-in conformation (type I inhibitors such as gefitinib), but can also extend to a neighboring hydrophobic pocket that is formed in kinase inactive DFG-out conformation (type II kinase inhibitors such as imatinib). Therefore, it is not surprising that in majority of docking studies ligand is placed into a highly conserved ATP binding pocket where it can form different bonding interactions (H-bond, electrostatic, hydrophobic, van der Waals, $\pi-\pi$ interactions) with functionally important amino acid residues of kinase active site. Hinge region, which connects the N- and C-terminal lobes and forms hydrogen bonds with the adenine ring of ATP, has been particularly identified as important for achieving key hydrogen bond interactions with H-bond donors and acceptors that are present in ligand heterocycles. Examples of heterocycles that have ability to bind the hinge region can be seen in FDA-approved kinase inhibitors and include pyridine (imatinib), indole (sunitinib), quinazoline (gefitinib), thiazole (dasatinib), pyridopyrazole (vemurafenib) (Ghosh & Gemma, 2014). Because the ATP binding site is conserved among kinases, they have the tendency to exert toxic reactions due to the low selectivity and this is where the use of in silico chemistry, especially docking methods, can guide medicinal chemist to design inhibitors that can be both potent and selective.

Low selectivity profile of type I and II kinase inhibitors has directed docking studies to also investigate binding of ligands to an allosteric and less conserved non-ATP binding site. This strategy has been shown successful not only in identification of more selective, less

toxic kinase inhibitors (type III) but also in overcoming mutation-related drug resistance (Gibbons et al., 2012). However, the success did not come without challenges such as identification of allosteric pockets that are often poorly accessible and susceptibility of allosteric ligands to switching receptor signaling to different pathways with only subtle structural changes (this is called mode switching or functional switching).

6.1 Targeting RAF/MEK/ERK Pathway

Mitogen-activated protein kinases (MAPKs) are a family of serine/threonine kinases that are involved in the RAS—RAF—MEK—ERK signaling pathway (Fig. 16.4); abnormal activity in this pathway is often found in many types of human cancer (McCubrey et al., 2007). Members of the MAPK family include ERK, p38, and c-Jun N-terminal protein kinases (JNKs) that are involved in cell proliferation and along with their activator MAPKK (mitogen activated protein kinase, also known as MEK) are frequently used target for anticancer drugs.

Vemurafenib, sorafenib, and dabrafenib act on B-RAF^{V600E}, the most common mutation of RAF that results in the RAS-independent upregulation of ERK pathway and is found in more than 50% human melanomas (Davies et al., 2002). However, they can induce dimerization of wild-type B-RAF and subsequent phosphorylation of MEK, leading to activation of ERK pathway and cell proliferation (Cox & Der, 2010; Durrant & Morrison, 2018). Overexpression of C-RAF is often found to contribute to amplification of this pathway (Johannessen et al., 2010). To avoid this paradox, Jung et al. have recently applied docking studies to develop a new class of inhibitors that selectively bind to B-RAF^{V600E} and C-RAF, but not to the wild-type B-RAF (Jung et al., 2019). They synthetized

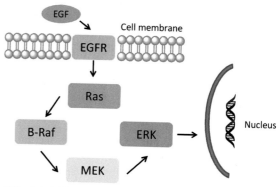

FIG. 16.4 The RAS—RAF—MEK—ERK signaling pathway.

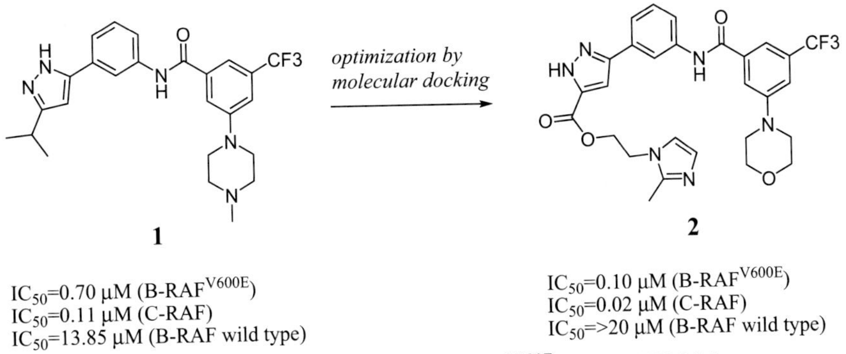

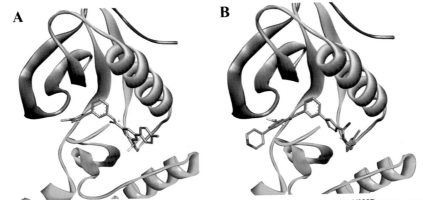

IC$_{50}$=0.70 μM (B-RAF^{V600E})
IC$_{50}$=0.11 μM (C-RAF)
IC$_{50}$=13.85 μM (B-RAF wild type)

IC$_{50}$=0.10 μM (B-RAF^{V600E})
IC$_{50}$=0.02 μM (C-RAF)
IC$_{50}$=>20 μM (B-RAF wild type)

FIG. 16.5 Optimization of selective B-RAF^{V600E} and C-RAF inhibitors.

A   **B**

FIG. 16.6 Docking mode of **(A)** compound **1** and **(B)** compound **2** on B-RAF^{V600E} (PDB: 4G9R).

a series of N-(3−(3-alkyl-1H-pyrazol-5-yl) phenyl)-aryl amide and urea analogs, with compound **1** (Fig. 16.5) exerting best potency and selectivity toward the B-RAF^{V600E} (IC$_{50}$ = 0.70 μM) and C-RAF (IC$_{50}$ = 0.11 μM) over other protein kinases and wild-type B-RAF (IC$_{50}$ = 13.85 μM). Investigation of compound **1** docking mode in B-RAF^{V600E} revealed H-bond interaction of pyrazole hydrogen with Thr529, π-cation interaction of piperazine with His574, and ionic interaction between protonated piperazine and Asp594, with piperazinyl-trifluoromethyl-phenyl moiety located in hydrophobic pocket formed by activation loop. In contrast, docking of **1** in wild-type B-RAF revealed only one H-bond interaction of pyrazole with Thr528, while piperazinyl-trifluoromethyl-phenyl moiety was located outside of the hydrophobic pocket, thus explaining the selectivity toward the B-RAF^{V600E}. In a follow-up study, authors further suggested that addition of carbonyl to pyrazole heterocycle could increase activity to B-RAF^{V600E} through additional interaction with Cys532 at the hinge region that was lacking in

compound **1** (Kim et al., 2019). They designed a new series of 3-carbonyl-5-phenyl-1H-pyrazole analogs (Fig. 16.5) and conducted docking studies on RAF kinases.

Docking scores were better, compared to compound **1** series, and more importantly, difference in docking scores between mutated and wild-type B-RAF was much greater, implicating improved selectivity. Binding mode confirmed intended interaction of carbonyl with hinge Cys532 and also additional π−π interaction with Phe595 of the activation loop (Fig. 16.6). Designed analogs were then synthesized and tested against RAF kinases and compound **2** showed excellent activity on B-RAF^{V600E} and C-RAF, suggesting a new direction toward development of efficient and selective RAF inhibitors that can avoid paradoxical ERK signaling activation.

Kaieda et al. employed high-throughput screening to identify lead compound **3** that showed moderate activity on p38 MAP kinase (IC$_{50}$ = 140 nM) (Kaieda et al., 2019). Co-crystallization and investigation of binding

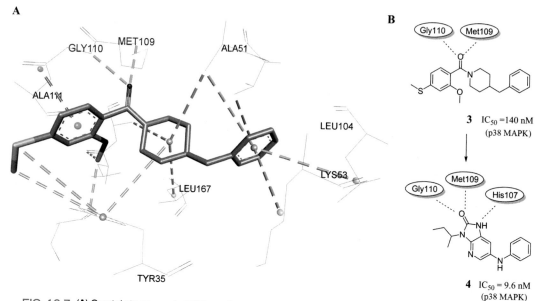

FIG. 16.7 **(A)** Crystal structure of p38 kinase in complex with compound **3** identified by HTS. **(B)** Discovery of new p38 inhibitor **4** by scaffold hopping.

mode showed that carbonyl group of **3** formed two hydrogen bonds with the Met109 and Gly110 at the hinge region (Fig. 16.7). They also noted a peptide bond flip between Met109 and Gly110, which is not usually seen in protein kinases, and hypothesized that this flip could be exploited by cyclization of carbonylpiperidine moiety present in compound **3**. After exploration of several scaffolds that contain hydrogen bond acceptors able to interact with Met109 and Gly110 and form additional interaction with His107, they finally designed and synthesized a series of imidazo [4,5-b]pyridin-2-one derivatives (Fig. 16.7). Compound **4** was identified as a potent inhibitor of p38 MAPK kinase ($IC_{50} = 9.6$ nM) and proposed as a lead candidate for further studies.

6.2 EGFR Kinase Inhibitors

The EGFR family of tyrosine kinases that consists of four structurally related members (erbB1-4 or HER1-4) is one of the earliest recognized and highly successful targets in the development of kinase inhibitors (Bach et al., 2020; Mendelsohn & Baselga, 2006). Overexpression or mutation of these kinases is often found in patients with non−small cell lung cancer (NSCLC), HER-2−positive breast cancer, colon cancer, and others (Gristina et al., 2020; Maennling et al., 2019; Thomas & Weihua, 2019). Majority of the approved EGFR inhibitors, such as gefitinib, erlotinib, lapatinib, and afatinib, share aminoquinazoline

scaffold that occupies the ATP binding pocket where it forms key H-bond with Met769, while the lipophilic moiety connected with amine linker extends to the neighboring selectivity pocket (Stamos et al., 2002). Modifications of this scaffold have been used as a successful strategy in structure-based design of many promising EGFR inhibitors (Alkahtani et al., 2020; Allam et al., 2020; Bathula et al., 2020; Le et al., 2020). A series of quinazoline derivatives were recently synthesized and subjected to molecular docking studies in order to investigate binding mode to EGFR kinase (Khodair et al., 2019). Based on the excellent binding affinities and ability to interact with Met769 hinge residue, which is essential for EGFR inhibitory activity, compounds **5−9** were selected for further in vitro studies (Fig. 16.8). Compound **9**, which formed two H-bonds with Met769, inhibited survival of MCF-7 and HepG2 cancer cells in low concentrations ($IC_{50} = 2.09$ and 2.08 μM, respectively) and was able to activate p53-related proapoptotic genes.

Recently synthesized derivatives of triazolo[4,3-c] quinazoline showed comparable EGFR inhibitory activity ($IC_{50} = 0.69-1.8$ μM) with gefitinib ($IC_{50} = 1.74$ μM). Docking of the most active compounds revealed key H-bond interactions of quinazoline N1 nitrogen with Lys745 and important hydrophobic interaction of quinazoline ring with Leu858. Good alignment was also observed with reference drug gefitinib (Ewes et al., 2020).

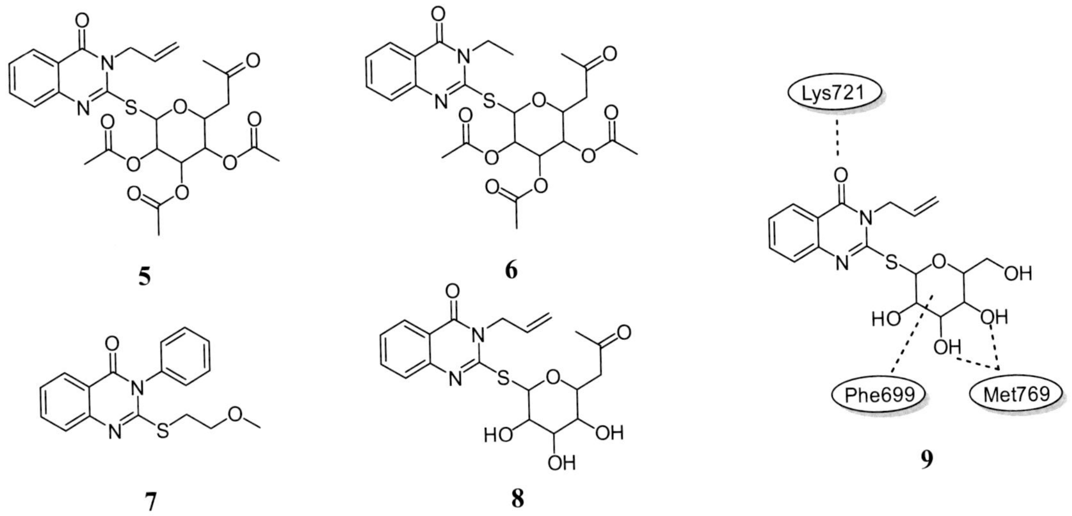

FIG. 16.8 Structures of quinazoline derivatives **5–9** and predicted interactions of **9** with EGFR binding site residues (PDB: 1M17).

George et al. reported that some 1,3,5-trisubstituted pyrazoline analogs display promising anticancer activity against breast cancer MCF7 cell line via EGFR inhibition (George et al., 2020). They performed molecular docking study in order to explain experimental results and found that the most active compound **10** also showed highest affinity to EGFR. Investigation of binding mode of **10** revealed key H-bond interaction of pyrazoline nitrogen with Met769, H-bond interaction of carboxamide substituent with Gln767, and hydrophobic interactions of methoxyphenyl moiety with Val702, Ala719, and Leu820 (Fig. 16.9).

Natural products isolated from plants, such as alkaloids, flavonoids, and polyphenols, have been proved as potential cancer agents, and in recent years, the number of newly discovered natural compounds has significantly increased (Lichota & Gwozdzinski, 2018). A group of researchers designed several analogs of chrysin, a flavone found in various plants and honey that has been derivatized and studied for anticancer activity (Debnath et al., 2019). They performed virtual screening of designed library using a 3D QSAR model developed from known EGFR inhibitors and then performed docking on the obtained hits. Three compounds were selected for synthesis based on their docking score, and in vitro testing of compound **11** (Fig. 16.9) showed selective binding and decreased phosphorylation of EGFR. Docking analysis indicated similar interactions as known EGFR inhibitors with residues such as Met793, Gln791, Leu718, Asp855, Lys745, and Phe856.

6.3 VEGFR Kinase Inhibitors

VEGFR family (VEGFR-1, 2, and 3) has been identified as the most important target for antiangiogenesis therapy in various types of cancers, including renal, gastrointestinal, hepatocellular, and leukemia. Several VEGFR inhibitors that are approved for clinical use (lenvatinib, sorafenib, sunitinib, etc.) are multitarget inhibitors, in which lack of selectivity might result in serious side effects (Kamba & McDonald, 2007). To achieve better safety, Jiang et al. discovered new highly selective VEGFR-2 inhibitor by screening of an in-house kinase inhibitor library (Jiang et al., 2020). Compound **12** displayed potent inhibitory activity against VEGFR-2 (Fig. 16.10) and high selectivity among structurally related kinases. It efficiently inhibited VEGF-induced angiogenesis in vitro using HUVEC cells and in vivo embryonic angiogenesis in zebrafish models, while exhibiting low acute toxicity in comparison to sunitinib. Results of docking study revealed that compound **12** binds to VEGFR-2 in a similar manner as type II kinase inhibitors. The indazole amine moiety formed H-bonds with Cys919 and Glu917 in the hinge region, while methyl benzamide moiety interacted with Glu885 and Asp1046.

Compound **13** was designed and synthesized as a part of novel class of 5-anilinoquinazoline-8-nitro derivatives that inhibit VEGFR-2 (Fig. 16.10). It exerted potent antiangiogenesis ability and the efficient cytotoxic activities in vitro against HUVEC ($IC_{50} = 0.58\ \mu M$) and HepG2 ($IC_{50} = 0.23\ \mu M$, same as ponatinib and superior to sorafenib) cells. Docking

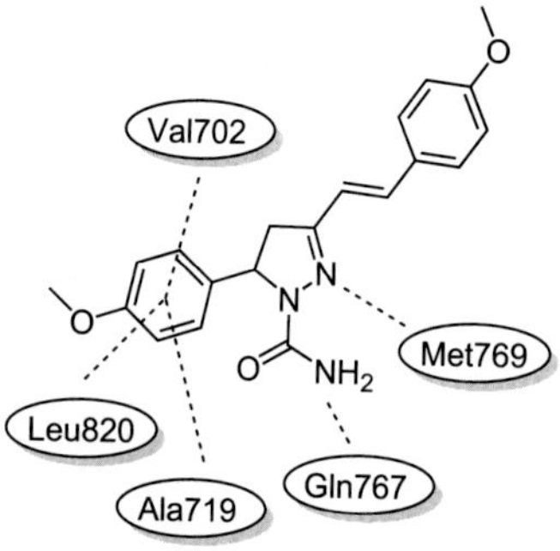

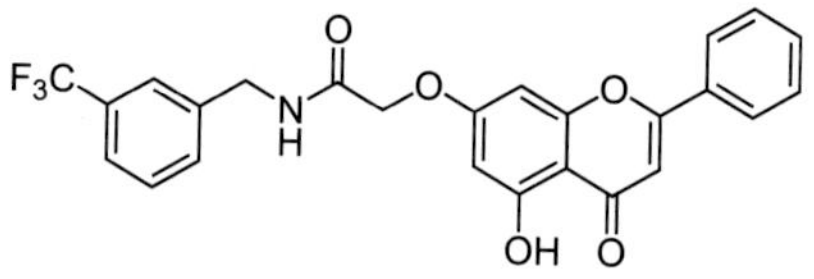

10

IC$_{50}$=0.33 µM (EGFR)

11

FIG. 16.9 Structures and binding interactions of EGFR inhibitors **10** and **11**.

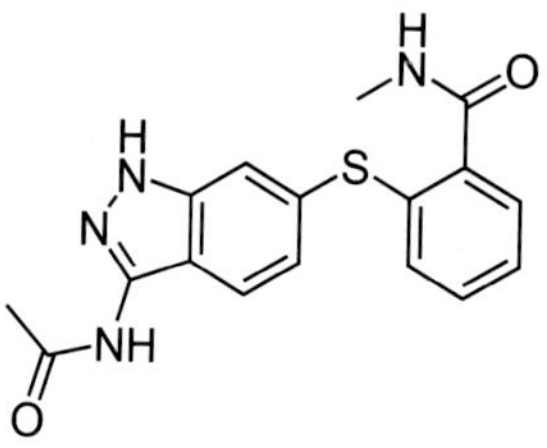

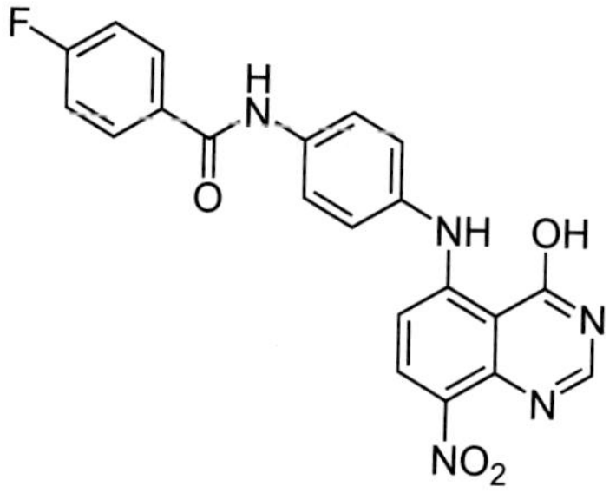

12

IC$_{50}$ = 66 nM (VEGFR-2)

13

IC$_{50}$ = 64.8 nM (VEGFR-2)

FIG. 16.10 Structures of VEGFR-2 inhibitors.

study revealed binding mode typical for type II kinase inhibitors. Quinazoline scaffold occupied the ATP binding region and formed three H-bonds with Cys919 and 4-fluorobenzoyl moiety formed π–π interactions with residues in the neighboring hydrophobic pocket (Zhao et al., 2012).

Simultaneous inhibition of structurally related angiokinases, such as VEGFR, PDGFR, and FGFR, creates opportunity to identify inhibitors with more potent antiangiogenesis action that is seen in clinically approved indoline analog nintedanib. Qin et al. in a recent study reported discovery of compound **14** (Fig. 16.11) bearing oxoindoline scaffold as a potent multikinase inhibitor (Qin et al., 2020). In addition to the strong inhibition of VEGFR, PDGFR, and FGFR, compound **14** displayed excellent activity on SRC and c-KIT kinases and moderate activity on LYN kinase,

while maintaining certain degree of favorable selectivity against B-RAF, cyclin-dependent kinase 2 (CDK2), CDK4, and CHK kinases. It was discovered by docking study that methoxycarbonyl indoline moiety formed three H-bonds with Glu917, Cys919, and Lys868 residues in the hinge region and that ionic interaction between piperazinyl moiety and Glu850 was crucial for high VEGFR-2 kinase affinity.

6.4 JAK Kinase Inhibitors

Janus kinases (JAK) family consists of four members (JAK1, JAK2, JAK3 and TYK2) that play pivotal in the downstream signaling of inflammatory cytokines and multiple growth factors. Currently approved JAK inhibitors (ruxolitinib, tofacitinib, fedratinib, etc.) are primarily used to treat hematopoietic malignancies, inflammatory, and autoimmune disorders that are linked

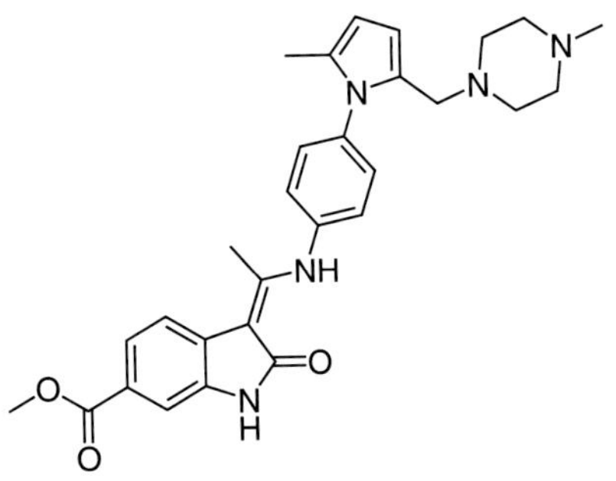

IC_{50} = 31 nM (VEGFR-1)
IC_{50} = 6.5 nM (VEGFR-2)
IC_{50} = 6.3 nM (VEGFR-3)
IC_{50} = 7 nM (PDGFRα)
IC_{50} = 9.9 nM (PDGFRβ)
IC_{50} = 23 nM (FGFR−1)
IC_{50} = 178 nM (SRC)
IC_{50} = 16 nM (LYN)
IC_{50} = 17 nM (c-KIT)

14

FIG. 16.11 Structure of indolinone multitarget kinase inhibitor.

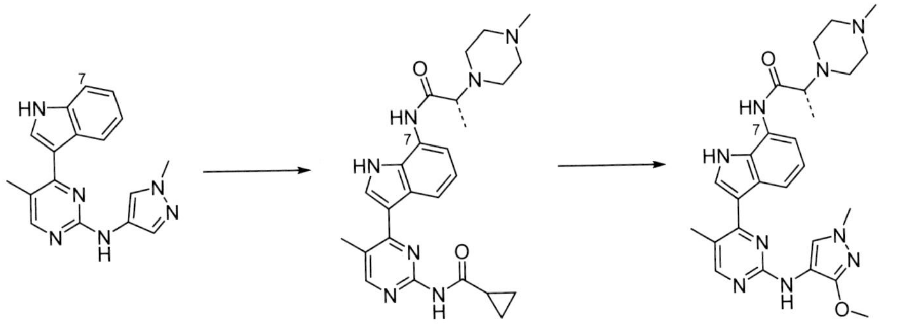

15

IC_{50} = 0.07 μM (JAK1)
IC_{50} = 0.38 μM (JAK2)

16

IC_{50} = 0.18 μM (JAK1)
IC_{50} = >30 μM (JAK2)

17

IC_{50} = 0.07 μM (JAK1)
IC_{50} = >15 μM (JAK2)

FIG. 16.12 Discovery of potent and highly selective JAK1 inhibitor **17**.

to abnormal cytokine stimulation and alteration of the JAK/STAT signaling pathway (Vainchenker et al., 2018). Undesirable side effects associated with the use of these inhibitors, such as thrombocytopenia and anemia, are believed to be the result of JAK2 inhibition (Kettle et al., 2017). Thus, Su et al. employed library screening and structure-based design in attempt to develop selective JAK1 kinase inhibitor (Su et al., 2020). They performed high-throughput screening of AstraZeneca compound collection and identified compound **15** as a potent JAK1 inhibitor with moderate selectivity over JAK2 (Fig. 16.12). Investigation of the docked pose of **15** onto the X-ray structure of JAK1 showed binding of indole NH with Asp1021 and pyridine with Leu959 in the hinge region. They concluded that through substitution of indole at position 7, a favorable π−π interaction with His885 in the JAK1 selectivity

pocket could be achieved, so they designed and synthesized a series of 7-substituted indole analogs.

Subsequent structural modifications and in vitro testing against kinase panel led to identification of compound **16** that had a favorable balance of JAK1 potency and selectivity. Looking into X-ray structure of **16** bound to JAK1, they discovered additional salt bridge interaction between the piperazine and Asp1003 that contributed to good inhibitory activity, so they kept piperazine moiety and continued optimization in order to further increase selectivity and potency (Fig. 16.13). Finally, compound **17** displayed high inhibitory activity and selectivity toward JAK1 (214 folds over JAK2) and also enhanced antitumor activity in combination with osimertinib in NSCLC xenograft model.

Structural optimization of a dual SYK/JAK inhibitor cerdulatinib (**18**) led to discovery of selective JAK3 inhibitor with diaminopyrimidine-carboxamide scaffold as a potential candidate for the treatment of rheumatoid arthritis (Bahekar et al., 2020). Docking studies of cerdulatinib showed favorable bonding interactions in the hinge region of JAK3, but not in the catalytic region. Thus, modifications were made in attempt to induce noncovalent hydrogen binding with Cys909 in the catalytic domain of JAK3 in order to achieve selectivity. Initial attempts included changes in position 2 of pyrimidine ring that led to discovery of compound **19** with potent activity, but moderate selectivity (Fig. 16.14). Further modification in position 4 of pyrimidine ring resulted in design of highly potent and selective compound **20** that showed additional H-bond interaction with Cys909 and promising efficacy and safety in animal models of rheumatoid arthritis.

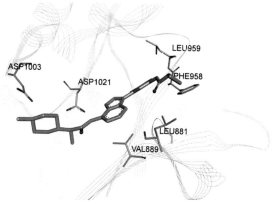

FIG. 16.13 X-ray structure of JAK1 (PDB: 6GGH) in complex with compound **16**.

7 STRUCTURE-GUIDED STRATEGIES FOR OVERCOMING KINASE INHIBITORS RESISTANCE

Several cell mechanisms can be involved in developing resistance to kinase inhibitors such as point mutation in kinase domain that impairs inhibitor binding, activation of downstream or parallel signaling pathways, and modification of drug transport and stability. Point mutations in the BCR-ABL kinase domain are the main reason for imatinib resistance in patients with CML. These mutations include different amino acid substitution in the active site or mutation of amino acids such as Tyr253 that destabilize inactive kinase conformation required for binding of imatinib. Substitution of Thr315 in the gatekeeper region of BCR-ABL with a bulkier and more hydrophobic isoleucine eliminates a critical hydrogen bonding interaction required for high-affinity binding and blocks entry to the hydrophobic pocket near Thr315. Information obtained from imatinib binding mode guided development of nilotinib, dasatinib, bosutinib, and ponatinib to overcome imatinib resistance and most are approved for the second-line treatment of CML.

Dasatinib is the second-generation BCR-ABL inhibitor developed shortly after approval of imatinib and currently only approved BCR-ABL inhibitor able to bind to the active kinase conformation (Hochhaus & Kantarjian, 2013; Shah et al., 2004). Nilotinib was designed to better exploit imatinib binding hydrophobic pocket by addition of trifluoromethyl group and substitution of piperazine with imidazole, while preserving benzamide pharmacophore to retain H-bond interactions with Glu286 and Asp381 (Weisberg

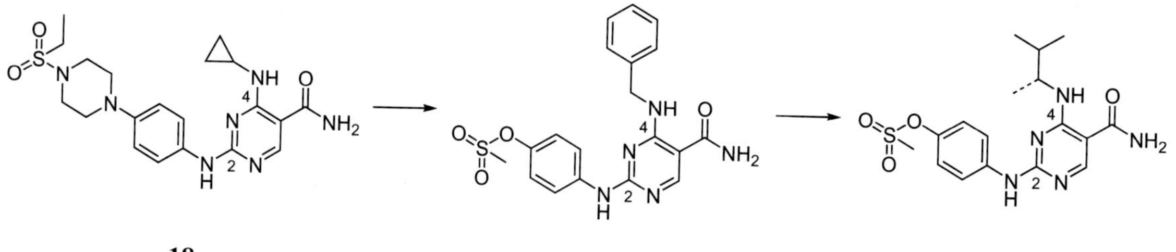

18

IC_{50} = 15 nM (JAK1)
IC_{50} = 7 nM (JAK2)
IC_{50} = 8 nM (JAK3)
IC_{50} = 5 nM (TYK2)

19

IC_{50} = 18 nM (JAK1)
IC_{50} = 42 nM (JAK2)
IC_{50} = 9.5 nM (JAK3)
IC_{50} = 45 nM (TYK2)

20

IC_{50} = 20 nM (JAK1)
IC_{50} = 171 nM (JAK2)
IC_{50} = 1.7 nM (JAK3)
IC_{50} = 186 nM (TYK2)

FIG. 16.14 Discovery of potent and selective JAK3 inhibitor **20**.

et al., 2006). Although nilotinib share similar binding mode to imatinib, because of the improved affinity, it is able to inhibit most mutant forms resistant to imatinib. However, nilotinib, dasatinib, and bosutinib were still not able to inhibit T315I mutant BCR-ABL, which confers resistance to all three second−generation drugs. Structure-based drug design led to discovery of ponatinib, the next-generation inhibitor that was able to combat T315I-related resistance (O'Hare et al., 2009; Zhou et al., 2011). Ponatinib was designed by combination of structural features present in imatinib (piperazine moiety) and nilotinib (trifluoromethyl group), which ensured binding to native BCR-ABL kinase, but the key to the success in overcoming resistance was replacement of amine linker between benzamide and pyridine with an extended alkyne moiety. This triple bond linker enabled access to specificity pocket that was restricted to imatinib and nilotinib by bulkier isoleucine gatekeeper residue while also producing favorable hydrophobic interaction with this residue, resulting in inhibition of T315I mutant (Fig. 16.15).

Similar mechanism of resistance is seen in EGFR kinases where the gatekeeper Thr790 is replaced with more hydrophobic methionine, causing the steric hindrance and increased affinity to ATP, consequently limiting efficacy of the first-generation ATP-competitive inhibitors such as gefitinib and erlotinib. Second-generation inhibitors afatinib and dacomitinib were designed to covalently target Cys797 via introduction of an acrylamide Michael acceptor, which resulted in improved potency. However, it failed to demonstrate clinical efficiency in patients harboring T790M mutation, due to simultaneous inhibition of wild-type EGFR and required high doses that led to severe adverse effects (Karachaliou et al., 2018). Third-generation EGFR inhibitor osimertinib was the first approved

T790M mutant-selective inhibitor (Butterworth et al., 2017). The rationale behind its design was to introduce groups that will selectively react with the more hydrophobic methionine of the T790M over the more hydrophilic threonine of the wild-type EGFR, while keeping the acrylamide moiety to ensure covalent binding to Cys797. However, patients develop resistance to osimertinib within the first year of treatment as a result of cysteine to serine C797S mutation that prevents formation of potency-essential covalent bond (Chen et al., 2018; Thress et al., 2015). Based on the structural analysis, Lategahn et al. developed a series of pyrrolopyrimidines that could interact reversibly with Cys797 and thus be less prone to the loss in activity toward the C797S mutant, as compared to osimertinib (Lategahn et al., 2019). Compounds **21** and **22** displayed high activity against triple mutated EGFR-L858R/T790M/C797S, which was revealed to be a result of the favorable interactions with Met790 and enhanced electron density of pyrrolopyrimidine ring due to the ether substitution (Fig. 16.16).

8 MOLECULAR DYNAMICS IN DESIGN AND DISCOVERY OF KINASE INHIBITORS

Numerous X-ray structures of protein kinases in their apo states and in complexes with diverse ligands in various conformations have been published in the last decades and are available in PDB. This represents valuable structural information on protein kinases and their complexes. However, it does not consider their dynamics, which could explain the affinity and selectivity of different ligands, and the thermodynamics of the solvated complexes (Caballero & Alzate-Morales, 2012). Protein kinases are flexible proteins; they go through significant conformational rearrangement in order to switch from inactive to active state. A key tool to describe the dynamics of macromolecular systems is molecular dynamic (MD) simulation, and it therefore represents the basis for structure-based drug discovery of protein kinase inhibitors.

MD is a computational method used for studying the physical movements of atoms/molecules, so that the motion of individual atoms can be tracked to reveal the information that could not be uncovered by in vitro methods (conformational changes in protein, protein folding, ligand binding, protein−protein interaction, mutations, etc.). Owing to improvements in computational resources, the usage of MD to understand protein structure-to-function relationships has advanced impressively (Beauchamp et al., 2012; Piana et al., 2014; Raval et al., 2012).

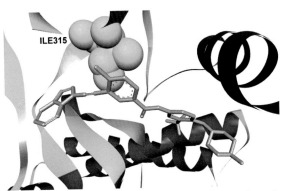

FIG. 16.15 Crystal structure of mutant T315I ABL kinase in complex with ponatinib.

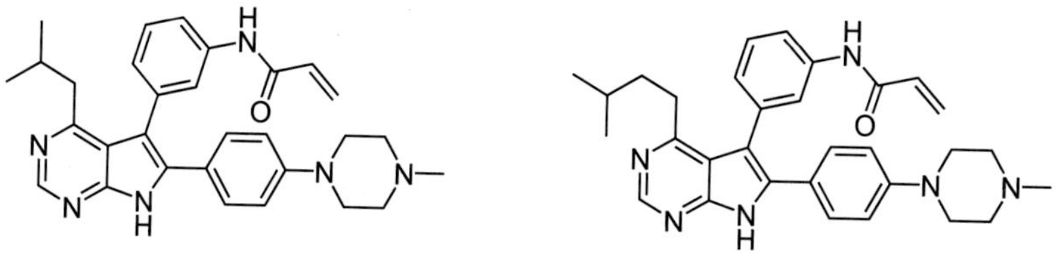

21   **22**

FIG. 16.16 Structure of compounds with activity against triple mutated EGFR-L858R/T790M/C797S.

The simulation system can be represented as protein (with or without ligand) in solution with additional membranes or other large macromolecules (such as DNA, nucleosomes, or ribosomes) if necessary. Experimental structures (obtained by X-ray crystallography, nuclear magnetic resonance spectroscopy, cryo-electron microscopy, etc.) or computational protein models (obtained by homology or ab initio modeling) are used to prepare an initial model of the system (Gelpi et al., 2015). When the system is prepared, atoms and molecules are allowed to interact with each other for a certain period of time. Afterward, their trajectories are determined by solving Newton's equations of motion for the whole system. Forces acting on each atom and their potential energies are calculated using force fields (Alder & Wainwright, 1959; Kukol, 2008; Rahman, 1964). Basically, force fields are describing the time evolution of bonded (bond stretching, angle bending, torsion potential, improper torsions) and nonbonded interactions (Lenard-Jones repulsion and dispersion and Coulomb electrostatics) (Leach, 2001). Force fields are represented by a set of equations used to calculate the potential energy and forces utilizing particle coordinates and parameters used in equations (Eq. 16.1).

$$E = \sum_{bonds} K_b(b - b_0)^2 + \sum_{angles} K_\theta(\theta - \theta_0)^2$$
$$+ \sum_{\substack{improper \\ dihedrals}} K_\varphi(\varphi - \varphi_0)^2 + \sum_{dihedrals} \sum_{n=1}^{6} K_{\phi,n}(1 + \cos(n\phi - \delta_n))$$
$$+ \sum_{\substack{nonbonded \\ pairs\ ij}} \frac{q_i\, q_j}{4\, \pi D r_{ij}}$$
$$+ \sum_{\substack{nonbonded \\ pairs\ ij}} \varepsilon_{ij}\left[\left(\frac{R_{min,ij}}{r_{ij}}\right)^{12} - 2\left(\frac{R_{min,ij}}{r_{ij}}\right)^6\right]$$

$$(16.1)$$

Eq. (16.1). The potential energy function is given by energy E, which is the sum of energies of bonded (bond,

angles, improper torsions) and nonbonded (electrostatic and Lenard-Johns) interactions (Leach, 2001).

Force fields differ in the way they are parameterized (Gelpi et al., 2015). A few force fields provide a precise depiction of the structure and dynamics of numerous proteins on the submicrosecond timescale (Piana et al., 2014). The most popular force fields nowadays are CHARMM (Yin & MacKerell, 1998), AMBER (Cornell et al., 1995; Weiner et al., 1984), GROMOS (Oostenbrink et al., 2004), and OPLS (Jorgensen et al., 1996). The most commonly used MD codes are CHARMM (Brooks et al., 2009), Desmond (Bowers et al., 2006), GROMACS (Berendsen et al., 1995; Lindahl et al., 2001), LAMMPS (Plimpton, 1995), NAMD (Phillips et al., 2005), and OpenMM (Eastman et al., 2017).

The time frame of typical unbiased atomistic MD simulation (100 ns −1 µs) is much shorter than the time necessary for the structural changes in proteins to take place (0.1–10 s). MD simulations are, therefore, unlikely to predict the complete structural rearrangements during these processes. In order to lower the computer costs and improve the efficiency of MD simulations, different approaches have been developed. First of all, advancements in computing hardware, the usage of graphical processing units, and parallelization have significantly improved the performance of MD simulations. Additionally, many efforts have been made to improve MD codes and develop enhanced MD sampling methods (Aci-Sèche et al., 2016; Gelpi et al., 2015; Singh & Li, 2019). Methods that introduce the artificial bias into the simulation (such as steered MD (Izrailev et al., 1999), metadynamics (Barducci et al., 2011), accelerated MD (Hamelberg et al., 2004), etc.) are the most frequently used methods for enhancing conformational sampling in MD simulations (Aci-Sèche et al., 2016). Alternatively, in order to accelerate conformational sampling, simplified coarse-grained models that reduce the degrees of freedom, while retaining the crucial molecular characteristics of the system, by converging several atoms into one big pseudoatom, can be used (Aci-Sèche et al., 2016; Singh & Li, 2019).

9 MOLECULAR DYNAMICS APPROACHES TO ENHANCE DOCKING

To obtain deeper insight into the ligand—protein binding process, molecular docking can be combined with MD. As aforementioned, the most commonly used in silico method for prediction of protein—ligand binding is molecular docking. However, one of the most significant drawbacks of molecular docking approaches is the lack of consideration for protein flexibility (Aci-Sèche et al., 2016; Gelpi et al., 2015). In nature, different ligands induce different structural rearrangements in the protein, but majority of docking programs work using rigid structures obtained from the PDB. Some flexibility had been effectively brought into docking by giving flexibility to several amino acid side chains in the binding pocket (Leach, 1994; Najmanovich et al., 2000; Rarey et al., 1996; Ravindranath et al., 2015). Nevertheless, in cases when large backbone movements are observed to obtain near-accurate results, it is necessary to allow the full receptor flexibility (De Paris et al., 2013; Sandak et al., 1998; Uehara & Tanaka, 2017).

There are several ways to overcome these problems that are based on a successive combination of molecular docking and MD. One way would be to use MD simulations after molecular docking to validate, refine, and/or rescore docking results. Another way would be to use multiple protein conformations (conformational ensembles), instead of a single one. These ensembles are a much more accurate approximation of real proteins, as they take into consideration flexibility and dynamics, which could improve the prediction rate of molecular docking. In the approach called "ensemble docking," a set of different protein conformations (i.e., snapshots from MD trajectory) is used as target for the docking. The specific, thermodynamically favored, protein conformation is then selected by a ligand of interest (Aci-Sèche et al., 2016; Evangelista Falcon et al., 2019; Gelpi et al., 2015; Gioia et al., 2017). The combination of MD and ensemble docking is widely used protocol in structure-based drug design of protein kinases.

In the last decade, researchers investigated diverse MD methods and conformational sampling techniques in combination with different docking programs on various protein kinases (Bajusz et al., 2016; Campbell et al., 2014; Chen et al., 2016; Patel, Athar, & Jha, 2020; Sharma et al., 2016; Tian et al., 2014; Zhao

et al., 2012). Asses et al. investigated cMet flexibility by comparing two different MD simulation strategies (1 long 1 μs MD simulation and 26 normal mode calculations followed by 10 ns MD simulations). For long simulation they used the apo-cMet, and for shorter MD simulations they used 26 available X-ray structures of c-Met in complexes with different ligands. Afterward, they clustered conformers and extracted five representative structures from long simulation and five from all short simulations. Four out of five conformers from long simulations could relate to those from short simulations. However, the combination of normal modes calculations and short MD simulation proved more efficient, especially in terms of computational cost (Asses et al., 2012). Later on, continuing their work, Bresso et al. used this protocol to identify novel potential inhibitors of the cMet kinase. Similarly as in their previous study, they obtained 45 X-ray structures of cMet kinase from PDB and submitted them to short MD simulations to prepare a conformation ensemble. They found that the conformation, which is halfway between classical type-I and type-II inhibitors, was forming additional interactions with studied ligand. This interesting finding could lead to design and discovery of novel type of cMet inhibitors (Bresso et al., 2020).

To take into account protein flexibility in virtual screening, Spyrakis et al. designed integrated MD-FLAP (molecular dynamics—fingerprint for ligand and proteins) (https://www.moldiscovery.com/) protocol and tested it on ABL1 kinase. After performing 100 ns MD simulation on ABL1 kinase, they clustered the entire trajectory based on the variability of the molecular interaction fields within the binding pocket using principal component analysis (Goodford, 1985). They used linear discriminant analysis approach to select the templates and FLAPscores able to discriminate between active compounds and decoys in training sets. Template-FLAPscore combinations were than validated using the test set (Spyrakis et al., 2015).

Uehara and Tanaka developed novel computationally inexpensive protocol by combining cosolvent-based molecular dynamics (CMD) and ensemble docking. By adding small molecule probes/fragments into solvent, CMD simulations induce conformational changes of a target proteins binding pocket. This way, diverse conformations able to accommodate diverse ligands can be obtained (Ghanakota & Carlson, 2016; Guvench & MacKerell, 2009). Additionally, for the selection of the conformers, they applied a method called the screening performance index, which selects good conformers through docking of known active compounds (Huang & Wong, 2016). They tested this protocol on several proteins, among others CDK2 (Uehara & Tanaka, 2017).

Additionally, MD simulation can be used to fully explore ligand–protein recognition, binding, and unbinding, both the mechanistically and energetically, in the approach called "dynamic docking" (Gioia et al., 2017). In this approach, MD is used to explore how the ligand enters the binding pocket, which binding position does it take, and how it leaves the pocket. Apart from calculation of the ligand–protein binding free energy, dynamic docking can give an insight into the kinetics of binding/unbinding. It can be used to calculate the association and dissociation rates of ligand binding. These two values can be used to estimate the residence time of ligand in the binding pocket. Ligands that have slow dissociation rate/long retention time are highly likely to bind target protein in vitro (De Vivo & Cavalli, 2017; Decherchi et al., 2018; Gioia et al., 2017; Spitaleri et al., 2018). The main advantage of dynamic docking is that it considers flexibility and solvation of the ligand–protein complex, as well as the temperature of the system. Trajectory of drug binding and unbinding reveals passing interactions between the ligand and the proximal amino acid residues during simulation time. This results in better prediction of ligand–protein binding, binding free energy and kinetics (De Vivo & Cavalli, 2017; Gioia et al., 2017).

Approaches such as ensemble docking and dynamic docking show clear advantages; however, as they are computationally expensive, they are not routinely used. For that reason, classical molecular docking still remains the method of choice for studying protein–ligand binding. With the rapid progresses in computer technology, it is highly likely that these methods will replace classical molecular docking, in the near future.

10 APPLICATION OF MOLECULAR DYNAMICS TO DESCRIBE ALLOSTERY

As aforementioned, the main advantages of allosteric kinase inhibitors (types III and IV) include high selectivity and low toxicity due to their binding to nonconserved allosteric sites (Fabbro, 2015; Roskoski, 2016). Additionally, design of allosteric kinase inhibitors becomes even more significant due to their ability to circumvent mutation-induced resistance (Aci-Sèche et al., 2016; Gibbons et al., 2012). This validates the reasoning behind investigating allosteric pockets and conformational shifts involved in allosteric transitions (Aci-Sèche et al., 2016; Gelpi et al., 2015). MD is currently a method of choice for exploring conformational rearrangements of proteins. This method offers

clear conclusions of allosteric mechanisms involving shifts among such conformations (Stolzenberg et al., 2016). Therefore, MD emerged as a viable tool to detect hidden allosteric pockets, explore conformational landscape, and obtain deeper insight into mechanisms underlining allosteric transition. In general, it would be possible to describe detailed mechanism of communication between allosteric and active sites by running long MD simulations (Guo & Zhou, 2016). However, this is still computationally very expensive. Consequently, targeted or biased MD, where the simulation is artificially forced into the preferred conformation, is often used (Gelpi et al., 2015).

The most common application of MD simulations in investigation of allostery involves determining the nature of an event (for example, binding of a ligand) at one site increasing or decreasing activity at a different site, with a goal to design additional ligands with matching or opposite effects. In other cases, the goal could be to find additional pockets that are coupled to the central binding site which might represent novel targets for drug design. In particular, some compounds can bind to "hidden" pockets, which are not clearly discernible in crystal structure. These pockets may appear during MD simulation, especially if ligands are included to facilitate pocket formation (Hertig et al., 2016). MD simulation has been broadly used in structure-based drug design of protein kinases for identification of allosteric pockets and design of novel allosteric inhibitors (Betzi et al., 2011; Gomez-Gutierrez et al., 2016; Gu et al., 2019; Schulze et al., 2016; Shukla et al., 2016).

Gkeka et al. used MD simulation in order to find allosteric binding pocket of PI3K kinase, which would help in design of novel selective inhibitors by running unbiased MD simulations of the wild-type human apo and holo forms, murine holo form, and human-mutant apo and holo forms, with an inhibitor placed in both ATP and non-ATP pockets. The allosteric site previously described in murine PI3K was confirmed in human PI3K. Additionally, by observing the dynamics of complexes and protein–ligand interactions, they discovered separate binding modes of studied ligand in murine, wild-type, and mutant forms of PI3K. Their finding could be further used for the design of potent selective PI3K inhibitors (Gkeka, Papafotika, Christoforidis, & Cournia, 2015). With the help of MD, Gomez-Gutierrez et al. discovered novel inhibitory allosteric site on p38 MAPK kinase. Starting from the X-ray structure of the p38α-MK2 (MAPK-activated protein kinase 2) heterodimer (White et al., 2007), they designed small peptide inhibitors (derived from the sequence of MK2 regulatory loop). To reveal the key protein–protein interactions and define a pharmacophore, they submitted these complexes to MD simulation. The obtained pharmacophore was used for virtual screening, which resulted in identification of few small molecule allosteric inhibitors (Gomez-Gutierrez et al., 2016).

Liu et al. investigated allosteric inhibition of 3-phosphoinositide-dependent protein kinase-1 (PDK1) by targeting the PDK1-interacting fragment (PIF) pocket. This pocket interacts with a region of protein kinase C−related kinase 2 (PRK2) called PIF. Activity of PDK1 can be modulated by ligand binding or mutations in amino acid in this site (Biondi et al., 2000). By means of molecular docking, MD simulations, and free energy calculations, Liu et al. identified novel allosteric inhibitor of PDK1 and elucidated allosteric conformational changes it causes (Liu et al., 2019). Li et al. performed 1 μs MD simulations of wild-type and mutant glycogen synthase kinase 3β (GSK3β) in order to obtain the mechanistic explanation of allosteric inactivation caused by the L343R mutation. MD trajectories were used for dynamic cross-correlation analysis and network analysis. The results revealed that the L343R mutation causes conformational changes in the activation loop and blocks the binding of substrates to the phosphate binding site (Li et al., 2019).

11 MACHINE LEARNING APPROACHES TO ENHANCE DOCKING

Scoring functions used in docking generalize and simplify physical aspects of ligand–protein interactions (for example, protein flexibility and solvation). This makes them computationally inexpensive, but it decreases their predictability. Moreover, classical scoring function presumes a predetermined functional form for the correlation between protein–ligand interactions and binding affinity, where parameters are fitted to experimental or simulation data, which leads to poor predictivity in those complexes that do not follow general rules (Ballester & Mitchell, 2010). Machine learning (ML) algorithms can be used to improve scoring functions for binding affinity predictions (Torres et al., 2019). All ML scoring functions are based on supervised nonparametric learning, meaning that the output variables are known and that the scoring function does not presume a predetermined functional form. A training set consisting of input and output variables is provided to the algorithm. The functional form is derived from the correlation between input and output data and can be further used to predict the unknown outputs for other datasets (Ballester & Mitchell, 2010; Torres et al., 2019). Numerous recently published studies confirmed that these scoring functions are more diverse and accurate than the classical docking scoring functions (Table 16.3) (Ashtawy & Mahapatra, 2015; Ballester & Mitchell, 2010; Hassan et al., 2018; Hsin et al., 2013; Kinnings et al., 2011; Ouyang et al., 2011; Pereira et al., 2016; Springer et al., 2005; Zhang et al., 2019).

ML scoring functions can be generic or family-specific. As the significance of variables that describe the data are different for different protein families, family-specific scoring functions better integrate these features, as a result of dealing with more sophisticated data. Consequently, scoring functions trained with

TABLE 16.3
Machine Learning Scoring Functions for Molecular Docking (Torres et al., 2019).

Scoring Function	Regression Algorithm	Type	Best Performance	References
B2BScore	Random forest	Generic	Rp = 0.746	Liu et al. (2013)
CScore	Neural network	Generic Family-specific	Rp = 0.7668 Rp = 0.8237	Ouyang et al. (2011)
DLScore	Deep neural network	Generic	Rp = 0.82	Hassan et al. (2018)
DeepVS	Deep neural network	Generic	ROC = 0.81	Pereira et al. (2016)
ID-Score	Support vector regression	Generic	Rp = 0.85	Li et al. (2013)
KDEEP	Deep neural network	Generic	Rp = 0.82	Jiménez et al. (2018)
NNScore	Neural network	Generic	EF = 10.3	Durrant & McCammon (2010)
PostDOCK	Random forest	Generic	92% accuracy	Springer et al. (2005)
RFScore	Random forest	Generic	Rp = 0.776	Ballester and Mitchell (2010)

ROC, Receiver operating curve; *Rp*, Pearson's correlation coefficient.

family-specific data outperformed most generic ones (Imrie et al., 2018; Tajbakhsh et al., 2016; Torres et al., 2019; Wang et al., 2015). As ML docking approaches are relatively novel, majority of published studies are still testing different regression models to improve predictiveness, and their widespread application in drug discovery experiments is yet to be seen (Ashtawy & Mahapatra, 2015; Ballester & Mitchell, 2010; Hassan et al., 2018; Hsin et al., 2013; Imrie et al., 2018; Kinnings et al., 2011; Ouyang et al., 2011; Pereira et al., 2016; Springer et al., 2005; Torres et al., 2019; Wang et al., 2015).

Imrie et al. investigated application of family-based convolutional neural network (CNN) scoring function to structure-based virtual screening by using a densely connected convolutional networks (DenseNet) architecture (Huang et al., 2018). CNN scoring function was previously described by Ragoza et al. A CNN uses a 3D structure of a protein–ligand interaction to learn the key features that correlate with binding affinity so that it can differentiate between known active and inactive compounds and find the correct binding poses (Ragoza et al., 2017). To avoid overfitting, family-specific models were created by fine-tuning of generic models on data that are specific for each target's family. The model was prepared for four specific protein families (kinases, proteases, nuclear proteins, and G-protein–coupled receptors). They found that fine-tuning significantly improved the performance of the scoring function, especially for kinases (up to 50%). Moreover, they noted that performance continued to rise with the increase of data in the training set (Imrie et al., 2018). De Avila et al. used an ML scoring function to predict binding affinity for CDK2. The dataset of high-resolution crystallographic structures of CDK2 with known binding affinities was used to train the model (de Azevedo, 2016). Their ML models outperformed other docking scoring functions (AutoDock4, AutoDock Vina, MolDock, and PLANTS) (de Ávila et al., 2017).

12 CONCLUSION

The growing body of known protein kinase structures along with rapid advances in computational chemistry offers enticing future prospective for the use of molecular docking tools in the design of new and potent kinase inhibitors. This chapter outlines the most recent structure-based drug design methods, with emphasis on molecular docking, which have been utilized in discovery of various kinase inhibitors. Analysis of the protein–ligand interactions enabled structural optimization of known kinase scaffolds, which led to further improvement of activity and selectivity. This approach also represents a valuable strategy to overcome mutation-associated drug resistance and has resulted in discovery of new generation of kinase inhibitors that have high clinical relevance.

ACKNOWLEDGMENTS

We acknowledge project of Ministry of Science and Technological Development of the Republic of Serbia for Faculty of Pharmacy, University of Belgrade, No. 451-03-68/2020- 14/200161.

REFERENCES

Aci-Sèche, S., Ziada, S., Braka, A., Arora, R., & Bonnet, P. (2016). Advanced molecular dynamics simulation methods for kinase drug discovery. *Future Medicinal Chemistry, 8*, 545–566.

Adams, J. A. (2001). Kinetic and catalytic mechanisms of protein kinases. *Chemical Reviews, 101*, 2271–2290.

Alder, B. J., & Wainwright, T. E. (1959). Studies in molecular dynamics. I. General method. *The Journal of Chemical Physics, 31*, 459–466.

Alkahtani, H. M., Abdalla, A. N., Obaidullah, A. J., Alanazi, M. M., Almehizia, A. A., Alanazi, M. G., Ahmed, A. Y., Alwassil, O. I., Darwish, H. W., Abdel-Aziz, A. A.-M., & El-Azab, A. S. (2020). Synthesis, cytotoxic evaluation, and molecular docking studies of novel quinazoline derivatives with benzenesulfonamide and anilide tails: Dual inhibitors of EGFR/HER2. *Bioorganic Chemistry, 95*, 103461.

Allam, H. A., Aly, E. E., Farouk, A. K. B. A. W., El Kerdawy, A. M., Rashwan, E., & Abbass, S. E. S. (2020). Design and Synthesis of some new 2,4,6-trisubstituted quinazoline EGFR inhibitors as targeted anticancer agents. *Bioorganic Chemistry, 98*, 103726.

Ashtawy, H. M., & Mahapatra, N. R. (2015). Machine-learning scoring functions for identifying native poses of ligands docked to known and novel proteins. *BMC Bioinformatics, 16*.

Asses, Y., Venkatraman, V., Leroux, V., Ritchie, D. W., & Maigret, B. (2012). Exploring c-Met kinase flexibility by sampling and clustering its conformational space. *Proteins: Structure, Function, and Bioinformatics, 80*, 1227–1238.

de Ávila, M. B., Xavier, M. M., Pintro, V. O., & de Azevedo, W. F. (2017). Supervised machine learning techniques to predict binding affinity. A study for cyclin-dependent kinase 2. *Biochemical and Biophysical Research Communications, 494*, 305–310.

de Azevedo, W. F. (2016). Opinion paper: Targeting multiple cyclin-dependent kinases (CDKs): A new strategy for molecular docking studies. *Current Drug Targets, 17*, 2.

Bach, D.-H., Kim, D., & Lee, S. K. (2020). Cancer chemopreventive potential of epidermal growth factor receptor inhibitors from natural products. In J. M. Pezzuto, & O. Vang

(Eds.), *Natural products for cancer chemoprevention* (pp. 469–488). Cham: Springer International Publishing.

Bahekar, R., Panchal, N., Soman, S., Desai, J., Patel, D., Argade, A., Gite, A., Gite, S., Patel, B., Kumar, J., Sachchidanand, S., Patel, H., Sundar, R., Chatterjee, A., Mahapatra, J., Patel, H., Ghoshdastidar, K., Bandyopadhyay, D., & Desai, R. C. (2020). Discovery of diaminopyrimidine-carboxamide derivatives as JAK3 inhibitors. *Bioorganic Chemistry, 99*, 103851.

Bajusz, D., Ferenczy, G. G., & Keserű, G. M. (2016). Discovery of subtype selective janus kinase (JAK) inhibitors by structure-based virtual screening. *Journal of Chemical Information and Modeling, 56*, 234–247.

Ballester, P. J., & Mitchell, J. B. O. (2010). A machine learning approach to predicting protein–ligand binding affinity with applications to molecular docking. *Bioinformatics, 26*, 1169–1175.

Barducci, A., Bonomi, M., & Parrinello, M. (2011). Metadynamics. *WIREs Computational Molecular Science, 1*, 826–843.

Bathula, R., Satla, S. R., Kyatham, R., & Gangarapu, K. (2020). Design, one pot synthesis and molecular docking studies of substituted-1H-Pyrido[2,1-b] quinazolines as apoptosis-inducing anticancer agents. *Asian Pacific Journal of Cancer Prevention, 21*, 411–421.

Beauchamp, K. A., Lin, Y.-S., Das, R., & Pande, V. S. (2012). Are protein force fields getting better? A systematic benchmark on 524 diverse NMR measurements. *Journal of Chemical Theory and Computation, 8*, 1409–1414.

Berendsen, H. J. C., van der Spoel, D., & van Drunen, R. (1995). GROMACS: A message-passing parallel molecular dynamics implementation. *Computer Physics Communications, 91*, 43–56.

Betzi, S., Alam, R., Martin, M., Lubbers, D. J., Han, H., Jakkaraj, S. R., Georg, G. I., & Schönbrunn, E. (2011). Discovery of a potential allosteric ligand binding site in CDK2. *ACS Chemical Biology, 6*, 492–501.

Biondi, R. M., Cheung, P. C. F., Casamayor, A., Deak, M., Currie, R. A., & Alessi, D. R. (2000). Identification of a pocket in the PDK1 kinase domain that interacts with PIF and the C–terminal residues of PKA. *The EMBO Journal, 19*, 979–988.

Bogoyevitch, M. A., Barr, R. K., & Ketterman, A. J. (2005). Peptide inhibitors of protein kinases—discovery, characterisation and use. *Biochimica et Biophysica Acta (BBA) - Proteins and Proteomics, 1754*, 79–99.

Bowers, K. J., Chow, D. E., Xu, H., Dror, R. O., Eastwood, M. P., Gregersen, B. A., Klepeis, J. L., Kolossvary, I., Moraes, M. A., Sacerdoti, F. D., Salmon, J. K., Shan, Y., & Shaw, D. E. (2006). Scalable algorithms for molecular dynamics simulations on commodity clusters. In *ACM/IEEE SC 2006 conference (SC'06). Presented at the SC 2006 proceedings supercomputing 2006, IEEE, Tampa, FL* (p. 43).

Bresso, E., Furlan, A., Noel, P., Leroux, V., Maina, F., Dono, R., & Maigret, B. (2020). Large-scale virtual screening against the MET kinase domain identifies a new putative inhibitor type. *Molecules, 25*, 938.

Brooks, B. R., Brooks, C. L., MacKerell, A. D., Nilsson, L., Petrella, R. J., Roux, B., Won, Y., Archontis, G., Bartels, C., Boresch, S., Caflisch, A., Caves, L., Cui, Q., Dinner, A. R., Feig, M., Fischer, S., Gao, J., Hodoscek, M., Im, W., … Karplus, M. (2009). CHARMM: The biomolecular simulation program. *Journal of Computational Chemistry, 30*, 1545–1614.

Butterworth, S., Cross, D. A. E., Finlay, M. R. V., Ward, R. A., & Waring, M. J. (2017). The structure-guided discovery of osimertinib: The first U.S. FDA approved mutant selective inhibitor of EGFR T790M. *Medchemcomm, 8*, 820–822.

Caballero, J., & H. Alzate-Morales, J. (2012). Molecular dynamics of protein kinase-inhibitor complexes: A valid structural information. *Current Pharmaceutical Design, 18*, 2946–2963.

Campbell, A. J., Lamb, M. L., & Joseph-McCarthy, D. (2014). Ensemble-based docking using biased molecular dynamics. *Journal of Chemical Information and Modeling, 54*, 2127–2138.

Carles, F., Bourg, S., Meyer, C., & Bonnet, P. (2018). PKIDB: A curated, annotated and updated database of protein kinase inhibitors in clinical trials. *Molecules, 23*, 908.

Chen, L., Fu, W., Zheng, L., Liu, Z., & Liang, G. (2018). Recent progress of small-molecule epidermal growth factor receptor (EGFR) inhibitors against C797S resistance in non-small-cell lung cancer. *Journal of Medicinal Chemistry, 61*, 4290–4300.

Chen, H., Li, S., Hu, Y., Chen, G., Jiang, Q., Tong, R., Zang, Z., & Cai, L. (2016). An integrated in silico method to discover novel Rock1 inhibitors: Multi- complex-based pharmacophore, molecular dynamics simulation and hybrid protocol virtual screening. *Combinatorial Chemistry and High Throughput Screening, 19*, 36–50.

Cohen, P., & Alessi, D. R. (2013). Kinase drug discovery – what's next in the field? *ACS Chemical Biology, 8*, 96–104.

Cormier, K. W., & Woodgett, J. R. (2016). *Protein kinases: Physiological roles in cell signalling* (pp. 1–9). Chichester, UK: John Wiley & sons Ltd.

Cornell, W. D., Cieplak, P., Bayly, C. I., Gould, I. R., Merz, K. M., Ferguson, D. M., Spellmeyer, D. C., Fox, T., Caldwell, J. W., & Kollman, P. A. (1995). A second generation force field for the simulation of proteins, nucleic acids, and organic molecules. *Journal of the American Chemical Society, 117*, 5179–5197.

Cowan-Jacob, S. W., Fendrich, G., Floersheimer, A., Furet, P., Liebetanz, J., Rummel, G., Rheinberger, P., Centeleghe, M., Fabbro, D., & Manley, P. W. (2007). Structural biology contributions to the discovery of drugs to treat chronic myelogenous leukaemia. *Acta Crystallographica. Section D, Biological Crystallography, 63*, 80–93.

Cox, A. D., & Der, C. J. (2010). The raf inhibitor paradox: Unexpected consequences of targeted drugs. *Cancer Cell, 17*, 221–223.

Davies, H., Bignell, G. R., Cox, C., Stephens, P., Edkins, S., Clegg, S., Teague, J., Woffendin, H., Garnett, M. J., Bottomley, W., Davis, N., Dicks, E., Ewing, R., Floyd, Y., Gray, K., Hall, S., Hawes, R., Hughes, J., Kosmidou, V., …

Futreal, P. A. (2002). Mutations of the BRAF gene in human cancer. *Nature, 417*, 949–954.

De Paris, R., Frantz, F. A., Norberto de Souza, O., & Ruiz, D. D. A. (2013). wFReDoW: A cloud-based web environment to handle molecular docking simulations of a fully flexible receptor model. *BioMed Research International, 2013*, 1–12.

De Vivo, M., & Cavalli, A. (2017). Recent advances in dynamic docking for drug discovery. *WIREs Computational Molecular Science, 7*.

Debnath, S., Kanakaraju, M., Islam, M., Yeeravalli, R., Sen, D., & Das, A. (2019). In silico design, synthesis and activity of potential drug-like chrysin scaffold-derived selective EGFR inhibitors as anticancer agents. *Computational Biology and Chemistry, 83*, 107156.

Decherchi, S., Bottegoni, G., Spitaleri, A., Rocchia, W., & Cavalli, A. (2018). BiKi life sciences: A new suite for molecular dynamics and related methods in drug discovery. *Journal of Chemical Information and Modeling, 58*, 219–224.

Dong, Q., Dougan, D. R., Gong, X., Halkowycz, P., Jin, B., Kanouni, T., O'Connell, S. M., Scorah, N., Shi, L., Wallace, M. B., & Zhou, F. (2011). Discovery of TAK-733, a potent and selective MEK allosteric site inhibitor for the treatment of cancer. *Bioorganic and Medicinal Chemistry Letters, 21*, 1315–1319.

Durrant, J. D., & McCammon, J. A. (2010). NNScore: A neural-network-based scoring function for the characterization of protein–ligand complexes. *Journal of Chemical Information and Modeling, 50*, 1865–1871.

Durrant, D. E., & Morrison, D. K. (2018). Targeting the raf kinases in human cancer: The raf dimer dilemma. *British Journal of Cancer, 118*, 3–8.

Eastman, P., Swails, J., Chodera, J. D., McGibbon, R. T., Zhao, Y., Beauchamp, K. A., Wang, L.-P., Simmonett, A. C., Harrigan, M. P., Stern, C. D., Wiewiora, R. P., Brooks, B. R., & Pande, V. S. (2017). OpenMM 7: Rapid development of high performance algorithms for molecular dynamics. *PLoS Computational Biology, 13*, e1005659.

Eldar-Finkelman, H., & Eisenstein, M. (2009). Peptide inhibitors targeting protein kinases. *Current Pharmaceutical Design, 15*, 2463–2470.

Evangelista Falcon, W., Ellingson, S. R., Smith, J. C., & Baudry, J. (2019). Ensemble docking in drug discovery: How many protein configurations from molecular dynamics simulations are needed to reproduce known ligand binding? *The Journal of Physical Chemistry B, 123*, 5189–5195.

Ewes, W. A., Elmorsy, M. A., El-Messery, S. M., & Nasr, M. N. A. (2020). Synthesis, biological evaluation and molecular modeling study of [1,2,4]-Triazolo[4,3-c]quinazolines: New class of EGFR-TK inhibitors. *Bioorganic and Medicinal Chemistry, 28*, 115373.

Fabbro, D. (2015). 25 years of small molecular weight kinase inhibitors: Potentials and limitations. *Molecular Pharmacology, 87*, 766–775.

Gagic, Z., Ruzic, D., Djokovic, N., Djikic, T., & Nikolic, K. (2020). In silico methods for design of kinase inhibitors as anticancer drugs. *Frontiers in Chemistry, 7*.

Gelpi, J., Hospital, A., Goñi, R., & Orozco, M. (2015). Molecular dynamics simulations: Advances and applications. *Advances and Applications in Bioinformatics and Chemistry, 37*.

George, R. F., Kandeel, M., El-Ansary, D. Y., & El Kerdawy, A. M. (2020). Some 1,3,5-trisubstituted pyrazoline derivatives targeting breast cancer: Design, synthesis, cytotoxic activity, EGFR inhibition and molecular docking. *Bioorganic Chemistry, 99*, 103780.

Ghanakota, P., & Carlson, H. A. (2016). Driving structure-based drug discovery through cosolvent molecular dynamics. *Journal of Medicinal Chemistry, 59*, 10383–10399.

Ghosh, A. K., & Gemma, S. (2014). *Structure-based design of drugs and other bioactive molecules: Tools and strategies*. Weinheim, Germany: Wiley-VCH Verlag GmbH & Co. KGaA.

Gibbons, D. L., Pricl, S., Kantarjian, H., Cortes, J., & Quintás-Cardama, A. (2012). The rise and fall of gatekeeper mutations? The *BCR-ABL1* T315I paradigm: Taming the T315I mutation. *Cancer, 118*, 293–299.

Gioia, D., Bertazzo, M., Recanatini, M., Masetti, M., & Cavalli, A. (2017). Dynamic docking: A paradigm shift in computational drug discovery. *Molecules, 22*, 2029.

Gkeka, P., Papafotika, A., Christoforidis, S., & Cournia, Z. (2015). Exploring a non-ΛTP pocket for potential allosteric modulation of PI3Kα. *The Journal of Physical Chemistry B, 119*(3), 1002–1016. https://doi.org/10.1021/jp506423e

Gomez-Gutierrez, P., Campos, P. M., Vega, M., & Perez, J. J. (2016). Identification of a novel inhibitory allosteric site in p38α. *PLoS One, 11*, Article e0167379.

Goodford, P. J. (1985). A computational procedure for determining energetically favorable binding sites on biologically important macromolecules. *Journal of Medicinal Chemistry, 28*, 849–857.

Gower, C. M., Chang, M. E. K., & Maly, D. J. (2014). Bivalent inhibitors of protein kinases. *Critical Reviews in Biochemistry and Molecular Biology, 49*, 102–115.

Gristina, V., Malapelle, U., Galvano, A., Pisapia, P., Pepe, F., Rolfo, C., Tortorici, S., Bazan, V., Troncone, G., & Russo, A. (2020). The significance of epidermal growth factor receptor uncommon mutations in non-small cell lung cancer: A systematic review and critical appraisal. *Cancer Treatment Reviews, 85*, 101994.

Guo, J., & Zhou, H.-X. (2016). Protein allostery and conformational dynamics. *Chemical Reviews, 116*, 6503–6515.

Guvench, O., & MacKerell, A. D. (2009). Computational fragment-based binding site identification by ligand competitive saturation. *PLoS Computational Biology, 5*, e1000435.

Gu, X., Wang, Y., Wang, M., Wang, J., & Li, N. (2019). Computational investigation of imidazopyridine analogs as protein kinase B (Akt1) allosteric inhibitors by using 3D-QSAR, molecular docking and molecular dynamics simulations. *Journal of Biomolecular Structure and Dynamics*, 1–16.

Hamelberg, D., Mongan, J., & McCammon, J. A. (2004). Accelerated molecular dynamics: A promising and efficient simulation method for biomolecules. *The Journal of Chemical Physics, 120*, 11919–11929.

Hanks, S. K., Quinn, A. M., & Hunter, T. (1988). The protein kinase family: Conserved features and deduced phylogeny of the catalytic domains. *Science, New Series, 241*, 42–52.

Hassan, M., Mogollon, D. C., Fuentes, O., & Sirimulla, S. (2018). *DLSCORE: A deep learning model for predicting protein-ligand binding affinities* (preprint).

Hertig, S., Latorraca, N. R., & Dror, R. O. (2016). Revealing atomic-level mechanisms of protein allostery with molecular dynamics simulations. *PLoS Computational Biology, 12*, e1004746.

Hilberg, F., Roth, G. J., Krssak, M., Kautschitsch, S., Sommergruber, W., Tontsch-Grunt, U., Garin-Chesa, P., Bader, G., Zoephel, A., Quant, J., Heckel, A., & Rettig, W. J. (2008). BIBF 1120: Triple angiokinase inhibitor with sustained receptor blockade and good antitumor efficacy. *Cancer Research, 68*, 4774−4782.

Hill, Z. B., Perera, B. G. K., Andrews, S. S., & Maly, D. J. (2012). Targeting diverse signaling interaction sites allows the rapid generation of bivalent kinase inhibitors. *ACS Chemical Biology, 7*, 487−495.

Hochhaus, A., & Kantarjian, H. (2013). The development of dasatinib as a treatment for chronic myeloid leukemia (CML): From initial studies to application in newly diagnosed patients. *Journal of Cancer Research and Clinical Oncology, 139*, 1971−1984.

Hsin, K.-Y., Ghosh, S., & Kitano, H. (2013). Combining machine learning systems and multiple docking simulation packages to improve docking prediction reliability for network pharmacology. *PLoS One, 8*, e83922.

Huang, G., Liu, Z., van der Maaten, L., & Weinberger, K. Q. (2018). *Densely connected convolutional networks*. arXiv: 1608.06993 [cs].

Huang, Z., & Wong, C. F. (2016). Inexpensive method for selecting receptor structures for virtual screening. *Journal of Chemical Information and Modeling, 56*, 21−34.

Imrie, F., Bradley, A. R., van der Schaar, M., & Deane, C. M. (2018). Protein family-specific models using deep neural networks and transfer learning improve virtual screening and highlight the need for more data. *Journal of Chemical Information and Modeling, 58*, 2319−2330.

Ivan, T., Enkvist, E., Viira, B., Manoharan, G. B., Raidaru, G., Pflug, A., Alam, K. A., Zaccolo, M., Engh, R. A., & Uri, A. (2016). Bifunctional ligands for inhibition of tight-binding protein-protein interactions. *Bioconjugate Chemistry, 27*, 1900−1910.

Izrailev, S., Stepaniants, S., Isralewitz, B., Kosztin, D., Lu, H., Molnar, F., Wriggers, W., & Schulten, K. (1999). Steered molecular dynamics. In P. Deuflhard, J. Hermans, B. Leimkuhler, A. E. Mark, S. Reich, & R. D. Skeel (Eds.), *Computational molecular dynamics: Challenges, methods, ideas, Lecture notes in computational science and engineering* (pp. 39−65). Springer Berlin Heidelberg.

Jahnke, W., Grotzfeld, R. M., Pellé, X., Strauss, A., Fendrich, G., Cowan-Jacob, S. W., Cotesta, S., Fabbro, D., Furet, P., Mestan, J., & Marzinzik, A. L. (2010). Binding or bending: Distinction of allosteric Abl kinase agonists from antagonists by an NMR-based conformational assay. *Journal of the American Chemical Society, 132*, 7043−7048.

Jiang, Z., Wang, L., Liu, X., Chen, C., Wang, B., Wang, W., Hu, C., Yu, K., Qi, Z., Liu, Q., Wang, A., Liu, J., Hong, G., Wang, W., & Liu, Q. (2020). Discovery of a highly selective VEGFR2 kinase inhibitor CHMFL-VEGFR2-002 as a novel anti-angiogenesis agent. *Acta Pharmaceutica Sinica B, 10*, 488−497.

Jiménez, J., Škalič, M., Martínez-Rosell, G., & De Fabritiis, G. (2018). KDEEP: Protein-ligand absolute binding affinity prediction via 3D-convolutional neural networks. *Journal of Chemical Information and Modeling, 58*, 287−296.

Johannessen, C. M., Boehm, J. S., Kim, S. Y., Thomas, S. R., Wardwell, L., Johnson, L. A., Emery, C. M., Stransky, N., Cogdill, A. P., Barretina, J., Caponigro, G., Hieronymus, H., Murray, R. R., Salehi-Ashtiani, K., Hill, D. E., Vidal, M., Zhao, J. J., Yang, X., Alkan, O., ... Garraway, L. A. (2010). COT drives resistance to RAF inhibition through MAP kinase pathway reactivation. *Nature, 468*, 968−972.

Jorgensen, W. L., Maxwell, D. S., & Tirado-Rives, J. (1996). Development and testing of the OPLS all-atom force field on conformational energetics and properties of organic liquids. *Journal of the American Chemical Society, 118*, 11225−11236.

Jung, H., Kim, J., Im, D., Moon, H., & Hah, J.-M. (2019). Design, synthesis, and in vitro evaluation of N-(3-(3-alkyl-1H-pyrazol-5-yl) phenyl)-aryl amide for selective RAF inhibition. *Bioorganic and Medicinal Chemistry Letters, 29*, 534−538.

Kaieda, A., Takahashi, M., Fukuda, H., Okamoto, R., Morimoto, S., Gotoh, M., Miyazaki, T., Hori, Y., Unno, S., Kawamoto, T., Tanaka, T., Itono, S., Takagi, T., Sugimoto, H., Okada, K., Snell, G., Bertsch, R., Nguyen, J., Sang, B., & Miwatashi, S. (2019). Structure-based design, synthesis, and biological evaluation of imidazo[4,5-*b*]pyridin-2-one-Based p38 MAP kinase inhibitors: Part 1. *ChemMedChem, 14*, 1022−1030.

Kamba, T., & McDonald, D. M. (2007). Mechanisms of adverse effects of anti-VEGF therapy for cancer. *British Journal of Cancer, 96*, 1788−1795.

Karachaliou, N., Fernandez-Bruno, M., Bracht, J. W. P., & Rosell, R. (2018). EGFR first- and second-generation TKIs— there is still place for them in EGFR-mutant NSCLC patients. *Translational Cancer Research, 8*, S23−S47.

Kettle, J. G., Åstrand, A., Catley, M., Grimster, N. P., Nilsson, M., Su, Q., & Woessner, R. (2017). Inhibitors of JAK-family kinases: An update on the patent literature 2013−2015, part 1. *Expert Opinion on Therapeutic Patents, 27*, 127−143.

Khodair, A. I., Alsafi, M. A., & Nafie, M. S. (2019). Synthesis, molecular modeling and anti-cancer evaluation of a series of quinazoline derivatives. *Carbohydrate Research, 486*, 107832.

Kim, J., Choi, B., Im, D., Jung, H., Moon, H., Aman, W., & Hah, J.-M. (2019). Computer-aided design and synthesis of 3-carbonyl-5-phenyl-1H-pyrazole as highly selective and potent BRAFV600E and CRAF inhibitor. *Journal of Enzyme Inhibition and Medicinal Chemistry, 34*, 1314−1320.

Kinnings, S. L., Liu, N., Tonge, P. J., Jackson, R. M., Xie, L., & Bourne, P. E. (2011). A machine learning-based method to improve docking scoring functions and its application to drug repurposing. *Journal of Chemical Information and Modeling, 51*, 408−419.

Knighton, D. R., Zheng, J. H., Ten Eyck, L. F., Ashford, V. A., Xuong, N. H., Taylor, S. S., & Sowadski, J. M. (1991). Crystal structure of the catalytic subunit of cyclic adenosine monophosphate-dependent protein kinase. *Science, 253,* 407–414.

Knight, J. D. R., Qian, B., Baker, D., & Kothary, R. (2007). Conservation, variability and the modeling of active protein kinases. *PLoS One, 2,* e982.

Kornev, A. P., Haste, N. M., Taylor, S. S., & Ten Eyck, L. F. (2006). Surface comparison of active and inactive protein kinases identifies a conserved activation mechanism. *Proceedings of the National Academy of Sciences United States of America, 103,* 17783–17788.

Kornev, A. P., & Taylor, S. S. (2015). Dynamics-driven allostery in protein kinases. *Trends in Biochemical Sciences, 40,* 628–647.

Kuglstatter, A., Ghate, M., Tsing, S., Villaseñor, A. G., Shaw, D., Barnett, J. W., & Browner, M. F. (2010). X-ray crystal structure of JNK2 complexed with the p38alpha inhibitor BIRB796: Insights into the rational design of DFG-out binding MAP kinase inhibitors. *Bioorganic and Medicinal Chemistry Letters, 20,* 5217–5220.

Kukol, A. (Ed.). (2008). *Molecular modeling of proteins.* Hatfield, UK: Humana Press.

La Sala, G., Riccardi, L., Gaspari, R., Cavalli, A., Hantschel, O., & De Vivo, M. (2016). HRD motif as the central hub of the signaling network for activation loop autophosphorylation in Abl kinase. *Journal of Chemical Theory and Computation, 12,* 5563–5574.

Lategahn, J., Keul, M., Klövekorn, P., Tumbrink, H. L., Niggenaber, J., Müller, M. P., Hodson, L., Flaßhoff, M., Hardick, J., Grabe, T., Engel, J., Schultz-Fademrecht, C., Baumann, M., Ketzer, J., Mühlenberg, T., Hiller, W., Günther, G., Unger, A., Müller, H., … Rauh, D. (2019). Inhibition of osimertinib-resistant epidermal growth factor receptor EGFR-T790M/C797S. *Chemical Science, 10,* 10789–10801.

Leach, A. R. (1994). Ligand docking to proteins with discrete side-chain flexibility. *Journal of Molecular Biology, 235,* 345–356.

Leach, A. R. (2001). *Molecular modelling principles and applications* (2nd ed.). Edinburgh, England: Glaxo Wellcome Research and Development, Pearson Education.

Le, Y., Gan, Y., Fu, Y., Liu, J., Li, W., Zou, X., Zhou, Z., Wang, Z., Ouyang, G., & Yan, L. (2020). Design, synthesis and in vitro biological evaluation of quinazolinone derivatives as EGFR inhibitors for antitumor treatment. *Journal of Enzyme Inhibition and Medicinal Chemistry, 35,* 555–564.

Lichota, A., & Gwozdzinski, K. (2018). Anticancer activity of natural compounds from plant and marine environment. *International Journal of Molecular Sciences, 19.*

Li, J., Fu, Q., Liang, Y., Cheng, B., & Li, X. (2019). Microsecond molecular dynamics simulations and dynamic network analysis provide understanding of the allosteric inactivation of GSK3β induced by the L343R mutation. *Journal of Molecular Modeling, 25.*

Lindahl, E., Hess, B., & van der Spoel, D. (2001). GROMACS 3.0: A package for molecular simulation and trajectory analysis. *Journal of Molecular Modeling, 7,* 306–317.

Liu, Q., Kwoh, C. K., & Li, J. (2013). Binding affinity prediction for protein-ligand complexes based on β contacts and B factor. *Journal of Chemical Information and Modeling, 53,* 3076–3085.

Liu, W., Li, P., & Mei, Y. (2019). Discovery of SBF1 as an allosteric inhibitor targeting the PIF-pocket of 3-phosphoinositide-dependent protein kinase-1. *Journal of Molecular Modeling, 25.*

Li, G.-B., Yang, L.-L., Wang, W.-J., Li, L.-L., & Yang, S.-Y. (2013). ID-score: A new empirical scoring function based on a comprehensive set of descriptors related to protein-ligand interactions. *Journal of Chemical Information and Modeling, 53,* 592–600.

Lovly, C. M., & Shaw, A. T. (2014). Molecular pathways: Resistance to kinase inhibitors and implications for therapeutic strategies. *Clinical Cancer Research, 20,* 2249–2256.

Maddox, S., Hecht, D., & Gustafson, J. L. (2016). Enhancing the selectivity of kinase inhibitors in oncology: A chemical biology perspective. *Future Medicinal Chemistry, 8,* 241–244.

Maennling, A. E., Tur, M. K., Niebert, M., Klockenbring, T., Zeppernick, F., Gattenlöhner, S., Meinhold-Heerlein, I., & Hussain, A. F. (2019). Molecular targeting therapy against EGFR family in breast cancer: Progress and future potentials. *Cancers, 11.*

Manning, G. (2002). The protein kinase complement of the human genome. *Science, 298,* 1912–1934.

McCubrey, J. A., Steelman, L. S., Chappell, W. H., Abrams, S. L., Wong, E. W. T., Chang, F., Lehmann, B., Terrian, D. M., Milella, M., Tafuri, A., Stivala, F., Libra, M., Basecke, J., Evangelisti, C., Martelli, A. M., & Franklin, R. A. (2007). Roles of the Raf/MEK/ERK pathway in cell growth, malignant transformation and drug resistance. *Biochimica et Biophysica Acta, 1773,* 1263–1284.

Mendelsohn, J., & Baselga, J. (2006). Epidermal growth factor receptor targeting in cancer. *Seminars in Oncology, 33,* 369–385.

Modi, V., & Dunbrack, R. L. (2019). Defining a new nomenclature for the structures of active and inactive kinases. *Proceedings of the National Academy of Sciences United States of America, 116,* 6818–6827.

Muckelbauer, J., Sack, J. S., Ahmed, N., Burke, J., Chang, C. Y., Gao, M., Tino, J., Xie, D., & Tebben, A. J. (2011). X-ray crystal structure of bone marrow kinase in the x chromosome: A tec family kinase. *Chemical Biology and Drug Design, 78,* 739–748.

Najmanovich, R., Kuttner, J., Sobolev, V., & Edelman, M. (2000). Side-chain flexibility in proteins upon ligand binding. *Proteins: Structure, Function, and Genetics, 39,* 261–268.

Ohren, J. F., Chen, H., Pavlovsky, A., Whitehead, C., Zhang, E., Kuffa, P., Yan, C., McConnell, P., Spessard, C., Banotai, C., Mueller, W. T., Delaney, A., Omer, C., Sebolt-Leopold, J., Dudley, D. T., Leung, I. K., Flamme, C., Warmus, J., Kaufman, M., … Hasemann, C. A. (2004). Structures of human MAP kinase kinase 1 (MEK1) and MEK2 describe novel noncompetitive kinase inhibition. *Nature Structural and Molecular Biology, 11,* 1192–1197.

Oostenbrink, C., Villa, A., Mark, A. E., & van Gunsteren, W. F. (2004). A biomolecular force field based on the free enthalpy of hydration and solvation: The GROMOS force-field parameter sets 53A5 and 53A6. *Journal of Computational Chemistry, 25*, 1656−1676.

Ouyang, X., Handoko, S. D., & Kwoh, C. K. (2011). CSCORE: A simple yet effective scoring function for protein−ligand binding affinity prediction using modified cmac learning architecture. *Journal of Bioinformatics and Computational Biology, 09*, 1−14.

O'Hare, T., Shakespeare, W. C., Zhu, X., Eide, C. A., Rivera, V. M., Wang, F., Adrian, L. T., Zhou, T., Huang, W.-S., Xu, Q., Metcalf, C. A., Tyner, J. W., Loriaux, M. M., Corbin, A. S., Wardwell, S., Ning, Y., Keats, J. A., Wang, Y., Sundaramoorthi, R., … Clackson, T. (2009). AP24534, a pan-BCR-ABL inhibitor for chronic myeloid leukemia, potently inhibits the T315I mutant and overcomes mutation-based resistance. *Cancer Cell, 16*, 401−412.

Park, J. H., Liu, Y., Lemmon, M. A., & Radhakrishnan, R. (2012). Erlotinib binds both inactive and active conformations of the EGFR tyrosine kinase domain. *Biochemical Journal, 448*, 417−423.

Patel, D., Athar, M., & Jha, P. C. (2020). Exploring ruthenium-based organometallic inhibitors against plasmodium calcium dependent kinase 2 (PfCDPK2): A combined ensemble docking, QM paramterization and molecular dynamics study (preprint). *Biorxiv*. https://doi.org/10.1101/2020.03.31.01754.

Pemovska, T., Johnson, E., Kontro, M., Repasky, G. A., Chen, J., Wells, P., Cronin, C. N., McTigue, M., Kallioniemi, O., Porkka, K., Murray, B. W., & Wennerberg, K. (2015). Axitinib effectively inhibits BCR-ABL1(T315I) with a distinct binding conformation. *Nature, 519*, 102−105.

Pereira, J. C., Caffarena, E. R., & dos Santos, C. N. (2016). Boosting docking-based virtual screening with deep learning. *Journal of Chemical Information and Modeling, 56*, 2495−2506.

Phillips, J. C., Braun, R., Wang, W., Gumbart, J., Tajkhorshid, E., Villa, E., Chipot, C., Skeel, R. D., Kalé, L., & Schulten, K. (2005). Scalable molecular dynamics with NAMD. *Journal of Computational Chemistry, 26*, 1781−1802.

Piana, S., Klepeis, J. L., & Shaw, D. E. (2014). Assessing the accuracy of physical models used in protein-folding simulations: Quantitative evidence from long molecular dynamics simulations. *Current Opinion in Structural Biology, 24*, 98−105.

Plimpton, S. (1995). Fast parallel algorithms for short-range molecular dynamics. *Journal of Computational Physics, 117*, 1−19.

Qin, M., Tian, Y., Han, X., Cao, Q., Zheng, S., Liu, C., Wu, X., Liu, L., Meng, Y., Wang, X., Zhang, H., & Hou, Y. (2020). Structural modifications of indolinones bearing a pyrrole moiety and discovery of a multi-kinase inhibitor with potent antitumor activity. *Bioorganic and Medicinal Chemistry, 115486*.

Qiu, C., Tarrant, M. K., Choi, S. H., Sathyamurthy, A., Bose, R., Banjade, S., Pal, A., Bornmann, W. G., Lemmon, M. A., Cole, P. A., & Leahy, D. J. (2008). Mechanism of activation and inhibition of the HER4/ErbB4 kinase. *Structure, 16*, 460−467.

Ragoza, M., Hochuli, J., Idrobo, E., Sunseri, J., & Koes, D. R. (2017). Protein−ligand scoring with convolutional neural networks. *Journal of Chemical Information and Modeling, 57*, 942−957.

Rahman, A. (1964). Correlations in the motion of atoms in liquid argon. *Physical Review, 136*, A405−A411.

Rarey, M., Kramer, B., Lengauer, T., & Klebe, G. (1996). A fast flexible docking method using an incremental construction algorithm. *Journal of Molecular Biology, 261*, 470−489.

Raval, A., Piana, S., Eastwood, M. P., Dror, R. O., & Shaw, D. E. (2012). Refinement of protein structure homology models via long, all-atom molecular dynamics simulations. *Proteins, 80*, 2071−2079.

Ravindranath, P. A., Forli, S., Goodsell, D. S., Olson, A. J., & Sanner, M. F. (2015). AutoDockFR: Advances in protein-ligand docking with explicitly specified binding site flexibility. *PLoS Computational Biology, 11*.

Renukuntla, J., Vadlapudi, A. D., Patel, A., Boddu, S. H. S., & Mitra, A. K. (2013). Approaches for enhancing oral bioavailability of peptides and proteins. *International Journal of Pharmaceutics, 447*, 75−93.

Rettenmaier, T. J., Fan, H., Karpiak, J., Doak, A., Sali, A., Shoichet, B. K., & Wells, J. A. (2015). Small-molecule allosteric modulators of the protein kinase PDK1 from structure-based docking. *Journal of Medicinal Chemistry, 58*, 8285−8291.

Roskoski, R. (2015). A historical overview of protein kinases and their targeted small molecule inhibitors. *Pharmacological Research, 100*, 1−23.

Roskoski, R. (2016). Classification of small molecule protein kinase inhibitors based upon the structures of their drug-enzyme complexes. *Pharmacological Research, 103*, 26−48.

Roskoski, R. (2019a). Properties of FDA-approved small molecule protein kinase inhibitors. *Pharmacological Research, 144*, 19−50.

Roskoski, R. (2019b). Cyclin-dependent protein serine/threonine kinase inhibitors as anticancer drugs. *Pharmacological Research, 139*, 471−488.

Sandak, B., Wolfson, H. J., & Nussinov, R. (1998). Flexible docking allowing induced fit in proteins: Insights from an open to closed conformational isomers. *Proteins: Structure, Function, and Genetics, 32*, 159−174.

Sawyers, C. L. (2003). Opportunities and challenges in the development of kinase inhibitor therapy for cancer. *Genes and Development, 17*, 2998−3010.

Schulze, J. O., Saladino, G., Busschots, K., Neimanis, S., Süß, E., Odadzic, D., Zeuzem, S., Hindie, V., Herbrand, A. K., Lisa, M.-N., Alzari, P. M., Gervasio, F. L., & Biondi, R. M. (2016). Bidirectional allosteric communication between the ATP-binding site and the regulatory PIF pocket in PDK1 protein kinase. *Cell Chemical Biology, 23*, 1193−1205.

Shah, N. P., Tran, C., Lee, F. Y., Chen, P., Norris, D., & Sawyers, C. L. (2004). Overriding imatinib resistance with a novel ABL kinase inhibitor. *Science, 305*, 399−401.

Sharma, V. K., Nandekar, P. P., Sangamwar, A., Pérez-Sánchez, H., & Agarwal, S. M. (2016). Structure guided design and binding analysis of EGFR inhibiting analogues of erlotinib and AEE788 using ensemble docking, molecular dynamics and MM-GBSA. *RSC Advances, 6,* 65725–65735.

Shi, Z., Resing, K. A., & Ahn, N. G. (2006). Networks for the allosteric control of protein kinases. *Current Opinion in Structural Biology, 16,* 686–692.

Shukla, D., Meng, Y., Roux, B., & Pande, V. S. (2014). Activation pathway of Src kinase reveals intermediate states as targets for drug design. *Nature Communications, 5.*

Singh, N., & Li, W. (2019). Recent advances in coarse-grained models for biomolecules and their applications. *International Journal of Molecular Sciences, 20.*

Smyth, L. A., & Collins, I. (2009). Measuring and interpreting the selectivity of protein kinase inhibitors. *Journal of Chemical Biology, 2,* 131–151.

Solca, F., Dahl, G., Zoephel, A., Bader, G., Sanderson, M., Klein, C., Kraemer, O., Himmelsbach, F., Haaksma, E., & Adolf, G. R. (2012). Target binding properties and cellular activity of afatinib (BIBW 2992), an irreversible ErbB family blocker. *Journal of Pharmacology and Experimental Therapeutics, 343,* 342–350.

Spitaleri, A., Decherchi, S., Cavalli, A., & Rocchia, W. (2018). Fast dynamic docking guided by adaptive electrostatic bias: The MD-binding approach. *Journal of Chemical Theory and Computation, 14,* 1727–1736.

Springer, C., Adalsteinsson, H., Young, M. M., Kegelmeyer, P. W., & Roe, D. C. (2005). PostDOCK: A structural, empirical approach to scoring protein ligand complexes. *Journal of Medicinal Chemistry, 48,* 6821–6831.

Spyrakis, F., Benedetti, P., Decherchi, S., Rocchia, W., Cavalli, A., Alcaro, S., Ortuso, F., Baroni, M., & Cruciani, G. (2015). A pipeline to enhance ligand virtual screening: Integrating molecular dynamics and fingerprints for ligand and proteins. *Journal of Chemical Information and Modeling, 55,* 2256–2274.

Stamos, J., Sliwkowski, M. X., & Eigenbrot, C. (2002). Structure of the epidermal growth factor receptor kinase domain alone and in complex with a 4-anilinoquinazoline inhibitor. *Journal of Biological Chemistry, 277,* 46265–46272.

Stolzenberg, S., Michino, M., LeVine, M. V., Weinstein, H., & Shi, L. (2016). Computational approaches to detect allosteric pathways in transmembrane molecular machines. *Biochimica et Biophysica Acta (BBA) - Biomembranes, 1858,* 1652–1662.

Su, Q., Banks, E., Bebernitz, G., Bell, K., Borenstein, C. F., Chen, H., Chuaqui, C. E., Deng, N., Ferguson, A. D., Kawatkar, S., Grimster, N. P., Ruston, L., Lyne, P. D., Read, J. A., Peng, X., Pei, X., Fawell, S., Tang, Z., Throner, S., … Kettle, J. G. (2020). Discovery of (2R)-N-[3-[2-[(3-Methoxy-1-methyl-pyrazol-4-yl)amino]pyrimidin-4-yl]-1H-indol-7-yl]-2-(4-methylpiperazin-1-yl)propenamide (AZD4205) as a potent and selective janus kinase 1 inhibitor. *Journal of Medicinal Chemistry, 63*(9), 4517–4527.

Tajbakhsh, N., Shin, J. Y., Gurudu, S. R., Hurst, R. T., Kendall, C. B., Gotway, M. B., & Liang, J. (2016). Convolutional neural networks for medical image analysis: Full training or fine tuning? *IEEE Transactions on Medical Imaging, 35,* 1299–1312.

Taylor, S. S., Keshwani, M. M., Steichen, J. M., & Kornev, A. P. (2012). Evolution of the eukaryotic protein kinases as dynamic molecular switches. *Philosophical Transactions of the Royal Society B: Biological Sciences, 367,* 2517–2528.

Taylor, S. S., & Kornev, A. P. (2011). Protein kinases: Evolution of dynamic regulatory proteins. *Trends in Biochemical Sciences, 36,* 65–77.

Thomas, R., & Weihua, Z. (2019). Rethink of EGFR in cancer with its kinase independent function on board. *Frontiers in Oncology, 9.*

Thress, K. S., Paweletz, C. P., Felip, E., Cho, B. C., Stetson, D., Dougherty, B., Lai, Z., Markovets, A., Vivancos, A., Kuang, Y., Ercan, D., Matthews, S., Cantarini, M., Barrett, J. C., Jänne, P. A., & Oxnard, G. R. (2015). Acquired EGFR C797S mediates resistance to AZD9291 in advanced non-small cell lung cancer harboring EGFR T790M. *Nature Medicine, 21,* 560–562.

Tian, S., Sun, H., Pan, P., Li, D., Zhen, X., Li, Y., & Hou, T. (2014). Assessing an ensemble docking-based virtual screening strategy for kinase targets by considering protein flexibility. *Journal of Chemical Information and Modeling, 54,* 2664–2679.

Torres, P. H. M., Sodero, A. C. R., Jofily, P., & Silva, F. P., Jr. (2019). Key topics in molecular docking for drug design. *International Journal of Molecular Sciences, 20,* 4574.

Uehara, S., & Tanaka, S. (2017). Cosolvent-based molecular dynamics for ensemble docking: Practical method for generating druggable protein conformations. *Journal of Chemical Information and Modeling, 57,* 742–756.

Vainchenker, W., Leroy, E., Gilles, L., Marty, C., Plo, I., & Constantinescu, S. N. (2018). JAK inhibitors for the treatment of myeloproliferative neoplasms and other disorders. *F1000Research, 7,* 82.

Vanderpool, D., Johnson, T. O., Ping, C., Bergqvist, S., Alton, G., Phonephaly, S., Rui, E., Luo, C., Deng, Y.-L., Grant, S., Quenzer, T., Margosiak, S., Register, J., Brown, E., & Ermolieff, J. (2009). Characterization of the CHK1 allosteric inhibitor binding site. *Biochemistry, 48,* 9823–9830.

Wang, Y., Guo, Y., Kuang, Q., Pu, X., Ji, Y., Zhang, Z., & Li, M. (2015). A comparative study of family-specific protein-ligand complex affinity prediction based on random forest approach. *Journal of Computer-Aided Molecular Design, 29,* 349–360.

Weiner, S. J., Kollman, P. A., Case, D. A., Singh, U. C., Ghio, C., Alagona, G., Profeta, S., & Weiner, P. (1984). A new force field for molecular mechanical simulation of nucleic acids and proteins. *Journal of the American Chemical Society, 106,* 765–784.

Weisberg, E., Manley, P., Mestan, J., Cowan-Jacob, S., Ray, A., & Griffin, J. D. (2006). AMN107 (nilotinib): A novel and selective inhibitor of BCR-ABL. *British Journal of Cancer, 94,* 1765–1769.

White, A., Pargellis, C. A., Studts, J. M., Werneburg, B. G., & Farmer, B. T. (2007). Molecular basis of MAPK-activated protein kinase 2:p38 assembly. *Proceedings of the National*

Academy of Sciences of the United States of America, 104, 6353–6358.

Wójcik, P., & Berlicki, Ł. (2016). Peptide-based inhibitors of protein-protein interactions. *Bioorganic and Medicinal Chemistry Letters, 26,* 707–713.

Wu, P., Nielsen, T. E., & Clausen, M. H. (2015). FDA-approved small-molecule kinase inhibitors. *Trends in Pharmacological Sciences, 36,* 422–439.

Wu, J., Tseng, Y. D., Xu, C.-F., Neubert, T. A., White, M. F., & Hubbard, S. R. (2008). Structural and biochemical characterization of the KRLB region in insulin receptor substrate-2. *Nature Structural and Molecular Biology, 15,* 251–258.

Yan, M., Wang, H., Wang, Q., Zhang, Z., & Zhang, C. (2016). Allosteric inhibition of c-Met kinase in sub-microsecond molecular dynamics simulations induced by its inhibitor, tivantinib. *Physical Chemistry Chemical Physics, 18,* 10367–10374.

Yin, D., & MacKerell, A. D. (1998). Combined ab initio/empirical approach for optimization of Lennard–Jones parameters. *Journal of Computational Chemistry, 19,* 334–348.

Yosaatmadja, S., McKeage, F., To be Published. 1.85 angstrom structure of EGFR kinase domain with gefitinib.

Zhang, J., Adrián, F. J., Jahnke, W., Cowan-Jacob, S. W., Li, A. G., Iacob, R. E., Sim, T., Powers, J., Dierks, C., Sun, F., Guo, G.-R., Ding, Q., Okram, B., Choi, Y., Wojciechowski, A., Deng, X., Liu, G., Fendrich, G., Strauss, A., ... Gray, N. S. (2010). Targeting Bcr-Abl by combining allosteric with ATP-binding-site inhibitors. *Nature, 463,* 501–506.

Zhang, H., Liao, L., Saravanan, K. M., Yin, P., & Wei, Y. (2019). DeepBindRG: A deep learning based method for estimating effective protein–ligand affinity. *PeerJ, 7,* e7362.

Zhao, H., Huang, D., & Caflisch, A. (2012). Discovery of tyrosine kinase inhibitors by docking into an inactive kinase conformation generated by molecular dynamics. *ChemMedChem, 7,* 1983–1990.

Zhou, T., Commodore, L., Huang, W.-S., Wang, Y., Thomas, M., Keats, J., Xu, Q., Rivera, V. M., Shakespeare, W. C., Clackson, T., Dalgarno, D. C., & Zhu, X. (2011). Structural mechanism of the pan-BCR-ABL inhibitor ponatinib (AP24534): Lessons for overcoming kinase inhibitor resistance. *Chemical Biology and Drug Design, 77,* 1–11.

Zuccotto, F., Ardini, E., Casale, E., & Angiolini, M. (2010). Through the "gatekeeper door": Exploiting the active kinase conformation. *Journal of Medicinal Chemistry, 53,* 2681–2694.

CHAPTER 17

Docking Approaches Used in Epigenetic Drug Investigations

YUDIBETH SIXTO-LÓPEZ • JOSÉ CORREA-BASURTO

1 EPIGENETICS

The term "epigenetics" was coined by Conrad Waddington in 1942, which refers to the mechanisms through which an organism adapts its phenotype to the environmental conditions; besides, it is also involved in the development and differentiation of superior organisms (Waddington, 2012). Currently, this term implies the way in which transcription is modified without modifications of the DNA sequence, which are also heritable. These modifications are more evident in multicellular organisms, as the cells possess the same genome and are able to differentiate and generate different tissues and organs with distinct functions from the others (Ganesan et al., 2019; Handy et al., 2011). But several definitions are proposed, from biological point of view: it is the regulation of eukaryotic gene expression by chromatin remodeling process, and the other definitions include the chemical perspective that implies covalent and reversible modifications on DNA and histone proteins (Deans & Maggert, 2015; Ganesan et al., 2019; Lu et al., 2018).

Chromatin is a complex of DNA and protein located in the nucleus of the eukaryotic cells. The DNA is wrapped around histone proteins forming the nucleosome (Berger, 2007; Handy et al., 2011). The nucleosome is composed by 147 base pairs (bp) of DNAs wrapped around an octameric core of histones, which is formed by two H3—H4 histone dimers surrounded by two H2A—H2B dimers, and 20—50 bp that link nucleosomes forming a structure that seems like beads and strings. The N-terminal histone tails protrude from the nucleosome toward the exterior; the protein histone H1 is located at the end of the nucleosome to form the chromosome which is condensed forming the chromatin (Hu et al., 2018; Robinson & Rhodes, 2006). Chromatin structure is divided into heterochromatin and euchromatin; the former is the closed state of the chromatin that is associated with transcriptional repression and the latter is the open state that favors the transcriptional process. This chromatin conformational changes are driven by posttranslational modifications on histone proteins and covalent modifications on DNA (Biswas & Rao, 2018). The epigenetic modifications are complex, dynamic, and highly regulated; at molecular level, the structural and chemical modifications are carried out on nucleic acids and also on the histone tails; these modifications are often referred to as "epigenetic marks," which conform the epigenetic code (Berger, 2007; Ganesan et al., 2019; Lu et al., 2018; Rutten & Mill, 2009).

This code confers the ability to adapt and respond to the environmental conditions during the life of an organism, and the proteins responsible to carry out the modifications are broadly divided into three groups: writers, readers, and erasers. Writers are enzymes that add small chemical groups such as a methyl group or a bigger ubiquitin molecule. These modifications influence the binding affinity between DNA and histone proteins and, also, in the recruitment of other molecules such as noncoding RNAs and chromatin remodelers. Readers are able to recognize the modifications added by writers and recruit chaperons. Finally, erasers are enzymes that efficiently catalyze the removal of these covalent modifications, making the process reversible (Biswas & Rao, 2018; Ganesan et al., 2019; Lu et al., 2018).

1.1 Writers in Epigenetics

Writers are enzymes which add chemical groups, such as methylation, acetylation, phosphorylation, ubiquitination, SUMOylation, and ADP-ribosylation, among others. This chemical addition could be carried onto DNA, histone proteins, or nonhistone substrates. But we are focused on the two most commonly studied epigenetic marks, methylation and acetylation from cofactor S-adenosyl-L-methionine and acetyl coenzyme

Molecular Docking for Computer-Aided Drug Design. https://doi.org/10.1016/B978-0-12-822312-3.00016-3

A (Ac-CoA), respectively. Methylation occurs on both DNA and histone proteins, whereas acetylation is associated only with histones (Biswas & Rao, 2018; Rossetto et al., 2012).

In turn, writers are divided into three groups: DNA methyltransferases (DNMTs), protein lysine/arginine methyltransferases, and histone acetyltransferases (HATs), whose activity affect the chromatin organization and also affect the gene transcription (Biswas & Rao, 2018; Lu et al., 2018) (Fig. 17.1).

1.2 Reader in Epigenetics

Readers are those enzymes that recognize the epigenetic marks that writers have added to further recruit downstream effectors to execute a wide variety of functions. Several epigenetic readers have been described, which include methyl-lysine readers, methyl-arginine readers, acetyl-lysine readers, and phospho-serine readers. They have a specialized domain that can identify and bind to modifications present on DNA and histones added by writers (Biswas & Rao, 2018; Lu et al., 2018). Methyl-lysine readers include PHD zinc finger domain, ankyrin repeats, MBT, WD40, Tudor, double/tandem Tudor, zf-CW, PWWP, and chromodomains. While methyl arginine readers, the primary readers of this kind of marks are Tudor domain containing proteins. Till now there are approximately 36, of this at least 10 bind methyl-lysine motifs as above stated and 8 bind methyl-arginine motifs. The Tudor domains that bind methyl-arginine motifs are survival motor neuron protein, which modulate the assembly of RNA–

protein complexes (small nuclear ribonucleoproteins) and bind the spliceosomal core proteins SmD1, SmD3, and SmB/B′ through its Tudor domain. Other examples are TDRD3, SND1, TDRD1-2, 6, and 9, the PAF1 complex, BRCT domains, and small nuclear RNA as a direct effector of methyl-arginine marks (Gayatri & Bedford, 2014; Greer & Shi, 2012). Finally, acetyl-lysine readers consist of bromodomains and the tandem PHD domains. Bromodomains (BRD) are domains that recognize ε-amine acetylated lysine and are composed by a great variety of proteins despite their BRD sharing a conserved folding (61 bromodomains). It is a left-handed bundle of four α-helices linked by loop regions with different sizes (ZA and BC loops) that house the acetylated lysine binding site. BRD are classified into eight families according to structural/sequence similarity. BRD include HATs (GCN5, PCAF), ATP-dependent chromatin-remodeling complexes (BAZ1B), helicases (SMARCA), methyltransferases (MLL, ASH1L), transcriptional coactivators (TRIM/TIF1, TAFs), transcriptional mediators (TAF1), nuclear scaffolding proteins (PB1), and the BET family (Filippakopoulos et al., 2012) (Fig. 17.1).

1.3 Erasers in Epigenetics

The covalent modifications previously added by writers can be removed by the eraser enzymes, depending on the requirement of the cell regarding the maintenance of the epigenetic marks with the end of modifying the expression states of the locus or modulating the transcriptional state of the cell. Based on the relative

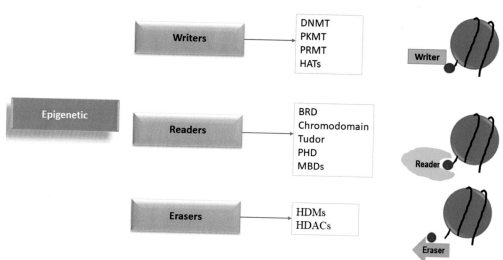

FIG. 17.1 "Epigenetic" readers, writers, and erasers. *BRD*, bromodomains; *DNMT*, DNA methyltransferase; *HATs*, histone acetyltransferases; *HDACs*, histone deacetylases; *HDMs*, histone demethyltransferases; *PKMT*, protein lysine methyltransferase; *PRMT*, protein arginine methyltransferase.

functions and substrate affinity, erasers are divided into several families including histone demethyltransferases (HDMs), DNA demethylases, histone deacetylases (HDACs), among others (Biswas & Rao, 2018; Lu et al., 2018) (Fig. 17.1).

2 EPIGENETICS AND DISEASES

There are several human diseases related to epigenetic modulation due to external factors such as nutrients, drugs, environment (air, water, etc.) pollution, etc. The mentioned factors could be prevented, however, due to the lifestyle of people; it is not possible sometimes to prevent the epigenetic modulation. The modulation of epigenetic factor decreases or increases the gene expression that dysregulates the cell function resulting in a complex disease. In case that the corresponding disease is in early stages, it is possible to be treated by modulating the epigenetic changes. Epigenetic dysregulation is associated with genesis, development, and progression of disease states. Epigenetic imbalance has been observed in some diseases such as cancer, metabolic diseases, cardiovascular diseases, chronic diseases, immune diseases, neurodegenerative diseases, diabetes, etc. (Ganesan et al., 2019; Lu et al., 2018; Lundstrom, 2017).

The identification and characterization of epigenetic modifications can also be exploited to establish biomarkers of diseases as potential diagnosis or prognosis parameter. In this sense, there are tests that allow to determine the status of DNA methylation of some genes such as septin 9 or NDGR4 and BMP3 for colorectal cancer that are approved by the FDA (Lamb & Dhillon, 2017). Many other epigenetic marks have been studied for other diseases that can be further commercialized (Garcia-Gimenez et al., 2017). The other side of epigenetic alterations is to identify them and find their role in the genesis or development of the disease in order to target them. As in some cases, the expression profile or activity of the writers, readers, or erasers is affected in disease condition in comparison with healthy state. As epigenetic alterations have been described in several diseases, such as cancer, neurological disorders, and metabolic disorders, we will discuss some epigenetics targeting drugs useful for treating those diseases.

3 DRUG DISCOVERY OF EPIGENETIC MODULATORS WITH DOCKING

Nowadays, writer, erasers, and readers in epigenetics are extensively under investigation, as they are considered as pharmacological targets to design molecules

(Ganesan et al., 2019; Lundstrom, 2017). The use of computers in drug design research field is essential, as it allows the use of novel and efficient techniques that reduce the time and cost involved in drug discovery (Yu & MacKerell, 2017). One of the computational tools is molecular docking that allows to predict the molecular binding between two molecules (small on macromolecule) as well as the molecular and energetics patterns that govern the recognition (Morris & Lim-Wilby, 2008). Several examples can be found, such as DNMT inhibitors, HAT inhibitors, histone lysine methyltransferase (HKMT) inhibitors, HDAC inhibitors (HDACi), histone lysine demethylase (HKDM) inhibitors, bromodomain inhibitors (Biswas & Rao, 2018) design using molecular docking, which are described below.

3.1 DNA Methyltransferases Inhibitors

DNMT inhibitors have been discovered from natural as well as synthetic sources employing molecular docking by either rational design or virtual screening from databases (Medina-Franco et al., 2014; Saldivar-Gonzalez et al., 2018). Medina-Franco et al. by virtual screening protocol using a multistep docking of natural products identified DNMT1 inhibitors. Most of these compounds have a coumarin scaffold; besides, they found that the side chains of the hit compounds have carboxylic group that formed hydrogen bonds with residues with important role in the DNA methylation (Medina-Franco et al., 2011) (Fig. 17.2).

Molecular docking has also been used to generate pharmacophore of DMNT inhibitors using 14 well-established DNTM1 inhibitors, among which are 5-azacitidine, decitabine, and parthenolide. The structure-based pharmacophore model obtained suggests that interactions with E1265, R1311, R1461, S1229, and G1230 are important, and interactions with one negative charge, hydrogen bond acceptor, aromatic ring, and two hydrogen bond donors are necessary (Yoo & Medina-Franco, 2011). In the same line, combination of molecular docking and molecular dynamics (MD) simulation was used by Yoo et al. and found that aurintricarboxylic acid (ATA) inhibits DNMT1 ($IC_{50} = 0.68\ \mu M$) and DNMT3a ($IC_{50} = 1.4\ \mu M$) in in vitro enzymatic inhibitory assay. Docking of ATA on DNTM1 (PDB code: 1MHT1) showed that it forms hydrogens bond with E1266, R1310, R1312, S1230, G1231, and K1535 by its hydroxybenzoic acid and its second carboxylate group makes interaction with N1578 (Yoo et al., 2012) (Fig. 17.3). Furthermore, Maldonado-Rojas et al. (2015) combined quantitative structure–activity

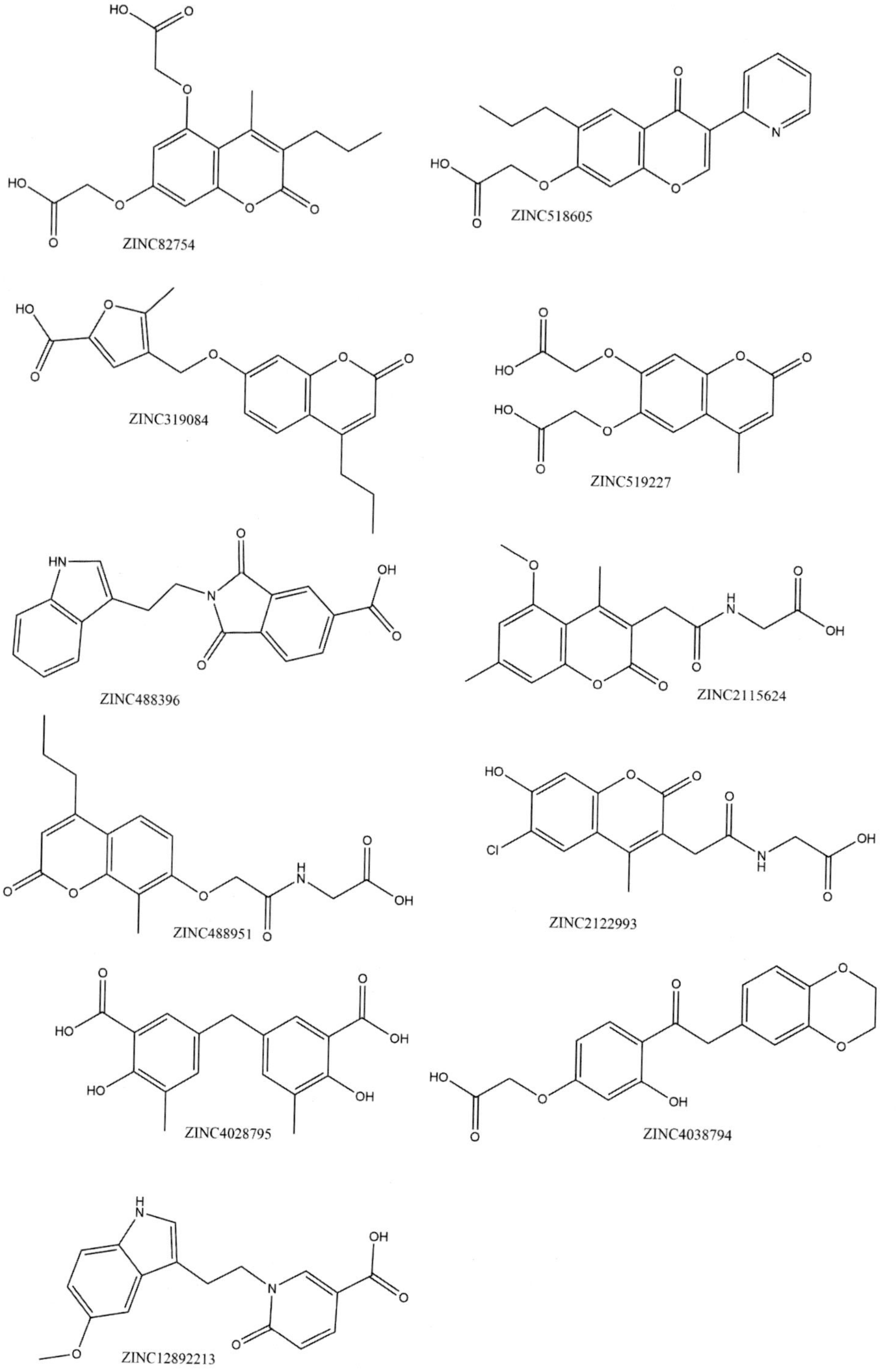

FIG. 17.2 Candidate DNMT1 inhibitors identified by multistep docking-based virtual screening approach (Medina-Franco et al., 2011).

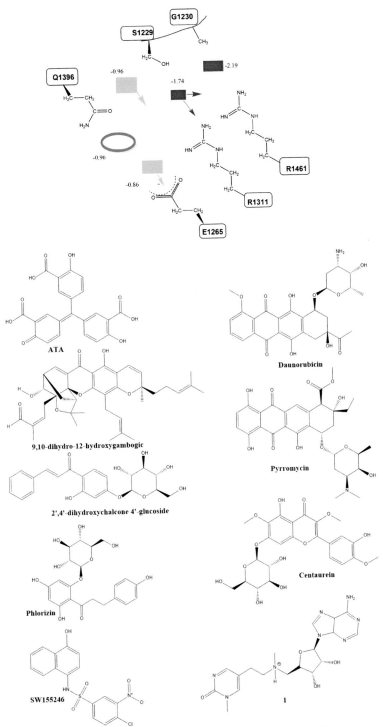

FIG. 17.3 DMNT1 and DNMT3A inhibitors discovered by molecular docking. Pharmacophore of DNMT1 inhibitors is depicted at the upper side of the figure. (Adapted from Yoo, J., & Medina-Franco, J. L. (2011). Homology modeling, docking and structure-based pharmacophore of inhibitors of DNA methyltransferase. *Journal of Computer-Aided Molecular Design, 25*(6), 555—567.)

relationship (QSAR) study with molecular docking, using AutoDock Vina and Surflex-Dock, to identify six natural product compounds as potential DNMT1 and DNMT3A inhibitors, that is, 9,10-dihydro-12-hydroxygambogic, phloridzin, 2′,4′-dihydroxychalcone 4′-glucoside, daunorubicin, pyrromycin, and centaurein (Maldonado-Rojas et al., 2015). Another example of combination is docking with MD simulations used to design DNMT1 inhibitors that reached both active site and cofactor site acting as transition state analog **1**. Besides, it is suggested that these compounds might target DNA-bond and DNA-free form of DNMT1 following a suicide inhibition step after bound to Cys1226 in the active site (Miletic et al., 2017) (Fig. 17.3).

Molecular docking helps to elucidate the mechanism of action of DNMT inhibitors that allow to establish the structure–activity relationships (SARs). This approach was used by employing induced fit docking to develop a model of the binding between SW155246 and DNMT1, as a result it was elucidated that SW155246 occupies part of the cofactor binding site as well as the catalytic site. SW15246 is a sulfonamide derivative that formed hydrogen bonds with N1267 in ENV motif at the catalytic site and hydrogen bond between the nitro group with R1312, an important residue that participates in the mechanism of cytosine-C5 methylation (Medina-Franco et al., 2014). DNMT1 was co-crystallized with its cofactor S-adenosyl-ʟ-homocysteine, and the interactions visualized; the interactions stabilized mainly by hydrogen bonds with S1400 and Q14455, π-π interactions, alkyl interactions and interactions of the residues with halogen groups (Zhang et al., 2015) (Fig. 17.4).

3.2 Histone Lysine Methyltransferase Inhibitors

For HKMT inhibitors (Fig. 17.5), several approaches have been used not only to target human HKMT such as G9a (Charles et al., 2020; Srimongkolpithak et al., 2014) but also some other organism like *Plasmodium falciparum* rendering potential antimalarial compounds (Sharma et al., 2014). Srimongkol pithak et al. designed and synthesized heterocyclic derivatives to obtain G9a inhibitors HKMI-1-248 (**41**) and HKMI-1-247 (**42**). They were obtained using molecular docking and found out that dimethoxy groups on the benzenoid ring and a basic nitrogen at position 1 were important for obtaining potent G9a inhibitory activity (Srimongkolpithak et al., 2014). The use of knowledge- and structure-based approaches assisted by molecular docking allowed the discovery of G9a and DNMT inhibitor **CMC-272** with a G9a IC$_{50}$ of 8 nM and DNMT1 IC$_{50}$ of 382 nM. This was identified by docking the hit **11** (G9a IC$_{50}$ of 0.7 nM and DNMT1 IC$_{50}$ of 619 nM) against DNMT1.

Previously, by structural studies in G9a, it was found that quinazoline-based inhibitor (**UNC-0638**) could be bound to histone binding pocket of G9, pointing out that the nitrogen atom of the N3 of the quinazoline

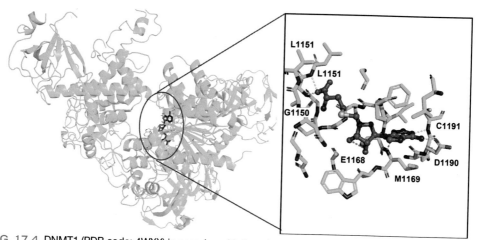

FIG. 17.4 DNMT1 (PDB code: 4WXX) in complex with its cofactor S-adenosyl-ʟ-homocysteine (SAH). In left panel is depicted the cofactor on the DNMT1 binding site. While in right panel, a zoom that shows the interaction is observed, only those residues with which SAH are forming hydrogen bond are labeled. DNMT1 residues are depicted as stick green color and SAH are depicted as magenta ball and stick. (Figure was generated using Pymol 2.0 DeLano, W. L. (2002). *PyMOL* (pp. 700). San Carlos, CA: DeLano Scientific.)

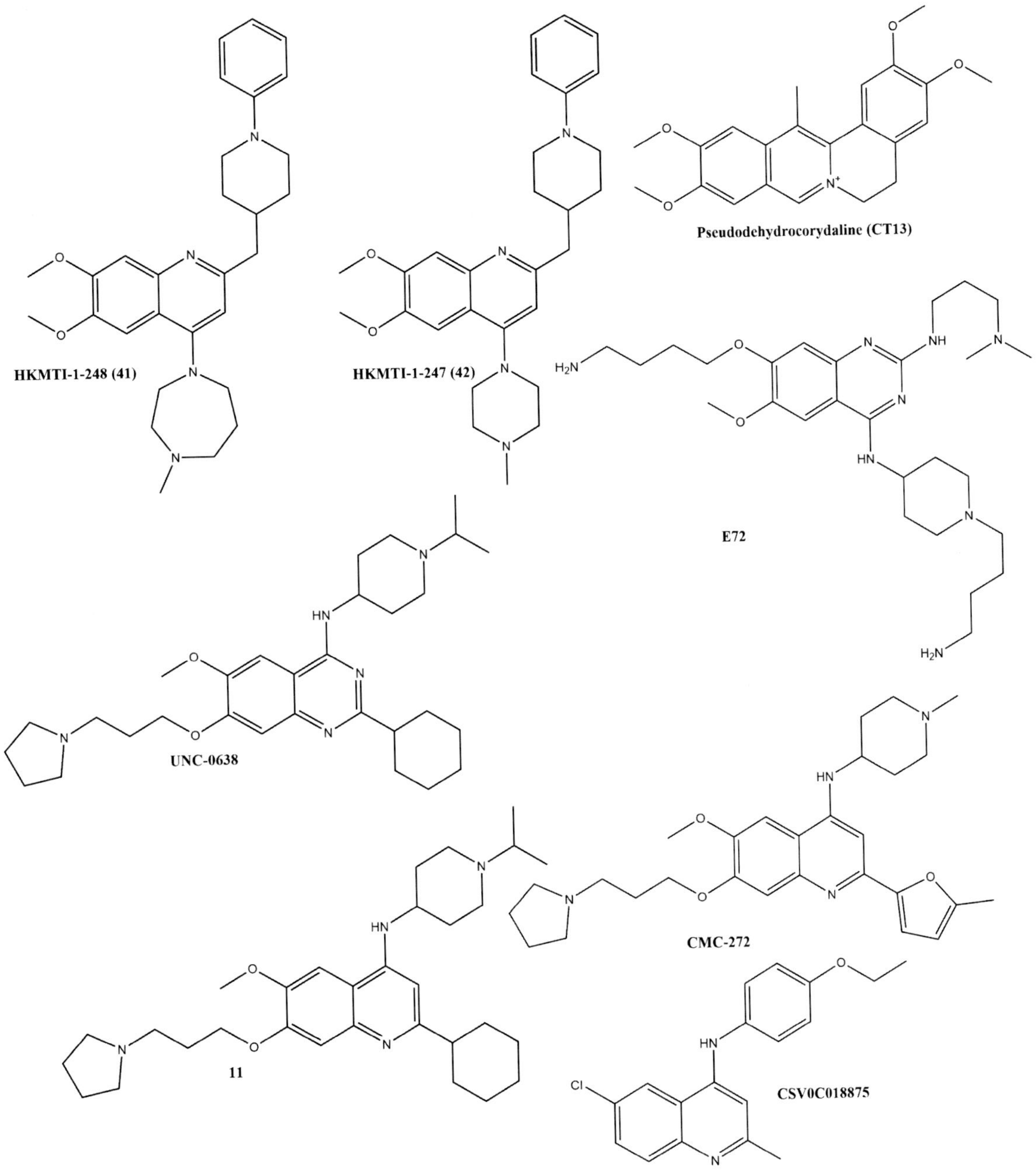

FIG. 17.5 Histone lysine methyltransferase inhibitors discovered by molecular docking.

does not participate in any direct key interaction with the protein; therefore, the quinazoline was replaced by quinoline scaffold, rendering the compound **11** with improved activity against G9 and DNMT1. Compound **11** is accommodated into the DNA binding pocket, overlapping on the hemi-methylated CpG.

The pyrrolidine ring is located at the cytidine-binding pocket through hydrogen bond with the carboxylate group of Q1269 (Q1266 in human DNMT1), and the oxygen is bound at the 7-position forming a hydrogen bond with R1315 (R1312, hDNMT1), while the amino group of the 4-position forms hydrogen bond with

S1233 (S1230 in hDNMT1). The isopropylpiperidyl group makes hydrophobic interactions with M1235 (M1232 in hDNMT1), and the cyclohexyl ring at the 2-position resides in the cavity bordered by T1530 (T1528, hDNMT1), Q12230 (Q1227, hDNMT1), Y1243 (Y1240, hDNMT1), and R1576 (R1574, hDNMT1) (Rabal et al., 2018) (Fig. 17.6).

There are other examples of HKMT inhibitors (Fig. 17.5) guided by molecular docking, such as the BIX-01294, an established G9a HKMT inhibitor. BIX-01294 was modified because by structural comparison of G9a-like protein (GLP), a closely related lysine methyltransferase, it was observed that the lysine binding channel was unoccupied; therefore, the O7-methoxy group was replaced by 5-aminopentyloxy moiety (**E72**), which increase the length of the aliphatic chain and added an amino group at the terminal. This derivative depicted an enhanced in vitro potency and lesser cell toxicity in vivo (Chang et al., 2010). The use of structure-based virtual screening identified protoberberine alkaloid pseudodehydrocorydaline (**CT13**) as a G9a inhibitor, and it depressed the level of H3K9me2 in MCF7 cell line. By molecular docking, it was observed that CT13 occupies the binding site of histone H3 substrate, interacting with acidic residues such as D1074, D1078, D1083, and D10888, particularly the quaternary nitrogen forms a salt bridge with D1083. This phenomenon was also observed with UNC0224 complex (experimentally determined), suggesting that interaction with D1083 displaying a key role for binding at the active site; in fact, CT13 shares a similar binding mode to the competitive G9a inhibitor UNC0224 (Chen et al., 2018).

Other example is the discovery of CSV0C018875 quinoline-based G9 inhibitor (Fig. 17.5) that was found by docking-based virtual screening; through the use of MD simulation, it was found that CSV0C018875 remains deeper at the active site cavity of G9a allowing a tighter binding and increasing the residence time that is reflected by G9a in vitro experimentally inhibition (Charles et al., 2020). Finally, López-López, E. et al., by SAR study, molecular docking, and MD simulation approach studied 251 experimentally reported G9a inhibitors identifying key interactions between G9a enzyme and their inhibitors, such as D1083, L1086, Y1154, and F1158 (Lopez-Lopez et al., 2020).

3.3 Histone Lysine Demethylase Inhibitors

For the identification of HKDM inhibitors (Fig. 17.7), different molecular docking approaches have been used, some of them are discussed below. In this sense, Zhou, C. et al. followed a pharmacophore-based virtual screening combined with molecular docking with 171,143 compounds, identifying nine LDS1 (KDM1) inhibitors (IC50 = 2.41−101 μM); finally, XZ09 compound was selected due to better selectivity for LSD1 than for MAO-A and B; both are homologous to LDS1. Both oxygen and nitrogen of the amide group formed a hydrogen bond with D56; XZ09 resembles the binding mode of the substrate-like peptide (PDB: 2V1D) (Zhou et al., 2015).

In another approach, Schmitt M. L. et al. developed a nonpeptidic propargylamine compound as LDS1 inhibitors, starting from the basis of the synthesis and in vitro LDS1 inhibition, of propargyl amines derived

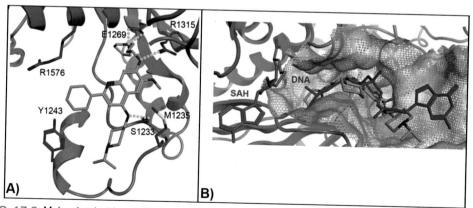

FIG. 17.6 Molecular docking of compound 11 with DNMT1. **(A)** Hydrogen bonds between **11** and DNMT1. **(B)** The compound **11** docked (*orange stick*) with DNMT1, DNA are superimposed (*pink stick*), and the cofactor S-adenosyl-L-homocysteine (SAH) (*green stick*). (Figure reprinted from the original publication with permission for reproduction Rabal, O., San Jose-Eneriz, E., Agirre, X., Sanchez-Arias, J. A., Vilas-Zornoza, A., Ugarte, A., et al. (2018). Discovery of reversible DNA methyltransferase and lysine methyltransferase G9a inhibitors with antitumoral in vivo efficacy. *Journal of Medicinal Chemistry*, 61(15), 6518−6545.)

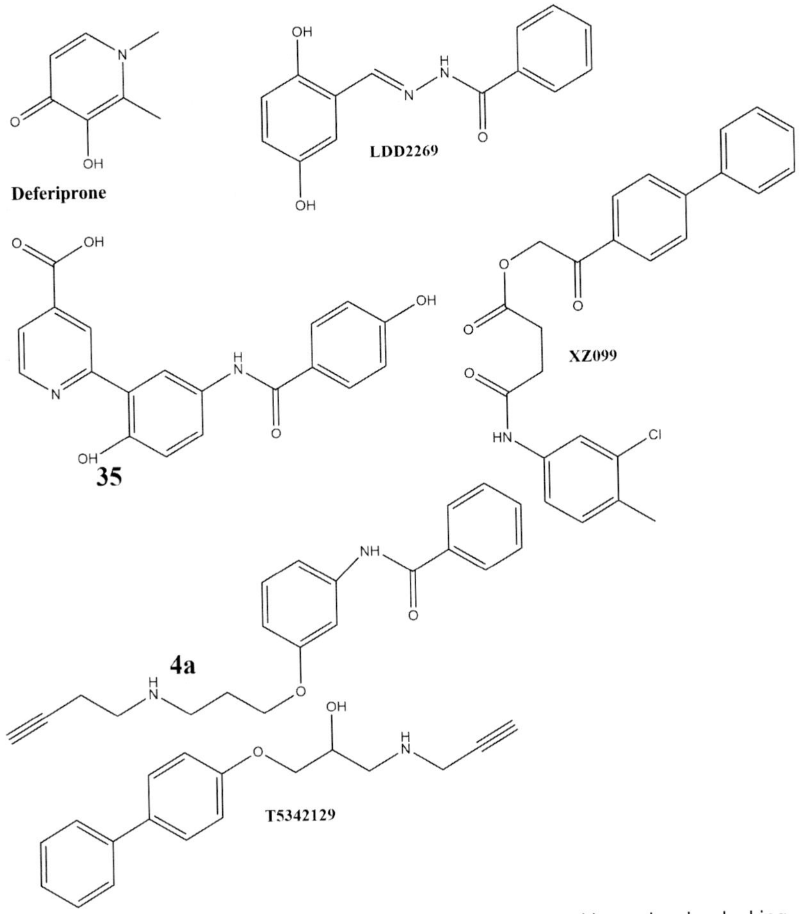

FIG. 17.7 Histone lysine demethylase inhibitors discovered by molecular docking.

from L-lysine. **4a** compound was obtained (IC$_{50}$ = 184.2 ± 16.0 μM), which was submitted to molecular docking. This revealed a hydrogen bond formation of the N-propargylamine amine with Y761 and the nitrogen of the benzamide group opposite to the side chain of the molecule also formed a hydrogen bond with D555. The terminal phenyl ring interacted hydrophobically (T-shaped aromatic interaction) with F5558, F560, Y807, and H815. Furthermore, these findings allow the identification and assay of a commercially available compound, **T5342129** (IC$_{50}$ = 44.0 ± 2.2 μM) with an N-propargyl group, as an LSD1 inhibitor. Docking study points out that the N-propargylamine group of this compound is located adjacent to the relative N5-nitrogen of the cofactor suggesting a covalent inhibitor, which was confirmed

in vitro. The hydroxyl group and the amine formed hydrogen bonds with A809 and Y761, respectively. And the hydrophobic biaryl group established an alkyl—aryl interaction with F560 and Y807 (Schmitt et al., 2013) (Fig. 17.8).

The repositioning of deferiprone (DFP; Fig. 17.7) as HKDM inhibitor used docking to guide the drug repositioning. DFP is a hydroxypyridinone derivative with antiproliferative activity; Khodavedian et al. report DFP as an HKDM (2A, 2B, 5C, 6A, 7A, and 7B) pan inhibitor, inhibiting at micromolar range and also inhibits downstream demethylation of H3K4me3 and H3K27me3. By molecular docking, it was shown that DFP chelates Fe^{2+} on KDM6A; besides, it forms a hydrogen bond with the phenolic group of Y1135 and the hydroxyl group of S1154, and the amino group

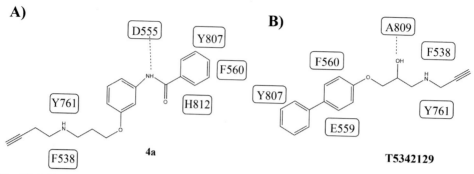

FIG. 17.8 Two-dimensional representation of the interactions found between LDS1 and **(A)** 4a, **(B)** **T5342129**. Hydrogen bonds are depicted as *dotted lines*.

protrudes toward the exit of the active site, opening an opportunity to make modifications (Khodaverdian et al., 2019). Also, the use of high-throughput screening of databases has guided the discovery of a Jumonji demethylase inhibitor, by the docking of 600,000 fragments against the structure of KDM4A, by using DOCK3.6. This allowed to identify molecules whose scaffold is 5-aminosalicylic nucleus, where carboxyl and hydroxyl moieties interacted with Fe^{2+} by coordination bond and the −NH− group formed a hydrogen bond with D135, yielding the compound **35** (Ki = 43 nM). Many other compounds were also discovered as selective KDM4C inhibitors over other enzymes such as FIH and KDM2A and KDM6B and nonselective for KDM3 and KDM5; furthermore, docking predictions were corroborated by crystallization of the complexes (Korczynska et al., 2016).

The discovery of a benzylidenehydrazine analog, **LDD2269** (Fig. 17.7), as a KDM4A (JMJD2A) inhibitor was done with an enzymatic in vitro assay (IC_{50} = 6.56 μM) and the rationalization of the binding mode was performed using docking in PDB: 3PDQ with Maestro software. The active site of KDM4A is characterized by the presence of polar charged amino acids (D135, D191, E190, H188, H276, and K241) by the presence of nickel and one aromatic residue (F185). The hydroxyl groups of the ligand interacted with D135 and D191 forming hydrogen bonds, while the benzyl group of the 2,5-dihydroxybenzyl moiety formed a π-cation interaction with K241, while the amine groups interacted with nickel; finally, the benzyl group forms a π−π interaction with F185 (Lee et al., 2017).

3.4 Bromodomain Inhibitors

In terms of bromodomains inhibitors (Fig. 17.9), several approaches have been explored, and one of

them involves the survey of natural products such as flavonoids. Virtual screening of flavonoid libraries through docking and MD simulation exploring bromodomain 1 (BD1) of BRD4 identified Fisetin as putative BRD4 ligand. Fisetin formed a hydrogen bond between the acetyl carbonyl oxygen and N140, which formed part of the hydrophobic cavity of acetylated lysine (K_{AC}), and hydrophobic interactions with P82, F83, V87, L92, L94, Y97, C136, Y139, I146, and W81. P82 and F83 belong to WPF shelf region, exclusively found on bromodomain and extra terminal domain (BET), and also interacted with E85. All of these interactions plus M105 remain stable along 50 ns of MD simulation (Raj et al., 2017). In another work, amentoflavone was also found as BDR4 ligand; this finding was supported by multistep approach using 2D and 3D similarity searching followed by molecular dockings (Prieto-Martinez & Medina-Franco, 2018a, b). Furthermore, Prieto-Martinez, F. D. et al. studied Fisetin and amentoflavone by molecular docking and MD simulations to identify the key interactions that define the binding mode of these compounds. Molecular docking was performed employing four docking programs with different algorithms (Autodock vina, MOE, Ledock, and PLANTS), and the results point out that amentoflavone established contact with ZA channel (L92) and hydrophobic contacts with residues of the WPF shelf (P81), which are regions frequently targeted by BDR inhibitors, as well as contacts with noncanonical residues (M105, Q135, C136, and D145). This behavior is explained by the atropisomerism of the molecule. In an in vitro assay, amentoflavone was found to inhibit BRD4 with an IC_{50} of 30−36 μM (Prieto-Martinez & Medina-Franco, 2018a, b).

On the other hand, Allen et al. reported the identification of N-[3-(2-oxo-pyrrolidinyl) phenyl]-benzenesulfonamide derivatives (Z115668302,

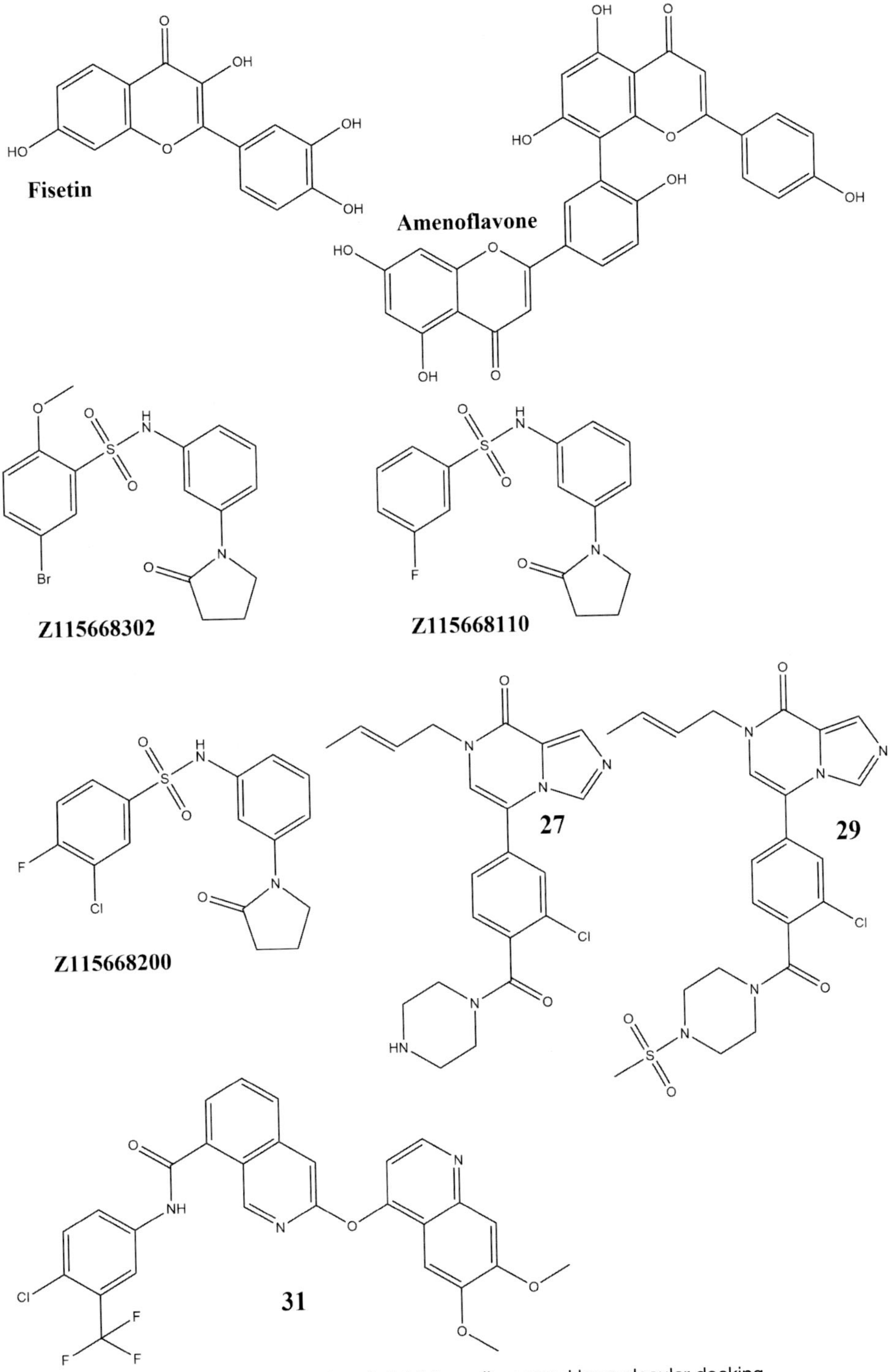

FIG. 17.9 Bromodomain inhibitors discovered by molecular docking.

Z115668110, Z115668200, Fig. 17.9) retrieved from enamine database using high-throughput docking-based virtual screening protocol, employing several BRD4 crystal structures in order to explore several conformations of the protein. Furthermore, IC$_{50}$ of benzenesulfonamide derivatives was determined and co-crystal complex of ligand enzyme was experimentally generated. Surprisingly, docking-binding mode prediction did not match with that obtained experimentally; therefore, authors took the binding mode predicted by docking and submitted to MD simulations; this led to the identification of binding mode conformations closely related to the experimental one. These findings highlight the limitations of docking because of the receptor flexibility and limited sampling without water environment (Allen et al., 2017).

In the chemical synthesis of novel compounds to target bromodomains, molecular docking has been used as a strategy to explain the binding mode and the SAR, such as in the case of the synthesis of imidazo [1,5-a]pyrazin-8(7H)-one derivatives as BRD9 inhibitors. For compounds **27** and **29** (IC$_{50}$ = 35 and 103 nM, respectively), molecular docking evidenced key interactions between BRD9 and the inhibitors, i.e., hydrogen bond with conserved N100, π–π interaction with Tyr106, and hydrogen bond with H42. Additionally, the (E)-but-2-en-1-yl portion favors the activity of the compound, because it extends to a small hydrophobic pocket unique to the BRD9 protein (Zheng et al., 2019) (Fig. 17.9).

Bromodomain PHD finger transcription factor, which participates in cancer, is also a promising epigenetic target, and to date there are not many ligands reported to inhibit this protein. Zhang, D. et al. using an integrated docking-based virtual screening and biochemical analysis identified DCB29 compound (IC$_{50}$ = 13.2 μM). In molecular docking, it was observed that DCB29 is located in the pocket of acetylated H4 peptide substrate (Zhang et al., 2019). As CREB is another bromodomain that is also involved in cancer progression, several authors have made efforts to design CREB inhibitors; Dash et al. used molecular docking, free energy calculations, and MD simulations to explore 9 naphthyl-based compounds, from which compound **31** (Fig. 17.9) was selected as the best because it demonstrate selectivity to CREB bromodomain in in silico. Compound **31** interacts with R1173 forming two hydrogen bonds due to the interaction between arginine guanidinium group and the acceptor oxygen of carboxamide group of **31**; R1173 is considered a key residue for the selectivity toward CREB. Also it formed a hydrogen bond with N1168 and halogen bond with L1109, whereas the naphthyl ring and hinge region interact hydrophobically with L1120 on ZA loop and V1174 via π-alkyl and π-σ bond; besides, Y1125 established nonbonded hydrophobic interaction and the MD simulations showed that the complex CREB-**31** was stable along the time, decreasing fluctuations of the binding site (Dash et al., 2018) (Fig. 17.10).

3.5 Histone Acetyltransferase Inhibitors
Several successful cases of HAT inhibitors (Fig. 17.11) have recently been reported which used docking studies. In the case of curcumin derivatives targeting

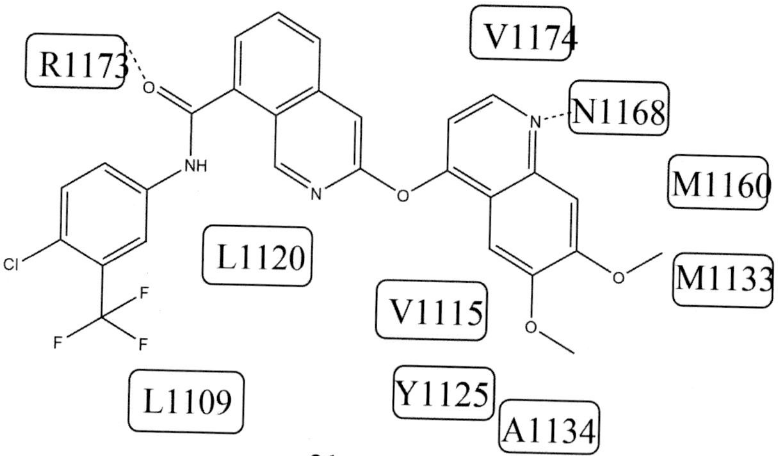

31

FIG. 17.10 Two-dimensional representation of the interactions found between CREB bromodomain and compound **31**.

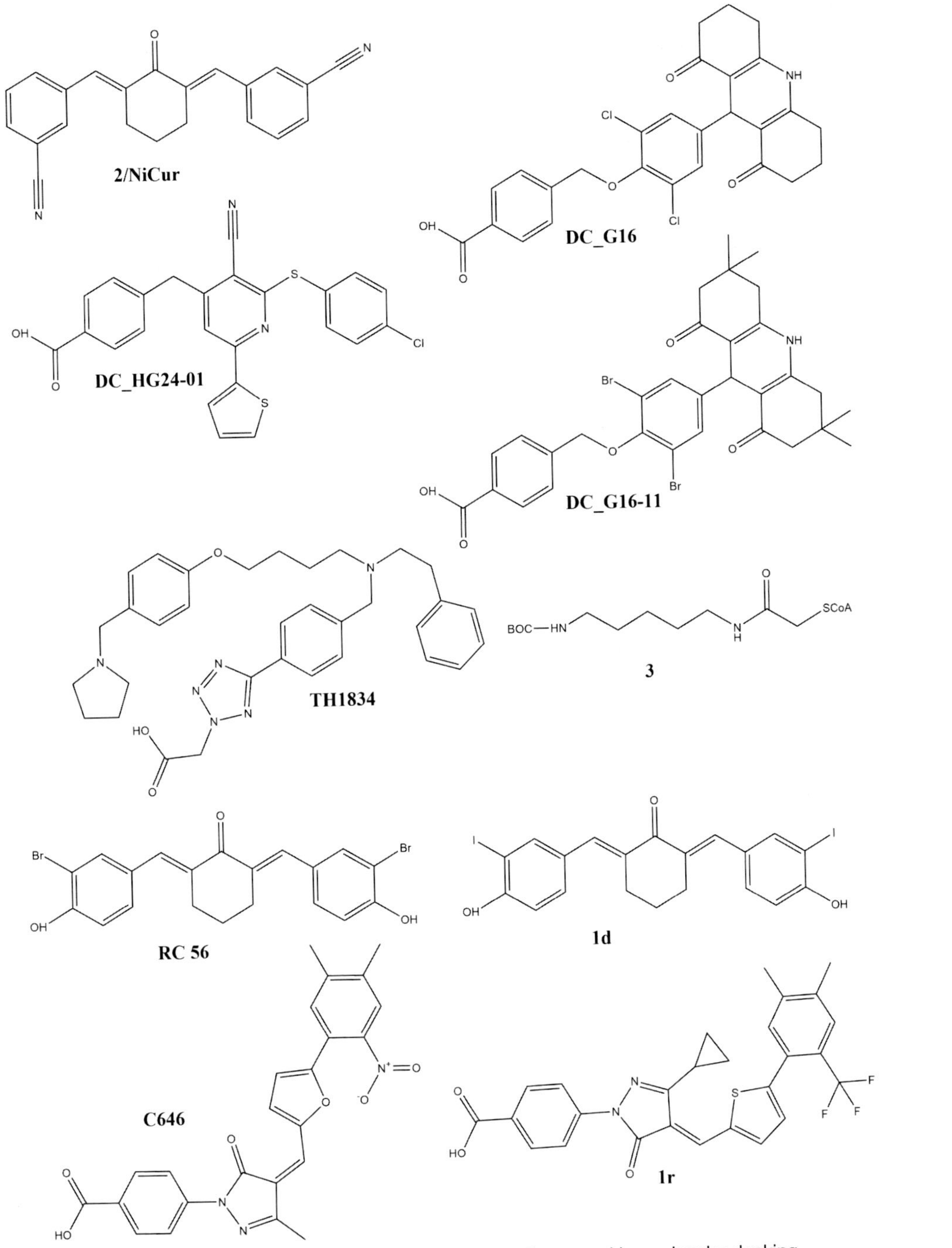

FIG. 17.11 Histone acetyltransferase inhibitors discovered by molecular docking.

CBP HAT, from 12 curcumin derivatives, **2/NiCur** demonstrates more that 99% of inhibition of CBP HAT at 500 nM with an $IC_{50} = 350$ nM. Molecular docking suggested that it reaches the active site of CBP HAT

and become extended along the active channel. 2/Nicur formed hydrogen bonds with Y1446, E1445, and Y1467 and hydrophobic contacts with P1458 and L1463, and due to the similar binding mode with Lys-

CoA substrate inhibitor, it is suggested that 2/NiCur might act as competitive inhibitor (Vincek et al., 2018).

By high-throughput screening of 21,980 compounds using AlphaScreen assay, DC_HG24-01 compound was discovered as GCN5 inhibitors (IC$_{50}$ 3.1 μM). Docking suggests that DC_HG24-01 occupied the binding pocket of Ac-CoA cofactor and its carboxyl group formed a hydrogen bond with V587 and G591 and nonbonded interactions with A618, F622, I576, A614, V577, L531, and F578, suggesting the importance of the carboxyl group, as other derivatives with different substituents in this position showed decreased activity (Tao et al., 2019). On the other hand, the compound DC_G16 (IC$_{50}$ = 34.7 μM) that possess a 1,8-acridinedione scaffold was discovered using the AlphaScreen assay as GCN5 inhibitor. Further structure optimization guided the design of more potent inhibitor DC_G16-11 (IC$_{50}$ = 6.8 μM), which occupied the substrate H3 peptide pocket where the carboxyl group formed a hydrogen bond with Y212, and it was experimentally observed that the absence of this group led to detrimental effect on the GNC5 inhibitory activity. Besides this, two other hydrogen bonds with K242 and Y135 were observed and also polar interactions, a hydrophobic environment pocket surrounded by V175, F176, Y244, Y212, Y240, and Y135 stabilized the complex. Also, the substitution of Br by Cl improves the activity of DC_G16-11 in comparison to the lead compound (DC_G16) because the strength of Br—O interaction is stronger than Cl—O (Xiong et al., 2018) (Fig. 17.11).

Gao, C. et al. suggest Tip60 as a candidate drug target for breast cancer because it mediates the DNA damage response and transcriptional coactivation. By molecular modeling, they elucidate that the binding pocket of Tip60 possesses opposite charges at each extreme of the pocket; this finding supports the development of TH1834 molecule as novel Tip60 inhibitor (Fig. 17.11). The design of TH1834 was guided based on molecular modeling studies between PNT compound and Tip60, resulting in the enhanced inhibitory activity of the compound. Docking indicates that the carboxylic group of TH1834 interacts with K331 and Q368, and the pyrrolidine group with E351 and H274; so, this approach offers a tool to generate Tip60 inhibitors with potential therapeutic applications (Gao et al., 2014).

Several efforts have been made in p300 HAT inhibitors using docking. One of them was the synthesis of CoA derivatives to target p300; the compound 3 present the best inhibitory activity resembling a bisubstrate-type compound (IC$_{50}$ of 40 nM). It was able to reach two binding sites, named pocket P1 and P2 at the active

site, because it possessed a C-5 spacing linker region that connects the CoA moiety to tert-butyloxycarbonyl (Boc) group; but the efficiency as p300 inhibitor could be attributed to the ability of 3 to reach pocket 2 (Kwie et al., 2011). In another approach, investigation of SARs of cinnamoyl derivatives of RC56 was carried out, which was identified as a hit compound that was chemically modified resulting in two new compounds 1d and 1i (IC$_{50}$ = 8.1 and 2.3 μM) as p300 HAT inhibitors (Fig. 17.11). Induced fit docking procedure, which takes into account receptor flexibility after ligand binding using Schrödinger Maestro, was used to investigate these derivatives. All docked positions were clustered into two distinct binding modes, A and B, independent of the ionization state of the ligands. Complementarily, MM-GBSA approach implemented in Schrödinger's Maestro suit was used to determine which one of the binding modes is most favored; this additional calculation indicates that binding pose A was the most favorable for both RC56 and 1d inhibitor, while for the compound 1i, both binding poses could be reliable. Furthermore, the stability of the complexes was tested by MD simulation; it was shown that all three complexes were stable, and ligands with the binding pose A interact with R1410 and interacted via hydrophobic interactions with Y1467, I1457, L1463, LK1398, W1436, Y1414, I1395, and I1435 and hydrogen bonds with S1396 and W1436. While with the binding pose B, polar interactions with R1410, Y1467, and W1436 and hydrophobic interactions with W146, Y1414, L1398, W1436, Y1446, I1435, and Y1467 were observed (Madia et al., 2017) (Fig. 17.12).

Finally, a well-known p300 inhibitor C646, which contains a toxic group, was taken as a scaffold in order to design new compound with better drug-like properties. Liu R. et al. used C646 as lead compound and modified its structure using the bioisosterism approach. The compound 1r (IC$_{50}$ 160 nM) was the best compound with p300 inhibitory activity, better than the lead compound with improved drug-like properties. By docking, it was observed that both compounds have similar binding mode, but the cyclopropyl group at the 3-position on the pyrazolone increases the interaction with the protein cavity due to their incremented volume in comparison to methyl group. Moreover, the replacement of nitro group by the trifluoromethyl group in 1r allows to form a new hydrogen bond with W1436 that could replace the ones formed by nitro group with Y1367. This reduced the toxicity associated with nitro group without affecting the p300 inhibitory activity of the compound (Liu et al., 2019) (Fig. 17.11). In another successful approach, Lasko et al. obtained a compound called A-485, a potent inhibitor of p300, from screening and molecular docking to identify hydantoin and conjugated thiazolidinedione as hits. Furthermore, hydantoin was optimized rendering A-485 compound with a potency superior by at least 1000-fold more than C646. A-485 was bound to the catalytic site of p300 (Lasko et al., 2017) (Fig. 17.13).

3.6 Histone Deacetylase Inhibitors

HDACs are the most explored molecular targets in the epigenetic field, and several molecules are discovered by molecular docking and also by others in silico techniques. HDACi share a common chemical structure scaffold that consist of zinc binding group (ZBG), linker part that connects ZBG with the final portion, known as cap portion. The ZBG consists of chemical group that

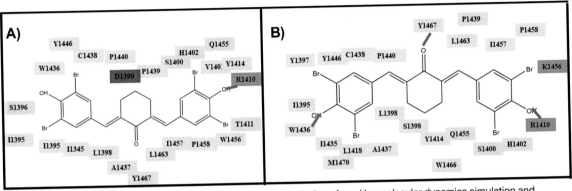

FIG. 17.12 Two-dimensional representation of the interactions found by molecular dynamics simulation and docking of the compound **1i**, **(A)** binding pose A, **(B)** binding pose B. Polar interactions are highlighted with blue color, hydrophobic with green, positive charged interaction with purple, and negative charged interaction with red color. Hydrogen bond interactions are specified with *purple line*.

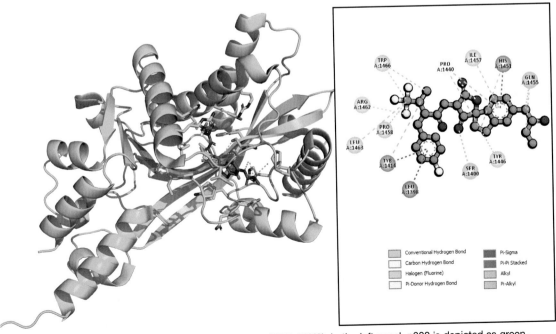

FIG. 17.13 P300 in complex with A-485 inhibitor (PDB: 5KJ2). In the left panel, p300 is depicted as green ribbon and A-485 as magenta stick. In the right panel, the interactions are established between A-485 and p300. (Figure was generated using Discovery Studio 2016 Dassault Systèmes BIOVIA, (2016). Discovery Studio modeling environment, release 2017. San Diego: Dassault Systèmes.)

confers the ability to chelate Zn^{2+} and also bind to residues around the zinc that play a key role in the inhibitory activity of the HDACi. Modifications in ZBD portion might modify the potency of the compound. The linker region binds to residues belonging to the active site and catalytic tunnel of the HDACs, while the cap portion usually has an aromatic character or heteroaromatic hydrophobic portion that participates in the interactions between residues at the rim of HDACs, and mainly the cap portion is responsible for the isoform selectivity and in few cases the linker region contributes to the selectivity (Micelli & Rastelli, 2015; Yang et al., 2019; Zhang et al., 2018) (Fig. 17.14).

The first HDACi were nonspecific, so-called pan inhibitors, such as trichostatin A (TSA), trapoxin, SAHA (vorinostat), FK228 (Istodax) (Seto & Yoshida, 2014). Later, more specific and selective compounds were designed due to the advance in the HDAC structure crystallization and the advance on molecular modeling techniques including molecular docking. Several HDACi (Fig. 17.15) will be discussed below.

The discovery and optimization of reported series of aroyl-pyrrolyl-hydroxyamides (APHAs) was a larger work guided by molecular docking and molecular modeling (Mai et al., 2003). At the beginning, docking study points out that the pyrrole portion serves as a linker region that interacts with the catalytic tunnel. Therefore, structure optimization was performed

modifying the pyrrole moiety or the length of the linker region of the APHA derivatives; these modifications guide to the discovery of more flexible ligands (Mai et al., 2004). On a third approach, compounds **3** and **4** were synthetized, and molecular docking predictions showed that the affinity with HDAC1 was up by 617-fold; this result was corroborated experimentally on both mouse HDAC1 and maize HD2 enzyme. By molecular docking, it was also elucidated that APHA derivatives reach the catalytic site and coordinate the Zn^{2+} atom using the hydroxamic group (Ragno et al., 2004). To achieve selectivity, specifically on HDACs of class I, 14 Å internal cavity adjacent to the catalytic site, which was postulated as the acetate release channel, was considered in the design and synthesis of compound **4** with a bis-(aryl)-type pharmacophore. This paved the way for new pharmacophore of HDACi, yielding in a new selective nonhydroxamic inhibitor, specifically benzamide inhibitor. The para-thienyl substituent depicted an improvement of a 27-fold and 5-fold improvement in activity comparison to HDAC1 and 2, respectively. Thus, the introduction of this thienyl group allows the ligand to introduce this portion into the 14 Å tunnel and improve the selectivity toward HDAC1 and HDAC2 isoforms, highlighting the importance of M30 and C151 on HDAC1 because it lines the wall of the 14 Å tunnel (Moradei et al., 2007) (Fig. 17.15).

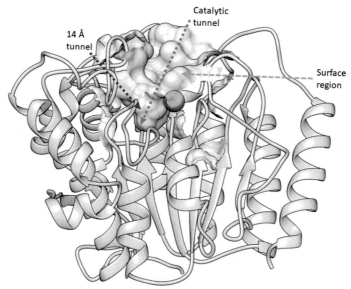

FIG. 17.14 Schematic representation of HDAC8, where catalytic tunnel (*dotted orange line*), 14 Å tunnel (*dotted red line*), and surface region can be observed. Zinc atom is represented as *cyan sphere*.

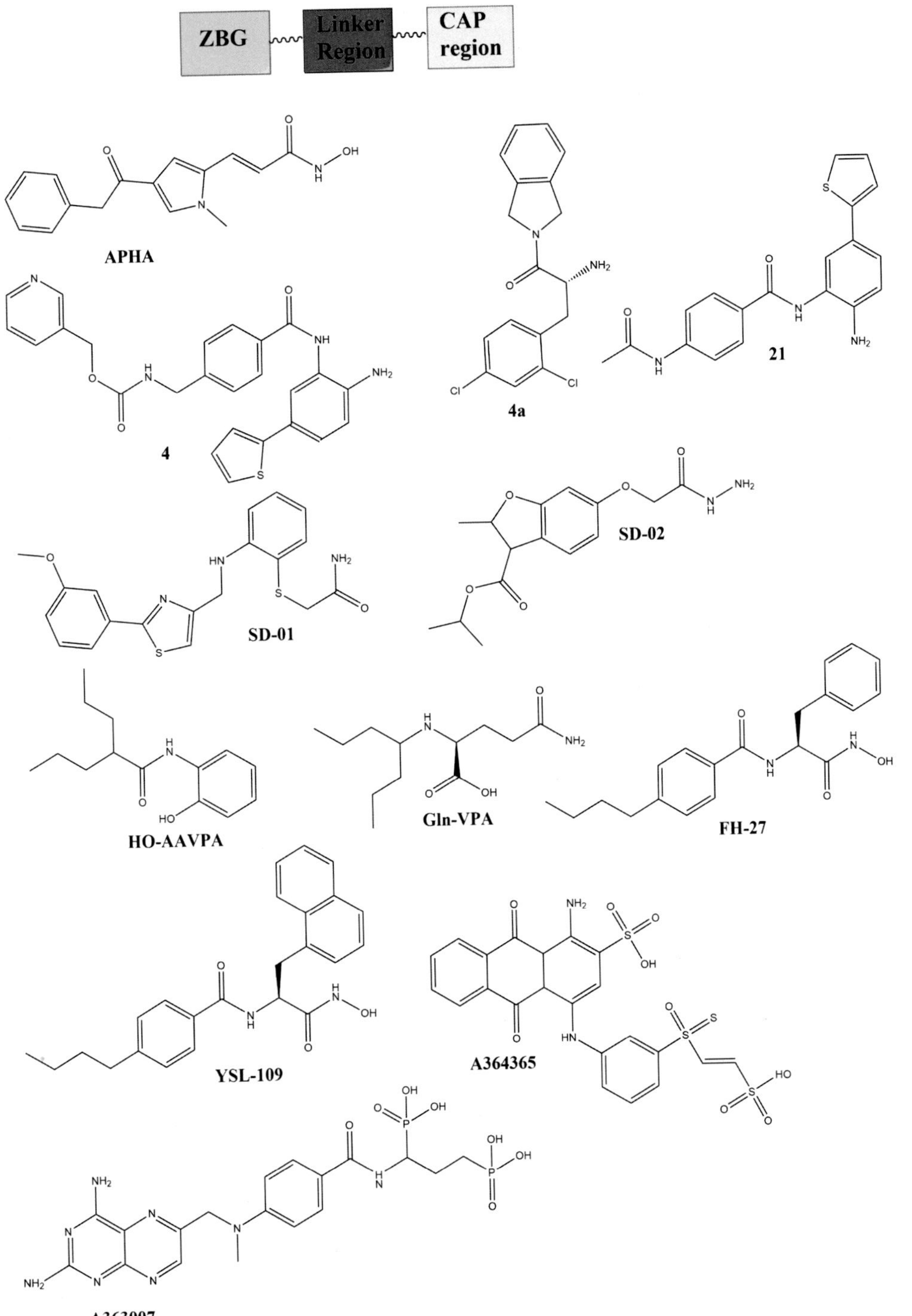

FIG. 17.15 Histone deacetylase inhibitors discovered by molecular docking. *ZBG*, zinc binding group.

Furthermore, two other reports also suggest the exploration of this internal cavity of the HDAC in order to achieve class I selectivity using molecular modeling studies, specifically molecular docking studies (Ganai et al., 2017; Methot et al., 2008; Whitehead et al., 2011). Methot et al. synthesized phenyl biaryl molecule, known as **21**, with an IC_{50} of 48 nM for HDAC1; this compound which is a benzamide derivative is bound to Zn^{2+} by carbonyl, while NH bond to G149, while phenyl ring is anchorage into the internal 14 Å cavity (Methot et al., 2008). Whitehead L. et al. synthetized a selective HDAC8 inhibitor molecule via exploitation of acetate release channel, hereafter referred as **4a** (IC_{50} = 90 nM). The activity is because the α-amino ketone group chelates zinc group, and the R chirality of the molecule is important; besides this, amine group also coordinated with D178, H180, and D267. While, the amino group of the compound donates a proton to H142 forming a hydrogen bond, and the carbonyl oxygen formed a weakened electrostatic interaction with zinc. This molecule was one of the first HDAC8 inhibitors with different binding modes because the ligand occupies space in both the 14 Å acetate release tunnel and the catalytic tunnel (Whitehead et al., 2011).

In late 2019, a 3D QSAR model was built and combined with XP glide docking studies employing HDAC1, 3, 4, 6, and 8 PDB structures available at the time to investigate the ligand binding affinity and used this to successfully find selective ligands. This led to the discovery of HDAC8 inhibitors with non−hydroxamic acid as ZBG. From this approach, compounds **SD-01** (HDAC8 IC_{50} = 9.0 nM and HDAC6 IC_{50} > 100 nM) and **SD-02** (HDAC8 IC_{50} = 0.67 nM and HDAC6 IC_{50} = 2.7 nM) were identified (Fig. 17.15). The compound SD-01 interacted with Zn^{2+} using the oxygen of the amide (2.01 Å) and Y306 (2.26 Å). Also, it interacted via π-π with F208 and F152 and hydrophobic interactions with L308, W141, P35, I34, H180, H142, and M274. However, SD-02 interacted with acetoxy hydrazide carbonyl oxygen which binds with Zn^{2+} (2.05 Å) of the active site and formed hydrogen bonds using NH_2 of hydrazine with H142 and G151 and π-π interactions with H180 and F208 and hydrophobic interactions with F252, Y306, and Y1000. Thus, this research is another example of non−hydroxamic specific HDACi discovered using docking (Debnath et al., 2019).

In another strategy, modification guided by docking of existing compound valproic acid (VPA) resulted in increase in the potency and decrease in the hepatotoxic effect of the compound. This strategy resulted in the discovery of the N-(2-hydroxyphenyl)-

2-propylpentanamide (**HO-AAVPA**); molecular docking showed that it might inhibit HDACs and depicted better cytotoxic effect on HeLa, breast cancer, and rhabdomyosarcoma cell lines in comparison to VPA (Prestegui-Martel et al., 2016). Other contribution guided by docking on HDAC8 was the development of (S)-5-amino-2-(heptan-4-ylamino)-5-oxopentanoic acid (**Gln-VPA**); docking studies show that Gln-VPA can reach the catalytic site of HDAC8 (Martinez-Ramos et al., 2016). Following this line, a virtual screening based on docking protocol using Autodock vina of hydroxamic acid derivatives guided to the discovery of **FH-27** (Figs. 17.15 and 17.16). The lead compound was further optimized by adding bulky and aromatic groups in order to increase the selectivity against HDAC6. This optimization gave eight derivatives, and by MD simulation and docking approach besides in vitro cytotoxic profile, **YSL-109** was selected as a potent selective HDAC6 inhibitor (IC_{50} = 0.537 nM) with moderate in vitro activity against HDAC1 (IC_{50} = 259.439 μM) and HDAC8 (IC_{50} = 2.24 μM) (Sixto-López et al., 2020b).

Regarding class IIa (HDAC4, 5, 7, and 9), Hsu K. C. et al. present an in silico virtual screening methodology using CDOCKER in Discovery Studio, by which they were able to identify nonhydroxamate inhibitors from National Cancer Institute database. **A364365** (IC_{50} class IIa: 1.33−14.59 μM) and **A363007** (IC_{50} class IIa: 1.27−16.16 μM) (Fig. 17.15) demonstrate potent inhibition against class IIa HDACs and weak inhibitory profile against class I HDACs. More detailed in silico study reveals that a single residue affects the cavity size between class I and IIa, and that is responsible for the selectivity of these new inhibitors. In fact, these inhibitors are capable to form key interactions within the lower pocket of the catalytic site of HDACs of class IIa and coordinate with the zinc ion. There is a difference of three nonconserved residues of the catalytic site between class I and IIa; HDACs of class 1 has Arg and Tyr lining the wall of the catalytic tunnel instead of Pro and His of class IIa, and Gly at the bottom of the tunnel instead of Glu; this difference is responsible for the selectivity of the new inhibitors (Hsu et al., 2017).

In the case of sirtuin inhibitors (Fig. 17.17), the search for selective inhibitors between isoforms has also been extensively explored. Two quercetin derivatives, diquercetin and 2-chloro-1,4-naphtoquinone-quercetin, were discovered as promising SIRT6 inhibitors with IC_{50} = 130 and 55 μM, respectively. Molecular docking studies suggested that diquercetin prefers the binding site of the nicotinamide moiety, whereas 2-chloro-1,4-naphtoquinone-quercetin prefers the substrate binding

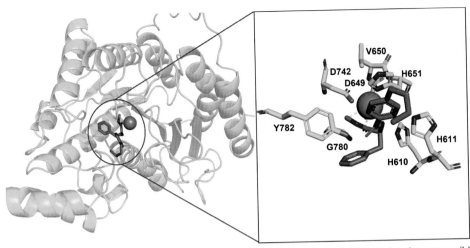

FIG. 17.16 FH27 in complex with CD2-HDAC6. In the left panel, CD2-HDAC6 is depicted as green ribbon and FH27 as magenta stick. In the right panel, the interactions are established between FH-27 and CD2-HDAC6. (Model obtained from Sixto-Lopez, Y., Bello, M., Rodriguez-Fonseca, R. A., Rosales-Hernandez, M. C., Martinez-Archundia, M., Gomez-Vidal, J. A., et al. (2017). Searching the conformational complexity and binding properties of HDAC6 through docking and molecular dynamic simulations. *Journal of Biomolecular Structure and Dynamics, 35*(13), 2794–2814; Figure was generated using Pymol 2.0 DeLano, W. L. (2002). *PyMOL* (pp. 700). San Carlos, CA: DeLano Scientific.)

site; furthermore, this result was proved by in vitro studies (Heger et al., 2019). A pan-SIRT1-3, **BZF9Q1**, with and $IC_{50} = 7.7$ µM for SIRT1, 5.6 µM for SIRT2, and 9.8 µM for SIRT3, was identified after screening a series of benzimidazole derivative based on the sirtuin inhibitor BZD9L1. Molecular docking indicates that BZD9Q1 occupies the adenosine binding pocket (Pocket A) at the active site of SIRT2 avoiding that NAD^+, the cofactor bound to the enzyme; this finding was experimentally corroborated by competition analysis. According to the binding mode predicted, BZD9Q1 forms a hydrogen bond with W262 and R97 and interactions with S98, K287, and C324, while the planar benzimidazole moiety is accommodated along the pocket A (Yeong et al., 2019).

Also, for identifying sirtuin inhibitors, virtual screening has been employed by docking of compound libraries against Sirt2, 5, and 6. This approach identified several molecules with inhibitory activity against the above isoform mentioned, but most of them act as pan-sirtuin inhibitors, except two compounds. **CSC8** (IC_{50}: 4.8 µM) and **CSC13** ($IC_{50} = 9.7$ µM) are Sirt-2 specific inhibitor (Fig. 17.17). The binding mode of both Sirt2-specific compounds indicates a pocket extending from the peptide-binding groove as target site allowing to achieve the isoform specificity. Both

compounds have steroid core as a common scaffold with 2-benzyl pyrimidine fused to the A ring. According to molecular docking, compound CSC8 interacted hydrophobically between the estradiol ring and F96, L107, F119, and I169 and also with residues belonging to hydrophobic cavity (A85, Y104, and I118), allowing the ligand to be extended toward the small sirtuin Zn^{2+} domain. Besides this, the hydroxyl group of CSC8 forms a hydrogen bond with N168 and the phosphate Q167 and the phosphate group of the cofactor ADP-ribose. However, CSC13 depicted two orientations: in the first one, it mainly interacted hydrophobically between the sterol ring and the hydrophobic pocket around F119, while the second one is in the same site but two additional hydrogen bonds were observed with N168 and Q167. Therefore, this work showed that the use of molecular docking allows to obtain potential sirtuin 2 inhibitors (Schlicker et al., 2011).

On the other hand, sirtuins are proposed to be suitable targets for the treatment of Leishmaniasis. Recently, compound 5 (Fig. 17.17) is described as *Leishmania infantum* LiSIR2rp1 inhibitor. Molecular docking studies indicated that it did not bind to the highly species-conserved pocket C and established interactions with residues such as V44, A45, R242, and E243 in pocket A, which are different in humans. Therefore,

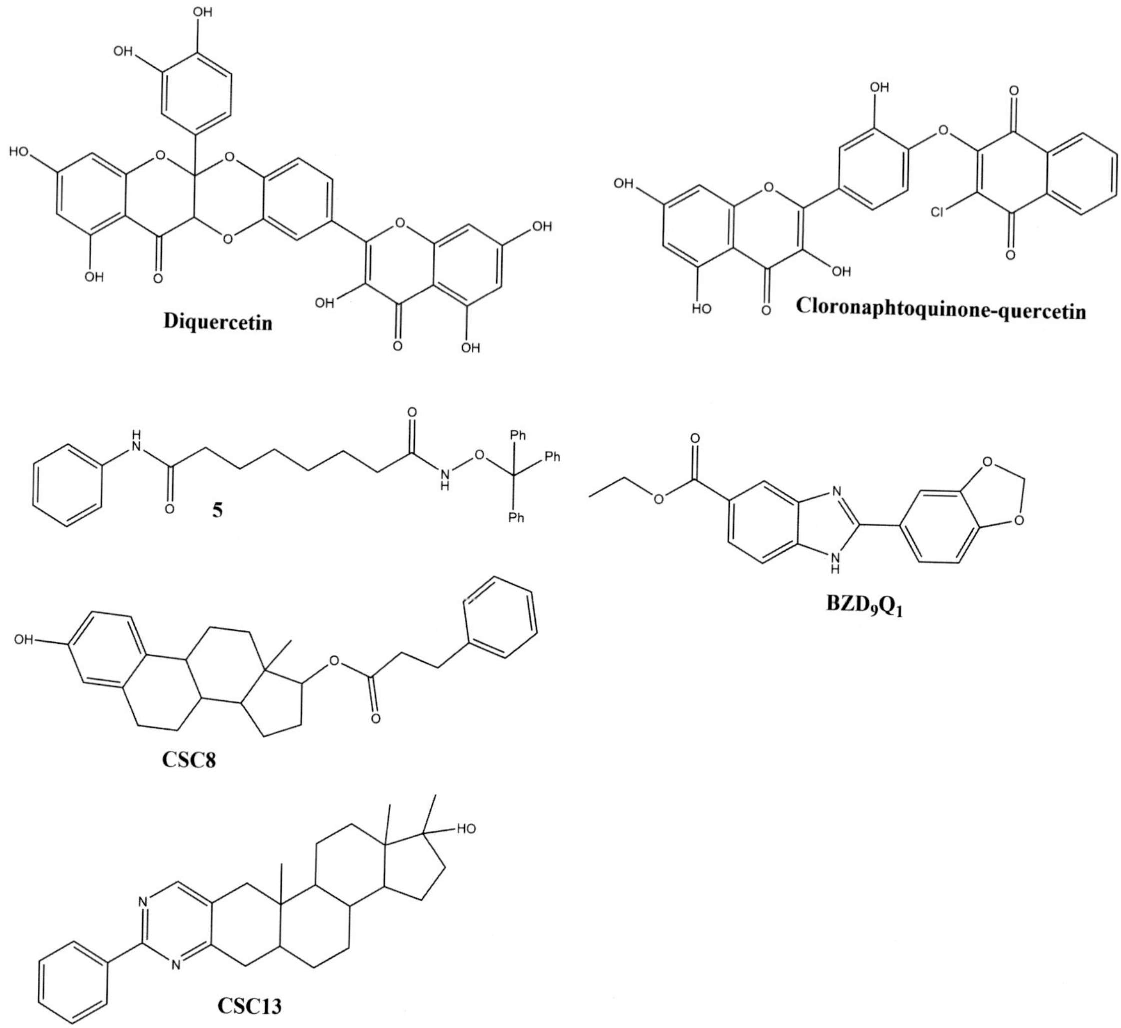

FIG. 17.17 Sirtuin inhibitors discovered by molecular docking.

this compound might represent a new selective ligand with antileishmanial activity (Corpas-Lopez et al., 2020). Thus, sirtuins represent a potential pharmacological target for the treatment of diverse disease not only in humans but also in other species. In this context also HDACi such as SAHA or vorinostat due to its HDACi activity and drug repositioning using techniques such as docking and MD simulation have been proposed for the treatment of amebiasis as an alternative to metronidazole, because SAHA has the capacity to inhibit the HDAC of the parasite (Montano et al., 2020). Besides, molecular docking also helps to elucidate the structural and energetics behaviors that are essential in the recognition of HDAC enzymes and their inhibitor, this elucidation allow to establish a better

knowledge on the molecule mechanism of action and support new HDAC drug design. Interested readers on this topic may refer to other publications elsewhere (Sixto-Lopez et al., 2014, 2017, 2019, 2020a; Tambunan et al., 2013; Uba & Yelekci, 2019).

4 CONCLUSION AND FUTURE PERSPECTIVES

Throughout this chapter we have shown how molecular docking is a very valuable tool for epigenetic drugs discovery. Imbalance in writers, readers, and erasers is involved in several diseases: cancer, neurodegenerative diseases, chronic diseases, etc. However, there are only few epigenetic drugs available in the market for treating these diseases. As investigations progress, new roles of epigenetic proteins are discovered as potential drug targets. Also, the epigenetic drugs that are now available need to be improved in order to increase the specificity as well as potency and decrease the side effects or improve other drug-like properties. Docking studies help to discover drugs and to improve them, by assisting in the understanding the binding properties (binding mode and binding energy) at the molecular level. Several examples are listed where docking was used to assist in epigenetic drug discovery (hit or lead compound), as well as in the optimization of hit compounds to generate drug candidates with improved properties. Docking studies can be used in virtual screening, pharmacophore design, and hit to drug optimization, used either alone or in combination with other in silico techniques such as MD simulation and QSAR approaches with several successful cases. However, in the field of epigenetic drug discovery, still much more research needs to be done in which docking could assist.

REFERENCES

Allen, B. K., Mehta, S., Ember, S. W. J., Zhu, J. Y., Schonbrunn, E., Ayad, N. G., et al. (2017). Identification of a novel class of BRD4 inhibitors by computational screening and binding simulations. *ACS Omega, 2*(8), 4760−4771.

Berger, S. L. (2007). The complex language of chromatin regulation during transcription. *Nature, 447*(7143), 407−412.

Biswas, S., & Rao, C. M. (2018). Epigenetic tools (the Writers, the Readers and the Erasers) and their implications in cancer therapy. *European Journal of Pharmacology, 837*, 8−24.

Chang, Y., Ganesh, T., Horton, J. R., Spannhoff, A., Liu, J., Sun, A., et al. (2010). Adding a lysine mimic in the design of potent inhibitors of histone lysine methyltransferases. *Journal of Molecular Biology, 400*(1), 1−7.

Charles, M. R. C., Mahesh, A., Lin, S. Y., Hsieh, H. P., Dhayalan, A., & Coumar, M. S. (2020). Identification of novel quinoline inhibitor for EHMT2/G9a through virtual screening. *Biochimie, 168*, 220−230.

Chen, J., Lin, X., Park, K. J., Lee, K. R., & Park, H. J. (2018). Identification of protoberberine alkaloids as novel histone methyltransferase G9a inhibitors by structure-based virtual screening. *Journal of Computer-Aided Molecular Design, 32*(9), 917−928.

Corpas-Lopez, V., Tabraue-Chavez, M., Sixto-Lopez, Y., Panadero-Fajardo, S., Alves de Lima Franco, F., Dominguez-Seglar, J. F., et al. (2020). O-alkyl hydroxamates display potent and selective antileishmanial activity. *Journal of Medicinal Chemistry, 63*(11), 5734−5751.

Dash, R., Mitra, S., Arifuzzaman, M., & Zahid Hosen, S. M. (2018). In silico quest of selective naphthyl-based CREBBP bromodomain inhibitor. *Silico Pharmacology, 6*(1), 1.

Dassault Systèmes BIOVIA. (2016). *Discovery Studio modeling environment, release 2017*. San Diego: Dassault Systèmes.

Deans, C., & Maggert, K. A. (2015). What do you mean, "epigenetic"? *Genetics, 199*(4), 887−896.

Debnath, S., Debnath, T., Bhaumik, S., Majumdar, S., Kalle, A. M., & Aparna, V. (2019). Discovery of novel potential selective HDAC8 inhibitors by combine ligand-based, structure-based virtual screening and in-vitro biological evaluation. *Scientific Reports, 9*(1), 17174.

DeLano, W. L. (2002). *PyMOL* (p. 700). San Carlos, CA: DeLano Scientific.

Filippakopoulos, P., Picaud, S., Mangos, M., Keates, T., Lambert, J. P., Barsyte-Lovejoy, D., et al. (2012). Histone recognition and large-scale structural analysis of the human bromodomain family. *Cell, 149*(1), 214−231.

Ganai, S. A., Abdullah, E., Rashid, R., & Altaf, M. (2017). Combinatorial in silico strategy towards identifying potential hotspots during inhibition of structurally identical HDAC1 and HDAC2 enzymes for effective chemotherapy against neurological disorders. *Frontiers in Molecular Neuroscience, 10*, 357.

Ganesan, A., Arimondo, P. B., Rots, M. G., Jeronimo, C., & Berdasco, M. (2019). The timeline of epigenetic drug discovery: From reality to dreams. *Clinical Epigenetics, 11*(1), 174.

Gao, C., Bourke, E., Scobie, M., Famme, M. A., Koolmeister, T., Helleday, T., et al. (2014). Rational design and validation of a Tip60 histone acetyltransferase inhibitor. *Scientific Reports, 4*, 5372.

Garcia-Gimenez, J. L., Seco-Cervera, M., Tollefsbol, T. O., Roma-Mateo, C., Peiro-Chova, L., Lapunzina, P., et al. (2017). Epigenetic biomarkers: Current strategies and future challenges for their use in the clinical laboratory. *Critical Reviews in Clinical Laboratory Sciences, 54*(7−8), 529−550.

Gayatri, S., & Bedford, M. T. (2014). Readers of histone methylarginine marks. *Biochimica et Biophysica Acta, 1839*(8), 702−710.

Greer, E. L., & Shi, Y. (2012). Histone methylation: A dynamic mark in health, disease and inheritance. *Nature Reviews Genetics, 13*(5), 343−357.

Handy, D. E., Castro, R., & Loscalzo, J. (2011). Epigenetic modifications: Basic mechanisms and role in cardiovascular disease. *Circulation, 123*(19), 2145−2156.

Heger, V., Tyni, J., Hunyadi, A., Horakova, L., Lahtela-Kakkonen, M., & Rahnasto-Rilla, M. (2019). Quercetin based derivatives as sirtuin inhibitors. *Biomedicine and Pharmacotherapy, 111*, 1326−1333.

Hsu, K. C., Liu, C. Y., Lin, T. E., Hsieh, J. H., Sung, T. Y., Tseng, H. J., et al. (2017). Novel class IIa-selective histone deacetylase inhibitors discovered using an in silico virtual screening approach. *Scientific Reports, 7*(1), 3228.

Hu, J., Gu, L., Ye, Y., Zheng, M., Xu, Z., Lin, J., et al. (2018). Dynamic placement of the linker histone H1 associated with nucleosome arrangement and gene transcription in early Drosophila embryonic development. *Cell Death and Disease, 9*(7), 765.

Khodaverdian, V., Tapadar, S., MacDonald, I. A., Xu, Y., Ho, P. Y., Bridges, A., et al. (2019). Deferiprone: Pan-selective histone lysine demethylase inhibition activity and structure activity relationship study. *Scientific Reports, 9*(1), 4802.

Korczynska, M., Le, D. D., Younger, N., Gregori-Puigjane, E., Tumber, A., Krojer, T., et al. (2016). Docking and linking of fragments to discover Jumonji histone demethylase inhibitors. *Journal of Medicinal Chemistry, 59*(4), 1580–1598.

Kwie, F. H., Briet, M., Soupaya, D., Hoffmann, P., Maturano, M., Rodriguez, F., et al. (2011). New potent bisubstrate inhibitors of histone acetyltransferase p300: Design, synthesis and biological evaluation. *Chemical Biology and Drug Design, 77*(1), 86–92.

Lamb, Y. N., & Dhillon, S. (2017). Epi proColon((R)) 2.0 CE: A blood-based screening test for colorectal cancer. *Molecular Diagnosis and Therapy, 21*(2), 225–232.

Lasko, L. M., Jakob, C. G., Edalji, R. P., Qiu, W., Montgomery, D., Digiammarino, E. L., Hansen, T. M., Risi, R. M., Frey, R., Manaves, V., Shaw, B., Algire, M., Hessler, P., Lam, L. T., Uziel, T., Faivre, E., Ferguson, D., Buchanan, F. G., Martin, R. L., Torrent, M., … Bromberg, K. D. (2017). Discovery of a selective catalytic p300/CBP inhibitor that targets lineage-specific tumours. *Nature, 550*(7674), 128–132.

Lee, H. J., Kim, B. K., Yoon, K. B., Kim, Y. C., & Han, S. Y. (2017). Novel inhibitors of lysine (K)-specific demethylase 4A with anticancer activity. *Investigational New Drugs, 35*(6), 733–741.

Liu, R., Zhang, Z., Yang, H., Zhou, K., Geng, M., Zhou, W., et al. (2019). Design, synthesis, and biological evaluation of a new class of histone acetyltransferase p300 inhibitors. *European Journal of Medicinal Chemistry, 180*, 171–190.

Lopez-Lopez, E., Rabal, O., Oyarzabal, J., & Medina-Franco, J. L. (2020). Towards the understanding of the activity of G9a inhibitors: An activity landscape and molecular modeling approach. *Journal of Computer-Aided Molecular Design, 34*(6), 659–669.

Lundstrom, K. (2017). Epigenetics: New possibilities for drug discovery. *Future Medicinal Chemistry, 9*(5), 437–441.

Lu, W., Zhang, R., Jiang, H., Zhang, H., & Luo, C. (2018). Computer-aided drug design in epigenetics. *Frontiers in Chemistry, 6*, 57.

Madia, V. N., Benedetti, R., Barreca, M. L., Ngo, L., Pescatori, L., Messore, A., et al. (2017). Structure-activity relationships on cinnamoyl derivatives as inhibitors of p300 histone acetyltransferase. *ChemMedChem, 12*(16), 1359–1368.

Mai, A., Massa, S., Cerbara, I., Valente, S., Ragno, R., Bottoni, P., et al. (2004). 3-(4-Aroyl-1-methyl-1H-2-pyrrolyl)-N-hydroxy-2-propenamides as a new class of synthetic histone deacetylase inhibitors. 2. Effect of pyrrole-C2 and/or -C4 substitutions on biological activity. *Journal of Medicinal Chemistry, 47*(5), 1098–1109.

Mai, A., Massa, S., Ragno, R., Cerbara, I., Jesacher, F., Loidl, P., et al. (2003). 3-(4-Aroyl-1-methyl-1H-2-pyrrolyl)-N-hydroxy-2-alkylamides as a new class of synthetic histone deacetylase inhibitors. 1. Design, synthesis, biological evaluation, and binding mode studies performed through three different docking procedures. *Journal of Medicinal Chemistry, 46*(4), 512–524.

Maldonado-Rojas, W., Olivero-Verbel, J., & Marrero-Ponce, Y. (2015). Computational fishing of new DNA methyltransferase inhibitors from natural products. *Journal of Molecular Graphics and Modelling, 60*, 43–54.

Martinez-Ramos, F., Luna-Palencia, G. R., Vasquez-Moctezuma, I., Mendez-Luna, D., Fragoso-Vazquez, M. J., Trujillo-Ferrara, J., et al. (2016). Docking studies of glutamine valproic acid derivative (S)-5- amino-2-(heptan-4-ylamino)-5-oxopentanoic acid (Gln-VPA) on HDAC8 with biological evaluation in HeLa cells. *Anticancer Agents in Medicinal Chemistry, 16*(11), 1485–1490.

Medina-Franco, J. L., Lopez-Vallejo, F., Kuck, D., & Lyko, F. (2011). Natural products as DNA methyltransferase inhibitors: A computer-aided discovery approach. *Molecular Diversity, 15*(2), 293–304.

Medina-Franco, J. L., Mendez-Lucio, O., & Yoo, J. (2014). Rationalization of activity cliffs of a sulfonamide inhibitor of DNA methyltransferases with induced-fit docking. *International Journal of Molecular Sciences, 15*(2), 3253–3261.

Methot, J. L., Chakravarty, P. K., Chenard, M., Close, J., Cruz, J. C., Dahlberg, W. K., et al. (2008). Exploration of the internal cavity of histone deacetylase (HDAC) with selective HDAC1/HDAC2 inhibitors (SHI-1:2). *Bioorganic and Medicinal Chemistry Letters, 18*(3), 973–978.

Micelli, C., & Rastelli, G. (2015). Histone deacetylases: Structural determinants of inhibitor selectivity. *Drug Discovery Today, 20*(6), 718–735.

Miletic, V., Odorcic, I., Nikolic, P., & Svedruzic, Z. M. (2017). In silico design of the first DNA-independent mechanism-based inhibitor of mammalian DNA methyltransferase Dnmt1. *PLoS One, 12*(4), e0174410.

Montano, S., Constantino-Jonapa, L. A., Sixto-Lopez, Y., Hernandez-Ramirez, V. I., Hernandez-Ceruelos, A., Romero-Quezada, L. C., et al. (2020). Vorinostat, a possible alternative to metronidazole for the treatment of amebiasis caused by Entamoeba histolytica. *Journal of Biomolecular Structure and Dynamics, 38*(2), 597–603.

Moradei, O. M., Mallais, T. C., Frechette, S., Paquin, I., Tessier, P. E., Leit, S. M., et al. (2007). Novel aminophenyl benzamide-type histone deacetylase inhibitors with enhanced potency and selectivity. *Journal of Medicinal Chemistry, 50*(23), 5543–5546.

Morris, G. M., & Lim-Wilby, M. (2008). Molecular docking. *Methods in Molecular Biology, 443*, 365–382.

Prestegui-Martel, B., Bermudez-Lugo, J. A., Chavez-Blanco, A., Duenas-Gonzalez, A., Garcia-Sanchez, J. R., Perez-Gonzalez, O. A., et al. (2016). N-(2-hydroxyphenyl)-2-propylpentanamide, a valproic acid aryl derivative designed in silico with improved anti-proliferative activity in HeLa, rhabdomyosarcoma and breast cancer cells. *Journal of Enzyme Inhibition and Medicinal Chemistry, 31*(Suppl. 3), 140–149.

Prieto-Martinez, F. D., & Medina-Franco, J. L. (2018a). Charting the bromodomain BRD4: Towards the identification of novel inhibitors with molecular similarity and receptor mapping. *Letters in Drug Design and Discovery, 15*(9), 1002–1011.

Prieto-Martinez, F. D., & Medina-Franco, J. L. (2018b). Flavonoids as putative epi-modulators: Insight into their binding mode with BRD4 bromodomains using molecular docking and dynamics. *Biomolecules, 8*(3).

Rabal, O., San Jose-Eneriz, E., Agirre, X., Sanchez-Arias, J. A., Vilas-Zornoza, A., Ugarte, A., et al. (2018). Discovery of reversible DNA methyltransferase and lysine methyltransferase G9a inhibitors with antitumoral in vivo efficacy. *Journal of Medicinal Chemistry, 61*(15), 6518−6545.

Ragno, R., Mai, A., Massa, S., Cerbara, I., Valente, S., Bottoni, P., et al. (2004). 3-(4-Aroyl-1-methyl-1H-pyrrol-2-yl)-N-hydroxy-2-propenamides as a new class of synthetic histone deacetylase inhibitors. 3. Discovery of novel lead compounds through structure-based drug design and docking studies. *Journal of Medicinal Chemistry, 47*(6), 1351−1359.

Raj, U., Kumar, H., & Varadwaj, P. K. (2017). Molecular docking and dynamics simulation study of flavonoids as BET bromodomain inhibitors. *Journal of Biomolecular Structure and Dynamics, 35*(11), 2351−2362.

Robinson, P. J., & Rhodes, D. (2006). Structure of the '30 nm' chromatin fibre: A key role for the linker histone. *Current Opinion in Structural Biology, 16*(3), 336−343.

Rossetto, D., Avvakumov, N., & Cote, J. (2012). Histone phosphorylation: A chromatin modification involved in diverse nuclear events. *Epigenetics, 7*(10), 1098−1108.

Rutten, B. P., & Mill, J. (2009). Epigenetic mediation of environmental influences in major psychotic disorders. *Schizophrenia Bulletin, 35*(6), 1045−1056.

Saldivar-Gonzalez, F. I., Gomez-Garcia, A., Chavez-Ponce de Leon, D. E., Sanchez-Cruz, N., Ruiz-Rios, J., Pilon-Jimenez, B. A., et al. (2018). Inhibitors of DNA methyltransferases from natural sources: A computational perspective. *Frontiers in Pharmacology, 9*, 1144.

Schlicker, C., Boanca, G., Lakshminarasimhan, M., & Steegborn, C. (2011). Structure-based development of novel sirtuin inhibitors. *Aging, 3*(9), 852−872.

Schmitt, M. L., Hauser, A. T., Carlino, L., Pippel, M., Schulz-Fincke, J., Metzger, E., et al. (2013). Nonpeptidic propargylamines as inhibitors of lysine specific demethylase 1 (LSD1) with cellular activity. *Journal of Medicinal Chemistry, 56*(18), 7334−7342.

Seto, E., & Yoshida, M. (2014). Erasers of histone acetylation: The histone deacetylase enzymes. *Cold Spring Harbor Perspectives in Biology, 6*(4), a018713.

Sharma, M., Dhiman, C., Dangi, P., & Singh, S. (2014). Designing synthetic drugs against plasmodium falciparum: A computational study of histone-lysine N-methyltransferase (PfHKMT). *Systems and Synthetic Biology, 8*(2), 155−160.

Sixto-Lopez, Y., Bello, M., & Correa-Basurto, J. (2019). Insights into structural features of HDAC1 and its selectivity inhibition elucidated by molecular dynamic simulation and molecular docking. *Journal of Biomolecular Structure and Dynamics, 37*(3), 584−610.

Sixto-Lopez, Y., Bello, M., Rodriguez-Fonseca, R. A., Rosales-Hernandez, M. C., Martinez-Archundia, M., Gomez-Vidal, J. A., et al. (2017). Searching the conformational complexity and binding properties of HDAC6 through docking and molecular dynamic simulations. *Journal of Biomolecular Structure and Dynamics, 35*(13), 2794−2814.

Sixto-Lopez, Y., Gomez-Vidal, J. A., & Correa-Basurto, J. (2014). Exploring the potential binding sites of some known HDAC inhibitors on some HDAC8 conformers by docking studies. *Applied Biochemistry and Biotechnology, 173*(7), 1907−1926.

Sixto-López, Y., Bello, M., & Correa-Basurto, J. (2020a). Exploring the inhibitory activity of valproic acid against the HDAC family using an MMGBSA approach. *Journal of Computer-Aided Molecular Design, 34*(8), 857−878. https://doi.org/10.1007/s10822-020-00304-2.

Sixto-López, Y., Gómez-Vidal, J. A., de Pedro, N., Bello, M., Rosales-Hernández, M. C., & Correa-Basurto, J. (2020b). Hydroxamic acid derivatives as HDAC1, HDAC6 and HDAC8 inhibitors with antiproliferative activity in cancer cell lines. *Scientific Reports, 18*.

Srimongkolpithak, N., Sundriyal, S., Li, F., Vedadi, M., & Fuchter, M. J. (2014). Identification of 2,4-diamino-6,7-dimethoxyquinoline derivatives as G9a inhibitors. *Medchemcomm, 5*(12), 1821−1828.

Tambunan, U. S., Bakri, R., Prasetia, T., Parikesit, A. A., & Kerami, D. (2013). Molecular dynamics simulation of complex histones deacetylase (HDAC) Class II Homo Sapiens with suberoylanilide hydroxamic acid (SAHA) and its derivatives as inhibitors of cervical cancer. *Bioinformation, 9*(13), 696−700.

Tao, H., Wang, J., Lu, W., Zhang, R., Xie, Y., Liu, Y.-C., et al. (2019). Discovery of trisubstituted nicotinonitrile derivatives as novel human GCN5 inhibitors through AlphaScreen-based high throughput screening. *RSC Advances, 9*(9), 4917−4924.

Uba, A. I., & Yelekci, K. (2019). Crystallographic structure versus homology model: A case study of molecular dynamics simulation of human and zebrafish histone deacetylase 10. *Journal of Biomolecular Structure and Dynamics, 1*−12.

Vincek, A., Patel, J., Jaganathan, A., Green, A., Pierre-Louis, V., Arora, V., et al. (2018). Inhibitor of CBP histone acetyltransferase downregulates p53 activation and facilitates methylation at lysine 27 on histone H3. *Molecules, 23*(8).

Waddington, C. H. (2012). The epigenotype. 1942. *International Journal of Epidemiology, 41*(1), 10−13.

Whitehead, L., Dobler, M. R., Radetich, B., Zhu, Y., Atadja, P. W., Claiborne, T., et al. (2011). Human HDAC isoform selectivity achieved via exploitation of the acetate release channel with structurally unique small molecule inhibitors. *Bioorganic and Medicinal Chemistry, 19*(15), 4626−4634.

Xiong, H., Han, J., Wang, J., Lu, W., Wang, C., Chen, Y., et al. (2018). Discovery of 1,8-acridinedione derivatives as novel GCN5 inhibitors via high throughput screening. *European Journal of Medicinal Chemistry, 151*, 740−751.

Yang, F., Zhao, N., Ge, D., & Chen, Y. (2019). Next-generation of selective histone deacetylase inhibitors. *RSC Advances, 9*(34), 19571−19583.

Yeong, K. Y., Nor Azizi, M. I. H., Berdigaliyev, N., Chen, W. N., Lee, W. L., Shirazi, A. N., et al. (2019). Sirtuin inhibition and anti-cancer activities of ethyl 2-benzimidazole-5-carboxylate derivatives. *MedChemComm, 10*(12), 2140–2145.

Yoo, J., Kim, J. H., Robertson, K. D., & Medina-Franco, J. L. (2012). Molecular modeling of inhibitors of human DNA methyltransferase with a crystal structure: Discovery of a novel DNMT1 inhibitor. *Advances in Protein Chemistry and Structural Biology, 87*, 219–247.

Yoo, J., & Medina-Franco, J. L. (2011). Homology modeling, docking and structure-based pharmacophore of inhibitors of DNA methyltransferase. *Journal of Computer-Aided Molecular Design, 25*(6), 555–567.

Yu, W., & MacKerell, A. D., Jr. (2017). Computer-aided drug design methods. *Methods in Molecular Biology, 1520*, 85–106.

Zhang, D., Han, J., Lu, W., Lian, F., Wang, J., Lu, T., et al. (2019). Discovery of alkoxy benzamide derivatives as novel BPTF bromodomain inhibitors via structure-based virtual screening. *Bioorganic Chemistry, 86*, 494–500.

Zhang, Z. M., Liu, S., Lin, K., Luo, Y., Perry, J. J., Wang, Y., & Song, J. (2015). Crystal structure of human DNA methyltransferase 1. *Journal of Molecular Biology, 427*(15), 2520–2531.

Zhang, L., Zhang, J., Jiang, Q., Zhang, L., & Song, W. (2018). Zinc binding groups for histone deacetylase inhibitors. *Journal of Enzyme Inhibition and Medicinal Chemistry, 33*(1), 714–721.

Zheng, P., Zhang, J., Ma, H., Yuan, X., Chen, P., Zhou, J., et al. (2019). Design, synthesis and biological evaluation of imidazo[1,5-a]pyrazin-8(7H)-one derivatives as BRD9 inhibitors. *Bioorganic and Medicinal Chemistry, 27*(7), 1391–1404.

Zhou, C., Kang, D., Xu, Y., Zhang, L., & Zha, X. (2015). Identification of novel selective lysine-specific demethylase 1 (LSD1) inhibitors using a pharmacophore-based virtual screening combined with docking. *Chemical Biology and Drug Design, 85*(6), 659–671.

CHAPTER 18

Molecular Docking for Natural Product Investigations: Pitfalls and Ways to Overcome Them

VERONIKA TEMML • DANIELA SCHUSTER

1 INTRODUCTION

The field of pharmacognosy, or pharmaceutical biology, has its roots in the wide variety of plant-based medicine systems around the globe. "Gegen jede Krankheit ist ein Kraut gewachsen" (there is an herb for every ailment) is a quote from Paracelsus, which is still referred to today. Before the rise of synthetic chemistry in the 20th century, nearly all medicines were plant-based. Dried herbs, teas, and plant extracts were used to cure ailments long before the knowledge of what made them a remedy was available. Modern natural product (NP) research aims to elucidate the individual compounds contained in a plant and to identify the active principles. Starting from dried plant material, scientists produce extracts with solvents of variable degrees of lipophilicity to separate different compound families from the plant. These extracts are then analyzed with thin layer chromatography or high-pressure liquid chromatography and further fractionated (e.g., with flash chromatography) to get down to the individual compounds, and their structures are elucidated with mass spectrometry and nuclear magnetic resonance spectroscopy. In search for a specific bioactivity and with a biological test system available, the fractionation process is guided by bioactivity results to focus on concentrating the active principles and isolating them (Atanasov et al., 2015; Weller, 2012).

The search for novel drug entities is a challenging task, where not only the therapeutic activity of a compound needs to be addressed but also all other biological effects, including toxicity they might exert on an organism, need to be explored. NPs continue to serve as sources for new drugs. Every year novel chemical entities that are either of plant origin or derived from NPs reach the market (Newman & Cragg, 2020). Due to the genetic relation, the plant metabolome is similar enough to that of mammalian organisms so that many secondary plant metabolites show bioactivity on animal proteins/targets (Marcus & Koch, 2004). Even though NPs promise and continue to deliver a wealth of novel bioactivities, realizing their potential into marketed drugs is challenging and many companies still prefer high-throughput screening of pure compound libraries (Amirkia & Heinrich, 2015). The main problems of NP drug development are the limited accessibility of NPs and their complex modes of action. Plant material is often not obtained locally and therefore subjected to the Nagoya Protocol to protect the exploitation rights of the region of origin (Soares, 2011). Correct plant material collection requires a prior knowledge on taxonomy and plant physiology to ensure the homogeneity of the plant material and avoid fluctuations in the concentrations of active principles. Individual NP concentrations may vary widely depending on the stage of physiological development and environmental factors, such as soil or temperature. This effect was, for example, investigated in 2010 for flavonoid concentrations in kale, which varied with climatic changes (Schmidt et al., 2010).

The process of isolating individual NPs from plant extracts is laborious and time-consuming and often leads to very small amounts of substance, consequently compelling researchers to prioritize biological experiments. Many plants contain several closely related compounds, which might contribute to the activity. They often consist of racemic mixtures of multiple stereoisomers.

For all these reasons, high-throughput screening is often not an option for NPs. NP activities are often driven by multi-target effects, and while it is becoming more clear that this might actually lead to efficient

Molecular Docking for Computer-Aided Drug Design. https://doi.org/10.1016/B978-0-12-822312-3.00027-8

drugs with a beneficial side effects profile (Koeberle & Werz, 2014; Morphy & Rankovic, 2005), it also renders the search for the mechanism of action an even more challenging task. Activities of plant extracts might also be caused by stacking or synergistic effects of different related components in the mixture that are hard to pin down on one isolated molecule (Ulrich-Merzenich et al., 2010).

Some of these challenges in NP research can be met with the aid of computational chemistry. One popular method to predict and rationalize the mode of action of NPs is molecular docking, where a molecule is fitted into the active site of a protein target, to determine if the molecule can bind to the target (Morris & Lim-Wilby, 2008). In drug discovery, this technique has been successfully used to rationalize and predict structure–activity relationships (SARs) (Ferreira et al., 2015) and to conduct large-scale virtual screening campaigns (Kontoyianni, 2017). Over the last decades, docking has also become more popular in NP chemistry, mainly to illustrate possible binding modes for observed bioactivities.

The aim of this chapter is to enlighten NP chemists on how to use and interpret molecular docking simulations in their field and on the other hand to inform computational chemists about the particularities they might encounter in the field of NP chemistry. Furthermore, we highlight high-quality studies that may serve as best practice examples from the field.

2 PECULIARITIES IN NATURAL PRODUCT INVESTIGATIONS

The most common strategy in present-day drug development is high-throughput screening of synthetic compounds developed by means of combinatorial chemistry. Consequently, computational chemistry is largely using activity data from these synthetic compounds to train and optimize their models. When working with NPs, it is vital to consider that there are overall statistical differences between the physicochemical properties of natural and synthetic products. While synthetic drugs predominantly possess a molecular weight between 200 and 500 Da, active NPs are found in a much wider range. NPs possess a larger number of chiral centers, as those are usually avoided in synthetic chemistry due to the complexity of separating racemic mixtures. NPs tend to be more rigid and they furthermore contain a higher number of rings. Synthetic molecules contain more nitrogen atoms on average, while NPs contain more oxygen atoms (Feher & Schmidt, 2003). Due to these differences in physicochemical

properties, computational chemistry focused on NPs should aim to validate its methods and workflows with data that at the very least includes NPs. NP likeness can be quantified in descriptor-based scoring functions that are used to evaluate the lead quality of novel structures (Ertl et al., 2008; Vanii Jayaseelan et al., 2012). Some of the most notable differences between NPs and synthetic drug–like molecules are highlighted in Table 18.1.

Compound series in NP research likely originate from a plant extract, which often contains several similar compounds. Finding the active principle among these closely related structures is a challenge (Balandrin et al., 1985) but sometimes derivative series can be used to elucidate the SAR of a compound class. In synthetic chemistry, derivatives are usually synthesized following an optimization scheme like the Topliss tree (Topliss, 1972) and are often aimed at elucidating the SAR. The structurally related compounds in plant extracts by contrast come from metabolic pathways, often interfering with the same protein targets and causing additive effects (Vue et al., 2015).

NPs are also known (and often derided) for their multi-target effects that can lead to unspecific activity. However, the potent and selective activity in a target-based assay is often followed by loss of efficacy in a clinical setting and ultimately resulting in drug failure. Furthermore, the vast majority of marketed drugs are found to be active on more than one target and multi-target mechanisms can also lead to more beneficial side effect profiles. The activity on multiple targets can avoid the formation of resistances and sometimes avoids pathway shunting (Morphy & Rankovic, 2005). In recent years, the approach of network pharmacology is gaining traction because it enables researchers to have better insight into the complex relationships between related structures and a wide variety of targets. A network of structures contained in an extract and their targets are mapped out to provide a complete picture of a plant extracts activities and their targets relation in the organism (Hopkins, 2007). This method mainly relies on data mining for already known activities for the contained compounds. While it is useful to gain an overview on plant extract activity, it cannot be used to find novel targets or activities. As many different species contain similar or even the same NPs, it is a valuable strategy not to replicate already known findings.

NPs also encompass structurally complex compound classes, such as macrocyclic compounds. Chen et al. investigated how many of these complex NPs they could find with models that were based exclusively on synthetic compounds. They found a hit rate

TABLE 18.1
The Mean of Several Molecular Properties That
Differ Significantly Between Compounds
Resulting From Combinatorial Chemistry
(n

Molecular Property	Combinatorial Compounds	Drugs	Natural Products
Number of chiral centers	0.4	2.3	6.2
Ratio of aromatic atoms to ring atoms	0.80	0.55	0.31
Normalized number of ring systems	0.85	0.75	0.47
Ring fusion degree	1.27	1.67	2.83
Number of nitrogen atoms	2.69	1.64	0.84
Number of oxygen atoms	2.77	4.03	5.9
Number of sulfur atoms	0.45	0.23	0.03
Number of halogen atoms	0.80	0.34	0.02
Ratio of nitrogen atoms to all heavy atoms	0.1	0.08	0.03
Ratio of oxygen atoms to all heavy atoms	0.1	0.16	0.19
Number of acceptor atoms	3.7	5.7	6.8
Number of donor atoms	1.0	1.9	2.6

Adapted from Feher, M., & Schmidt, J. M. (2003). Property distributions: Differences between drugs, natural products, and molecules from combinatorial chemistry. *Journal of Chemical Information and Computer Sciences*, 43(1), 218-227. doi: 10.1021/ci0200467.

of 30%–40% for 28 distinct targets in a similarity-based virtual screening with the shape-based comparison tool vROCS (Hawkins et al., 2007). This shows that even though structurally largely different NPs can be found with models based on synthetic compounds, a large portion of them is likely to be overlooked, if only synthetic compounds are used to build a model. Including NPs in the training sets makes the model more suited to retrieve other NPs (Chen et al., 2020).

3 MOLECULAR DOCKING FOR NATURAL PRODUCT INVESTIGATIONS

Molecular docking first described in 1982 (Kuntz et al., 1982) summarizes computational methods that aim to simulate the interaction of molecules with macromolecular targets and to quantify their binding affinity. In macromolecule-ligand docking, the binding of a small molecule ligand to a protein (or ribonucleic acid) binding pocket is simulated and scoring functions are used to quantify the energetic stability of the resulting binding poses. Protein–protein docking is used to find and simulate the binding of protein interaction sites (Chen et al., 2003). For both forms, a variety of different software solutions and algorithms are available (Pagadala et al., 2017).

Docking can be used to calculate binding modes of NPs within a protein, but unfortunately this method is sometimes used to merely create pictures that have little or no base in reality. The aim of this chapter is therefore to introduce readers from noncomputational chemistry fields to the method and pitfalls associated with it.

3.1 Docking Algorithms and Scoring

There is a wide range of freeware and commercial software programs available for docking. These programs use different approaches to predict binding affinity and their performance is compared in several reviews (Cross et al., 2009; Pagadala et al., 2017). Comparative studies on NPs are rare, but in 2017, Castro-Alvarez et al. evaluated the performance of Glide and different Autodock versions for their ability to replicate macrolide protein complexes. Depending on the quality measure, Glide and AutoDock Vina showed the best results, respectively (Castro-Alvarez et al., 2017). Depending on the research question, it can be beneficial to compare different software and select the one best suited to the task.

Even within one software solution, there is often the option to use different scoring functions. Scoring functions aim to quantify how well a ligand fits into the binding pocket. Ligand binding depends on factors such as steric fit, electrostatic interactions, hydrophobic contacts, and entropic contributions from replacing water in the binding site. There are multiple approaches to weigh different contributions to ligand binding, so there is a wide array of scoring functions and the topic of scoring still remains controversial within the field (Xu et al., 2015).

If a number of active compounds of a similar scaffold are available, researches may turn to customized scoring functions that reflect their knowledge from experiments, e.g., by rewarding poses that interact with known key binding residues. The process of customizing scoring functions has also been computationally optimized by employing machine learning approaches (Ragoza et al., 2017).

NPs inhibit proteins with a wide variety of binding modes and mechanisms of action. Many NPs interact via direct reaction with the protein target, binding with covalent bonds, e.g., via Michael addition (Caprioglio et al., 2020; Drahl et al., 2005). These covalent mechanisms of action can lead to irreversible inhibition, which can be not only advantageous because only a low dose is needed for a long-lasting effect but also disadvantageous because the target protein is often irrevocably damaged. Standard docking programs are not designed to take covalent binding mechanisms into account. Specific workflows can be designed, e.g., with constraints that place reactive groups in the reaction vicinity, but the information about the reaction mechanism and reactive groups has to come from the user.

Over time, several docking programs were designed or adapted from noncovalent docking programs to tackle the specific challenges of covalent docking. Usually a covalent reaction is characterized by two steps. In the initial step, a noncovalent complex is formed, where the electrophilic group is positioned in an orientation that enables the following reaction with its nucleophile counterpart. The reaction site geometry is therefore the cornerstone of covalent docking.

Scarpino et al. provide a comparative evaluation of six different software solutions for covalent docking. In this work, they evaluated the performance of these tools in regards to seven different mechanisms of action: Michael addition, nucleophilic substitution, ring opening, addition to nitrile, addition to ketone, addition to aldehyde, and disulfide formation. In this study, ICM-Pro (http://www.molsoft.com/icm_pro.html) outperformed the other tools, although the performance differed widely depending on which reaction type was investigated (Scarpino et al., 2018).

3.2 Databases

There are several database resources for molecular modeling in the context of NPs available. Unfortunately, many NP databases do not contain the necessary

information for molecular modeling approaches. It is vital that the correct stereochemistry is specified. Lagunin et al. reviewed 46 NP databases for their suitability to use in a molecular modeling context. They chronicled the quality of the contained data and the public availability of the databases (Lagunin et al., 2014). More recently, Chen et al. reviewed 25 virtual and 31 physical NP libraries in the context of chemoinformatics applications such as docking. The review also examined the overlap between the NPs contained within the databases, showing that a large part of the available NPs is contained within the freely available databases (Chen, de Bruyn Kops, & Kirchmair, 2017). In this review, we would like to highlight a few highly valuable data resources in Table 18.2.

3.3 Targets and Bioassay Considerations for Successful Docking

3.3.1 Does the biological test system really verify the target?

Before undertaking an in silico project with the goal to elucidate an SAR, it is vital to familiarize oneself with the biological test systems available for determining the potency of the compounds. The requirements of a bioassay for NP research and target-based computational chemistry differ. While a bioassay should always be reliable, inexpensive, and sensitive, it is especially important in NP research that the assay requires little material and covers a spectrum of activities, for example, general antiinflammatory assays (Atanasov et al., 2015; Banerjee et al., 2014; Colegate & Molyneux, 2008). Such general assays enable us to identify activity for small samples of plant material and increase the likelihood of translating this activity to the organism at large; they are, however, not well suited for molecular modeling purposes. Structure-based modeling techniques, such as docking, pharmacophore modeling, or molecular dynamics simulations, all simulate the interaction between a single-target protein and its direct interaction partner. Off-target effects or even physicochemical factors such as solubility and cell permeability are not part of the simulation. To optimize a target-based computational workflow, based on a known structure activity dataset, it is vital that the biological activity is measured as direct target–ligand interaction, for example, radioligand binding assays, receptor binding assays or enzyme inhibition assay.

This allows the modeler to fine-tune their workflow for maximum activity within the experimental setup. When outside factors (e.g., cell permeability in cell-based assays) play a role in the assay, it becomes nearly

TABLE 18.2
The Largest Natural Product (NP) Databases Currently Available.

Database	Number of Compounds (K)	Molecular Library	Free of Charge	Bioactivities	Literature
Dictionary of NPs (DNP)	153	No	No	Yes	No
Super Natural II	325	No	No	Yes	Banerjee et al. (2015)
3DMET	18	No	Yes	No	No
Ambinter-Greenpharma Natural Compound Library (GPNCL)	150	No	No	No	No
AntiMarin	60	Yes	Yes	No	No
CMAUP	47	Yes	Yes	Yes	Zeng et al. (2019)
CNPD (Chinese Natural Products Database)	57	No	Unknown	Unknown	Shen et al. (2003)
ZINC natural products catalog	85	Yes	Yes	No	Sterling and Irwin (2015)
DMNP (Dictionary of Marine NPs)	30	Yes	Yes	No	No
DFC (Dictionary of Food Compounds)	41	Yes	Yes	No	No
MarinLIT	29	Yes	Yes	No	Blunt et al. (2018)
NPASS	30	Yes	Yes	No	Zeng et al. (2018)
NPAtlas	20	Yes	Yes	No	No
UNDP	229	Yes	Yes	No	Gu et al. (2013)
TCMID	13	Yes	Yes	Yes	Xue et al. (2013)
TCM Taiwan	58	Yes	Yes	Yes	Chen (2011)

impossible to determine if a compound is inactive because it does not fit the target (which should prompt refinement of the model) or because it cannot reach the target (an effect that is outside the scope of the model). Assay interference (e.g., from a strong coloring of a compound) can lead to false positive results, which also distract the optimization process (Fig. 18.1).

In some protein targets, there is more than one modulating binding site available. A prominent example for this would be the androgen receptor, with more than five distinct and druggable binding sites (Ban et al., 2017). For the purpose of structure-based molecular modeling, each distinct binding site has to be treated as a separate protein target. Attempting to build a "one-size-fits-all" workflow would be akin to trying to assemble two different jigsaw puzzles as one. In many biological assays, the possibility of

different binding sites is not taken into account, even though novel binding sites are often discovered for well-known established targets. This was the case for 5-lipoxygenase (5-LOX), which was long assumed to be only inhibited by molecules interacting with the substrate binding site. Only recently, an alternative allosteric binding site was verified. This new site is likely the target of many known 5-LOX inhibitors that should be reevaluated (Chan et al., 2019; Pein et al., 2018).

Surprisingly in many NP publications, the target is not sufficiently verified. There might be a general activity (e.g., antiinflammatory) and a docking study on a specific target that is associated with that activity (e.g., cyclooxygenase), but other targets that could cause the activity in the assay are often not experimentally verified. Such studies should be regarded with scrutiny and ideally be removed in peer review.

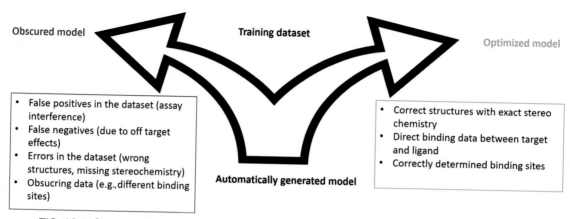

Obscured model

Training dataset

Optimized model

- False positives in the dataset (assay interference)
- False negatives (due to off target effects)
- Errors in the dataset (wrong structures, missing stereochemistry)
- Obsucring data (e.g., different binding sites)

- Correct structures with exact stereo chemistry
- Direct binding data between target and ligand
- Correctly determined binding sites

Automatically generated model

FIG. 18.1 Observed effects that are related to direct ligand binding can be used to improve the model, while effects occurring outside the model will obscure the structure–activity relationship of NPs.

3.3.2 Target fishing with docking?

In the context of high-throughput screening of large synthetic compound libraries, computational filters are often used to narrow down testing to the most promising drug candidate molecules. NPs, by contrast, have usually been isolated in small amounts from plants or plant extracts with a known general activity. Instead of identifying active molecules, the question posed to the computational chemist is usually to find a target for the isolated NPs.

Such a virtual target fishing approach can be used to narrow down experimental tests to the most promising candidate targets (Lee & Kim, 2019). In this setup, a given structure is docked into the binding sites of multiple proteins and the targets that "fit best" are experimentally tested. This process is sometimes also called reverse docking (Kharkar et al., 2014). The advantage of virtual target fishing is that it allows the user to virtually screen a large number of target structures. However, novel targets without three-dimensional structural data are not available cannot be accessed by this method.

How well a molecule fits into a binding pocket is quantified by a scoring function. There are different mathematical concepts that scoring functions use to calculate binding affinity between protein and ligand conformations (Li et al., 2019). In a recent review, the reliability of docking as a method for target fishing was evaluated. A dataset of 60 targets and 600 known active ligands was assembled. The authors used 13 distinct docking workflows with different docking algorithms and scoring functions to assign the ligands to their corresponding targets. Around 25%–35% of ligand–target combinations were predicted correctly. The best performance within the setup was achieved by a workflow in Glide SP and by a consensus approach, combining the results from the different docking setups (Lapillo et al., 2019).

3.3.3 Finding the binding site in the target

In case that the binding site of the ligand on the protein structure is unknown, there are multiple tools available to identify binding pockets on a proteins surface from a crystallographic structure. A variety of software solutions for pocket finding were collected in a review by Dukka in 2013 (Dukka, 2013). In the meantime, the problem has also been tackled by other research groups. In 2019, Jendele et al. introduced a web server for pocket finding and visualization, called PrankWeb (Jendele et al., 2019).

The program BsiteFinder employs structural alignment of known target–ligand complexes to propose novel binding sites on proteins based on similarity (Gao et al., 2016). DeepDrug3D uses a neural network to predict and evaluate binding sites (Pu et al., 2019). One tool, cryptoSite, even offers to predict binding sites that are closed off by allosteric shifts or otherwise hidden. Such crypto pockets are under discussion for their therapeutic potential (Cimermancic et al., 2016).

3.4 Validation of Docking Workflows

Creating a reliable docking workflow requires in-depth analysis of the available structure and activity data. These data can then be used to optimize and validate docking settings to ensure reliable performance.

3.4.1 Validation by redocking

The first step of validating a docking workflow is a redocking of a co-crystallized compound. This compound is prepared with the same workflow as the potential ligands that are later docked in the prospective simulation. It is vital not to use the co-crystallized conformation as a starting point but to draw and calculate the starting conformation the same way it is executed for the other molecules. The resulting poses are then compared with the original pose from the crystal structure and a root mean square deviation (RMSD) value is calculated to quantify the difference between the docking pose and the bioactive conformation (Fig. 18.2).

This setup is used to optimize docking settings like the exact binding site position and size or scoring function. Water co-crystallized in the binding pocket is sometimes relevant for ligand binding and should be examined and taken into account for a docking simulation. Water molecules stabilized in the binding pocket by multiple hydrogen bonds with the protein residues of the binding site most likely play a role in ligand binding (Huang & Shoichet, 2008). As the RMSD value is size-dependent, it is difficult to define an exact cutoff. The redocking settings should be aimed to minimize the RMSD, ideally below 2 Å. A fast way to calculate RMSD values online is DOCK RMSD (Bell & Zhang, 2019).

3.4.2 Validation by cross-docking

An advanced validation strategy is cross-docking, which can be pursued if crystallographic data of the same protein co-crystallized with multiple ligands are available. In this case, the ligands of other crystal structures (prepared as in a prospective docking) are docked into the investigated structure and their binding pose is then compared to the one they assumed in co-crystallization (Rueda et al., 2010). This approach improves upon redocking because it verifies the ability of the workflow to also find binding conformations with novel molecules and it takes into account protein flexibility (Mukherjee et al., 2010). A protein sometimes assumes adaptive conformations to accommodate different ligand molecules. For docking, all possible states of the protein should be sampled. If they differ widely, it can also be feasible to use more than one crystal structure and workflow (Rueda et al., 2012). Online tools like Protein-Ligand Interaction Profiler (PLIP) offer a quick way to gain an overview of key interactions in a large number of crystal structures (Salentin et al., 2015).

3.4.3 Validation by using test and training set data

Especially for novel, highly innovative targets, crystallographic data are often scarce and there might not be any structures for cross-docking available. In some cases, only an apo-protein structure was crystallized and therefore not even a validation with redocking is feasible.

An alternative approach in such a setting is the use of a test set comprised of high-quality, target-based activity data. The workflow is then optimized to discriminate between known active and inactive molecules. Validation is key for any reliable docking workflow (Deligkaris et al., 2014; Friedrich et al., 2017). Ideally, a theoretical validation test set should contain NPs with a similar size as the investigated compounds.

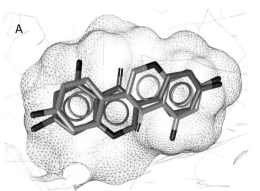
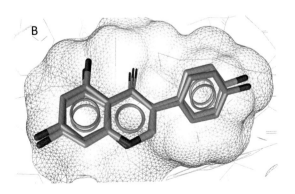

FIG. 18.2 Docking poses of genistein in the binding pocket of the estrogen receptor. The co-crystallized conformation is shown in blue. (A) Depicts a failed redocking, where the ligand is flipped in the binding pocket. (B) Shows a successful redocking, where the ligand is orientated to form the same interactions as in the co-crystallized complex (Manas, Xu, Unwalla, & Somers, 2004).

Modelling

- Workflow validation
- High quality dataset
- Optimized methods

Experiment

- Reliable target-based assay
- Pure compounds
- Defined stereochemistry

FIG. 18.3 Collaboration between molecular modeling and biological activity testing team.

Careful workflow validation, based on reliable experimental data, leads to a robust and widely applicable predictive workflow (Fig. 18.3).

4 OTHER COMPUTATIONAL METHODS FOR NATURAL PRODUCT INVESTIGATIONS

Other computational tools can even further enhance the predictive ability of a molecular docking study. Quantitative structure–activity relationship (QSAR) methods can be used to filter down hit lists or to refine docking models. Similarity searches can bring up related compounds with known activities. They can also help to recognize pan-assay interference compounds (PAINS), which might give misleading results in bioassays (Baell & Holloway, 2010). PAINS are a controversial topic in NP chemistry because several well-investigated NP classes are contained in PAINS list (Baell, 2016; Baell & Nissink, 2018; Capuzzi et al., 2017). As it is paramount to discuss PAINS compounds if they appear as hits in a biological screening, the benefits of fast computational recognition are obvious.

QSAR models are also used to predict properties outside direct ligand binding, such as bioavailability, distribution, drug metabolism, and potential toxicity. Also, other computational methods, such as pharmacophore modeling, can successfully adapt for being applied to NPs (Seidel et al., 2020). Molecular dynamics simulations are in use to further refine docking poses and elucidate binding modes. There are numerous studies aiming to synthetically optimize NP lead structures. Here docking is used to prioritize synthetic proposals.

5 RECENT EXAMPLES OF DOCKING IN NATURAL PRODUCT RESEARCH

Docking has been used in a variety of studies to enhance NP-related activity studies. The transient receptor

potential ankyrin 1 (TRPA1) channel is a therapeutic target to alleviate pain, itching, and other irritant reactions. TRPA1 can either be activated via a reaction with electrophilic compounds or by noncovalent agonists. A series of piperidine–carboxamide analogs was used to explore the different states and binding mechanisms. Homology modeling and molecular docking were combined to propose a binding site for the compounds and to rationalize their SAR. The predictions made in the molecular modeling efforts were then experimentally confirmed in detail. The most potent compound, PIPC1, is shown in Fig. 18.4 (Chernov-Rogan et al., 2019).

In 2017, a review by Dhiman et al. summarized a variety of in silico studies of NPs and related derivatives as monoamine oxidase (MAO) inhibitors. MAO isoforms play a role in the metabolism of neurotransmitters and are targets for depression and neurodegenerative diseases. Authors discussed docking studies on a range of NPs, such as chalcones, flavonoids, coumarins, xanthones, and a variety of alkaloids (Dhiman et al., 2018). In a study by Pein et al., docking was used to rationalize the binding mode of vitamin E–derived NPs on 5-LOX and other targets. Experiments had shown that the most active compound garcinoic acid was not acting in competition with the 5-LOX substrate arachidonic acid. It seemed therefore inevitable that the compound series was binding to an alternative binding site. Such a site was located with a pocket finder software, and docking in the proposed site could indeed be used to rationalize the observed SAR (Pein et al., 2018). Molecular docking was then also used to guide the synthetic optimization of the lead structure in a follow-up publication (Dinh et al., 2020). Recently, a crystal structure of another noncompetitive 5-LOX inhibitor, the boswellic acid AKBA (Fig. 18.4), showed this molecule co-crystallized in vicinity of the proposed binding site for garcinoic acid (Gilbert et al., 2020) (see Figs. 18.4 and 18.5).

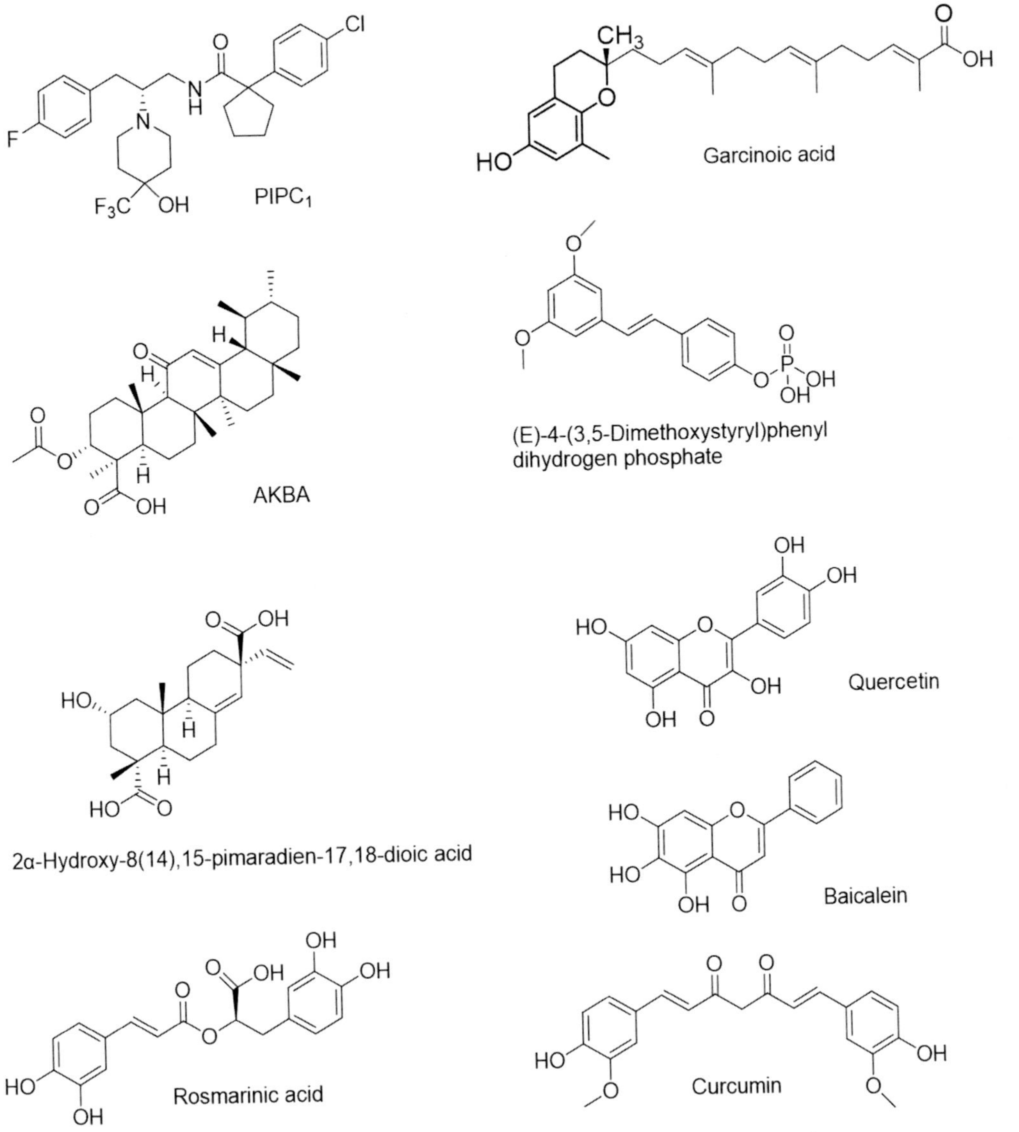

FIG. 18.4 Structures of natural products examined in docking studies.

Mizuno et al. employed docking in the optimization of pterostilbene, a natural PPARα activator. The most active compound from the series (E)-4-(3,5-dimethoxystyryl) phenyl dihydrogen phosphate (Fig. 18.4) outperformed both pterostilbene and the control drug ciprofibrate. The docking study revealed that there is a key set of polar amino acids in the binding pocket that are crucial for the activation of the receptor. This part of the pocket could not be reached by the cis-isomers of the series which were uniformly inactive, while the active trans-isomers formed these critical interactions

(Mizuno et al., 2008). A recent study by Thawai et al. investigated diterpene derivatives isolated from *Micro-bispora hainanensis* strain CSR-4 and their in vitro and in silico inhibition effects on acetylcholine esterase enzyme (AChE). The most active compound 2α-hy-droxy-8(14),15-pimaradien-17,18-dioic acid (see Fig. 18.4) inhibited AChE with an IC_{50} of 96.87 ± 2.31 μg/mL (Thawai et al., 2020). In 2020, Redford and Abbott reported that the well-known flavonoid quercetin (see Fig. 18.4) from capers, known for multiple activities, based on a variety of targets

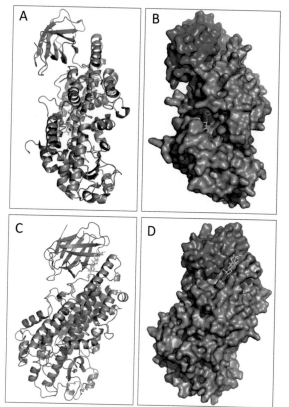

FIG. 18.5 The crystal structures of 5-lipoxygenase co-crystallized with NDGA in the substrate binding site **(A)** and **(B)** and co-crystallized with AKBA within the alternate binding site **(C and D)** (Gilbert et al., 2020).

(Anand David et al., 2016), also acted as an atypical KCNQ potassium channel activator. They proposed interaction modes of quercetin on different states of the protein and supported the findings from the simulations by experimentally confirming the proposed key interaction residues (Redford & Abbott, 2020).

In a target fishing approach, Chen et al. employed reverse docking to predict target proteins for marine NPs with antitumor activity. The method was validated by comparing the results to known bioactivities of the compounds (Chen, Wang, et al., 2017). Target fishing by reverse docking was also conducted by Park and Cho and proposed several anticancer targets for a series of 26 ginsenosides (Park & Cho, 2017). Gupta et al. provided a comprehensive review of the wide variety of protein targets that have been found and verified for curcumin. Many of the 50+ known targets have been investigated with docking to propose a mode of

action for curcumin (Gupta et al., 2011). Zhao et al. used metabolite docking to map out metabolic pathways. Genetic pathway context and docking workflows for multiple metabolic enzymes (or homology models) were used to predict the metabolism of compounds. The predicted pathways and intermediates were then experimentally validated by in vitro assays and metabolomics (Zhao et al., 2013).

In 2016, Corral-Lugo et al. investigated NPs with the ability to affect quorum sensing via binding to the quorum sensing regulator RhlR of *Pseudomonas aeruginosa* PAO1. They found rosmarinic acid acting as a homoserine lactone mimic. Docking was successfully used to propose a binding mode of rosmarinic acid to RhlR (Corral-Lugo et al., 2016).

In a 2020 study, Liang et al. used an approach termed targeted covalent NP design to search for NP inhibitors of polo-like kinase (PLK) 1. They were looking for covalent binders by using consensus (noncovalent) docking with four different programs. Only hits that were positioned similarly in three out of four programs were considered for hit selection. Five herbs that contained the virtual hits were then experimentally evaluated. The most potent activity was exhibited by *Scutellaria baicalensis*. Two of the main constituents baicalein (see Fig. 18.4) and baicalin were shown to covalently bind to PLK1 via Cys133 (Liang et al., 2020).

Molecular docking is an approach that allows us to quickly integrate the available data into a predictive workflow to find novel ligands. Reacting to the current COVID-19 pandemic, Battisti et al. used a combined approach of docking, molecular dynamics simulation, and pharmacophore modeling to determine compounds with a high likelihood of inhibiting SARS-CoV-2. While at this point, no experimental validation of the proposed antiviral activities was possible, this study shows that molecular docking can produce a fast and straightforward list of promising test compounds when time is pressing (Battisti et al., 2020).

6 CONCLUSIONS

High-quality docking studies can do their part in enhancing research on NPs. They can aid in elucidating the often complex and unusual binding mechanisms of NPs. Prerequisite for this is a workflow based on strictly target-based activity data and an intimate understanding of the corresponding experimental work. Over the last decades, more databases focusing on NPs have become available, providing us with a plethora of data that can be accessed and processed

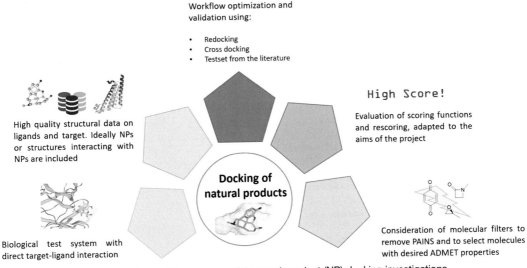

FIG. 18.6 Key points of successful natural product (NP) docking investigations.

computationally. Some of the important points to be considered for a successful NP docking investigations are shown in Fig. 18.6.

Every year more specialized tools are developed, which allow us to handle the challenges of NP drug development, such as multitarget effects or covalent binding modes. This fast growth of ever improving methods should inspire scientists of NP research fields to explore and evaluate new tools and apply them, where they can propel their research ahead. Vice versa, computational chemists should aim to meet the challenges posed by NP chemistry to explore this diverse and extraordinary part of chemical space.

REFERENCES

Amirkia, V., & Heinrich, M. (2015). Natural products and drug discovery: A survey of stakeholders in industry and academia. *Frontiers in Pharmacology, 6*, 237. https://doi.org/10.3389/fphar.2015.00237.

Anand David, A. V., Arulmoli, R., & Parasuraman, S. (2016). Overviews of biological importance of quercetin: A bioactive flavonoid. *Pharmacognosy Reviews, 10*(20), 84–89. https://doi.org/10.4103/0973-7847.194044.

Atanasov, A. G., Walterberger, B., Pferschy-Wenig, E.-M., Linder, T., Wawrosch, C., Uhrin, P., Temml, V., Wang, L., Schwaiger, S., Heiss, E. H., Rollinger, J. M., Schuster, D., Breuss, J. M., Bochkov, V., Mihovilovic, M. D., Kopp, B., Bauer, R., Dirsch, V. M., & Stuppner, H. (2015). Discovery and resupply of pharmacologically active plant-derived natural products: A review. *Biotechnology Advances, 33*(8), 1582–1614. https://doi.org/10.1016/j.biotechadv.2015.08.001.

Baell, J. B. (2016). Feeling nature's PAINS: Natural products, natural product drugs, and pan assay interference compounds (PAINS). *Journal of Natural Products, 79*(3), 616–628. https://doi.org/10.1021/acs.jnatprod.5b00947.

Baell, J. B., & Holloway, G. A. (2010). New substructure filters for removal of pan assay interference compounds (PAINS) from screening libraries and for their exclusion in bioassays. *Journal of Medicinal Chemistry, 53*(7), 2719–2740. https://doi.org/10.1021/jm901137j.

Baell, J. B., & Nissink, J. W. M. (2018). Seven year itch: Pan-assay interference compounds (PAINS) in 2017—utility and limitations. *ACS Chemical Biology, 13*(1), 36–44. https://doi.org/10.1021/acschembio.7b00903.

Balandrin, M. F., Klocke, J. A., Wurtele, E. S., & Bollinger, W. H. (1985). Natural plant chemicals: Sources of industrial and medicinal materials. *Science, 228*(4704), 1154. https://doi.org/10.1126/science.3890182.

Ban, F., Dalal, K., Li, H., LeBlanc, E., Rennie, P. S., & Cherkasov, A. (2017). Best practices of computer-aided drug discovery: Lessons learned from the development of a preclinical candidate for prostate cancer with a new mechanism of action. *Journal of Chemical Information and Modeling, 57*(5), 1018–1028. https://doi.org/10.1021/acs.jcim.7b00137.

Banerjee, S., Chanda, A., Adhikari, A., Das, A., & Biswas, S. (2014). Evaluation of phytochemical screening and anti inflammatory activity of leaves and stem of *Mikania scandens* (L.) wild. *Annals of Medical and Health Sciences Research, 4*(4), 532–536. https://doi.org/10.4103/2141-9248.139302.

Banerjee, P., Erehman, J., Gohlke, B. O., Wilhelm, T., Preissner, R., & Dunkel, M. (2015). Super natural II—a database of natural products. *Nucleic Acids Research, 43*(Database issue), D935–D939. https://doi.org/10.1093/nar/gku886.

Battisti, V., Wieder, O., Garon, A., Seidel, T., Urban, E., & Langer, T. (2020). A computational approach to identify potential novel inhibitors against the coronavirus SARS-CoV-2. *Molecular Informatics*. https://doi.org/10.1002/minf.202000090.

Bell, E. W., & Zhang, Y. (2019). DockRMSD: An open-source tool for atom mapping and RMSD calculation of symmetric molecules through graph isomorphism. *Journal of Cheminformatics*, 11(1), 40. https://doi.org/10.1186/s13321-019-0362-7.

Blunt, J. W., Carroll, A. R., Copp, B. R., Davis, R. A., Keyzers, R. A., & Prinsep, M. R. (2018). Marine natural products. *Natural Product Reports*, 35(1), 8−53. https://doi.org/10.1039/C7NP00052A.

Caprioglio, D., Minassi, A., Avonto, C., Taglialatela-Scafati, O., & Appendino, G. (2020). Thiol-trapping natural products under the lens of the cysteamine assay: Friends, foes, or simply alternatively reversible ligands? *Phytochemistry Reviews*. https://doi.org/10.1007/s11101-020-09700-w.

Capuzzi, S. J., Muratov, E. N., & Tropsha, A. (2017). Phantom PAINS: Problems with the utility of alerts for pan-assay interference compounds. *Journal of Chemical Information and Modeling*, 57(3), 417−427. https://doi.org/10.1021/acs.jcim.6b00465.

Castro-Alvarez, A., Costa, A. M., & Vilarrasa, J. (2017). The performance of several docking programs at reproducing protein-macrolide-like crystal structures. *Molecules (Basel, Switzerland)*, 22(1), 136. https://doi.org/10.3390/molecules22010136.

Chan, H. C. S., Li, Y., Dahoun, T., Vogel, H., & Yuan, S. (2019). New binding sites, new opportunities for GPCR drug discovery. *Trends in Biochemical Sciences*, 44(4), 312−330. https://doi.org/10.1016/j.tibs.2018.11.011.

Chen, C. Y.-C. (2011). TCM Database@Taiwan: The world's largest traditional Chinese medicine database for drug screening in silico. *PLoS One*, 6(1), e15939. https://doi.org/10.1371/journal.pone.0015939.

Chen, Y., de Bruyn Kops, C., & Kirchmair, J. (2017). Data resources for the computer-guided discovery of bioactive natural products. *Journal of Chemical Information and Modeling*, 57(9), 2099−2111. https://doi.org/10.1021/acs.jcim.7b00341.

Chen, Y., Mathai, N., & Kirchmair, J. (2020). Scope of 3D shape-based approaches in predicting the macromolecular targets of structurally complex small molecules including natural products and macrocyclic ligands. *Journal of Chemical Information and Modeling*, 60(6), 2858−2875. https://doi.org/10.1021/acs.jcim.0c00161.

Chen, R., Mintseris, J., Janin, J., & Weng, Z. (2003). A protein−protein docking benchmark. *Proteins: Structure, Function, and Bioinformatics*, 52(1), 88−91. https://doi.org/10.1002/prot.10390.

Chen, F., Wang, Z., Wang, C., Xu, Q., Liang, J., Xu, X., Yang, J., Wang, C., Jiang, T., & Yu, R. (2017). Application of reverse docking for target prediction of marine compounds with anti-tumor activity. *Journal of Molecular Graphics and Modelling*, 77, 372−377. https://doi.org/10.1016/j.jmgm.2017.09.015.

Chernov-Rogan, T., Gianti, E., Liu, C., Villemure, E., Cridland, A. P., Hu, X., Ballini, E., Lange, W., Deisemann, H., Li, T., Ward, S. I., Hackos, D. H., Magnuson, S., Safina, B., Klein, M. L., Volgraf, M., Carnevale, V., & Chen, J. (2019). TRPA1 modulation by piperidine carboxamides suggests an evolutionarily conserved binding site and gating mechanism. *Proceedings of the National Academy of Sciences*, 116(51), 26008. https://doi.org/10.1073/pnas.1913929116.

Cimermancic, P., Weinkam, P., Rettenmaier, T. J., Bichman, L., Keedy, D. A., Woldeyes, R. A., Schneidman-Duhovny, D., Demerdash, O. N., Mitchell, J. C., Wells, J. A., Fraser, J. S., & Sali, A. (2016). CryptoSite: Expanding the druggable proteome by characterization and prediction of cryptic binding sites. *Journal of Molecular Biology*, 428(4), 709−719. https://doi.org/10.1016/j.jmb.2016.01.029.

Colegate, S. E., & Molyneux, R. (Eds.). (2008). *Bioactive natural products*. Boca Raton: CRC Press.

Corral-Lugo, A., Daddaoua, A., Ortega, A., Espinosa-Urgel, M., & Krell, T. (2016). Rosmarinic acid is a homoserine lactone mimic produced by plants that activates a bacterial quorum-sensing regulator. *Science Signaling*, 9(409), ra1. https://doi.org/10.1126/scisignal.aaa8271.

Cross, J. B., Thompson, D. C., Rai, B. K., Baber, J. C., Fan, K. Y., Hu, Y., & Humblet, C. (2009). Comparison of several molecular docking programs: Pose prediction and virtual screening accuracy. *Journal of Chemical Information and Modeling*, 49(6), 1455−1474. https://doi.org/10.1021/ci900056c.

Deligkaris, C., Ascone, A. T., Sweeney, K. J., & Greene, A. J. Q. (2014). Validation of a computational docking methodology to identify the non-covalent binding site of ligands to DNA. *Molecular BioSystems*, 10(8), 2106−2125. https://doi.org/10.1039/C4MB00239C.

Dhiman, P., Malik, N., & Khatkar, A. (2018). 3D-QSAR and in-silico studies of natural products and related derivatives as monoamine oxidase inhibitors. *Current Neuropharmacology*, 16(6), 881−900. https://doi.org/10.2174/1570159X15666171128143650.

Dinh, C. P., Ville, A., Neukirch, K., Viault, G., Temml, V., Koeberle, A., Werz, O., Schuster, D., Stuppner, H., Richomme, P., Helesbeux, J.-J., & Séraphin, D. (2020). Structure-based design, semi-synthesis and anti-inflammatory activity of tocotrienolic amides as 5-lipoxygenase inhibitors. *European Journal of Medicinal Chemistry*, 112518. https://doi.org/10.1016/j.ejmech.2020.112518.

Drahl, C., Cravatt, B. F., & Sorensen, E. J. (2005). Protein-reactive natural products. *Angewandte Chemie International Edition*, 44(36), 5788−5809. https://doi.org/10.1002/anie.200500900.

Dukka, B. K. (2013). Structure-based methods for computational protein functional site prediction. *Computational and Structural Biotechnology Journal*, 8(11), e201308005. https://doi.org/10.5936/csbj.201308005.

Ertl, P., Roggo, S., & Schuffenhauer, A. (2008). Natural product-likeness score and its application for prioritization of compound libraries. *Journal of Chemical*

Information and Modeling, 48(1), 68–74. https://doi.org/10.1021/ci700286x.

Feher, M., & Schmidt, J. M. (2003). Property distributions: Differences between drugs, natural products, and molecules from combinatorial chemistry. *Journal of Chemical Information and Computer Sciences, 43*(1), 218–227. https://doi.org/10.1021/ci0200467.

Ferreira, L. G., Dos Santos, R. N., Oliva, G., & Andricopulo, A. D. (2015). Molecular docking and structure-based drug design strategies. *Molecules (Basel, Switzerland), 20*(7), 13384–13421. https://doi.org/10.3390/molecules200713384.

Friedrich, N.-O., Meyder, A., de Bruyn Kops, C., Sommer, K., Flachsenberg, F., Rarey, M., & Kirchmair, J. (2017). High-quality dataset of protein-bound ligand conformations and its application to benchmarking conformer ensemble generators. *Journal of Chemical Information and Modeling, 57*(3), 529–539. https://doi.org/10.1021/acs.jcim.6b00613.

Gao, J., Zhang, Q., Liu, M., Zhu, L., Wu, D., Cao, Z., & Zhu, R. (2016). bSiteFinder, an improved protein-binding sites prediction server based on structural alignment: More accurate and less time-consuming. *Journal of Cheminformatics, 8*, 38. https://doi.org/10.1186/s13321-016-0149-z.

Gilbert, N. C., Gerstmeier, J., Schexnaydre, E. E., Börner, F., Garscha, U., Neau, D. B., Werz, O., & Newcomer, M. E. (2020). Structural and mechanistic insights into 5-lipoxygenase inhibition by natural products. *Nature Chemical Biology, 16*(7), 783–790. https://doi.org/10.1038/s41589-020-0544-7.

Gu, J., Gui, Y., Chen, L., Yuan, G., Lu, H.-Z., & Xu, X. (2013). Use of natural products as chemical library for drug discovery and network pharmacology. *PLoS One, 8*(4), e62839. https://doi.org/10.1371/journal.pone.0062839.

Gupta, S. C., Prasad, S., Kim, J. H., Patchva, S., Webb, L. J., Priyadarsini, I. K., & Aggarwal, B. B. (2011). Multitargeting by curcumin as revealed by molecular interaction studies. *Natural Product Reports, 28*(12), 1937–1955. https://doi.org/10.1039/c1np00051a.

Hawkins, P. C. D., Skillman, A. G., & Nicholls, A. (2007). Comparison of shape-matching and docking as virtual screening tools. *Journal of Medicinal Chemistry, 50*(1), 74–82. https://doi.org/10.1021/jm0603365.

Hopkins, A. L. (2007). Network pharmacology. *Nature Biotechnology, 25*(10), 1110–1111. https://doi.org/10.1038/nbt1007-1110.

Huang, N., & Shoichet, B. K. (2008). Exploiting ordered waters in molecular docking. *Journal of Medicinal Chemistry, 51*(16), 4862–4865. https://doi.org/10.1021/jm8006239.

Jendele, L., Krivak, R., Skoda, P., Novotny, M., & Hoksza, D. (2019). PrankWeb: A web server for ligand binding site prediction and visualization. *Nucleic Acids Research, 47*(W1), W345–W349. https://doi.org/10.1093/nar/gkz424.

Kharkar, P. S., Warrier, S., & Gaud, R. S. (2014). Reverse docking: A powerful tool for drug repositioning and drug rescue. *Future Medicinal Chemistry, 6*(3), 333–342. https://doi.org/10.4155/fmc.13.207.

Koeberle, A., & Werz, O. (2014). Multi-target approach for natural products in inflammation. *Drug Discovery Today, 19*(12), 1871–1882. https://doi.org/10.1016/j.drudis.2014.08.006.

Kontoyianni, M. (2017). Docking and virtual screening in drug discovery. In I. M. Lazar, M. Kontoyianni, & A. C. Lazar (Eds.), *Proteomics for drug discovery: Methods and protocols* (pp. 255–266). New York, NY: Springer New York.

Kuntz, I. D., Blaney, J. M., Oatley, S. J., Langridge, R., & Ferrin, T. E. (1982). A geometric approach to macromolecule-ligand interactions. *Journal of Molecular Biology, 161*(2), 269–288. https://doi.org/10.1016/0022-2836(82)90153-X.

Lagunin, A. A., Goel, R. K., Gawande, D. Y., Pahwa, P., Gloriozova, T. A., Dmitriev, A. V., Ivanov, S. M., Rudik, A. V., Konova, V. I., Pogodin, P. V., Druzhilovsky, D. S., & Poroikov, V. V. (2014). Chemo- and bioinformatics resources for in silico drug discovery from medicinal plants beyond their traditional use: A critical review. *Natural Product Reports, 31*(11), 1585–1611. https://doi.org/10.1039/C4NP00068D.

Lapillo, M., Tuccinardi, T., Martinelli, A., Macchia, M., Giordano, A., & Poli, G. (2019). Extensive reliability evaluation of docking-based target-fishing strategies. *International Journal of Molecular Sciences, 20*(5), 1023. https://doi.org/10.3390/ijms20051023.

Lee, A., & Kim, D. (2019). CRDS: Consensus reverse docking system for target fishing. *Bioinformatics, 36*(3), 959–960. https://doi.org/10.1093/bioinformatics/btz656.

Liang, H., Liu, H., Kuang, Y., Chen, L., Ye, M., & Lai, L. (2020). Discovery of targeted covalent natural products against PLK1 by herb-based screening. *Journal of Chemical Information and Modeling.* https://doi.org/10.1021/acs.jcim.0c00074.

Li, J., Fu, A., & Zhang, L. (2019). An overview of scoring functions used for protein–ligand interactions in molecular docking. *Interdisciplinary Sciences: Computational Life Sciences, 11*(2), 320–328. https://doi.org/10.1007/s12539-019-00327-w.

Manas, E. S., Xu, Z. B., Unwalla, R. J., & Somers, W. S. (2004). Understanding the selectivity of genistein for human estrogen receptor-β using X-ray crystallography and computational methods. *Structure, 12*(12), 2197–2207. https://doi.org/10.1016/j.str.2004.09.015.

Marcus, A., & Koch, H. W. (2004). Natural product-derived compound libraries and protein structure similarity as guiding principles for the discovery of drug candidates. In *Chemogenomics in drug discovery* (pp. 377–403).

Mizuno, C. S., Ma, G., Khan, S., Patny, A., Avery, M. A., & Rimando, A. M. (2008). Design, synthesis, biological evaluation and docking studies of pterostilbene analogs inside PPARα. *Bioorganic & Medicinal Chemistry, 16*(7), 3800–3808. https://doi.org/10.1016/j.bmc.2008.01.051.

Morphy, R., & Rankovic, Z. (2005). Designed multiple ligands. An emerging drug discovery paradigm. *Journal of Medicinal Chemistry, 48*(21), 6523–6543. https://doi.org/10.1021/jm058225d.

Morris, G. M., & Lim-Wilby, M. (2008). Molecular docking. *Methods in Molecular Biology, 443*, 365–382. https://doi.org/10.1007/978-1-59745-177-2_19.

Mukherjee, S., Balius, T. E., & Rizzo, R. C. (2010). Docking validation resources: Protein family and ligand flexibility experiments. *Journal of Chemical Information and Modeling, 50*(11), 1986–2000. https://doi.org/10.1021/ci1001982.

Newman, D. J., & Cragg, G. M. (2020). Natural products as sources of new drugs over the nearly four decades from 01/1981 to 09/2019. *Journal of Natural Products, 83*(3), 770–803. https://doi.org/10.1021/acs.jnatprod.9b01285.

Pagadala, N. S., Syed, K., & Tuszynski, J. (2017). Software for molecular docking: A review. *Biophysical Reviews, 9*(2), 91–102. https://doi.org/10.1007/s12551-016-0247-1.

Park, K., & Cho, A. E. (2017). Using reverse docking to identify potential targets for ginsenosides. *Journal of ginseng research, 41*(4), 534–539. https://doi.org/10.1016/j.jgr.2016.10.005.

Pein, H., Ville, A., Pace, S., Temml, V., Garscha, U., Raasch, M., Alsabil, K., Viault, G., Dinh, C.-P., Guilet, D., Troisi, F., Neukirch, K., König, S., Bilancia, R., Waltenberger, B., Stuppner, H., Wallert, M., Lorkowski, S., Weinigel, C., … Koeberle, A. (2018). Endogenous metabolites of vitamin E limit inflammation by targeting 5-lipoxygenase. *Nature Communications, 9*(1), 3834. https://doi.org/10.1038/s41467-018-06158-5.

Pu, L., Govindaraj, R. G., Lemoine, J. M., Wu, H.-C., & Brylinski, M. (2019). DeepDrug3D: Classification of ligand-binding pockets in proteins with a convolutional neural network. *PLoS Computational Biology, 15*(2), e1006718. https://doi.org/10.1371/journal.pcbi.1006718.

Ragoza, M., Hochuli, J., Idrobo, E., Sunseri, J., & Koes, D. R. (2017). Protein–ligand scoring with convolutional neural networks. *Journal of Chemical Information and Modeling, 57*(4), 942–957. https://doi.org/10.1021/acs.jcim.6b00740.

Redford, K. E., & Abbott, G. W. (2020). The ubiquitous flavonoid quercetin is an atypical KCNQ potassium channel activator. *Communications Biology, 3*(1), 356. https://doi.org/10.1038/s42003-020-1089-8.

Rueda, M., Bottegoni, G., & Abagyan, R. (2010). Recipes for the selection of experimental protein conformations for virtual screening. *Journal of Chemical Information and Modeling, 50*(1), 186–193. https://doi.org/10.1021/ci9003943.

Rueda, M., Totrov, M., & Abagyan, R. (2012). ALiBERO: Evolving a team of complementary pocket conformations rather than a single leader. *Journal of Chemical Information and Modeling, 52*(10), 2705–2714. https://doi.org/10.1021/ci3001088.

Salentin, S., Schreiber, S., Haupt, V. J., Adasme, M. F., & Schroeder, M. (2015). PLIP: Fully automated protein-ligand interaction profiler. *Nucleic Acids Research, 43*(W1), W443–W447. https://doi.org/10.1093/nar/gkv315.

Scarpino, A., Ferenczy, G. G., & Keserű, G. M. (2018). Comparative evaluation of covalent docking tools. *Journal of Chemical Information and Modeling, 58*(7), 1441–1458. https://doi.org/10.1021/acs.jcim.8b00228.

Schmidt, S., Zietz, M., Schreiner, M., Rohn, S., Kroh, L. W., & Krumbein, A. (2010). Genotypic and climatic influences on the concentration and composition of flavonoids in kale (*Brassica oleracea* var. sabellica). *Food Chemistry, 119*(4), 1293–1299. https://doi.org/10.1016/j.foodchem.2009.09.004.

Seidel, T., Wieder, O., Garon, A., & Langer, T. (2020). Applications of the pharmacophore concept in natural product inspired drug design. *Molecular Informatics.* https://doi.org/10.1002/minf.202000059.

Shen, J., Xu, X., Cheng, F., Liu, H., Luo, X., Shen, J., Chen, K., Zhao, W., Shen, X., & Jiang, H. (2003). Virtual screening on natural products for discovering active compounds and target information. *Current Medicinal Chemistry, 10*(21), 2327–2342. https://doi.org/10.2174/0929867033456729.

Soares, J. (2011). The Nagoya protocol and natural product-based research. *ACS Chemical Biology, 6*(4), 289. https://doi.org/10.1021/cb200089w.

Sterling, T., & Irwin, J. J. (2015). ZINC 15 – ligand discovery for everyone. *Journal of Chemical Information and Modeling, 55*(11), 2324–2337. https://doi.org/10.1021/acs.jcim.5b00559.

Thawai, C., Bunbamrung, N., Pittayakhajonwut, P., Chongruchiroj, S., Pratuangdejkul, J., He, Y.-W., Tadtong, S., Sareedenchai, V., Prombutara, P., & Qian, Y. (2020). A novel diterpene agent isolated from *Microbispora hainanensis* strain CSR-4 and its in vitro and in silico inhibition effects on acetylcholine esterase enzyme. *Scientific Reports, 10*(1), 11058. https://doi.org/10.1038/s41598-020-68009-y.

Topliss, J. G. (1972). Utilization of operational schemes for analog synthesis in drug design. *Journal of Medicinal Chemistry, 15*(10), 1006–1011. https://doi.org/10.1021/jm00280a002.

Ulrich-Merzenich, G., Panek, D., Zeitler, H., Vetter, H., & Wagner, H. (2010). Drug development from natural products: Exploiting synergistic effects. *Indian Journal of Experimental Biology, 48*(3), 208–219.

Vanii Jayaseelan, K., Moreno, P., Truszkowski, A., Ertl, P., & Steinbeck, C. (2012). Natural product-likeness score revisited: An open-source, open-data implementation. *BMC Bioinformatics, 13*(1), 106. https://doi.org/10.1186/1471-2105-13-106.

Vue, B., Zhang, S., & Chen, Q.-H. (2015). Synergistic effects of dietary natural products as anti-prostate cancer agents. *Natural Product Communications, 10*(12). https://doi.org/10.1177/1934578X1501001241.

Weller, M. G. (2012). A unifying review of bioassay-guided fractionation, effect-directed analysis and related techniques. *Sensors (Basel), 12*(7), 9181–9209. https://doi.org/10.3390/s120709181.

Xue, R., Fang, Z., Zhang, M., Yi, Z., Wen, C., & Shi, T. (2013). TCMID: Traditional Chinese medicine integrative database for herb molecular mechanism analysis. *Nucleic Acids Research*, *41*(Database issue), D1089–D1095. https://doi.org/10.1093/nar/gks1100.

Xu, W., Lucke, A. J., & Fairlie, D. P. (2015). Comparing sixteen scoring functions for predicting biological activities of ligands for protein targets. *Journal of Molecular Graphics and Modelling*, *57*, 76–88. https://doi.org/10.1016/j.jmgm.2015.01.009.

Zeng, X., Zhang, P., He, W., Qin, C., Chen, S., Tao, L., Wang, Y., Tan, Y., Gao, D., Wang, B., Chen, Z., Chen, W., Jiang, Y. Y., & Chen, Y. Z. (2018). NPASS: Natural product activity and species source database for natural product research, discovery and tool development. *Nucleic Acids Research*, *46*(D1), D1217–D1222. https://doi.org/10.1093/nar/gkx1026.

Zeng, X., Zhang, P., Wang, Y., Qin, C., Chen, S., He, W., Tao, L., Tan, Y., Gao, D., Wang, B., Chen, Z., Chen, W., Jiang, Y. Y., & Chen, Y. Z. (2019). CMAUP: A database of collective molecular activities of useful plants. *Nucleic Acids Research*, *47*(D1), D1118–D1127. https://doi.org/10.1093/nar/gky965.

Zhao, S., Kumar, R., Sakai, A., Vetting, M. W., Wood, B. M., Brown, S., Bonanno, J. B., Hillerich, B. S., Seidel, R. D., Babbitt, P. C., Almo, S. C., Sweedler, J. V., Gerlt, J. A., Cronan, J. E., & Jacobson, M. P. (2013). Discovery of new enzymes and metabolic pathways by using structure and genome context. *Nature*, *502*(7473), 698–702. https://doi.org/10.1038/nature12576.

CHAPTER 19

Advances in Docking-Based Drug Design for Microbial and Cancer Drug Targets

DIVYA GUPTA • ASAD U. KHAN

1 INTRODUCTION

Drug discovery and development is a very tedious process as it takes 10—15 years of research for a compound to transform into a drug molecule. This process also incorporates the expenditure of significant capital running from millions to billions of dollars for every effective drug molecule (Paul et al., 2010). The conventional drug discovery and development is a complex multistep process involving the identification of lead candidate, preclinical development studies, clinical trials, and pharmacovigilance. Modern drug discovery approach relies more upon computer-aided drug design (CADD) for exploration of novel drug candidates, which not only lowers the cost but also decreases the time required for discovering the drugs (Eweas et al., 2014). CADD utilizes in silico methods to mimic ligand receptor binding to confirm the binding of lead candidate to the target, simultaneously depicting its binding affinity (Parvu, 2003). It can further help to reduce the number of choices of drug candidates to just 10 or fewer by employing various tools and techniques of bioinformatics, as thousands of drug molecules are usually retrieved from a chemical compound database for screening.

Molecular docking is an automated computer algorithm employed to determine binding orientation and binding ability of ligand to its target protein, which in turn is used to calculate the affinity and activity of the provided ligand molecule. It is used to investigate whether the binding orientation between two molecular structures (such as ligand and target protein) is perfect or not (Gane & Dean, 2000; Hakes et al., 2007). Molecular docking is used to visualize 3D structure of the final complex that was composed based on the binding properties of lead candidate and targeted protein. Most docking software creates various conceivable conformation of the complex that is ranked using the scoring techniques

of the software. The scoring techniques of the software correctly rank the docking conformations based on the calculated energy (McConkey et al., 2002).

Virtual screening is one of the important tools of bioinformatics used for the screening of thousands of lead candidates available in the database against a targeted macromolecule. It is used to select few molecules to turn them into drug candidate. It does not consume much resources, time, and money (Doss et al., 2014). During virtual screening process, it is also necessary to consider Lipinski's "rule of five," also known as Pfizer's rule of five, to differentiate between drug-like molecules and non—drug-like molecules (Lipinski, 2000). Lipinski's "rule of five" consists of filters which help to eliminate selection of wrong drug candidates at early stage and prevent preclinical and clinical failure. The filters are physiochemical properties, which are as follows:

- Molecular weight ≤ 500 Da
- cLogP <5 (lipophilicity)
- Hydrogen bond donors <5
- Hydrogen bond acceptors <10

The list of the drug molecules developed through virtual screening is very long. Some cases involve the screening of drug molecule to make it work as an inhibitor to block the function of a protein. However, some cases involve binding of drug molecule to the target protein to induce its function, ultimately providing aid to combat disease. Either of the case relies upon the binding ability of the ligand to the protein that can be determined using docking studies (Gulati et al., 2013).

Virtual screening can be of two types: structure-based virtual screening (SBVS) and ligand-based virtual screening (LBVS). LBVS takes advantage of the information from known active and inactive ligands. It is the best option to adopt for screening of lead candidates, when the complete information of target is not available.

Molecular Docking for Computer-Aided Drug Design. https://doi.org/10.1016/B978-0-12-822312-3.00020-5

While SBVS is usually used to dock multiple molecules against a known target, it requires the complete knowledge of the structure of the target protein (Batool et al., 2019).

Other computational tools are also required before or in conjunction with docking applications. Foremost step requires 3D structure of the target protein, which can be easily acquired from RCSB website. In the case of nonavailability of the Protein Data Bank (PDB) structure of the protein, one has to predict protein structure, using some software or online servers such as Robetta, SWISS-MODEL, PEPstr, and QUARK. It is then followed by finding active site of the protein that assists in defining the search area in the protein for docking algorithms (Roche et al., 2015; Xie & Hwang, 2015; Zhu et al., 2015). Active site of most of the proteins is known or available in the literature; if not, software such as metapocket, SCFBio, COACH, and others can be employed for the prediction of active site. Lead candidate can be acquired from databases such as ZINC, NCI, Maybridge, and so on. Preceding to the docking, some computational methods such as pharmacophore modeling and quantitative structure–activity relationship (QSAR) models can be employed to cut down computational burden and time (Abdolmaleki et al., 2017; Desaphy et al., 2012; Kumari et al., 2016).

This approach of molecular docking has become more advanced and can be utilized in various phases of the drug discovery paradigm. Also, the different approaches of in silico drug design can be applied in different ways. The execution of the project relies upon the data that are accessible or obtainable. However, each approach is related to one another. If one has complete information regarding protein and ligand structures, it simply requires docking and virtual screening to be carried out. But the case of unknown protein structure and known ligand must be started with similarity searching, pharmacophore, or QSAR modeling. In the interim, it would obviously be useful to attempt to inquire about the structure of a protein, which itself assists to quicken the drug design venture. It is than followed by virtual screening step.

There are numerous successful stories available that have effectively used the aforementioned approaches. However, the effective stories have not been referenced regularly and are not broadly known, but this chapter will throw light on some important applications.

2 STRUCTURE-BASED DRUG DESIGN

Structure-based drug design (SBDD) refers to the efficient utilization of structural information, such as

protein targets, and chemical information, such as ligand, which are generally acquired either through experimentally or computational tools and techniques (Mandal & Mandal, 2009). SBVS has turned out into an imperative technique for the drug design and development nowadays (Jorgensen, 2004; Lavecchia & Di Giovanni, 2013; Park et al., 2012). It is not only specific and relevant but also saves money and time. Application of genomics and proteomics in bioinformatics has opened doors for future development of novel drugs by contributing to providing thousands of different therapeutically vital protein targets. SBDD is an iterative procedure which continues over multiple cycles, ultimately assisting scientists to identify diverse lead candidates, which can become drugs and then introduced into the market (Wang et al., 2018).

The initial step is the determination of protein target structure followed by virtual screening of the lead candidates available in the database to recognize potential ligands. The potential ligands are scrutinized based on docking score or binding energy, which helps to determine potency, binding affinity, and efficacy of the drug molecule with the target protein (Fang, 2012). The selected ligand–receptor complexes are then resolved to throw light on the intermolecular interactions between them. Basic depictions of ligand–receptor complexes are valuable to deduce some important features, such as ligand conformation after binding, explanation of intermolecular forces (hydrogen bonds, hydrophobic interactions, etc.), portrayal of ligand-prompted conformational changes in protein structure, and many more (Kahsai et al., 2011; Wilson & Lill, 2011). The data obtained through in silico studies are then compared with the data of experimental studies (Shoichet & Kobilka, 2012). Selected ligand molecules may also undergo structural modifications, if required. Following this, the SBDD procedure begins once again with structurally modified ligand molecule to confirm enhanced binding ability of ligand in the active site of the target protein (Ferreira et al., 2015). Selected ligand molecules after incorporation of modification are finally evaluated for pharmacological properties to assess, ADMET properties (absorption, distribution, metabolism, excretion, and toxicity), as well as biological activity (Lipinski et al., 1997) (Fig. 19.1).

SBDD can by employed for drug design in a variety of ways. Some examples are briefly described:

- Ligands as inhibitors of target protein
- Protein–protein interaction (PPI) inhibitors
- Application for covalent drugs

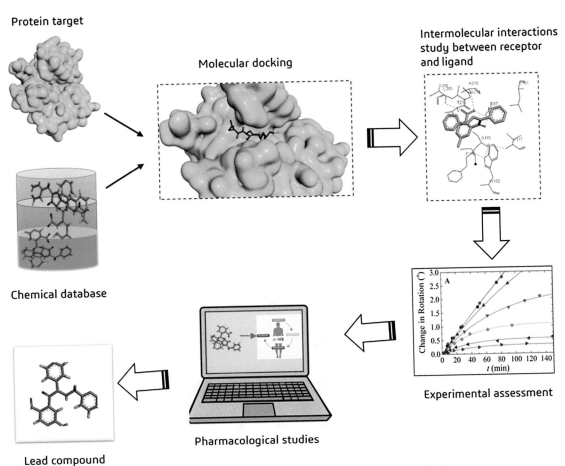

FIG. 19.1 Schematic representation of structure-based drug designing.

2.1 Ligands as Inhibitors of Target Protein

SBDD finds its application in the identification of lead molecules to cure many severe diseases by researchers as it provides extraordinary understanding into protein–ligand interactions with their binding mechanism and information on the ideal direction of the ligand bound to its receptor to determine the most stable complex. SBDD is used to identify lead molecule that can be used to block the enzymatic activity to the targeted enzyme, ultimately assisting in combating diseases. Some examples of the applications of SBDD to identify lead molecules as inhibitors are provided in Table 19.1.

2.2 Protein–Protein Interaction Inhibitors

Most of the biochemical and cellular processes involved in the interactions of two diverse class of proteins (Gonzalez & Kann, 2012). The impaired PPIs can lead to the emergence of many severe disease. Hence, the inhibition of PPIs is an emerging concept to produce novel therapeutics (Leung et al., 2011; Petta et al., 2016; Ryan & Matthews, 2005; Wilson, 2009). The huge and undefined interfaces of the proteins represent major hurdle in research to focus on PPIs. In most of the cases, the binding sites of PPIs are consisting of flat surface and do not possess a solitary huge and very much characterized pocket; in contrast, they comprise of numerous tiny pockets. Hence, the major concern to handle in SBDD for PPIs is the determination of their binding cavities and to locate their capacity for interaction with lead compounds (Fuller et al., 2009). The potential binding pockets having druggable properties can be identified using some web-based tool such as Q-SiteFinder and AN-CHOR (Laurie & Jackson, 2005; Meireles et al., 2010).

Recently, Liu et al. used SBDD in conjunction with interaction-based pharmacophore (IBP) model and stepwise molecular docking to identify Mcl-1 inhibitors

TABLE 19.1

Some Examples of Applications of Structure-Based Drug Design to Identify Lead Candidates as Inhibitors of Target Protein.

S. No	Target	Lead molecule/ database	Software	Application	Reference
CANCER DRUG TARGETS					
1	20S proteasome	PI-083	GLIDE version3.0	For cancer therapeutics	Kazi et al., 2009
2	20S proteasome	Peptide aldehyde derivatives	GOLD version 4.0	For cancer therapeutics	Ma et al., 2011
3	20S proteasome	Porphyrins	AutoDock Vina	For cancer therapeutics	Santoro et al., 2012
4	Carbonic anhydrase IX	Five potential CA IX inhibitors	Autodock 4.2	To target cancer cell lines and tumor tissues	Amresh et al., 2013
5	Carbonic anhydrase IX	ZINC database	GLIDE/HTVS and GLIDE/XP protocols	To target cancer cell lines and tumor tissues	Salmas et al., 2016
6	Carbonic anhydrase II	Leadquest and Maybridge database	FlexX and Drug score	To target cancer cell lines and tumor tissues	Grüneberg et al., 2002
7	EGFR+HER2+HSP90	Mcule database	Autodock Vina	To target breast cancer	Yousuf et al., 2017
8	EGFR	50000 lead molecules	GLIDE module →SP docking → XP docking	To identify anticancer molecule	Mahajan et al., 2017
9	COX-2	Five carborane containing derivatives of rofecoxib	Autodock	Drug design for human melanoma and colon cancer cells	Buzharevski et al., 2020
MICROBIAL DRUG TARGETS					
10	HIV reverse transcriptase	3000 molecules from ZINC database	AutoDock Vina	To identify anti-HIV agent	Hosseini et al., 2016
11	BlaC enzyme of *Mycobacterium tuberculosis*	Selected compounds from ChEMBL/MolPort/ ZINC databases	AutoDock Vina	To cure tuberculosis	Gonzalez et al., 2018
12	*Klebsiella pneumoniae* KPC-2 β-lactamases	MySQL-database	DOCK3.6	To combat antibiotic resistance	Klein et al., 2018a,b
13	CTX-M-15 type β-lactamase	Maybridge HitFinder™ database	GOLD 5.0 version	To identify novel anti −β-lactamase agents to combat antibiotic resistance	Ali et al., 2018
14	NS5B RNA-dependent RNA polymerase as anti −hepatitis C virus (HCV) agent	In-house library	GLIDE module →SP docking → XP docking	To identify anti-HCV agent	Barreca et al., 2013

TABLE 19.1
Some Examples of Applications of Structure-Based Drug Design to Identify Lead Candidates as Inhibitors of Target Protein.—cont'd

S. No	Target	Lead molecule/ database	Software	Application	Reference
15	LasR, transcription factor that controls quorum sensing in *Pseudomonas aeruginosa*	2603 compounds from ZINC database	Autodock and AutodockVina	To control biofilm development	Kalia et al., 2017
16	Shikimate kinase	ligand library commercially obtainable at Acros Organics+ZINC database	MOLDOCK program	Identification of new lead candidates to cure tuberculosis	Vianna and de Azevedo, 2012
17	NDM-1 and IMP-1 β-lactamases	Maybridge HitFinder™ database	GOLD 5.0 version	To identify novel anti −β-lactamase agents to combat antibiotic resistance	Khan et al., 2017

that can further be used as lead compound to cure cancer. Virtual screening was carried out on a commercial database based on the joint use of selected IBP model and docking studies. Out of 210,000 compounds, 30 hit compounds were selected to evaluate their binding abilities with Mcl-1 and other Bcl-2 proteins. Specifically, one of the compounds showed anticancer activity in Jurkat cells. In conclusion, this research provided a useful methodology for looking through small lead compounds to target PPIs related with Mcl-1 (Du et al., 2020).

Another example of the use of in silico studies to find inhibitors for protein–protein interfaces was cited by Betzi et al. in 2007 to specifically target HIV-1 Nef protein. They have successfully identified two lead drug compounds having binding ability with the HIV-1 Nef SH3 binding surface and can functionally compete for SH3-Hck interaction, both in vitro and in cell-based assays. These lead compounds were identified utilizing a primary step of in silico screening followed by their validation in a cell-based screening method for the given database of compounds. The outcomes of the research give the premise to a groundbreaking disclosure process that ought to be appropriate to larger libraries of compounds by either analogy searching or docking. It will not only assist in accelerating identification of PPI inhibitors but also open roads for new class of antiviral molecules (Betzi et al., 2007).

2.3 Application for Covalent Drugs

Covalent ligands are such drug molecules that bind to their target in an irreversible manner and hence exhibit high affinity for their target proteins. Once covalent drug gets bound with its target protein; it completely hinders protein function. Covalent inhibitors prompt an everlasting pharmacological reaction and subsequently require low quantity of the drug to get effective (Singh et al., 2011). About 33% of the commercial enzyme activity regulators are covalent inhibitors (Kumalo et al., 2015).

In 2009, Zhang et al. employed covalent docking in combination with 5 ns molecular dynamic simulation method to deduce the complex structure MG132-proteasome complex. Covalent docking was done using GOLD software, and MD simulation was done using Amber platform. MG132 (Z-Leu-Leu-Leu-al) is a peptide aldehyde, which is essentially progressively powerful against the 20S proteasome. In most of the cellular processes, 20S proteasome plays very crucial role and hence attracted broad intrigue in tumor research. This research employs covalent docking and MD simulation method to understand the binding mechanism and structure–activity relationship of the peptide aldehyde inhibitor with proteasome and may give valuable data for further drug design and development (Zhang et al., 2009).

A recent study published by Gupta et al. also elucidates the role of covalent docking to identify covalent interactions between proposed inhibitors and the OXA variants. OXA β-lactamases are the enzymes that have ability to degrade antibiotics such as penicillin, cephalosporins, and carbapenems and hence the bacteria become resistant to these antibiotics. This study includes the identification of non−β-lactam inhibitor molecules against

OXA variants through virtual screening using GOLD 5.0 version and their validation using 50 ns MD simulation that was performed using GROMACS v5.038,39 assigning GROMOS9640,41 43a1 force field. They have further examined covalent interactions between proposed inhibitors and the OXA after employing Discovery Studio software. Root-mean-square deviation value below 2 Å is generally considered adequate for effective covalent docking. Covalent docking was also performed using GOLD software (Gupta et al., 2020).

3 LIGAND-BASED VIRTUAL SCREENING

In the event of unavailability of 3D structure of macromolecule target, LBVS becomes primary alternative for drug development and optimization. It utilizes the information present in known active ligands to find new ligands from databases (Dai & Guo, 2019; Banegas-Luna et al., 2018). Hence, all known active ligands of a target are gathered to generate a pharmacophore model, molecular descriptors, or QSAR model. It might employ pharmacophore modeling for virtual screening of the database to discover new drug molecule against the macromolecule target (Pedretti et al., 2019; Tresadern et al., 2009). LBVS method is based on the way that ligands having identical properties to active ligand would be more appropriate in comparison to random ligands. LBVS method is basically dependent on the information available of the known active ligands (Hamza et al., 2012). Ligand-based screening method incorporates three strategies: similarity searching, pharmacophore modeling, and machine learning methods (Fig. 19.2). Some commercial and freely available software for pharmacophore modeling are HipHop, DISCO, HypoGen, LigandScout, Pharmer, MolSign, Catalyst, UNITY, Zinc-Pharmer, and PHASE (Malathi & Ramaiah, 2018; Martin, 2000). COMFA and COMSIA models by sybyl can be used for 3D QSAR modeling and LQTAgrid can be used for 4D QSAR (Malathi & Ramaiah, 2018).

Case study 1: One of the earlier uses of LBVS was explained by Wang et al. in 2008. They have used pharmacophore modeling method for the identification of a novel class of azetidinone CB1 antagonists. In the treatment of obesity and related disorders, CB1 receptor antagonists are clinically viable approach. They have produced a pharmacophore model after comparing the conformational space of individual lead candidate in the training set of known active ligands of CB1 antagonists and determine the general 3D configurations of functional features. It was the first ever 3D pharmacophore model of CB1 antagonist. This model assists in the recognition of the important chemical features involved in ligand–receptor interactions, which further helps to demonstrate the crucial role of hydrogen bonding and aromatic interactions in the binding of ligand to the target. The pharmacophore model was used for the screening of a database of about a half million Schering-Plough compounds. They have also used some filters to reduce the total number of lead compounds to a sensible scale. It was concluded with the finding of novel structural class of CB1 antagonists, which can be further optimized to cure obesity and related diseases (Wang et al., 2008).

Case study 2: The farnesoid X receptor (FXR) played crucial role in regulating glucose and lipid metabolism and hence can be used as protein target in the treatment of dyslipidemia, atherosclerosis, and type 2 diabetes. D. Schuster et al. in 2011 developed structure-based pharmacophore models to screen out novel FXR agonists. The models were evaluated against ChEMBL database and best models were employed to identify novel active compounds from the NCI database (Schuster et al., 2011).

Case study 3: Another successful application of pharmacophore modeling was carried out for screening of 11β-hydroxysteroid dehydrogenase (11β-HSD) inhibitors, which was published in 2006 (Schuster et al., 2006), and further work on the similar receptor was accounted in 2010 (Rollinger et al., 2010).

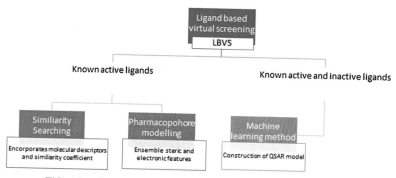

FIG. 19.2 Different approaches of ligand-based virtual screening.

Glucocorticoid-associated diseases such as obesity, diabetes, wound healing, and muscle atrophy can be treated by blocking the enzymatic activity of 11β-HSD1. Hence, ligand-based multifeature pharmacophore models for 11β-HSD1 were constructed using known active ligands. The model was further employed for virtual screening and in vitro evaluation of lead candidates resulted in the identification of promising 11β-HSD1 inhibitors (Prieto-Martínez et al., 2019).

Case study 4: The dysregulation of PI3K/Akt/mTOR pathway can lead to emergence of some serious diseases such as cancer, metabolic disorders, neurological diseases, and inflammation. Initially, rapamycin and its analogs were used to block mTOR receptor; however, their clinical uses demonstrate serious side effects induced by ATP-competitive inhibitors. In 2017, Kist et al. utilized LBVS in an effort to identify non−ATP-competitive inhibitors of the mTOR complex. They have constructed a pharmacophore model followed by molecular docking, pharmacokinetic properties analysis, and molecular dynamics simulation to identify eight novel potential mTOR inhibitors having preferable properties over the typical inhibitor, rapamycin (Kist et al., 2018).

Case study 5: Tropomyosin receptor kinase A (TrkA) inhibitors are at present studied as an emerging approach to cure many cancers. About 161 TrkA inhibitors were analyzed to examine their physicochemical properties and pharmacophoric binding modes, which in turn used to prepare optimal QSAR model. The optimal QSAR model was analyzed and further employed for virtual screening of NCI database to find TrkA inhibitors. Out of 41 lead candidates, exclusively 21 lead candidates possess the ability to block TrkA enzymes (Shahin et al., 2018).

Case study 6: Another application of ligand-based drug design was demonstrated by McKay et al. in 2011 for the identification of plasmepsin (a family of malarial parasitic aspartyl protease) inhibitors as antimalarial agents. They have constructed pharmacophore models, and the best pharmacophore models were explored for virtual screening of databases to identify novel drug molecule having the ability to block plasmepsin (McKay et al., 2011).

Case study 7: Tong et al. in 2015 have prepared a QSAR model using topomer CoMFA for 38 HIV-1 protease inhibitors to determine the biochemical interactions between the protease inhibitors and protease target. The prepared model was further used for virtual screening from ZINC database to identify novel HIV-1 protease inhibitors that can be employed as novel HIV/AIDS drugs. They have identified 60 novel molecules that possess better activity results compared to the originals. Eventually, Surflex-Dock was applied for the analysis of binding mode of training molecules and identified inhibitors independently (Tong et al., 2016).

4 IN SILICO FRAGMENT-BASED DRUG DESIGN

Fragment-based drug design (FBDD) is an emerging technique for the drug designing that initiates with the screening of fragment hits followed by their optimization to convert them into lead molecule with enhanced potency and efficacy. In silico screening has become an essential tool in FBDD, and various computational methods and techniques have been created to help FBDD (Kumar et al., 2012). There are two frequently utilized approaches for the optimization of fragment hits into lead compounds: (1) fragment growing and (2) fragment linking.

Fragment-linking approach initiates with the identification of the binding cavity of the target protein to depict the possible interacting points for different functional groups available in the scrutinized fragments, bound independently in proximity. Multiple fragments can be recognized to attach with the target binding cavity, with each fragment occupying a specific subarea. This was followed by the covalent linking of the functional group to conceive a lead compound. On the other hand, fragment growing is the stepwise expansion or multiple substitution of functional groups to the fragment core to amplify the possible binding interactions. Fragment growing is a widely recognized and mainstream approach for the fragment optimization. It is a repetitive procedure, and at each progression, multiple substitutions are performed to the fragment core with the objective of augmenting potency and pharmacological properties (Bian & Xie, 2018). The schematic representation of both approaches is illustrated in Fig. 19.3.

Like the traditional in vitro FBDD, there are three fundamental strides during in silico FBDD:

- Fragment library generation,
- Virtual screening of fragment libraries, and
- Optimization of fragment hits.

4.1 Fragment Library Generation

One imperative thing to take into account during fragment library generation is to select the low molecular weight compounds having defined and required chemical properties. These compounds are preferred to be chosen in light of the acknowledged physicochemical properties avowed as the Astex Rule of Three (Ro3) (Congreve et al., 2003):

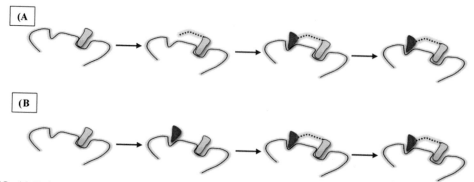

FIG. 19.3 Approaches employed for fragment-based drug designing. **(A)** Fragment growing, **(B)** fragment linking. *Color scheme: Fragment 1 (*red triangle shaped*), fragment 2 (*yellow rod shaped*) and fragment linker (*red dotted line*).

- Molecular weight ≤300,
- Hydrogen bond donor and acceptor ≤3,
- cLogP ≤3,
- Number of rotatable bonds ≤3, and
- Polar surface area ≤60 Å^2

Generally, the compound suppliers these days promote a ready-to-screen fragment library; a thorough analysis gives an outline of the available databases. The databases can be chosen are Asinex, ChemBridge, Enamine, Key Organics, Life Chemicals, Maybridge, Specs, and ZINC database (Mortier et al., 2012).

4.2 Virtual Screening of Fragment Libraries

Molecular docking has been utilized in FBDD for the assortment of compounds available in fragment libraries. In any case, the common conviction is that outcomes from docking fragment libraries are not truly dependable (Chen & Shoichet, 2009). The explanation might be that fragments may be more promiscuous in their binding modes. Also, scoring methods of docking might get erroneous in any event, even for the high molecular weight compounds for which they have been originally defined (Hubbard et al., 2007). Albeit most docking programs were not intended for fragment screening, it is the most frequently utilized strategy, as it is rapid and cost-effective method compared to experimental one. Examples of some software that can be used as a docking approach for the virtual screening of fragment libraries are AutoDock, UCSF Dock, Glide, GOLD, Surflex-Dock, etc.

4.3 Optimization of Fragment Hits

It is one of the critical steps of FBDD that initiates from crude low molecular weight compounds that completely satisfy steric and electrostatic prerequisites forced by the complex structure of a binding cavity (Sheng & Zhang, 2013). It at last prompts the lead compound with high efficiency against target proteins. Similarity searching or pharmacophore screening method is one of the commonly used methods to optimize the scrutinized fragments hits. These methods are usually used to search for similar compounds in the databases (Murray et al., 2010). Other strategy is de novo synthesis of new ligands either by developing from leading fragment or by connecting at least two diverse fragments. Some examples of de novo design programs are LUDI, GROWMOL, PRO_LIGAND, LigBuilder, SMoG, Allegrow, GANDI, CONFIRM, Autogrow, FOG, etc. (Hartenfeller & Schneider, 2010).

In silico FBDD has some advantages:

- Efficient,
- Work sparing, and
- Material sparing.

In spite of the numerous points of interest in computational part—based screening, there are a few downsides too. One of the most genuine is the generally little certainty of predictions and other is the quick aggregation of mistakes. This could be because of the defects in the scoring methods employed to assess protein—ligand interactions. Additionally, there could be troubles in accomplishing complete sampling of the conformational space (Kumar et al., 2012).

4.4 Application of In Silico Fragment-Based Drug Design

In silico FBDD in combination with experimental techniques has been used in many studies to facilitate the compound development processes. An overview of the application of in silico FBDD is provided in Table 19.2. Albeit in silico FBDD does not produce reliable results and is less accurate in contrast to in vitro studies, yet they can create impressively great outcomes whenever utilized in conjunction with experimental techniques.

TABLE 19.2
List of Some Recent Application of In Silico Fragment-Based Drug Design in Infectious Disease and Cancer.

S. No	Target	Fragment libraries	Software	Application	Reference
CANCER DRUG TARGETS					
1	Signal transducer and activator of transcription 3 (STAT3)	Fragment libraries were categorized from the known STAT3 dimerization inhibitors based on their binding modes	AutoDock	STAT3 has been validated as an attractive therapeutic target for different carcinomas, including breast cancer and sarcomas	Yu et al., 2013
2	Lysine-specific demethylase enzyme LSD1	Maybridge Ro3 2000 Diversity Fragment Library	Ludi-based protocols and MCSS protocol	To discover novel LSD1 inhibitors: LSD1 get overexpressed in various cancers	Alnabulsi et al., 2019
3	Human bromodomains ATAD2, BAZ2B, BRD4(1), and CREBBP	Chemical Library of the Lausanne Bioscreening Facility and anchor-based library tailoring approach for virtual screening (ALTA-VS)	In-house docking software SEED	Anticancer activity	Marchand et al., 2017
4	Hsp90 molecular chaperone	In-house corporate library	FlexX	For the treatment of cancer	Buchstaller et al., 2012
MICROBIAL DRUG TARGETS					
5	6-phosphogluconate dehydrogenase, of parasitic protozoan *Trypanosoma brucei*	Available chemicals and screening compounds directories (ACD—SCD)	DOCK 3.5.54	To cure human African trypanosomiasis	Ruda et al., 2010
6	Subfamily B1 metallo-β-lactamase (MBL) including NDM-1, IMP-1, and VIM-2 enzymes	De novo molecular design program SPROUT1	AutoDock 4.2	To produce specific MBL inhibitors to combat antibiotic resistance	Cain et al., 2018
7	AmpC β-lactamase	ZINC database	DOCK 3.5.54	To combat antibiotic resistance	Teotico et al., 2009

5 IN SILICO DRUG REPURPOSING

Drug repurposing (or drug repositioning) depends on the possibility of the drug which have been approved by FDA as a medicine against a specific disease or drugs which have succeeded in the early steps of assessment but terminated in subsequent steps of development, might be utilized to cure the disease aside from the first target. Drug repurposing is picking up prominence because of the critical decrease in the quantity of safe and viable drugs launched by pharmaceutical sector in the last few years (Phillips et al., 2018, p. 141). The repurposing of the drugs can be done using two approaches: experimental technology or computational technology. A multitude of drug molecules requires to get assembled in the experimental methodology. Also, many screening tests need to build up, prompting significant time and work costs. On the contrary, in silico drug repurposing utilizes virtual screening and docking methodology to recognize drug molecules for repurposing and hence viewed as a quicker methodology (Serçinoğlu & Sarica, 2019). Various databases are available to prepare screening library for drug repurposing, such as BindingDB, DrugBank, ChEMBL, SuperDRUG2, and many more. Some examples of the case studies for utilization of computational technology for drug repurposing are provided in Table 19.3.

In silico strategies are demonstrated to have a crucial role in the fight against a numbers of world's most dangerous diseases. Recently, an outbreak of COVID-19 has emerged from the Wuhan city of China and researchers are continuously trying to find remedy against COVID-19 using in silico drug repurposing methodology.

6 REVERSE OR INVERSE DRUG DESIGNING

A molecular docking method, which is used to identify protein target against a query ligand, is known as reverse or inverse drug designing. In inverse docking strategy, virtual screening of protein database has been carried out against a query ligand (Xu et al., 2018). The steps involved in this approach are like those of normal docking techniques, assembling libraries, scanning for binding sites, and scoring and ranking of the complex structures (Lee et al., 2016). Converse or reverse docking is an important approach for drug repositioning and drug redemption (Kharkar et al., 2014).

There are some freely accessible databases containing the data about targetable proteins, such as the PDB, SuperTarget, BindingDB, Potential Drug Target database, Therapeutic Target Database, and many more. The knowledge of phytomedicine or phytotherapy has been collected for many years, and because of that, products of natural origin have turned into a plentiful asset for new drug identification (Ji et al., 2009). Identification of the target protein related to infectious disease, against naturally obtained ligands, is now become a new approach of modern drug design and discovery. Several examples of using inverse docking for the identification of protein targets against natural query ligand have been demonstrated by many researchers (Table 19.4).

7 ANTIBIOTIC RESISTANCE AND DRUG DESIGNING THROUGH DOCKING

Antibiotic resistance poses serious health risk to human beings worldwide. Antibiotic resistance could emerge in microbes through diverse type of mechanisms. It might emerge through the selection of preexisting types, species, and variants; modifying antibiotic target; decreasing cell permeability for antibiotics; enhancing antibiotic efflux; and increasing the expression of antibiotic-inactivating enzymes such as β-lactamases enzymes (Mulvey & Simor, 2009). On account of the extreme antibiotic resistance, researchers have been searching for a long time to conquer the hurdle of drug resistance and investigate new options that can be used in conjunction with antibiotics (Danishuddin & Khan, 2015). The most significant and frequent cause of resistance, especially in gram-negative bacteria, is the expression of β-lactamases enzymes, causing hydrolysis of β-lactam ring of antibiotic (Watkins & Bonomo, 2017). Virtual screenings using docking methodology permitted quick determination of target ligands for β-lactamases from large databases or screening libraries (Table 19.5).

8 CONCLUSION

Molecular docking approach can be employed on various steps of the drug design and identification for infectious diseases and cancer. It can be employed to identify a lead molecule against a protein target or in contrary it can be employed to identify protein target against a query ligand. It can further be employed for de novo design of the lead molecule against a disease-causing protein target. Molecular docking also helps in rejuvenation of drug molecule for a different protein target, ultimately helping reuse of the drugs. The incorporation of docking methodology in the beginning of the research helps to reduce the time and cost. However, the outcomes of the docking methodology require to be confirmed with experimental approach before turning them into practical use. Table 19.6 lists approved drugs developed using computational methods in combination with rational drug design approach. There are still a lot of issues that need attention in using this approach, and hence many scientists in research labs or pharmaceutical organizations become suspicious in adopting this methodology for drug design and development. However, the outcomes of the research depend on using docking methods in the right place at the right time.

TABLE 19.3

Few Case Studies That Have Used Docking for Drug Repurposing.

S. No	Drug	Known pharmaco-logical target	Predicted drug target	Predicted treatment	Software used	Reference
CANCER DRUG TARGET						
1	Bromocriptine	It is an ergot alkaloid and dopamine D2 receptor agonist, which is used to treat Parkinson's disease	NF-kB pathway proteins	Inhibition of NF-kB sensitizes cancer cells toward anticancer drugs, cytotoxic phytochemicals, biological agents, and radiation	Molecular docking was done using Autodock	Seo et al., 2018
2	Dasatinib	Multi-BCR/ABL and Src family tyrosine kinase inhibitor approved for chronic myelogenous leukemia	ACK1, a novel cancer target	To cure breast and prostate cancers	Docking was conducted with the Glide software	Phatak and Zhang, 2013
3	Carvedilol	Beta-1 adrenergic receptor	Cytochrome P450 1B1 (CYP1B1)	For prevention and therapy of cancer	Glide software used to perform the molecular docking–based virtual screening	Wang et al., 2020a
MICROBIAL DRUG TARGET						
4	Cefmenoxime, ceforanide, cefotetan, cefonicid sodium, and cefpiramide	Antibacterial drugs bind with penicillin-binding proteins or transpeptidases and inhibits cell wall synthesis of bacteria	E1-E2 interface of chikungunya virus (CHIKV) envelope protein complex	To cure CHIKV	Molecular docking and structure-based drug design were conducted using the FlexX/LeadIT software	Agarwal et al., 2019
5	Guanethidine	An antihypertensive agent	Influenza A virus (IAV) matrix protein 2 (M2)	Antiinfluenza drug	Molecular docking was carried in Autodock Vina	Radosevic et al., 2019
	Methamphetamine	Psychostimulant and sympathomimetic drug				
	Cycrimine	To treat Parkinson's disease				

Continued

TABLE 19.3
Few Case Studies That Have Used Docking for Drug Repurposing.—cont'd

S. No	Drug	Known pharmaco-logical target	Predicted drug target	Predicted treatment	Software used	Reference
6	Velpatasvir and ledipasvir	Anti--hepatitis C virus drug	3-Chymotrypsin-like protease (3CL^{pro}) inhibitor of SARS-CoV-2	To cure COVID-19	MTiOpenScreen web service used for virtual screening and AutoDock Vina was used for docking	Chen et al., 2020
7	Apixaban	Anticoagulant administered by the oral route	Main protease of SARS-CoV-2 (M^{pro})	To cure COVID-19	GOLD 2020.1 software	da Silva Hage-Melim et al., 2020
8	Ribavirin	To treat hepatitis C and viral hemorrhagic fevers	Main protease of SARS-CoV-2 (M^{pro})	To cure COVID-19	Schrodinger glide docking module was used for virtual Screening	Kandeel and Al-Nazawi, 2020
	Telbivudine	To treat hepatitis B virus				
	Vitamin B12	Vitamin				
	Nicotinamide	Vitamin				
9	Itraconazole	Antifungal agent	Plasmodium falciparum lactate dehydrogenase enzyme	To discover new antimalarial drugs	Molegro Virtual Docker software	Penna-Coutinho et al., 2011
	Atorvastatin	Lipid-lowering drug				
	Posaconazole	Antifungal agent				

TABLE 19.4
Some Case Studies Involving Application of Docking to Identify Potential Target Molecule Against the Natural Ligands to Combat Diseases.

Natural ligand	Identified target	Software	References
Curcumin	CDK2	Schrödinger Glide	Lim et al., 2014
ε-Viniferin	Cyclic nucleotide phosphodiesterase 4 (PDE4)	FlexX docking program	Do and Bernard, 2004 and Do et al., 2005
Meranzin	COX1, COX2, and PPAR$_\Upsilon$		
Danshensu, an active compound of TCM Danshen (*Salvia miltiorrhiza*)	GTPase HRas	idTarget server and a ligand-based IVS server PharmMapper	Chen and Ren, 2014
Tea polyphenol, epigallocatechin gallate (EGCG)	Leukotriene A4 hydrolase	Autodock software and TarFisDock	Zheng et al., 2011
Xanthohumol	PDK1protein kinases	Autodock-Vina32 software	Lauro et al., 2012
Isoxanthohumol	PKC protein kinases		

TABLE 19.5
List of Docking-Based Virtual Screening for Ligands Against β-Lactamases to Combat Antibiotic Resistance.

Type of β-lactamases	Databases used	Software used	References
KPC-2	In-house MySQL database	DOCK3.6	Klein et al., 2018a,b
CTX-M-15, KPC-2, NDM-1, VIM-2, AmpC	Specs database	FLAPdock implemented in FLAP (Fingerprint for Ligands and Proteins)	Francesca et al., 2020
NDM-1	ZINC database	AutoDock Vina	Wang et al., 2020b
CTXM-15	ZINC database	Pyrx 0.8 software	Farhadi et al., 2018
KPC-2	ZINC database	GOLD 5.0 version	Danishuddin et al., 2014
TEM-1 and SHV-1	ZINC database	GOLD 5.0 version	Baig et al., 2015
TEM-1	ZINC database	Glide tool of Maestro	Avci et al., 2018
OXA β-lactamases	Maybridge database	GOLD 5.0 version	Gupta et al., 2020

TABLE 19.6

Approved Drugs Developed Using Computational Methods in Combination With Rational Drug Design Approach.

S. No	Approved drugs	Drug target/action
1	Saquinavir Indinavir Ritonavir Nelfinavir Lopinavir Fosamprenavir Atazanavir Tipranavir Darunavir	HIV protease inhibitor
2	Lapatinib Erlotinib Gefitinib	EGFR kinase inhibitor for the treatment of cancer
3	Oseltamivir Zanamivir	Neuraminidase inhibitor to treat influenza A and influenza B
4	Crizotlnib	ALK inhibitor
5	Abiraterone	Androgen synthesis inhibitor to cure metastatic castration-resistant prostate cancer or hormone refractory prostate cancer
6	Sorafenib	VEGFR kinase inhibitor to treat renal cancer, liver cancer, and thyroid cancer

ALK, anaplastic lymphoma kinase; *EGFR*, epidermal growth factor receptor; *HIV*, human immunodeficiency virus; *VEGFR*, vascular epidermal growth factor receptor.

Some of the examples are taken from (Prada-Gracia et al., 2016). Application of computational methods for anticancer drug discovery, design, and optimization. *Boletín Médico Del Hospital Infantil de México (English Edition)*, 73(6), 411–423.

REFERENCES

Abdolmaleki, A., Ghasemi, J. B., & Ghasemi, F. (2017). Computer aided drug design for multi-target drug design: SAR/QSAR, molecular docking and pharmacophore methods. *Current Drug Targets*, 18(5), 556–575.

Agarwal, G., Gupta, S., Gabrani, R., Gupta, A., Chaudhary, V. K., & Gupta, V. (2019). Virtual screening of inhibitors against envelope glycoprotein of chikungunya virus: A drug repositioning approach. *Bioinformation*, 15(6), 439.

Ali, A., Danishuddin, Maryam, L., Srivastava, G., Sharma, A., & Khan, A. U. (2018). Designing of inhibitors against CTX-M-15 type β-lactamase: Potential drug candidate against β-lactamases-producing multi-drug-resistant bacteria. *Journal of Biomolecular Structure and Dynamics*, 36(7), 1806–1821.

Alnabulsi, S., Al-Hurani, E. A., & El-Elimat, T. (2019). Aminocarboxamide benzothiazoles as potential LSD1 hit inhibitors. Part I: Computational fragment-based drug design. *Journal of Molecular Graphics and Modelling*, 93, 107440.

Amresh, P., Kumar, K., Islam, A., Hassan, I., & Ahmad, F. (2013). Receptor chemoprint derived pharmacophore model for development of CAIX inhibitors. *Journal of Carcinogenesis and Mutagenesis*, S8, 17–20.

Avci, F. G., Altinisik, F. E., Karacan, I., Karagoz, D. S., Ersahin, S., Eren, A., Sayar, N. A., Ulu, D. V., Ozkirimli, E., & Akbulut, B. S. (2018). Targeting a hidden site on class A beta-lactamases. *Journal of Molecular Graphics and Modelling*, 84, 125–133.

Baig, M. H., Balaramnavar, V. M., Wadhwa, G., & Khan, A. U. (2015). Homology modeling and virtual screening of inhibitors against TEM-and SHV-type-resistant mutants: A multilayer filtering approach. *Biotechnology and Applied Biochemistry*, 62(5), 669–680.

Banegas-Luna, A. J., Cerón-Carrasco, J. P., & Pérez-Sánchez, H. (2018). A review of ligand-based virtual screening web tools and screening algorithms in large molecular databases in the age of big data. *Future Medicinal Chemistry*, 10(22), 2641–2658.

Barreca, M. L., Manfroni, G., Leyssen, P., Winquist, J., Kaushik-Basu, N., Paeshuyse, J., Krishnan, R., Iraci, N., Sabatini, S., Tabarrini, O., Basu, A., Danielson, U. H., Neyts, J., & Cecchetti, V. (2013). Structure-based discovery of pyrazolobenzothiazine derivatives as inhibitors of hepatitis C virus replication. *Journal of Medicinal Chemistry*, 56(6), 2270–2282.

Batool, M., Ahmad, B., & Choi, S. (2019). A structure-based drug discovery paradigm. *International Journal of Molecular Sciences*, 20(11), 2783.

Betzi, S., Restouin, A., Opi, S., Arold, S. T., Parrot, I., Guerlesquin, Morelli, X., & Collette, Y. (2007). Protein-protein interaction inhibition (2P2I) combining high throughput and virtual screening: Application to the HIV-1 Nef protein. *Proceedings of the National Academy of Sciences*, 104(49), 19256–19261.

Bian, Y., & Xie, X. Q. S. (2018). Computational fragment-based drug design: Current trends, strategies, and applications. *The AAPS Journal*, 20(3), 59.

Buchstaller, H. P., Eggenweiler, H. M., Sirrenberg, C., Grädler, U., Musil, D., Hoppe, E., Zimmermann, A., Schwartz, H., März, J., Bomke, J., Wegener, A., & Wolf, M. (2012). Fragment-based discovery of hydroxy-indazole-carboxamides as novel small molecule inhibitors of Hsp90. *Bioorganic and Medicinal Chemistry Letters*, 22(13), 4396–4403.

Buzharevski, A., Paskaš, S., Sárosi, M. B., Laube, M., Lönnecke, P., Neumann, W., Murganić, B., Mijatović, S., Maksimović-Ivanić, D., Pietzsch, J., & Hey-Hawkins, E. (2020). Carboranyl Derivatives of Rofecoxib with Cytostatic Activity against Human Melanoma and Colon Cancer Cells. *Scientific Reports*, 10(1), 1–13.

Cain, R., Brem, J., Zollman, D., McDonough, M. A., Johnson, R. M., Spencer, J., Makena, A., Abboud, M. I., Cahill, S., Lee, S. Y., McHugh, P. J., Schofield, C. J., &

Fishwick, C. W. G. (2018). In silico fragment-based design identifies subfamily B1 metallob-lactamase inhibitors. *Journal of Medicinal Chemistry, 61*(3), 1255–1260.

Chen, S. J., & Ren, J. L. (2014). Identification of a potential anti-cancer target of danshensu by inverse docking. *Asian Pacific Journal of Cancer Prevention, 15*(1), 111–116.

Chen, Y., & Shoichet, B. K. (2009). Molecular docking and ligand specificity in fragment-based inhibitor discovery. *Nature Chemical Biology, 5*(5), 358–364.

Chen, Y. W., Yiu, C. P. B., & Wong, K. Y. (2020). *Protease (3CL) structure: virtual screening reveals velpatasvir, ledipasvir, and other drug repurposing candidates.*

Congreve, M., Carr, R., Murray, C., & Jhoti, H. (2003). A'rule of three'for fragment-based lead discovery? *Drug Discovery Today, 19*(8), 876–877.

Dai, W., & Guo, D. (2019). A ligand-based virtual screening method using direct quantification of generalization ability. *Molecules, 24*(13), 2414.

Danishuddin, M., & Khan, A. U. (2015). Virtual screening strategies: A state of art to combat with multiple drug resistance strains. *MOJ Proteomics Bioinform, 2*(2), 00042.

Danishuddin, M., Khan, A., Faheem, M., Kalaiarasan, P., Hassan Baig, M., Subbarao, N., & Khan, A. U. (2014). Structure-based screening of inhibitors against KPC-2: Designing potential drug candidates against multidrug-resistant bacteria. *Journal of Biomolecular Structure and Dynamics, 32*(5), 741–750.

Desaphy, J., Azdimousa, K., Kellenberger, E., & Rognan, D. (2012). Comparison and druggability prediction of protein–ligand binding sites from pharmacophore-annotated cavity shapes. *Journal of Chemical Information and Modeling.*

Do, Q. T., & Bernard, P. (2004). Pharmacognosy and reverse pharmacognosy: A new concept for accelerating natural drug discovery. *Idrugs: The Investigational Drugs Journal, 7*(11), 1017–1027.

Do, Q. T., Renimel, I., Andre, P., Lugnier, C., Muller, C. D., & Bernard, P. (2005). Reverse pharmacognosy: Application of selnergy, a new tool for lead discovery. The example of ε-viniferin. *Current Drug Discovery Technologies, 2*(3), 161–167.

Doss, C. G. P., Chakraborty, C., Narayan, V., & Kumar, D. T. (2014). Computational approaches and resources in single amino acid substitutions analysis toward clinical research. In *Advances in protein chemistry and structural biology* (Vol. 94, pp. 365–423). Academic Press.

Du, J., Liu, L., Liu, B., Yang, J., Hou, X., Yu, J., & Fang, H. (2020). Structure-based virtual screening, biological evaluation and biophysical study of novel Mcl-1 inhibitors. *Future Medicinal Chemistry, 12*(14), 1293–1304.

Eweas, A. F., Maghrabi, I. A., & Namarneh, A. I. (2014). Advances in molecular modeling and docking as a tool for modern drug discovery. *Der Pharma Chemica, 6*(6), 211–228.

Fang, Y. (2012). Ligand–receptor interaction platforms and their applications for drug discovery. *Expert Opinion on Drug Discovery, 7*(10), 969–988.

Farhadi, T., Fakharian, A., & Ovchinnikov, R. S. (2018). Virtual screening for potential inhibitors of CTX-M-15 protein of *Klebsiella pneumoniae. Interdisciplinary Sciences: Computational Life Sciences, 10*(4), 694–703.

Ferreira, L. G., Dos Santos, R. N., Oliva, G., & Andricopulo, A. D. (2015). Molecular docking and structure-based drug design strategies. *Molecules, 20*(7), 13384–13421.

Francesca, S., Matteo, S., Lorenzo, M., Cross, S., Eleonora, G., Filomena, S., Federica, V., Filomena, D. L., Jean-Denis, D., Laura, C., Donatella, T., Alberto, V., Gabriele, C., & Maria, P. C. (2020). Virtual screening identifies broad-spectrum b-lactamase inhibitors with activity on clinically relevant serine-and metallocarbapenemases. *Scientific Reports, 10*(1), 1–15.

Fuller, J. C., Burgoyne, N. J., & Jackson, R. M. (2009). Predicting druggable binding sites at the protein–protein interface. *Drug Discovery Today, 14*(3–4), 155–161.

Gane, P. J., & Dean, P. M. (2000). Recent advances in structure-based rational drug design. *Current Opinion in Structural Biology, 10,* 401–404.

Gonzalez, M. W., & Kann, M. G. (2012). Protein interactions and disease. *PLoS Computational Biology, 8*(12), e1002819.

Gonzalez, J., Lendebol, E., Shen, A., Philipp, M., & Clement, C. (2018). In silico-mediated virtual screening and molecular docking platforms for discovery of non β-lactam inhibitors of Y-49 β-lactamase from Mycobacterium tuberculosis. *MOJ Proteomics Bioinform, 7*(1), 1-14: 00207.

Grüneberg, S., Stubbs, M. T., & Klebe, G. (2002). Successful virtual screening for novel inhibitors of human carbonic anhydrase: Strategy and experimental confirmation. *Journal of Medicinal Chemistry, 45*(17), 3588–3602.

Gulati, S., Cheng, T. M., & Bates, P. A. (August 2013). Cancer networks and beyond: Interpreting mutations using the human interactome and protein structure. *Seminars in Cancer Biology, 23*(4), 219–226. Academic Press.

Gupta, D., Singh, A., Somvanshi, P., Singh, A., & Khan, A. U. (2020). Structure-based screening of non-β-lactam inhibitors against class D β-lactamases: An approach of docking and molecular dynamics. *ACS Omega, 5*(16), 9356–9365.

Hakes, L., Lovell, S. C., Oliver, S. G., & Robertson, D. L. (2007). Specificity in protein interactions and its relationship with sequence diversity and coevolution. *Proceedings of the National Academy of Sciences of the United States of America, 104*(19), 7999–8004.

Hamza, A., Wei, N. N., & Zhan, C. G. (2012). Ligand-based virtual screening approach using a new scoring function. *Journal of Chemical Information and Modeling, 52*(4), 963–974.

Hartenfeller, M., & Schneider, G. (2010). De novo drug design. In *Chemoinformatics and computational chemical biology* (pp. 299–323). Totowa, NJ: Humana Press.

Hosseini, Y., Mollica, A., & Mirzaie, S. (2016). Structure-based virtual screening efforts against HIV-1 reverse transcriptase to introduce the new potent non-nucleoside reverse transcriptase inhibitor. *Journal of Molecular Structure, 1125,* 592–600.

Hubbard, R. E., Chen, I., & Davis, B. (2007). Informatics and modeling challenges in fragment-based drug discovery.

Current Opinion in Drug Discovery and Development, 10(3), 289–297.

Ji, H. F., Li, X. J., & Zhang, H. Y. (2009). Natural products and drug discovery: Can thousands of years of ancient medical knowledge lead us to new and powerful drug combinations in the fight against cancer and dementia? *EMBO Reports, 10*(3), 194–200.

Jorgensen, W. L. (2004). The many roles of computation in drug discovery. *Science, 303*(5665), 1813–1818.

Kahsai, A. W., Xiao, K., Rajagopal, S., Ahn, S., Shukla, A. K., Sun, J., G Oas, T., & Lefkowitz, R. J. (2011). Multiple ligand-specific conformations of the b 2-adrenergic receptor. *Nature Chemical Biology, 7*(10), 692–700.

Kalia, M., Singh, P. K., Yadav, V. K., Yadav, B. S., Sharma, D., Narvi, S. S., Mani, A., & Agarwal, V. (2017). Structure based virtual screening for identification of potential quorum sensing inhibitors against LasR master regulator in Pseudomonas aeruginosa. *Microbial Pathogenesis, 107*, 136–143.

Kandeel, M., & Al-Nazawi, M. (2020). Virtual screening and repurposing of FDA approved drugs against COVID-19 main protease. *Life Sciences*, 117627.

Kazi, A., Lawrence, H., Guida, W. C., McLaughlin, M. L., Springett, G. M., Berndt, N., Yip, R. M. L., & Sebti, S. M. (2009). Discovery of a novel proteasome inhibitor selective for cancer cells over non-transformed cells. *Cell Cycle, 8*(12), 1940–1951.

Khan, A. U., Ali, A., Srivastava, G., & Sharma, A. (2017). Potential inhibitors designed against NDM-1 type metallo-β-lactamases: An attempt to enhance efficacies of antibiotics against multi-drug-resistant bacteria. *Scientific Reports, 7*(1), 1–14.

Kharkar, P. S., Warrier, S., & Gaud, R. S. (2014). Reverse docking: A powerful tool for drug repositioning and drug rescue. *Future Medicinal Chemistry, 6*(3), 333–342.

Kist, R., Timmers, L. F. S. M., & Caceres, R. A. (2018). Searching for potential mTOR inhibitors: Ligand-based drug design, docking and molecular dynamics studies of rapamycin binding site. *Journal of Molecular Graphics and Modelling, 80*, 251–263.

Klein, R., Linciano, P., Celenza, G., Bellio, P., Papaioannou, S., Blazquez, J., Cendron, L., Brenk, R., & Tondi, D. (2018a). In silico identification and experimental validation of hits active against KPC-2 β-lactamase. *PLoS One, 13*(11), e0203241.

Klein, R., Linciano, P., Celenza, G., Bellio, P., Papaioannou, S., Blazquez, J., Cendron, L., Brenk, R., & Tondi, D. (2018b). *silico identification and experimental validation of novel KPC-2 b-lactamase inhibitors* (p. 396283). bioRxiv.

Kumalo, H. M., Bhakat, S., & Soliman, M. E. (2015). Theory and applications of covalent docking in drug discovery: Merits and pitfalls. *Molecules, 20*(2), 1984–2000.

Kumari, M., Chandra, S., Tiwari, N., & Subbarao, N. (2016). 3D QSAR, pharmacophore and molecular docking studies of known inhibitors and designing of novel inhibitors for M18 aspartyl aminopeptidase of Plasmodium falciparum. *BMC Structural Biology, 16*(1), 12.

Kumar, A., Voet, A., & Zhang, K. Y. J. (2012). Fragment based drug design: From experimental to computational

approaches. *Current Medicinal Chemistry, 19*(30), 5128–5147.

Laurie, A. T., & Jackson, R. M. (2005). Q-SiteFinder: An energy-based method for the prediction of protein–ligand binding sites. *Bioinformatics, 21*(9), 1908–1916.

Lauro, G., Masullo, M., Piacente, S., Riccio, R., & Bifulco, G. (2012). Inverse virtual screening allows the discovery of the biological activity of natural compounds. *Bioorganic and Medicinal Chemistry, 20*(11), 3596–3602.

Lavecchia, A., & Di Giovanni, C. (2013). Virtual screening strategies in drug discovery: A critical review. *Current Medicinal Chemistry, 20*(23), 2839–2860.

Lee, A., Lee, K., & Kim, D. (2016). Using reverse docking for target identification and its applications for drug discovery. *Expert Opinion on Drug Discovery, 11*(7), 707–715.

Leung, C. H., Chan, D. S. H., Yang, H., Abagyan, R., Lee, S. M. Y., Zhu, G. Y., Fong, W. F., & Ma, D. L. (2011). A natural product-like inhibitor of NEDD8-activating enzyme. *Chemical Communications, 47*(9), 2511–2513.

Lim, T. G., Lee, S. Y., Huang, Z., Chen, H., Jung, S. K., Bode, A. M., Lee, W. K., & Dong, Z. (2014). Curcumin suppresses proliferation of colon cancer cells by targeting CDK2. *Cancer Prevention Research, 7*(4), 466–474.

Lipinski, C. A. (2000). Drug-like properties and the causes of poor solubility and poor permeability. *Journal of Pharmacological and Toxicological Methods, 44*(1), 235–249.

Lipinski, C. A., Lombardo, F., Dominy, B. W., & Feeney, P. J. (1997). Experimental and computational approaches to estimate solubility and permeability in drug discovery and development settings. *Advanced Drug Delivery Reviews, 23*(1–3), 3–25.

Mahajan, P., Suri, N., Mehra, R., Gupta, M., Kumar, A., Singh, S. K., & Nargotra, A. (2017). Discovery of novel small molecule EGFR inhibitory leads by structure and ligand-based virtual screening. *Medicinal Chemistry Research, 26*(1), 74–92.

Malathi, K., & Ramaiah, S. (2018). Bioinformatics approaches for new drug discovery: A review. *Biotechnology and Genetic Engineering Reviews, 34*(2), 243–260.

Mandal, S., & Mandal, S. K. (2009). Rational drug design. *European Journal of Pharmacology, 625*(1–3), 90–100.

Marchand, J. R., Dalle Vedove, A., Lolli, G., & Caflisch, A. (2017). Discovery of inhibitors of four bromodomains by fragment-anchored ligand docking. *Journal of Chemical Information and Modeling, 57*(10), 2584–2597.

Martin, Y. C. (2000). Disco: What we did right and what we missed. In *Pharmacophore perception, development, and use in drug design, 2* pp. 49–68).

Ma, Y., Xu, B., Fang, Y., Yang, Z., Cui, J., Zhang, L., & Zhang, L. (2011). Synthesis and SAR study of novel peptide aldehydes as inhibitors of 20S proteasome. *Molecules, 16*(9), 7551–7564.

McConkey, B. J., Sobolev, V., & Edelman, M. (2002). The performance of current methods in ligand-protein docking. *Current Science, 83*(7), 845–856.

McKay, P. B., Peters, M. B., Carta, G., Flood, C. T., Dempsey, E., Bell, A., Berry, C., Lloyd, D. G., & Fayne, D. (2011). Identification of plasmepsin inhibitors as selective anti-malarial agents using ligand based drug design. *Bioorganic and Medicinal Chemistry Letters, 21*(11), 3335–3341.

Meireles, L. M., Dömling, A. S., & Camacho, C. J. (2010). ANCHOR: A web server and database for analysis of protein–protein interaction binding pockets for drug discovery. *Nucleic Acids Research, 38*(Suppl. l_2), W407–W411.

Mortier, J., Rakers, C., Frederick, R., & Wolber, G. (2012). Computational tools for in silico fragment-based drug design. *Current Topics in Medicinal Chemistry, 12*(17), 1935–1943.

Mulvey, M. R., & Simor, A. E. (2009). Antimicrobial resistance in hospitals: How concerned should we be? *Canadian Medical Association Journal, 180*(4), 408–415.

Murray, C. W., Carr, M. G., Callaghan, O., Chessari, G., Congreve, M., Cowan, S., Coyle, J. E., Downham, R., Figueroa, E., Frederickson, M., Graham, B., McMenamin, R., O'Brien, M. A., Patel, S., Phillips, T. R., Williams, G., Woodhead, A. J., & Woolford, A. J. A. (2010). Fragment-based drug discovery applied to Hsp90. Discovery of two lead series with high ligand efficiency. *Journal of Medicinal Chemistry, 53*(16), 5942–5955.

Park, H., Chien, P. N., & Ryu, S. E. (2012). Discovery of potent inhibitors of receptor protein tyrosine phosphatase sigma through the structure-based virtual screening. *Bioorganic and Medicinal Chemistry Letters, 22*(20), 6333–6337.

Parvu, L. (2003). QSAR-a piece of drug design. *Journal of Cellular and Molecular Medicine, 7*(3) (Online).

Paul, S. M., Mytelka, D. S., Dunwiddie, C. T., Persinger, C. C., Munos, B. H., Lindborg, S. R., & Schacht, A. L. (2010). How to improve R&D productivity: The pharmaceutical industry's grand challenge. *Nature Reviews Drug Discovery, 9*(3), 203–214.

Pedretti, A., Mazzolari, A., Gervasoni, S., & Vistoli, G. (2019). Rescoring and linearly combining: A highly effective consensus strategy for virtual screening campaigns. *International Journal of Molecular Sciences, 20*(9), 2060.

Penna-Coutinho, J., Cortopassi, W. A., Oliveira, A. A., França, T. C. C., & Krettli, A. U. (2011). Antimalarial activity of potential inhibitors of Plasmodium falciparum lactate dehydrogenase enzyme selected by docking studies. *PLoS One, 6*(7), e21237.

Petta, I., Lievens, S., Libert, C., Tavernier, J., & De Bosscher, K. (2016). Modulation of protein–protein interactions for the development of novel therapeutics. *Molecular Therapy, 24*(4), 707–718.

Phatak, S. S., & Zhang, S. (2013). A novel multi-modal drug repurposing approach for identification of potent ACK1 inhibitors. In *Biocomputing 2013* (pp. 29–40).

Phillips, M. A., Stewart, M. A., Woodling, D. L., & Xie, Z. R. (2018). *Has molecular docking ever brought us a medicine*. London, UK: Molecular Docking. IntechOpen Limited.

Prada-Gracia, D., Huerta-Yépez, S., & Moreno-Vargas, L. M. (2016). Application of computational methods for anti-cancer drug discovery, design, and optimization. *Boletín Médico Del Hospital Infantil de México (English Edition), 73*(6), 411–423.

Prieto-Martínez, F. D., Norinder, U., & Medina-Franco, J. L. (2019). Cheminformatics explorations of natural products. In *Progress in the chemistry of organic natural products 110* (pp. 1–35). Cham: Springer.

Radosevic, D., Sencanski, M., Perovic, V., Veljkovic, N., Prljic, J., Veljkovic, V., Mantlo, E., Bukreyeva, N., Paessler, S., & Glisic, S. (2019). Virtual screen for repurposing of drugs for candidate influenza a M2 ion-channel inhibitors. *Frontiers in Cellular and Infection Microbiology, 9*, 67.

Roche, D. B., Brackenridge, D. A., & McGuffin, L. J. (2015). Proteins and their interacting partners: An introduction to protein–ligand binding site prediction methods. *International Journal of Molecular Sciences, 16*(12), 29829–29842.

Rollinger, J. M., Kratschmar, D. V., Schuster, D., Pfisterer, P. H., Gumy, C., Aubry, E. M., Brandstötter, S., Stuppner, H., Wolber, G., & Odermatt, A. (2010). 11b-Hydroxysteroid dehydrogenase 1 inhibiting constituents from Eriobotrya japonica revealed by bioactivityguided isolation and computational approaches. *Bioorganic and Medicinal Chemistry, 18*(4), 1507–1515.

Ruda, G. F., Campbell, G., Alibu, V. P., Barrett, M. P., Brenk, R., & Gilbert, I. H. (2010). Virtual fragment screening for novel inhibitors of 6-phosphogluconate dehydrogenase. *Bioorganic and Medicinal Chemistry, 18*(14), 5056–5062.

Ryan, D. P., & Matthews, J. M. (2005). Protein–protein interactions in human disease. *Current Opinion in Structural Biology, 15*(4), 441–446.

Salmas, R. E., Senturk, M., Yurtsever, M., & Durdagi, S. (2016). Discovering novel carbonic anhydrase type IX (CA IX) inhibitors from seven million compounds using virtual screening and in vitro analysis. *Journal of Enzyme Inhibition and Medicinal Chemistry, 31*(3), 425–433.

Santoro, A. M., Lo Giudice, M. C., D'Urso, A., Lauceri, R., Purrello, R., & Milardi, D. (2012). Cationic porphyrins are reversible proteasome inhibitors. *Journal of the American Chemical Society, 134*(25), 10451–10457.

Schuster, D., Markt, P., Grienke, U., Mihaly-Bison, J., Binder, M., Noha, S. M., Rollinger, J. M., Stuppner, H., Bochkov, V. N., & Wolber, G. (2011). Pharmacophore-based discovery of FXR agonists. Part I: Model development and experimental validation. *Bioorganic and Medicinal Chemistry, 19*(23), 7168–7180.

Schuster, D., Maurer, E. M., Laggner, C., Nashev, L. G., Wilckens, T., Langer, T., & Odermatt, A. (2006). The discovery of new 11β-hydroxysteroid dehydrogenase type 1 inhibitors by common feature pharmacophore modeling and virtual screening. *Journal of Medicinal Chemistry, 49*(12), 3454–3466.

Seo, E. J., Sugimoto, Y., Greten, H. J., & Efferth, T. (2018). Repurposing of bromocriptine for cancer therapy. *Frontiers in Pharmacology, 9*, 1030.

Serçinoğlu, O., & Sarica, P. O. (2019). In silico databases and tools for drug repurposing. In *In silico drug design* (pp. 703–742). Academic Press.

Shahin, R., Mansi, I., Swellmeen, L., Alwidyan, T., Al-Hashimi, N., Al-Qarar'h, Y., & Shaheen, O. (2018). Ligand-based computer aided drug design reveals new tropomycin receptor kinase a (TrkA) inhibitors. *Journal of Molecular Graphics and Modelling, 80*, 327–352.

Sheng, C., & Zhang, W. (2013). Fragment informatics and computational fragment-based drug design: An overview and update. *Medicinal Research Reviews, 33*(3), 554–598.

Shoichet, B. K., & Kobilka, B. K. (2012). Structure-based drug screening for G-protein-coupled receptors. *Trends in Pharmacological Sciences, 33*(5), 268–272.

da Silva Hage-Melim, L. I., Federico, L. B., de Oliveira, N. K. S., Francisco, V. C. C., Correa, L. C., de Lima, H. B., Gomes, S. Q., Barcelos, M. P., Francischini, I. A. G., & de Paula da Silvab, C. H. T. (2020). *Virtual screening, ADME/Tox predictions and the drug repurposing concept for future use of old drugs against the COVID-19* (p. 117963). Life Sciences.

Singh, J., Petter, R. C., Baillie, T. A., & Whitty, A. (2011). The resurgence of covalent drugs. *Nature Reviews Drug Discovery, 10*(4), 307–317.

Teotico, D. G., Babaoglu, K., Rocklin, G. J., Ferreira, R. S., Giannetti, A. M., & Shoichet, B. K. (2009). Docking for fragment inhibitors of AmpC β-lactamase. *Proceedings of the National Academy of Sciences of the United States of America, 106*(18), 7455–7460.

Tong, J. B., Bai, M., & Zhao, X. (2016). 3D-QSAR and docking studies of HIV-1 protease inhibitors using R-group search and Surflex-dock. *Medicinal Chemistry Research, 25*(11), 2619–2630.

Tresadern, G., Bemporad, D., & Howe, T. (2009). A comparison of ligand based virtual screening methods and application to corticotropin releasing factor 1 receptor. *Journal of Molecular Graphics and Modelling, 27*(8), 860–870.

Vianna, C. P., & de Azevedo, W. F. (2012). Identification of new potential *Mycobacterium tuberculosis* shikimate kinase inhibitors through molecular docking simulations. *Journal of Molecular Modeling, 18*(2), 755–764.

Wang, H., Duffy, R. A., Boykow, G. C., Chackalamannil, S., & Madison, V. S. (2008). Identification of novel cannabinoid CB1 receptor antagonists by using virtual screening with a pharmacophore model. *Journal of Medicinal Chemistry, 51*(8), 2439–2446.

Wang, X., Song, K., Li, L., & Chen, L. (2018). Structure-based drug design strategies and challenges. *Current Topics in Medicinal Chemistry, 18*(12), 998–1006.

Wang, Y., He, X., Li, C., Ma, Y., Xue, W., Hu, B., Wang, J., Zhang, T., & Zhang, F. (2020a). Carvedilol serves as a novel CYP1B1 inhibitor, a systematic drug repurposing approach through structurebased virtual screening and experimental verification. *European Journal of Medicinal Chemistry, 193,* 112235.

Wang, X., Yang, Y., Gao, Y., & Niu, X. (2020b). Discovery of the novel inhibitor against New Delhi metallo-β-lactamase based on virtual screening and molecular modelling. *International Journal of Molecular Sciences, 21*(10), 3567.

Watkins, R. R., & Bonomo, R. A. (2017). β-Lactam antibiotics. In *Infectious diseases* (pp. 1203–1216). Elsevier.

Wilson, A. J. (2009). Inhibition of protein–protein interactions using designed molecules. *Chemical Society Reviews, 38*(12), 3289–3300.

Wilson, G. L., & Lill, M. A. (2011). Integrating structure-based and ligand-based approaches for computational drug design. *Future Medicinal Chemistry, 3*(6), 735–750.

Xie, Z. R., & Hwang, M. J. (2015). Methods for predicting protein–ligand binding sites. In *Molecular modeling of proteins* (pp. 383–398). New York, NY: Humana Press.

Xu, X., Huang, M., & Zou, X. (2018). Docking-based inverse virtual screening: Methods, applications, and challenges. *Biophysics reports, 4*(1), 1–16.

Yousuf, Z., Iman, K., Iftikhar, N., & Mirza, M. U. (2017). Structure-based virtual screening and molecular docking for the identification of potential multi-targeted inhibitors against breast cancer. *Breast Cancer: Targets and Therapy, 9,* 447.

Yu, W., Xiao, H., Lin, J., & Li, C. (2013). Discovery of novel STAT3 small molecule inhibitors via in silico site-directed fragment-based drug design. *Journal of Medicinal Chemistry, 56*(11), 4402–4412.

Zhang, S., Shi, Y., Jin, H., Liu, Z., Zhang, L., & Zhang, L. (2009). Covalent complexes of proteasome model with peptide aldehyde inhibitors MG132 and MG101: Docking and molecular dynamics study. *Journal of Molecular Modeling, 15*(12), 1481.

Zheng, R., Chen, T. S., & Lu, T. (2011). A comparative reverse docking strategy to identify potential antineoplastic targets of tea functional components and binding mode. *International Journal of Molecular Sciences, 12*(8), 5200–5212.

Zhu, X., Xiong, Y., & Kihara, D. (2015). Large-scale binding ligand prediction by improved patch-based method Patch-Surfer2. 0. *Bioinformatics, 31*(5), 707–713.

CHAPTER 20

Role of Bioinformatics in Subunit Vaccine Design

HEMANT ARYA • TARUN KUMAR BHATT

1 INTRODUCTION

Vaccine is an inactivated form of a microorganism (that causes disease/s), which improves immunity against the targeted pathogen and guards human health (He et al., 2010; Leroux-Roels et al., 2011; Prugnola, 1994). The vaccine is the safest medical artifact which protects healthy human from deadly pathogen (microorganisms such as virus, bacteria, etc.) infection. In vaccination, a dead or weak biological substance (microorganism) is injected in healthy humans, which leads to triggering of an immune response against the microorganism and also keeping memories to eradicate the actual pathogen infection in the future (Bartlett et al., 2009). In short, the vaccine helps the human immune system to recognize and devastate the pathogen during future infection. As per WHO, "Vaccine helps to trigger the body's immune system to recognize and fight against the pathogen infection and keep the human population safe from the diseases (https://www.who.int/topics/vaccines/en/)" (Prugnola, 1994). Vaccine plays a crucial role in human health by protecting from life-threatening diseases such as polio, influenza, meningitis, measles, rubella, tetanus, typhoid, cervical cancer, diphtheria, smallpox, pertussis, etc. (Plotkin, 2005).

Vaccine may be divided into live attenuated vaccine, inactivated vaccine, subunit vaccine, toxoid vaccine, recombinant vector vaccine, DNA vaccine, RNA vaccine, conjugate vaccine, and biosynthetic vaccine (He et al., 2010; Huang et al., 2017; Leroux-Roels et al., 2011; Plotkin, 2005). In this chapter, we will discuss mostly the designing of the subunit vaccine only. Interested readers may refer to useful article for the designing of other type of vaccines (Abraham Peele et al., 2020; Bhamarapravati & Sutee, 2000; Bonten et al., 2015; Ferraro et al., 2011; Pardi et al., 2017; Smith et al., 2020). Few examples of the different type of vaccines along with their uses and trade names are shown in Table 20.1.

2 BASICS OF VACCINE

Vaccine is an inactivated or weak form of a pathogen or it is a kind of a pathogen imposter which looks similar to a virus or bacteria or pathogen but does not make the body sick. During vaccination, the inactivated form is introduced to human that generates immunity against the pathogen (Prugnola, 1994). Once a vaccine is delivered to human, antigen-presenting cells (macrophages or dendritic cells) digest the pathogen, break it apart, and display the foreign particles (antigen) via major histocompatibility complex (MHC) on the surface of the cells. In the presence of interleukin 12 (IL12), T cell receptor of the Th_0/T naïve cell recognizes the antigen and gets converted into T helper (Th_1) cell. Th1 cells further activate Tc (cytotoxic cell) and Th_2 in the presence of IL2 and IL10, respectively. Tc cell kills the pathogen containing cells, whereas Th_2 cells activate B cell. The antigen-specific activated B cells can undergo cell division and produce more specific B cells. Some of these will transform into plasma B cell and others develop into B memory cells. The plasma B cells which is generally called as antibody factories produce "Y"-shaped antibodies specific to the vaccine antigen, and each antibody tightly binds with target antigen (like lock and key) and kills the pathogen (Giese, 2016; Kennedy, 2010). If the same pathogen or antigen attacks, the memory B cells produce antibodies immediately and prevent human from the pathogen infection. Fig. 20.1 represents how vaccines trigger immune response and prevent humans from any pathogen infection.

3 PROPERTIES OF AN IDEAL VACCINE AND ITS COMPONENTS

An ideal vaccine should stimulate or boost the immune response and prevent pathogen infection and provide immunity from life-threatening diseases. A

TABLE 20.1
Vaccine Types Along With Their Popular Examples and Trade Names.

Sl. no.	Type of Vaccine	Vaccine Name	Disease	Trade Name
1	Live attenuated vaccine	MMR vaccine	Measles mumps rubella	M-M-R II
		Rotavirus vaccine	Rotavirus infection	Rotarix, RotaTeq
		Oral polio vaccine	Poliomyelitis	Poliovax
		Influenza vaccine (nasal spray)	Influenza	FluMist
		FluMist	Influenza	FluMist Quadrivalent
		Varicella vaccine	Varicella (chickenpox)	Varivax, ACAM2000
		Shingles vaccine	Herpes zoster	Zostavax
		Yellow fever vaccine	Yellow fever	YF-Vax
		Adenovirus oral vaccine	Adenovirus infection	Adenovirus Type 4 Type 7
		BCG vaccine	Tuberculosis	TheraCys, TICE BCG
		Cholera vaccine	Cholera	Vaxchora, Dukoral
		Typhoid vaccine	Typhoid	Vivotif
2	Inactivated vaccine	Polio vaccine (IPV)	Poliomyelitis	Ipol
		Hepatitis A vaccine	Hepatitis A	Havrix, Vaqta
		Flu (shot only)	Influenza	Afluria, Fluarix, Fluzone
		Rabies vaccine	Rabies	RabAvert, Imovax
		Japanese encephalitis vaccine	Japanese encephalitis	Ixiaro
		Anthrax vaccine	Anthrax	BioThrax
3	Toxoid vaccine	Diphtheria vaccine	Diphtheria	Daptacel, Infanrix
		Tetanus vaccine	Tetanus	Boostrix, Adacel
		Pertussis vaccine	Whooping cough	Tenivac
4	Subunit, recombinant, polysaccharide, and conjugate vaccines	Hepatitis B vaccine	Hepatitis B	Engerix-B, Recombivax HB
		Influenza (injection)	Influenza	Flublok
		Hib vaccine	Haemophilus influenzae type b	ActHIB, Hiberix, PedvaxHIB
		Pertussis vaccine	Whooping cough	Tenivac
		Pneumococcal vaccine	S. pneumonia	Pneumovax 23, Prevnar 13
		Meningococcal vaccine	Meningococcemia	Trumenba, Bexsero
		HPV	Human papillomavirus	Gardasil 9
		PPV	S. pneumonia	Pneumovax 23
		Shingles vaccine	Herpes zoster	Shingrix
5	DNA-based vaccine	Hepatitis B vaccine	Hepatitis B	Engerix-B, Recombivax HB

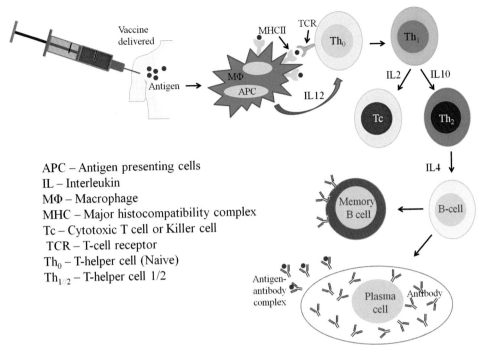

FIG. 20.1 A workflow depicting how vaccine works.

APC – Antigen presenting cells
IL – Interleukin
MΦ – Macrophage
MHC – Major histocompatibility complex
Tc – Cytotoxic T cell or Killer cell
TCR – T-cell receptor
Th_0 – T-helper cell (Naive)
$Th_{1/2}$ – T-helper cell 1/2

perfect vaccine should be safe (does not harm a healthy human), protective (prevent from pathogen infection), competent, target-specific, capable of invoking an immune response, and provide long-lasting protection, easily deliverable, cost-effective, stable at high temperature, multivalent, and high immunogenicity invoking both the humoral and cell-mediated immune responses and should not induce hypersensitivity or autoimmunity.

Mostly a subunit vaccine is made up of epitope/s, linker/s, and adjuvant that are also known as essential components of an ideal vaccine. Epitope or antigenic determinant is a molecular region or part of an antigen or foreign protein that is capable of triggering or stimulates an immune response (Liang, 1998). It interacts with a specific antigen receptor protein of the B cell surface and leads to antibody production. An adjuvant is an essential ingredient of vaccines that is attached to the antigen region and enhances the immune response (Pasquale et al., 2015). Linkers are short and flexible amino acid sequences which join B- and T cell epitopes to form an ideal biological/vaccine construct (Srivastava et al., 2018).

4 BIOINFORMATICS IN VACCINE DESIGN

Bioinformatics plays a vital role in drug discovery and vaccine design program. Bioinformatics or computational biology includes structural bioinformatics (SB), sequence analysis, pathway analysis, next-generation sequencing, proteomics, genomics, big data analysis, artificial intelligence, structural biology, algorithms, etc. Among all, SB is one of the vital sections of bioinformatics, which explores the three-dimensional structure of biomolecules such as proteins, lipids, and nucleic acids (Altman & Dugan, 2005; Bourne & Weissig, 2003). Besides, the SB helps in understanding the protein structure information such as secondary structural components (α-helices and β-sheets), tertiary structure, protein domain, active and binding site information, protein folding, the relation between 3D structure and amino acids sequence, etc. Besides, SB also deals with sequence analysis (protein/nucleic acid), structure prediction via molecular modeling, molecular docking (computer-aided drug design), molecular dynamic (MD) simulations, protein–protein interactions (PPIs), network analysis, etc. (Kapetanovic, 2008; Sliwoski et al., 2014).

SB plays an essential role in the modern vaccine design and development program. Computational prediction of the essential elements (B and T epitopes, suitable linkers, and adjuvant) of an ideal vaccine that is capable of stimulating the human immune response which eliminates the pathogen infection is known as a biological construct or vaccine candidate, and this computational prediction process is called as vaccine design (Davies & Flower, 2007; Flower, 2009; Zagursky & Russell, 2001). In general, vaccine design and development takes around 5–10 years, but with the help of computational approaches, the timeline downs to 1–2 years, and this is the beauty of bioinformatics. The following sections explain how computational tools could be used in the designing of a subunit vaccine.

5 COMPUTATIONAL DESIGN OF VACCINE CONSTRUCTS

In silico design of vaccine is a step-by-step process which includes potential antigen (target) selection from the pathogen, B- and T cell epitope prediction, linker and adjuvant selection, vaccine construct design, computational prediction of antigenicity, allergenicity, toxicity, solubility and stability, 3D structure prediction of the construct, protein (vaccine)–protein (Toll-like receptors [TLRs]) interactions, and protein–protein stability analysis (Fig. 20.2) (Ada, 1991). The aforementioned steps are essential for an ideal vaccine design process and are discussed below.

5.1 Antigen Selection

The whole vaccine design process depends upon the identification and selection of a potential vaccine target or antigen (Zagursky & Russell, 2001). In other words, the antigen selection is a primary, crucial, and tough step in vaccine development. The antigen needs to be a unique and essential component of the microorganism/pathogen that should have the capability of inducing a defensive immune response in humans (Huang et al., 2017; Monterrubio-López et al., 2015). A suitable target should have unique properties such as an effective antigen present at the cell surface, secretary or membrane protein, less sequence similarity with any of the human protein, less cross-reactivity, and excellent adhesive property (Monterrubio-López

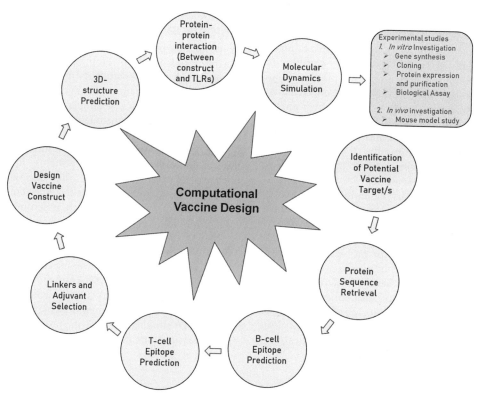

FIG. 20.2 Computational vaccine design: a step-by-step process.

et al., 2015). Identification of potential antigens among thousands of vaccine targets in a particular pathogen is not easy, and there is no straightforward method for choosing potential antigens with specific immune functions. The best way to select potential antigens is a literature survey. Besides, few protein databases such as LOCATE (http://locate.imb.uq.edu.au/), LocDB (https://www.rostlab.org/services/locDB/), TargetP (http://www.cbs.dtu.dk/services/TargetP/), CELLO (http://cello.life.nctu.edu.tw/), and eSLDB (http://gpcr.biocomp.unibo.it/esldb/) are available, which helps in providing information about protein subcellular localization (María et al., 2017).

5.2 Epitope Prediction

An epitope is composed of 7−10 amino acids and is a part of a foreign protein which is capable of triggering the immune response and producing antibody against the specific pathogen infection. Generally, it's of two types: one is continuous epitope (amino acids sequence is linear) and another is discontinuous epitope (appear when protein is folded and the sequence of amino acid is not linear). Nowadays, epitope-based vaccine design is a practical approach due to its successful outcomes, such as EMD640744 (cancer vaccine), EP HIV-1090 (HIV vaccine), Lu AF20513 (Alzheimer's disease), etc., which are in phase 1 clinical trials (Davtyan et al., 2013; Nilvebrant & Rockberg, 2018; Wilson et al., 2008).

In vaccine design, the epitope prediction is an essential and crucial step as the presence of the potential epitope leads to stimulation of the immune response after interacting with the antigen-specific receptor on the surface of B cell or T cell. For the interaction, the structures of epitope and receptor should be complementary to each other (Liang, 1998; Sanchez-Trincado et al., 2017). The B cell epitope interacts with immunoglobulin, and the T cell epitope binds with MHC, CD4, and CD8. An ideal subunit vaccine should have B and T cell (Tc and Th) epitopes, which could initiate an effective response against the pathogen (Patronov & Doytchinova, 2013; Regenmortel, 2009).

The B cells are the essential elements of the adaptive immune system as they are capable of providing long-term defense against microorganisms (Cornaby et al., 2015). After interacting with a specific antigen, the cell surface of the B cell produces an antigen-specific antibody (unique antibody), called B cell epitope (Regenmortel, 2009; Sanchez-Trincado et al., 2017; Srivastava et al., 2018). Generally, crystallography, ELISA, and peptide chip methods are being used for epitope determination/identification, which is expensive and time-consuming. The scientists from

bioinformatics and computational biology background have developed sequence-based computational algorithms that could help in predicting B cell epitope with high accuracy. Few examples which are routinely used for B cell epitope predictions are BepiPred (http://www.cbs.dtu.dk/services/BepiPred/), IBDB (http://tools.iedb.org/main/bcell/), Disco Tope (http://www.cbs.dtu.dk/services/DiscoTope/), LBtope (http://crdd.osdd.net/raghava/lbtope/), ABCpred (http://crdd.osdd.net/raghava/abcpred/), SEPPA (http://www.badd-cao.net/seppa3/index.html), CBtope (http://crdd.osdd.net/raghava/cbtope/), Epitopia Server (http://epitopia.tau.ac.il/), iBCE-EL (http://www.thegleelab.org/iBCE-EL/), etc. (El-Manzalawy & Honavar, 2010; Kringelum et al., 2013; Potocnakova et al., 2016; Regenmortel, 2009; Sanchez-Trincado et al., 2017; Yang & Yu, 2009).

Similarly, the T cell epitope, which includes cytotoxic T lymphocyte (CTL) and helper T lymphocyte (HTL), is responsible for triggering T cell response (Liang, 1998; Patronov & Doytchinova, 2013; Sanchez-Trincado et al., 2017). T cell epitopes are identified via costly and time-consuming experimental methods that determine epitope interaction with MHC. Bioinformaticians developed several computational methods such as sequence motifs, QSAR, artificial neural networks, support vector machines, quantitative matrices, and molecular docking simulations, which could help in predicting T cell epitopes with high accuracy. Few publicly available T cell epitope (MHC I and MHC II) prediction tools are EpiJen (http://www.ddg-pharmfac.net/epijen/EpiJen/EpiJen.htm), NetTepi (http://www.cbs.dtu.dk/services/NetTepi/), IEDB (https://www.iedb.org/), SYFPEITHI (http://www.syfpeithi.de/), Propred (http://crdd.osdd.net/raghava/propred/), NetMHC (http://www.cbs.dtu.dk/services/NetMHC/), MMBPred (https://bio.tools/mmbpred), MHC2Pred (http://crdd.osdd.net/raghava/mhc2pred/), NetMHCcons (http://www.cbs.dtu.dk/services/NetMHCcons/), TepiTool (http://tools.iedb.org/tepitool/), NetMHCIIpan (http://www.cbs.dtu.dk/services/NetMHCIIpan-4.0/), and Rankpep (http://imed.med.ucm.es/Tools/rankpep.html); whereas CTL could be predicated using CTL-Pred (http://crdd.osdd.net/raghava/ctlpred/), PAComplex (http://pacomplex.life.nctu.edu.tw/), NetCTL (http://www.cbs.dtu.dk/services/NetCTL/), etc. (Desai & Kulkarni-Kale, 2014; Lafuente & Reche, 2009; Salimi et al., 2010; Sanchez-Trincado et al., 2017; Yang & Yu, 2009).

The selection of B and T cell epitopes is dependent upon the software/s. Computational tools will be having all the information like how the epitope will be predicated, algorithm details, user-defined parameters, how to select the epitope, etc., so read the tutorial and choose the epitopes carefully.

5.3 Linkers Selection

The linker is a small flexible amino acid sequence (two to five), which connects predicted B and T cell epitopes (Reddy Chichili et al., 2013). Few experimentally proved linkers such as KK, AAY, G_4S, GGGS, KFERQ, EAAAK, GPGPG, and ACELGT are available, which are being used in vaccine designing (Gallagher et al., 2019; Nezafat et al., 2017). To date, no such bioinformatics tools exist for linker prediction; only one linker database VU (IBIVU) (http://www.ibi.vu.nl/programs/linkerdbwww/) is available online where experimentally reported linkers information is present.

5.4 Adjuvant

An adjuvant is a key component of any vaccine, which helps in stimulating and enhancing the immune response (Gupta & Siber, 1995). Also, adjuvant increases the immunogenicity of a weak or poor antigen, accelerates the immune response, persuades cell-mediated immunity, and modulates humoral response (Marciani, 2003; Mastelic et al., 2010). Generally, an adjuvant is mixed with the purified protein/antigen before infecting the model organism. Examples of few experimentally proven adjuvants are AS01 (Shingrix), monophosphoryl lipid A (hepatitis B), aluminum (hepatitis A and B, anthrax, IPV, HPV), virosomes (hepatitis and influenza), AS03 (influenza—pandemic), AS04 (HPV, hepatitis B), CpG 1018 (Heplisav-B), MF59 (influenza—seasonal and pandemic), thermoreversible oil-in-water (influenza—pandemic), ISA51 (lung cancer), etc. (Gupta & Siber, 1995; Olafsdottir et al., 2015; Pasquale et al., 2015). Moreover, saponins (steroid molecule which activates immune response), minerals (aluminum salt that stimulates Th2 response), synthetic products (helps in pattern recognition receptors [PRRs] and TLR receptor activation), microbial products (polysaccharide chain induced immune reaction), oil emulsion (persuade Th2 immune response and slow release effect of antigen), and cytokine (interferon and interleukin; released by an immune cell to activate each other) could also be used as adjuvants in the vaccine development process (Marciani, 2003; Mastelic et al., 2010; Pasquale et al., 2015).

Recently, it was shown that an adjuvant sequence might be joined at the beginning of the vaccine construct. These are the peptide sequences known to elicit the immune response, and therefore, the addition of these sequences during vaccine construct design leads to self-adjuvant vaccine construct (Rudra et al., 2010; Xin et al., 2012). Similar to linker selection, only experimentally reported adjuvant sequences could be used in vaccine design as there are no computational tools or standard recipes available for adjuvant selection (Marciani, 2003; Olafsdottir et al., 2015). Scientists should study and then select proper adjuvant for their designed vaccine construct for better outcomes.

5.5 Design of Vaccine Construct/s

After the successful selection of antigen, B- and T cell (CTL and HTL) epitopes, linkers and adjuvant, and the amino acid sequence will be arranged appropriately (by randomly placing epitopes and joining them using appropriate linkers) that leads to design a potential vaccine construct (Lafuente & Reche, 2009; Leroux-Roels et al., 2011; Monterrubio-López et al., 2015; Potocnakova et al., 2016; Reddy Chichili et al., 2013). In other words, vaccine construct design is a crucial and essential step where the principal components (epitope/s, linker/s, and adjuvant) will be organized to form a protein sequence (Rueckert & Guzmán, 2012). For vaccine design, no such standards for arranging epitopes and linkers in a particular order are there. It is essential that only adjuvant be placed at N-terminal end, whereas epitopes and linkers could be added at any place, and several constructs may also be designed by placing the epitope/s in different combinations (by changing epitope and linker positions) (Fig. 20.3). Next, the designed construct should be checked for their physicochemical properties.

6 COMPUTATIONAL PREDICTION OF PHYSICOCHEMICAL PROPERTIES OF DESIGNED VACCINE CONSTRUCT/S

The designed vaccine construct should be safe. For this, the designed vaccine construct/s must be checked for their antigenicity, toxicity, allergenicity, solubility, and stability via computational tools because the designed construct should be a suitable antigen, trigger, and stimulate the immune response, nonallergic, nontoxic, highly stable, and soluble as these properties make it an appropriate vaccine candidate for detailed study and further investigations.

6.1 Antigenicity Prediction

The designed construct should be a pathogen-specific antigen and needs to show higher antigenicity, which is required to provoke a protective immune response against the pathogen (Shey et al., 2019). In the absence of suitable and potential antigen, the vaccine construct

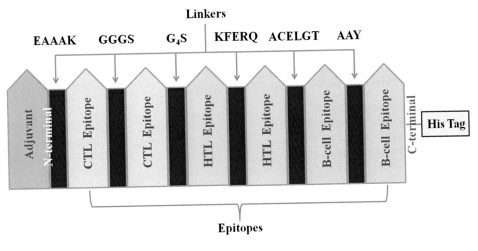

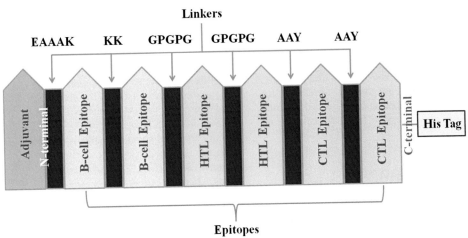

FIG. 20.3 Schematic representation of two general multiepitope vaccines constructs.

will be nonantigenic and futile. Bioinformatics tools such as ANTIGENPro (http://scratch.proteomics.ics.uci.edu/), Protegen (http://www.violinet.org/protegen/), VaxiJen (http://www.ddg-pharmfac.net/vaxijen/VaxiJen/VaxiJen.html), EpiToolKit (https://bio.tools/epitoolkit), SVMTriP (http://sysbio.unl.edu/SVMTriP/), etc., are broadly used for antigenicity predictions (Magnan et al., 2010; María et al., 2017).

6.2 Allergenicity Prediction

It is essential to check that the newly designed construct is safe and does not cause allergic reactions after vaccination in a healthy human. Any essential component of vaccines such as epitopes, linkers, and adjuvant should not lead to creating any type of allergy, as all of them are capable of triggering an allergic reaction.

The designed construct should process through computational tools such as AllerTOP (https://www.ddg-pharmfac.net/AllerTOP/), AlgPred (http://crdd.osdd.net/raghava/algpred/), AllergenFP (https://ddg-pharmfac.net/AllergenFP/), AllerHunter (http://tiger.dbs.nus.edu.sg/AllerHunter), Allerdictor (http://allerdictor.vbi.vt.edu/), etc., or their allergenicity prediction (Dimitrov et al., 2013; Muh et al., 2009) to make sure that the vaccine construct is nonallergic.

6.3 Toxicity Prediction

Toxicity is the study of the adverse effects of any chemical/natural/synthesized component on human health. In other words, any chemical or physical agent which causes toxic effect in a living organism that may lead to death of the organism is considered toxic. The designed vaccine construct/s (which synthesized) should not show any toxic effect or side effect on human health. ToxinPred, a widely used bioinformatics tool, is developed to differentiate between toxic and nontoxic peptides through machine learning technique and quantitative matrix (Gupta et al., 2013).

6.4 Solubility and Stability Prediction

The designed vaccine construct needs to be soluble and stable (Dar et al., 2019). Computational tools for solubility prediction are ProsoII, SODA, CamSol, Protein-sol, etc., and the publicly available ProtParam for stability prediction (Hebditch et al., 2017; Smialowski et al., 2012).

Each computational tool has their own standard value to determine the antigenicity, allergenicity, toxicity, stability, and solubility. The computational tools which help in physicochemical properties predications use experimental datasets to determine whether the designed vaccine construct is allergen or nonallergen/antigenic or nonantigenic/toxic or nontoxic/stable or unstable/soluble or insoluble. Once the novel vaccine constructs are designed, checked for their physicochemical properties, they are further processed for structure prediction and docking studies to determine its suitability for development (Soria-Guerra et al., 2015).

7 STRUCTURE PREDICTION, PROTEIN–PROTEIN INTERACTIONS, AND COMPLEX STABILITY ANALYSIS

Till now, the vaccine sequence is designed, and the sequence-based screening was done. Furthermore, the 3D structure of the protein/construct should be predicted as structure helps in predicting its function. Molecular modeling, a powerful computational technique, helps in the prediction of a 3D dimensional structure of the designed vaccine construct (Kalita et al., 2019; Kalita et al., 2020; Qin et al., 2010), which provides a clear insight to the molecule behavior and structure. Later, the predicted structure should be optimized and validated through Galaxy Refine Server and Ramachandran plot analysis, respectively. I-TASSER, an online tool, could also help in 3D structure determination.

Next, the designed vaccine constructs should be docked with TLRs using the PPI/docking approach (an essential event in several biological activities, including immune system response) to see binding between them (Bakail & Ochsenbein, 2016; Mooney et al., 2013).

Molecular docking is a computational program that is used to predict the interactions between biomolecules

(protein–protein/ligand/nucleic acids) (Morris & Lim-Wilby, 2008). Through docking investigations, binding energy, hydrophobic and hydrogen bond, electrostatic, Van der Waals interactions, etc., could be determined that helps in deciding how tightly the biomolecules interacted with each other (Ferreira et al., 2015; Pagadala et al., 2017).

TLRs localized on the immune cells such as neutrophils, macrophages, dendritic cells, B lymphocytes, mast cells, endothelium, eosinophils, epithelial cells, etc. (Vasselon, 2002), and 10 human TLRs (TLR 1 to TLR 10; Mol. Wt. of 90–115 kDa) are identified. The TLRs represent the first-line defense against the pathogen, which identify and respond against the conserved region (peptide sequence) of the pathogen and stimulate the immune response (Werling & Jungi, 2003). The TLRs play an important role in antibody secretion by sensing conserved molecular patterns for early immune recognition of a pathogen (Cook et al., 2004; De Nardo, 2015; Werling & Jungi, 2003). The TLRs are capable of perceiving pathogen infection using PRRs, which leads to instigating adaptive and innate immune response to fight against the pathogen. Overall, the designed construct (the combination of B- and T cell epitopes) should interact with TLRs to signal antibody production (Norman, 1995; Schjetne et al., 2003). PPI analysis will suggest that how efficiently TLRs bind with the designed construct. If the binding energy between TLRs and the designed vaccine construct/s are more, there is potential interactions with each other that may trigger good immune response. Computational tools such as ZDOCK (http://zdock.umassmed.edu/), HADDOCK (https://alcazar.science.uu.nl/services/HADDOCK2.2/haddockserver-easy.html), ClusPro (https://cluspro.bu.edu), RosettaDOCK (http://rosettadock.graylab.jhu.edu/), etc., are available online, which helps in identifying interactions between TLRs and vaccine construct.

Next, the protein–protein complex (which has an excellent binding score and more interactions) will be subjected to MD studies to check the stability of the protein–protein complex and analyze the stable hydrogen bonds between proteins using MD simulations (Arya & Coumar, 2014; Arya et al., 2018; Martin Karplus & McCammon, 2002). MD is a powerful computational tool that helps to understand the structure and dynamic behavior of biological complexes (Durrant & McCammon, 2011; Karplus & Kuriyan, 2005). MD uses the force field to describe the atomic-level properties of biomolecules and is a physics-based approach where time-dependent conformations of biomolecules or bimolecular complexes will be recorded and analyzed. The ultimate aim of MD in bioinformatics is to study the protein folding, protein stability, protein–protein or protein–ligand interactions, fluctuation, and conformational changes in biomolecule's structure. Fig. 20.4 shows various in silico steps involved in the design and evaluation of a vaccine construct.

Upto now, bioinformatics and computational biology techniques played crucial role in the design of potential vaccine construct/s, and the tools, as mentioned earlier, help in predicting physicochemical properties, which is essential for an ideal vaccine candidate. Next, the designed vaccine construct (with positive bioinformatics analysis) could be synthesized and cloned in the bacterial system (Lessard, 2013). Later, the designed protein (vaccine candidate) needs to be expressed and purified in the laboratory (Růčková et al., 2014). The purified protein should be subjected to biological assay such as enzyme inhibition, effect on pathogen culture, animal model study, etc. Once a

positive in vitro and in vivo results are obtained for a vaccine construct, the vaccine candidate could be progressed to clinical studies.

8 AN OVERVIEW OF SUCCESSFUL VACCINE DESIGN USING BIOINFORMATICS

Several research groups are working on computational vaccinology and try to develop novel vaccine candidate using bioinformatics and later validate the in silico outcomes through experimental studies. Few vaccines constructs such as EMD640744 (cancer vaccine), EP HIV-1090 (HIV vaccine), Lu AF20513 (Alzheimer's disease), etc., are designed and developed using bioinformatics and are now in clinical trials.

Kalita and her research group (Kalita et al., 2019) in 2019 have designed a multiepitope subunit vaccine against a helminth disease fascioliasis, which is caused by *Fasciola gigantica* and *Fasciola hepatica*. In their study, researchers did a literature survey and selected seven

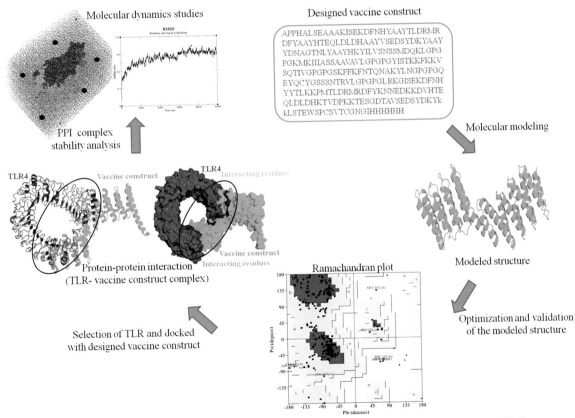

FIG. 20.4 Steps in vaccine evaluation using in silico methods. Molecular modeling, structure validation, protein–protein docking, and molecular dynamics study are used to investigate the potential vaccine construct.

important proteins that are responsible for the parasite survival and retrieved their protein sequence from Uni-Prot database. Later, B cell, CTL, and HTL epitopes were predicted using ABCPred, NetCTL, and IEDB computational servers, respectively, and selected potential epitopes (seven each) based on their scores. The researchers have checked the epitope toxicity using the ToxinPred server and later the predicted epitopes were joined with the help of linkers (35), and an adjuvant (lipoprotein LprA) was added at the N-terminal. Next, the physicochemical properties of the designed vaccine construct were calculated to check its antigenicity and allergenicity, stability, and solubility using bioinformatics tools. Furthermore, the 3D structure of the designed vaccine construct was predicted and docked with TLR-2 (PDB ID: 5D3I) using the ClusPro server, and a complex model that has lower binding energy (more stable) was selected and processed for MD studies to check its stability and binding with TLR-2. DIMPLOT program was used to see the amino acids involved in the interactions between the vaccine construct and TLR-2. Moreover, in silico cloning was done using Java Codon Adaptation Tool to know the expressibility of the vaccine construct (Fig. 20.5). The researcher concluded that the designed construct against *F. gigantica* could be processed for the experimental evaluation of its immunogenic behavior (Kalita et al., 2019).

Recently, the same research group has designed a novel peptide-based corona vaccine (Kalita et al., 2020). Severe acute respiratory syndrome coronavirus 2 (SARS-CoV-2) belongs to the Coronaviridae family and is responsible for the coronavirus disease 2019 (COVID-19). COVID-19 is declared as epidemic and as public health emergency by WHO (Khan & Naushad, 2020). Researchers have sequenced the viral genome and are trying to develop medicine or vaccine against COVID-19. Few drugs such as hydroxychloroquine, chloroquine, and remdesivir are clinically tested against the corona virus infection. According to WHO, more than 140 vaccines for SARS-CoV-2 are under investigation and around 34 vaccines such as ChAdOx1 nCov-19, Adenovirus Type 5 Vector, Gam-COVID-Vac Lyo, mRNA-1273, CoronaVac, BTN162, GRAde-4800, NVX CoV2373, etc., are on human clinical trial (Abraham Peele et al., 2020; Folegatti et al., 2020; Pandey et al., 2020; Smith et al., 2020; Wang et al., 2020; Zhu et al., 2020).

Timir tripathi and his research team have designed peptide-based novel subunit vaccine against SARS-CoV-2. In this, researcher selected three proteins (nucleocapsid protein, membrane glycoprotein, and surface spike glycoprotein) and found that they are antigenic in nature. A total of 33 epitopes (B cells—9, HTLs—6, and CTLs—18) were predicted using computational tools and designed a subunit vaccine construct with the help of linkers. Next, the construct was processed to check its physicochemical properties and found as a potential antigen, nonallergen, and stable in nature. Later, 3D structure of the subunit vaccine construct was predicted using molecular modeling technique and validated the structure through the Ramachandran plot.

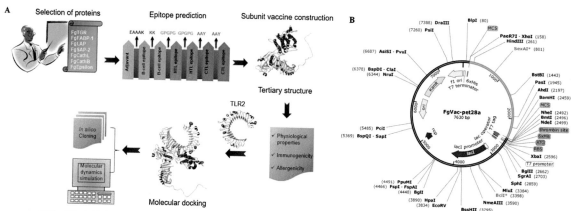

FIG. 20.5 Schematic representation of the multiepitope subunit vaccine candidate designed by Kalita et al. where they have computationally predicted **(A)** B cell, cytotoxic T lymphocyte (CTL), and helper T lymphocyte (HTL) epitopes and performed molecular docking, dynamics simulation, and **(B)** in silico cloning. (This figure is reproduced with permission from Elsevier Kalita, P., Lyngdoh, D. L., Padhi, A. K., Shukla, H., & Tripathi, T. (2019). Development of multi-epitope driven subunit vaccine against *Fasciola gigantica* using immunoinformatics approach. *International Journal of Biological Macromolecules, 138*, 224–233. https://doi.org/10.1016/j.ijbiomac.2019.07.024.)

The vaccine receptor docking was performed to evaluate the binding efficiency (binding energy: −1491 kJ/mol) between the construct and TLR3. The protein−protein complex was further subjected for 40 ns MD studies to check the complex stability using root mean square deviation, root mean square fluctuation, radius of gyration (Rg), hydrogen bond, Principal component analysis (PCA), and contact energy analysis. The MD result suggests that the complex is stable in nature and formed 327 hydrogen bonds (intermolecular interactions) between the vaccine construct and TLR3. In silico cloning was also performed using SnapGene 1.1.3 restriction cloning tool. Computational investigations suggest that the designed multiepitope vaccine construct may show good protective efficacy and safety against SARS-CoV-2 infection, and authors suggest the synthesis and experimental studies of the designed subunit vaccine construct to determine its immunogenic potential (Kalita et al., 2020).

He et al. (2018) in 2018 designed and developed vaccine candidates for diagnostic and therapies for the lectin allergy caused by *Phaseolus vulgaris* L. (black turtle bean). In this investigation, the lectin protein sequence was obtained from NCBI and checked its physicochemical properties and later predicted 3D structure and validated through the Ramachandran plot. Next, seven B cell epitopes and three T cell epitopes were predicted using computational tools. The overlapped epitope regions were clubbed together, and detailed structural analysis led to finalize the potential epitopes selection for further studies. Later, the predicted epitopes were synthesized with >95% purity; an ELISA was performed where four B cell epitopes and one T cell epitope were significantly interacting with IgE. Moreover, cytokine profile and lymphocyte proliferation assay with identified T cell epitopes and the lectin were performed and found that two T cell epitopes and the lectin show significantly higher peripheral blood mononuclear cells proliferation in relation to control. Next, a QSAR study was done to see the relation between potential antigenicity and the physicochemical characteristics of the amino acids, which were present in the B- and T cell epitopes. The QSAR analysis result showed that B- and T cell epitopes were antigenic in nature, which could trigger the immune response. In conclusion, a total of four B cell epitopes and two T cell epitopes were combined and designed a vaccine candidate. In the epitope regions of the lectin, positively charged amino acids with high hydrophobicity were present, and that could be potentially used for diagnostics and treatment for the lectin allergy (He et al., 2018).

Similarly, Shey et al. (2019) have designed a novel vaccine candidate against a parasitic nematode *Onchocerca volvulus*, which causes human onchocerciasis (river blindness). In their investigation, researchers have selected six proteins that were highly expressed in the L3 larval stage of the parasite. With the help of computational tools, scientists have predicted linear B cell epitopes, CTL, and HTL. Later, they designed a multiepitope vaccine candidate sequence and joined via linkers and added an adjuvant (TLR-4 agonist) and then predicted their physicochemical properties. With the successful outcomes, the 3D structure of the vaccine candidate was predicted and docked with TLR-4 to show that the construct was potentially bound with TLR-4 and could trigger an immune response against the pathogen. Later, they also checked the protein−protein complex (vaccine construct−TLR-4) stability via MD studies and computational cloning optimization was also performed. Furthermore, it was predicted to show immune stimulation with significantly high levels of IgG_1, HTL, CTL, INF-γ, and IL-2. The engineered recombinant peptide showed better antigenicity than the current vaccine candidates and could also be used for other neglected tropical disease control programs such as lymphatic filariasis, loiasis, etc. (Shey et al., 2019).

9 CONCLUSION

Bioinformatics is being extensively used in most of the life science laboratories, which deals with proteomics, genomics, artificial intelligence, next-generation sequencing, RNA-seq analysis, drug discovery, vaccine development, big data analysis, structural biology, etc. (Flower, 2009; Huang et al., 2010). The focus of the book chapter is to brief the role of bioinformatics in the subunit vaccine construct design. To design a biological construct with the essential components such as epitopes (B and T cells), linkers and adjuvant (with the help of computational tools), which is capable of triggering an immune response to eliminate the pathogen infection, is called a vaccine design. The book chapter provides information about the basics of how vaccines work, components of a potential subunit vaccine (B cell, T cell, TLRs, linkers, and adjuvants), the available computational tools for the prediction of B cell and T cell epitopes, 3D structure prediction tools, PPI techniques, and MD studies. Also few successful case studies are provided in this book chapter. The successful outcome of these computational tools in vaccine design saves time, money, and manpower.

ACKNOWLEDGMENT

Hemant Arya thanks the Indian Council of Medical Research (ICMR), Government of India, for the Research Associate Fellowship (Project ID: 2019-6285; File No. ISRM/11(35)/2019).

REFERENCES

Abraham Peele, K., Srihansa, T., Krupanidhi, S., Ayyagari, V. S., & Venkateswarulu, T. C. (2020). Design of multi-epitope vaccine candidate against SARS-CoV-2: A in-silico study. *Journal of Biomolecular Structure and Dynamics*, 1–9. https://doi.org/10.1080/07391102.2020.1770127

Ada, G. L. (1991). The ideal vaccine. *World Journal of Microbiology and Biotechnology*, 7(2), 105–109. https://doi.org/10.1007/BF00328978

Altman, R. B., & Dugan, J. M. (2005). Defining bioinformatics and structural bioinformatics. In *Structural bioinformatics* (pp. 1–14). https://doi.org/10.1002/0471721204.ch1

Arya, H., & Coumar, M. S. M. S. (2014). Virtual screening of Traditional Chinese Medicine (TCM) database: Identification of fragment-like lead molecules for filariasis target asparaginyl-tRNA synthetase. *Journal of Molecular Modeling*, 20(6), 2266. https://doi.org/10.1007/s00894-014-2266-9

Arya, H., Syed, S. B., Singh, S. S., Ampasala, D. R., & Coumar, M. S. (2018). In silico investigations of chemical constituents of clerodendrum colebrookianum in the anti-hypertensive drug targets: ROCK, ACE, and PDE5. *Interdisciplinary Sciences*, 10(4), 792–804. https://doi.org/10.1007/s12539-017-0243-6

Bakail, M., & Ochsenbein, F. (2016). Targeting protein–protein interactions, a wide open field for drug design. *Comptes Rendus Chimie*, 19(1–2), 19–27. https://doi.org/10.1016/j.crci.2015.12.004

Bartlett, B. L., Pellicane, A. J., & Tyring, S. K. (2009). Vaccine immunology. *Dermatologic Therapy*, 22(2), 104–109. https://doi.org/10.1111/j.1529-8019.2009.01223.x

Bhamarapravati, N., & Sutee, Y. (2000). Live attenuated tetravalent dengue vaccine. *Vaccine*, 18(Suppl. 2), 44–47. https://doi.org/10.1016/s0264-410x(00)00040-2

Bonten, M. J. M., Huijts, S. M., Bolkenbaas, M., Webber, C., Patterson, S., Gault, S., & Grobbee, D. E. (2015). Polysaccharide conjugate vaccine against pneumococcal pneumonia in adults. *New England Journal of Medicine*, 372(12), 1114–1125. https://doi.org/10.1056/NEJMoa1408544

Bourne, P. E., & Weissig, H. (2003). *Structural bioinformatics*. Hoboken, NJ, USA: John Wiley & Sons, Inc. https://doi.org/10.1002/0471721204

Cook, D. N., Pisetsky, D. S., & Schwartz, D. A. (2004). Toll-like receptors in the pathogenesis of human disease. *Nature Immunology*, 5(10), 975–979. https://doi.org/10.1038/ni1116

Cornaby, C., Gibbons, L., Mayhew, V., Sloan, C. S., Welling, A., & Poole, B. D. (2015). B cell epitope spreading: Mechanisms and contribution to autoimmune diseases. *Immunology Letters*, 163(1), 56–68. https://doi.org/10.1016/j.imlet.2014.11.001

Dar, H. A., Zaheer, T., Shehroz, M., Ullah, N., Naz, K., Muhammad, S. A., Zhang, T., & Ali, A. (2019). Immunoinformatics-aided design and evaluation of a potential multi-epitope vaccine against *Klebsiella pneumoniae*. *Vaccines*, 7(3), 88. https://doi.org/10.3390/vaccines7030088

Davies, M., & Flower, D. (2007). Harnessing bioinformatics to discover new vaccines. *Drug Discovery Today*, 12(9–10), 389–395. https://doi.org/10.1016/j.drudis.2007.03.010

Davtyan, H., Ghochikyan, A., Petrushina, I., Hovakimyan, A., Davtyan, A., Poghosyan, A., Marleau, A. M., Movsesyan, N., Kiyatkin, A., Rasool, S., & Larsen, A. K. (2013). Immunogenicity, efficacy, safety, and mechanism of action of epitope vaccine (Lu AF20513) for Alzheimer's disease: Prelude to a clinical trial. *Journal of Neuroscience*, 33(11), 4923–4934. https://doi.org/10.1523/JNEUROSCI.4672-12.2013

De Nardo, D. (2015). Toll-like receptors: Activation, signalling and transcriptional modulation. *Cytokine*, 74(2), 181–189. https://doi.org/10.1016/j.cyto.2015.02.025

Desai, D. V., & Kulkarni-Kale, U. (2014). T-cell epitope prediction methods: An overview. *Methods in Molecular Biology*, 333–364. https://doi.org/10.1007/978-1-4939-1115-8_19

Dimitrov, I., Flower, D. R., & Doytchinova, I. (2013). AllerTOP - a server for in silico prediction of allergens. *BMC Bioinformatics*, 14(Suppl. 6), S4. https://doi.org/10.1186/1471-2105-14-S6-S4

Durrant, J. D., & McCammon, J. A. (2011). Molecular dynamics simulations and drug discovery. *BMC Biology*, 9(1), 71. https://doi.org/10.1186/1741-7007-9-71

El-Manzalawy, Y., & Honavar, V. (2010). Recent advances in B-cell epitope prediction methods. *Journal of Immunology Research*, 6(Suppl. 2), S2. https://doi.org/10.1186/1745-7580-6-S2-S2

Ferraro, B., Morrow, M. P., Hutnick, N. A., Shin, T. H., Lucke, C. E., & Weiner, D. B. (2011). Clinical applications of DNA vaccines: Current progress. *Clinical Infectious Diseases*, 53(3), 296–302. https://doi.org/10.1093/cid/cir334

Ferreira, L. G., Dos Santos, R. N., Oliva, G., & Andricopulo, A. D. (2015). Molecular docking and structure-based drug design strategies. *Molecules*, 20(7), 13384–13421. https://doi.org/10.3390/molecules200713384

Flower, D. R. (2009). *Bioinformatics for vaccinology*. Chichester, UK: John Wiley & Sons, Ltd. https://doi.org/10.1002/9780470699836

Folegatti, P. M., Ewer, K. J., Aley, P. K., Angus, B., Becker, S., Belij-Rammerstorfer, S., & Oxford COVID Vaccine Trial Group. (2020). Safety and immunogenicity of the ChAdOx1 nCoV-19 vaccine against SARS-CoV-2: A preliminary report of a phase 1/2, single-blind, randomised controlled trial. *Lancet*, 396(10249), 467–478. https://doi.org/10.1016/S0140-6736(20)31604-4

Gallagher, T. B., Mellado-Sanchez, G., Jorgensen, A. L., Moore, S., Nataro, J. P., Pasetti, M. F., & Baillie, L. W. (2019). Development of a multiple-antigen protein fusion vaccine candidate that confers protection against *Bacillus anthracis* and *Yersinia pestis*. *PLoS Neglected Tropical Diseases*, 13(8), e0007644. https://doi.org/10.1371/journal.pntd.0007644

Giese, M. (2016). *Introduction to molecular vaccinology*. Cham: Springer International Publishing. https://doi.org/10.1007/978-3-319-25832-4

Gupta, S., Kapoor, P., Chaudhary, K., Gautam, A., Kumar, R., & Raghava, G. P. S. (2013). In silico approach for predicting toxicity of peptides and proteins. *PLoS One*, 8(9), e73957. https://doi.org/10.1371/journal.pone.0073957

Gupta, R. K., & Siber, G. R. (1995). Adjuvants for human vaccines—current status, problems and future prospects. *Vaccine*, 13(14), 1263—1276. https://doi.org/10.1016/0264-410X(95)00011-O

Hebditch, M., Carballo-Amador, M. A., Charonis, S., Curtis, R., & Warwicker, J. (2017). Protein—sol: A web tool for predicting protein solubility from sequence. *Bioinformatics*, 33(19), 3098—3100. https://doi.org/10.1093/bioinformatics/btx345

He, Y., Rappuoli, R., De Groot, A. S., & Chen, R. T. (2010). Emerging vaccine informatics. *Journal of Biomedicine and Biotechnology*, 2010, 1—26. https://doi.org/10.1155/2010/218590

He, S., Zhao, J., Elfalleh, W., Jemaà, M., Sun, H., Sun, X., & Lang, F. (2018). In silico identification and in vitro analysis of B and T-cell epitopes of the black turtle bean (*Phaseolus vulgaris* L.) lectin. *Cellular Physiology and Biochemistry*, 49(4), 1600—1614. https://doi.org/10.1159/000493496

Huang, Z., Wang, X., & Liu, Q. (2017). Vaccine development. In *The norovirus* (pp. 187—206). Elsevier. https://doi.org/10.1016/B978-0-12-804177-2.00013-0.

Huang, H.-J., Yu, H. W., Chen, C.-Y., Hsu, C.-H., Chen, H.-Y., Lee, K.-J., & Chen, C. Y.-C. (2010). Current developments of computer-aided drug design. *Journal of the Taiwan Institute of Chemical Engineers*, 41(6), 623—635. https://doi.org/10.1016/j.jtice.2010.03.017

Kalita, P., Lyngdoh, D. L., Padhi, A. K., Shukla, H., & Tripathi, T. (2019). Development of multi-epitope driven subunit vaccine against *Fasciola gigantica* using immunoinformatics approach. *International Journal of Biological Macromolecules*, 138, 224—233. https://doi.org/10.1016/j.ijbiomac.2019.07.024

Kalita, P., Padhi, A. K., Zhang, K. Y. J., & Tripathi, T. (2020). Design of a peptide-based subunit vaccine against novel coronavirus SARS-CoV-2. *Microbial Pathogenesis*, 145, 104236. https://doi.org/10.1016/j.micpath.2020.104236

Kapetanovic, I. M. (2008). Computer-aided drug discovery and development (CADDD): In silico-chemico-biological approach. *Chemico-Biological Interactions*, 171(2), 165—176. https://doi.org/10.1016/j.cbi.2006.12.006

Karplus, M., & Kuriyan, J. (2005). Molecular dynamics and protein function. *Proceedings of the National Academy of Sciences of the United States of America*, 102(19), 6679—6685. https://doi.org/10.1073/pnas.0408930102

Karplus, M., & McCammon, J. A. (2002). Molecular dynamics simulations of biomolecules. *Nature Structural and Molecular Biology*, 9(9), 646—652. https://doi.org/10.1038/nsb0902-646

Kennedy, M. A. (2010). A brief review of the basics of immunology: The innate and adaptive response. *Veterinary Clinics of North America: Small Animal Practice*, 40(3), 369—379. https://doi.org/10.1016/j.cvsm.2010.01.003

Khan, N., & Naushad, M. (2020). Effects of corona virus on the world community. *SSRN Electronic Journal*. https://doi.org/10.2139/ssrn.3532001

Kringelum, J. V., Nielsen, M., Padkjær, S. B., & Lund, O. (2013). Structural analysis of B-cell epitopes in antibody:protein complexes. *Molecular Immunology*, 53(1—2), 24—34. https://doi.org/10.1016/j.molimm.2012.06.001

Lafuente, E., & Reche, P. (2009). Prediction of MHC-peptide binding: A systematic and comprehensive overview. *Current Pharmaceutical Design*, 15(28), 3209—3220. https://doi.org/10.2174/138161209789105162

Leroux-Roels, G., Bonanni, P., Tantawichien, T., & Zepp, F. (2011). Vaccine development. *Perspectives in Vaccinology*, 1(1), 115—150. https://doi.org/10.1016/j.pervac.2011.05.005

Lessard, J. C. (2013). Molecular cloning. In *Methods enzymol* (pp. 85—98). https://doi.org/10.1016/B978-0-12-418687-3.00007-0

Liang, T. C. (1998). Epitopes. In *Encyclopedia of immunology* (pp. 825—827). Elsevier. https://doi.org/10.1006/rwei.1999.0219.

Magnan, C. N., Zeller, M., Kayala, M. A., Vigil, A., Randall, A., Felgner, P. L., & Baldi, P. (2010). High-throughput prediction of protein antigenicity using protein microarray data. *Bioinformatics*, 26(23), 2936—2943. https://doi.org/10.1093/bioinformatics/btq551

Marciani, D. J. (2003). Vaccine adjuvants: Role and mechanisms of action in vaccine immunogenicity. *Drug Discovery Today*, 8(20), 934—943. https://doi.org/10.1016/S1359-6446(03)02864-2

María, R. R., Arturo, C. J., Alicia, J. A., Paulina, M. G., & Gerardo, A. O. (2017). The impact of bioinformatics on vaccine design and development. In *Vaccines*. InTech. https://doi.org/10.5772/intechopen.69273.

Mastelic, B., Ahmed, S., Egan, W. M., Del Giudice, G., Golding, H., Gust, I., & Lambert, P.-H. (2010). Mode of action of adjuvants: Implications for vaccine safety and design. *Biologicals*, 38(5), 594—601. https://doi.org/10.1016/j.biologicals.2010.06.002

Monterrubio-López, G. P., González-Y-Merchand, J. A., & Ribas-Aparicio, R. M. (2015). Identification of novel potential vaccine candidates against tuberculosis based on reverse vaccinology. *Biomed Research International*, 2015, 1—16. https://doi.org/10.1155/2015/483150

Mooney, M., McWeeney, S., Canderan, G., & Sékaly, R.-P. (2013). A systems framework for vaccine design. *Current Opinion in Immunology*, 25(5), 551—555. https://doi.org/10.1016/j.coi.2013.09.014

Morris, G. M., & Lim-Wilby, M. (2008). Molecular docking. In *Methods mol biol* (pp. 365—382). https://doi.org/10.1007/978-1-59745-177-2_19

Muh, H. C., Tong, J. C., & Tammi, M. T. (2009). AllerHunter: A SVM-pairwise system for assessment of allergenicity and allergic cross-reactivity in proteins. *PLoS One*, 4(6), e5861. https://doi.org/10.1371/journal.pone.0005861

Nezafat, N., Eslami, M., Negahdaripour, M., Rahbar, M. R., & Ghasemi, Y. (2017). Designing an efficient multi-epitope oral vaccine against *Helicobacter pylori* using immunoinformatics and structural vaccinology approaches. *Molecular BioSystems*, 13(4), 699—713. https://doi.org/10.1039/C6MB00772D

Nilvebrant, J., & Rockberg, J. (2018). An introduction to epitope mapping. In *Methods mol biol* (pp. 1—10). https://doi.org/10.1007/978-1-4939-7841-0_1

Norman, P. (1995). Immunobiology: The immune system in health and disease. *The Journal of Allergy and Clinical Immunology, 96*(2), 274. https://doi.org/10.1016/S0091-6749(95)70025-0

Olafsdottir, T., Lindqvist, M., & Harandi, A. M. (2015). Molecular signatures of vaccine adjuvants. *Vaccine, 33*(40), 5302–5307. https://doi.org/10.1016/j.vaccine.2015.04.099

Pagadala, N. S., Syed, K., & Tuszynski, J. (2017). Software for molecular docking: A review. *Biophysical Reviews, 9*(2), 91–102. https://doi.org/10.1007/s12551-016-0247-1

Pandey, S. C., Pande, V., Sati, D., Upreti, S., & Samant, M. (2020). Vaccination strategies to combat novel corona virus SARS-CoV-2. *Life Sciences, 256*, 117956. https://doi.org/10.1016/j.lfs.2020.117956

Pardi, N., Hogan, M. J., Pelc, R. S., Muramatsu, H., Andersen, H., DeMaso, C. R., Dowd, K. A., Sutherland, L. L., Scearce, R. M., Parks, R., & Wagner, W. (2017). Zika virus protection by a single low-dose nucleoside-modified mRNA vaccination. *Nature, 543*(7644), 248–251. https://doi.org/10.1038/nature21428

Pasquale, A., Preiss, S., Silva, F., & Garçon, N. (2015). Vaccine adjuvants: From 1920 to 2015 and beyond. *Vaccines, 3*(2), 320–343. https://doi.org/10.3390/vaccines3020320

Patronov, A., & Doytchinova, I. (2013). T-cell epitope vaccine design by immunoinformatics. *FEBS Open Bio, 3*(1), 120139. https://doi.org/10.1098/rsob.120139

Plotkin, S. A. (2005). Vaccines: Past, present and future. *Nature Medicine, 11*(S4), S5–S11. https://doi.org/10.1038/nm1209

Potocnakova, L., Bhide, M., & Pulzova, L. B. (2016). An introduction to B-cell epitope mapping and in silico epitope prediction. *Journal of Immunology Research, 2016*, 1–11. https://doi.org/10.1155/2016/6760830

Prugnola, A. (1994). Vaccine design. *Trends in Biotechnology, 12*(6), 251. https://doi.org/10.1016/0167-7799(94)90132-5

Qin, J., Lei, B., Xi, L., Liu, H., & Yao, X. (2010). Molecular modeling studies of Rho kinase inhibitors using molecular docking and 3D-QSAR analysis. *European Journal of Medicinal Chemistry, 45*(7), 2768–2776. https://doi.org/10.1016/j.ejmech.2010.02.059

Reddy Chichili, V. P., Kumar, V., & Sivaraman, J. (2013). Linkers in the structural biology of protein-protein interactions. *Protein Science, 22*(2), 153–167. https://doi.org/10.1002/pro.2206

Regenmortel, M. H. V. (2009). What is a B-cell epitope?. In *Methods mol biol* (pp. 3–20). https://doi.org/10.1007/978-1-59745-450-6_1

Rudra, J. S., Tian, Y. F., Jung, J. P., & Collier, J. H. (2010). A self-assembling peptide acting as an immune adjuvant. *Proceedings of the National Academy of Sciences United States of America, 107*(2), 622–627. https://doi.org/10.1073/pnas.0912124107

Rueckert, C., & Guzmán, C. A. (2012). Vaccines: From empirical development to rational design. *PLoS Pathogens, 8*(11), e1003001. https://doi.org/10.1371/journal.ppat.1003001

Růčková, E., Müller, P., & Vojtěšek, B. (2014). Protein expression and purification. *Klinická Onkologie, 27*(Suppl. 1), S92–S98. https://doi.org/10.14735/amko20141S92

Salimi, N., Fleri, W., Peters, B., & Sette, A. (2010). Design and utilization of epitope-based databases and predictive tools. *Immunogenetics, 62*(4), 185–196. https://doi.org/10.1007/s00251-010-0435-2

Sanchez-Trincado, J. L., Gomez-Perosanz, M., & Reche, P. A. (2017). Fundamentals and methods for T- and B-cell epitope prediction. *Journal of Immunology Research, 2017*, 1–14. https://doi.org/10.1155/2017/2680160

Schjetne, K. W., Thompson, K. M., Nilsen, N., Flo, T. H., Fleckenstein, B., Iversen, J.-G., & Bogen, B. (2003). Cutting edge: Link between innate and adaptive immunity: Toll-like receptor 2 internalizes antigen for presentation to CD4+ T cells and could be an efficient vaccine target. *The Journal of Immunology, 171*(1), 32–36. https://doi.org/10.4049/jimmunol.171.1.32

Shey, R. A., Ghogomu, S. M., Esoh, K. K., Nebangwa, N. D., Shintouo, C. M., Nongley, N. F., & Souopgui, J. (2019). In-silico design of a multi-epitope vaccine candidate against onchocerciasis and related filarial diseases. *Scientific Reports, 9*(1), 4409. https://doi.org/10.1038/s41598-019-40833-x

Sliwoski, G., Kothiwale, S., Meiler, J., & Lowe, E. W. (2014). Computational methods in drug discovery. *Pharmacological Reviews, 66*(1), 334–395. https://doi.org/10.1124/pr.112.007336

Smialowski, P., Doose, G., Torkler, P., Kaufmann, S., & Frishman, D. (2012). PROSO II - a new method for protein solubility prediction. *FEBS Journal, 279*(12), 2192–2200. https://doi.org/10.1111/j.1742-4658.2012.08603.x

Smith, T. R. F., Patel, A., Ramos, S., Elwood, D., Zhu, X., Yan, J., & Broderick, K. E. (2020). Immunogenicity of a DNA vaccine candidate for COVID-19. *Nature Communications, 11*(1), 2601. https://doi.org/10.1038/s41467-020-16505-0

Soria-Guerra, R. E., Nieto-Gomez, R., Govea-Alonso, D. O., & Rosales-Mendoza, S. (2015). An overview of bioinformatics tools for epitope prediction: Implications on vaccine development. *Journal of Biomedical Informatics, 53*, 405–414. https://doi.org/10.1016/j.jbi.2014.11.003

Srivastava, S., Kamthania, M., Singh, S., Saxena, A., & Sharma, N. (2018). Structural basis of development of multi-epitope vaccine against Middle East respiratory syndrome using in silico approach. *Infection and Drug Resistance, 11*, 2377–2391. https://doi.org/10.2147/IDR.S175114

Vasselon, T. (2002). Toll receptors: A central element in innate immune responses. *Infection and Immunity, 70*(3), 1033–1041. https://doi.org/10.1128/IAI.70.3.1033-1041.2002

Wang, F., Kream, R. M., & Stefano, G. B. (2020). An evidence based perspective on mRNA-SARS-CoV-2 vaccine development. *Medical Science Monitor, 26*, e924700. https://doi.org/10.12659/MSM.924700

Werling, D., & Jungi, T. W. (2003). TOLL-like receptors linking innate and adaptive immune response. *Veterinary Immunology and Immunopathology, 91*(1), 1–12. https://doi.org/10.1016/S0165-2427(02)00228-3

Wilson, C. C., Newman, M. J., Livingston, B. D., MaWhinney, S., Forster, J. E., Scott, J., Schooley, R. T., & Benson, C. A. (2008). Clinical phase 1 testing of the safety and immunogenicity of an epitope-based DNA vaccine in human immunodeficiency virus type 1-infected subjects receiving highly active antiretroviral therapy. *Clinical and Vaccine Immunology, 15*(6), 986–994. https://doi.org/10.1128/CVI.00492-07

Xin, H., Cartmell, J., Bailey, J. J., Dziadek, S., Bundle, D. R., & Cutler, J. E. (2012). Self-adjuvanting glycopeptide conjugate vaccine against disseminated candidiasis. *PLoS One, 7*(4), e35106. https://doi.org/10.1371/journal.pone.0035106

Yang, X., & Yu, X. (2009). An introduction to epitope prediction methods and software. *Journal of Medical Virology, 19*(2), 77–96. https://doi.org/10.1002/rmv.602

Zagursky, R. J., & Russell, D. (2001). Bioinformatics: Use in bacterial vaccine discovery. *Biotechniques, 31*(3), 636–659. https://doi.org/10.2144/01313dd02

Zhu, F. C., Li, Y. H., Guan, X. H., Hou, L. H., Wang, W. J., Li, J. X., Wu, S. P., Wang, B. S., Wang, Z., Wang, L., & Jia, S. Y. (2020). Safety, tolerability, and immunogenicity of a recombinant adenovirus type-5 vectored COVID-19 vaccine: A dose-escalation, open-label, non-randomised, first-in-human trial. *The Lancet, 395*, 1845–1854. https://doi.org/10.1016/S0140-6736(20)31208-3

Computational Approaches Toward Development of Topoisomerase I Inhibitor: A Clinically Validated Target

ARINDAM TALUKDAR[a] • SOURAV PAL[b]

1 INTRODUCTION

DNA topoisomerases are ubiquitous enzymes which are responsible for relaxing of topological strain in DNA due to supercoiling during replication, transcription, and chromatin remodeling process. The supercoiling of DNA does not allow the polymerases to act on DNA during cell division (Bansal et al., 2017; Champoux, 2001; Pommier, 2006; Wang, 2002). The topoisomerases relax the torsional strain by nicking the phosphate backbone of either one or both the DNA strands. Therefore, these enzymes play vital roles during cell division through regulation of DNA supercoiling.

Expression of DNA topoisomerase enzymes happens in both prokaryotic and eukaryotic cells. As DNA Top1 enzyme relaxes the torsional stress which is significant for cell division process, Top1 enzyme could be one of the significant targets for the cancer chemotherapy. Notably, DNA replication, repairing process, and gene transcription in most of the mitotic cells are regulated transcendently by Top1 enzyme. Therefore, controlling the biomolecular procedure due to cancer cell growth regulation through restraining of the Top1 enzyme can be an attractive therapeutic intervention (Dev et al., 2016; Drwal et al., 2011; Keszthelyi et al., 2016). DNA topoisomerase enzymes are divided into two types based on the number of strands topoisomerase breaks. TopI cleaves single strand of DNA whereas type II topoisomerases cleaves both the DNA strands. TopI includes four mammalian enzymes: nuclear Top1, mitochondrial TopI, topoisomerase IIIα, and topoisomerase IIIβ. Type II topoisomerase group can be divided into type IIA and type IIB enzymes.

Top1 enzyme performs its activity through three significant steps: firstly, binding of the enzyme with the supercoiled DNA; secondly, enzyme through nucleophilic attack of tyrosine 723 cleaves the single DNA strand by hydrolyzing the phosphodiesterase bond between the nucleotide base pairs forming a covalent "cleavage complex" of the DNA strand and the enzyme (Bomgaars et al., 2001; Sheng-Yong, 2010; Tsao et al., 1989). Thirdly, the primary hydroxyl group of the ribose unit attacks the cleavage complex leading to religation of the DNA strand and releases topoisomerase enzyme (Denny, 2007). The process releases the torsional strain by strand uncoiling and religation of the broken strand. The rate of religation procedure of the split DNA strands by Top1 happens much faster in contrast with the DNA cleavage process (Kizisar et al., 2016). Though topoisomerase types IA and IB both initiate momentary break of single DNA strand, their mechanism of action is significantly different. Topoisomerase type IA enzymes release the torsional stress of both positive and negatively supercoiled DNA by nicking one of the strands in presence of ATP molecules. However, type IB enzymes cut the single strand in absence of ATP molecules and facilitate the broken DNA strand to make a turnaround the cleavage complex of DNA-Top1 (Pommier, 2006).

2 STRUCTURE OF HUMAN TOPOISOMERASE I

Human topoisomerase I (hTop1) belongs to the type IB topoisomerase families, which is present in combination with 100 kDa DNA strands having endonuclease and ligase activities. The hTop1 consists of 765 amino acids and composed 4 different domains in their crystal structures (PDB ID: 1A36) (Baker et al., 2009; Staker et al., 2005; Stewart et al., 1998) (Fig. 21.1); the N-terminal

---

[a]Academic pediatrics.
[b]SP wants to thank ICMR, New Delhi for the fellowship.

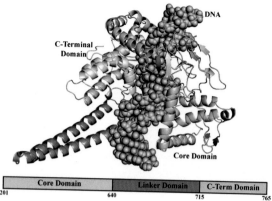

FIG. 21.1 Schematic representation of hTop1-DNA duplex crystal structure (PDB ID: 1A36). Highly conserved core domain is signified by the orange cartoon representation which makes a complex with the double stranded DNA. The DNA double helices are represented with deep green sphere representation. Poorly conserved linker domain is depicted by light brown cartoon representation. The C-terminal domain is portrayed by light yellow cartoon representation.

and C-terminal domains, the central core domain which contains the active site, surrounded with the DNA strands and the linker region associating the C-terminal and central core (Champoux, 2001; Kato & Kikuchi, 1998). Highly charged N-terminal domain plays a pivotal role for interactions with other cellular components (Haluska et al., 1998; Lebel et al., 1999) and processes several nuclear localization signals (Alsner et al., 1992). Enzymatic activity of the hTop1 is maintained due to the presence of the poorly conserved linker domain having cleavage complex with the DNA strands (Stewart et al., 1997). There is also a subdivision of the core domain, namely subdomains I, II, and III. Subdomains II and III process a new region entitled as "lips" that is open at the time of DNA binding and unbinding. Important residue tyrosine (Tyr723), which is present on the C-terminal domain, attaches on the supercoiled DNA and cleaves the phosphodiester bond.

Many structural studies have been carried out on the human Top1-DNA cleavage complex for understanding the structural basis of Top1 functionality. Redinbo and co-authors describe that the central DNA-binding pore has been formulated by combining with the core domain and C-terminal domain (Redinbo et al., 1998). Furthermore, it has been shown that positively charged residues along with the catalytic residues take part for the assembly of the central pore, which has a direct interaction with the backbone phosphate groups of the DNA double strands (Redinbo et al., 1998; Stewart et al., 1998).

3 INHIBITORS OF HUMAN TOPOISOMERASE I

DNA topoisomerases are validated clinical targets for cancer chemotherapy (Hevener et al., 2018; Pommier, 2009; Staker et al., 2002). Top1 enzyme can be inhibited generally through two mechanisms (Tselepi et al., 2011). Catalytic inhibitors do not allow the formation of Top1−DNA complex by reacting either with DNA or with Top1 enzyme and consequently enzymatic turnover. The second class is the most clinically relevant topoisomerase poison (Pommier et al., 2018), which traps the Top1 cleavage complex (Top1cc) through formation of stabilized locked ternary complex of Top1-DNA and the drug resulting in cytotoxic effect (Lindsey et al., 2014). The natural product camptothecin (CPT) obtained from *Camptotheca acuminata* by Monroe Wall and co-workers (Wall and Wani, 1995; Wall et al., 1966) was found to selectively inhibit Top1 by trapping Top1cc (Fig. 21.2) (Dexheimer & Pommier, 2008; Hsiang et al., 1985). CPT went into clinical trial but failed due to issues related to efficacy and toxicity such as severe diarrhea, hepatotoxicity, and neutropenia (Pommier, 2009). The lactone E-ring of CPT opens up to form inactive carboxylate analog spontaneously within minute at physiological pH. The conversion of carboxylate and the active lactone form is reversible with the inactive carboxylate form being favored. Moreover, the carboxylate form has higher affinity for serum albumin which further reduces the concentration of the active lactone form in the blood (Fig. 21.2) (Asano et al., 1994).

Subsequently, many CPT-inspired agents were developed to overcome the issues related to CPT. Water-soluble CPT derivatives, topotecan (Hycamtin) and irinotecan, have been approved by the FDA. Topotecan is being used for the treatment of ovarian cancer and small cell lung cancer in combination with the chemotherapeutic regimen. Among the many side effects of topotecan, the most common is hematological toxicity causing loss of white blood cells due to the damaged bone marrow progenitors; as a result, the patients are prone to infections. Irinotecan, a prodrug, is used for the treatment of colorectal tumors. Carboxylesterase enzyme converts irinotecan to its active metabolite SN-38. The most severe side effect associated with irinotecan therapy is diarrhea. Most of the side effects observed are similar to that caused by topotecan. Apart from topotecan and irinotecan, many CPT derivatives went into clinical trials such as rubitecan investigated for the treatment of pancreatic cancer, melanoma, and ovarian cancer; belotecan investigated for the treatment of epithelial ovarian cancer, and exatecan investigated

Active and Inactive form of Camptothecin

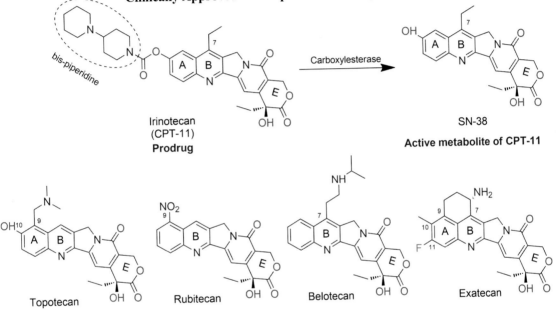

FIG. 21.2 Development of camptothecin derivatives as anticancer agents.

for the treatment of sarcoma, leukemia, lymphoma, and lung cancer (Fig. 21.2) (Bax et al., 2019; Bjornsti & Kaufmann, 2019; Pommier, 2009). To address the instability of the lactone E-ring, the first strategy adopted was to increase the ring size of lactone E-ring by one carbon to limit the opening of the E-ring and at the same time does not allow the recloser of the opened ring (Chen et al., 2005; Lavergne et al., 1998; Tangirala et al., 2006).

These compounds are classified as homocamptothecins with diflomotecan as the clinical analog. Unlike in CPT derivatives where the opening and closing of the lactone ring is reversible, with homocamptothecins, the E-ring opening is irreversible and it inactivates homocamptothecins (Urasaki et al., 2000). An added advantage that homocamptothecins have over CPT derivatives is that they are not substrate of P-glycoprotein (P-gp) drug efflux pump, which is responsible for multidrug resistance to CPTs (Bates et al., 2004; Brangi et al., 1999; Liao et al., 2008).

In another approach the size of the E-ring was shortened to five-membered ring with removal of the lactone ring oxygen. The strategy successfully led to complete stabilization of the E-ring in the keto derivative S39625 (Takagi et al., 2007). The derivative was found to be very potent with selective inhibition of Top1 (Fig. 21.2). The development of topoisomerase poison S39625 suggested the lactone E-ring once thought to be necessary is not important for trapping the cleavage complex of Top1-DNA.

Several limitations of the CPT and its derivatives have prompted the development of "non-CPT" Top1 poisons as anticancer agents (Fig. 21.3). Recent times have seen several non-CPT Top1 poisons as drug leads which are undergoing clinical trials. Indenoisoquinoline NSC 314622 was first reported by Cushman and Pommier group in 1998 (Kohlhagen et al., 1998). Thereafter, development of indenoisoquinolines led to NSC 725776 and NSC 724998, for which phase II clinical trials are over. Other than indenoisoquinolines, indolocarbazole and phenanthridine derivatives are under clinical development.

Unlike CPT derivatives, indenoisoquinolines are chemically stable and trap the cleavage complex at differential sites. Also, they have the ability to overcome drug resistance as they are not substrate of P-gp. Recently, discovery and mechanistic study of quinoline derivatives (Fig. 21.3, TBK-1-53) has been reported as Top1 poison with potent anticancer activity (Kundu et al., 2019). The study describes strategically placed substitution of five-membered oxadiazole at C3 position of quinoline ring, which forms an intramolecular hydrogen bond with nitrogen at C4 position of the quinoline ring to provide the necessary geometry essential to fit into the narrow cavity and trap the top1cc. The quinoline derivatives have all the features of indenoisoquinolines which are under clinical development.

4 COMPUTATIONAL APPROACHES FOR THE DISCOVERY OF TOPOISOMERASE I INHIBITORS

4.1 Quantitative Ligand-Based Pharmacophore Methods

A pharmacophore is defined as a theoretical depiction of molecular features that are essential for recognition of ligand by biological macromolecules; the abstract description has been effectively applied in the field of drug development (Langer & Hoffmann, 2008; Leach et al., 2010; Sheng-Yong, 2010). A pharmacophore model describes how structurally different ligands along with its features can bind to its target receptor sites. IUPAC characterizes a pharmacophore to be "an ensemble of steric and electronic features that is necessary to ensure the optimal supramolecular interactions with a specific biological target and to trigger (or block) its biological response" (Daisy et al., 2011). The model can be built solely on ligand data which is defined as ligand-based pharmacophore modeling or the pharmacophore model; the model can also be built based on protein three-dimensional (3D) structure and is designated as structure-based pharmacophore (SBP). Sometimes, pharmacophore model can also be built with both the data from protein and ligand complex (complex-based).

Ligand-based drug design (LBDD) is performed through mapping of molecular structure, electronic properties, and biological activities of the ligands for building the model without the receptor 3D structural information. The structural insights are significant for building chemometric model which connects molecular properties with their pharmacokinetic as well as pharmacodynamic parameters (Dudek et al., 2006). Molecular descriptors or molecular properties are mathematical representations of the target molecules generated through quantitative structure–activity and structure–property relationships (QSAR and QSPR, separately) methodology (Fig. 21.4) (John et al., 2011; Mati Karelson, 2000; Yousefinejad & Hemmateenejad, 2015). Information regarding the ligand binding on the target protein and the protein inhibition assay is one of the important parameters to construct the ligand-based pharmacophore models that can correlates the structural features of the participating ligands

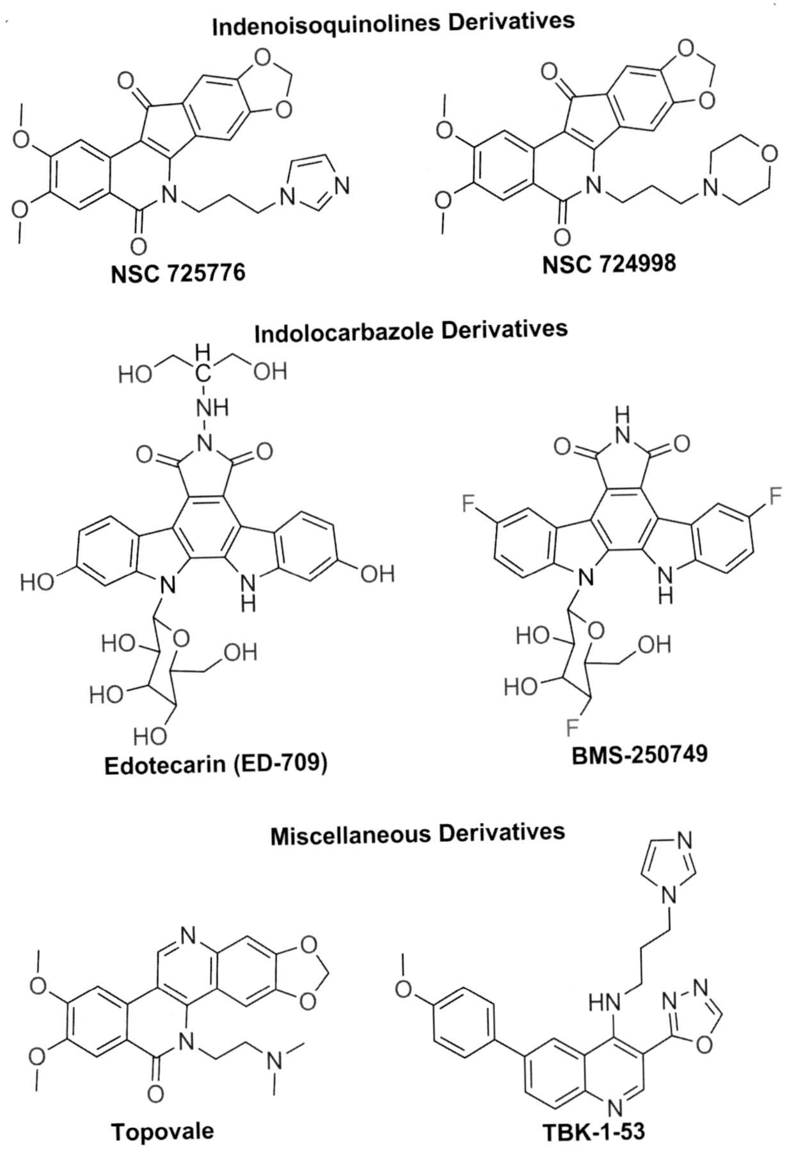

Indenoisoquinolines Derivatives

NSC 725776

NSC 724998

Indolocarbazole Derivatives

Edotecarin (ED-709)

BMS-250749

Miscellaneous Derivatives

Topovale

TBK-1-53

FIG. 21.3 Development of noncamptothecin topoisomerase poison from different structural scaffolds.

along with their necessary binding or inhibition of target protein. At least 16 structurally diverse training set compounds with their activities uniformly spanning over four orders of magnitude is considered to produce statistically legitimate ligand-based pharmacophore models (Li et al., 2000). HypoGen and HypeGen Refine algorithms in Discovery Studio are one of the prime platforms for developing the ligand-based pharmacophore models (Fang et al., 2011). HypoGen modules work for developing the pharmacophore hypotheses

in three different phases, which are constructive, subtractive, and optimization phase. The lead compounds of the training set in the constructive phase are distinguished and are utilized for the development of the pharmacophore hypotheses. The activities (A) of the training set compounds as well as the uncertainty value (U) set by the user play pivotal roles for deciding the number of the training set molecules to be taken into consideration as lead molecules. A compound will be considered as a lead if it fulfills the following condition:

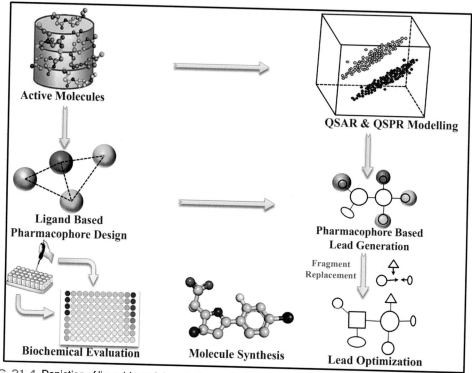

FIG. 21.4 Depiction of ligand-based drug design and quantitative structure–activity relationship (QSAR) and quantitative structure–property relationship (QSPR) modeling. Used for the design of novel chemical entity and to predict their pharmacodynamics and pharmacokinetics properties. The validated experimental data of the newly designed molecules can facilitate further design through generation of enriched models.

$$A_{MA} * U_{MA} - \frac{Ax}{Ux} > 0,$$

In the training set, MA represents the most active compound and x is the compound being referred to. The algorithm differentiates all pharmacophore hypotheses that are shared within the most active compounds of the training set. Usually pharmacophore hypotheses consist of maximum five features that are representative of the spatial orientation of conformational space of the molecule of a particular target (Klebe, 2013). Mapping of the rest of the lead compounds is utilized for filtrations of the generated pharmacophore hypotheses. Hypotheses that do not get mapped with a minimum subset of features are not considered. In the subtractive stage, their respective activity values of the inactive compounds of the training set play crucial roles for their recognition.

Recently, a diverse LBDD strategy has been adopted to design novel Top1 inhibitors (Drwal et al., 2011). Combination of structure-based drug design (SBDD) methodology and LBDD approaches along with experimental validations has been implemented for the

investigation of the novel Top1 inhibitors. The predicted activity and the ADMET parameters of the screened hit compounds can assist exploration of novel molecules with better profile by incorporating the QSAR and QSPR models into ligand-based virtual screening (LBVS) (Kandakatla & Ramakrishnan, 2014; Lee et al., 2011).

A similar approach reports the discovery of non–CPT-based hit compounds as a potent novel Top1 inhibitor (Pal et al., 2019) developed through the ligand-based pharmacophore model based on 3D QSAR pharmacophore generation followed by virtual screening techniques with drug-like molecules from ZINC databases (Fei et al., 2013). The criterion for selection of the best quantitative pharmacophore is based on the highest cost difference, best correlation coefficient, and the lowest total cost value. The selected Hypo1 model comprises of two hydrogen bond acceptors (HBAs) and one ring aromatic feature (Fig. 21.5A). The Hypo1 pharmacophore was used for the virtual screening of more than 1 million drug-like molecules from ZINC database. The structurally diversified molecular libraries were subjected to various constraints. The

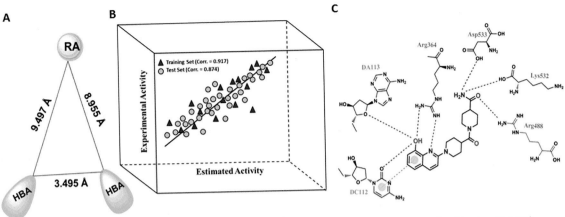

FIG. 21.5 **(A)** The resultant ligand-based pharmacophore, where spatial arrangements of hydrogen bond acceptor (HBA) and ring aromatic (RA) features are depicted in green and brown vectors, respectively. **(B)** Training set and test set correlation graph. **(C)** Docking result of one of the hit compounds onto the active site domain of the human Top1-DNA cleavage complex (PDB ID: 1T8I). The conventional hydrogen bonding interactions are shown in blue dotted line and hydrophobic π–π interactions are depicted in brown *dotted line.*

extensive screening finally provided possible six hit molecules. The final selection of these six molecules is based on the combination of features specific to selective Top1 poison, such as molecular planarity, structural rigidity along with molecular interactions with specific residues, and thereafter extensive molecular docking of these ligands on the hTop1 protein (PDB ID: 1T8I) (Fig. 21.5C) (Arthur & Uzairu, 2018). Based on this high correlation ($R^2 = 0.917678$) (Fig. 21.5B), this QSAR model predicted the activities of the identified hit molecules which is at μM range. Authors also demonstrated the toxicity assessment and pharmacokinetic behavior of the identified six non-CPT compounds under TOPKAT program (Prival, 2001). Molecular dynamics studies revealed the probable binding nature of the hit molecules in the active site (Koch et al., 2013). Such studies will enable identification of the structural features and molecular geometry in the non-CPT top1 inhibitors required for inhibition of the human Top1 enzyme.

Kathiravan and co-authors demonstrated another drug discovery program against Top1 inhibition through 3D QSAR ligand-based methodology (Kathiravan, Khilare, Chothe, & Nagras, 2013) with the help of QSAR modeling (Dixon et al., 2006; Golbraikh et al., 2003; Kubinyi, 1997). They reported in silico discoveries of 25 numbers of 2,4,6-tri-substituted pyridine derivatives as selective topoisomerase inhibitors. The developed atom-based 3D QSAR model gives the knowledge into the structural prerequisite of novel 2,4,6-tri-substituted pyridine derivatives as selective

topoisomerase inhibitors. The library of 63 molecular datasets belonging to the 2,4,6-tri-substituted pyridine derivatives demonstrating anticancer activities against HCT-15 cell lines was gathered for the development of ligand-based pharmacophore hypothesis and atom-based 3D QSAR model (Basnet et al., 2007, 2010; Karki et al., 2010; Thapa, Karki, Choi, et al., 2010; Thapa, Karki, Thapa, et al., 2010). The top scoring pharmacophore hypothesis among the other hypotheses consists of one hydrophobic group (H4) and four aromatic groups rings (R5, R6, R7, R8) based on the training set data points (Fig. 21.6B). The selected pharmacophore hypothesis also yields 3D QSAR model with good PLS statistics ($r^2_{training} = 0.7892$, SD = 0.2948, F = 49.9, P = 1.379).

These models offer a valuable structural insight for the future endeavors on the enhancement of this series of compounds. The favorable and unfavorable interactive regions based on the molecular dataset were calculated with the volume occlusion maps by using the atom-based 3D QSAR strategy (Fig. 21.6A). Enhancements in the formal charged regions on the dataset which are expected to increase the activities are marked in red; whereas the decrease in the formal charge to improve the biological activities on the hydrophobic volume occlusion maps of the 3D QSAR model is marked in blue. QSAR results (Fig. 21.6A) demonstrate that five-membered ring (thienyl or furyl) at second and fourth positions, a phenyl ring at sixth position, and hydrophobic group (2′), methyl (5′) at the second position are significant for exhibiting better topoisomerase inhibitory activities. Newly designed novel 2,4,6-tri-substituted pyridine derivatives developed from ligand-based pharmacophore hypothesis (Fig. 21.6B) and their binding patterns convey the new strategy for the discovery of novel topoisomerase inhibitors.

With the aid of QSAR modeling, Zhi and co-authors (Zhi et al., 2012) described the structural motifs and geometry of 48 indenoisoquinoline derivatives which exhibited Top1 inhibitory activities on SN12C (human renal cell carcinoma cell line). In this study, a group of 48 indenoisoquinoline derivatives (Morrell et al., 2007; Nagarajan et al., 2006) were collected as data set points for building comparative molecular field analysis

(CoMFA) and comparative molecular similarity indices analysis (CoMSIA) models, which have a good predictability that can efficiently guide further modification for the generation of more potent novel Top1 inhibitors.

The CoMFA contour maps (Fig. 21.7) signify that the steric fields (indicated by green and yellow contours) at the end of the flexible chain connected with the nitrogen of isoquinoline ring did not play significant roles toward their anticancer activities. However, the requirement of bulky and electron withdrawing substituent at position 3 of the isoquinoline rings for the activity has been evaluated by observing the high density of two large green and two small red contour maps around

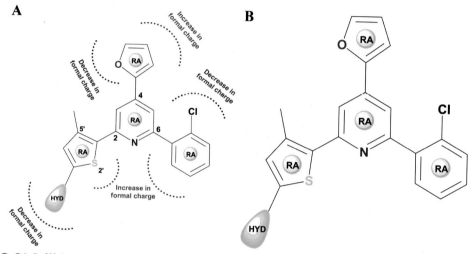

FIG. 21.6 **(A)** Atom-based 3D QSAR model of 2,4,6-tri-substituted pyridine derivatives. Blue regions indicate favorable regions, while red zones indicate unfavorable region for the activity. **(B)** Pharmacophore mapping upon the 2,4,6-tri-substituted pyridine derivatives. Yellow sphere indicates the ring aromaticity (RA) and green elongated sphere directs the hydrophobicity (HYD) toward the target protein/DNA.

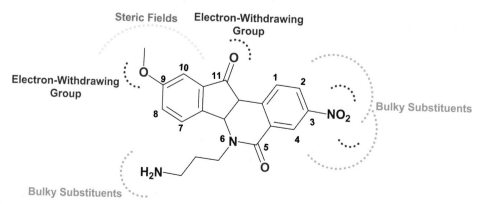

FIG. 21.7 Schematic representation of the comparative molecular field analysis contour maps of the highest active indenoisoquinoline derivatives.

this position. This study also revealed the importance of protonated hydrogen at position 1 by monitoring a large yellow contour around them.

Authors also scrutinized the ligand-based CoMSIA by alignment-dependent 3D QSAR methodology on active compounds. This CoMSIA is a modified version of CoMFA model (Klebe et al., 1994). The approaches toward the structural analysis, CoMFA, and CoMSIA are almost similar, except for the similarity analysis which is performed by CoMSIA. The results obtained from CoMFA model are nicely correlated with steric and electrostatic contour plots calculated by CoMSIA contour maps. CoMSIA results indicate that the position 3 of the isoquinoline ring is more susceptible to HBA functional groups, such as nitro and methoxy for better Top1 inhibitory activities, as on the contour map one large magenta polyhedron is visible around that position. Also, large cyan polyhedral contour maps at CoMSIA model (Fig. 21.8) suggested that the HBAs at positions 2 and 4 on the isoquinoline ring do not have any significant role around the flexible side chain attached to the isoquinoline ring toward their anticancer activity (Basith et al., 2017; Srivastava et al., 2005). Authors have also designed and synthesized a series of 15 new compounds based on the results obtained from the 3D QSAR model, which showed potent anticancer activities as comparable to the most active compounds belonging to the training set.

In another study, Drwal and co-authors performed both the ligand-based and structure-based methodology to explore the ligand space required for DNA TopI inhibition (Drwal et al., 2011). Twenty novel inhibitors were identified based on LBVs and structure-based virtual screening followed by molecular docking and manual investigation. Best two 3D QSAR pharmacophore hypotheses were selected among 10 hypotheses which maintained a high correlation (0.96 and 0.94, respectively) with the biological activities data upon the DNA cleavage assay. The models were built with various CPT derivatives as training and test set data points. The resultant hypotheses indicate the significance of stereochemistry at the 20th position on the CPT structure with hydroxyl group contributing toward hydrogen bond donor (HBD) feature at that position. The study states that the distribution of the cyclic π-interactive (CYPI) features on the pyridine ring and the hydrophobic (HYD) features at the ethyl group at 20th position on CPT are necessary for better inhibitory activities (Fig. 21.10). This CYPI feature on the aromatic ring of CPT sandwiched between nucleotide base pairs DT10, TGP11, DC112, and DA113 (Fig. 21.9) stabilized through hydrophobic π–π interactions.

The combined pharmacophore model based on both ligand-based and structure-based design was taken as a 3D query for virtual screening against the National Cancer Institute (NCI) databases containing 240,000 compounds. To filter out the number of hit molecules, screening methods were restricted to only the drug-like molecules. This screening was able to identify 15 hit molecules which were subjected to biological evaluation. Out of the 15 hit molecules, 6 molecules exhibit potent inhibitory activities against the human tumor cell lines. Interestingly, the cytotoxicity of those compounds was modest and could not be correlated with their Top1 inhibitory activity. The authors have explained that maybe the cytotoxicity of the drug-like molecules is controlled by various other properties such as solubility of the molecule, off-target activity, selectivity, and cellular uptake, which needed further modulation (Chivere et al., 2020). The probable reason for that exception is that the chemical space and physicochemical properties around the anticancer agent are different in comparison to that of the other drug-like compounds (Lloyd et al., 2006). As a future direction the combined pharmacophore can be utilised for the designing of new Top1 inhibitors.

4.2 Structure-Based Drug Design Methods

The design of SBPs represents drug development strategy based on the protein structural information of the 3D coordinates of the binding pocket that utilizes ligand–receptor interactions (Anderson, 2003; Van Montfort & Workman, 2017). Information about 3D architecture of the target binding pockets obtained from X-ray crystallography techniques has played pivotal roles for the SBDD programs. Studies based on SBDD can provide ligands with better interactions and high affinity toward the target molecules by revealing the binding site features, such as the shape and electronic distributions of the active site (Batool et al., 2019; Ferreira et al., 2015). Structure–activity relationship of the designed molecules can be predicted by correlating the molecular docking result with the experimental activity results (Wang et al., 2018). These results are also important to understand the concept regarding the ligand–receptor affinity and other structural properties with correlation to their activity (Fig. 21.11) (Reddy et al., 2007; Docking et al., 2018).

One of the promising macromolecular targets for anticancer drug development is TopI enzyme (Danks, 2001; Kathiravan, Khilare, Nikoomanesh, et al., 2013; Kundu et al., 2019). In the binary cleavage complex, the TopI enzyme is covalently bonded with the cleaved DNA. The crystal structure of human DNA-TopI

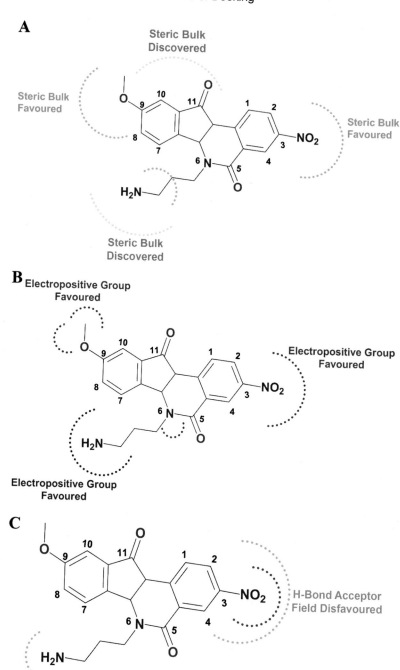

FIG. 21.8 Ligand-based comparative molecular similarity indices analysis fields are depicted with active indenoisoquinoline molecule. **(A)** Green contours indicate favorable steric bulk and yellow contour indicates disfavored steric bulk; **(B)** blue contour indicates favorable electropositive groups and red contour fields indicate favorable electronegative groups; **(C)** magenta contour indicates favorable H-bond acceptor fields and cyan indicates disfavored field.

cleavage complex (PDB ID: 1T8I) revealed the structural integrity of the ternary complex of Top1-DNA and inhibitor molecules of CPT (Cretaio et al., 2007). So, the drug design approaches based on the crystal structure of the target maintain a specific workflow (Fig. 21.11), where potential ligand interactive sites and pharmacophoric location vectors play crucial roles for defining the HBA and HBD features, and the hydrophobic interaction points features on the SBP.

De novo drug design approach is also similar with this SBDD techniques, where the designing aspect of new lead molecules is depicted based on the active

site domain of the experimentally solved target protein structure along with varied distances and angles of the altered functional groups attached with the co-crystal ligands (Böhm, 1992a). Another techniques associated with the SBDD is LUDI techniques. It is a rule-based technique, where small suitable fragments are arranged properly onto the active site pocket of the target protein in order to form hydrogen bonding interactions with the active site residue as well as in the hydrophobic pockets through hydrophobic interactions (Kawai et al., 2014). The strength of the LUDI program acknowledges the structural features of the binding site which rationally process the idea about the selection of those fragments that could occupy the binding domain and should be connected with the main scaffold (Böhm, 1992b).

Design and exploration of novel ligands has been accomplished by following the SBDD approaches for well-established Top1 drug targets. Drwal and coauthors (Drwal et al., 2011) have selected structurally similar CPT derivatives (PDB ID: 1T8I) (Staker et al., 2005) and topotecan (PDB ID: 1K4T) (Staker et al., 2002) bound to human Top1-DNA forming ternary complex structure for the development of CPT-based Top1 inhibitory pharmacophore. The protein drugs are interconnected by hydrogen bonds and π−cation interactions; while the DNA counterparts of the target protein interact with the drug through π−π interactions. The final pharmacophore has three CYPI features, two HBAs features, and one HBD feature. Most interestingly, the CYPI and HBD features are similarly mapped

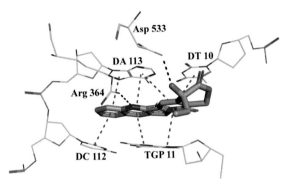

FIG. 21.9 Binding of camptothecin (CPT) onto the active site of human topoisomerase I (Top1) DNA cleavage complex (PDB ID: 1T8I). Orange stick representation depicts the CPT ligand; whereas the representation of green lines portrays the Top1 active site residues and DNA nucleotide base pairs with proper labeling.

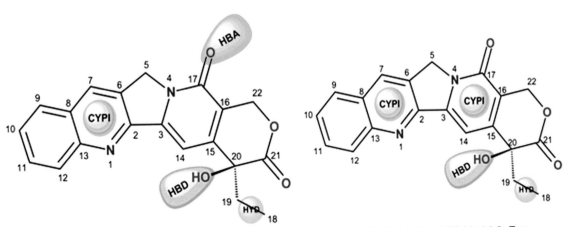

FIG. 21.10 Two developed ligand-based pharmacophore models: **(A)** Model 1 and **(B)** Model 2. Two-dimensional (2D) representation of pharmacophore mapping upon camptothecin (CPT). Orange color represents cyclic π-interaction (CYPI); green color represents hydrogen bond acceptor (HBA); pink color represents hydrogen bond donor (HBD); blue color represents hydrophobic (HYD). The 2D representation of CPT is color-coded (carbon: black, nitrogen: blue, oxygen: red).

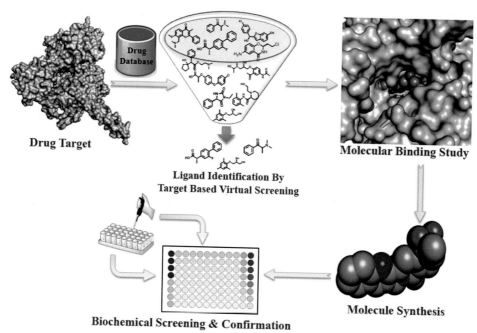

FIG. 21.11 Strategy and flowchart for drug discovery using structure-based drug design, virtual screening, and molecular docking.

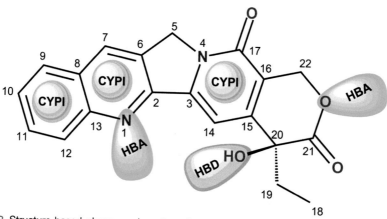

FIG. 21.12 Structure-based pharmacophore two-dimensional mapping to camptothecin (CPT) indicating location of the features and their interactions. Orange: cyclic π-interaction (CYPI); green: hydrogen bond acceptor (HBA); pink: hydrogen bond donor (HBD).

with those of the ligand-based pharmacophore. The excluded volumes were adjusted by considering the shape of the binding pockets (Fig. 21.12).

Previously described ligand-based and SBPs have been taken together for subsequent virtual screening against NCI databases. Apart from the Top1 inhibitors showing anticancer activity, studies by Jiang and co-authors (Jiang et al., 2018) discovered that DNA Top1

enzyme could be a probable target for designing new insecticides. SBDD was performed against Top1 (*Sf*Top1)-DNA cleavage complex from *Spodoptera frugiperda* species. Binding mode analysis was done with the help of interactions of CPT in the active site of homology build *Sf*Top1-DNA cleavage complex. The binding conformation of CPT on the *Sf*Top1-DNA cleavage complex guided rational synthesis of 17 novel 7-amide CPT derivatives. Fifteen compounds among them expressed potent antiproliferative activity with IC$_{50}$ values ranging from 2.01 to 6.78 µM with respect to that of the CPT with 29.47 µM against cultured Sf9 cell line. The binding conformational analysis of the CPT on the *Sf*Top1-DNA-CPT ternary complex unravels that the planer geometry of the CPT interacts with the *Sf*Top1-DNA cleavage complex with hydrophobic π–π stacking interactions (Hunter & Sanders, 1990) with the nucleic acid base pairs (A965, G941, T940, and C964). As observed from the crystal structure of the ternary complex of topotecan-hTop1-DNA, the hydroxyl group at 20th position and the nitrogen atom at N-1 position on the CPT skeleton play a crucial role for making two essential hydrogen bond interactions with active residue Arg530 and Asp698, respectively (Fig. 21.13) (Staker et al., 2002).

Interactive maps of the CPT portray the SBP modeling consisting of chemical features with HBA and HBD contributing at around the position 7 of the CPT and on two base pairs (A965 and G941). Thus, the substitution at the seventh position with the group containing both HBA and HBD features such as acyl amino has the potential to interact with the active site base pairs and exhibits potent activities. However, the structure–activity relationship (SAR) of the diverse CPT derivatives (Martino et al., 2017; Sharma et al., 2016) has helped to improve the stability of the CPTs through direct substitution of the lipophilic groups at the seventh position resulted in improvement of their antitumor activity (Li et al., 2009; Liu et al., 2011; Zhang et al., 2012). The structure-based modifications and SAR development based on CPT are useful for further design and development of potentially effective insecticide.

A recent study has demonstrated that Top1 enzyme could also be a probable drug target for the development of antibacterial drugs toward tuberculosis (Sandhaus et al., 2018). The author utilized the active site of *mycobacterium tuberculosis* topoisomerase I (MtbTopI) crystal structure (PDB ID: 5D5H) (Cao et al., 2018) for target-based virtual screening against the Asinex elite library containing 104,000 compounds and Chembridge library subsequently. Active site domain of the MtbTopI resides on the DNA binding region in between the D1 and D4 domains. Molecular docking–based *in silico* screening program resulted in 1000 piperidine amide derivatives as initial hits. Subsequent molecular dynamics study of the MtbTop1-DNA-compound ternary complex structure opened up the DNA binding domain which can facilitate the compound to bind with greater affinity.

Pan-assay interference compounds (PAINS) are those chemical entity that can bind to nonspecific target providing a false positive result on the high-throughput screening (Baell & Walters, 2014; Dahlin & Walters, 2016). Prof. Jonathan Baell has proposed a group of disruptive functional groups/compounds

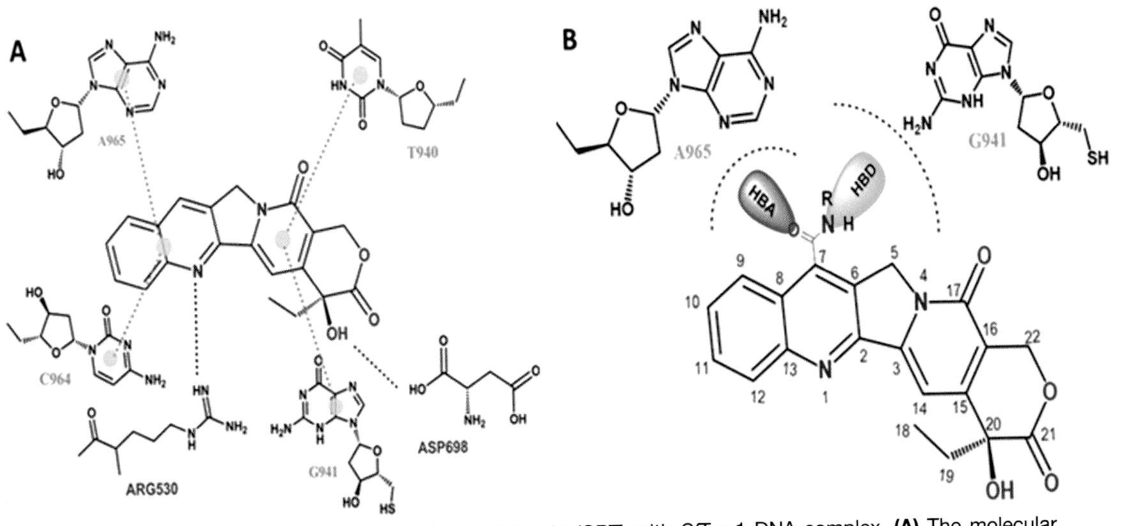

FIG. 21.13 Binding mode analysis of camptothecin (CPT) with *Sf*Top1-DNA complex. **(A)** The molecular docking model of CPT with *Sf*Top1-DNA complex. The hydrogen bond interactions between the ligand and the protein residues are indicated as red *dashed lines*; green *dashed lines* indicate π–π stacked interactions. **(B)** The pharmacophore model describing the binding model of the 7-C of CPT in the center of the *Sf*Top1-DNA complex. The hydrogen bond donor (HBD) is represented as green elongated spheres and the hydrogen bond acceptor (HBA) is represented by smaller purple elongated sphere.

(isothiazolones, curcumin, hydroxyphenylhydrazones, enones, quinones, catechols, etc.) that can be recognized by the PAINS electronic filter through which they should be excluded from further analysis and biological evaluation (Baell & Nissink, 2018). FAF-Drugs3 program can also be used to filter out the PAINS which can interact with the nonspecific target to avoid false positive results. Six compounds among the other hits exhibited inhibitory activities with $IC_{50} \geq 500\ \mu M$ along with the fourfold selectivity against MtbTop1 versus the hTop1. In order to improve potency of the inhibition, Sandhaus and co-authors have identified 2000 cyclic tertiary amide derivatives which were biologically evaluated against purified MtbTopI to find out their inhibitory activities on the MtbTop1 relaxation assay. Among them, the five compounds showed

similar inhibitory activities with IC_{50} value of 62.5 μM (Fig. 21.14). Their selectivity for type IA topoisomerase inhibition versus *Escherichia coli* DNA gyrase (bacterial type IIA topoisomerase) and hTop1 inhibition was measured. This structure-based molecular scaffold generation and its modifications will enable improvement of antibacterial activities that can penetrate into the mycobacteria active site cavity.

4.3 Scaffold Hopping Methods

Scaffold hopping is a kind of ligand-based technique that can design new ligands starting from the existing ligands (Hessler & Baringhaus, 2010; Sun et al., 2012). Fingerprint methodology and shape-based methodology have been applied in scaffold hopping technique (Hu et al., 2017). Among the new drugs which were

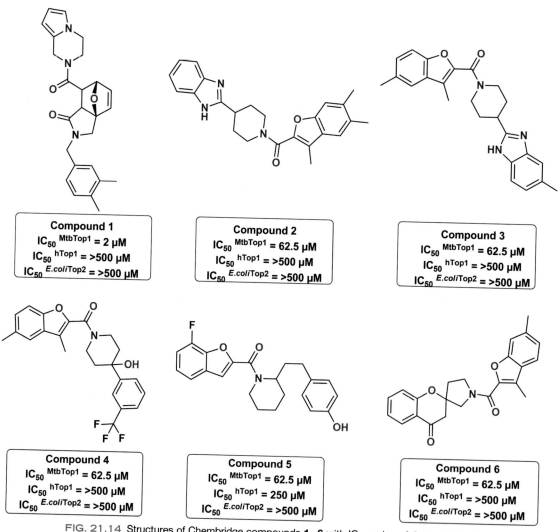

Compound 1
$IC_{50}^{\ MtbTop1}$ = 2 μM
$IC_{50}^{\ hTop1}$ = >500 μM
$IC_{50}^{\ E.coliTop2}$ = >500 μM

Compound 2
$IC_{50}^{\ MtbTop1}$ = 62.5 μM
$IC_{50}^{\ hTop1}$ = >500 μM
$IC_{50}^{\ E.coliTop2}$ = >500 μM

Compound 3
$IC_{50}^{\ MtbTop1}$ = 62.5 μM
$IC_{50}^{\ hTop1}$ = >500 μM
$IC_{50}^{\ E.coliTop2}$ = >500 μM

Compound 4
$IC_{50}^{\ MtbTop1}$ = 62.5 μM
$IC_{50}^{\ hTop1}$ = >500 μM
$IC_{50}^{\ E.coliTop2}$ = >500 μM

Compound 5
$IC_{50}^{\ MtbTop1}$ = 62.5 μM
$IC_{50}^{\ hTop1}$ = 250 μM
$IC_{50}^{\ E.coliTop2}$ = >500 μM

Compound 6
$IC_{50}^{\ MtbTop1}$ = 62.5 μM
$IC_{50}^{\ hTop1}$ = >500 μM
$IC_{50}^{\ E.coliTop2}$ = >500 μM

FIG. 21.14 Structures of Chembridge compounds **1–6** with IC_{50} value of 2 and 62.5 μM.

approved in between 2002 (Graul, 2003) and 2003 (Graul, 2004), many drugs were discovered with little variations of the existing drug structures. This lack of structural diversity can be an impediment toward discovery of new chemotypes. However, in scaffold hopping technique, the newly designed molecules are constructed upon modification of the central core of the known active compounds (Grisoni et al., 2018; Jürgen Bajorath, 2017), so it is expected those compounds to have a high chance to exhibit good activity. In addition to better activities, scaffold hopping techniques are also significant for the modification of the affinities/selectivity toward the receptors and also responsible for the improvement of the physicochemical and ADMET properties of the lead compounds (Grisoni et al., 2018; Schuffenhauer, 2012; Wang et al., 2020). Wang and co-authors introduced this scaffold hopping technique on the previously published type I and type II topoisomerase dual inhibitor evodiamine, a natural product (Wang et al., 2020), in order to develop novel indolopyrazinoquinazolinone derivatives. The newly designed evodiamine derivatives were developed by modulating the ring connectivity between the B-ring and C-ring (Fig. 21.15).

This study describes SAR of the indolopyrazinoquinazolinone antitumor scaffold (Fig. 21.16). Introduction of methyl or methyl substituent at C1 position is very significant, whereas the halogen atom (Cl, Br, and F) at same position is not important for its activity. However, fluoro substitution at C21 position is very important. The study describes the expansion of the six-membered C-ring, and alteration of thiocarbonyl group from the D-ring carbonyl groups does not influence the antitumor activity. Substitution at C9 position has a negative impact upon the activities.

The best active compound **15j** has been developed based on optimizing the main indolopyrazinoquinazolinone scaffolds according to their SARs, which exhibited potent selective TopI inhibitory activities at nanomolar ranges against HCT116 cell lines. The lead compound also exhibited anticancer activities by tubulin inhibition. The binding studies of **15j** in the active site domain of the human Top1-DNA cleavage complex (PDB ID: 1T8I) as well as colchicine binding domain of the tubulin (PDB ID: 1SA0) were done to understand structural basis of inhibition (Bueno et al., 2018). Expectedly, compound **15j** maintains the planner conformation (similar with CPT) and stacking interactions with the TGP11 nucleotide base pairs. In the colchicine binding domain on tubulin, the oxygen atom present in the D-ring carbonyl interacts with Ala250 through hydrogen bond and the interactions between E-ring and Lys254 through salt bridge (Fig. 21.17). Further exploration to improve the drugability of the template drug evodiamine by application of the scaffold hopping techniques is underway.

Mamidala and co-workers (Mamidala et al., 2016) explored Top1 of *Leishmania donovani* (*Ld*Top1) as a molecular target for the SBDD study to discover new drug molecules to cure leishmaniasis. Leishmaniasis is a vector-borne parasitic disease infected by flagellated protozoans, Leishmania parasite (van Griensven & Diro, 2012; Herwaldt, 1999; Villa et al., 2003). Out of 20 different species of Leishmania, *L. donovani* is the most significant among them. The intracellular protozoan parasite lives in the infected sand fly (Dostálová & Volf, 2012). Scaffold hopping and bioisosteric manipulation of known Top1 inhibitors has been used for the discovery of series of furopyridinedione derivatives as new *Ld*Top1 inhibitors (Hu et al., 2018). The molecular design started based on CPT and edotecarin, known Top1 inhibitors, as a starting point by using molecular docking against the X-ray structure of *Ld*Top1 and hTop1 orthologs. Six compounds exhibit selective *Ld*Top1 inhibitory activity having EC$_{50}$ in the range of 1–30 μM (Fig. 21.18).

One of the compounds (compound **5**) showed potent activity (EC$_{50}$ = 3.51 μM) against *L. donovani*

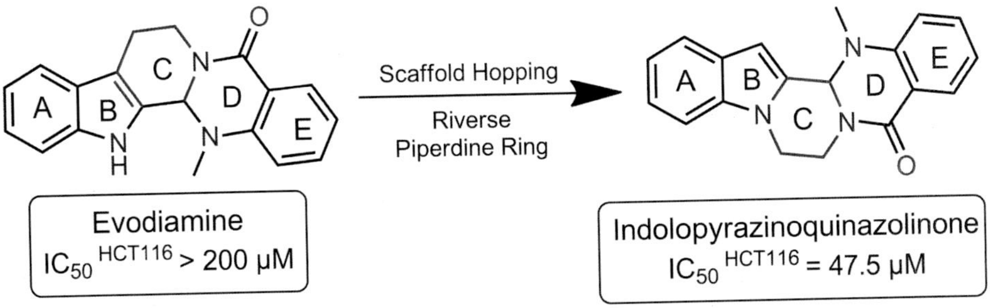

FIG. 21.15 Design of indolopyrazinoquinazolinone derivatives.

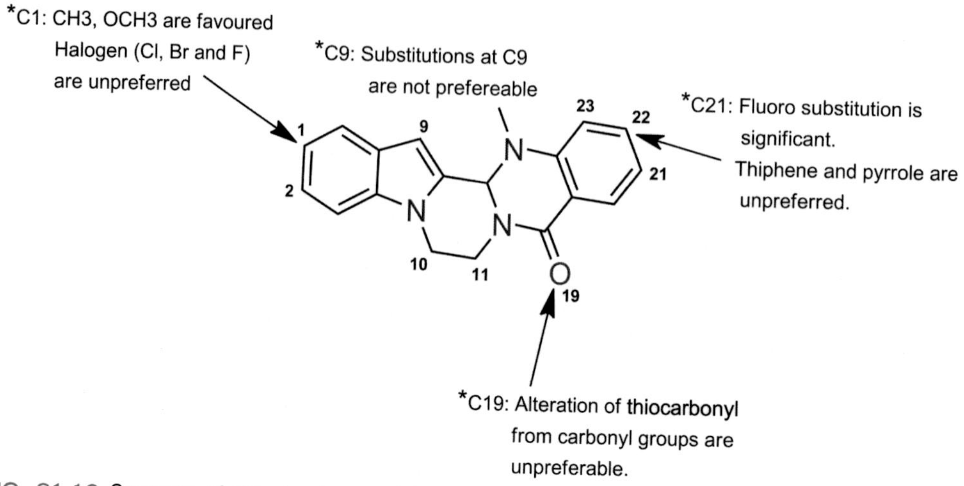

FIG. 21.16 Summary of the structure–activity relationships of indolopyrazinoquinazolinone derivatives (compound **15j**).

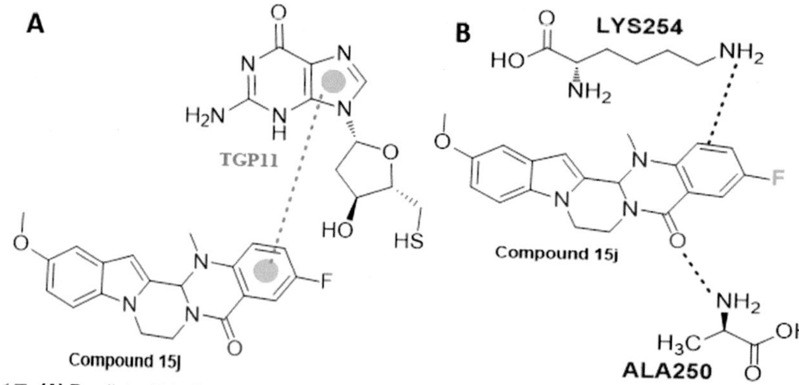

FIG. 21.17 **(A)** Predicted binding interaction for compound **15j** in the template Top1-DNA complex with PDB code: 1T8I. **(B)** Predicted binding interaction for compound **15j** in the template colchicine binding domain of tubulin with PDB code: 1SA0.

promastigotes with no cytotoxicity against mammalian cell line (COS7). Authors explored the pharmacophore modeling and structural requirement for the better biological activity. This study depicts the required molecular topology and SAR through pharmacophore modeling and X-ray crystal structure of the best active molecules. The binding mode analysis of the designed potent compound was based on the *Ld*Top1-DNA cleavage complex crystal structure (PDB ID: 2B9S)

(Roy et al., 2011). The pharmacophore-driven SAR study reveals that functionalized pyridinedione should be present at the central scaffold; polar protic functionality at the C3 and C4 of the benzene ring attached to pyridinedione moiety is required for better activity, and electron donating groups at C3 position of the benzene ring close to the pyridinedione moiety are important for optimum *Ld*Top1 inhibitory activities (Fig. 21.19).

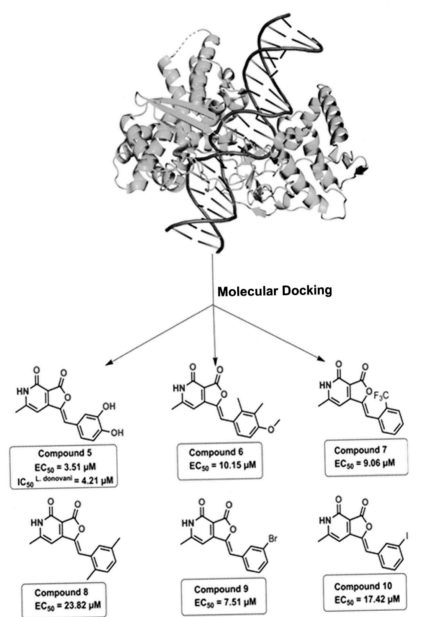

FIG. 21.18 *Leishmania donovani* topoisomerase 1 (*Ld*Top1) inhibitors through structure-based drug design approach. The docking analysis provided identification of lead compounds with in vitro antiparasitic activity.

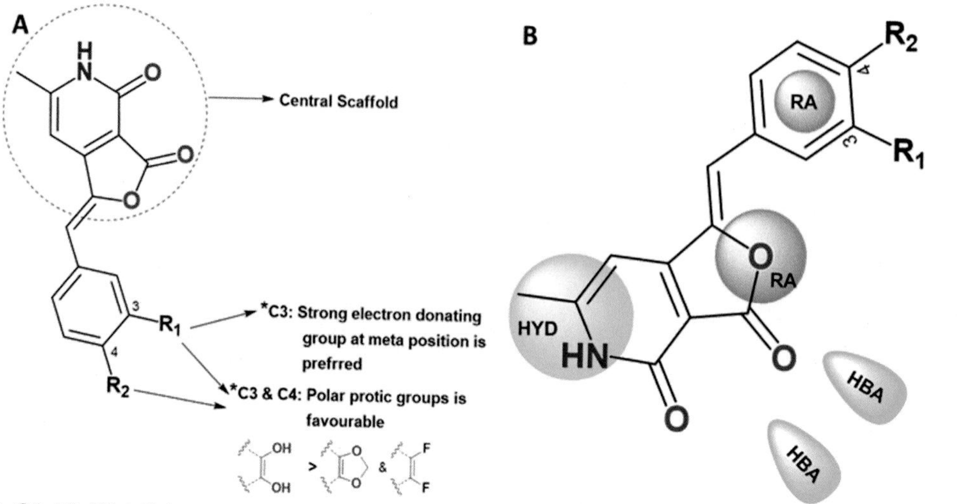

FIG. 21.19 (A) Initial structure–activity relationship studies. (B) Two-dimensional pharmacophore model mapping.

Scaffold hopping technique–driven molecular modeling studies in correlation with their crystal structure can enable the researchers to understand the proper binding conformation of the inhibitors on the LdTop1 active site for designing potent molecules.

5 CONCLUSION

TopI has been clinically validated as a drug target for designing of the new drug molecules with anticancer activity. Irinotecan, a CPT derivative, was the first clinically approved drug targeting TopI enzyme. CTPs have various limitations related to metabolic stability and safety. This paved the way for development of many non-CPTs which are currently under clinical development. The development of several new drug candidate molecules from both CPT and non-CPT scaffold was based on various computational approaches. SBDD approaches utilize the spatial arrangement and intermediate interactions between the crystal structures of Top1 along with their co-crystal inhibitors. The information from active site residues and neighboring environment helps to build the pharmacophores and subsequently facilitates identification of probable candidate through target-based virtual screening. Similarly, ligand-based methodology has been adopted where no crystal structure is present. The physicochemical characteristics and molecular fingerprints of the reported active molecules are taken as initial queries for designing and development of identified hit molecules. Molecular features of the active set of the molecules guide to build the

ligand-based pharmacophore as a lead identification process. More strong and potent drug discovery has been facilitated by performing the scaffold hopping as a lead optimization technique. This chapter summarizes all such computational approaches which have contributed toward rational design and development of potent Top1 inhibitors.

REFERENCES

Alsner, J., Svejstrup, J. Q., Kjeldsen, E., Sorensen, B. S., & Westergaard, O. (1992). Identification of an N-terminal domain of eukaryotic DNA topoisomerase I dispensable for catalytic activity but essential for in vivo function. *Journal of Biological Chemistry, 267*(18), 12408–12411.

Anderson, A. C. (2003). The process of structure-based drug design. *Chemistry & Biology, 10*(9), 787–797.

Arthur, D. E., & Uzairu, A. (2018). Molecular docking study and structure-based design of novel camptothecin analogues used as topoisomerase I inhibitor. *Journal of the Chinese Chemical Society, 65*(10), 1160–1178.

Asano, N., Oseki, K., Kizu, H., & Matsui, K. (1994). Nitrogen-in-the-ring pyranoses and furanoses: Structural basis of inhibition of mammalian glycosidases. *Journal of Medicinal Chemistry, 37*(22), 3701–3706.

Baell, J. B., & Nissink, J. W. M. (2018). Seven year itch: Pan-assay interference compounds (PAINS) in 2017 — Utility and limitations. *ACS Chemical Biology, 13*(1), 36–44.

Baell, J., & Walters, M. A. (2014). Chemistry: Chemical con artists foil drug discovery. *Nature, 513*(7519), 481–483.

Bajorath, J. (2017). Computational scaffold hopping: Cornerstone for the future of drug design? *Future Medicinal Chemistry, 9*(7), 629–631.

Baker, N. M., Rajan, R., & Mondragón, A. (2009). Structural studies of type I topoisomerases. *Nucleic Acids Research, 37*.

Bansal, S., Bajaj, P., Pandey, S., & Tandon, V. (2017). Topoisomerases: Resistance versus sensitivity, how far we can go? *Medicinal Research Reviews, 37*(2), 404−438.

Basith, S., Cui, M., Macalino, S. J. Y., & Choi, S. (2017). Expediting the design, discovery and development of anticancer drugs using computational approaches. *Current Medicinal Chemistry, 24*(42), 4753−4778.

Basnet, A., Thapa, P., Karki, R., Choi, H., Choi, J. H., Yun, M., Jeong, B. S., Jahng, Y., Na, Y., Cho, W. J., Kwon, Y., Lee, C. S., & Lee, E. S. (2010). 2,6-Dithienyl-4-furyl pyridines: Synthesis, topoisomerase I and II inhibition, cytotoxicity, structure-activity relationship, and docking study. *Bioorganic and Medicinal Chemistry Letters, 20*(1), 42−47.

Basnet, A., Thapa, P., Karki, R., Na, Y., Jahng, Y., Jeong, B. S., Jeong, T. C., Lee, C. S., & Lee, E. S. (2007). 2,4,6-Trisubstituted pyridines: Synthesis, topoisomerase I and II inhibitory activity, cytotoxicity, and structure-activity relationship. *Bioorganic and Medicinal Chemistry, 15*(13), 4351−4359.

Bates, S. E., Medina-Pérez, W. Y., Kohlhagen, G., Antony, S., Nadjem, T., Robey, R. W., & Pommier, Y. (2004). ABCG2 mediates differential resistance to SN-38 (7-ethyl-10-hydroxycamptothecin) and homocamptothecins. *Journal of Pharmacology and Experimental Therapeutics, 310*(2), 836−842.

Batool, M., Ahmad, B., & Choi, S. (2019). A structure-based drug discovery paradigm. *International Journal of Molecular Sciences, 20*(11).

Bax, B. D., Murshudov, G., Maxwell, A., & Germe, T. (2019). DNA topoisomerase inhibitors: Trapping a DNA-cleaving machine in motion. *Journal of Molecular Biology, 431*(18), 3427−3449.

Bjornsti, M. A., & Kaufmann, S. H. (2019). Topoisomerases and cancer chemotherapy: Recent advances and unanswered questions. *F1000Research, 8*, 1−18.

Böhm, H.-J. (1992a). LUDI: Rule-based automatic design of new substituents for enzyme inhibitor leads. *Journal of Computer-Aided Molecular Design, 6*(6), 593−606.

Böhm, H.-J. (1992b). The computer program LUDI: A new method for the de novo design of enzyme inhibitors. *Journal of Computer-Aided Molecular Design, 6*(1), 61−78.

Bomgaars, L., Berg, S. L., & Blaney, S. M. (2001). The development of camptothecin analogs in childhood cancers. *The Oncologist, 6*(6), 506−516.

Brangi, M., Litman, T., Ciotti, M., Nishiyama, K., Kohlhagen, G., Takimoto, C., Robey, R., Pommier, Y., Fojo, T., & Bates, S. E. (1999). Camptothecin resistance: Role of the ATP-binding cassette (ABC), mitoxantrone-resistance half-transporter (MXR), and potential for glucuronidation in MXR-expressing cells. *Cancer Research, 59*(23), 5938−5946.

Bueno, O., Estévez Gallego, J., Martins, S., Prota, A. E., Gago, F., Gómez-Sanjuan, A., & Priego, E. M. (2018). High-affinity ligands of the colchicine domain in tubulin based on a structure-guided design. *Scientific Reports, 8*(1), 1−17.

Cao, N., Tan, K., Annamalai, T., Joachimiak, A., & Tse-Dinh, Y. C. (2018). Investigating mycobacterial topoisomerase I mechanism from the analysis of metal and DNA substrate interactions at the active site. *Nucleic Acids Research, 46*(14), 7296−7308.

Champoux, J. J. (2001). DNA topoisomerases: Structure, function and mechanism. *Annual Review of Biochemistry, 70*, 369−413.

Chen, A. Y., Shih, S. J., Garriques, L. N., Rothenberg, M. L., Hsiao, M., & Curran, D. P. (2005). Silatecan DB-67 is a novel DNA topoisomerase I-targeted radiation sensitizer. *Molecular Cancer Therapeutics, 4*(2), 317−324.

Chivere, V. T., Kondiah, P. P. D., Choonara, Y. E., & Pillay, V. (2020). Nanotechnology-based biopolymeric oral delivery platforms for advanced cancer treatment. *Cancers, 12*(2).

Cretaio, E., Pattarello, L., Fontebasso, Y., Banedetti, P., & Losasso, C. (2007). Human DNA topoisomerase IB: Structure and functions. *Italian Journal of Biochemistry, 56*(2), 91−102.

Dahlin, J. L., & Walters, M. A. (2016). How to triage PAINS-full research. *Assay and Drug Development Technologies, 14*(3), 168−174.

Daisy, P., Singh, S. K., Vijayalakshmi, P., Selvaraj, C., Rajalakshmi, M., & Suveena, S. (2011). A database for the predicted pharmacophoric features of medicinal compounds. *Bioinformation, 6*(4), 167−168.

Danks, M. K. (2001). Topoisomerase enzymes as drug targets BT. In M. Schwab (Ed.), *Encyclopedic reference of cancer* (pp. 900−903). Berlin, Heidelberg: Springer Berlin Heidelberg.

Denny, W. A. (2007). In J. B. Taylor, & D. J. B. T.-C. M. C. I. I. Triggle (Eds.), *7.05 − Deoxyribonucleic acid topoisomerase inhibitors* (pp. 111−128). Oxford: Elsevier.

Dev, S., Dhaneshwar, S., & Mathew, B. (2016). Discovery of camptothecin based topoisomerase I inhibitors: Identification using an atom based 3D-QSAR, pharmacophore modeling, virtual screening and molecular docking approach. *Combinatorial Chemistry & High Throughput Screening, 19*(9), 752−763.

Dexheimer, T. S., & Pommier, Y. (2008). DNA cleavage assay for the identification of topoisomerase I inhibitors. *Nature Protocols, 3*(11), 1736−1750.

Dixon, S. L., Smondyrev, A. M., Knoll, E. H., Rao, S. N., Shaw, D. E., & Friesner, R. A. (2006). PHASE: A new engine for pharmacophore perception, 3D QSAR model development, and 3D database screening: 1. Methodology and preliminary results. *Journal of Computer-Aided Molecular Design, 20*(10−11), 647−671.

Docking, M., Santos, R. N., Ferreira, L. G., & Andricopulo, A. D. (2018). Practices in molecular docking and structure-based virtual screening. In *Computational drug discovery and design* (Vol. 1762, pp. 31−50). New York, NY: Humana Press.

Dostálová, A., & Volf, P. (2012). Leishmania development in sand flies: Parasite-vector interactions overview. *Parasites & Vectors, 5*, 276.

Drwal, M. N., Agama, K., Wakelin, L. P. G., Pommier, Y., & Griffith, R. (2011). Exploring DNA topoisomerase I ligand space in search of novel anticancer agents. *PLoS One, 6*(9), 1−12.

Dudek, A., Arodz, T., & Galvez, J. (2006). Computational methods in developing quantitative structure-activity relationships (QSAR): A review. *Combinatorial Chemistry & High Throughput Screening, 9*(3), 213–228.

Fang, C., Xiao, Z., & Guo, Z. (2011). Generation and validation of the first predictive pharmacophore model for cyclin-dependent kinase 9 inhibitors. *Journal of Molecular Graphics and Modelling, 29*(6), 800–808.

Fei, J., Zhou, L., Liu, T., & Tang, X. Y. (2013). Pharmacophore modeling, virtual screening, and mo-lecular docking studies for discovery of novel Akt2 inhibitors. *International Journal of Medical Sciences, 10*(3), 265–275.

Ferreira, L. G., Dos Santos, R. N., Oliva, G., & Andricopulo, A. D. (2015). Molecular docking and structure-based drug design strategies. *Molecules, 20*.

Golbraikh, A., Shen, M., Xiao, Z., De Xiao, Y., Lee, K. H., & Tropsha, A. (2003). Rational selection of training and test sets for the development of validated QSAR models. *Journal of Computer-Aided Molecular Design, 17*(2–4), 241–253.

Graul, A. I. (2003). The year's new drugs. *Drug News and Perspectives, 16*(1), 22–39.

Graul, A. I. (2004). The year's new drugs. *Drug News and Perspectives, 17*(1), 43.

van Griensven, J., & Diro, E. (2012). Visceral leishmaniasis. *Infectious Disease Clinics of North America, 26*(2), 309–322.

Grisoni, F., Merk, D., Consonni, V., Hiss, J. A., Tagliabue, S. G., Todeschini, R., & Schneider, G. (2018). Scaffold hopping from natural products to synthetic mimetics by holistic molecular similarity. *Communications Chemistry, 1*(1).

Haluska, P., Saleem, A., Edwards, T. K., & Rubin, E. H. (1998). Interaction between the N-terminus of human topoisomerase I and SV40 large T antigen. *Nucleic Acids Research, 26*(7), 1841–1847.

Herwaldt, B. L. (1999). Leishmaniasis. *Lancet, 354*(9185), 1191–1199.

Hessler, G., & Baringhaus, K. H. (2010). The scaffold hopping potential of pharmacophores. *Drug Discovery Today: Technologies, 7*(4), e263–e269.

Hevener, K. E., Verstak, T. A., Lutat, K. E., Riggsbee, D. L., & Mooney, J. W. (2018). Recent developments in topoisomerase-targeted cancer chemotherapy. *Acta Pharmaceutica Sinica B, 8*(6), 844–861.

Hsiang, Y. H., Hertzberg, R., Hecht, S., & Liu, L. F. (1985). Camptothecin induces protein-linked DNA breaks via mammalian DNA topoisomerase I. *The Journal of Biological Chemistry, 260*(27), 14873–14878.

Hu, W., Huang, X. S., Wu, J. F., Yang, L., Zheng, Y. T., Shen, Y. M., Li, Z. Y., & Li, X. (2018). Discovery of novel topoisomerase II inhibitors by medicinal chemistry approaches. *Journal of Medicinal Chemistry, 61*(20), 8947–8980.

Hunter, C. A., & Sanders, J. K. M. (1990). The nature of π-π interactions. *Journal of the American Chemical Society, 112*(14), 5525–5534.

Hu, Y., Stumpfe, D., & Bajorath, J. (2017). Recent advances in scaffold hopping. *Journal of Medicinal Chemistry, 60*(4), 1238–1246.

Jiang, Z., Zhang, Z., Cui, G., Sun, Z., Song, G., Liu, Y., & Zhong, G. (2018). DNA Topoisomerase 1 structure-BASED design, synthesis, activity evaluation and molecular simulations study of new 7-amide camptothecin derivatives against *Spodoptera frugiperda*. *Frontiers in Chemistry, 6*(Sep), 1–14.

John, S., Thangapandian, S., Arooj, M., Hong, J. C., Kim, K. D., & Lee, K. W. (2011). Development, evaluation and application of 3D QSAR Pharmacophore model in the discovery of potential human renin inhibitors. *BMC Bioinformatics, 12*(14), S4.

Kandakatla, N., & Ramakrishnan, G. (2014). Ligand based pharmacophore modeling and virtual screening studies to design novel HDAC2 inhibitors. *Advances in Bioinformatics, 2014*, 1–11.

Karelson, M. (2000). *Molecular descriptors in QSAR/QSPR*. John Wiley & Sons Inc.

Karki, R., Thapa, P., Kang, M. J., Jeong, T. C., Nam, J. M., Kim, H. L., Na, Y., Cho, W. J., Kwon, Y., & Lee, E. S. (2010). Synthesis, topoisomerase I and II inhibitory activity, cytotoxicity, and structure-activity relationship study of hydroxylated 2,4-diphenyl-6-aryl pyridines. *Bioorganic and Medicinal Chemistry, 18*(9), 3066–3077.

Kathiravan, M. K., Khilare, M. M., Chothe, A. S., & Nagras, M. A. (2013). Design and development of topoisomerase inhibitors using molecular modelling studies. *Journal of Chemical Biology, 6*(1), 25–36.

Kathiravan, M. K., Khilare, M. M., Nikoomanesh, K., Chothe, A. S., & Jain, K. S. (2013). Topoisomerase as target for antibacterial and anticancer drug discovery. *Journal of Enzyme Inhibition and Medicinal Chemistry, 28*(3), 419–435.

Kato, S., & Kikuchi, A. (1998). DNA topoisomerase: The key enzyme that regulates DNA super structure. *Nagoya Journal of Medical Science, 61*(1–2), 11–26.

Kawai, K., Nagata, N., & Takahashi, Y. (2014). De novo design of drug-like molecules by a fragment-based molecular evolutionary approach. *Journal of Chemical Information and Modeling, 54*(1), 49–56.

Keszthelyi, A., Minchell, N. E., & Baxter, J. (2016). The causes and consequences of topological stress during DNA replication. *Genes, 7*(12), 1–13.

Kizisar, D., Subasi, N. T., Eroksuz, S., Demir, A. S., & Mert, O. (2016). Investigation of the stabilization of camptothecin anticancer drug via PSA-PEG polymeric particles. *Anadolu University Journal of Science and Technology-A Applied Sciences and Engineering, 17*(1), 221–231.

Klebe, G. (2013). Pharmacophore hypotheses and molecular comparisons. In G. Klebe (Ed.), *Drug design: Methodology, concepts, and mode-of-action* (pp. 349–370). Springer International Publishing.

Klebe, G., Abraham, U., & Mietzner, T. (1994). Molecular similarity Indices in a comparative analysis (CoMSIA) of drug molecules to correlate and predict their biological activity. *Journal of Medicinal Chemistry, 37*(24), 4130–4146.

Koch, O., Cappel, D., Nocker, M., Jäger, T., Flohé, L., Sotriffer, C. A., & Selzer, P. M. (2013). Molecular dynamics reveal binding mode of glutathionylspermidine by trypanothione synthetase. *PLoS One, 8*(2), 1–10.

Kohlhagen, G., Paull, K. D., Cushman, M., Nagafuji, P., & Pommier, Y. (1998). Protein-linked DNA strand breaks induced by NSC 314622, a novel noncamptothecin topoisomerase I poison. *Molecular Pharmacology, 54*(1), 50−58.

Kubinyi, H. (1997). QSAR and 3D QSAR in drug design Part 1: Methodology. *Drug Discovery Today, 2*(11), 457−467.

Kundu, B., Das, S. K., Paul Chowdhuri, S., Pal, S., Sarkar, D., Ghosh, A., Mukherjee, A., Bhattacharya, D., Das, B. B., & Talukdar, A. (2019). Discovery and mechanistic study of tailor-made quinoline derivatives as topoisomerase 1 poison with potent anticancer activity. *Journal of Medicinal Chemistry, 62*(7), 3428−3446.

Langer, T., & Hoffmann, R. D. (2008). Pharmacophore modelling. *Encyclopedia of Molecular Pharmacology*, 960.

Lavergne, O., Lesueur-Ginot, L., Rodas, F. P., Kasprzyk, P. G., Pommier, J., Demarquay, D., Prévost, G., Ulibarri, G., Rolland, A., Schiano-Liberatore, A. M., Harnett, J., Pons, D., Camara, J., & Bigg, D. C. (1998). Homocamptothecins: Synthesis and antitumor activity of novel E-ring-modified camptothecin analogues. *Journal of Medicinal Chemistry, 41*(27), 5410−5419.

Leach, A. R., Gillet, V. J., Lewis, R. A., & Taylor, R. (2010). Three-dimensional pharmacophore methods in drug discovery. *Journal of Medicinal Chemistry, 53*(2), 539−558.

Lebel, M., Spillare, E. A., Harris, C. C., & Leder, P. (1999). The Werner syndrome gene product co-purifies with the DNA replication complex and interacts with PCNA and topoisomerase I. *Journal of Biological Chemistry, 274*(53), 37795−37799.

Lee, C. H., Huang, H. C., & Juan, H. F. (2011). Reviewing ligand-based rational drug design: The search for an ATP synthase inhibitor. *International Journal of Molecular Sciences, 12*(8), 5304−5318.

Liao, Z., Robey, R. W., Guirouilh-Barbat, J., To, K. K. W., Polgar, O., Bates, S. E., & Pommier, Y. (2008). Reduced expression of DNA topoisomerase I in SF295 human glioblastoma cells selected for resistance to homocamptothecin and diflomotecan. *Molecular Pharmacology, 73*(2), 490−497.

Li, M., Jin, W., Jiang, C., Zheng, C., Tang, W., You, T., & Lou, L. (2009). 7-Cycloalkylcamptothecin derivatives: Preparation and biological evaluation. *Bioorganic and Medicinal Chemistry Letters, 19*(15), 4107−4109.

Lindsey, R. H., Pendleton, M., Ashley, R. E., Mercer, S. L., Deweese, J. E., & Osheroff, N. (2014). Catalytic core of human topoisomerase IIα: Insights into enzyme-DNA interactions and drug mechanism. *Biochemistry, 53*(41), 6595−6602.

Li, H., Sutter, J., & Hoffmann, R. (2000). Pharmacophore perception, development, and use in drug design, ch. Hypogen: An automated system for generating 3D predictive pharmacophore models. *International University Line*, 49−68.

Liu, W., Zhu, L., Guo, W., Zhuang, C., Zhang, Y., Sheng, C., Cheng, P., Yao, J., Wang, W., Dong, G., Wang, S., Miao, Z., & Zhang, W. (2011). Synthesis and biological evaluation of novel 7-acyl homocamptothecins as Topoisomerase I inhibitors. *European Journal of Medicinal Chemistry, 46*(6), 2408−2414.

Lloyd, D. G., Golfis, G., Knox, A. J. S., Fayne, D., Meegan, M. J., & Oprea, T. I. (2006). Oncology exploration: Charting cancer medicinal chemistry space. *Drug Discovery Today, 11*(3−4), 149−159.

Mamidala, R., Majumdar, P., Jha, K. K., Bathula, C., Agarwal, R., Chary, M. T., Mazumdar, H., Munshi, P., & Sen, S. (2016). Identification of *Leishmania donovani* topoisomerase 1 inhibitors via intuitive scaffold hopping and bioisosteric modification of known Top 1 inhibitors. *Scientific Reports, 6*(26603), 1−13.

Martino, E., Della Volpe, S., Terribile, E., Benetti, E., Sakaj, M., Centamore, A., Sala, A., & Collina, S. (2017). The long story of camptothecin: From traditional medicine to drugs. *Bioorganic and Medicinal Chemistry Letters, 27*(4), 701−707.

Morrell, A., Placzek, M., Parmley, S., Antony, S., Dexheimer, T. S., Pommier, Y., & Cushman, M. (2007). Nitrated indenoisoquinolines as topoisomerase I inhibitors: A systematic study and optimization. *Journal of Medicinal Chemistry, 50*(18), 4419−4430.

Nagarajan, M., Morrell, A., Ioanoviciu, A., Antony, S., Kohlhagen, G., Agama, K., Hollingshead, M., Pommier, Y., & Cushman, M. (2006). Synthesis and evaluation of indenoisoquinoline topoisomerase I inhibitors substituted with nitrogen heterocycles. *Journal of Medicinal Chemistry, 49*(21), 6283−6289.

Pal, S., Kumar, V., Kundu, B., Bhattacharya, D., Preethy, N., Reddy, M. P., & Talukdar, A. (2019). Ligand-based pharmacophore modeling, virtual screening and molecular docking studies for discovery of potential topoisomerase I inhibitors. *Computational and Structural Biotechnology Journal, 17*, 291−310.

Pommier, Y. (2006). Topoisomerase I inhibitors: Camptothecins and beyond. *Nature Reviews Cancer, 6*(10), 789−802.

Pommier, Y. (2009). DNA topoisomerase I inhibitors: Chemistry, biology, and interfacial inhibition. *Chemical Reviews, 109*(7), 2894−2902.

Pommier, Y., Leo, E., & Zhang, H.,M. C. (2018). DNA topoisomerases and their poisoning by anticancer and antibacterial drugs. *Physiology & Behavior, 176*(1), 139−148.

Prival, M. J. (2001). Evaluation of the TOPKAT system for predicting the carcinogenicity of chemicals. *Environmental and Molecular Mutagenesis, 37*(1), 55−69.

Reddy, A. S., Pati, S. P., Kumar, P. P., Pradeep, H. N., & Sastry, G. N. (2007). Virtual screening in drug discovery - a computational perspective. *Current Protein & Peptide Science, 8*(4), 329−351.

Redinbo, M. R., Stewart, L., Kuhn, P., Champoux, J. J., & Hol, W. G. J. (1998). Crystal structures of human topoisomerase I in covalent and noncovalent complexes with DNA. *Science, 279*(5356), 1504−1513.

Roy, A., Chowdhury, S., Sengupta, S., Mandal, M., Jaisankar, P., D'Annessa, I., Desideri, A., & Majumder, H. K. (2011). Development of derivatives of 3, 3′-diindolylmethane as potent Leishmania donovani Bi-subunit topoisomerase IB poisons. *PLoS One, 6*(12).

Sandhaus, S., Chapagain, P. P., & Tse-Dinh, Y. C. (2018). Discovery of novel bacterial topoisomerase i inhibitors by use of in silico docking and in vitro assays. *Scientific Reports, 8*(1), 1−9.

Schuffenhauer, A. (2012). Computational methods for scaffold hopping. *Wiley Interdisciplinary Reviews: Computational Molecular Science, 2*(6), 842−867.

Sharma, P., Xu, S., Schaubel, D. E., Sung, R. S., & Magee, J. C. (2016). Perspectives on biologically active camptothecin derivatives. *Liver Transplantation, 3*(10), 973−982.

Sheng-Yong, Y. (2010). Pharmacophore modeling and applications in drug discovery: Challenges and recent advances. *Drug Discovery Today, 15*(11), 444−450.

Srivastava, V., Negi, A. S., Kumar, J. K., Gupta, M. M., & Khanuja, S. P. S. (2005). Plant-based anticancer molecules: A chemical and biological profile of some important leads. *Bioorganic and Medicinal Chemistry, 13*(21), 5892−5908.

Staker, B. L., Feese, M. D., Cushman, M., Pommier, Y., Zembower, D., Stewart, L., & Burgin, A. B. (2005). Structures of three classes of anticancer agents bound to the human topoisomerase I-DNA covalent complex. *Journal of Medicinal Chemistry, 48*(7), 2336−2345.

Staker, B. L., Hjerrild, K., Feese, M. D., Behnke, C. A., Burgin, A. B., & Stewart, L. (2002). The mechanism of topoisomerase I poisoning by a camptothecin analog. *Proceedings of the National Academy of Sciences of the United States of America, 99*(24), 15387−15392.

Stewart, L., Ireton, G. C., & Champoux, J. J. (1997). Reconstitution of human topoisomerase I by fragment complementation. *Journal of Molecular Biology, 269*(3), 355−372.

Stewart, L., Redinbo, M. R., Qiu, X., Hol, W. G. J., & Champoux, J. J. (1998). A model for the mechanism of human topoisomerase I. *Science, 279*(5356), 1534−1541.

Sun, H., Tawa, G., & Wallqvist, A. (2012). Classification of scaffold-hopping approaches. *Drug Discovery Today, 17*(7−8), 310−324.

Takagi, K., Dexheimer, T. S., Redon, C., Sordet, O., Agama, K., Lavielle, G., Pierré, A., Bates, S. E., & Pommier, Y. (2007). Novel E-ring camptothecin keto analogues (S38809 and S39625) are stable, potent, and selective topoisomerase I inhibitors without being substrates of drug efflux transporters. *Molecular Cancer Therapeutics, 6*(12), 3229−3238.

Tangirala, R. S., Antony, S., Agama, K., Pommier, Y., Anderson, B. D., Bevins, R., & Curran, D. P. (2006). Synthesis and biological assays of E-ring analogs of camptothecin and homocamptothecin. *Bioorganic and Medicinal Chemistry, 14*(18), 6202−6212.

Thapa, P., Karki, R., Choi, H., Choi, J. H., Yun, M., Jeong, B. S., Jung, M. J., Nam, J. M., Na, Y., Cho, W. J., Kwon, Y., & Lee, E. S. (2010). Synthesis of 2-(thienyl-2-yl or -3-yl)-4-furyl-6-aryl pyridine derivatives and evaluation of their topoisomerase I and II inhibitory activity, cytotoxicity, and structure-activity relationship. *Bioorganic and Medicinal Chemistry, 18*(6), 2245−2254.

Thapa, P., Karki, R., Thapa, U., Jahng, Y., Jung, M. J., Nam, J. M., Na, Y., Kwon, Y., & Lee, E. S. (2010). 2-Thienyl-4-furyl-6-aryl pyridine derivatives: Synthesis, topoisomerase I and II inhibitory activity, cytotoxicity, and structure-activity relationship study. *Bioorganic and Medicinal Chemistry, 18*(1), 377−386.

Tsao, Y.-P., Wu, H.-Y., & Liu, L. F. (1989). Transcription-driven supercoiling of DNA: Direct biochemical evidence from in vitro studies. *Cell, 56*(1), 111−118.

Tselepi, M., Papachristou, E., Emmanouilidi, A., Angelis, A., Aligiannis, N., Skaltsounis, A. L., Kouretas, D., & Liadaki, K. (2011). Catalytic inhibition of eukaryotic topoisomerases i and II by flavonol glycosides extracted from vicia faba and lotus edulis. *Journal of Natural Products, 74*(11), 2362−2370.

Urasaki, Y., Takebayashi, Y., & Pommier, Y. (2000). Activity of a novel camptothecin analogue, homocamptothecin, in camptothecin-resistant cell lines with topoisomerase I alterations. *Cancer Research, 60*(23), 6577−6580.

Van Montfort, R. L. M., & Workman, P. (2017). Structure-based drug design: Aiming for a perfect fit. *Essays in Biochemistry, 61*(5), 431−437.

Villa, H., Marcos, A. R. O., Reguera, R. M., Balaña-Fouce, R., García-Estrada, C., Pérez-Pertejo, Y., Tekwani, B. L., Myler, P. J., Stuart, K. D., Bjornsti, M. A., & Ordóñez, D. (2003). A novel active DNA topoisomerase I in Leishmania donovani. *Journal of Biological Chemistry, 278*(6), 3521−3526.

Wall, M. E., & Wani, M. C. (1995). Camptothecin and taxol: Discovery to clinic—thirteenth Bruce F. Cain Memorial Award Lecture. *Cancer Research, 55*(4), 753−760.

Wall, M. E., Wani, M. C., Cook, C. E., Palmer, K. H., McPhail, A. T., & Sim, G. A. (1966). Plant antitumor agents. I. The isolation and structure of camptothecin, a novel alkaloidal leukemia and tumor inhibitor from *Camptotheca acuminata*. *Journal of the American Chemical Society, 88*(16), 3888−3890.

Wang, J. C. (2002). Cellular roles of DNA topoisomerases: A molecular perspective. *Nature Reviews Molecular Cell Biology, 3*(6), 430−440.

Wang, L., Fang, K., Cheng, J., Li, Y., Huang, Y., Chen, S., Dong, G., Wu, S., & Sheng, C. (2020). Scaffold hopping of natural product evodiamine: Discovery of a novel antitumor scaffold with excellent potency against colon cancer. *Journal of Medicinal Chemistry, 63*(2), 696−713.

Wang, X., Song, K., Li, L., & Chen, L. (2018). Structure-based drug design strategies and challenges. *Current Topics in Medicinal Chemistry, 18*(12), 998−1006.

Yousefinejad, S., & Hemmateenejad, B. (2015). Chemometrics tools in QSAR/QSPR studies: A historical perspective. *Chemometrics and Intelligent Laboratory Systems, 149*, 177−204.

Zhang, L., Zhang, Y., He, W., Ma, D., & Jiang, H. (2012). Effects of camptothecin and hydroxycamptothecin on insect cell lines Sf21 and IOZCAS-Spex-II. *Pest Management Science, 68*(4), 652−657.

Zhi, Y., Yang, J., Tian, S., Yuan, F., Liu, Y., Zhang, Y., & Chen, Z. (2012). Quantitative structure-activity relationship studies on indenoisoquinoline topoisomerase I inhibitors as anticancer agents in human renal cell carcinoma cell line SN12C. *International Journal of Molecular Sciences, 13*(5), 6009−6025.

CHAPTER 22

Docking-Based Virtual Screening Using PyRx Tool: Autophagy Target Vps34 as a Case Study

SREE KARANI KONDAPURAM • SAILU SARVAGALLA • MOHANE SELVARAJ COUMAR

1 INTRODUCTION

Drug discovery is an attractive area of research that enables the application of cutting-edge biomedical findings to improve the health of people (Ng, 2009). In ancient times, people have begun to use naturally occurring resources, including plant products, for the treatment of various diseases. Based on the fact that plants contain valuable products which possess large medicinal benefits, researchers have started extraction and purification of the medicinally active components and marketed them as commercial medicines (Mohsa & Greig, 2017; Pan et al., 2013). A recent analysis of the approved drugs has identified that approximately 50% of them are either natural products or designed based on them (Newman & Cragg, 2020; Rankovic & Morphy, 2010); i.e., the natural product could be a good source of lead molecules for drug discovery.

A lead is a prototype molecule with essential biological activity. The discovery of the lead molecule is an important step in drug discovery (Hughes et al., 2011). Lead molecules can be obtained from different sources, and as aforementioned, natural products are a good place to look for them. Additionally, lead molecules can be obtained from already approved drugs, i.e., drug repurposing, where new use for old drugs is identified (Pushpakom et al., 2018; Sarvagalla et al., 2019). Alternatively, lead can be obtained from synthetic molecules such as those obtained from combinatorial synthesis or analog synthesis (Liu et al., 2017). One of the caveats of identifying a suitable lead molecule is that a large number of compounds need to be experimentally tested, which is challenging and can be a time-consuming and costly affair. Therefore, as an alternative to cut both the time and cost, computational method for screening these compounds has become a

promising approach in the drug discovery field. Virtual screening (VS) is a computational approach (Forli, 2015; Lionta et al., 2014; Scior et al., 2012; Wang et al., 2020) where a large number of molecules are evaluated for their potential using a set of criteria. In docking-based virtual screening (DBVS), the ability of molecules to interact with a drug target is used as the criteria to evaluate the database compounds. Based on the criterion, which is docking score in DBVS, molecules are prioritized for testing experimentally. Hence, VS methods will help in reducing the number of molecules that are experimentally evaluated to find a lead molecule. Fig. 22.1 shows various options available for lead identification in a drug discovery program.

To carry out DBVS, several free and commercial software/tools were developed and made available for researchers. Interested readers may refer to Chapter 10 in this book for a nonexhaustive list of such docking software. Also, chapter 10 discusses various databases of ligands and targets that are useful in VS experiments.

In this chapter, we demonstrate in a step-by-step manner the use of the PyRx 0.8 tool to perform DBVS of database compounds for autophagy target Vps34 (vacuolar protein sorting 34), to identify potential molecules for testing in the lab.

2 PYRX 0.8 AS A DOCKING-BASED VIRTUAL SCREENING TOOL

To date, several molecular docking cum VS tools/software have been developed and made freely available for researchers. In this work, we use PyRx 0.8 tool (https://pyrx.sourceforge.io/; https://sourceforge.net/projects/pyrx/) to virtually screen the database compounds to identify molecules for biological testing. So

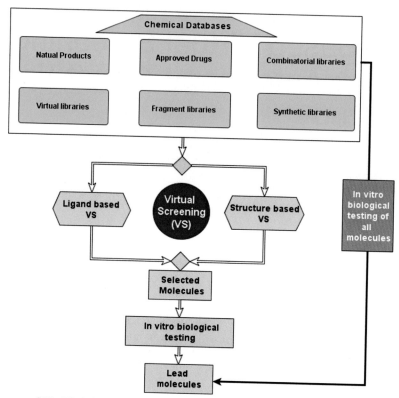

FIG. 22.1 Lead identification strategies in a drug discovery program.

a brief introduction to the tool and its recent use is provided below.

PyRx 0.8 (Dallakyan & Olson, 2015) is a VS tool developed by Sargis Dallakyan and Arthur J. Olson. The software can run on Windows, Linux, and Mac OS and is written in Python language. For the convenience of the users and to improve the performance, PyRx incorporates AutoDock 4 (Forli et al., 2016), Autodock Vina (Trott & Olson, 2010), AutoDock tools, and Open Babel (O'Boyle et al., 2011), so that DBVS could be performed seamlessly. PyRx 0.8 version is freely available, but the higher versions are available at a nominal cost.

PyRx uses both AutoDock 4 and AutoDock Vina to carry out the docking process. AutoDock tool is used for preparing the input files and acts as a visualization tool to analyze the docking results, while Open Babel provides an option for importing ligand SDF files and cleaning them. The AutoDock 4 uses Assisted Model Building with Energy Refinement (AMBER) force field and a large set of diverse protein—ligand complexes with known inhibition constants for scoring the docked poses. Even though AutoDock Vina inherits some of the ideas from AutoDock, it functions differently. Autodock Vina's new search algorithm and hybrid scoring function combining empirical and knowledge-based scoring function offers a full range of speed for DBVS. The computational efficiency is enhanced by its ability to use multiple CPUs or CPU cores. This provides improvements in binding mode predictions and performs the screening up to two orders of magnitude faster than AutoDock.

Several research groups have used PyRx in their work to carry out both docking and VS to understand drug—target interaction as well as to identify leads. Some of the recent investigations are highlighted to show the usefulness of the tool. By applying AutoDock Vina in the PyRx tool, researchers have shown antileishmanial activity with the target leishmanolysin against biflavonoids. The study has shown lanaroflavone as a potential agent for experimental binding affinity testing (Mercado-Camargo et al., 2020). Adeniji et al. (2020) utilized the PyRx tool to investigate the ligand—receptor interactions between quinoline derivatives and the target DNA gyrase of *Mycobacterium tuberculosis*. They conclude that the quinoline derivative 10 (−18.8 kcal/

mol) could be a good structural template for the development of the antitubercular drug as it shows better binding to the target than isoniazid (−14.6 kcal/mol).

Kalia et al. (2019) in 2019 retrieved 75 natural compounds from the zinc database and screened them using the PyRx tool against Fabl, an enzymatic target in the fatty acid biosynthesis pathway of *Pseudomonas aeruginosa*. They have identified potential molecules that could be evaluated as quorum sensing inhibitors against *P. aeruginosa*. Gandhi, Rupareliya, Shukla, Donga, & Acharya (2020) investigated 160 plant constituents from different plants used in Ayurvedic formulation against the COVID-19 3clpro/Mpro (Protein Data Bank [PDB] ID: 6LU7) target using PyRx tool. From the in silico results and Ayurvedic concept of disease etiology, they conclude that Nagaradi Kashayam, which includes Sunthi (*Zingiber officinale* Roscoe), Guduchi (*Tinospora cordifolia* Miers.), Pushkarmool (*Inula racemosa* Hook.F.), and Kantakari (*Solanum virginianum* L.), may have an appreciable effect in combating SARS-CoV-2.

Abdelfatah et al. (2019) retrieved 30,793 chemically modified natural products from the ZINC database and utilized the PyRx tool to screen against PLK1. Based on in silico screening results, 25 compounds were purchased and tested experimentally to identify ZINC20503376 as a potent PLK1 inhibitor. This compound could act as a novel scaffold for selective PLK1 inhibitor design. Singh and Mohanty (2018) screened 617 compounds from PubChem, ZINC, and ChemEMBL database using PyRx and iGEMDOCK tools against PPAR-γ, for identifying possible leads for metabolic disorder treatment. Among them, 30 compounds showed good binding affinity; they were further filtered based on ADMET and molecular dynamics simulation to get three potential compounds (ZINC IDs 00181552, 00276456, 00298314) for experimental testing (PPAR-γ modulation). Hassaan et al. (2016) used two different docking programs, AutoDock Vina in PyRx and MOE, to mine the ZINC database for novel PDE9 inhibitors. Based on consensus scoring, they selected compounds for experimental testing in cancer cell lines, which identified three novel chemical scaffolds that inhibited the growth of breast tumor cell lines, MCF-7 and MDA-468.

3 CASE STUDY: DBVS TO IDENTIFY SMALL MOLECULE INHIBITORS OF VPS34

In this section, we discuss in a step-by-step manner DBVS protocol to identify potential small molecule inhibitors of Vps34. For this purpose, PyRx was used. Phosphatidylinositol 3-kinase, Vps34 (PDB ID:

4UWH), is chosen as the target and the compound library used for the VS experiment was obtained from Selleckchem (https://www.selleckchem.com/).

3.1 Vps34—a Crucial Regulator of the Autophagy Process

Macroautophagy (hereafter referred to as autophagy) is a highly conserved catabolic process and stress response, protecting cells from unfavorable condiions, such as nutrient starvation (Morel et al., 2017). Moreover, autophagy removes aggregated proteins, damaged organelles, and intracellular pathogens. The dysfunction of various signaling components of the autophagy pathway has been implicated in cancer, liver disease, heart disease, and several neurodegenerative diseases such as Alzheimer's, Parkinson's, and Huntington's diseases (Saha et al., 2018). The autophagy process begins with the nucleation of phagophores. These phagophores mature by recruiting lipids and form a double-membrane structure referred to as autophagosomes. Furthermore, these autophagosomes carry cell's cargo to lysosomes and fuse with them, forming autolysosomes. Cellular components are degraded in autolysosomes and get recycled (Kondapuram et al., 2019).

Autophagy is a complex process and is accomplished by the involvement of numerous signaling components; among them, class III phosphatidylinositol 3-kinase, Vps34, is a critical regulator of autophagy initiation (Stjepanovic et al., 2017). Vps34 is a class III phosphoinositide 3-kinase (PI3K) that phosphorylates phosphatidylinositol to generate phosphatidylinositol 3-phosphate [PI(3)P], a phospholipid central for membrane trafficking processes and autophagosome formation during the autophagy process. Accumulating pieces of evidence suggest that Vps34 may be involved in the development of several human cancers (Marinkovi et al., 2018). A more recent study has reported that Vps34 acts as a transcriptional activator of p62 in the process of cancer progression (Jiang et al., 2017). Besides, Vps34 expression has shown a positive correlation with the tumorigenic signature of human breast cancer. It has also been reported that inhibition of Vps34 resulted in the reduction of tumor growth in melanoma and colorectal cancer (Noman et al., 2020). Therefore, developing specific inhibitors of Vps34 will help in better understanding autophagy and its association with cancer progression. Moreover, such inhibitors will be in addition to our cache of cancer therapeutics for the future. So, in this case study, we choose Vps34 as the target to carry our DBVS to help in identifying suitable molecules for testing in the laboratory.

3.2 Docking-Based Virtual Screening Using PyRx 0.8

Here, DBVS is used for identifying suitable hits for the target Vps34. The DBVS protocol has three steps: (1) protein preparation, (2) ligand preparation, and (3) docking and analysis. An experimentally (crystallographic, NMR spectroscopic, and cryo-EM) determined or predicted (homology modeling, fold recognition, and threading) 3D structure sometimes not only appears with missing atoms, side chains, or even complete residues but also may contain unwanted water molecules. These are potential factors that interfere with the molecular docking process and affect the accuracy of the results. Thus, repairing the structures or fixing the errors by using suitable tools or servers such as WHAT IF (https://swift.cmbi.umcn.nl/servers/html/index.html) is widely followed and is an essential activity for getting reliable docking results. Once the structure refinement task is completed, loading the structure into Protein Preparation Wizard, an embedded module of many docking packages, is a key step for preparing the structure by removing unwanted water molecules, adding hydrogen atoms, and correcting bond orders. Moreover, structure minimization is also a necessary action to reach fruitful outcomes. After all the above protein preparation steps, the structure needs to be saved in the docking format (PDBQT for AutoDock).

Also, the ligands used for docking need to be prepared by optimizing it and generate possible isomers, tautomers, and ionization states. This process is done by the Open Babel program within the PyRx environment. Furthermore, the site of macromolecule for ligand interaction is termed as the active site and is used for defining the "grid box," which needs to be generated before actually starting the docking process. The grid box is defined by the active site residues of the target, and within this grid box, various conformations of the ligands will be docked using a genetic algorithm (Morris et al., 1999). After this step, the docked complexes will be evaluated via binding energy using protein–ligand interactions such as hydrogen bond, electrostatic, and hydrophobic interactions. The following steps demonstrate how to retrieve Vps34 structure from PDB, prepare the protein and ligands, carry out docking, and then analyze the results. The workflow of the DBVS protocol is shown in Fig. 22.2 and discussed in a step-by-step manner below.

3.2.1 Step 1: Retrieving 3D structure of Vps34

- Open the PDB website (https://www.rcsb.org/) and enter the target protein name or PDB ID of the target protein. For instance, 4UWH is entered in the PDB browser search box for retrieving the Vps34 protein structure. In the case of searching with protein name, more than one structure for query name may appear;

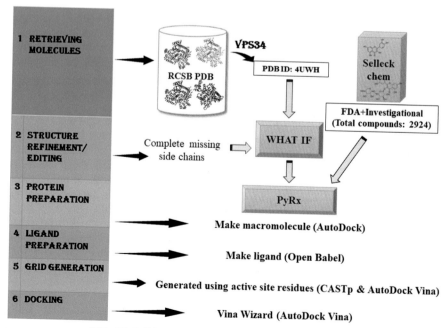

FIG. 22.2 Workflow for docking-based virtual screening.

a better structure can be chosen following the criteria such as domains of interest, the length of the solved structure, normal or mutated form, atomic resolution of the structure, and R-factor. Download the selected structure in PDB format from the dropdown menu (Fig. 22.3).

3.2.2 Step 2: Protein structure refinement

- Open WHAT IF web server (https://swift.cmbi. umcn.nl/servers/html/index.html) and choose the option "Build/check/repair model." In the "Complete a structure" window, upload the PDB file (modeled or downloaded; in this case PDB ID: 4UWH) and run the program. Download the fixed .pdb as a result file (refined and complete structure) (Fig. 22.4).

3.2.3 Step 3: PyRx installation and basic operations

- Download PyRx (https://sourceforge.net/projects/ pyrx/) and install the software.
PyRx has three panels for operation: Navigator, View, and Controls panels. "Navigator panel" is used for navigating the macromolecule and ligands. It has two tabs "Molecules" and "AutoDock," which will be used here. In the "View panel," the macromolecules/ligands can be graphically visualized and also the results are visualized as graphs/tables. Under this panel, there

is the "3D Scene" and "Tables" tab, which will be useful in this work. The "Control panel" provides the ability to execute jobs such as ligand preparation and docking. Under this panel, "Vina Wizard," "AutoDock Wizard," and "Open Babel" tabs are used in this work (Fig. 22.5).

3.2.4 Step 4: Protein preparation

- Create a folder (in desktop) and store the refined target protein as .pdb and ligands as .sdf files.
- Open PyRx and load the macromolecule from the "File" option. On successful launch, protein (Vps34; PDB ID: 4UWH) appears on the View panel under the "3D Scene" tab. As protein preparation can be performed in a single step in PyRx software, in the Navigator panel of the "Molecules" tab, right-click on the protein PDB ID followed by "AutoDock" and "Make Macromolecule" option. Now, the prepared protein structure is automatically saved into a docking format .pdbqt in the Navigator panel of the "AutoDock" tab (Fig. 22.6).

3.2.5 Step 5: Ligand preparation

Ligand structures of interest that are being screened against a target protein either need to be downloaded or can be drawn using various tools including MarvinSketch (https://chemaxon.com/products/marvin/download) or ChemDraw (https://www.perkinelmer.com/category/chemdraw), etc. The ligand files may be saved in either

FIG. 22.3 RCSB Protein Data Bank (PDB) webpage, showing how to download a PDB file.

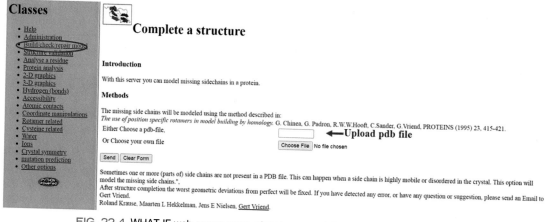

FIG. 22.4 WHAT IF web server page, showing completing a protein structure option.

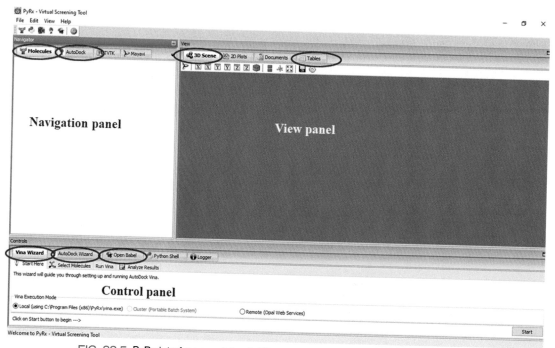

FIG. 22.5 PyRx interface consists of different panels for executing virtual screening.

.sdf or .pdb formats for use by PyRx. Here, the ligand molecules used for VS are "approved and Investigational compounds" from Selleckchem (https://www.selleckchem.com/screening/fda-approved-drug-library.html) downloaded in .sdf format.

- Ligand preparation starts by importing the molecules into PyRx work environment using the "File" option (Fig. 22.7). The list of imported ligand structures will appear under the Control panel of the "Open Babel" tab

- Next, all the imported ligands can be minimized using the "Open Babel" tab in the Control panel, by right-clicking and executing the "Minimize All" option.

- Then, from the Control panel of the "Open Babel" tab, right-click and execute "Convert All to

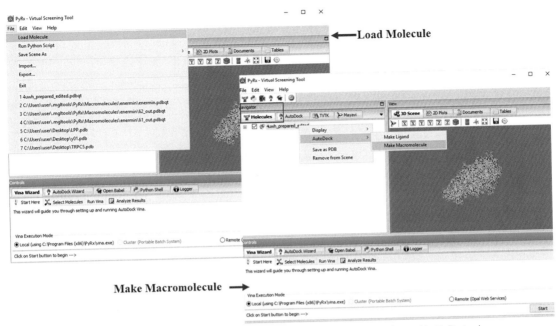

FIG. 22.6 Loading and protein preparation (Make Macromolecule) panel in PyRx tool.

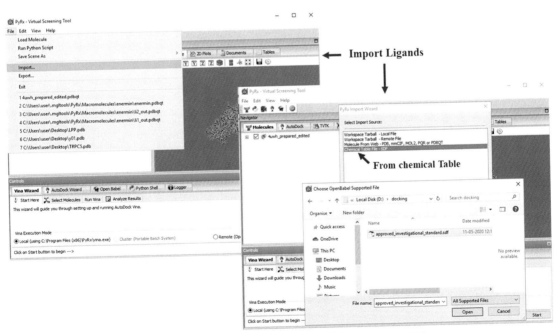

FIG. 22.7 Importing ligands into PyRx 3D Scene (workspace).

AutoDock ligand (.pdbqt)" format. The ligands get converted and stored in .pbdqt format under ligands and can be accessed using the Navigator panel of the "AutoDock" tab ready for docking (Fig. 22.8).

3.2.6 Step 6: Receptor grid generation

In most of the docking programs, directing the ligands to dock into a particular site of the macromolecule is required. A large number of molecules target the substrate binding site, but docking against the allosteric site can also be performed. Therefore, defining a site of macromolecule for ligand interaction is termed as "grid box" generation and is an essential step in docking. Within the grid box region, the ligands are docked and scored. In PyRx, the grid box can be generated by using "Vina Wizard."

- The active site residues were obtained from the binding site prediction tool CASTp (http://sts.bioe. uic.edu/castp/index.html?4jii) (Tian et al., 2018). Once a protein 3D structure is loaded in the browser, by choosing the window "Calculation," select "File," browse from the local system, and "Submit" the .pdb file. Users can interactively interrogate the protein structure and choose the largest pocket residues for grid generation (Fig. 22.9).

- For the grid box generation, choose the active site residues of Vps34 (Phe612, Lys613, Ser614, Leu616, Pro618, Ile634, Lys636, Asp639, Asp640, Leu641, Asp644, Tyr670, Met682, Gln683, Phe684, Ile685, Ser687, Pro689, Glu692, His745, Asp747, Asn748, Leu750, Phe758, Ile760, Asp761, Gly763)

under the Navigator panel "Molecules" tab. This can be done using the toggle selection sphere option, which results in highlighting the selected protein residues in the "3D Scene" tab in the scene panel (Fig. 22.10).

- Next, in the Control panel "Vina Wizard" under the "Start Here" tab, use the "Start" button to execute the grid generation. This will automatically bring up the selection option of the prepared ligands and protein under the Navigator panel "AutoDock" tab. Choose all the ligands and proteins and use the Control panel to move to the next step by using the "Forward" button. This will result in the placement of a grid box over the active site residues in the "3D Scene" window. The box size can be adjusted manually to encompass all the selected (highlighted) residues of the protein active site. In the case of Vps34, the generated grid box size had a dimension of 41, 41, 25 Å (Fig. 22.11).

3.2.7 Step 7: Docking-based virtual screening

Most of the algorithms used in docking programs calculate the binding affinity of a ligand to the target by using a scoring function. Using the scoring function, the docked ligands are ranked and the best-ranked ligands could be chosen for further analysis and experimental testing.

- In PyRx, after the grid generation, docking can be started by using the "Forward" button in the Control panel of "Vina Wizard" under the "Run Vina" tab (Fig. 22.11).

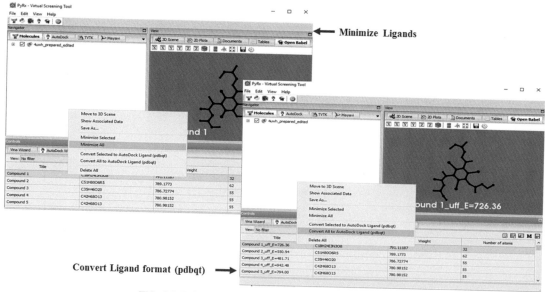

FIG. 22.8 Ligand preparation in PyRx using Open Babel.

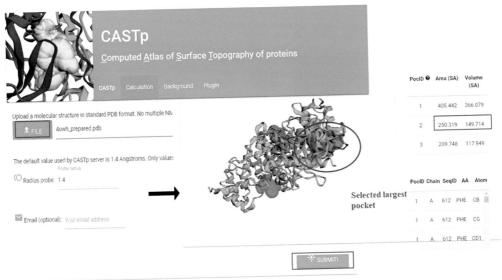

FIG. 22.9   CASTp webpage server, showing active site residues prediction.

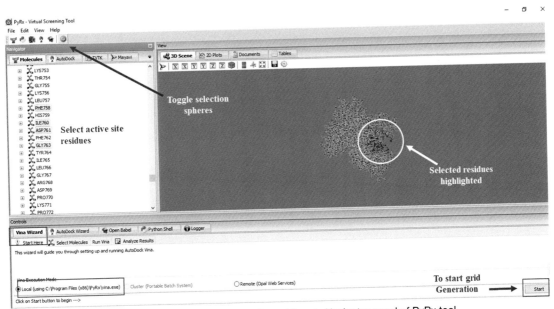

FIG. 22.10   Selection of active site residues in Navigator panel of PyRx tool.

- Progress of the job can be monitored both in the control and scene panels (Fig. 22.12).
- Once the job is completed, the results are displayed under the View panel of the "Tables" tab as "Docking Results" in a rank order manner based on the ligand binding energy (Fig. 22.12).

In our screening exercise, compound number 16 was the topmost hit with a binding energy of −12.7 kcal/mol.

3.2.8 Step 8: Result analysis

The results are stored and are available for analysis from the Navigator panel "AutoDock" tab under the "Macromolecules" section as the "4uwh_prepared_edited" folder (in this study, protein name is 4uwh).

- Open the "4uwh_prepared_edited" folder, where the protein and ligand output file can be accessed. Select both the protein "4uwh_prepared_edited.pdqt" and the ligand of interest (e.g.,

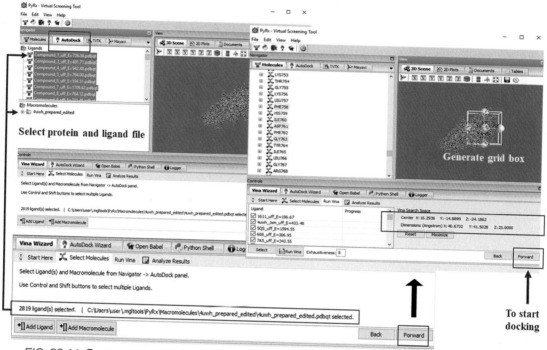

FIG. 22.11 Receptor grid generation by encompassing all the active site residues. This is done by adjusting the grid box dimension.

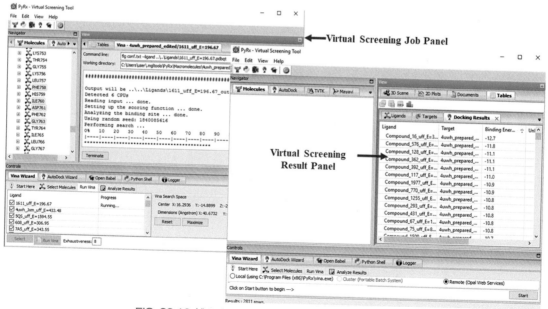

FIG. 22.12 Virtual screening job and result panel in PyRx.

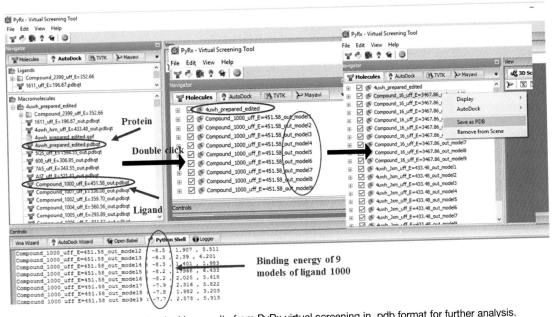

FIG. 22.13 Saving the docking results from PyRx virtual screening in .pdb format for further analysis.

compound_1100_out.pdbqt) and double-click to move these files to the "Molecules" tab (Fig. 22.13).

- In the "Molecules" tab, protein and nine poses (model 1–9) of the selected ligand appear. Right-click on the particular ligand pose and save as a .pdb file in a place of your convenience. The binding energies of all the nine ligand models are displayed in the "Python Shell" of the Control panel (Fig. 22.13). However, in the virtual screening View panel "Docking Results" tab, only model 1 with the highest binding energy is displayed for each ligand (Fig. 22.12).
- Similarly, save the protein as .pdb file by right-clicking on the file (4uwh_prepared_edited.pdbqt) from the "Molecule" tab in the Navigator panel.
- For visualizing protein–ligand interaction, open the saved protein and ligand .pdb files simultaneously in Pymol (https://pymol.org/2/) or Chimera (https://www.cgl.ucsf.edu/chimera/download.html) or LIGPLOT (https://www.ebi.ac.uk/thornton-srv/software/LigPlus/download.html) or other visualization software and analyze the displayed results visually.

Table 22.1 shows the PyRx VS results for the target Vps34. The top-scored 20 ligands are listed, along with their binding energy and H-bond interacting residues.

4 DISCUSSION

In the drug discovery process, identifying a suitable lead molecule for development is of paramount importance.

In this respect, there are several strategies to identify lead (Guido et al., 2011; Holenz et al., 2016; Rankovic & Morphy, 2010). One of the methods commonly used nowadays is the computer-aided drug design (CADD), among which DBVS of a large database has provided leads for several drugs (Berry et al., 2015; Ferreira et al., 2015). The DBVS can rank molecules based on the predicted binding affinity to a target macromolecule and can help to pick up suitable molecules for testing. The cost of running a VS experiment is minimal compared to high-throughput screening (wet lab) experiments. With the advances in computer software and hardware, and also with the increasing number of publicly available data, VS will continue to aid the discovery of new drugs.

Recently, the importance of autophagy in several pathological states such as cancer, inflammation, aging, neurodegeneration, and heart diseases is being investigated intensively. This has suggested Vps34 as an important signaling molecule in the initiation and sustenance of the autophagy process. Moreover, few Vps34 inhibitors have shown a promising anticancer effect, both in vitro and in vivo. In this chapter, we have attempted to screen a large number of compounds against Vps34 using DBVS to identify possible molecules for testing in the lab. For this purpose, "Approved and investigational compounds" from Selleckchem were chosen, as we can purchase the compounds once we identify suitable hits for testing. Moreover, these compounds are already either approved drugs or in

TABLE 22.1

Top 20 Hits From Docking-Based Virtual Screening Using PyRx for Vps34 Protein.

Sl. no.	Compound name	Binding energy (kcal/mol)	H-bond forming residues
1	Paritaprevir (ABT-450)	−12.7	SER614, GLN620
2	LY333531	−11.8	LYS636, SER687
3	Cepharanthine	−11.1	ASP761
4	LY2090314	−11.1	SER687
5	Adavivint (SM04690)	−11.1	LYS613, GLN683, ILE685, SER687, ASP761
6	KPT-9274	−11	THR610, LEU611, LYS636
7	Cyproheptadine (Periactin)	−10.9	–
8	Sotrastaurin (AEB071)	−10.9	LYS636, TYR670
9	Go6976	−10.8	LYS636
10	Dutasteride	−10.8	–
11	Conivaptan (Vaprisol)	−10.8	LYS613
12	Tariquidar	−10.8	ASP747
13	Evacetrapib (LY2484595)	−10.8	ILE685, LYS613
14	RHPS 4	−10.7	ILE685
15	S55746 (BCL201)	−10.7	ILE685, LYS613
16	Lonafarnib (SCH66336)	−10.7	ASP644
17	Mifepristone (Mifeprex)	−10.7	ILE685, SER687
18	Fangchinoline	−10.6	LYS636
19	Azatadine	−10.6	–
20	Anacetrapib (MK-0859)	−10.6	ILE685
21	Co-crystal ligand JXM from PDB ID: 4UWH	−9.0	ILE685, ASP761

advanced stages of development and hence their toxicity profiles and ADME properties will be well established.

Before actually running the DBVS experiment, the docking protocol was validated to make sure that the AutoDock Vina (docking module in PyRx) could actually pick up/identify the active compounds. For this purpose, the co-crystal ligand JXM from PDB ID: 4UWH was redocked (self-docking) onto the protein. The conformation of the top pose (−9.0 kcal/mol) generated by the docking program matched well with the co-crystal conformation with root mean square deviation of <2 Å. This confirms that the docking program could identify compounds that interact with the target and hence it is suitable for screening large databases to identify potential Vps34 inhibitors. The results from the redocking are shown in Fig. 22.14. Such validation of docking protocol using redocking in VS experiments is essential and increases the confidence in the results. Readers may also refer to Chapter 4 in this book for other validation methods for DBVS. If the redocking experiments do not provide satisfactory results, i.e., the conformation generated by the docking program is deviant from that of the co-crystal conformation, it is advisable to either change the parameters of the docking (such as search and scoring parameters) or choose an alternate docking program.

As the redocking experiment was successful, we initiated the DBVS experiment to identify lead from the compound library. While preparing the screening library, we also included the known Vps34 inhibitor (co-crystal ligands JXM from the PDB ID: 4UWH) in the compound library. Incorporation of known inhibitors to the screening library would help us to identify a threshold binding score, above which to choose the hit compounds for analysis/testing. The DBVS showed that the known inhibitor JXM had a binding energy of −0.90 kcal/mol. Moreover, the Vps34 inhibitor maintained a hydrogen bond with ILE685 residue, which is a key interaction observed in the Vps34 co-crystal structure. In addition to this interaction, JXM made an H-bond interaction with ASP761 residue (Fig. 22.14B).

In the DBVS, a total of 300 compounds were identified to have a binding energy greater (higher negative score) than −0.9 kcal; i.e., 300 compounds were identified to possess stronger interaction with the target than the Vps34 co-crystal ligand JXM. Among them, the top 20 hits were retrieved and their protein−ligand interactions are analyzed using Pymol and listed in Table 22.1. Among the 20 hits, 6 compounds adavivint

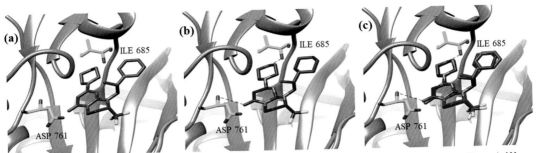

FIG. 22.14 Validation of docking protocol by redocking experiment in PDB ID: 4UWH using PyRx tool. **(A)** Co-crystal ligand JXM conformation in PDB ID: 4UWH, **(B)** docked conformation of ligand JXM in PDB ID: 4UWH, and **(C)** superimposition of co-crystal pose (red color) and docking pose (blue color) of ligand JXM.

(-11.1 kcal/mol), evacetrapib (-10.8 kcal/mol), RHPS 4 (-10.7 kcal/mol), S55746 (-10.7 kcal/mol), mifepristone (-10.7 kcal/mol), and anacetrapib (-10.6 kcal/mol) maintain a hydrogen bond with ILE685, similar to the known Vps34 inhibitor JXM. Adavivint is a selective and potent inhibitor of Wnt signaling at low concentrations (EC_{50} value is 19.5 nM) in colon cancer cells (Deshmukh et al., 2018). Evacetrapib is known to have a key role in preventing cardiovascular events via involvement in lipid profiling where it has shown a decrease in low-density lipoprotein (LDL) and an increase in high-density lipoprotein (HDL) cholesterol (Filippatos & Elisaf, 2017). Studies have reported RHPS 4 to contain anticancer activity. This compound primarily inhibits telomerase in medulloblastoma and glioblastoma, while another study reported that this compound induces apoptosis in cancer cells and sensitizes cancer cells to radiation resistance (Berardinelli et al., 2015; Lagah et al., 2014). S55746 compound is a well-known potent apoptosis inducer and selectively inhibits Bcl-2 with low toxicity and suppresses tumor growth in several cancers (Casara et al., 2018). Researchers have shown that mifepristone lowers tumor growth by inhibiting progesterone and glucocorticoid in triple-negative breast cancer and suggested that this compound may fight against cancer (Tieszen et al., 2011). Many studies have shown that anacetrapib can be used for treating cardiovascular diseases (Di Bartolo & Nicholls, 2017). This compound inhibits cholesteryl ester transfer protein (CETP) and decreases LDL cholesterol and increases HDL cholesterol.

From the DBVS, we identified hits, which show key interactions with the target protein and could be further considered for in silico assessment such as molecular dynamics simulation experiments to determine if these compounds indeed make stable interactions with the Vps34 protein. Based on such assessment, the promising compounds could be tested in wet lab for Vps34 inhibition. As autophagy and apoptosis processes often cross-talk with each other, some of the hits identified could be interesting as they are known to be involved in apoptotic signaling pathways. Therefore, these hits if proven to modulate the autophagy pathway by targeting Vps34 would be a useful addition to already known anticancer agents.

5 CONCLUSION

Searching a new drug for a particular disease is challenging and demands continuous improvements in the available drugs, as acquired drug resistance and toxic effects of existing drugs increase upon prolonged usage. In drug discovery, developing a novel drug molecule typically takes $10-15$ years. To reduce the time and cost, researchers have been developing and implementing new strategies. Particularly, the use of CADD techniques has enabled cost and time reductions in drug discovery.

For researchers who are new to the field of CADD, this chapter demonstrates in a step-by-step manner to perform DBVS of database compounds using PyRx 0.8 tool to select few molecules for biological testing. As a case study, we have used autophagy target Vps34 and approved drugs as ligands for the DBVS. We believe that the step-by-step protocol provided in the chapter will help the novice to practice and implement CADD in their research work. Moreover, researchers could also benefit from the YouTube video about PyRx, which is available at https://www.youtube.com/watch?v=2t12UlI6vuw and https://www.youtube.com/watch?v=gtAur4Rj7Lo.

REFERENCES

Abdelfatah, S., Berg, A., Böckers, M., & Efferth, T. (2019). A selective inhibitor of the polo-box domain of polo-like kinase 1 identified by virtual screening. *Journal of Advanced Research*, 16, 145–156. https://doi.org/10.1016/j.jare.2018.10.002.

Adeniji, S. E., Adamu Shallangwa, G., Ebuka Arthur, D., Abdullahi, M., Mahmoud, A. Y., & Haruna, A. (2020). Quantum modelling and molecular docking evaluation of some selected quinoline derivatives as anti-tubercular agents. *Heliyon*, 6(3), Article e03639. https://doi.org/10.1016/j.heliyon.2020.e03639.

Berardinelli, F., Siteni, S., Tanzarella, C., Stevens, M. F., Sgura, A., & Antoccia, A. (2015). The G-quadruplex-stabilising agent RHPS4 induces telomeric dysfunction and enhances radiosensitivity in glioblastoma cells. *DNA Repair*, 25, 1–12. https://doi.org/10.1016/j.dnarep.2014.10.009.

Berry, M., Fielding, B., & Gamieldien, J. (2015). Practical considerations in virtual screening and molecular docking. *Emerging Trends in Computational Biology, Bioinformatics, and Systems Biology*, 21(1), 487–502. https://doi.org/10.1155/2010/706872.

Casara, P., Davidson, J., Claperon, A., Le Toumelin-Braizat, G., Vogler, M., Bruno, A., Chanrion, M., Lysiak-Auvity, G., Le Diguarher, T., Starck, J. B., Chen, I., Whitehead, N., Graham, C., Matassova, N., Dokurno, P., Pedder, C., Wang, Y., Qiu, S., & Girard, A. M. (2018). S55746 is a novel orally active BCL-2 selective and potent inhibitor that impairs hematological tumor growth. *Oncotarget*, 9(28), 20075–20088. https://doi.org/10.18632/oncotarget.24744.

Dallakyan, S., & Olson, A. J. (2015). Small-molecule library screening by docking with PyRx. *Chemical Biology*, 1263, 243–249. https://doi.org/10.1007/978-1-4939-2269-7.

Deshmukh, V., Hu, H., Barroga, C., Bossard, C., KC, S., Dellamary, L., Stewart, J., Chiu, K., Ibanez, M., Pedraza, M., Seo, T., Do, L., Cho, S., Cahiwat, J., Tam, B., Tambiah, J. R. S., Hood, J., Lane, N. E., & Yazici, Y. (2018). A small-molecule inhibitor of the Wnt pathway (SM04690) as a potential disease modifying agent for the treatment of osteoarthritis of the knee. *Osteoarthritis and Cartilage*, 26(1), 18–27. https://doi.org/10.1016/j.joca.2017.08.015.

Di Bartolo, B. A., & Nicholls, S. J. (2017). Anacetrapib as a potential cardioprotective strategy. *Drug Design, Development and Therapy*, 11, 3497–3502. https://doi.org/10.2147/DDDT.S114104.

Ferreira, L. G., Dos Santos, R. N., Oliva, G., & Andricopulo, A. D. (2015). Molecular docking and structure-based drug design strategies. *Molecules*, 20. https://doi.org/10.3390/molecules200713384.

Filippatos, T. D., & Elisaf, M. S. (2017). Evacetrapib and cardiovascular outcomes: Reasons for lack of efficacy. *Journal of Thoracic Disease*, 9(8), 2308–2310. https://doi.org/10.21037/jtd.2017.07.75.

Forli, S. (2015). Charting a path to success in virtual screening. *Molecules*, 20, 18732–18758. https://doi.org/10.3390/molecules201018732.

Forli, S., Huey, R., Pique, M. E., Sanner, M. F., Goodsell, D. S., & Olson, A. J. (2016). Computational protein–ligand docking and virtual drug screening with the AutoDock suite.

Nature Protocols, 11(5), 905–918. https://doi.org/10.1038/nprot.2016.051.

Gandhi, A. J., Rupareliya, J. D., Shukla, V. J., Donga, S. B., & Acharya, R. (2020). An ayurvedic perspective along with in silico study of the drugs for the management of SARS-CoV-2. *Journal of Ayurveda and Integrative Medicine*, (In Press), 1–6. https://doi.org/10.1016/j.jaim.2020.07.002.

Guido, R. V. C., Oliva, G., & Andricopulo, A. D. (2011). Modern drug discovery technologies: Opportunities and challenges in lead discovery. *Combinatorial Chemistry and High Throughput Screening*, 14(10), 830–839. https://doi.org/10.2174/138620711797537067.

Hassaan, E. A., Sigler, S. C., Ibrahim, T. M., Lee, K. J., Cichon, L. K., Gary, B. D., Canzoneri, J. C., Piazza, G. A., & Abadi, A. H. (2016). Mining ZINC database to discover potential phosphodiesterase 9 inhibitors using structure-based drug design approach. *Medicinal Chemistry*, 12(5), 472–477. https://doi.org/10.2174/1573406412666151204002836.

Holenz, J., Mannhold, R., Kubinyi, H., & Folkers, G. (2016). Lead generation: Methods and strategies, 2 volume set. *Pharmaceutical & Medicinal Chemistry*, 2, 1–824. https://doi.org/10.1017/CBO9781107415324.004.

Hughes, J. P., Rees, S. S., Kalindjian, S. B., & Philpott, K. L. (2011). Principles of early drug discovery. *British Journal of Pharmacology*, 162(6), 1239–1249. https://doi.org/10.1111/j.1476-5381.2010.01127.x.

Jiang, X., Bao, Y., Liu, H., Kou, X., Zhang, Z., Sun, F., Qian, Z., Lin, Z., Li, X., Liu, X., Jiang, L., & Yang, Y. (2017). VPS34 stimulation of p62 phosphorylation for cancer progression. *Oncogene*, 36(50), 6850–6862. https://doi.org/10.1038/onc.2017.295.

Kalia, M., Yadav, V. K., Singh, P. K., Dohare, S., Sharma, D., Narvi, S. S., & Agarwal, V. (2019). Designing quorum sensing inhibitors of *Pseudomonas aeruginosa* utilizing FabI: An enzymic drug target from fatty acid synthesis pathway. *3 Biotech*, 9(40), 1–10. https://doi.org/10.1007/s13205-019-1567-1.

Kondapuram, S. K., Sarvagalla, S., & Coumar, M. S. (2019). Targeting autophagy with small molecules for cancer therapy. *Journal of Cancer Metastasis and Treatment*, 5, 1–26. https://doi.org/10.20517/2394-4722.2018.105.

Lagah, S., Tan, I. L., Radhakrishnan, P., Hirst, R. A., Ward, J. H., O'Callaghan, C., Smith, S. J., Stevens, M. F., Grundy, R. G., & Rahman, R. (2014). RHPS4 G-quadruplex ligand induces anti-proliferative effects in brain tumor Cells. *PLoS One*, 9(1), e86187. https://doi.org/10.1371/journal.pone.0086187.

Lionta, E., Spyrou, G., Vassilatis, D., & Cournia, Z. (2014). Structure-based virtual screening for drug discovery: Principles, applications and recent advances. *Current Topics in Medicinal Chemistry*, 14(16), 1923–1938. https://doi.org/10.2174/1568026614666140929124445.

Liu, R., Li, X., & Lam, K. S. (2017). Combinatorial chemistry in drug discovery. *Current Opinion in Chemical Biology*, 176(3), 117–126. https://doi.org/10.1016/j.cbpa.2017.03.017.

Marinkovi, M., Matilda, Š., Buljuba, M., & Novak, I. (2018). Autophagy modulation in cancer : Current knowledge on action and therapy. *Oxidative Medicine and Cellular Longevity*, 8023821. https://doi.org/10.1155/2018/8023821.

Mercado-Camargo, J., Cervantes-Ceballos, L., Vivas-Reyes, R., Pedretti, A., Serrano-García, M. L., & Gómez-Estrada, H. (2020). Homology modeling of leishmanolysin (gp63) from *Leishmania panamensis* and molecular docking of flavonoids. *ACS Omega, 5*(24), 14741−14749. https://doi.org/10.1021/acsomega.0c01584.

Mohsa, R. C., & Greig, N. H. (2017). Drug discovery and development: Role of basic biological research. *Alzheimer's and Dementia: Translational Research and Clinical Interventions, 3*(4), 651−657. https://doi.org/10.1016/j.trci.2017.10.005.

Morel, E., Mehrpour, M., Botti, J., Dupont, N., Hamaï, A., Nascimbeni, A. C., & Codogno, P. (2017). Autophagy: A druggable process. *Annual Review of Pharmacology and Toxicology, 57*(1), 375−398. https://doi.org/10.1146/annurev-pharmtox-010716-104936.

Morris, G. M., Goodsell, D. S., Halliday, R. S., Huey, R., Hart, W. E., Belew, R. K., & Olson, A. J. (1999). Automated docking using a Lamarckian genetic algorithm and an empirical binding free energy function. *Journal of Computational Chemistry, 19*(14), 1639−1662. https://doi.org/10.1002/(SICI)1096-987X(19981115)19:14<1639::AID-JCC10>3.0.CO;2-B.

Newman, D. J., & Cragg, G. M. (2020). Natural products as sources of new drugs over the nearly four decades from 01/1981 to 09/2019. *Journal of Natural Products, 83*(3), 770−803. https://doi.org/10.1021/acs.jnatprod.9b01285.

Ng, R. (2009). *Drugs from discovery to approval* (Vol. 2). Wiley-BlackWell.

Noman, M. Z., Parpal, S., Van Moer, K., Xiao, M., Yu, Y., Viklund, J., De Milito, A., Hasmim, M., Andersson, M., Amaravadi, R. K., Martinsson, J., Berchem, G., & Janji, B. (2020). Inhibition of Vps34 reprograms cold into hot inflamed tumors and improves anti-PD-1/PD-L1 immunotherapy. *Science Advances, 6*(18), eaax7881. https://doi.org/10.1126/sciadv.aax7881.

O'Boyle, N. M., Banck, M., James, C. A., Morley, C., Vandermeersch, T., & Hutchison, G. R. (2011). Open babel: An open chemical toolbox. *Journal of Cheminformatics, 3*(33), 1−14. https://doi.org/10.1186/1758-2946-3-33.

Pan, S. Y., Zhou, S. F., Gao, S. H., Yu, Z. L., Zhang, S. F., Tang, M. K., Sun, J. N., Ma, D. L., Han, Y. F., Fong, W. F., & Ko, K. M. (2013). New perspectives on how to discover drugs from herbal medicines: CAM'S outstanding contribution to modern therapeutics. *Evidence-Based Complementary and Alternative Medicine*, 1−25. https://doi.org/10.1155/2013/627375.

Pushpakom, S., Iorio, F., Eyers, P. A., Escott, K. J., Hopper, S., Wells, A., Doig, A., Guilliams, T., Latimer, J., McNamee, C., Norris, A., Sanseau, P., Cavalla, D., &

Pirmohamed, M. (2018). Drug repurposing: Progress, challenges and recommendations. *Nature Reviews Drug Discovery, 18*(1), 41−58. https://doi.org/10.1038/nrd.2018.168.

Rankovic, Z., & Morphy, R. (2010). *Lead generation approaches in drug discovery*. John Wiley & Sons, Inc. https://doi.org/10.1002/9780470584170.

Saha, S., Panigrahi, D. P., Patil, S., & Bhutia, S. K. (2018). Autophagy in health and disease: A comprehensive review. *Biomedicine and Pharmacotherapy, 104*, 485−495. https://doi.org/10.1016/j.biopha.2018.05.007.

Sarvagalla, S., Syed, S. B., & Coumar, M. S. (2019). An overview of computational methods, tools, servers, and databases for drug repurposing. In *In silico drug design*. https://doi.org/10.1016/b978-0-12-816125-8.00025-0.

Scior, T., Bender, A., Tresadern, G., Medina-Franco, J. L., Martínez-Mayorga, K., Langer, T., Cuanalo-Contreras, K., & Agrafiotis, D. K. (2012). Recognizing pitfalls in virtual screening: A critical review. *Journal of Chemical Information and Modeling, 52*(4), 867−881. https://doi.org/10.1021/ci200528d.

Singh, S., & Mohanty, A. (2018). In silico identification of potential drug compound against Peroxisome proliferator-activated receptor-gamma by virtual screening and toxicity studies for the treatment of diabetic nephropathy. *Journal of Biomolecular Structure and Dynamics, 36*(7), 1776−1787. https://doi.org/10.1080/07391102.2017.1334596.

Stjepanovic, G., Baskaran, S., Lin, M. G., & Hurley, J. H. (2017). Vps34 kinase domain dynamics regulate the autophagic PI 3-kinase complex. *Molecular Cell, 67*(3), 528−534. https://doi.org/10.1016/j.molcel.2017.07.003.

Tian, W., Chen, C., Lei, X., Zhao, J., & Liang, J. (2018). CASTp 3.0: Computed atlas of surface topography of proteins. *Nucleic Acids Research, 46*(W), 363−367. https://doi.org/10.1093/nar/gky473.

Tieszen, C. R., Goyeneche, A. A., Brandhagen, B. A. N., Ortbahn, C. T., & Telleria, C. M. (2011). Antiprogestin mifepristone inhibits the growth of cancer cells of reproductive and non-reproductive origin regardless of progesterone receptor expression. *BMC Cancer, 11*(207), 1−12. https://doi.org/10.1186/1471-2407-11-207.

Trott, O., & Olson, A. J. (2010). AutoDock vina: Improving the speed and accuracy of docking. *Journal of Computational Chemistry, 31*(2), 455−461. https://doi.org/10.1002/jcc.21334.AutoDock.

Wang, Z., Sun, H., Shen, C., Hu, X., Gao, J., Li, D., Cao, D., & Hou, T. (2020). Combined strategies in structure-based virtual screening. *Physical Chemistry Chemical Physics, 22*(6), 3149−3159. https://doi.org/10.1039/c9cp06303j.

CHAPTER 23

Molecular Docking: A Contemporary Story About Food Safety

FRANCESCA CAVALIERE • GIULIA SPAGGIARI • PIETRO COZZINI

1 INTRODUCTION

What is the link between medicinal chemistry and food safety, and could we apply to food problems the same "in silico" approach used in medicinal chemistry in the last 30 years?

"Questa o quella per me pari sono …" ("Neither is any different…") sang the Duke of Mantova from Rigoletto by Giuseppe Verdi. In the opera, the meaning is regarding women—all women are equal for the duke of Mantova, no difference among them. In this manuscript, it has no negative facet, but it is just referred to molecules. From a chemistry point of view, all the molecules are "molecules," independently from the research field. Then we can apply the same computational methods to different molecules considered -drugs or lead compound or food contact chemicals. The main difference between medicinal chemistry and food science is shown in Fig. 23.1.

Molecular docking is a well-know approach in medicinal chemistry widely used to study the interaction between a receptor and a possible lead compounds, after a screening of a huge number of compounds. While docking in medicinal chemistry is a technique applied for several decades, in food science, it was born 15 years ago, more or less. It could be considered as a new promising application in food science for the discovery of new possible food contaminants, acting as endocrine disruptors, or to understand a mechanism of binding to activate a flavor (umami, sweet, salty, etc.) or to decipher the activity of a dimer against a monomer.

Food safety refers to handling, cooking, and storing food in order to reduce the risk and protect people from foodborne illnesses caused by microbes, chemicals, and other food contact chemicals. A very high number of substances can contaminate food causing a possible risk to the people. An important milestone for screening/docking approaches is the availability of a three-dimensional (3D) database to collect the huge

amount of food contact chemicals in order to make possible testing these compounds otherwise unfeasible with traditional in vitro tests. (To give an idea of the huge chemicals that can interact with food, the most collection of substance information is CAS REGISTRY. It contains more than 163 million unique organic and inorganic chemical substances and more than 68 million biosequences.) The application of computational methods, such as repository or database design, screening, and molecular docking, in food safety, could be applied to predict the interaction between food contact chemicals and different receptors/targets involved in human diseases and/or to decipher their mechanism of binding.

2 FOOD SAFETY

How often do we ask ourselves if the food we are eating is safe? Do we know if it is free from bacteria, viruses, chemicals, and other contaminants? Over the years, food safety is becoming one of the major issues of public concern, food policy, industry, and research. There is no uniform/standard definition of food safety, but anyway in 1993, OECD, the Organisation for Economic Co-operation and Development, gave it a working definition, namely "a reasonable certainty that no harm will result from intended uses under the anticipated conditions of consumption". Food safety can be defined as the probability of not contracting a disease as a consequence of consuming food. In a broad sense, food safety refers to the scientific process to deal with, manufacture, and store food in order to prevent foodborne diseases. The concept of food safety is closely related to the concept of food security: it is not enough to ensure that the food is safe from a health point of view, but it is necessary to delete the obstacles to food such as the supply, the poverty, and the climate changes. In 1970, the World Food

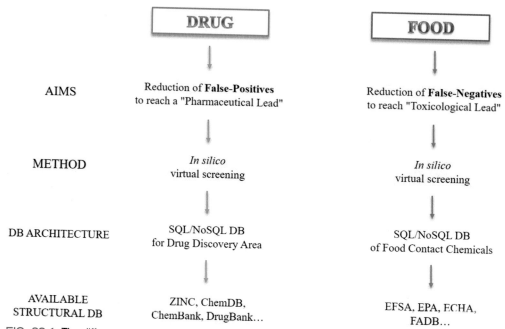

FIG. 23.1 The different approaches to screen compounds in drug discovery and food safety' areas. The same in silico methods, widely used in drug discovery, can be applied in the food field. The unique difference between them is the aim of the screening process: in drug discovery, it is important to retrieve compounds that strongly bind target protein, avoiding false positives; in food safety, the aim is to retrieve all possible food contaminant molecules that have the capacity to bind the target protein, also with low binding affinity, avoiding to exclude true negatives.

Conference defined food security in terms of food supply; it was the Word Food Summit to provide the final definition: "Food security exists when all people, at all times, have physical and economic access to sufficient, safe and nutritious food that meets their dietary needs and food preferences for an active and healthy life". These two elements do not take into account a third important factor: food quality. Let us say that food safety is obtained when everyone has access to food guaranteed as healthy from a hygienic and a nutritional point of view. Therefore, in order to fully understand what food safety means, it is required to define the other two terms: *hazard* and *risk*. These two words are often used interchangeably or confused with each other, but they have a different meaning. A hazard is the capacity of a thing to cause harm and in particular referred to food safety. It is any agent (biological, chemical, or physical) or substance in food with the potential to cause adverse consumer health effects. A risk is the probability of an adverse effect in an organism caused by exposure to an agent. For example,

salmonella, a biological agent that can contaminate different food such as raw eggs, is considered a biological hazard for the consumer. The risk of getting salmonella food poisoning is minimal when the egg is cooked, but, otherwise, if the eggs are eaten raw, the health risk from salmonella will be higher as a result of the higher likelihood that the hazard will be present and consumed.

Ensuring food safety is a significant challenge to protect public health in both developing and developed countries. For this reason, the food safety risk analysis was introduced: it is a fundamental food safety aspect that wants to reduce foodborne illness. This approach aimed at producing high-quality goods and products to ensure safety and protect consumers' health and comply with international and national standards and market regulations; this consists of three components: risk assessment, risk management, and risk communication. In a typical instance, a food safety problem is identified, and risk managers initiate a risk management process, which they then see through to completion.

Risk management is defined as "the process of weighing policy alternatives to accept, minimize, or reduce assessed risks and to select and implement appropriate options". The risk assessment process consists of hazard identification and hazard characterization. Fundamental for all these processes and in general for food safety is the Hazard Analysis Critical Control Points (HACCP), an internationally recognized system, composed of seven points, used to identify, evaluate, and control hazards to food safety. These principles are included in the international standard ISO 22000, a complete food safety management system. Apart from this, the presence of regulations established by national and international organizations (such as the European Food Safety Authority [EFSA] in the European Union, which provides scientific advice and information on existing and emerging risks related to the food chain, and the Environmental Protection Agency [EPA] in the United States, which is in charge of environmental protection and that of human health) ensures that consumers are more protected from health risks.

The World Health Organization (WHO) estimated that 600 million (almost 1 in 10 people in the world) fall ill after eating and/or drinking contaminated food or water resulting in around 420,000 death every year. In recent years with the movement of the people, the increase of globalization, the modernization of industries, and the international trade, people and/or consumers are exposed on a daily basis to chemical substances, and consequently, the risk of foodborne diseases has increased. Therefore, the control of contaminates and the prevention of foodborne diseases have become one of the most public and private health problems in the contemporary world involving the cooperation of all stages of the food chain: from the field to the table.

In the last years, with the increase of diversity and complexity of contaminants and foodborne diseases, not only researchers but also industries and consumers are urged to discover new rapid, sensitive, and selective methods to quantify and qualify damaging substances in food products. Therefore, in vitro and in vivo techniques, such as colorimetric detection, fluorescence sensing (using high quantum yields, narrow and symmetric size-tunable emission, and pronounced photostability, quantum dots, and high signal-to-background ratio and sensitivity as a result of large anti-Stokes shifts, UCNPs), electrochemical sensing, chromatographic separation (high-performance liquid chromatography), immunoassays (enzyme-linked immunosorbent assay), and real-time and in situ analytical methods, have been joined by in silico methods (Liu et al., 2018). These methods, as well as being quick and inexpensive, make up the alternative to animal testing, described by the principles of three Rs (3Rs): replacement, reduction, and refinement (https://www.nc3rs.org.uk/).

3 DATABASES AND BIG DATA IN FOOD SAFETY

Evaluating the effects of food contaminant chemicals is a challenging task. Human exposure can derive from different sources, such as molecules that are naturally present in food products (mycotoxins produced by fungi, flavonoids, etc.), intentionally added molecules (additives, flavorings, etc.), or unintentionally added to food. Some examples are pesticides, biocides that are in contact with the food product, or molecules derived from its packaging and storage such as bisphenols, polycarbonates, etc. The exposure to one chemical can occur via different sources, but rarely humans come in contact with just one single chemical. Instead, we are exposed to a mixture of contaminants. The scenario becomes more complicated if one also considers environmental chemicals and food contaminant metabolites. Thus, the number of molecules that require risk assessment analysis is very high. Moreover, considering the new molecules that are produced every year and that could accidently be released in food, environment, etc., the number of chemicals that should be investigated increases rapidly. Risk assessment of these huge amounts of chemicals using standard toxicological in vitro methods is unthinkable, although, with the advent of high-throughput screening (HTS), toxicological data can be retrieved quickly. However, considering chemical mixtures exposure, it is physically impossible to test all combinations. Thus, in silico methods can be applied to screen this amount of chemicals in a very fast and economic way. To speed up these analyses, it is fundamental to have access to databases that store all food contact chemicals containing information regarding their physical/chemical properties, the 3D structures, their bioactivity, etc.

The huge amount of data produced has raised the need for efficient methods that allow the collection, storage, and processing of data. In this scenario, the big data methods are emerging and becoming an increasingly popular term. Big data is a relatively recent word that has become a ubiquitous term in different sectors of society: business, health care, government, etc.

The term is seldom used in the food safety field. The principal reason is that toxicological data were produced very slowly due to laboratory experiment time limitations. However, after the advent of techniques that allow laboratory automation and HTS, toxicological data are produced very rapidly and at a low cost for many molecules. Moreover, with the advances in data mining and deep learning, more chemical information can be also retrieved from various online sources, including scientific articles and patent documents. Thus, from a lack of data, it has been passed to "data overload" (Richarz, 2020). Many definitions of big data exist and the majority of them refer to the characteristics that a database should have, named versus attributes. Currently, there are more or less 10 different attributes for big data, but the 3 common versus are volume, velocity, and variety. Volume refers to the amount of data generated, velocity refers to the speed at which these data are produced, and variety refers to the types of data. Based on the context and the use, big data can include other attributes, such as variability, veracity, value, etc.

The European Commission (EC) has defined big data as "the large amounts of different types of data produced with high velocity from a number of various types of sources" (European Commission, 2014). Because of the complexity of data generation and curation, big data requires a high-performing computer (HPC) infrastructure. HPCs are very helpful not only to store and manage this high-velocity flow of data but also to make possible the collection of new insight, solutions, and decisions based on this information. The EC definition has also stated: "Handling today's highly variable and real-time data sets requires new tools and methods, such as powerful processors, software and algorithms" (European Commission, 2014). We thought that this definition could be the best one in the context of food safety. Data and information are scattered across food, health, and agriculture sectors for food assessment. As the information is derived from different assays and techniques, many different types of data are produced and should be stored and processed. Moreover, considering in silico assessment, it is also mandatory to store chemical information and 3D structures. Thus, different types of sources and data are used (variety). Although data are not yet generated in real time as in other big data fields, the speed by which they are produced is increased in the last years with the advent of HTS, omics technologies, and (bio) monitoring (velocity).

The first requirement of big data in food science is the collection of information from different sources considering different aspects of the food toxicology and food safety fields. Thus, a database should be storing and making accessible information regarding the physical/chemical properties, the 3D structures of molecules along with toxicological data, derived from different assays, and regulatory information. With the free access, online databases, chemical structures, and data are available for their use in cheminformatics, bioinformatics, systems biology, drug discovery, and food science. From the computational point of view, different public databases store important information, which is currently used in drug discovery and design. Just to cite some of them, PubChem is a large public repository containing information on chemical substances, their biological activities, and their chemical structures. Another chemical database is ChemSpider, a free chemical structure database providing fast text and structure search access to 85 million chemical structures from 275 different data sources. ZINC is a free database that contains the 3D formats of over 230 million purchasable compounds in a ready-to-dock format and over 750 million purchasable compounds allowing the possibility to search for analogs in a very fast way. Moreover, 3D databases are also present in literature that are specific for in silico screening in food toxicology. For example, Ginex et al. have released a 3D version of the EAFUS (Everything Added to Food in the United States) list, a sum of WHO, FAO food additive databases (Ginex, Spyrakis, & Cozzini, 2014).

Data stored in these databases contain important toxicological information and comprise a variety of different types of data: in vitro and in vivo assay results, in silico predictions, gene arrays and omics read-outs, regulatory data, 3D and 2D chemical structures, physical/chemical information, etc. All these information represent a big data set (volume) containing several different types of data (variety and variability) and data can be collected in a single repository or otherwise connected.

Retrieving information from different sources highlights the importance of uniform data to avoid incongruence among them. In fact, some efforts should be made in the direction of database data quality to enforce the utility of big data in drug design and food safety fields. For example, an important point in chemical toxicity data is the identity name of the chemical used. Each molecule must be having an unambiguous name linked to a unique 3D structure. This issue should be guaranteed by the use of CAS numbers, but it is not uncommon to find some errors in public databases. Moreover, errors in chemical structures are not so rare. Williams and Ekins (Williams & Ekins, 2011) estimated that around 5% −10% of molecule structures have errors in their stereochemistry, valency, and charge. Thus, an important issue is the data curation to improve data quality. There is also a great data variability in terms of differences in data measurement and types of assay across different laboratories. Therefore, data could not be comparable. Data standardization should be desirable. The use of nonrelational databases is becoming more common, as they are open source and horizontally scalable and they are referred to as NoSQL databases.

Why big data is becoming so popular? How could it be useful in food safety? Correlated with the concept of the term big data, there are techniques such as text mining and machine learning methods. These methodologies, in some cases, allow us to use the big amount of data to find new knowledge from already available information in a perspective manner. Using information from human cell lines, HTS assays, in vivo animal models could allow the building of predictive models for different applications, such as computer-aided drug design (for the development of new drugs), food toxicology, and/or predictive toxicology (for safety assessment and decision-making). Moreover, the use of big data databases also allows to reduce unnecessary in vivo studies. Hartung et al. (Hartung, 2019) have reported that, on average, every assay was carried out three times and sometimes more than this value. For example, they have reported that two chemicals have been tested more than 90 times in the Draize rabbit eye. Moreover, having a database that stores all chemical information of food contact chemicals, such as the 3D structures, can increase the velocity of in silico methods results. Virtual screening, molecular docking, and molecular dynamics can take a great advantage by the usage of these data.

4 IN SILICO METHODS

In silico methods are computer methods (computing hardware, algorithms, programming, databases, and other domain-specific knowledge) used to study molecular systems in the fields of computational chemistry, computational biology, and material sciences.

Computational methods developed since the 1950s with the increase of computers used for predicting and studying the physical-chemical proprieties, the interactions, and the structures of molecules. Molecular modeling includes all those theoretical methods and computational techniques, such as homology modeling, molecular docking, and molecular dynamics, which are used to represent and/or simulate the behavior of molecules. It, therefore, allows the use of innovative in silico methods, based on the use of computers and information technology, to predict the behavior of biological molecules. Molecular modeling, by studying the energy state of molecules and exploiting calculation algorithms and force fields, or a set of parameters that expresses the potential energy of a particle system, is able to predict and determine quickly and at a low cost the final structure of a molecule.

The sources of starting data for the molecular modeling come from experimental determinations (X-ray, nuclear magnetic resonance, and cryogenic electron microscopy) or computational structure prediction, based on homology modeling, in the event that the 3D structure is not present. 3D databases, such as the Protein Data Bank (PDB) for protein structures (https://www.rcsb.org/) and the Cambridge Structural Database (CSD) (https://www.ccdc.cam.ac.uk/solutions/csd-system/components/csd/) for small organic molecules, contain experimental data. The three parameters we have to consider to understand the quality limits of structural data are (1) resolution (Å), which is a statement of the accuracy in data collection and not a measure of the accuracy in refinement, (2) R-factor, which is a measure of how well the refined structure explains the observed data, and (3) temperature factor, which models the effects of static and dynamic disorder in the crystal. All these parameters are fundamental to choose the best "starting point" for the following computational prediction. A schema of in silico approaches in food safety is shown in Fig. 23.2.

4.1 Molecular Docking

Molecular docking is a complex and simple multistep computational technique used to predict and evaluate the structural chemical—physical interaction between two molecules. The method aims to identify the correct positions of the ligands in the binding pocket of a protein and to predict the affinity between the ligand

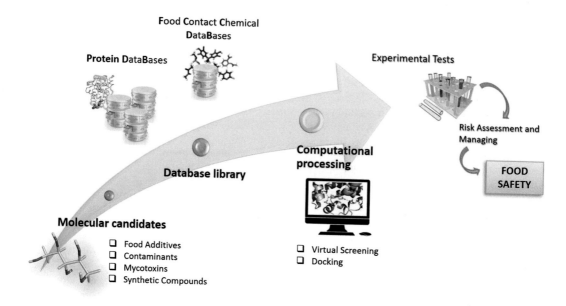

FIG. 23.2 The in silico approach in food safety schema.

and the protein. Ligand-based and structure-based are the two approaches for virtual screening. Structure-based virtual screening is based on the protein cavity shape, while ligand-based virtual screening refers to the shape of the natural ligand.

At the basis of the docking, there is the molecular recognition between the two molecules that interact according to the "lock and key" model developed by Emil Fischer in 1890. In this model the protein has a conformation where the ligand "fits" perfectly, just as it happens for a key inside a lock. The highly specific molecular complementarity between key (ligand) and lock (receptor) plays a fundamental role in biological processes. The receptor's ability to bind to its ligand with high specificity and affinity is due to the formation of a series of weak bonds and favorable interactions. Usually, the interaction between the ligand and its receptor involves the formation of weaker and reversible forces such as (1) hydrogen bonds (10–40 kJ/mol); (2) hydrophobic interactions that constitute the "driving force" capable of promoting bond formation; (3) van der Waals forces (0.03–0.1 kcal/mol); (4) electrostatic interactions (0.3–4 kcal/mol); (5) π–π interactions; and (6) coordination with metals. Electrostatic interactions and hydrogen bonds provide specificity to the protein–ligand interaction and determine its complementarity.

During the formation of the complex, a series of enthalpic and entropic interactions are established between protein and ligand, which are mutually concerted. There is, therefore, a variation of enthalpy (due to the formation of intra- and intermolecular noncovalent bonds) and entropy (due to desolvation) in the system, with consequent variation of free energy.

The binding affinity between the molecules, a ligand (L) and a protein (P), is characterized by the dissociation constant (K_d):

$$K_d = [L][P]/[LP]$$

corresponding to the process $LP \leftrightarrow L + P$.

The fundamental equation that governs everything is

$$\Delta G = \Delta H - T\Delta S$$

where ΔG is the change in free energy of a reaction, ΔH and ΔS are the corresponding changes in enthalpy and entropy, and T is the temperature of the system. The binding affinity can be expressed in terms of the equilibrium constant (K) for the formation of the complex between the two molecules:

$$\Delta G^0 = RT \ln K_d$$

where R is the universal gas constant and T is the absolute temperature.

Even while the interactions between protein and ligand are important for generating a positive enthalpy of binding, we must also consider the presence of the water. In fact, molecular recognition takes place in an aqueous medium. Both the protein and the ligand are solvated before complexation; the formation of the intermolecular bond requires the desolvation of the ligand and the macromolecule with simultaneous breakage of the hydrogen water–receptor and water–ligand bonds (Murcko & Murcko, 1995). The water molecules are organized in such a way as to form as many hydrogen bonds as possible and thus decrease the entropic contribution of the interaction. The clear difference in free energy is often close to zero as many of these breaking bonds are reformed between the ligand and the receptor and the water molecules reorganize around the newly formed complex (Fersht, 1987; Salari & Chong, 2010). For this reason, in order to obtain a reliable energetic estimation of the overall binding process (the total free energy of binding, ΔG°_{bind}), we must use an equation like this:

$$\Delta G^\circ_{bind} = \Delta G^{\circ\,compl}_{solv} - \Delta G^{\circ\,prot}_{solv} - \Delta G^{\circ\,lig}_{solv} + \Delta G^\circ_{int} - T\Delta S^\circ + \Delta\lambda$$

where (ΔG°_{int}) is the interaction free energy of the complex, the solvation energy of the ligand ($\Delta G^{\circ\,lig}_{solv}$), the protein ($\Delta G^{\circ\,prot}_{solv}$), and the complex ($\Delta G^{\circ\,compl}_{solv}$),

and the entropic ($T\Delta S^\circ$) and conformational ($\Delta\lambda$) changes (Spyrakis et al., 2010). However, as Dill said, "Biological interactions are concerted events, not neat sum of terms where each represents an ingredient of the overall process." Thus, even the best practices in treating each of these disparate interaction types individually will not necessarily yield an accurate and reliable ΔG°_{bind} in the end (Dill, 1997).

Docking software can be differentiated based on their two main components: the sampling algorithm, which searches the possible molecule position, and the scoring function, which evaluates the interaction energy of each position. The first is an "easy" geometrical aspect that "mixes" two or more bodies (molecules) in the same Cartesian space without superpositions and obeying to some elementary chemical rules. In fact, anybody (or molecule) is composed by several or many solid spheres and springs. One of the most difficult part of the molecular docking is due to the fact that it involves many degrees of freedom, the high dimensionality of the energy surface where the search for the global minimum is performed by a docking program. Each algorithm generates poses (where and how a ligand binds a protein); a series of conformations result from rotation about single bonds (Fig. 23.3). For a molecule with n rotatable bonds, if each torsion angle is rotated in increments of x degrees, the number of conformations is $(360^\circ/x)^n$.

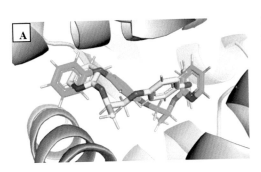
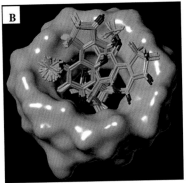

FIG. 23.3 Different poses generated by a generic docking software (GOLD). **(A)** A protein-ligand complex. In grey the protein (PDB ID: 1FM6) and in yellow and green two different poses of a pesticide (pyriproxyfen) are shown. As we can see the two ligand poses are opposite to each other. **(B)** The most important poses of a mycotoxins (Aflatoxin) within the cavity of a beta-cyclodextrin. All the poses identify the same position.

We can classify the different search algorithms and consequently the different docking software according to the degrees of freedom that they consider:

1. Rigid docking: This type of molecular docking ignores the flexibility of the molecules, both for the ligand and for the protein, and treats them like rigid objects. In this case, the side chains and the backbone of the two molecules are kept fixed with no torsion angles or distance between two atoms allowed to change upon the docking simulation.

2. Semiflexible docking: During this docking, the receptor remains unchanged, while the conformation of the ligand changes. It focuses on the changes in the ligand structure and it is usually used for the docking between small ligands and macromolecules.

3. Flexible docking: This docking, the most common today, considers every conformational change of both the protein and the ligand.

We can summarize scoring functions in three classes: (1) force field—based, where the binding affinity is estimated by the sum of the strength of intermolecular (van der Waals and electrostatic interactions) interactions between all atoms of the two molecules; (2) empirical, where the binding affinity between the two molecules is estimated by the number of various types of interactions; (3) knowledge-based, based on a statistical analysis of observed pairwise distributions. More than 60 different docking software have been reported in the literature, such as AutoDock (Morris et al., 2009), DOCK (Allen et al., 2015), GOLD (Verdonk et al., 2003), FlexX (Schellhammer & Rarey, 2004), Glide (Friesner et al., 2004), Surflex (Jain, 2003), distinguished by the algorithms, the evaluation methods, the docking types (rigid, semiflexible, or flexible docking), and more. One or more scoring functions can be associated with each scoring program. There is no docking—scoring combination valid for each type of analysis, but these combinations must be evaluated based on the characteristics of the target. Most docking scoring functions use very simplified models for hydrophobic interactions; then simulating the binding (or docking) process with explicit terms for entropy has proven to be an elusive goal. To get around this, in 1991, Abraham and Kellogg developed HINT (hydropathic interactions), a scoring function that simulates and quantifies all of the subtle effects contributing to entropy in the docking process (Shoichet & Kuntz, 1993). HINT uses a force field that allows it to evaluate both the entropic

aspect (due to desolvation) and the enthalpic aspect (due to interactions). The fact that you also evaluate the entropic aspect is what differentiates HINT from other scoring functions. The function also includes the computational titration method for predicting and optimizing the protonation state of ionizable residues at a complex interface and the Rank algorithm for rationalizing the role of structural water molecules in protein binding pockets (Amadasi et al., 2008; Cozzini et al., 2004).

Deciding which program is the best one is a challenging task. Docking software is normally validated using a training set of protein—ligand complexes with the known crystal structure and known binding affinity. The 3D complex is used to validate the internal algorithm of the docking package to predict the correct binding pose based on the crystallographic one. The correlation is usually assessed using the root-main-square deviation between the docked and the crystal ligand pose: the lower the value, the better is the reliability of the docking algorithm. Binding affinity value is used to test the ability of scoring functions to discriminate between compounds having strong-, medium-, and lower binding affinity, or in alternative to test their ability to discriminate among a library of true, false, and decoy compounds. There is no general rule for choosing the best docking program, but it is advisable to utilize a software that was validated against the same class of protein under investigation or with proteins sharing common physical-chemical and shape characteristics. However, the goal of any docking program is to be used for every protein—ligand system. It has been estimated that the averaged success rate in predicting the correct poses and top scores is in the range of 54.0%—67.8% for commercial programs and in the range of 47.4%—68.4% for the academic ones (Pagadala et al., 2017). Even if the performances are quite high, a certain grade of uncertainty and error can occur. Thus, the best practice is to apply a consensus score prediction. The concept was introduced to enhance the performances of docking protocols. Multiple scoring functions are simultaneously used rather than a single one. Compounds are then ranked based on the consensus existing among them and only the top scored compounds common to the scoring functions will be used for further in vitro/in vivo assays (Wang & Wang, 2001). The concept of consensus scoring could be seen as weather forecasts: if many of them agree that, during the weekend, there will be the sun, and just one predicts a thunderstorm, then it is more probably there will be a sunny weekend. Compared to a single scoring procedure, it has been shown that the combination of different scoring functions reduces false positives and hence improves the hit rates (Wang & Wang, 2001). It has also been reported that the use of three scoring methods is enough to enhance the capability hit rates of ~50% (Bissantz et al., 2000).

5 CASE STUDIES

In food science, molecular docking approach is applied for different needs: to study the interaction of a food chemical with a protein receptor understanding the mechanism of binding or the competition between a natural ligand of a protein and a food molecule or to design chemosensors able to include a toxin in a cavity to take away the dangerous molecule from water or food. Hereinafter, we illustrate a few real cases of docking applications. As the majority of our key studies

are focused on nuclear receptors (NRs) and how food contact chemicals may act as endocrine disruptors, Fig. 23.4 shows a schematic view of the perturbation induced by endocrine disruptor compounds (EDCs) activity.

5.1 Mycotoxins Detection

Mycotoxins are important because of possible danger to humans; depending on the intake dose, they can act as endocrine disruptors binding mostly to NRs. Two well-known NRs are recognized as responsible for breast cancer in women and prostate cancer for men: estrogen receptor (ER) and androgen receptor (AR), respectively.

5.1.1 Aflatoxins and ochratoxins

Cyclodextrins are cheap and relatively easy to manage, and they show a lipophilic cavity; mycotoxins are small molecules with a hydrophobic and a hydrophilic side, present in plants but dangerous for humans. They could be included in a cyclodextrins cavity, the lipophilic side, and the complex could be detected using spectroscopy fluorescence (Cozzini et al., 2008). Moreover, the

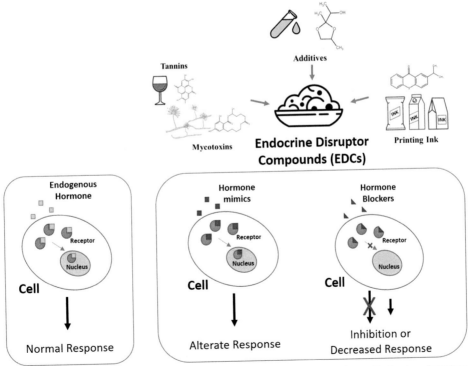

FIG. 23.4 How food contact chemicals can affect the nuclear receptors pathway. Endocrine disruptors are chemicals that can interfere with endocrine or hormonal systems. These disruptions can cause adverse effects such as tumors, birth deffects, and several other disorders. In fact, these molecules can decrease or increase normal hormone levels altering their normal production.

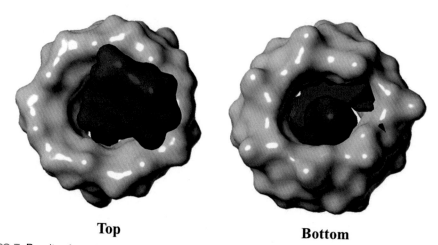

Top   **Bottom**

FIG. 23.5 Results of a molecular docking simulation between beta-cyclodextrin and a mycotoxin. Molecules are depicted using occupancy volume. In red is the volume occupied by the mycotoxin, in gray the cyclodextrin volume, and in blue the empty volume that could be filled by water molecules (GRID analysis).

mechanism of action (MOA) understanding allows designing specific cyclodextrins customized for different toxin structures (Fig. 23.5) (Amadasi et al., 2007). Afla-toxins and ochratoxins (Fig. 23.7) affect a large number of mays and grains production in Italy (more or less 5%). They can be detected using cyclodextrins and fluo-rescence spectroscopy. In particular, in this case, the modeling allowed us to understand why the same beta-cyclodextrin can include aflatoxin and not ochra-toxin. The latter requires a specifically designed cyclodextrin.

5.1.2 Zearalenone

Zearalenone (ZEN) (Fig. 23.7) and its metabolites that are known to act through activation of the estrogen receptor alpha (ER alpha) has been studied (Cozzini & Dellafiora, 2012) against ER to understand if they can bind competitively with the endogenous ligand, estradiol. A molecular docking–based study demon-strates that it is possible to discriminate between *cis*- and *trans*-isomers for ZEN using the same docking approach: for the *cis* isomer, a stronger interaction has been predicted (Dellafiora, Galaverna, et al., 2015). Moreover, ZEN and its reduced metabolites have been used within the framework of reduction, refinement, and replacement of animal experiments (Ehrlich et al., 2015). Mixed methods, docking/scoring, and toxicolog-ical methods for identification and characterization of chemical hazards have been developed. The results suggest that activation of ER alpha may play a role in the molecular initiating event and be predictive of adverse effects. The investigation of receptor–ligand

interactions through docking simulation showed the suitability of the model to address estrogenic potency for this group of compounds. Therefore, the model was further applied to biologically uncharacterized, commercially unavailable, oxidized ZEN metabolites (6 alpha-, 6 beta-, 8 alpha-, 8 beta-, and 13- and 15-OH-ZEN). The main conclusion is that, except for 15-OH-ZEN, the data indicate that in general, the oxidized metabolites would be considered of a lower estrogenic concern than ZEN and reduced metabolites.

5.1.3 Alternariol

Another mycotoxin, a widespread microfungi second-ary metabolites that may accumulate in crops and enter in contact with some foods, is from *Alternaria* species (Dellafiora, Dall'Asta, et al., 2015). The whole corn production in Italy is affected by mycotoxin alternariol every year, depending on tem-perature. Thus, the comprehension of the MOA of alternariol and its derivatives against some proteins is crucial to understand if toxic potency may drastically be reduced by metabolic modifications. Alternariol (Fig. 23.7) and alternariol methyl ether show evidence of toxicity binding to topoisomerases but it is not enough. Too many compounds and its derivatives are candidates to be endocrine disruptors because of binders of several proteins. Because wet-lab tests are expensive and require long times, it is really chal-lenging to have a fast and cheap method to discrimi-nate among possible poisons and no poisons as in silico methods. In this work, the methods have been applied for the topoisomerase case.

5.2 Ellagitannin Metabolites

Dellafiora et al. have applied the same in silico approach to ellagitannins and their metabolites (glucuronidation, sulfation, and methylation, occurring in vivo) (Fig. 23.7) (Dellafiora et al., 2013). Urolithin metabolites could act as phytoestrogens able to interact with the ER binding cavity. These hydroxylation patterns are presented in our models coming from berries, walnuts, pomegranate, and oak-aged red wines. They are well known as "natural drugs" that can contribute to decreasing the risk of some ER-dependent diseases. Ones again, the in silico approach to study the MOA suggested that hydroxylation can play an important role in the agonistic behavior of these derivates.

5.3 Printing Inks

As stated, another tumor marker is the AR, involved in prostate cancer, able to interact with many food contact chemicals. Thioxanthones are analogs of xanthone and are largely used as photoinitiators (TX) by printing industry to promote ink polymerization. However, a certain level of contamination by isopropyl thioxanthone (ITX) and 2-ethylhexyl-4-dimethylamino-benzoate has been found in food products, especially in infant formulas (as reported by the European Food Safety Authority in 2005). Ginex et al. (Ginex, Dall'Asta, & Cozzini, 2014) have reported an in silico approach to predict the binding affinity of thioxanthone derivates and thioxanthone metabolites against AR. In fact, it is well known by in vitro analyses that this class of compounds is able to bind to AR. Using the in vitro affinity values of some TX compounds as validation test of in silico procedure, different metabolites have been computationally analyzed to predict their binding affinity for the ligand binding cavity of AR. The authors have found that different metabolites have the same or higher binding affinity of 2-ITX, 4-ITX, and 2-Chloro-TX, which are the three well-known AR-mediated endocrine disrupting compounds.

5.4 Food Additives

More than 3000 substances could be added to the food depending on the different countries' laws. In the search for xenoestrogens within food additives, the Joint FAO-WHO expert committee database, containing 1500 compounds, was checked using an integrated in silico and in vitro approaches (Amadasi et al., 2009). The main question was, are we confident about the safety of food additives allowed? Docking and screening could assume the same meaning but, usually, screening is reserved to "screen" a huge number of molecules against one or more receptors based on ligand structure or receptor cavity structure. Both techniques can be applied in a pipeline to extract a smaller set of data from a big database (screening) to be docked within a receptor cavity. Wet lab tests applied to predicted molecules identified propyl gallate as an antagonist and 4-hexylresorcinol as a potent transactivator (nanomolar concentration) based to in silico prediction. The final meaning is to consider these two compounds as probable ER interactors but not certified as poison.

5.5 Bisphenols in Food

The bisphenol case is another example used to demonstrate that docking methods could be a valid approach to screen estrogenic and androgenic activity of food contact materials (Cavaliere et al., 2020). One of the most common bisphenols is bisphenol A (BPA) or 4,4′-isopropylidenediphenol (Fig. 23.7). This plastic, used to make many food containers, has been classified by the European Chemical Agency (ECHA) as a substance of very high concern for its toxicological effect on reproduction and its endocrine disrupting properties. EDCs can exert their adverse effects binding directly with the ligand binding domain of NRs interfering with the normal hormone response. Thus, a lot of efforts is made to find alternative molecules that can exert the same plasticizing effects in polycarbonate materials (Fig. 23.6) with no or lower adverse effects for human health. The estrogenic and androgenic effects of 26 different bisphenols (including 7 BPA metabolites) have been evaluated using a mix of molecular docking and consensus scoring methods to evaluate the activity of some BPA alternatives and BPA metabolites.

Six different NRs have been included in the analysis: three NRs for the estrogenic pathway and three NRs for the androgenic one. The ligand binding pockets of these NRs have different physicochemical properties. Thus, two different molecular docking software and four different scoring functions have been applied to overcome the possible limitations derived by molecular docking package and to reduce the number of false positive across different targets. The results have shown that (1) some BPA metabolites could lower the harmful effects of BPA exposure; (2) bisphenol S, a BPA′ substitute, turned out a lower interactor for all NRs, except for AR, for which its binding activity is found similar to a pharmacological antiandrogen; (3) only 2,2-bis(4-hydroxyphenyl) propanol (BPAol), a BPA metabolite, was predicted as a lower interactor for all NRs considered.

BPA and BPA alternatives/metabolites

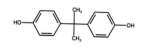

Estrogenic/Androgenic pathways

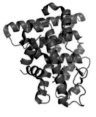

❑ Virtual Screening
❑ Docking

Likely safer BPA substitutes

FIG. 23.6 The case of docking/scoring application on food contact materials: Bisphenols.

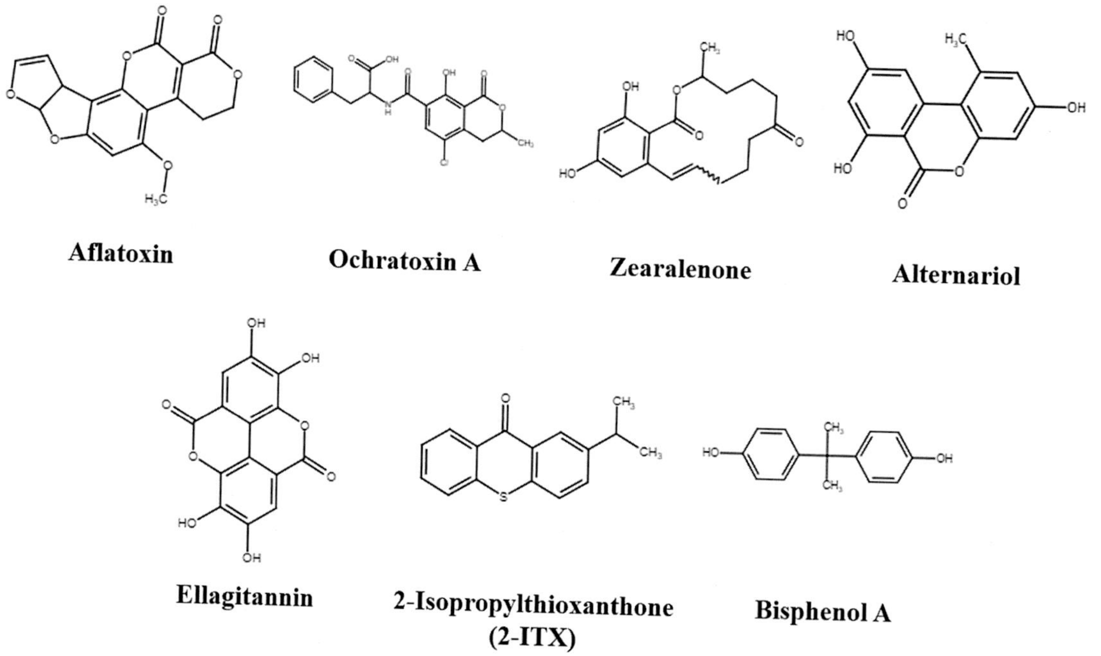

Aflatoxin

Ochratoxin A

Zearalenone

Alternariol

Ellagitannin

2-Isopropylthioxanthone (2-ITX)

Bisphenol A

FIG. 23.7 The chemical structures of case studies compounds.

6 CONCLUSIONS

The lesson learning from medicinal chemistry suggests we can use computational simulations in food safety, in particular molecular modeling and molecular dynamics. The possibility to screen a huge number of chemicals to find endocrine disruptors in a reasonable time is, to date, a real low-cost opportunity, allowing to apply wet lab test only to the chemicals predicted as most probable interactors. From this chapter we got few take-home messages: (1) be careful with starting structural data (check the structural parameters); (2) be careful in choosing the software, there is not a general package able to solve all modeling problems; (3) a complete analysis should include a lot of factors: waters, protons, metals, cofactors, etc.; and (4) do not trust docking results blindly without a discussion; the software is not a wizard able to predict exactly the future.

REFERENCES

Allen, W. J., Balius, T. E., Mukherjee, S., Brozell, S. R., Moustakas, D. T., Lang, P. T., Case, D. A., Kuntz, I. D., & Rizzo, R. C. (2015). DOCK 6: Impact of new features and current docking performance. *Journal of Computational Chemistry, 36*(15), 1132–1156. https://doi.org/10.1002/jcc.23905

Amadasi, A., Dall'Asta, C., Ingletto, G., Pela, R., Marchelli, R., & Cozzini, P. (2007). Explaining cyclodextrin-mycotoxin interactions using a "natural" force field. *Bioorganic & Medicinal Chemistry, 15*(13), 4585–4594. https://doi.org/10.1016/j.bmc.2007.04.006

Amadasi, A., Mozzarelli, A., Meda, C., Maggi, A., & Cozzini, P. (2009). Identification of xenoestrogens in food additives by an integrated in silico and in vitro approach. *Chemical Research in Toxicology, 12*, 52–63.

Amadasi, A., Surface, J. A., Spyrakis, F., Cozzini, P., Mozzarelli, A., & Kellogg, G. E. (2008). Robust classification of "relevant" water molecules in putative protein binding sites. *Journal of Medicinal Chemistry, 51*(4), 1063–1067. https://doi.org/10.1021/jm701023h

Bissantz, C., Folkers, G., & Rognan, D. (2000). Protein-based virtual screening of chemical databases. 1. Evaluation of different docking/scoring combinations. *Journal of Medical Chemistry, 43*, 4759–4767. https://doi.org/10.1021/jm001044l

Cavaliere, F., Lorenzetti, S., & Cozzini, P. (2020). Molecular modelling methods in food safety: Bisphenols as case study. *Food and Chemical Toxicology, 137*. https://doi.org/10.1016/j.fct.2020.111116

Cozzini, P., & Dellafiora, L. (2012). In silico approach to evaluate molecular interaction between mycotoxins and the estrogen receptors ligand binding domain: A case study

on zearalenone and its metabolites. *Toxicology Letters.* https://doi.org/10.1016/j.toxlet.2012.07.023

Cozzini, P., Fornabaio, M., Marabotti, A., Abraham, D. J., Kellogg, G. E., & Mozzarelli, A. (2004). Free energy of ligand binding to protein: Evaluation of the contribution of water molecules by computational methods. *Current Medicinal Chemistry, 11*(23), 3093–3118. https://doi.org/10.2174/0929867043363929

Cozzini, P., Ingletto, G., Singh, R., & Dall'Asta, C. (2008). Mycotoxin detection plays "cops and robbers": Cyclodextrin chemosensors as specialized police? *International Journal of Molecular Sciences, 9*(12), 2474–2494. https://doi.org/10.3390/ijms9122474

Dellafiora, L., Dall'Asta, C., Cruciani, G., Galaverna, G., & Cozzini, P. (2015). Molecular modelling approach to evaluate poisoning of topoisomerase I by alternariol derivatives. *Food Chemistry, 189*, 93–101. https://doi.org/10.1016/j.foodchem.2015.02.083

Dellafiora, L., Galaverna, G., Dall'Asta, C., & Cozzini, P. (2015). Hazard identification of cis/trans-zearalenone through the looking-glass. *Food and Chemical Toxicology, 86*, 65–71. https://doi.org/10.1016/j.fct.2015.09.009

Dellafiora, L., Mena, P., Cozzini, P., Brighenti, F., & Del Rio, D. (2013). Modelling the possible bioactivity of ellagitannin-derived metabolites. In silico tools to evaluate their potential xenoestrogenic behavior. *Food & Function, 4*(10), 1442–1451. https://doi.org/10.1039/c3fo60117j

Dill, K. A. (1997). Additivity principles in biochemistry. *The Journal of Biological Chemistry, 272*(2), 701–704. https://doi.org/10.1074/jbc.272.2.701

Ehrlich, V. A., Dellafiora, L., Mollergues, J., Asta, C. D., & Serrant, P. (2015). Hazard assessment through hybrid in vitro/in silico approach : The case of zearalenone. *ALTEX: Alternativen Zu Tierexperimenten, 32*(4), 275–286.

European Commission. (2014). *Towards a thriving data-driven economy.* Brussels: Communication from the Commission to the European Parliament, The Council, The European Economic and Social Committee and The Committee of the Regions.

Fersht, A. R. (1987). The hydrogen bond in molecular recognition. *Trends in Biochemical Sciences, 12*(C), 301–304. https://doi.org/10.1016/0968-0004(87)90146-0

Friesner, R. A., Banks, J. L., Murphy, R. B., Halgren, T. A., Klicic, J. J., Mainz, D. T., Repasky, M. P., Knoll, E. H., Shelley, M., Perry, J. K., Shaw David, E., Francis, P., & Shenkin, P. S. (2004). Glide: A new approach for rapid, accurate docking and scoring. 1. Method and assessment of docking accuracy. *Journal of Medicinal Chemistry, 47*(7), 1739–1749. https://doi.org/10.1021/jm0306430

Ginex, T., Dall'Asta, C., & Cozzini, P. (2014). Preliminary hazard evaluation of androgen receptor-mediated endocrine-disrupting effects of thioxanthone metabolites through structure-based molecular docking. *Chemical Research in Toxicology, 27*(2), 279–289. https://doi.org/10.1021/tx400383p

Ginex, T., Spyrakis, F., & Cozzini, P. (2014). FADB: A food additive molecular database for in silico screening in food toxicology. *Food Additives & Contaminants Part A, Chemistry, Analysis, Control, Exposure & Risk Assessment, 31*(5), 792–798. https://doi.org/10.1080/19440049.2014.888784

Hartung, T. (2019). Predicting toxicity of chemicals: Software beats animal testing. *EFSA Journal, 17*(S1), e170710. https://doi.org/10.2903/j.efsa.2019.e170710

Jain, A. N. (2003). Surflex: fully automatic flexible molecular docking using a molecular similarity-based search engine. *Journal of Medicinal Chemistry, 46*(4), 499–511. https://doi.org/10.1021/jm020406h

Liu, J.-M., Wang, Z.-H., Ma, H., & Wang, S. (2018). Probing and quantifying the food-borne pathogens and toxins: From in vitro to in vivo. *Journal of Agricultural and Food Chemistry, 66*(5), 1061–1066. https://doi.org/10.1021/acs.jafc.7b05225

Morris, G. M., Huey, R., Lindstrom, W., Sanner, M. F., Belew, R. K., Goodsell, D. S., & Olson, A. J. (2009). AutoDock4 and AutoDockTools4: Automated docking with selective receptor flexibility. *Journal of Computational Chemistry, 30*(16), 2785–2791. https://doi.org/10.1002/jcc.21256

Murcko, A., & Murcko, M. A. (1995). Computational methods to predict binding free energy in ligand-receptor complexes. *Journal of Medicinal Chemistry, 38*(26), 4953–4967. https://doi.org/10.1021/jm00026a001

Pagadala, N. S., Syed, K., & Tuszynski, J. (2017). Software for molecular docking: A review. *Biophysical Reviews, 9*(2), 91–102. https://doi.org/10.1007/s12551-016-0247-1

Richarz, A.-N. (2020). Big data in predictive toxicology: Challenges, opportunities and perspectives. In *Big data in predictive toxicology* (pp. 1–37). The Royal Society of Chemistry. https://doi.org/10.1039/9781782623656-00001.

Salari, R., & Chong, L. T. (2010). Desolvation costs of salt bridges across protein binding interfaces: Similarities and differences between implicit and explicit solvent models. *The Journal of Physical Chemistry Letters, 1*(19), 2844–2848. https://doi.org/10.1021/jz1010863

Schellhammer, I., & Rarey, M. (2004). FlexX-scan: Fast, structure-based virtual screening. *Proteins, 57*(3), 504–517. https://doi.org/10.1002/prot.20217

Shoichet, B. K., & Kuntz, I. D. (1993). Matching chemistry and shape in molecular docking. *Protein Engineering Design and Selection, 6*(7), 723–732. https://doi.org/10.1093/protein/6.7.723

Spyrakis, F., Cozzini, P., & Kellogg, G. E. (2010). Docking and scoring in drug discovery. In *Burger's medicinal chemistry and drug discovery* (pp. 601–684). https://doi.org/10.1002/0471266949.bmc140

Verdonk, M. L., Cole, J. C., Hartshorn, M. J., Murray, C. W., & Taylor, R. D. (2003). Improved protein-ligand docking using GOLD. *Proteins, 52*(4), 609–623. https://doi.org/10.1002/prot.10465

Wang, R., & Wang, S. (2001). How does consensus scoring work for virtual library screening? An idealized computer experiment. *Journal of Chemical Information and Computer Sciences, 41*(5), 1422–1426. https://doi.org/10.1021/ci010025x

Williams, A. J., & Ekins, S. (2011). A quality alert and call for improved curation of public chemistry databases. *Drug Discovery Today, 16*(17), 747–750. https://doi.org/10.1016/j.drudis.2011.07.007

Index

Note: Page numbers followed by "f" indicate figures and "t" indicate tables.